Diterpenes of Flowering Plants

F. Seaman   F. Bohlmann
C. Zdero   T.J. Mabry

Diterpenes of Flowering Plants

Compositae (Asteraceae)

Springer-Verlag
New York Berlin Heidelberg
London Paris Tokyo Hong Kong

Fred Seaman
Drug Dynamics Institute
College of Pharmacy
The University of Texas
Austin, Texas 78712, USA

Christa Zdero
Institute of Organic Chemistry
Technical University
D-1000 Berlin 12
Federal Republic of Germany

Ferdinand Bohlmann
Institute of Organic Chemistry
Technical University
D-1000 Berlin 12
Federal Republic of Germany

Tom J. Mabry
Department of Botany
The University of Texas
Austin, Texas 78712, USA

Cover: The plant illustrated is *Ageratina gypsophylla* B. Turner, a localized endemic species of Nuevo Leon, Mexico. *Ageratina* species produce both the acyclic geranylnerol-type and bicyclic *ent*-labdane diterpenes shown on the cover (in the background). The illustration is provided by B. Turner.

Library of Congress Cataloging-in-Publication Data
Diterpenes of flowering plants/F. Seaman ... [et al.].
   p. cm.
  Contents: Compositae (Asteraceae)
  Includes bibliographical references.
  ISBN-13:978-1-4612-7945-7   e-ISBN-13:978-1-4612-3274-2
  DOI: 10.1007/978-1-4612-3274-2
  1. Diterpenes. 2. Angiosperms—Composition. 3. Angiosperms—Classification. 4. Botanical chemistry. 5. Plant chemotaxonomy.
  I. Seaman, F. (Fred)
  QK898.T4D57 1989
  582.13′04192—dc20   89-21877

Printed on acid-free paper

Typeset by Technical Typesetting Incorporated, Baltimore, Maryland.

9 8 7 6 5 4 3 2 1

ISBN-13:978-1-4612-7945-7

Contents

Chapter 1. Introduction 1

Chapter 2. Compound Names, Structures, and Sources 3

    II.1. Names of Reported Compounds 3
    II.2. Structures of Reported Compounds 3
    II.3. Plant Sources and References for Diterpenes 6

Chapter 3. Biogenesis 385

    III.1. *ent*- and Normal-labdanes 385
    III.2. Tricyclic Diterpenes 388
    III.3. Tetracyclic Diterpenes 389
    III.4. Biosynthetic Evidence 389

Chapter 4. Diterpene Distribution: Compositae 431

    IV.1. Subfamilial Distribution Patterns 431
    IV.2. Tribal Affinities 432
    IV.3. Infrageneric Distribution Patterns 434

Chapter 5. Biological Activity of Diterpenes 485

    V.1. Introduction 485
    V.2. Linear Diterpenes 485
    V.3. Bicyclic Diterpenes 485
    V.4. Tricyclic Diterpenes 486
    V.5. Tetracyclic Diterpenes 487

Chapter 6. Diterpene Analysis with Emphasis on Clerodanes 493

    VI.1. Clerodanes 493
    VI.2. Other Bicarbocyclics 505

VI.3. Tricarbocyclics 505
VI.4. Tetracarbocyclics and Related Skeletons 506

Chapter 7. References 551

Appendix   Alphabetical Listing of Molecular Substituents
            including Ester Sidechains 579

Indexes 591

CHAPTER 1

Introduction

More than 1200 diterpenes have been identified from approximately 550 Compositae taxa (Figure 1 [pp. 219–384], Tables 1–3 [pp. 7–218]). The annual output of such reports rose sharply during the past several years, a trend that was also reflected by other major lipophilic constituents of the Compositae: acetylenes (*63, 465*), sesquiterpene lactones (*266, 458*), and benzofurans and chromenes (*429*). Unlike these other natural products, no review of the diterpene chemistry of the Compositae has been published. Given the bulk of data currently on hand, a review of this topic is clearly overdue.

Several major goals influenced the preparation of this review. The first was to provide a useful reference to guide terpenoid chemists to the literature of known compounds. The published reports for 1200 structures provide a sizable body of spectroscopic data applicable to the investigation of new compounds.

The study of diterpenes has been hampered by the absence of an all-inclusive nomenclature and classification system for the various skeletal types. Consequently, a second goal was to organize structural data into a classification scheme designed to group compounds by biogenetic homology, and to standardize such features as diterpene nomenclature and numbering. Pursuant to this goal, the published names (Table 1) and numbering systems (Table 2) for all compounds are included.

The third goal was to compare the distribution of diterpenes to the classification scheme used to subdivide the Compositae into its subfamilies, tribes, subtribes, genera, etc., and to seek chemical distribution patterns that paralleled the plant taxonomic boundaries. As diterpenes constitute a substantial body of raw taxonomic data, they may hold some promise as novel characters for use in plant taxonomy.

The fourth goal was to critically review the quality of published structure-elucidation arguments. As will be discussed in detail (Chapter six), eagerness to describe novel compounds in absolute stereochemical terms has led to much confusion, particularly in the description of new clerodane structures. The lengthy review of this topic hopefully will aid in the proper assessment of stereochemical assignments.

The fifth goal was to develop a computerized diterpene database from which this review's tables could be extracted. Major topics dealt with in each chapter represent some facet of programmed access to the diterpene database. Accomplishing this final goal simplifies the future updating of this review by automating the process of compiling up-to-date tabular information (e.g., Tables 1–3).

Diterpene data for the Compositae are included in Tables 1–3 (Chapter 2). In Table 1, the compound names and numbers are grouped according to membership in one of 103 different skeletal types and listed alphabetically within the skeletal-type group. Table 2 lists for each skeleton the compound numbers and associated molecular descriptions, while Figure 1 displays the corresponding structures. Table 3 lists all compound numbers and the Compositae taxa from which the compounds were reported together with the reference numbers corresponding to these reports. Distributional data were extracted from the literature and from the authors' unpublished results. Alphabetized lists of compound names and species appear in the Index together with citation page numbers.

In the discussion of diterpenoid biogenesis (Chapter 3), each skeletal type is described in terms of its likely mode of origin along one of two routes, the *ent*-labdane-based route or the normal-labdane-based route. Hypothetical biogenetic schemes for most skeletons are illustrated. Next, compound distribution is discussed in terms of patterns of structural variability found at different plant taxonomic levels (Chapter 4). Interspecific variation is detailed using six genera (*Brickellia*, *Stevia*, *Baccharis*, *Helichrysum*, *Helianthus* and *Montanoa*) representing the major diterpene-producing tribes of the Compositae. The next chapter (five) describes the biological activities associated with different diterpene skeletal types. The last chapter reviews the analytical methods used to identify diterpenes and provides an interpretation of spectroscopic data for several types, principally the clerodanes.

CHAPTER 2

Compound Names, Structures, and Sources

II.1. Names of Reported Compounds

The names (Table 1 [pp. 7–59]) and structures (Figure 1 [pp. 219–384]) of diterpenes reported from the Compositae are organized according to skeletal type following the sequence indicated in Figure 2 (p. 392). If a name was assigned to a structure at the time the compound was reported, that name was included in Table 1. Otherwise, the structure was given a name compatible with the names previously assigned to related compounds. If more than one name was applied in the literature to the same structure, usually the first-published name was included in Table 1, followed by additional names listed within brackets. The compound number listed with each name was used throughout this treatment when reference to the compound was made. Comments about a reported compound were also included in brackets on a line below the compound name. Usually these comments describe situations in which the name applied to the compound was incompatible with either the illustrated structure or the published spectroscopic data.

II.2. Structures of Reported Compounds

An alternative database-generated method of diterpene structure representation introduced in Table 2 (pp. 60–149) departs from the traditional approach using the "R-group" system for differentiating related compounds. The rationale for developing an alternative method is that by using it, a computer-generated set of structures (e.g. an expanded Table 2 contained within a future update of this review) can easily be produced from the diterpene database, even when the database has grown to many times its present size. The alternative insertion of many "R-group"-style structure illustrations into an updated version of Figure 1 would be prohibitively time consuming. Thus, illustrating structures in updates of this review will be handled by this alternative method.

II.2.1. Structure Description Codes

In Table 2, diterpene structures are identified by compound number and grouped with related structures belonging to the same skeletal type. The structural information given for each molecule (Table 2), together with the accompanying carbocyclic (or in some cases heterocyclic) skeleton, are sufficient to derive the compound's constitution and stereochemistry. By superimposing on the skeleton the substituents listed in the molecular description code, a molecule can be constructed with as much structural information as was included in the original report. In the molecular description, each substituent code is separated by semicolons within a string that begins with double bond substituents, proceeds through various oxygen substituents, and ends with hydrogens:

I. Double Bonds
Double bonds are always indicated by a two-center code (e.g., 3(4), 9(11)) that identifies the two carbons linked by the double bond. If double bonds occur within the ring system, the steric relations around the double bond are defined by the skeleton illustrated with the description. Thus, for each cyclic double bond the steric relationship (Z or E) of the ring carbons attached to the doubly bonded carbons is defined by the relationship indicated in the illustrated skeleton.

For linear diterpenes or the linear segments of cyclic compounds with ambiguous double bond stereochemistries, the steric relationship about the double bond is indicated by a Z or E suffix (e.g., 13(14)E, 2(3)Z). The Z or E assignment results from the comparison of the sequence-rule-preference of the four atoms attached to the doubly bonded carbons (as specified in Rule E-2.2. of the IUPAC Rules for Nomenclature of Organic Chemistry). Several examples of Z and E assignment are discussed below in order to familiarize the reader with this method.

II. Oxygen Functionalities
A. Carboxyls: If present, the first-listed oxygen substituent is the carboxyl function (e.g., 19(=O,OH)). This code assigns a carbonyl group (=O) and a hydroxyl group (OH) to a specific center, carbon number 19. In compounds containing a lactone function, this carboxyl code is replaced by a single carbonyl group code (e.g. 19(=O)) and a separate string-entry indicating the position and orientivity of the oxido linkage (e.g., 19(—O—)6β).

B. Ketones and Aldehydes: Ketones and aldehydes are denoted by the carbonyl group code, (=O), associated with a secondary (ketone) or primary (aldehyde) carbon of the skeleton (e.g. 3(=O)).

C. Hydroxyls, Esters, etc.: Hydroxyls are indicated by a code, (OH), that is preceded by the number of the carbon to which it is attached and the orientation (α or β) of the attachment (e.g., 6β(OH)). Esters and other attached groups are specified in the same fashion, using approximately 80 different codes corresponding to the attached organic acid, etc. (e.g. 7α(OAng), 2β(OAc)). The Appendix lists the codes and structures for these attached groups. If attachment is through an oxygen, this is indicated in the code. For example, the codes for an acetate group and an acetyloxy substituent are respectively (Ac) and (OAc).

D. Epoxides, Ether- or Oxido-Linkages, etc.: Epoxides and other oxido-bridges are indicated by a code (e.g. 3β(—O—)4β) specifying the two centers of attachment and, if known, the orientivity of attachment.

III. Hydrogens

Hydrogens are included in order to define the orientivity (R or S) at a chiral center. For example, the entry, 8β(H), in reference to a labdane diterpene defines the orientation of the attached ring-methyl at C-8 as being alpha (α):

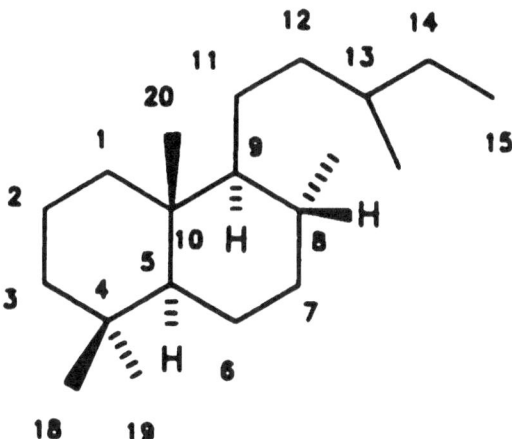

If no substituent is indicated for an available point of attachment at a skeletal carbon, then the presence of hydrogen(s) at that position is implicit.

2.2.2. Examples of Molecular Descriptions

I. Linear diterpene example: 15,19-dihydroxy-14-oxo-geranylnerol Skeleton (No. 1. Geranylnerol):

Molecular description:

1004   2(3)Z;6(7)Z;10(11)E;14(=O);1(OH);15(OH);19(OH)

Because the skeleton immediately preceding the compound number 1004 description lacks double bonds, it is necessary to include in the molecular description the steric relationships about each of the three double bonds. The first double bond occurs at C-2-C-3 and according to sequence-rule-preference, the "Z" suffix stipulates that C-1 and C-4 occur on the same side of the double bond. The "Z" suffix of the C-6-C-7 double bond requires that the two sequence-rule-preferred groups on this double bond, the hydroxymethylene function (C-19) and C-5 (and its associated ligands), occupy the same side of the double bond. The "E" suffix of the 10,11-double bond indicates that, because C-9 and C-12 and their associated ligands are the sequence-rule-preferred groups, they must be on opposite sides of the double bond.

The remainder of the description indicates that a carbonyl group is located at C-14 and hydroxyl groups are located at C-1, C-15, and C-19.

II.   Bicyclic example: 14β-Carterothamnotriol-15-O-acetate
      Skeleton (No. 16. "Normal"-labdane):

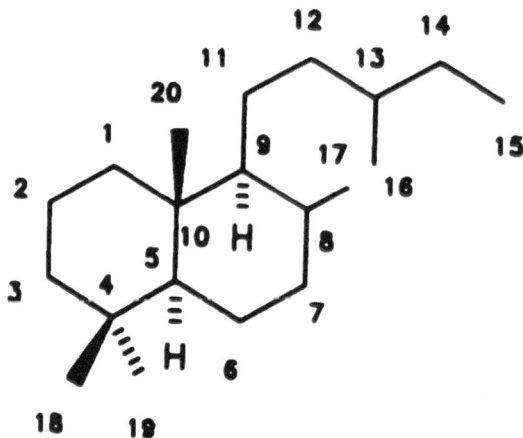

Molecular description:

95   12(13)E;9α(OH);14β(OH);15(OAc);8α(H)

The skeleton immediately above this description of compound number 95 possesses a six-carbon sidechain into which the single double bond must be inserted. Because this sidechain is not part of the cyclic portion of the molecule, the molecular description code, 12(13)E, includes an "E" defining the steric relationship around the double bond. Thus, sequence rule-preferred C-11 and C-14 are situated on opposite sides of the double bond.

The next three entries, 9α(OH), 14β(OH), and 15(OAc), indicate the placement of an α-oriented hydroxyl group at C-9, a β-oriented hydroxyl group at C-14 and an acetate at C-15. The presence of an α-oriented H-14 is implicit in the placement of a β-hydroxyl at C-14.

The code, 8α(H), defines the orientivity about C-8. In this case, it dictates β-orientation for the C-8 methyl group.

II.3. Plant Sources and References for Diterpenes

Table 3 (pp. 150–218) lists all diterpene compound numbers, the species of the Compositae from which they were reported and the reference numbers for the reports. Species names complete with authorities are listed alphabetically in the Species Index.

Table 1. Compound Numbers and Names   7

TABLE 1. Compound numbers and names of diterpenes from the Compositae (Asteraceae).

Compound:

No.   Name

I. Linear or Unicarbocyclic

1. Geranylnerol

1   Geranylnerol

2   Geranylnerol, 18-hydroxy-

3   Geranylnerol, 20-hydroxy-

4   Geranylnerol, 17,20-dihydroxy-

5   Geranylneral

6   Wyethic acid

7   Geranylnerol, 19-hydroxy-13-oxo-

8   Geranylnerol, 12,20-dihydroxy-16-oxo-

14   Viguieric acid [(2Z,6Z,10E)-3,15-dimethyl-7-carboxy-11-formyl-2,6,10,14-hexadecatetraen-1-ol]

15   Geranylnerol, 12,20-dihydroxy-19-acetoxy-

984   Geranylnerol, 19-acetoxy-20-hydroxy-

948   Geranylnerol, 8,12,19-trihydroxy-

957   Geranylnerol, 16,18,19-trihydroxy-

949   Geranylnerol-6,7-epoxide, 8,12,19-trihydroxy-

978   Geranylnerol, 19-hydroxy-12-oxo-

979   Geranylnerol, 6,7-epoxy-19-hydroxy-12-oxo-6,7-dihydro-

1002 Geranylnerol-19-oic acid

1003 Geranylnerol-19-oic acid, 1-acetyl-

1004 Geranylnerol, 15,19-dihydroxy-14-oxo-

1005 Geranylnerol acetate, 19-acetoxy-15-hydroxy-14-oxo-

1020 Geranylnerol, 20-acetoxy-10,11-epoxy-10,11-dihydro-9-hydroxy-

1095 Zoapatanol, pre-

1096 Tomexanthol, pre-

1109 Geranylnerol, 17,18-dihydroxy-

1110 Geranylnerol, 17,18,20-trihydroxy-

1111 Geranylnerol, 12-oxo-

TABLE 1. (*contd.*)

Compound:
No. Name

1112 Geranylnerol, 18,19-diacetoxy-17,20-dihydroxy-

1113 Geranylnerol, 18,19,20-triacetoxy-17-hydroxy-

1114 Geranylnerol, 18-acetoxy-17,20-dihydroxy-

1115 Geranylnerol, 19-acetoxy-17,20-dihydroxy-

1116 Geranylnerol, 9,17-dihydroxy-

1117 Geranylnerol, 9,17,19-trihydroxy-

2. Geranylgeraniol

11   Geranylgeraniol, 12,19-dihydroxy-

12   Geranylgeraniol acetate, 12,19-dihydroxy-

13   Geranylgeraniol, 16-hydroxy-

16   Geranylgeraniol, 14,
     15-dihydro-4,14,15-trihydroxy-

17   Geranylgeranial

18   Geranylgeranial, 4-hydroxy-

19   Koanoadmantic acid

20   Geranylgeraniol,13-oxo-2,3-dihydro-

21   Geranylgeraniol, 19-hydroxy-12-oxo-10,11-dihydro-

22   Geranylgeraniol, 12,18-dihydroxy-6,7Z-

954 Hexa-deca-2E,6E,10E,14-tetraene, 1-hydroxy-11-carboxy-3,
     7,15-trimethyl-

1118 Geranylgeraniol, 17-hydroxy-18-acetoxy-

1119 Geranylgeranal, 12-oxo-

1120 Aspiliaparviol

3. Geranyllinalool

23   Geranyllinalool

24   Gerranyllinalool, 13-hydroxy-

25   Geranyllinalool, 13-acetoxy-

26   Geranyllinalool, 9-hydroxy-

27   Geranyllinalool, 9-acetoxy-

28   Geranyllinalool, 5-hydroxy-

Table 1. Compound Numbers and Names   9

TABLE 1. *(contd.)*

Compound:

No.   Name

29   Geranyllinalool, 9-acetoxy-5-hydroxy-

30   Geranyllinalool, 5,9-diacetoxy-

31   Geranyllinalool, 13-acetoxy-5-hydroxy-

32   Geranyllinalool, 12-oxo-

33   Geranyllinalool, 13-oxo-

34   Geranyllinalool, 13-oxo-14,15-epoxy-

35   Geranyllinalool, 15-hydroxy-13,14-dehydro-14,15-dihydro-

36   Geranylinalool, 14,15-dihydro-13,14,15,16-teradehydro-

37   Geranyllinalool, 14,15-dihydro-14,15-dihydroxy-

38   Geranyllinalool, 5,14,15-trihydroxy-14,15-dihydro-

39   Geranyllinalool, 14,15-dihydroxy-9-acetoxy-14,15-dihydro-

1121 Geranyl linlol, 13-acetoxy-

1122 Geranyl linalol, 12-oxo-10,11-dihydro-

1123 Geranyl linalol, 12-oxo-15-hydroxy-13,14E-dehydro-
      10,11,14,15-tetrahydro-

4. Phytol-Derived

59   Neophytadiene

80. 14-Methylgeranylnerol

9   Geranylnerol, 12,19,20-trihydroxy-14-methylene-

10   Geranylnerol, 12,20-dihydroxy-19-acetoxy-14-methylene-

980   Smallantha-2Z,6Z,10E,13E-tetraene, 1,19-dihydroxy-

981   Smallantha-2Z,10E,13E-triene,
      1,19-dihydroxy-6,7-epoxy-12-oxo-

982   Smallantha-2Z,6Z,10E,14(21)-tetraene, 1,
      19-dihydroxy-12-oxo-

983   Smallantha-2Z,10E,14(21)-triene, 1,
      19-dihydroxy-6,7-epoxy-12-oxo-

1094 Tomentol, pre-

5. Acanthoaustralane

TABLE 1. *(contd.)*

Compound:
No.   Name

40   Acanthoaustralide

41   Acanthoaustralide-1-O-acetate

42   Acanthoaustralide, 17-hydroxy-

43   Acanthoaustralide, 17-acetoxy-

1124 Acanthoaustralide, 1,6-diacetyl-

6. Acanthoaustralane-6,11-Epoxide

44   Melfusanolide, 1,10-dihydroxy-17-acetoxy-

45   Melfusanolide, 1,10,17-trihydroxy-

7. Isoacanthoaustralane

46   Isoacanthoaustralide, 17-hydroxy-

47   Isoacanthoaustralide, 17-acetoxy-

48   Isoacanthoaustralide-1-O-acetate

      [According to illustrated structures, **46** and **47** were
      assigned an orientation at 2(3) opposite to the one assigned
      earlier to **48**. No revision of **48** was indicated in the later
      publication.]

8. Geranylgeraniol-18,8-Lactone

49   Ichthyouleolide

50   Ichthyouleolide, 19-acetoxy-

51   Ichthyouleolide, 15-peroxy-13,14t-dehydro-14-15-dihydro-

52   Ichthyouleolide, 15-hydroxy-13,14t-dehydro-14-15-dihydro-

53   Ichthyouleolide, 14-peroxy-15,17-dehydro-14,15-dihydro-

54   Ichthyouleolide, 14-hydroxy-15,17-dehydro-14,15-dihydro-

9. Oxepane

55   Zoapatanol

56   Tomexanthin

57   Montanol

1097 Tomentol

Table 1. Compound Numbers and Names   11

TABLE 1. *(contd.)*

Compound:
No.   Name

1098 Tomexanthol

1099 Tomentanol

86. Melcanthane

1006 Melcantholide, 1,6,7-trihydroxy-17-acetoxy-

10. Mikanofurane

58   Mikanifuran

11. Centipedane

60   Centipedic acid

61   Conypododiol

1125 Furosolidagonone

1126 Furosolidagonol

1127 Furosolidagonone, 13-hydroxy-

1128 Furosolidagonone, 15-peroxy-13,14E-dehydro-14,15-dihydro-

12. Geranylgeraniol-1,20-Lactone

62   Dimeroperatic acid

63   Dimeroaperatate, methyl-

64   Dimerobrasiolide

1129 Furosolidagonone, 1-oxo-2,3-dehydro-1,2,3,20-tetrahydro-

1130 Gutiesolbriolide, 17-hydroxy-

1131 Gutiesolbriolide, 10E-17-hydroxy-

1132 Gutiesolbriolide, 17-acetoxy-

1133 Geranylgeraniol-20-acid lactone, 17,19-dihydroxy-18-acetoxy-
      6,7,10,11-tetrahydro-

1134 Geranylgeraniol-20-acid lactone, 17-hydroxy-18,19-diacetoxy-
      6,7,10,11-tetrahydro-

1135 Gutiesolbriolide, 17-hydroxy-iso-

1136 Gutiesolbriolide, 17-acetoxy-iso-

TABLE 1. (*contd.*)

Compound:

No.   Name

1137 Furosolidagonone, 20-oxo-1,2-epoxy-1,2,3,20-tetrahydro-

13. Geranylterpinene

65   Helicallen-16-ol

66   Helicallen-16-al

67   Helicallen-16-oic acid, 14,15-dihydro-

68   Geranyl-α-terpinene, 9-

69   Geranyl-α-terpinene, 10,11H-10,11-dihydroxy-9-

70   Geranylcurcumen,9-

71   Geranylcurcumen, 10,11H-10,11-dihydroxy-9-

      [Compounds **69** and **70** were named "11,12H-11,12-dihydroxy-..."
      derivatives, but this numbering was incompatible with
      illustrated structures and previously published homologous
      compounds.]

72   Geranylascaridol, β-10,11H-10,11-dihydroxy-9-

1101 Geranylascaridol, α-10,11H-10,11-dihydroxy-9-

1138 Geranyl-α-terpineol, 9-

14. Isocembrene

73   Isocembrene, 15-hydroxy-13,18H-

15. Geranylnerol-11,14-α-Epoxide

74   Ligantrol

75   Ligantrol-monoacetate

II. Normal-Bicarbocyclics and Derivatives

16. Normal-Labdane

76   Labda-8(17)-13-dien-15-ol

1103 Labd-13-ene, 8α,15-dihydroxy-

77   Acanthospermol-β-galactosidopyranoside

78   Labda-8(17),13-diene

Table 1. Compound Numbers and Names 13

TABLE 1. *(contd.)*

Compound:
No. Name

80   Gymnospermin

79   Labdan-8α,15-diol

81   Labdan-8α,15-diol, 13-epi-

     [In published structure, C-5-H was shown as β-oriented.]

82   Carterothaminotriol

83   Carterothaminotriol-15-O-acetate

84   Carterothaminotriol-14,15-O-diacetate

85   Labd-8(17)-ene, 15-hydroxy-13,14-epoxy-

     [In the published name, this compound was called "...14,15-epoxy-...".]

86   Labd-7-ene, 13R,14R,15-trihydroxy-

87   Labd-(8,9)-ene, 13R,14R,15-trihydroxy-

88   Labd-8(17)-ene, 14,15-dihydroxy-

89   Labd-8(17)-ene, 13,14,15-trihydroxy-

90   Labd-8(17)-ene, 15-acetoxy-14-hydroxy-

993 Labd-7-ene, 3α,15-dihydroxy-

994 Labd-7-ene, 15-[2-methylbutyryloyloxy]-3α-hydroxy-

995 Labd-8(17)-ene, 3α,7α,15-trihydroxy-

91   Labd-8(17)-ene, 15-acetoxy-13,14-dihydroxy-

92   Labd-8(17)-ene, 14-acetoxy-15-hydroxy-

93   Labd-8(17)-ene, 15-acetoxy-13-hydroxy-14-propionyloxy-

94   Carterothamnotriol, 14β-

97   Brickellidiffusic acid angelate

98   Cativic acid, 2α,3α-[angeloyloxy-&
     2-hydroxy-2-methylbutyryloxy]-7α-trans-cinnamoyloxy-

99   Cativic acid, 2α,3α-[angeloyloxy- &
     2-hydroxy-2-methylbutyryloxy]-7α-cis cinnamoyloxy-

100 Cativic acid, 2-angeloyloxy-13,14Z-dehydro-

116 Labda-7,13Z-15-oic acid, 3α-hydroxy-

101 Cativic acid, 18-angeloyloxy-13,14-dehydro-

102 Cativic acid, 18-tigloyloxy-13,14-dehydro-

TABLE 1. *(contd.)*

Compound:
No.   Name

103   Cativic acid, 3α-angeloyloxy-2α-hydroxy-13,14Z-dehydro-

104   Cativic acid,3α-trans-cinnamoyloxy-2α-hydroxy-13,
      14Z-dehydro-

105   Cativic acid, 3α-cis-cinnamoyloxy-2α-hydroxy-13,
      14Z-dehydro-

106   Cativic acid, 3α-angeloyloxy-2α-hydroxy-(13Z)-13,
      14-didehydro-

      [Compound 106 is identical to 103.]

107   Cativic acid, 2α-angeloyloxy-3α-hydroxy-(13Z)-13,
      14-didehydro-

108   Cativic acid, 3α-angeloyloxy-2α-hydroxy-(13Z)-13,
      14-didehydro-8(17)-

109   Cativic acid, 2α-angeloyloxy-3α-hydroxy-(13Z)-13,
      14-didehydro-8(17)-

110   Batudioic acid

111   Labda-8(17),13E-dien-15-oic acid

112   Labda-7,13E-dien-15-oic acid, 2α-tigloyloxy-

113   Sempervirenic acid [3β-acetoxy-labda-7,13-dien-15-oic acid]

114   Labda-[7.13E-dien]-15-oic acid, 3α-hydroxy-

115   Labda-[7.13E-dien]-15-oic acid, 3β-hydroxy-

958   Labdan-15-oic acid, 8α-hydroxy-

117   Labda-7,13E-dien-15-oic acid, 3-oxo-

118   Cativic acid

991   Cativic acid, 3α-hydroxy-

992   Cativic acid methyl ester, 3α-hydroxy-

119   Labda-7-en-15-oic acid, 3α-hydroxy-

120   Labda-7-en-15-oic acid, 3β-hydroxy-

360   Labdane, (-)-7β,8α,15-trihydroxy-   [Gymnospermin]

121   Labdanol acid

871   Labd-7-en-2α,15-diol, 13 -

872   Labd-7-ene, 2α-hydroxy-15-[3,4-dihydroxycinnamoyloxy]-13 -

873   Labd-7-ene,2α-hydroxy-15-
      [4-hydroxy-3-methoxy-cinnamoyloxy]-13 -

Table 1. Compound Numbers and Names   15

TABLE 1. *(contd.)*

Compound:

No. Name

874 Labd-7-ene, 2α-hydroxy-15-[4-hydroxy-cinnamoyloxy]-13 -

122 Cativic acid, 18-angeloyloxy-

123 Cativic acid, 18-tigloyloxy-

[Published names for **122** and **123** indicated that they were C-19-substituted, but the illustrated structures and reference to the substituents' axial orientation are compatible only with a C-18-substituted structure.]

124 Cativic acid, 2α,3α-dihydroxy-

125 Cativic acid, 2α-hydroxy-3α-[2-hydroxy-2-methylbutyryloxy]-

126 Cativic acid, 3α-angeloyloxy-2α-hydroxy- [Brickellia acid C]

127 Cativic acid, 3α-trans-cinnamoyloxy-2α-hydroxy-

128 Cativic acid, 3α-cis-cinnamoyloxy-2α-hydroxy-

129 Cativic acid, 3-oxo-

870 Labd-7-en-15-oic acid, 3β-hydroxy-2α-senecioyloxy-13 -

130 Agathenic acid, 8,17H-7,8-dehydro-

977 Labd-8(17)-en-15-al-19-oic acid

131 Agathenic acid, 8,17,13,14H-7,8-dehydro-

132 Biformene, 12,13E-

133 Biformene, 3α-hydroxy-12,13E-

134 Biformene, 3β-hydroxy-12,13E-

135 Biformene, 3β-acetoxy-12,13E-

136 Biformene, 3-oxo-12,13E-

959 Labd-8(17)-en-15-oic acid, 2β-acetoxy-

137 Communic acid

1104 Labd-8(17),12E,14-trien-19-oic acid

138 Communic acid, 7α-acetoxy-trans-

139 Abienol

140 Abienol, 7β-hydroxy-

141 Nidorellol

142 Nidorellol, 6α-hydroxy-

143 Nidorellol, 6α-angeloyloxy-

TABLE 1. (*contd.*)

Compound:

No.   Name

144   Abienol, 6β-hydroxy-12E-

145   Abienol, 7β-acetoxy-12E-

146   Abienol, 6β,7β-dihydroxy-12E-

147   Abienol, 6β,18-dihydroxy-12E-

148   Abienol, 7β-acetoxy-6β-hydroxy-12E-

149   Aristin-19-O-arachinate, 1β-acetoxy-

150   Aristin-19-O-coumarate, 1β-acetoxy-

151   Aristin-19-O-cis-coumarate, 1β-acetoxy-

152   Labda-12(E),14-diene, 9α,19-dihydroxy-

153   Sclareol, 6α-angeloyloxy-

      [Published structure contained a C-6β-oriented angelate, contradicting the published name.]

1102 Sclareol

996   (+)-Manool

1105 Manool, iso-

1106 Manool, 3β-hydroxy-iso-

1107 Manool, 3-oxo-iso-

154   Salicifoliol

      [Absolute stereochemistry not indicated.]

155   Abienol, iso-

156   Abienol, 7β-acetoxy-6β-hydroxy-12β-peroxiiso-

157   Abienol, 7β-acetoxy-6β-hydroxy-12α-peroxiiso-

158   Solidagenone

159   Lambertianic acid, 17-hydroxyiso-

160   Gutierrezianol acid-[2'-methylbutyrate]

161   Gutierrezianol acid-isobutyrate

162   Gutierrezianol acid-isovalerate

163   Lambertianic acid, 7α-hydroxy-

1021 Austroinulin-7-O-acetate, (12Z)-6-desoxy-

1022 Abienol, 6α-hydroxy-7β-acetoxy-12E-

Table 1. Compound Numbers and Names   17

TABLE 1. (*contd.*)

Compound:
No.   Name

1023 Austroinulin-7-O-acetate, (12Z)-

1024 Labdenol acid, (13Z)-

1139 Labdan-8α,15-diol, 13β-H-

1140 Cordobic acid [7α,18-Dihydroxylabd-8(17)-en-oic acid]

1141 Cordobic acid 18-acetate

1142 Cordobic acid, 7-epi-

1143 Salvic acid

1144 Salvic acid, 7α-acetyl-

1145 Discoidic acid [4β-hydroxymethyllabd-7-en-15-oic acid]

1147 Labd-8(17),12E,14-trien-19-oic acid

96   Gutierrezia sphaerocephala labdane-α-L-arabinopyranoside 1

97   Gutierrezia sphaerocephala labdane-α-L-arabinopyranoside 2

1148 Gutierrezia sphaerocephala labdane-β-D-arabinopyranoside 3

1149 Gutierrezia sphaerocephala labdane-β-D-arabinopyranoside 5

1150 Gutierrezia sphaerocephala labdane-β-D-arabinopyranoside 6

1151 Abienol, 3β-acetoxy-iso-

1152 Gnaphala-13(16),14-diene, 8α-hydroxy-

17. Grindelane

165 Grindelic acid, 7α-hydroxy-7,8-dihydro-8(17)-dehydro-

166 Grinde-6,8(17)-dienic acid

167 Grindelic acid

168 Grindelic acid, 1α-hydroxy-

169 Grindelic acid, 3β-hydroxy-

170 Grindelic acid, 6α-hydroxy-

172 Grindelic acid, 6β-hydroxy-

173 Grindelic acid, 17-hydroxy- [Oxygrindelic acid]

171 Grindelic acid, 6α-formyloxy-

174 Grindelic acid, 17-acetoxy-

175 Grindelic acid, 17-propionyloxy-

TABLE 1. *(contd.)*

Compound:	
No.	Name

176 Grindelic acid, 17-isobutyryloxy-

177 Grindelic acid, 17-isovaleryloxy-

178 Grindelic acid, 17-[2-methylbutyryloxy]-

179 Grindelic acid, 17-methoxy-

180 Grindelic acid, 19-hydroxy-

181 Grindelic acid, 19-acetoxy-

182 Grindelic acid, 19-isovaleryloxy-

183 Grindelic acid, 19-isobutyryloxy-

184 Grindelic acid, 19-[2-methylbutyryloxy]-

185 Grindelic acid, 19-succinyloxy-

186 Grindelic acid, 19-oxo-

187 Grindelic acid, 19-oic-

188 Grindelic acid, 18-hydroxy-

189 Grindelic acid, 6-oxo-

190 Grindelic acid, 19-acetoxy-6-oxo-

191 Grindelic acid, 7β,8β-epoxy-7,8-dihydro-

192 Grindelic acid, 7α,8α-epoxy-7,8-dihydro-

955 Grindelic acid, iso-

956 Grindelic acid, 17-grindeloxy-

193 Solidagenone, 9-dehydroxy-9α,13α-epoxy-13,16-dihydro-

194 Solidagenone, 13-epi-dehydroxy-9α,13α-epoxy-13,16-dihydro-

1035 Grindelic acid, 3α-hydroxy-

1036 Grindelic acid, 7β-hydroxy-8(17)-dehydro-7,8-dihydro-

1060 Grindelic acid-15-O-arabinoside

1153 Grindelate, methyl-

1187 Grindelic acid 15-O-arabinoside

     [Incorrectly illustrated as an <u>ent</u>-labdane]

18. Normal-Labdane-8,13-Epoxide

272 Manoyl oxide

Table 1. Compound Numbers and Names   19

TABLE 1. (*contd.*)

Compound:

No. Name

195 Jhanol

196 Jhanol acetate

197 Jhanidiol

198 Jhanidiol-19-monoacetate

199 Jhanidiol-diacetate

200 Manoyl oxide, 3β-hydroxy- [18-deoxylazarcardenasol]

201 Manoyl oxide, 18-benzoyloxy-3β-hydroxy-
     [18-benzoyloxylarcardenasol]

964 Manoyl oxide, 15-oxo-14,15-dihydro-

273 Manoyl oxide, 13-epi-

202 Manoyl oxide, 18-hydroxy-13-epi-

     [Published structure was shown with a C-18 substituent but
     given a name indicating C-19 substitution. Reference to the
     substituent's axial orientation indicates a C-18
     substituent.]

19. Normal-Labdane-8,12-Epoxide

203 Carterochaetol, 16-hydroxy-

204 Carterochaetol, 16-acetoxy-

205 Carterochaetol acetate, 16-acetoxy-

206 Carterochaetol, 16-oxo-

207 Carterochaetal, 16-oxo-

208 Carterochaetol, 16-hydroxy-13,14H-13α,14α-epoxy-

209 Carterochaetol, 16-hydroxy-13,14H-13β,14β-epoxy-

210 Carterochaetol

211 Carterochaeta acid

212 Carterochaetal, 16-oxo-13,14H-12,13-dehydro-

213 Carterochaetal, 16-acetoxy-14-oxo-13,14H-12,13-dehydro-

20. Normal-Labdane-15,16-Lactone

164 Agathenic acid-16-lactone, 8,17H-7,8-dehydro-

965 Labda-7,13-dien-15-acid-16-lactone, 17-hydroxy-

TABLE 1. (*contd.*)

Compound:

No.   Name

966   Labda-7,13-dien-15-acid-16-lactone, 17-oxo-

21. Normal-Labdane-20,12-Lactone

214   Nidorella lactone

22. Normal-Labdane Spiroketal

215   Brickellidiffusic acid spiro ketallactone,
      2α-angeloyloxy-17,O-dihydro-

216   Brickellidiffusic acid spiro ketal lactone,
      2α-angeloyloxy-17α-hydroxy-17,O-dihydro-

217   Brickellidiffusic acid spiro ketal lactone,
      2α-angeloyloxy-17β-hydroxy-17,O-dihydro-

218   Brickellidiffusic acid spiro ketal lactone, 2α-hydroxy-17,
      O-dihydro-

88. Normal-Labdane-15,12-Lactone

938   Evillosin

23. Normal-Labdane, 7,8-Seco-

219   Koanolabda-12E,14-diene, seco-

24. Normal-Labdane, 3,4-Seco-

220   Cativin-3,15-dioic acid, 3,4-seco-

25. Grindelane, 8,9-Seco-

223   Chrysothane   [Strictanonic acid]

89. Grindelane, 19-Nor-

221   Grindelic acid, 4α-hydroxy-19-nor-

222   Grindelic acid, 4α-formyloxy-19-nor-

1154 Grindelate, methyl-4β-hydroxy-6-oxo-19-nor-

Table 1. Compound Numbers and Names 21

TABLE 1. *(contd.)*

Compound:
No.   Name

26. Grindelane, 8,17-Bis-Nor-8,9-Seco-

224   Grindelistrictic acid

27. Normal-Labdane-20,12-Lactone, 5,10-Seco-

225   Nidorella lactone, seco-

226   Nidorella lactone-6-O-angelate, seco-

227   Nidorella lactone-6-O-isobutyrate, seco-

228   Nidorella lactone-6-O-[2-methylbutyrate], seco-

229   Nidorella lactone-6-O-isovalerate, seco-

230   Nidorella lactone-6-O-angelate, 5-deoxy-5-hydroxy-5H-seco-

100. Normal-Labdane, 14,15-Bis-Nor-

1155 Labden-13-one, 14,15-bisnor-8-hydroxy-11E-

1157 Sterebin A

1158 Sterebin B

1159 Sterebin C

1160 Sterebin D

101. Normal-Labdane, 13,14,15,16-Tetra-Nor-

1156 Norambreinolide

87. Hexanorlabdane

941   Jhanilactone

28. Relhaniane

231   Relhania acid

232   Koanophyllic acid B

233   Koanophyllic acid A

234   Koanophyllic acid C

TABLE 1. (*contd.*)

Compound:
No.   Name

235   Koanophyllic acid D

[Absolute stereochemistry for the koanophyllic acid series
was unproven.]

29. trans-Normal-Clerodane

(No proven structures of this type)

30. Clerodane, 5,10-Seco-

255   Nidoreseda acid, seco-   [Conyzic acid, strictic acid]
[Revised to 9α-methyl, = 573]

256   Nidoreseda acid methyl ester, seco-

[Revised to 9α-methyl, = 573 methyl ether]

31. cis-Normal-Clerodane

238   Solidago alcohol, 8-epi-

[Reported as 5β,10α-trans-clerodane]

253   Nidorellalactone, iso-

[Reported as 5β,10α-trans-clerodane]

254   Nidorellalactone

[Reported as 5β,10α-trans-clerodane]

257   Clerodan-15-oic acid, 3,4-dehydro-cis-

258   Solidagolactone II   [Elongatolide C]

259   Solidagolactone III

260   Elongatolide E   [Solidagolactone VII]

261   Solidagolactone III, 3α,4α-epoxy-3,4-dihydro-

262   Solidagolactone V

263   Solidagolactone II, 2α-hydroxy-   [Solidago virgaurea
compound 4a]

264   Solidagolactone III, 2α-hydroxy-   [Solidago virgaurea
compound 4c]

265   Solidagolactone V, 2β-hydroxy-

266   Solidagolactone II, 3β,4α-dihydroxy-3,4-dihydro-

267   Solidagolactone III, 3β,4α-dihydroxy-3,4-dihydro-

Table 1. Compound Numbers and Names   23

TABLE 1. (*contd.*)

Compound:

No. Name

268 Solidagolactone II, 3-oxo-3,4-dihydro-

269 Solidagolactone III, 3-oxo-3,4-dihydro-

545 Elongatolide A [SolidagolactoneIV;
     6β-hydroxy-solidagolactone]

546 Elongatolide B

551 Kolavenic-15-acid lactone, 2α,16-dihydroxy-3α,4α-epoxy-
     [Revised from trans-ent-clerodane]

552 Kolavenic-15-acid lactone, 6β-angeloyloxy-16-hydroxy-3α,
     4α-epoxy-
     [Revised from trans-ent-clerodane]

553 Elongatolide D [Solidagolactone VI(structure no.555)]

988 Solidagolactone VIII

239 Solidagoic acid A

240 Solidago aldehyde

241 Solidago alcohol

242 Solidago glycol

243 Solidago dialdehyde

244 Solidagoic acid B

246 Solidago lactol

247 Solidago epoxylactol

270 Guteriolide

556 Kolavenic acid lactone, 16-hydroxy-3,4-epoxy-
     [Structure was revised to 5α,10α-cis]

557 Kolavenic acid lactone, 7β-angeloyloxy-16-hydroxy-3,
     4-epoxy-

     [Revised to 5α,10α-cis]

1025 Kolavenool, 6β-acetoxy-2-oxo-

1062 Solidago gigantea compound I

1063 Solidago gigantea compound III

1064 Solidago gigantea compound IV

1065 Solidago gigantea compound V

1066 Solidago gigantea compound II

1067 Solidago gigantea compound VI

TABLE 1. *(contd.)*

Compound:
No.   Name

1068 Solidago gigantea compound VII

271 Chrysolic acid

1083 Cleroda-3,13(14)-diene-15,16-olide, 6α,18-dihydroxy-cis-

1084 Cleroda-3,13(14)-diene-15,16-olide, 18,19-dihydroxy-cis-

1085 Cleroda-3,13(14)-diene-15,16:18,19-diolide, cis-

1086 Cleroda-3,13(14)-diene-15,16-olide,
     18,19-epoxy-19α-hydroxy-cis-

1087 Cleroda-13(14)-ene-15,16-olide, 3α,4:18,19-diepoxy-18β,
     19α-dihydroxy-cis-

1088 Cleroda-13(14)-ene-15,16:18,19-diolide,
     3α,4-epoxy-19α-hydroxy-cis-

1089 Cleroda-13(14)-ene-15,16-olide,
     3α,4:18,19-diepoxy-19α-hydroxy-cis-

1090 Cleroda-13(14)-ene-15,16-olide, 3α,4β,19α-trihydroxy-18,
     19-epoxy-cis-

1091 Cleroda-3,13(14)-diene-15,16-olide-19-oic ester,
     19-O-α-L-arabinopyranosyl-cis-

1092 Cleroda-3,13(14)-diene-15,16:18,6α-diolide,
     2β,6α-dihydroxy-cis-

90. Tricycloclerodane

947 Tricyclosolidagolactone

III. Normal-Tricarbocyclics and Derivatives

32. Sandaracopimarane (Normal-Pimarane)

274 Sandaracopimaradiene

927 Sandaracopimar-8(14),15-diene, 7α-hydroxy-

928 Sandaracopimar-8(14),15-diene, 1β,7α-dihydroxy-

929 Sandaracopimar-8(14),15-diene, 7α-hydroxy-1β-acetoxy-

930 Sandaracopimar-8(14),15-diene, 1β-hydroxy-7α-acetoxy-

931 Sandaracopimar-8(14),15-diene, 1β,11α-dihydroxy-

932 Sandaracopimar-8(14),15-diene, 1β,11α-diacetoxy-

Table 1. Compound Numbers and Names 25

TABLE 1. *(contd.)*

Compound:
No. Name

933 Sandaracopimar-8(14),15-diene, 1β,
     11α-dihydroxy-7α-acetoxy-

934 Sandaracopimar-8(14),15-diene, 7α-hydroxy-1β,
     11α-diacetoxy-

935 Sandaracopimar-8(14),15-diene, 1β,7α,11α-triacetoxy-

936 Sandaracopimar-8(14),15-diene, 1β,11α-diacetoxy-6,7-epoxy-

937 Sandaracopimar-8(14),15-diene, 1β,11α-diacetoxy-7-oxo-

275 Sandaracopimaradiene, iso-

276 Sandaracopimar-15-ene, 8β-hydroxy-

277 Sandaracopimar-15-en-6β,8β-diol

278 Sandaracopimar-15-en-8β-ol, 6β-acetoxy-

279 Sandaracopimar-15-en-8β,11α-diol

280 Sandaracopimar-15-ene, 8β-hydroxy-11α-senecioyloxy-

281 Sandaracopimar-15-ene, 8β-hydroxy-11α-tiglinoyloxy-

282 Sandaracopimar-15-ene, 8β-hydroxy-11α-acetoxy-

283 Sandaracopimar-15-en-11-one, 8β-hydroxy-

284 Sandaracopimar-15-ene, 8β-hydroxy-12β-acetoxy-

285 Sandaracopimar-15-ene, 3β,18-diol

286 Sandaracopimar-15-ene, 8β,20-dihydroxy-

287 Sandaracopimar-15-ene, 20-hydroxy-8β,20-oxido-

288 Sandaracopimar-15-ene, 8β,11α,12β-trihydroxy-

289 Sandaracopimar-15-en-8β,12β-diol, 11α-acetoxy-

290 Sandaracopimar-15-ene, 8β,12β-dihydroxy-11α-senecioyloxy-

291 Sandaracopimar-15-ene, 8β,12β-dihydroxy-11α-tiglinoyloxy-

292 Sandaracopimar-15-en-8β,11α-diol, 12β-acetoxy-

293 Sandaracopimar-15-en-8β,11α-diol,
     12β-(p-hydroxycinnamoyloxy)-

294 Sandaracopimar-15-en-8β-ol, 11α,12β-diacetoxy-

295 Sandaracopimar-15-en-8β,
     12α-diol, 11β-acetoxy-

296 Sandaracopimar-15-en-11-one. 8β,
     12α-dihydroxy-

TABLE 1. (*contd.*)

Compound:

No. Name

297 Sandaracopimar-15-en-11-one, 12β-acetoxy-8β-hydroxy-

298 Sandaracopimar-15-en-6β,8β,11α-triol

299 Sandaracopimar-15-en-8β,11α-diol, 6β-acetoxy-

300 Sandaracopimar-8(14),15-dien-18-oic acid

301 Sandaracopimar-7,15-dien-18-oic acid

302 Sandaracopimar-8(14),15-dien-18-oic acid, 3β-acetoxy-

807 Trinervinol [Isopimara-8(14)-en-3β,15,16,17-tetrol]

303 Sandaracopimar-7,15-dien-18-oic acid, 3β-acetoxy-

921 Sandaracopimar-8(14)-en-19-oic acid, 6β,15,16-trihydroxy-

304 Palarosan, 3α-hydroxy-13-epi-

305 Palarosan, 9α-hydroxy-13-epi-

891 Pimar-15-en-8β-ol, 13-epi-

306 Sandaracopimar-7-ene, 3α,15,16-trihydroxy-13-epi-
     [Revised to rosane structure, = 313; 168]

307 Sandaracopimar-7-ene, 9α,15,16-trihydroxy-
     [Revised to rosane structure, = 314; 168]

308 Julslimtetrol

922 Sandaracopimar-8(14)-en-6β-D-glucoside, 15,16-dihydroxy-

923 Sandaracopimar-8(14)-ene, 6α,15,16,18-tetrahydroxy-

     [Compounds 922 and 923 were illustrated with β-oriented C-10
     methyl groups.]

34. Cassane

310 Osteomuricone, 2α-hydroxy-

311 Osteomuricone, 1β,2α-dihydroxy-

312 Osteomuricone, 1β,2α-dihydroxy-15,16-dihydro-

35. Rosane

306 Palarosane, 3α-hydroxy

313 Palarosane, 3β-hydroxy-

314 Palarosane, 3β,18-dihydroxy-   [Jesromotetrol]

Table 1. Compound Numbers and Names   27

TABLE 1. (*contd.*)

Compound:

No.   Name

315   Palarosane, 3-oxo-

881   Rimuen-5β-ol, ent-

882   Rimuen-5β,18-diol, ent-

883   Rimuen-18-yl acetate, 5β-hydroxy-ent-

884   Rimuen-3α,5β-diol, ent-

885   Rimuen-18-yl tiglate, 5β-hydroxy-ent-

886   Rimuen-3α,5β,18-triol, ent-

887   Rosa-5,15-dien-3α,18-diol

888   Rosa-5,15-dien-3α-yl acetate, 18-hydroxy-

889   Rosa-5,15-dien-18yl acetate, 3α-hydroxy-

890   Rosa-1(10),15-dien-18-oic acid

      [Published structure indicated a C-19 acid, while the
      published name and other data indicated a C-18 substituent.]

1161 Jesromotrol, 3β,19-diacetoxy-

1162 Jesromotrol, 3β-acetoxy-

36. Sandaracopimarane, 19-Nor-13-Epi-

309. Julslimdiolone, nor-

37. Normal-Abietane

316   Abietadien-2α-ol acetate

317   Abietadien-3β-ol, 7,13-

318   Abietadien-3-one, 7,13-

      [Compounds **316-317** are probably artefacts of 8(9),13(14)-
      abietadiene-type compounds.]

319   Ferruginol

320   Inuroyleanol

321   Royleanone

322   Royleanone, dehydro-

323   Royleanone, 7-acetoxy-

324   Royleanone, 7-keto-

TABLE 1. *(contd.)*

Compound:

No.   Name

1072 Junceanol W

1073 Junceanol X

1074 Junceanol Y

1163 Turbinatone

94. Stevisalane

1026 Stevisalicinone

IV. Normal-Tetracarbocyclics

73. Normal-Stachane

325   Cupressen-18-oic acid
      [= 774?]

V. ent-Bicarbocyclics and Derivatives

38. ent-Labdane

326   ent-Labda-8(17),13-dien-15-ol

848   ent-Labda-8(17),13-dien-15-ol, 2α-hydroxy-

327   ent-Labda-8(17),13-dien-15-O-bernsteinoate

849   ent-Labda-8(17),13-dien-15-ol, 2-oxo-

328   ent-Labda-8(17),13,14E-dien-15-ol, 7β-acetoxy-

844   ent-Labda-8(17),13E-diene, 3α-angeloyloxy-15,18-dihydroxy-

329   ent-Labda-8(17),13E-dien-15-ol-18-oic acid
      [Viscidic acid A]

330   ent-Labda-8(17),13E-dien-15-acetoxy-18-oic acid
      [Viscidic acid B]

331   ent-Labda-8(17),13E-dien-15-al, 2α-hydroxy-

332   ent-Labda-8(17),13Z-dien-15-al, 2α-hydroxy-

333   ent-Labda-8(17),13E-dien-15-al, 2-oxo-

334   ent-Labda-8(17),13Z-dien-15-al, 2-oxo-

335   ent-Labda-8(17),13E-dien-15-oic acid   [E-Copalic acid]

336   ent-Labda-8(17),13Z-dien-15-oic acid   [Z-Copalic acid]

337   ent-Labda-8(17),13Z-dien-15-oic acid, 3α-hydroxy-

Table 1. Compound Numbers and Names   29

TABLE 1. (*contd.*)

Compound:
No.   Name

338   ent-Labda-8(17),13Z-dien-15-oic acid, 3α,7α-dihydroxy-

339   ent-Labda-13Z-en-15-oic acid, 3α-hydroxy-17-oxo-8β-H-

340   ent-Labda-13Z-en-15-oic acid, 3α-hydroxy-17-oxo-8α-H-

341   ent-Labda-13-en-15-oic acid, 3,17-dioxo-8β-H-

342   ent-Labda-13Z-en-15-oic acid, 3,17-dioxo-8α-H-

343   ent-Labda-13Z-en-15-oic acid, 3α,8α-dihydroxy-17-oxo-8β-H-

      [Published name, "3α,9β-dihydroxy-17-oxo-8β-H-ent-labda-13Z-
      en-oic acid", was inconsistent with the structure and
      published data.]

344   ent-Labda-13Z-en-15,17-dioic acid, 8β-H-

345   ent-Labda-13E-en-15,17-dioic acid, 3β-H-

346   ent-Labda-13Z-en-15-oic acid, 3-oxo-8,17-epoxy-

347   ent-Labda-8(17),13-dien-15-O-bernsteinoate, 8,17-dihydro-8β,
      17-epoxy-

348   ent-Labda-13(14)E-en-15-ol, 8-hydroxy-

349   ent-Labda-13(14)E-en-15-al, 8-hydroxy-

350   ent-Labda-13(14)Z-en-15-al, 8-hydroxy-

351   ent-Labda-7,8,13,14(Z)-diene, 6β,15-dihydroxy-

352   ent-Labda-7,8,13,14(Z)-dien-6-one, 15-hydroxy-

353   ent-Labda-7,13-diene, 3α,15-diacetoxy-11-hydroxy-

354   ent-Labda-7,13E-dien-15-oic acid

355   ent-Labda-7,13-dien-15-oic acid, 2β-angeloyloxy-
      [Dendroidinic acid]

356   ent-Labda-6,8(17)-dien-15-oic acid

357   ent-Labda-8(17)-ene, 2α,15-dihydroxy-

358   Eperuic acid, 2α-iso-valeroyloxy-

      [C-13 orientivity conflicts with earlier assigned
      stereochemistry for this center in eperuic acid.]

857   ent-Labda-8(17)-en-15-oic acid, 2α-hydroxy-

858   ent-Labda-8(17)-en-15-oic acid, 2α-angeloyloxy-

859   ent-Labda-8(17)-en-15-oic acid, 2α-tigloyloxy-

860   ent-Labda-8(17)-en-15-oic acid, 2α-[2-methylbutyryloxy]-

TABLE 1. (*contd.*)

Compound:

No.   Name

861   ent-Labda-8(17)-en-15-oic acid, 2-oxo-

862   ent-Labdan-15-oic acid, 2α-angeloyloxy-8β,17-epoxy-

863   ent-Labdan-15-oic acid, 2α-[2-methylbutyryloxy]-8β,17-epoxy-

359   ent-Labda-8(17)-en-15-ol, 2-oxo-

360   Haplopappus pectinatus ent-labdane 16

361   ent-Labda-7-en-15-oic acid, 2β-angeloyloxy-
       [13,14-Dihydrodendroidinic acid]

362   ent-Labda-7(8),13E-diene, 15,16,17-trihydroxy-

363   ent-Labda-6,13-diene, 15,16-diacetoxy-

364   ent-Labda-6,13-diene, 15,16,17-triacetoxy-

365   ent-Labda-6,8(17),13-trien-15-oic acid, 16-hydroxy-

366   ent-Labda-7,13E-dien-15-oic acid, 16-hydroxy

367   ent-Labda-7,13E-dien-15-oic acid, 16,17-dihydroxy-

368   ent-Labda-7,13Z-dien-15-oic acid, 12-hydroxy-16-oxo-

369   ent-Labda-7,13E-dien-15-oic acid, 16-hydroxy-17-oxo-

370   ent-Labda-7,13E-dien-15-oic acid, 17-hydroxy-16-acetoxy-

371   ent-Labda-7,13-dien-15-oic acid, 16-acetoxy-17-oxo-

372   ent-Labda-7,13E-dien-15-oic acid,
       16-acetoxy-6α-hydroxy-17-oxo-

373   ent-Labda-7,13E-dien-15-oic acid,
       16-acetoxy-6β-hydroxy-17-oxo-

374   ent-Labda-7,13Z-dien-15-oic acid, 16-acetoxy-6-oxo-

375   ent-Labda-7,13Z-dien-15-oic acid, 6,16-dioxo-

376   ent-Labda-7,13E-dien-15-oic acid, 16-hydroxy-6-oxo-

377   ent-Labda-7,13E-dien-15-oic acid, 16-acetoxy-6-oxo-

378   ent-Labda-7,13E-dien-15-oic acid, 16-methoxy-6-oxo-

379   ent-Labda-6,13E-dien-15-oic acid, 8α,16-dihydroxy-

380   ent-Labda-6,13E-dien-15-oic acid, 16-acetoxy-17-oxo-

381   ent-Labda-6,13E-dien-15,17-dioic acid, 16-acetoxy-

382   ent-Labda-8(17),13-diene, 7β,15,16-trihydroxy-

383   ent-Labda-8(17)-en-15-oic acid, 16-oxo-

Table 1. Compound Numbers and Names   31

TABLE 1. (*contd.*)

Compound:

No. Name

384 ent-Labda-6,8(17),13-trien-15-oic acid lactone, 16-hydroxy-

385 ent-Labda-8(17),13-dien-15-oic acid lactone,
     16,19-dihydroxy-

386 ent-Labda-7,13-dien-15-oic acid lactone, 16-hydroxy-

387 ent-Labda-7,13-dien-15-oic acid lactone, 12,16-dihydroxy-

388 ent-Labda-7,13-dien-15-oic acid lactone, 16,16α-dihydroxy-

389 ent-Labda-7,13-dien-15-oic acid lactone, 16,16β-dihydroxy-

390 ent-Labda-7,13-dien-15-oic acid lactone,
     16,12,16-trihydroxy-

391 ent-Labda-7,13-dien-15-oic acid lactone, 16,12,
     16α-trihydroxy-

392 ent-Labda-7,13-dien-15-oic acid lactone, 16,12,
     16β-trihydroxy-

393 ent-Labda-7,13-dien-15-oic acid lactone, 16-hydroxy-6-oxo-

394 ent-Labda-7,13-dien-15-oic acid lactone,
     16,16-dihydroxy-6-oxo-

395 ent-Labda-7,13-dien-15-oic acid lactone, 16,
     16α-dihydroxy-17-oxo-   [Acritolongifolide A]

396 ent-Labda-7,13-dien-15-oic acid lactone, 16,
     16β-dihydroxy-17-oxo-   [Acritolongifolide B]

397 ent-Labda-8(17),13(14)-dien-16-oic acid lactone, 7β,
     15-dihydroxy-

843 Nivenolide

398 ent-Labda-7,13-dien-16-oic acid lactone, 15-hydroxy-

399 ent-Labda-7,13-dien-16-oic acid lactone,
     15,12,15β-trihydroxy-

400 ent-Labda-7,13-dien-16-oic acid lactone, 15,12,
     15α-trihydroxy-

401 Polyalthine, 7β-hydroxy-   [Austrochaparol]

402 Austrochaparol acetate

403 Daniellol

404 Psiadiol

839 Daniellic acid

840 Daniellic acid, 3α-hydroxy-

841 Daniellol, 18-hydroxy-

TABLE 1. (*contd.*)

Compound:

No.   Name

842   Daniellol, 3α,18-dihydroxy-19-deoxy-

974   ent-Labda-8(17)-ene, 3α-angeloyloxy-18-hydroxy-13-furyl-

975   ent-Labda-8(17)-ene, 18-angeloyloxy-3α-hydroxy-13-furyl-

830   Polyalthic acid

831   Polyalthic acid, 3α-hydroxy-

832   Polyalthic acid, 3α-angeloyloxy-

833   Polyalthic acid, 3α-[2-methylbutyryloxy]-

834   Polyalthic acid, 3α-isobutyryloxy-

835   Polyalthic acid, 19-hydroxy-

836   Polyalthic acid, 8α,17-epoxy-8,17-dihydro-

837   Polyalthic acid, 8β,17-epoxy-8,17-dihydro-

838   Polyalthic acid, 17-oxo-8β,17-dihydro-

405   Conyzic acid

406   Austrofolin-12-ol, 12-desoxo-

407   Austrofolin

408   Austrochaparol acetate, 18-hydroxy-

409   Austrochaparol acetate, 19-acetoxy-

410   Austrochaparol, 19-oxo-

411   Austrochaparol acetate, 18-oxo-

412   Austrofolin, 3α-hydroxy-

413   Austrofolin, 3α,12-dihydroxy-12-desoxo-

414   Polyalthine, 17-hydroxy-iso-

415   ent-Labda-7,13(16),14-trien-15,16-epoxy-18-oic acid,
      5β,9βH,10α-

416   ent-Labda-13(16),14-diene, 3α-angeloyloxy-8-hydroxy-15,
      16-epoxy-

417   Austrofolin-15α-ol, 15,16-dihydro-

418   Austrofolin-15β-ol, 15,16-dihydro-

421   Pumiloxide, 6β-hydroxy-

      [Published structure indicated a C-7β-hydroxyl group, but
      name and data contradicted structure.]

Table 1. Compound Numbers and Names   33

TABLE 1. *(contd.)*

Compound:

No. Name

422 ent-Biformene

423 ent-Biformene, 18-oxo-

424 Ozic acid, cis-

425 Ozic acid, 12,13E-
     [ent-Labda-8(17),12E,14-trien-18-oic acid]

426 Ozic acid, 2α-hydroxy-12,13Z-

427 Ozic acid, 2-oxo-12,13Z-

428 Austroinulin

429 Austroinulin, 6-O-acetyl-   [Stevinsol]

808 Austroinulin, 6-angelyl-7-acetyl-

430 ent-Biformene, 12,13E-

431 ent-Labda-8(17),12E,14-triene, 19-hydroxy-

432 ent-Labda-8(17),12E,14-triene, 3α,19-dihydroxy-

433 ent-Labda-8(17),12E,14-triene, 18-hydroxy-

434 ent-Labda--8(17),12E,14-triene, 18,19-dihydroxy-

435 Ozic acid, 1β-acetoxy-trans-   [1β-acetoxy-ent-Labda-8(17),
     12E,14-trien-19-oic acid]

     [Published structure contained a C-1α-oriented acetate,
     while the name indicated that the acetate was β-oriented.
     Although the compound was named as an ozic acid derivative,
     the published structure contained an α-oriented carboxyl
     function rather than the β-oriented carboxyl group typical
     of ozic acid.]

436 ent-Abienol, 3β-hydroxy-

809 ent-Abienol, 6β,7α-dihydroxy-

810 ent-Abienol, 6β-acetoxy-7α-hydroxy-

811 ent-Abienol, 6β-angeloyloxy-7α-acetoxy-

437 ent-Labda-8(17),14-dien-18-oic acid, 12α,13αZ-epoxy-

438 ent-Labda-8(17),14-dien-18-oic acid, 12β,13βZ-epoxy-

439 ent-Labda-8(17),14-dien-18-oic acid, 12α,13αE-epoxy-

440 ent-Labda-8(17),14-dien-18-oic acid, 12β,13βE-epoxy-[12,
     13-Dehydro-12,13α-epoxy-ozic acid]

441 ent-Labda-13(16),14-diene, 8α-hydroxy-

442 ent-Manool

TABLE 1. *(contd.)*

Compound:

No.   Name

443   ent-Manool, 2β-hydroxy-

444   ent-Manool, 2α-hydroxy-

445   ent-Manool, 2β-succinyloxy-

446   ent-Manool, 2-oxo-

447   ent-Labda-7,14-diene, 13-hydroxy-

448   ent-Labda-7,14-diene, 3,13-dihydroxy-

449   ent-Labda-7,14-diene, 13-hydroxy-3-oxo-

450   ent-Sclareol, 8-epi-

460   ent-Labda-13-Z-en-15-oic acid, 6,16-dioxo-8,12-oxido-

461   Erigerol [1,6β-dihydroxy-7,8β-epoxydihydrogrendelyl ester
      of 3,3-dimethylacrylic acid]

1012 ent-Labda-7(8),13E-diene, 15,18-diacetoxy-2β-hydroxy-
      [or enantiomer]

1013 ent-Labda-7(8),13E-diene, 18-acetoxy-2β,15-dihydroxy-
      [or enantiomer]

1014 ent-Labda-8(17),13E-diene,3α-acetoxy-15-hydroxy-18-
      tigloyloxy- [or enantiomer]

1015 Abienol, 18-hydroxy-

1016 ent-Labd-8,14-diene, 13,18-dihydroxy-7-oxo-ent-

1017 ent-Labd-14-ene, 18-hydroxy-7-oxo-9β,13β-epoxy-

1018 ent-Labd-14-ene, 7,18-dioxo-9β,13β-epoxy-

1019 ent-Labd-14-en-18-oic acid, 7-oxo-9β,13β-epoxy-

1037 ent-Labda-8(17),13(16),14-triene,
      18-hydroxy-3α-angeloyloxy-15,16-epoxy-

1038 ent-Labda-8(17),13(16),14-triene,2β,18-dihydroxy-
      3α-angeloyloxy-15,16-epoxy-

1039 ent-Labda-8(17),13Z-diene, 2β,15,16,
      18-tetrahydroxy-3α-angeloyloxy-

1040 ent-Labda-8(17),13E-dien-15-oic acid, [E-Copalic acid]

      [Structure is the same as 335.]

1041 ent-Labda-8(17),13E-dien-15-oic acid, 18-hydroxy-

1042 ent-Labda-7(8),13E-dien-15-oic acid, 6β-hydroxy-18-oxo-

1043 ent-Labda-7(8),13E-dien-15,
      18-dioic acid, 6β-isovaleryloxy-

Table 1. Compound Numbers and Names 35

TABLE 1. (contd.)

Compound:
No.   Name

1044 ent-Labda-13E-en-15-oic acid, 5β,8β-peroxy-

1045 ent-Labda-7,13E-dien-15-oic acid, 6β,18-dihdroxy-

1046 ent-Labda-7-en-15-oic acid, 17-hydroxy-

1047 ent-Labda-7-en-15-oic acid, 17-acetoxy-

1048 ent-Labda-7-en-15-oic acid, 17-oxo-

1049 ent-Labda-7-en-15-oic acid, 2α,17-dihydroxy-

1050 ent-Labda-7-en-15-oic acid, 2α-hydroxy-17-acetoxy-

1051 ent-Labda-7-en-15-oic acid, 2α-hydroxy-17-oxo-

1052 ent-Labda-7-en-15,17-dioic acid

1053 ent-Labda-7-en-15-oic acid, 19-hydroxy-17-acetoxy-

1164 ent-Labd-8(17),13E-dien-15,18-dioic acid   [Guamaic acid]

1165 ent-Labd-7,13E-dien-2-O-β-fucopyranoside, 3α-angeloyloxy-
     2β,15-dihydroxy-

1166 ent-Labd-7,13E-dien-2-O-β-[fucopyranoside-4'-O-acetate],
     3α-angeloyloxy-2β,15-dihydroxy-

1167 ent-Labd-7,13E-dien-2-O-β-[fucopyranoside-3'-O-acetate], 3α-
     angeloyloxy-2β,15-dihydroxy-

1168 ent-Labd-7,13E-dien-2-O-β-[fucopyranoside-4'-O-acetate],
     3α-angeloyloxy-2β,15,16-trihydroxy-

1169 ent-Labd-7,13(16)-dien-2-O-β-[fucopyranoside-4'-O-acetate],
     3α-angeloyloxy-2β,14,15-trihydroxy-

1170 ent-Labd-7,13E-dien-2-O-[rhamnopyranoside-4'-O-acetate],
     3α-angeloyloxy-2β,15-dihydroxy-

1171 ent-Labd-7,13E-dien-2-O-β-xylopyranoside,
     3α-angeloyloxy-2β,15-dihydroxy-

1172 ent-Labd-13E-ene-8-O-β-xylopyranoside, 8β,15-dihydroxy-

1173 ent-Labd-8(17)-en-19-oic acid, 15-hydroxy-

1174 ent-Labd-13E-ene, 2β,8β,15-trihydroxy-

1175 Haplopappus pectinatus ent-labdane 13

1176 Haplopappus pectinatus ent-labdane 14

1177 Haplopappus pectinatus ent-labdane 15

1178 Haplopappus pectinatus ent-labdane 16

1179 ent-Labd-7-en-15-oic acid, 2-hydroxy-17-methoxy-

TABLE 1. (*contd.*)

Compound:
No.   Name

1180 ent-Labd-7,13(14)-diene-15-oic acid, 16-hydroxy-17-methoxy-

1181 ent-Labd-8(17),13-dien-15,16-olide, 2β,18-dihydroxy-

1182 ent-Labd-8(17),13-dien-15,16-olide, 3α-angeloyloxy-18-
      hydroxy-

1183 ent-Labd-8(17),13-dien-15,16-olide, 3α-angeloyloxy-18β-
      glucopyranosyloxy-

1184 ent-Labd-8(17),13-dien-15,16-olide, 3α-angeloyloxy-2β,18β-
      dihydroxy-

1185 ent-Labd-8(17),12E,14-triene, 18,19-diacetoxy-

1186 ent-Biformene, 3α,14,15-trihydroxy-12E,14,15-dihydro-

1188 ent-Labda-7,13(16),14-trien-2β,3α-diol, 15,16-epoxy-

1189 ent-Labd-8(17),13E-dien-15-oic acid, 3α,18-dihydroxy-

1190 ent-Labd-8(17),13E-dien-15-oic acid, 3α-angeloyloxy-18-
      hydroxy-

1191 ent-Labd-8(17),13Z-diene, 2β,15,16,18-tetrahydroxy-3α-
      angeloyloxy-

1192 ent-Labd-7,13-dien-2β,15-diol, 3α-angeloyloxy-

1193 ent-Labd-7,13E-diene, 2β,3β,15-trihydroxy-

1194 Gutierrezia spathulata ent-labdane 4

1195 Gutierrezia spathulata ent-labdane 5

1196 ent-Cativic acid, 2β,3β-dihydroxy-

1197 ent-Labd-7-ene, 2β,3β,15-trihydroxy-

39. ent-Labdane-19,6-Lactone

462   Hartwrightia acid

40. ent-Labdane-8,13-Epoxide-15,16-Hemiacetal

463   Schkuhriadiol, cyclo-

464   Schkuhriadiol-16-O-acetate, cyclo-

97. ent-Labdane Dimer

419   Acritopappus lactone A

Table 1. Compound Numbers and Names   37

TABLE 1. *(contd.)*

Compound:
No.   Name

420   Acritipappus lactone B

98. ent-Manoyl Oxide, 13-Epi-

451   ent-Manoyl oxide, 13-epi- [ent-8,13β-Epoxy-14-labdene]

452   ent-Manoyl oxide, 3α-hydroxy-

453   ent-Manoyl oxide, 3-oxo-13-epi- [ent-8,
      13β-epoxy-14-labden-3-one]

454   ent-Manoyl oxide, 2-oxo-13-epi-

455   ent-Manoyl oxide, 19-hydroxy-13-epi-

456   ent-Manoyl oxide, 2-oxo-19-hydroxy-13-epi-

457   Schkuhrianol acetate [16-acetoxy-14,15-epoxy-14,
      15-dihydro-13-epi-ent-Manoyl oxide]

1198 ent-Manoyl oxide, 3α-acetoxy-

1199 ent-Manoyl oxide, 19-cinnamoyloxy-3α-hydroxy-

99. ent-Manoyl Oxide

458   ent-Manoyl oxide

459   ent-Manoyl oxide, 3β-hydroxy-

      [This structure is apparently identical to a compound from
      Coespeletia lutescens named 3β-Hydroxy-13-epi-ent-manoyl
      oxide for which the published structure did not indicate a
      13-epi-orientation.]

41. ent-Labdane, 17-Nor-

465   ent-Labda-6,13E-dien-15-oic acid,
      16-acetoxy-8β-formyl-8-desmethyl-

466   ent-Labda-6,13E-dien-15-oic acid,
      16-acetoxy-8-oxo-8-desmethyl-

467   Ayapanone acid, 3α-hydroxy-

468   Ayapanone acid, 3-oxo-

1201 ent-Labda-13Z-en-15-oic acid, 8-oxo-8-desmethyl-

1202 ent-Labda-13Z-en-15-oic acid, 3,8-dioxo-

TABLE 1. (*contd.*)

Compound:
No.   Name

42. ent-Labdane, 15-Nor-

469   ent-Labda-8(17),12E-diene, 14-oxo-nor-15-

43. ent-Labdane, 14,15-Bis-Nor-

470   ent-Labda-8(17)-ene, 13-oxo-bis-nor-14,15-

1203 ent-Labd-7-en-2-O-[fucopyranoside-4'-O-acetate],
      14,15-nor-

44. ent-Labdane, 13,14,15,16-Tetra-Nor-

471   ent-Labda-7-en-aldehyde, nor-   [Tetra-nor-13,14,15,
      16-12-oxo-ent-labda-7-ene]

472   Acritoconfertic acid   [Tetra-nor-13,14,15,
      16-17-hydroxy-ent-labda-7-en-12-oic acid]

45. ent-Labdane, 7,8-Seco-

473   Athrixic acid, seco-

74. ent-Labdane, 18-Nor-

828   ent-Labdane triol 1a, nor-18-

829   ent-Labdane triol 2a, nor-18-

1205 Austroeupatorione, 12-desoxo-

1206 Austroeupatorione, 12-desoxo-2-desacetyl-

1207 Austroeupatorione

95. ent-Friedolabdane

1059 ent-Labda-5,13E-dien-15-oic acid,
      18-hydroxy-10-desmenthyl-9α-methyl-8β(H)-

1208 [= 1059]

1209 ent-Friedolabd-1(10),13E-dien-15-ol, 2-oxo-

1210 ent-Friedolabd-1(10),13E-diene, 2α,15-dihydroxy-

1211 Friedolabdaturbinic acid

Table 1. Compound Numbers and Names 39

TABLE 1. (*contd.*)

Compound:
No.   Name

1212 Chiliolide, 3α,5α-dihydroxy-

1213 Chiliolide, 3α-hydroxy-5,6-dehydro-

1214 Chiliolide, 3α-hydroxy-5β,10β-epoxy-

102. ent-Friedolabdane, 3,4-Seco-

1215 Chiliolide aldehyde, seco-

1216 Chiliolide acid, seco-

1217 Chiliolide, 19-hydroxy-seco-

1218 Chiliolide acid methyl ester, 19-hydroxy-

1219 Chiliolide lactone, seco-

1220 Chiliotrin, seco-

103. ent-Friedolabdane, Rearranged 3,4-Seco-

1221 Chiliolide lactone, iso-

46. trans-ent-Clerodane

236   Bedfordia diterpene alcohol

248   Solidago-18-oic acid, 8-epi-1,2-dehydro- [Reported as
      5β,10α-trans-clerodane]

249   Solidago-18-oic acid, 8-epi- [Reported as 5β,10α-trans-
      clerodane; probably (-)-hardwickiic acid]

250   Solidago-18-oic acid, 1,2-dehydro- [Reported as 5β,10α-
      trans-clerodane with C-8β-methyl;probably = 248]

251   Solidago-18-oate, methyl 8-epi-1,2-dehydro- [Nidoreseda
      acid methyl ester; Reported as 5β,10α-trans- clerodane]

245   Solidago-18-oic acid [Reported as 5β,10α-trans-clerodane
      with C-8β-methyl; probably the same as 249]

475   Kolavenol

476   Kolavenol, 17-hydroxy-13E-

477   Kolavenol acetate, 2α-hydroxy-3β,4β-epoxy-3,4-dihydro-

478   Kolavenol acetate, 2α,16-dihydroxy-3β,4β-epoxy-3,4-dihydro-

TABLE 1. *(contd.)*

Compound:	
No.	Name

1010 Kolavenol arachidate, 17-hydroxy-13E-

1011 Kolavenol behenate, 17-hydroxy-13E-

479   Kolavelool   [Kolavenool]

480   Kolavelool, 6β-angeloyloxy-   [6β-Angeloyloxy-kolavenool]

481   Kolavenol, 17-hydroxy-13,14-dihydro-

482   Kolavenol arachidate, 17-hydroxy-13,14-dihydro-

483   Kolavenol behenate, 17-hydroxy-13,14-dihydro-

484   Kolavenol, 18-succinyloxy-13,14-dihydro-

485   Kolavenol, 2α-hydroxy-3α,4α-epoxy-13,14-dihydro-

486   Kolavenol, 17-oxo-13,14-dihydro-

487   Gochnatol-17-O-acetate

488   Kolav-3-en-15-ol, 6α,18-dihydroxy-17-acetoxy-

489   Kolav-3-en-15-ol, 6α,18-dihydroxy-17-phenylacetoxy-

490   Kolav-3-en-15-ol, 6α-hydroxy-17-phenylacetoxy-18-oxo-

491   Kolav-3-en-15-ol, 6α-hydroxy-17-acetoxy-18-oxo-

492   Kolaven-15-al, 17-hydroxy-13,14-dihydro-

493   Kolavenoic acid, 13,14Z-

494   Kolavenoic acid, 2β-acetoxy-13,14Z-

495   Kolavenoic acid [13,14E-Kolavenoic acid]

496   Kolavenoic acid, 6-acetoxy- [6β-Acetoxy-13,
14E-kolavenoic acid]

497   Kolavenoic acid, 6-angeloyloxy- [6β-Angeloyloxy-13,
14E-kolavenoic acid]

498   Kolavenic acid, 6-angeloyloxy- [6α-Angeloyloxy-13,
14E-kolavenoic acid]

499   Kolavenic acid, 6-tigloyloxy- [6α-Tigloyloxy-13,
14E-kolavenoic acid]

500   Solidagonic acid [7α-Acetoxy-13,
14E-kolavenoic acid]

962   Tucumanoic acid

501   Kolavenoic acid, 6α,18-dihydroxy-17-acetoxy-

502   Kolavenoic acid, 6α,18-dihydroxy-17-phenylacetoxy-

Table 1. Compound Numbers and Names   41

TABLE 1. *(contd.)*

Compound:

No. Name

503 Kolav-3-en-15-oic acid, 13,14-dihydro-

504 Kolav-3-en-15-oic acid, 18-acetoxy-

961 Kolavenic acid, 2-oxo-

505 Haplociliatic acid

506 Kolavan-2-on-15-oic acid, 3β,4β-epoxy-

507 Kolav-3,13(16),14-triene, 15,16-epoxy-

508 Bacchalineol malonate

509 Bacchotricuneatin D

846 Kingidiol [Barticulidiol]

510 Barticulidiol, 18-acetoxy-19-malonyloxy-

967 Cleroda-3,13(16),14-trien-15,16-oxide, 17-oxo-

511 Hardwickiic acid, (-)-
     [= 517]

953 Hardwickiic acid, 12α-[2-methylbutyryloxy]-(-)-

512 ent-cleroda-1,3,13(16),14-tetraen-18-oic acid, 15,
     16-epoxy-

911 Hautriwaic acid, 19-O-acetyl-1,2-dehydro-

513 ent-Kolav-3,13(16),14-trien-18-oic acid, 2β-hydroxy-15,
     16-epoxy-

514 Imbricatol-isovalerate, 3-hydroxy-

515 Imbricatol-α-methylbutyrate, 3-hydroxy-

516 Imbricatol-angelate, 3-hydroxy-

518 Kolav-3,13(16),14-trien-18-oic acid, 15,
     16-epoxy-19-hydroxy-ent- [Hautriwaic acid]

519 Kolav-3,13(16),14-trien-18-oic acid, 15,16-epoxy-2α,
     19-dihydroxy-ent-

900 Hautriwaic acid, 19-acetyl-

901 Hautriwaic acid, 10-O-methyl

902 Hautriwaic acid, 19-O-angelyl-

903 Hautriwaic acid, 19-O-isovaleryl-

904 Hautriwaic acid, 19-oxo-

524 ent-Kolav-3,13(16),14-trien-20-oic acid, 15,
     16-epoxy- [Junceic acid]

TABLE 1. *(contd.)*

Compound:	
No.	Name

520	Hautriwaic acid lactone
950	Hautriwaic acid-19-lactone, 12α-hydroxy-
951	Hautriwaic acid-19-lactone, 7α,12α-dihydroxy-
521	Bacchofertin
522	Gochnatoic acid-17-O-phenylacetate
523	Hardwickiic acid lactone, 7-oxo-6α-hydroxy
525	Conycephaloide
526	Conycephaloide, 7-hydroxy-17-oxo-7,8-dehydro-8,17-dihydro-
527	Bacchotricuneatin A
528	Bacchotricuneatin C
969	Conycephaloide, 12-epi-
970	Conycephaloide, 12-epi-8β,17-dihydro-
971	Salviarin, 6α-angeloyloxy-1-oxo-2,3-dihydro-
972	Bacchotricuneatin A, 6β-hydroxy-7,8-dehydro-
973	Salviarin
529	Bacchotricuneatin B
968	Bacchotricuneatin B, 2α-angeloyloxy-
530	Bacchotricuneatin B, 7β-angeloyloxy-
531	Bacchotricuneatin, 2α-angeloyloxy-3α-senecioyloxy-3,4βH-
532	Bacchotricuneatin, 2α-senecioyloxy-3α-angeloyloxy-3,4βH-
533	Bacchotricuneatin, 2α-angeloyloxy-3α-[2-methylbutyryloxy]-3,4βH-
534	Kolaven-15-al, 17-hydroxy-13Z-
535	Kolaven-15-al, 17-hydroxy-13E-
536	Symphyoreticulic acid, 13Z-
537	Symphyoreticulic acid, 13E-
865	Kolava-3,13E-dien-15-oic acid, 16-hydroxy-
962	Solidagonic acid
538	Kolavenic acid, 16,18-dihydroxy-
539	Kolavenic acid, 16-acetoxy-18-hydroxy-

Table 1. Compound Numbers and Names   43

TABLE 1. *(contd.)*

Compound:

No.   Name

540   Kolavenic acid, 16-hydroxy-18-oxo-

541   Kolavenic acid, 16-acetoxy-18-oxo-

542   Kolav-3-en-15-oic acid, 16-oxo-

543   Solidagolactone   [Soldagolactone I; 16-Hydroxy-kolavenic acid lactone]

544   Kolavenic acid lactone, 2α,16-dihydroxy-

864   Kolava-3,13Z-dien-15-oic acid lactone, 16,16-dihydroxy-

550   Kolavenic acid lactone, 16,18-dihydroxy-

551   Kolavenic-15-acid lactone, 2α,16-dihydroxy-3α,4α-epoxy-

      [Revised to <u>cis</u>-normal-clerodane]

552   Kolavenic-15-acid lactone, 6β-angeloyloxy-16-hydroxy-3α, 4α-epoxy-
      [Revised to <u>cis</u>-normal-clerodane]

558   Hardwickiic acid, 15,16H-15-oxo-

      [Compound was named as "16-oxo-...", but data and structure indicate otherwise.]

909   Hautriwaic acid, 16α-acetoxy-19-O-acetyl-15-oxo-15, 16-dihydro-

910   Hautriwaiic acid, 16β-acetoxy-19-O-acetyl-15-oxo-15, 16-dihydro-

559   Kolaven-15, 18-dioic acid dilactone, 7α,16,19-trihydroxy-

560   Olearin

561   Kolaven-15,18-dioic acid dilactone,7-oxo-16, 19-dihydroxy-13,14-dihydro-

562   Kolaven-15,18-dioic acid dilactone, 7α,16,19-trihydroxy-13, 14-dihydro-

563   Kolaven-15,18-dioic acid dilactone, 16,19-dihydroxy-13, 14-dihydro-

564   Kolaven-15-oic acid lactone, 16-hydroxy-13, 14-dihydro-

565   Kolav-3-en-15-oic acid lactone, 16,16-dihydroxy-

963   Kolaven-15-oic acid lactone, 16-hydroxy-16-methoxy-18-acetoxy-

566   Rugosolide

905   Hautriwaic acid, 15α-hydroxy-16-oxo-15,16-dihydro-

TABLE 1. *(contd.)*

Compound:

No.   Name

906   Hautriwaic acid, 15β-hydroxy-16-oxo-15,16-dihydro-

907   Hautriwaic acid, 14α,15α-epoxy-16-oxo-13,14,15,
      16-tetrahydro-

908   Hautriwaic acid, 14β,15β-epoxy-16-oxo-13,14,15,
      16-tetrahydro-

567   Kolaven-16-oic acid lactone, 2β,15-dihydroxy-

568   Kolaven-16-oic acid lactone, 15-hydroxy-2-oxo-

569   Kolav-3-ene, 18-malonyloxy-15-hydroxy-15,16-epoxy-

570   Stephalic acid

1027 Kolavenic acid, 6α-isobutyryloxy-7β-acetoxy-13-Z-

1028 Kolavenic acid, 6α-angeloyloxy-7β-acetoxy-13Z-

1029 Kolavenic acid, 6α,7β-diacetoxy-13Z-

1030 Kolavenic acid, 6α,7β-dihydroxy-2-oxo-13E-

1031 Kolavenic acid, 6α-isobutyryloxy-7β-angeloyloxy-13,
      14-dihydro-

1033 Hautriwaic Acid

      [= 518]

1034 Hautriwaic acid, 2β-hydroxy-

1069 Bacchotricuneatin A, 1α-hydroxy-

1070 Hautriwaic acid, 1,2-dehydro-

1071 Junceic acid epoxide

1075 Kolav-13Z-en-15-al, 3α,4α-epoxy-

1076 Kolav-13(16),14-diene, 3α,4α,15,16-diepoxy-

1077 Kolav-13E-en-15-al, 3α,4α-epoxy-

1078 Kolav-13Z-en-15-al, 3-oxo-

1079 Kolav-13E-en-15-al, 3-oxo-

1080 Kolav-13-en-15-oic acid lactone, 16-hydroxy-3α,4α-epoxy-

1081 Kolav-13-en-15-oic acid lactone, 16-hydroxy-3-oxo-

1082 Kolav-13-en-15-oic acid lactone, 16,3α,4β-trihydroxy

1222 Bincatriol

1223 Salvicin [19-Carboxy-6α,15-dihydroxy-trans-cleroda-
      3,13(14)-diene]

Table 1. Compound Numbers and Names   45

Table 1. *(contd.)*

Compound:
No.   Name

1224 Bacchasalicyclic acid

1225 Bacchasalicyclic acid-15-O-acetate, 16-hydroxy-

1226 Bacchasalicyclic acid-15-O-[1'-β-xylopyranoside]

1227 Baccharis incarum neoclerodane 1

1228 Kolavenol acetate, 16-hydroxy-2α-acetoxy-3β,4β-epoxy-3,4-dihydro-

1229 Baccharis rhomboidalis neoclerodane 5

1230 Baccharis rhomboidalis neoclerodane 6

1231 Kingidiol, 19-O-malonyl-

1232 Bacchalineol

1233 Barticulidiol

1234 Salvin [15,16-Epoxy-trans-cleroda-3,13(16),14-trien-19,6β-olide]

1235 Salvinin [7-Hydroxy-15,16-epoxy-trans-cleroda-3,13(16),14-trien-19,6β-olide]

1236 Baccharis incarum neoclerodane 2

1237 Articulin

1238 Articulin acetate

1239 Baccharis grandicapitulata neoclerodane 16

1240 Baccharis incarum neoclerodane 3

1241 Baccharis rhomboidalis neoclerodane 1

1242 Baccharis rhomboidalis neoclerodane 3

1243 Baccharis rhomboidalis neoclerodane 4

1244 Bacchomagellin A

1245 Bacchomagellin B

1257 Bacchotricuneatin A, 1α-acetyl-

1258 Bacchotricuneatin A, 7α-hydroxy-

47. Printziane

571   Printzianic acid

TABLE 1. *(contd.)*

Compound:	
No.	Name

572 Printzianic acid, iso-

913 Printziaic acid, 5α-hydroxy-5,10-dihydro-

914 Printziaic acid, 5α-methoxy-5,10-dihydro-

915 Printziaic acid, 5α-hydroxy-1,2-dehydro-5,10-dihydro-

916 Printziaic acid lactone, 5α-hydroxy-5,10-dihydro-

76.Conyscabrane

917 Conyscabraic acid, 5,6-dehydroiso-

918 Conyscabraic acid, 5,6-dehydro-

919 Conyscabraic acid, 5α-hydroxyiso-

920 Conyscabraic acid, 5α-hydroxy-

48. ent-Clerodane, 5,10-Seco-

573 Strictic acid

912 Nidoresedaic acid, 19-acetoxyseco-

985 Pulicaria angustifolia compound 1a

986 Pulicaria angustifolia compound 2a

987 Pulicaria angustifolia compound 3a

952 Strictic acid, 12α-[2-methylbutyryloxy]-

96. trans-ent-Clerodane, 17-Nor-

1061 Bacchascoparone

49. trans-ent-Clerodane, 13,14,15,16-Tetra-Nor-

574 Acritopappusol

575 Acritopappus acid

1032 Kolavenic acid, 6α,7β-dihydroxy-nor-

1093 Hautriwaic acid, 13,14,15,16-tetranor-1,2-dehydro-

Table 1. Compound Numbers and Names 47

TABLE 1. (*contd.*)

Compound:
No.   Name

Undetermined

237   Bedfordia diterpene alcohol, iso-   [Clerodane of unknown
      absolute stereochemistry]

50. cis-ent-Clerodane

578   Cleroda-3,13(16),14-triene, 15,16-epoxy-cis-

579   Cleroda-3,13(16),14-triene, 18-hydroxy-15,16-epoxy-cis-

580   Cleroda-3,13(16),14-triene, 18-acetoxy-15,16-epoxy-cis-

576   Haplopappic acid

577   Haplopappate, monomethyl-

581   Cleroda-3,13(16),14-triene, 6α,18-dihydroxy-15,16-epoxy-cis-

582   Cleroda-3,13(16),14-triene, 6α-hydroxy-18-acetoxy-15,
      16-epoxy-cis-

583   Cleroda-3,13(16),14-trien-18,6α-olide, 15,16-epoxy-cis-

584   Cleroda-3-en-15,16,18,6α-diolide, cis-

585   Cleroda-3-en-15,16,18,6α-diolide, 2β-hydroxy-cis-

1055 Cleroda-3,13E-dien-15-oic acid, 18-acetoxy-cis-

1056 Cleroda-3,13E-dien-15-oic acid, 18-hydroxy-cis-

1057 Cleroda-3-en-15-oic acid, 18-hydroxy-cis-

1058 Cleroda-3-en-15-oic acid, 19-hydroxy-

1246 Chiliomarin

82. Solidagonane

960   Solidagonal acid

VI. ent-Tricarbocyclics and Derivatives

51. ent-Labdane, Cyclopropyl-

586   Gnaphalene, 8α,13α-dihydroxy-

77. Gutierreziane

845   Gutierrezial

TABLE 1. *(contd.)*

Compound:

No. Name

52. ent-Pimarane

587 Sandaracopimar-15-en-8β-ol [Published structure incorrectly drawn = Cmpd No. 276]

588 Sandaracopimar-15-en-8β,11α-diol, 12β-acetoxy- [Structure incorrectly drawn = Cmpd No. 292]

589 Kirenol

590 Pimara-8(14),15-dien-19-oic acid, ent-

820 Pimara-7,15-dien-19-oic acid, ent-

821 Pimara-8,15-dien-19-oic acid, ent-7-oxo-

823 Pimara-9(11)-en-19-oic acid, ent-12β-acetoxy-

591 Darutigenol

592 Darutigenol-3α-glucoside

593 Pimara-8(14)-en-6β-O-glucoside, 15β,16-dihydroxy-

594 Pimara-8(14)-ene, 6β,15β,16-trihydroxy-

595 Pimara-8(14)-ene, 2β,15β,16,19-trihydroxy-

944 Pimara-8(14)-ene, ent-6β,15,16,18-tetrahydroxy-

599 Pimar-9,11(15)-dien-19-ol, (-)-

600 Pimar-9,11(15)-dien-19-oic acid, (-)-

1247 ent-8(14),15-Pimaradiene-3β,19-diol

1248 ent-8(14),15-Pimaradiene-3β,18-diol

1249 ent-Pimar-8(9),15-diene, 20-hydroxy-7-oxo-

1250 ent-Pimara-9(11),15-dien-19-ol

1251 ent-Pimara-9(11),15-diene, 18-hydroxy-

1252 ent-Pimar-9(11),15-diene malonate, 18-hydroxy-

1253 ent-Pimar-9(11),15-diene malonate, 19-hydroxy-

1254 ent-Pimar-9(11),15-diene, 3α,19-dihydroxy-

53. ent-Pimarane-8,15-Tetrahydrofuran

596 Pimara-3β,12α,16-triol C-15-epimer 9a, ent-8,15-epoxy-

597 Pimara-3β,12α,16-triol C-15 epimer 10a, ent-8,15-epoxy-

Table 1. Compound Numbers and Names   49

TABLE 1. (*contd.*)

Compound:

No.   Name

824   Pimar-9(11)-en-19-oic acid, ent-8(R),
      15(S)-epoxy-12α-acetoxy-

598   Pimara-12α,16-diol, ent-8,15R-epoxy-3-oxo-

54. Cleistanthane (Cleistanthol)

601   Cleistanth-12-en-11-one, 15,16-epoxy-

602   Cleistanth-12-en-17-al, 15,16-epoxy-11-oxo-

603   Cleistanth-12-en-11-one, 17-acetoxy-15,16-epoxy-

604   Cleistanth-12-en-11-one, 3α-angeloyloxy-15,16-epoxy-

605   Cleistanth-12-en-11-one, 17-acetoxy-14β-hydroxy-15,16-epoxy-

606   Cleistanth-12-en-11-one,
      17-acetoxy-3α-angeloyloxy-15,16-epoxy-

607   Cleistanth-12-en-17-al, 11-oxo-8,9,15,16-diepoxy-

608   Cleistanth-12-en-11-one, 14β,15β-dihydroxy-16,17-oxido-

85. Cleistanthane, Iso-

609   Cleistanth-12-en-11-one, 17-acetoxy-15,16-epoxy-iso-

610   Cleistanth-12-en-11-one, 17-acetoxy-3,4,15,16-diepoxy-iso-

55. ent-Abietane

611   Missourienol A [initially isolated as dehydration product:
      3-oxo-abieta-7,13-diene]

612   Missourienol B [initially isolated as dehydration
      product:3β-hydroxy-abieta-7,13-diene]

613   Missourienol C [initially isolated as dehydration
      product:2β-acetoxy-abieta-7,13-diene]

614   Abieta-7,13-diene

615   Abieta-7,13(14)-diene, 5-acetoxy-

616   Abieta-7,13(14)-dien-18-oic acid

1054 Abieta-7,13-diene, 19-succinyloxy-

56. Acritoconfertane

617   Acritoconfert-7-en-15-oic acid, 16-acetoxy-17-hydroxy-

TABLE 1. (*contd.*)

Compound:

No.   Name

57. Acritoconfertane-Acetal

1100 Acritoconfertic acid, 8,16,16,17-bisoxido-

83. Erythroxane

976 Erythroxa-3,15-dien-18-oic acid

VII. ent-Tetracarbocyclics and Derivatives

58. ent-Kaurane

618 Kaur-16-ene, ent-

619 Kaur-16-ene, 19-hydroxy-ent-

997 Kauranol, ent-   [(-)-kauranol]

620 Kaur-16-ene, 19-acetoxy-ent-

621 Kaurene, 19-formyloxymethylenoxy-ent-

622 Kaurene, 19-[p-hydroxyhydrocinnamoyloxy]-ent-

623 Kaur-16-en-3α,19-diol, (-)-

626 Kaur-16-ene, 18-acetoxy-ent-

624 Kaur-16-ene, 19-oxo-ent-

625 Kauren-19-oic acid, ent-

926 Kauren-18-oic acid, ent-

     [Although referred to as kaurenic acid, the published
     structure of this compound from Melampodium perfoliatum
     contained a C-4β-oriented carboxyl group.]

939 Kaurenic acid, 18-angeloyloxy-
     [Although referred to as a kaurenic acid derivative, the
     published structure of this compound from Melampodium
     perfoliatum contained a C-4β-oriented carboxyl group and a
     C-4α-oriented ester function.]

627 Kaur-16-en-19-oic acid glycol ester, ent-

628 Kaur-16-en-19-oic acid, 3α-hydroxy-ent-

629 Kaur-16-en-19-oic acid, 3α-acetoxy-ent-

630 Kaur-16-en-19-oic acid, 3α-angeloyloxy-ent-

631 Kaur-16-en-19-oic acid, 3α-tigloyloxy-ent-

632 Kaur-16-en-19-oic acid, 3α-cinnamoyloxy-ent-

Table 1. Compound Numbers and Names 51

TABLE 1. (*contd.*)

Compound:	
No.	Name

633 Kaur-16-en-19-oic acid, 3α-senecioyloxy-ent-

634 Kaur-16-en-19-oic acid, 3α-isovaleryloxy-ent-

635 Kaur-16-en-18-al, ent-

636 Kaur-16-en-19-oic acid, 9β,13β-dihydroxy-15α-angeloyloxy-ent-

637 Kaur-16-en-19-oic acid, 11β,15α-dihydroxy-ent-

638 Kaurane, 18-hydroxy-16α,17-epoxy-ent-

1000 Kaur-16-en-19-oic acid; 7β-hydroxy-(-)-

639 Stenolobin

640 Kaur-16-en-19-oic acid, 11β-acetoxy-ent-

641 Kaur-16-en-19-oic acid, 12α-hydroxy-ent-

642 Kaur-16-en-19-oic acid, 12α-acetoxy-ent-

1108 [= **642**]

643 Kaur-16-en-19-oic acid, 18-angeloyloxy-ent-

644 Kaur-16-en-19-oic acid, 18-senecioyloxy-ent-

645 Kaur-16-en-19-oic acid, 18-isovaleryloxy-ent-

646 Grandifloric acid [15α-hydroxy-ent-kaur-16-en-19-oic acid]

647 Kaur-16-en-19-oic acid, 15β-hydroxy-ent-

648 Kaur-16-en-19-oic acid, 15α-acetoxy-ent-

649 Kaur-16-en-19-oic acid, 15α-angeloyloxy-ent-

1007 Stenolobin, 15α-angeloyloxy-
[methyl-15α-angeloyloxy-ent-kaur-16-en-19-oate]

650 Kaur-16-en-19-oic acid, 15α-tigloyloxy-ent-

651 Kaur-16-en-19-oic acid, 15α-senecioyloxy-ent-

652 Kaur-16-en-19-oic acid, 15α-isovaleryloxy-ent-

653 Kaur-16-en-19-oic acid, 15α-[2-methylacryloyloxy]-ent-

654 Kaur-16-en-19-oic acid, 15α-isobutyroyloxy-ent-

655 Kaur-16-en-19-oic acid, 15α-cinnamoyloxy-ent-
[cinnamoyloxygrandifloric acid]

656 Kauren-16-en-19-oic acid, 15α-[2,
3-epoxy-2-methylbutyloxy]-ent-

657 Kaur-16-en-19-oic acid, 15α-benzoyloxy-ent-

TABLE 1. *(contd.)*

Compound:

No. Name

818 Kauren-16-en-19-oic acid,15α-[2',
     3'-dihydroxy-2-methylbutyryloxy]-ent-

892 Kaur-16-en-19-oic acid, 15α[2'-methylbutyryloxy]-ent-

658 Kaur-16-en-19-oic acid, 15β-angeloyloxy-ent-

659 Kaur-16-en-19-oic acid, 15β-tigloyloxy-ent-

660 Kaur-16-en-19-oic acid, 15β-senecioyloxy-ent-

661 Kaur-16-en-19-oate, methyl-15α-hydroxy-ent-   [grandifloric
     acid methylester]

662 Kaur-16-en-19-oic acid, 3α-angeloyloxy-9β-hydroxy-ent-

663 Kaur-16-en-19-oic acid, 3α-cinnamoyloxy-9β-hydroxy-ent-

664 Kaur-16-en-19-oic acid, 9β,15α-dihydroxy-ent-
     [grandifloric acid, 9β-hydroxy]

665 Kaur-16-en-19-oic acid, 9β-hydroxy-15α-angeloyloxy-ent-

666 Kaur-16-en-19-oic acid, 9β-hydroxy-15α-tigloyloxy-ent-

667 Kaur-16-en-19-oic acid, 9β-hydroxy-15α-acetoxy-ent-

668 Kaur-16-en-19-oic acid, 9β-hydroxy-15α-senecioyloxy-ent-

669 Kaur-16-en-19-oic acid, 9β-hydroxy-15α-isovaleryloxy-ent-

670 Kaur-16-en-19-oic acid, 9β-hydroxy-15α-cinnamoyloxy-ent-

819 Kaur-16-en-19-oic acid, 9β-hydroxy-
     15α-[3'-hydroxy-2'-methylbutyryloxy]-ent-

671 Kaur-16-en-19-oic acid, 11β,15β-dihydroxy-ent-

812 Kaur-16-en-19-oic acid, 17-carbomethoxy-ent-

672 Kaur-16-en-19-oic acid, 11β-hydroxy-15β-acetoxy-ent-

673 Kaur-16-en-19-oic acid, 15β-hydroxy-11-oxo-ent-

674 Kaur-16-en-19-oic acid, 11β-hydroxy-15-oxo-ent-

1260 Kaur-16-en-19-oic acid, 11β,12β,15β-trihydroxy-ent-

675 Kaur-16-en-19-oic acid, 3β,9β-dihydroxy-15α-angeloyloxy-ent-

869 Kaur-16-en-19-oic acid,
     9β,11β-dihydroxy-15α-angeloyloxy-ent-

676 Kaur-16-en-19-oic acid, 3β,
     9β-dihydroxy-15α-senecioyloxy-ent-

677 Kaur-16-en-19-oic acid, 3β,9β-dihydroxy-15α-tigloyloxy-ent-

Table 1. Compound Numbers and Names 53

TABLE 1. *(contd.)*

Compound:
No. Name

945 Doronicoside D

678 Carboxyatractyloside

679 Wedeloside

680 Wedeloside, L-rhamnopyranosyl-

681 **Atractyloside**

895 Paniculoside I

896 Paniculoside II

897 Paniculoside III

898 Paniculoside IV

899 Paniculoside V

682 Steviolbioside

683 Stevioside

684 Rebaudioside B

685 Rebaudioside A

686 Dulcoside A

687 Dulcoside B

688 Rebaudioside D

689 Rebaudioside E

690 Kaur-9(11),16-dien-19-oic acid, ent-

691 Kaur-9(11),16-dien-19-oic acid, 2β-hydroxy-ent-

692 Kaur-9(11),16-dien-19-oic acid, 3β-hydroxy-ent-

693 Kaur-9(11),16-dien-19-oic acid, 7β-hydroxy-ent-

694 Kaur-9(11),16-dien-19-oic acid, 12β-hydroxy-

998 Kaur-9(11),16-dien-19-oic acid, 12β-ethoxy-ent-

695 Kaur-9(11),16-dien-19-oic acid, 12-oxo-ent-

696 Kaur-9(11),16-dien-19-oic acid, 15α-hydroxy-ent-

697 Kaur-15-ene, 17-hydroxy-ent-   [17-hydroxy-isokaurene]

698 Kaur-15-ene, 17,19-dihydroxy-ent-

699 Kaur-15-ene, 17-hydroxy-19-[p-hydroxy-hydrocinnamoyloxy]-
     [17-hydroxy-19-[p-hydroxy-hydrocinnamoyloxy]-isokaurene]

TABLE 1. *(contd.)*

Compound:
No.   Name

946   Kaur-15-en-19-al, (-)-

700   Kaur-15-en-19-oic acid, ent-

701   Kaur-15-en-19-oic acid, 17-hydroxy-ent-
      [17-hydroxy-ent-isokaur-15(16)-en-19-oic acid]

702   Kaur-15-en-19-oic acid, 17-oxo-ent-

703   Kaur-15-en-17,19-dioic acid, ent-

704   Kaur-9(11),15(16)-dien-19-oic acid, 3α-isovaleryloxy-
      [3α-isovaleryloxy-polymnia acid]

705   Kaur-9(11),15(16)-dien-19-oic acid, 3α-isobutyryloxy-ent-
      [3α-isobutyryloxy-polymnia acid]

706   Kaur-9(11),15(16)-dien-19-oic acid, 3α-tiglyloxy-ent-
      [3α-tiglyloxy-polymnia acid]

707   Kaur-15(16)-en-19-oic acid, 3α-isovaleryloxy-9β-hydroxy-ent-

708   Kaur-15(16)-en-19-oic acid, 3α-isobutyryloxy-9β-hydroxy-ent-

733   Kaurane, 16α,17-epoxy-ent-

709   Kaurane, 3α-angeloyloxy-16α,17-epoxy-ent-

734   Kauran-19-ol, 16α,17-epoxy-ent-

710   Kaurane, 19-angeloyloxy-16α,17-epoxy-ent-

711   Kaurane, 3α-angeloyloxy-19-oxo-16α,17-epoxy-ent-

735   Kauran-19-al, 16α,17-epoxy-ent-

712   Kauran-19-al, 16β,17-epoxy-ent-

713   Kauran-19-oic acid, 16α,17-epoxy-ent-

714   Kauran-19-oic acid, 16β,17-epoxy-ent-

715   Kauran-19-oic acid, 15α-angeloyloxy-16,
      17-epoxy-ent-   [Perymenium acid]

716   Kauran-19-oic acid, 15α-acetoxy-16,17-epoxy-ent-

717   Kauran-19-oic acid, 15β-angeloyloxy-16,17-epoxy-ent-

718   Kauran-19-oic acid, 15β-tigloyloxy-16,17-epoxy-ent-

719   Kauren-19-oic acid, 12β-hydroxy-16α,17-epoxy-16,
      17-dihydro-9(11)-dehydro-ent-

720   Kaurane, 16α-hydroxy-ent-   [(-)-kauranol]

1001 Kauran-16β-ol, (-)-

721   Kaurane, 3β-acetoxy-16α-hydroxy-ent-

Table 1. Compound Numbers and Names   55

TABLE 1. *(contd.)*

Compound:

No. Name

722 Kaurane, 16α-hydroxy-19-acetoxy-ent-

723 Kauran-19-oic acid, 16α-hydroxy-ent-

724 Kaur-11-en-19-oic acid, 16α-hydroxy-ent-

725 Kaurane, 16α,17-dihydroxy-ent-   [17-hydroxy-ent-kauranol]

726 Kaurane, 16α,17,19-trihydroxy-ent-

727 Kaurane, 16α,17-dihydroxy-3-oxo-ent-   [Abbeokutone]

851 Kaurane, 16β-hydroxy-ent- [ent-kauran-16β-ol]

852 Kaurane, 16β,19-dihydroxy-ent- [ent-kauran-16β,19-diol]

728 Kauran-19-oic acid; 16α,17-dihydroxy-ent-

924 Kauran-19-oic acid, 17-hydroxy-16α-H-ent-

925 Kauran-17,19-dioic acid, 16α-H-ent-

822 Kauran-19-oic acid, 16β,17-dihydroxy-ent-

729 Kauran-17-al, 16β-H-ent-

730 Kauran-17-al, 19-hydroxy-16β-H-ent-

731 Kauran-17-oic acid, 16β-H-ent-

732 Kauran-17-oic acid, 19-hydroxy-16β-H-ent-

736 Kaurane, 16α-acetyl-19-hydroxy-16-desmethyl-ent-

737 Kauran-17,19-dial, 16β-H-ent-

738 Kauran-19-oic acid, 17-hydroxy-16β-H-ent-

739 Kaur-9(11)-en-19-oic acid, 16α,17-dihydroxy-ent-

989 Kauran-19-al-17-oic acid, 16α(-)-

990 Kauran-17,19-dioic acid, 16α(-)-

740 Kauran-17-al, 18-[p-hydroxyhydrocinnamoyloxy]-ent-

741 Kauran-19-oic acid, 17-isobutyrloyloxy-16β-H-ent-

742 Kauran-19-oic acid, 17-[n-dodecanoyloxy]-16β-H-ent-

743 Kauran-19-oic acid, 17-[n-tetradodecanoyloxy]-16β-H-ent-

744 Kauran-19-oic acid, 17-[n-hexadecanoyloxy]-16β-H-ent-

745 Kauran-19-oic acid, 17-[n-octadecanoyloxy]-16β-H-ent-

746 Kauran-19-oic acid, 15α,16α-epoxy-17-hydroxy-ent-

747 Kauran-19-oic acid, 11β-hydroxy-15-oxo-16α-H-ent-

TABLE 1. (*contd.*)

Compound:	
No.	Name

748 [= 815 = 747]

942 Kaurenal, 9,11-dehydro-ent-

943 Kaur-16-en-19-oic acid, 9(11)-dehydro-15α-cinnamoyloxy-ent-

850 Kaur-16-en-19-oic acid thujanol ester, ent-

1255 Kauran-19-al, 16α-hydroxy-

1256 Ruilopezia acid [16α-Hydroxy-kaur-9(11)-en-19-oic acid]

59. ent-Kaurane, 17-Nor-

749 Kauran-19-oic acid, 17-nor-16-oxo-ent-

60. ent-Kaurane, 18-Nor-

752 Kaur-16-ene, 18-nor-ent-

753 Kaur-16-ene, 4β-hydroxy-18-nor-ent- [4-epi-ruilopeziol]

754 Athrixianone, 4-hydroxymethyl-

755 Athrixianone, 4-formyl-

756 Athrixianone, 4-carbomethoxy-

757 Athrixianone, 16,17-dihydro-16,17-epoxy-

758 Atractyligenin

813 Kaurene, 4β,19-epoxy-18-nor-ent-

61. ent-Kaurane, 19-Nor-

750 Kaur-16-ene, 19-nor-ent-

751 Kaur-16-ene, 4α-hydroxy-19-nor-ent- [Ruilopeziol]

62. ent-Kaurane, 9,10-Seco-

759 Terminaloic acid

866 Wederegiolide, 15α-hydroxy-

867 Wederegiolide, 15β-hydroxy-

868 Wederegiolide, 15β-acetoxy-

760 Wedelia-seco-kaurenolide, 9-oxo-
     [wedelia-seco-kaurenolide]

Table 1. Compound Numbers and Names   57

TABLE 1. *(contd.)*

Compound:

No.   Name

761   Wedelia-seco-kaurenolide, 9α-hydroxy-9-desoxo-

762   Wedelia-seco-kaurenolide, 9β-hydroxy-9desoxo-

763   Wedelia-seco-kaurenolide, 15β-hydroxy-

64. ent-Stachane (ent-Beyerane)

764   Beyer-15-en-3α-ol, ent-

765   Beyer-15-en-19-ol, ent-   [Erythroxylol-A]

766   Beyer-15-en-18-ol, ent-

767   Beyer-15-en-3α,19-diol, ent-

768   Beyer-15-en-3α,12β-diol, ent-

769   Erythroxylol A-malonate

770   Beyer-15-en-3α,17-diol, ent-

817   Beyer-15-en-12β,19-diol, ent-

771   Beyer-15-en-17,19-diol, ent-   [17-hydroxymonogynol]

772   Beyer-15-en-18,19-diol, ent-

774   Beyer-15-en-19-oic acid, ent-   [stach-15-en-19-oic acid; beyerenic acid]

773   Beyer-15-ene, 19-oxo-ent-   [20-oxostachene]

775   Beyer-15-en-19-oic acid; 3α-hydroxy-ent-

776   Beyer-15-en-19-oic acid, 3α-tigloyloxy-ent-

777   Beyer-9(11),15-dien-19-oic acid, ent-

778   Beyer-9(11),15-dien-19-oic acid, 3α-tigloyloxy-ent-

779   Beyeran-18-ol, 15β,16β-epoxy-ent-

780   Beyeran-19-ol, 15β,16β-epoxy-ent-   [erythroxylol A-14, 15-oxide]

781   Nidoanomalin

93. ent-Stachane, 19-Nor-

782   Beyer-15-ene, 4α-hydroxy-19-nor-ent-

65. Atisirane (Atisane)

TABLE 1. *(contd.)*

Compound:

No.   Name

783   Atis-16-en-19-oic acid

853   Atisan-16α-ol, ent-

854   Atisan-16β-ol, ent-

814   Atis-16-en-19-oic acid, 13-hydroxy-

816   Atis-13-en-3α,16β-diol, ent-

784   Atis-16-en-19-oic acid, 7α-hydroxy-ent-   [occidentalic acid]

785   Atisirene acid, 11α-acetoxy-
      [11α-acetoxy-atisirene-20-acid]

786   Atisirene acid, 13α-angeloyloxy-

787   Atisirene acid, 13α-isovaleryloxy-

788   Atisirene acid, 13α-isobutyryloxy-

789   Atisirene acid, 7α-acetoxy-9,11-didehydro-   [7α-acetoxy-9,
      11-didehydroatisirene-20-acid]

790   Atisirene acid, 14β-acetoxy-9,11-didehydro- [14β-acetoxy-9,
      11-didehydro-atisirene-20-acid]

66. Trachylobane

855   Trachyloban-19-al, ent-

791   Trachyloban-19-oic acid

825   Trachyloban-19-oic acid, 15α-acetoxy-

826   Trachyloban-19-oic acid, 15α-angeloyloxy-

827   Trachyloban-19-oic acid; 15α-isovaleryloxy-

893   Trachyloban-19-oic acid, 15α-isobutyryloxy-

792   Trachyloban-19-oic acid, 7α-hydroxy-   [Ciliaric acid]

793   Trachyloban-19-oic acid, 11-oxo-

794   Trachylobanoic acid, 9,11-dehydro-

795   Trachyloban-19-oic acid,
      11α-[16'α-hydroxy-ent-kaur-11'-en-19'-oyloxy]-

856   Trachyloban-19-oic acid thujanol ester, ent-

1204 Tachyloban-19-oic acid methyl ester

1259 Trachyloban-19-oic acid, 7β-hydroxy-

67. Helifulvane

796   Helifulvan-19-ol

Table 1. Compound Numbers and Names   59

TABLE 1. (*contd.*)

Compound:

No.   Name

797   Helifulvan-19-oic acid

798   Helifulvan-19-oic acid, 11α-hydroxy-

799   Helifulvan-19-oic acid, 11α-acetoxy-

68. Tetrachyrane

800   Tetrachyrin   [Zoapatlin]

847   Zoapatlin, 15-oxo-

805   Eupatalbin

806   Eupatoralbin

78. Villanovane

875   Villanovane, 13α,17-dihydroxy-19-isovaleryloxy-

876   Villanovane, 17-acetoxy-13α-hydroxy-19-isovaleryloxy-

877   Villanovane, 13α,17-dihydroxy-19-(3-methylvaleryloxy)-

878   Villanovane, 17-acetoxy-13α-hydroxy-19-(3-methylvaleryloxy)-

879   Villanovane, 17-acetoxy-13α-hydroxy-3α-isovaleryloxy-

880   Villanovane, 17-acetoxy-13α-hydroxy-3α-(3-methylvaleryloxy)-

894   Villanovane, 3α,15α-dihydroxy-

69. Lycoctonine

801   Lycaconitine, methyl-

802   Lycoctonine   [royline; roylene]

803   Lycoctonine, anthranoyl-

VIII. Undetermined Stereochemistry

79. Inulaefane

804   Inulaefol

84. Hebeclinane (9,10-Seco-Labdane)

940   Jhanic acid

999   Hebeclinolide

TABLE 2. The Substitutional Patterns of the Diterpenes Reported from the Compositae (Asteraceae).

Compound: No.	Substituents

I. Linear or Unicarbocyclic

1. Geranylnerol

1	1(OH)
2	1(OH);18(OH)
3	1(OH);20(OH)
4	1(OH);17(OH);20(OH)
5	1(=O)
6	18(=O,OH);1(OH);16(OH)
7	13(=O);1(OH);19(OH)
8	16(=O);1(OH);19(OH);20(OH)
15	1(OH);12(OH);20(OH);19(OAc)
984	1(OH);20(OH);19(OAc)
948	1(OH);8(OH);12(OH);19(OH)
957	1(OH);16(OH);18(OH);19(OH)
978	12(=O);1(OH);19(OH)
1002	19(=O,OH);1(OH)
1003	19(=O,OH);1(OAc)
1109	1(OH);17(OH);18(OH)
1110	1(OH);17(OH);18(OH);20(OH)
1111	1(=O);12(=O)
1112	1(OH);16(OH);20(OH);18(OAc);19(OAc)
1113	1(OH);16(OH);18(OAc);19(OAc);20(OAc)
1114	1(OH);16(OH);20(OH);18(OAc)

Table 2. Substitutional Patterns   61

TABLE 2. (*contd.*)

Compound: No.	Substituents
1115	1(OH);16(OH);20(OH);19(OAc)
1116	1(OH);9(OH);17(OH)
1117	1(OH);9(OH);17(OH);19(OH)

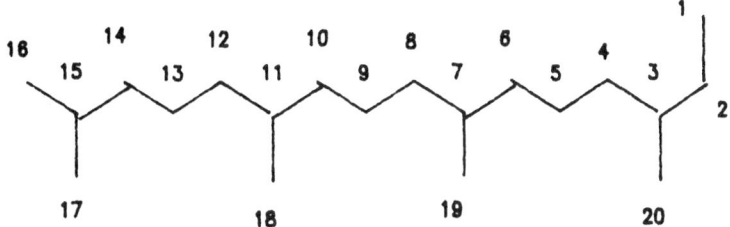

14	2(3)Z;6(7)Z;10(11)E;14(15);19(=O,OH);18(=O);1(OH)
949	2(3)Z;10(11)E;14(15);6(-O-)7;1(OH);8(OH); 12(OH);19(OH)
979	2(3)Z;10(11)E;14(15);12(=O);1(OH);19(OH);6(-O-)7
1004	2(3)Z;6(7)Z;10(11)E;14(=O);1(OH);15(OH);19(OH)
1005	2(3)Z;6(7)Z;10(11)E;14(=O);15(OH);1(OAc);19(OAc)
1095	2(3)E;14(15);12(=O);1(OH);20(OH);6(-O-)7
1096	2(3)E;13(14)E;12(=O);1(OH);15(OH);20(OH);6(-O-)7

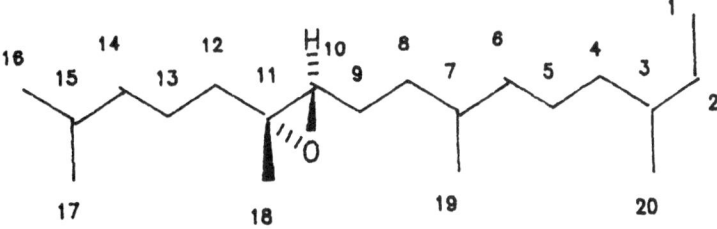

1020	2(3)E;6(7)E;14(15);1(OH);9(OH);20(OAc)

2. Geranylgeraniol

TABLE 2. (*contd.*)

Compound: No.	Substituents
11	1(OH);12(OH);19(OH)
12	12(OH);19(OH);1(OAc)
13	1(OH);16(OH)
17	1(=O)
18	1(=O);4α(OH)
19	17(=O,OH);1(OAc)
954	18(=O,OH);1(OH)
1118	1(OH);17(OH);18(OAc)
1119	1(=O);12(=O)

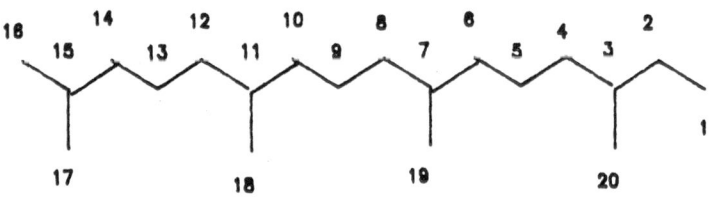

16	2(3)E;6(7)E;10(11)E;1(OH);4(OH);14(OH);15(OH)
20	6(7)E;10(11)E;13(=O);1(OH)
21	2(3)E;6(7)Z;14(15);12(=O);1(OH);19(OH)
22	2(3)E;6(7)Z;10(11)E;14(15);1(OH);12(OH);18(OH)
1120	2(3)E;6(7)E;10(11)E;14(15);18(=O);19(=O);1(OH)

3. Geranyllinalool

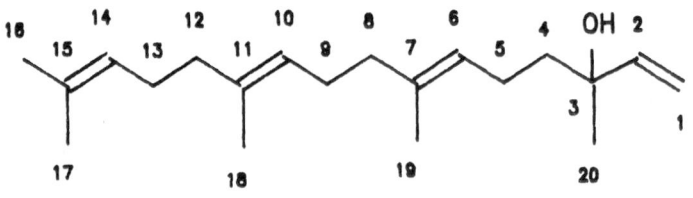

23	
24	13(OH)
25	13(OAc)

Table 2. Substitutional Patterns 63

TABLE 2. *(contd.)*

Compound: No.	Substituents
26	9(OH)
27	9(OAc)
28	5(OH)
29	5(OH);9(OAc)
30	5(OAc);9(OAc)
31	5(OH);13(OAc)
32	12(=O)
33	13(=O)
1121	13(OAc)

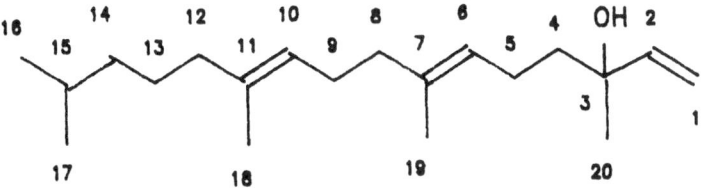

34	13(=O);14(-O-)15
35	13(14)E;15(OH)
36	13(14)E;15(17)
37	14(OH);15(OH)
38	5(OH);14(OH);15(OH)
39	14(OH);15(OH);9(OAc)

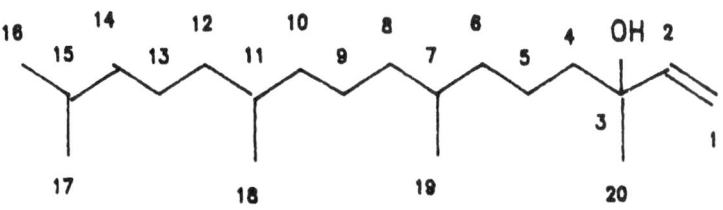

1122	6(7)Z;10(11)Z;14(15);12(=O)
1123	6(7)Z;13(14)E;12(=O);15(OH)

TABLE 2. *(contd.)*

Compound:
No.   Substituents

4. **Phytol-Derived**

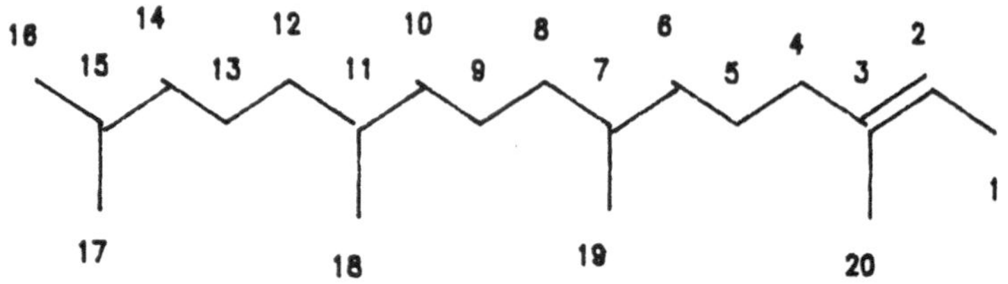

59   1(2);3(20)

80. **14-Methylgeranylnerol**

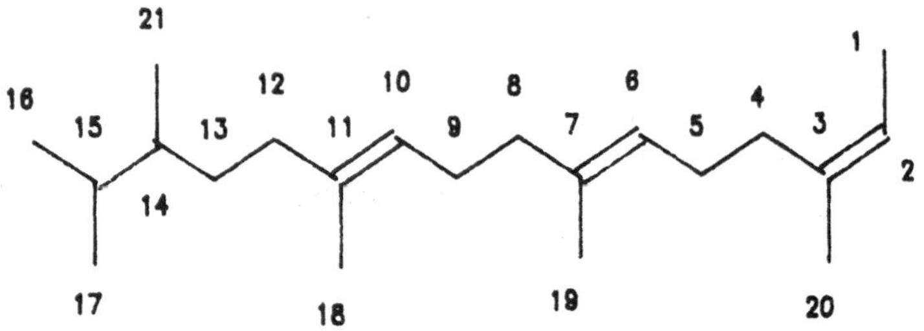

9   6(7)Z;10(11)E;14(21);1(OH);12(OH);19(OH);
          20(OH)

10   6(7)Z;10(11)E;14(21);1(OH);12(OH);20(OH);
          19(OAc)

980   6(7)Z;10(11)E;13(14)E;12(=O);1(OH);19(OH)

981   10(11)E;13(14)E;12(=O);6(-O-)7;1(OH);
          19(OH)

982   6(7)Z;10(11)E;14(21);12(=O);1(OH);19(OH)

983   10(11)E;14(21);12(=O);1(OH);19(OH);6(-O-)7

1094   14(21);12(=O);1(OH);20(OH);6(-O-)7

5. **Acanthoaustralane**

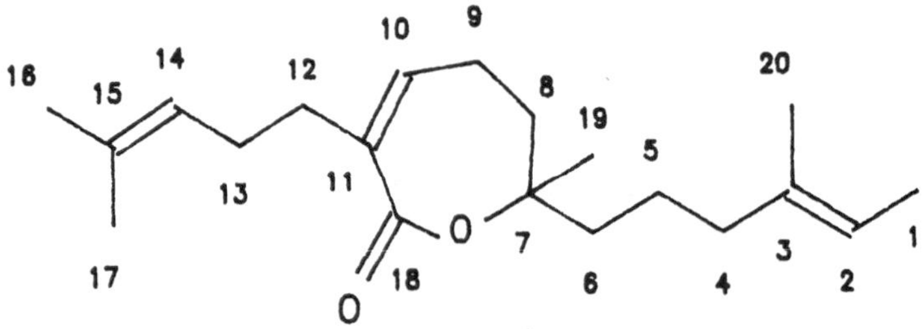

40   1(OH);6(OH)

Table 2. Substitutional Patterns   65

TABLE 2. (*contd.*)

Compound:No.	Substituents
41	6(OH);1(OAc)
42	1(OH);6(OH);17(OH)
43	1(OH);6(OH);17(OAc)
1124	1(OAc);6(OAc)

6. Acanthoaustralane-6,11-Epoxide

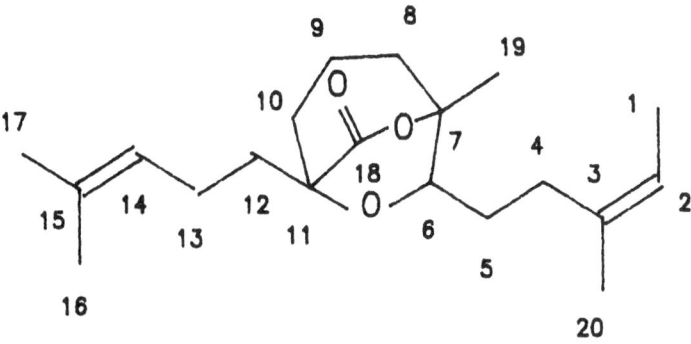

44	1(OH);10(OH);17(OAc)
45	1(OH);10(OH);17(OH)

7. Isoacanthoaustralane

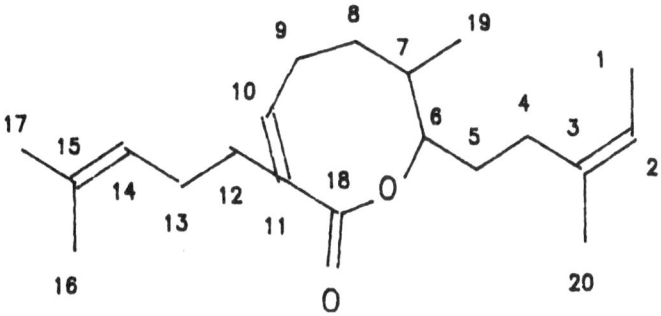

46	1(OH);7(OH);17(OH)
47	1(OH);7(OH);17(OAc)
48	7(OH);1(OAc)

TABLE 2. *(contd.)*

Compound:

No.   Substituents

8. **Geranylgeraniol-18,8-Lactone**

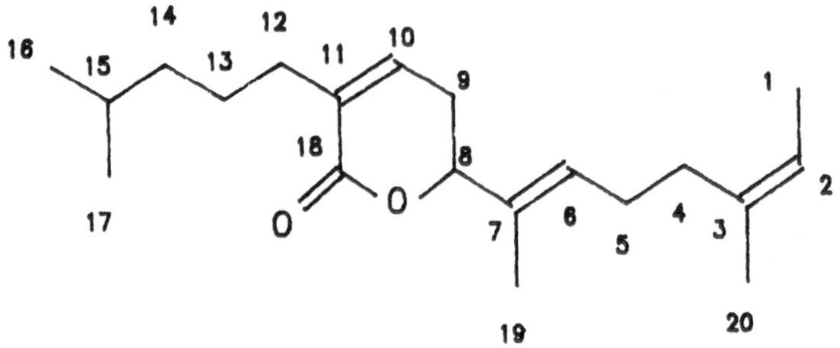

49   14(15);1(OH)

50   14(15);1(OH);19(OAc)

51   13(14)E;1(OH);15(OOH)

52   13(14)E;1(OH);15(OH)

53   15(17);1(OH);14(OOH)

54   15(17);1(OH);14(OH)

9. **Oxepane**

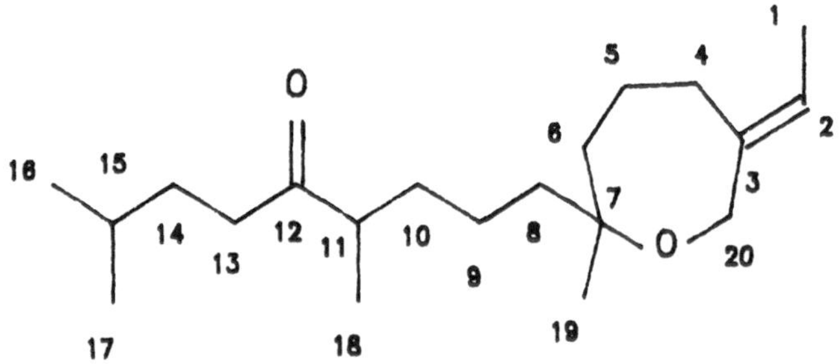

55   14(15);1(OH);6α(OH)

56   13(14)E;6(OH);15(OH);1(OAc)

57   13(14)E;1(OH);6α(OH);14(Me)

1097   1(OH);6α(OH);14(CH2)

1098   13(14)E;1(OH);6α(OH);15(OH)

1099   15(16);1(OH);6α(OH);14(CH3)

Table 2. Substitutional Patterns   67

TABLE 2. *(contd.)*

Compound:
No.   Substituents

86. Melcanthane

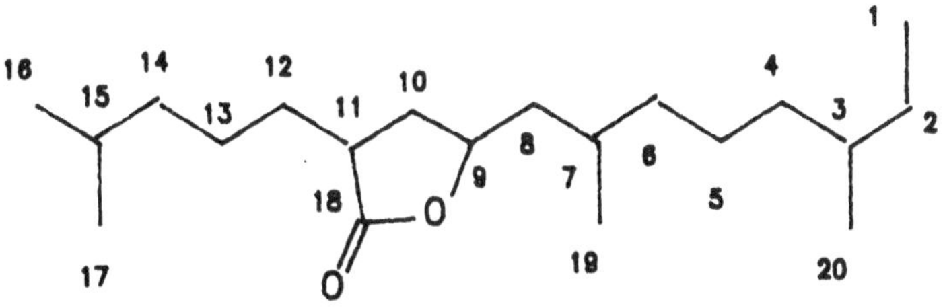

1006   2(3)Z;10(11)Z;14(15);1(OH);6(OH);7(OH);17(OAc)

10. Mikanofurane

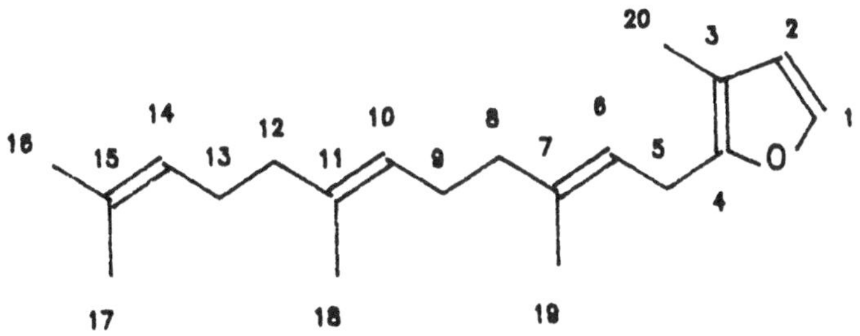

58

11. Centipedane

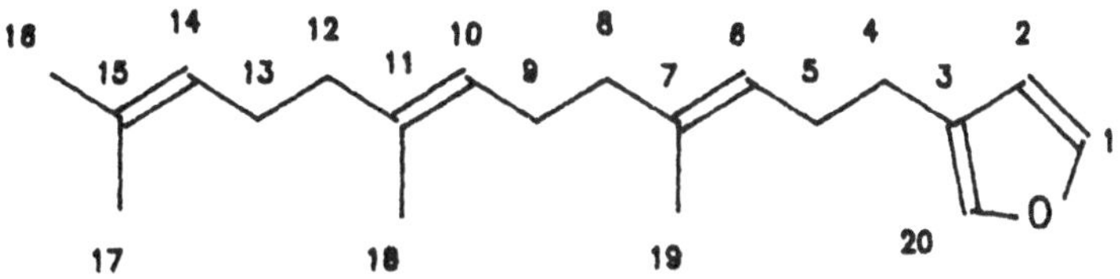

60   18(=O,OH)

61   18(OH);19(OH)

1125   12(=O)

1126   12(OH)

1127   12(=O);13(OH)

TABLE 2. *(contd.)*

Compound:
No.   Substituents

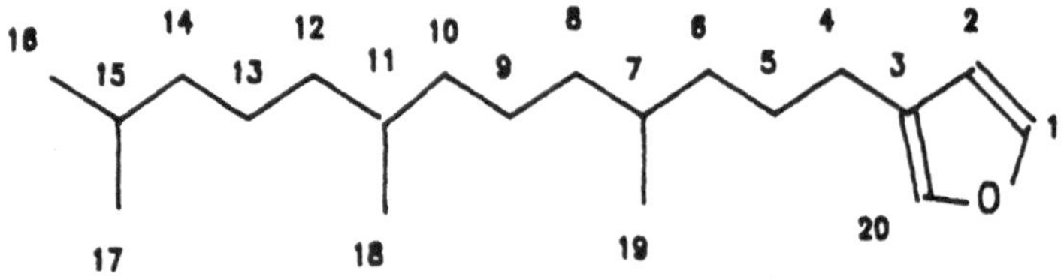

1128   6(7)E;10(11)E;13(14)E;12(=O);15(OOH)

12. Geranylgeraniol-1,20-Lactone

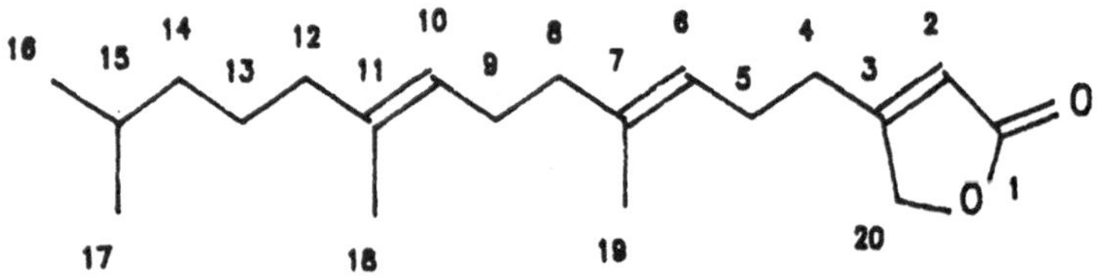

62   14(15);17(=O,OH)

63   14(15);17(=O,OMe)

64   14(OH);15(OH)

1129   14(15);12(=O)

1130   14(15);18(=O,OH);17(OH)

1132   14(15);18(=O,OH);17(OAc)

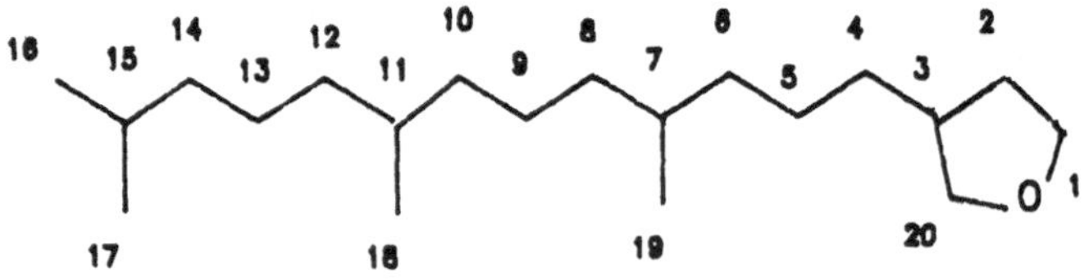

1131   2(3);6(7)E;10(11)E;14(15);18(=O,OH);1(=O);17(OH)

1133   2(3);14(15);20(=O);17(OH);19(OH);18(OAc)

1134   2(3);14(15);20(=O);17(OH);18(OAc);19(OAc)

1135   2(3);6(7)E;10(11)Z;14(15);18(=O,OH);20(=O);17(OH)

1136   2(3);6(7)E;10(11)Z;14(15);18(=O,OH);20(=O);17(OAc)

Table 2. Substitutional Patterns 69

TABLE 2. *(contd.)*

Compound:	
No.	Substituents
1137	6(7)E;10(11)E;14(15);12(=O);1β(-O-)2β;20(=O)

13. Geranylterpinene

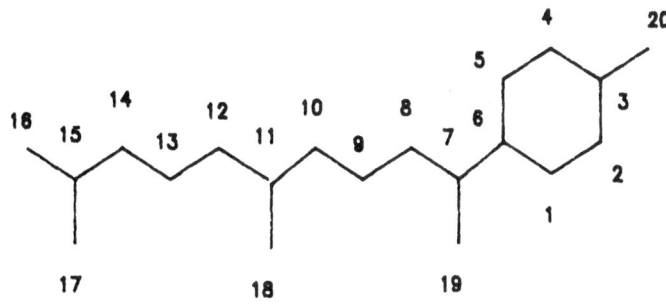

65	2(3);7(8)E;10(11)E;14(15)E;16(OH)
66	2(3);7(8)E;10(11)E;14(15)E;16(=O)
67	2(3);7(8)E;10(11)E;16(=O,OH)
68	1(6);2(3);10(11)E;14(15)
69	1(6);2(3);14(15);10α(OH);11(OH)
71	1(6);2(3);4(5);14(15);10α(OH);11(OH)
70	1(6);2(3);4(5);10(11)E;14(15)
72	14(15);3β(-O-O-)6β;10α(OH);11(OH)
1101	14(15);3α(-O-O-)6α;10α(OH);11(OH)
1138	2(3);10(11)E;14(15);7(OH)

14. Isocembrene

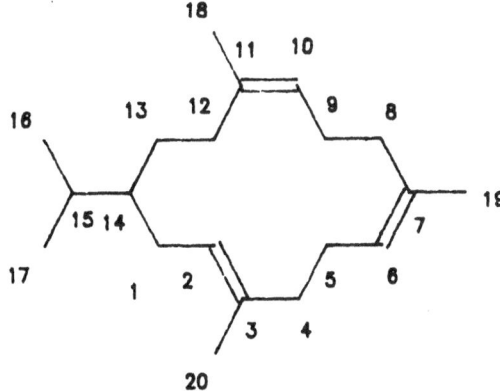

73	15(OH)

TABLE 2. *(contd.)*

Compound:
No.   Substituents

15. Geranylnerol-11,14-α-Epoxide

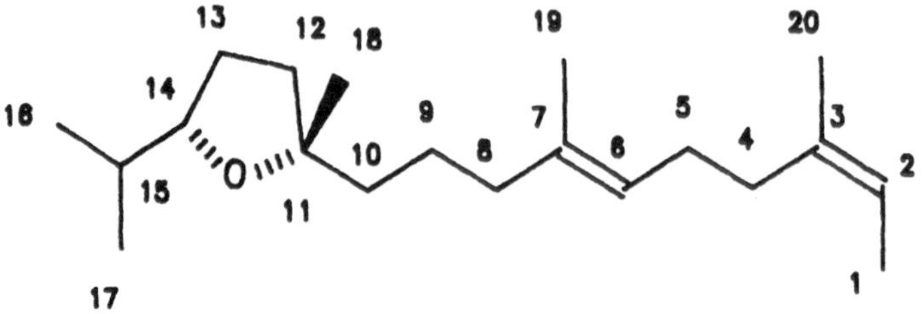

74   1(OH);10α(OH);15(OH);19(OH)

75   1(OH);10α(OH);15(OH);19(OAc)

II. Normal-Bicarbocyclics and Derivatives

16. Normal-Labdane

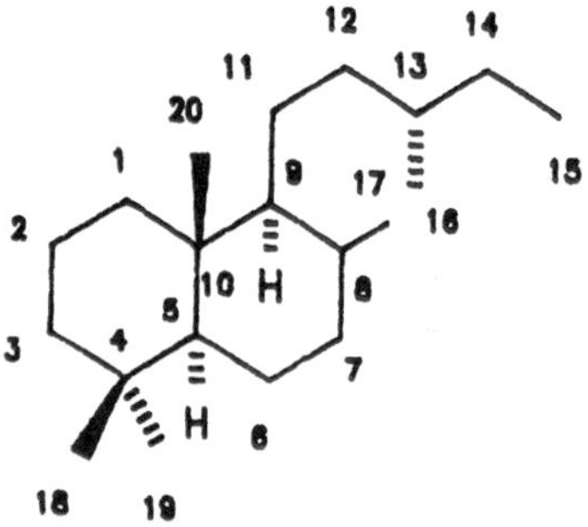

79   8α(OH);15(OH)

80   7β(OH);8α(OH);15(OH)

958   15(=O,OH);8α(OH)

1139   8α(OH);15(OH)

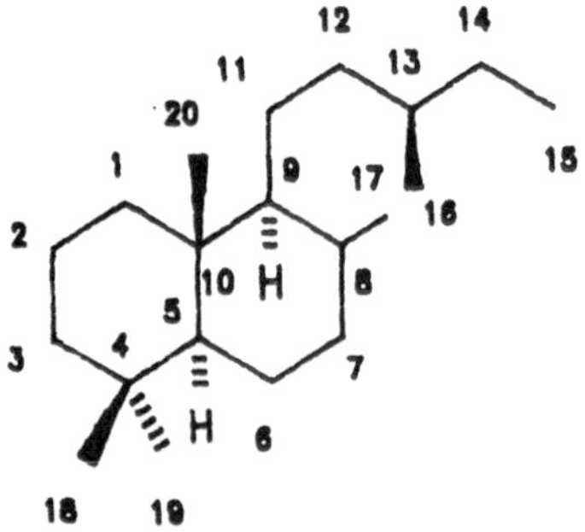

81   8α(OH);15(OH)

Table 2. Substitutional Patterns   71

TABLE 2. *(contd.)*

Compound: No.	Substituents
86	7(8);13β(OH);14β(OH);15(OH);14α(H)
87	8(9);13β(OH);14β(OH);15(OH);14α(H)
124	7(8);15(=O,OH);2α(OH);3α(OH)
125	7(8);15(=O,OH);2α(OH);3α(OMebut-2-OH)
129	7(8);15(=O,OH);3(=O)
1140	8(17);15(=O,OH);7α(OH);19(OH)
1141	8(17);15(=O,OH);7α(OH);19(OAc)
1142	8(17);15(=O,OH);7β(OH);19(OH)

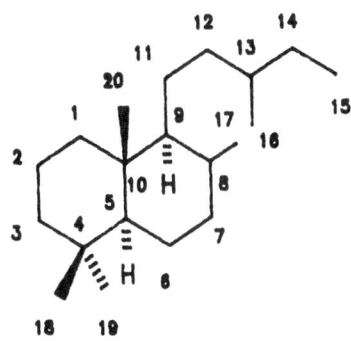

76	8(17);13(14)E;15(OH)
77	8(17);13(14)E;3α(OH);15(OH);7α(Oβ-Gal)
78	8(17);13(14)Z;2α(OH);15(OH)
1103	13(14)E;8α(OH);15(OH)
82	12(13)E;9α(OH);14α(OH);15(OH);8α(H)
83	12(13)E;9α(OH);14α(OH);15(OAc);8α(H)
84	12(13)E;9α(OH);14α(OAc);15(OAc);8α(H)
85	8(17);13(-O-)14;15(OH)
88	8(17);14(OH);15(OH)
89	8(17);13(OH);14(OH);15(OH)
90	8(17);14(OH);15(OAc)
993	7(8);3α(OH);15(OH);13β(H)
994	7(8);15(O2-Mebut);3α(OH);13β(H)
995	8(17);3α(OH);7α(OH);15(OH);13β(H)

TABLE 2. (*contd.*)

Compound: No.	Substituents
91	8(17);13(OH);14(OH);15(OAc)
92	8(17);15(OH);14(OAc)
93	8(17);13(OH);14(OPro);15(OAc)
94	12(13)E;9α(OH);14β(OH);15(OH);8α(H)
97	7(8);13(14)Z;15(=O,OH);12(=O);17(=O);2α(OAng);3α(OAng)
98	13(14)Z;15(=O;OH);8α(OH);7α(O-trans-Cinn);2(OAng or OMebut-2-OH);3(OAng or OMebut-2-OH)
99	13(14)Z;15(=O,OH);8α(OH);7α(O-cis-Cinn);2(OAng or OMebut-2-OH); 3(OAng or OMebut-2-OH)
100	7(8);13(14)Z;15(=O,OH);2(OAng)
116	7(8);13(14)Z;15(=O,OH);3α(OH)
101	7(8);13(14)Z;15(=O,OH);18(OAng)
102	7(8);13(14)Z;15(=O,OH);18(OTig)
103	7(8);13(14)Z;15(=O,OH);2α(OH);3α(OAng)
104	7(8);13(14)Z;15(=O,OH);2α(OH);3α(O-trans-Cinn)
105	7(8);13(14)Z;15(=O,OH);2α(OH);3α(O-cis-Cinn)
106	7(8);13(14)Z;15(=O,OH);2α(OH);3α(OAng)
107	7(8);13(14)Z;15(=O,OH);3α(OH);2α(OAng)
108	8(17);13(14)Z;15(=O,OH);2α(OH);3α(OAng)
109	8(17);13(14)Z;15(=O,OH);3α(OH);2α(OAng)
110	7(8);13(14)Z;15(=O,OH);19(=O,OH)
111	8(17);13(14)E;15(=O,OH)
112	7(8);13(14)E;15(=O,OH);2α(OTig)
113	7(8);13(14)E;15(=O,OH);3β(OAc)
114	7(8);13(14)E;15(=O,OH);3α(OH)
115	7(8);13(14)E;15(=O,OH);3β(OH)
117	7(8);13(14)E;15(=O,OH);3(=O)
118	7(8);15(=O,OH)
991	7(8);15(=O,OH);3α(OH);13β(H)
992	7(8);15(=O,OMe);3α(OH);13β(H)
119	7(8);15(=O,OH);3α(OH)

Table 2. Substitutional Patterns 73

TABLE 2. (*contd.*)

Compound: No.	Substituents
120	7(8);15(=O,OH);3β(OH)
121	15(=O,OH);8α(OH)
871	7(8);2α(OH);15(OH)
872	7(8);2α(OH);15(OCoum)
873	7(8);2α(OH);15(O3-Me-Coum)
874	7(8);2α(OH);15(p-OH-Cinn)
122	7(8);15(=O,OH);18(OAng)
123	7(8);15(=O,OH);18(OTig)
126	7(8);15(=O,OH);2α(OH);3α(OAng)
127	7(8);15(=O,OH);2α(OH);3α(O-trans-Cinn)
128	7(8);15(=O,OH);2α(OH);3α(O-cis-Cinn)
870	7(8);15(=O,OMe);3β(OAc);2α(OSen)
130	7(8);13(14)E;15(=O,OH);18(=O,OH)
977	8(17);19(=O,OH);15(=O);13α(H)
131	7(8);15(=O,OH);18(=O,OH)
132	8(17);12(13)E;14(15)
133	8(17);12(13)E;14(15);3α(OH)
134	8(17);12(13)E;14(15);3β(OH)
135	8(17);12(13)E;14(15);3β(OAc)
136	8(17);12(13)E;14(15);3(=O)
959	8(17);15(=O,OH);2β(OAc)
137	8(17);12(13)E;14(15);18(=O,OH)
1104	8(17);12(13)E;14(15);19(=O,OH)
138	8(17);12(13)E;14(15);18(=O,OH);7α(OAc)
139	12(13)Z;14(15);8α(OH)
140	12(13)Z;14(15);7β(OH);8α(OH)
141	12(13)E;14(15);7β(OH);8α(OH)
142	12(13)E;14(15);6α(OH);7β(OH);8α(OH)
143	12(13)E;14(15);7β(OH);8α(OH);6α(OAng)

TABLE 2. (*contd.*)

Compound: No.	Substituents
144	12(13)E;14(15);6β(OH);8α(OH)
145	12(13)E;14(15);8α(OH);7β(OAc)
146	12(13)E;14(15);6β(OH);7β(OH);8α(OH)
147	12(13)E;14(15);6β(OH);8α(OH);19(OH)
148	12(13)E;14(15);6β(OH);8α(OH);7β(OAc)
149	12(13)E;14(15);8α(OH);1β(OAc);19(OArac)
150	12(13)E;14(15);8α(OH);1β(OAc);19(OCoum)
151	12(13)E;14(15);8α(OH);1β(OAc);19(O-cis-Coum)
152	12(13)E;14(15);9α(OH);19(OH);8α(H)
153	14(15);8α(OH);13β(OH);6α(OAng)
1102	14(15);8α(OH);13β(OH)
996	8(17);14(15);13β(OH)
154	7(8);14(15);2(OH);13(OH)
1105	7(8);14(15);13β(OH)
1106	7(8);14(15);3β(OH);13β(OH)
1107	7(8);14(15);3(=O);13β(OH)
155	13(16);14(15);8α(OH)
156	13(16);14(15);6β(OH);8α(OH);7β(OAc);12β(O-OH)
157	13(16);14(15);6β(OH);8α(OH);7β(OAc);12α(O-OH)
158	7(8);13(16);14(15);6(=O);9α(OH);15(-O-)16
159	7(8);13(16);14(15);18(=O,OH);17(OH);15(-O-)16
160	7(8);13(16);14(15);17(=O,OH);6α(O2-Mebut);15(-O-)16
161	7(8);13(16);14(15);17(=O,OH);6α(Oi-But);15(-O-)16
162	7(8);13(16);14(15);17(=O,OH);6α(Oi-Val);15(-O-)16
163	8(17);13(16);14(15);19(=O,OH);7α(OH);15(-O-)16
1021	12(13)Z;14(15);8α(OH);7β(OAc)
1022	12(13)E;14(15);6α(OH);8α(OH);7β(OAc)
1023	12(13)Z;14(15);6α(OH);8α(OH);7β(OAc)
1024	13(14)Z;15(=O,OH);8α(OH)

Table 2. Substitutional Patterns   75

Tᴀʙʟᴇ 2. *(contd.)*

Compound:
No.   Substituents

1143   8(17);15(=O,OH);7α(OH)

1144   8(17);15(=O,OH);7α(OAc)

1145   7(8);15(=O,OH);19(OH)

1146   7(8);15(=O,OH);3β(OH)

1147   8(17);12(13)E;14(15);19(=O,OH)

1151   13(16);14(15);8α(OH);3β(OAc)

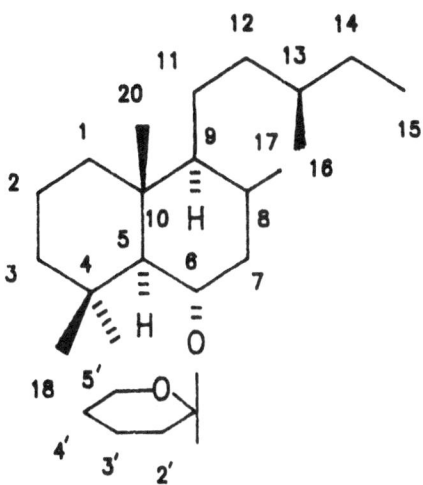

95   7(8);14(15);19(=O);13β(OH);2'α(OH);4'β(OH);3'β(OTig)

96   7(8);14(15);19(=O);13β(OH);2'α(OH);3'β(OH);4'β(OH)

1148   7(8);14(15);19(=O);13β(OH);3'β(OH);4'α(OH);2'α(OAc)

1149   7(8);14(15);19(=O,OMe);13β(OH);3'β(OH);4'α(OH);2'α(OAc)

1150   7(8);14(15);19(=O,OMe);13β(OH);2'α(OH);3'β(OH);4'α(OH)

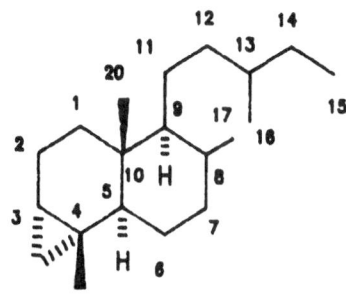

1152   13(16);14(15);8α(OH)

TABLE 2. (*contd.*)

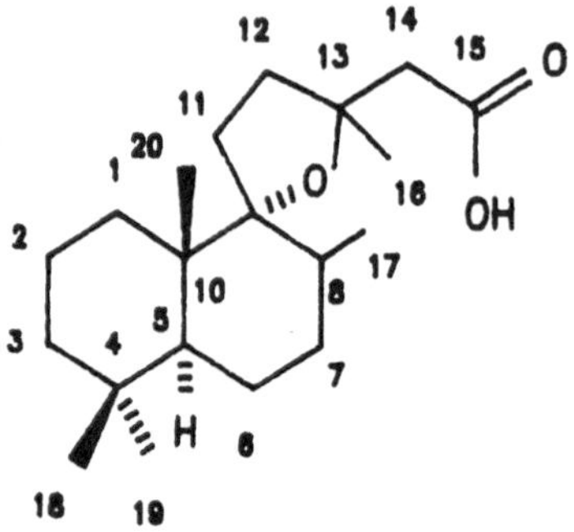

Compound:
No.   Substituents

17. Grindelane

165	8(17);7α(OH)
166	6(7);8(17)
167	7(8);13(S)
955	7(8);13(R)
168	7(8);1α(OH)
169	7(8);3β(OH)
170	7(8);6α(OH)
172	7(8);6β(OH)
173	7(8);17(OH)
171	7(8);6α(OCHO)
174	7(8);17(OAc)
175	7(8);17(OPro)
176	7(8);17(Oi-But)
177	7(8);17(Oi-Val)
178	7(8);17(O2-Mebut)
179	7(8);17(OMe)
180	7(8);19(OH)
181	7(8);19(OAc)
182	7(8);19(Oi-Val)
183	7(8);19(Oi-But)

Table 2. Substitutional Patterns 77

TABLE 2. *(contd.)*

Compound: No.	Substituents
184	7(8);19(O2-Mebut)
185	7(8);19(OSucc)
186	7(8);19(=O)
187	7(8);19(=O,OH)
188	7(8);18(OH)
189	7(8);6(=O)
190	7(8);6(=O);19(OAc)
191	7β(-O-)8β
192	7α(-O-)8α
1035	7(8);3α(OH)
1036	8(17);7β(OH)
1153	7(8);19(=O,OH);15(=O,OMe)

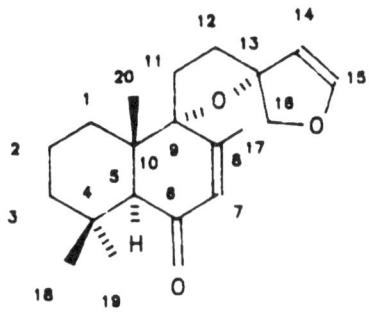

193

194

TABLE 2. (*contd.*)

Compound:
No.   Substituents

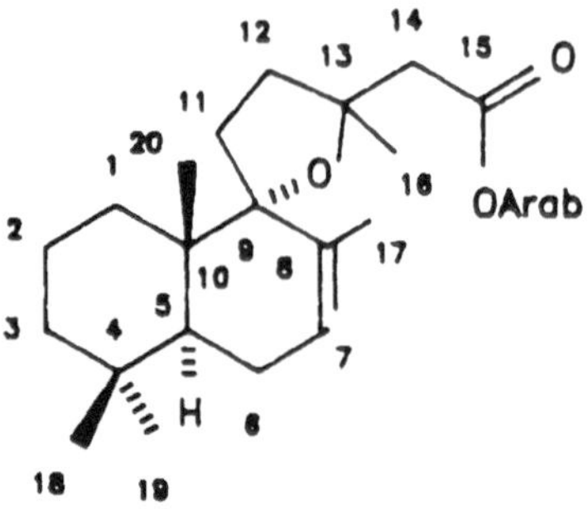

1060

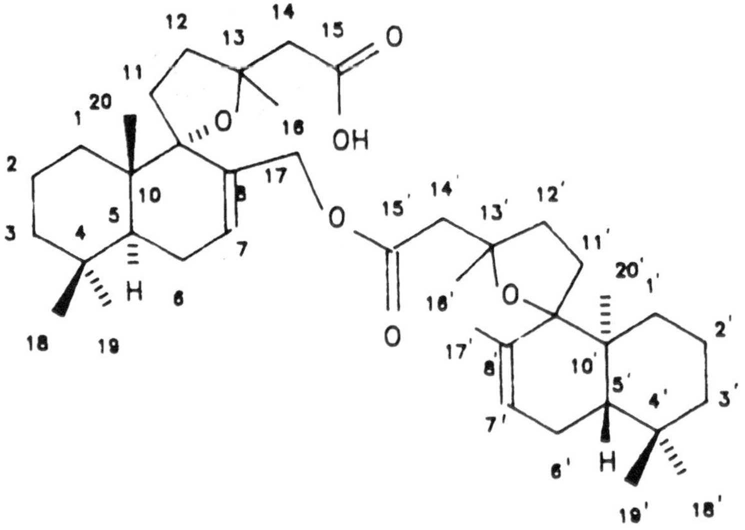

956

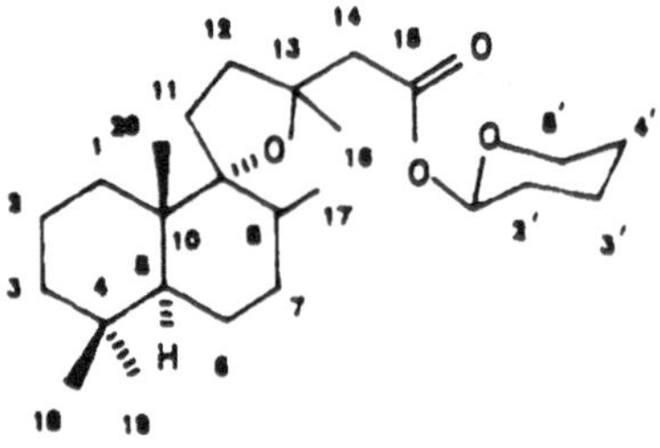

1187   7(8);2'α(OH);3'β(OH);4'β(OH)

Table 2. Substitutional Patterns 79

Table 2. (*contd.*)

Compound:
No.   Substituents

18. Normal-Labdane-8,13-Epoxide

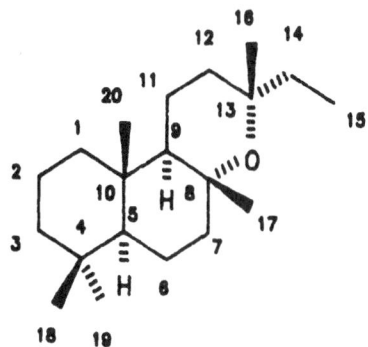

272	14(15)
195	14(15);19(OH)
196	14(15);19(OAc)
197	14(15);1β(OH);19(OH)
198	14(15);1β(OH);19(OAc)
199	14(15);1β(OAc);19(OAc)
200	14(15);3β(OAc)
201	14(15);3β(OH);18(OBenz)
964	15(=O)

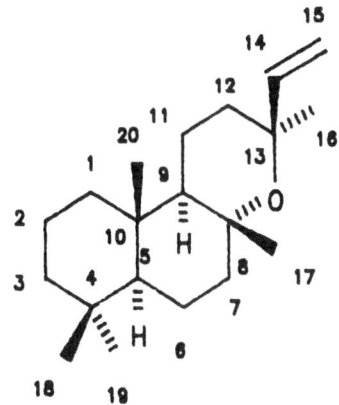

| 273 | |
| 202 | 19(OH) |

TABLE 2. (*contd.*)

Compound:	
No.	Substituents

19. Normal-Labdane-8,12-Epoxide

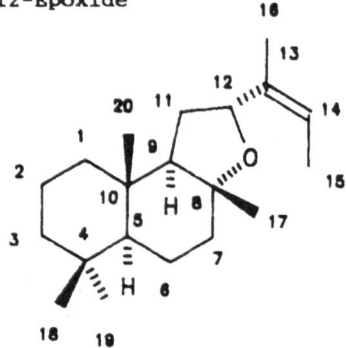

203	15(OH);16(OH)
204	15(OH);16(OAc)
205	15(OAc);16(OAc)
206	16(=O);15(OH)
207	15(=O);16(=O)
210	15(OH)
211	15(=O,OH)

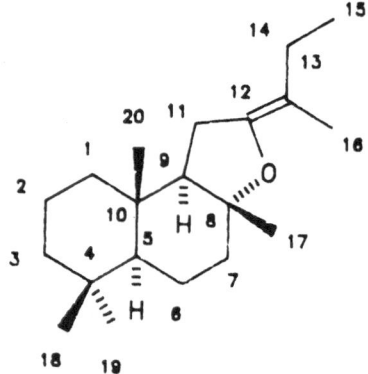

| 212 | 15(=O);16(=O) |
| 213 | 14(=O);15(=O);16(OAc) |

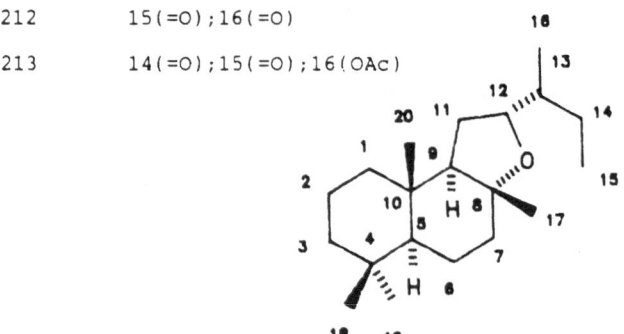

Table 2. Substitutional Patterns 81

TABLE 2. (*contd.*)

Compound:

No.	Substituents
208	15(OH);16(OH);13α(-O-)14α;14β(H)
209	15(OH);16(OH);13β(-O-)14β;14α(H)

20. Normal-Labdane-15,16-Lactone

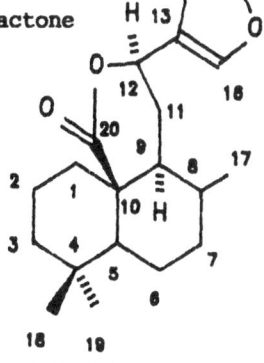

164	7(8);18(=O,OH)
965	7(8);17(OH)
966	7(8);17(=O)

21. Normal-Labdane-20,12-Lactone

214	1(=O);5α(OH);9α(OH);8β(H)

22. Normal-Labdane Spiroketal

215	7(8);13(14);2α(OAng);3α(OAng)

TABLE 2. (*contd.*)

Compound: No.	Substituents
216	7(8);13(14);17α(OH);2α(OAng);3α(OAng)
217	7(8);13(14);17β(OH);2α(OAng);3α(OAng)

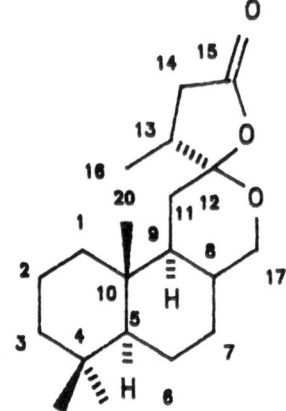

218	7(8);13(14);2α(OH);3α(OAng)

88. Normal-Labdane-15,12-Lactone

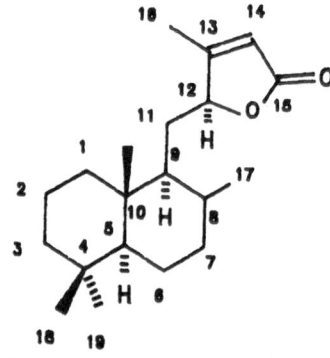

938	3α(OH);7α(OAc);8β(H)

23. Normal-Labdane, 7,8-Seco-

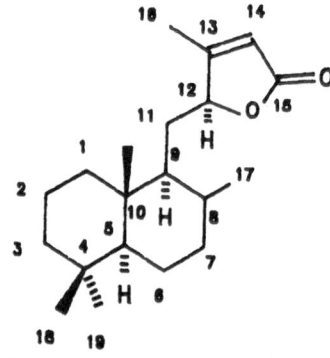

219	6(7);14(15);12(13)E;19(OH)

Table 2. Substitutional Patterns 83

TABLE 2. *(contd.)*

Compound:
No.   Substituents

24. Normal-Labdane, 3,4-Seco-

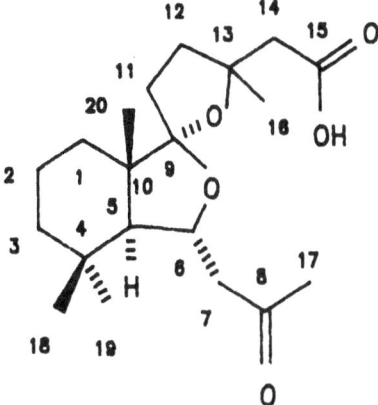

220   7(8);15(=O,OH)

25. Grindelane, 8,9-Seco-

223

89. Grindelane, 19-Nor-

221   7(8);15(=O,OH);4α(OH)

222   7(8);15(=O,OH);4α(OCHO)

1154   7(8);15(=O,OMe);6(=O);4α(OH)

TABLE 2. *(contd.)*

Compound:

No.   Substituents

26. Grindelane, 8,17-Bis-Nor-8,9-Seco-

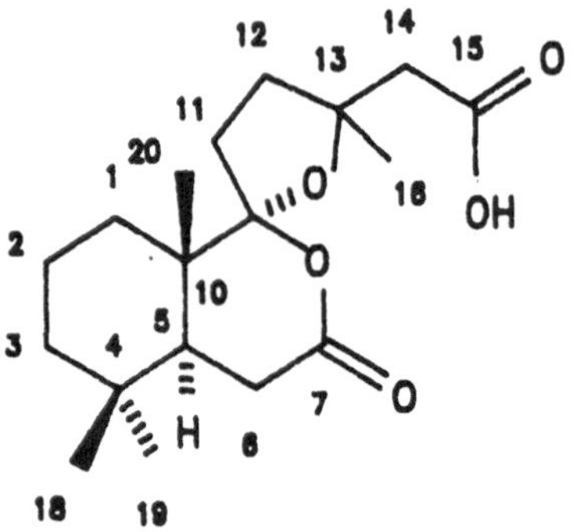

224

27. Normal -Labdane-20,12-Lactone, 5,10-Seco-

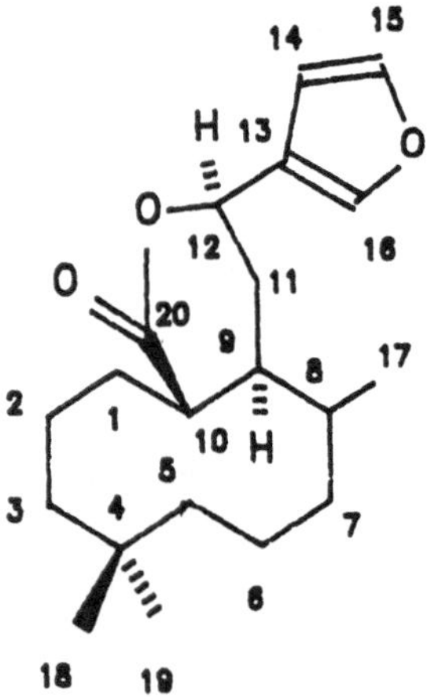

225	5(=O);6α(OH);1(-O-)10
226	5(=O);6α(OAng);1(-O-)10
227	5(=O);6α(Oi-But);1(-O-)10
228	5(=O);6α(O2-Mebut);1(-O-)10
229	5(=O);6(Oi-Val);1(-O-)10
230	5(OH);6α(OAng);1(-O-)10

Table 2. Substitutional Patterns   85

TABLE 2. (*contd.*)

Compound:
No.   Substituents

100. Normal-Labdane, 14,15-Bis-Nor-

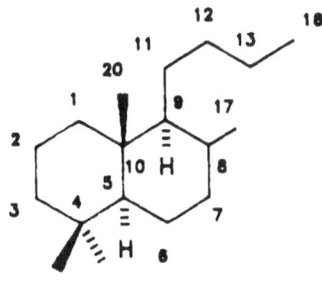

1155   11(12)E;13(=O);8α(OH)

1157   11(12)E;13(=O);6α(OH);7β(OH);8α(OH)

1158   11(12)E;13(=O);7β(OH);8α(OH);6α(OAc)

1159   11(12)E;13(=O);6α(OH);8α(OH);7β(OAc)

1160   11(12)E;13(=O);7β(OH);8α(OH)

101. Normal-Labdane, 13,14,15,16-Tetranor-

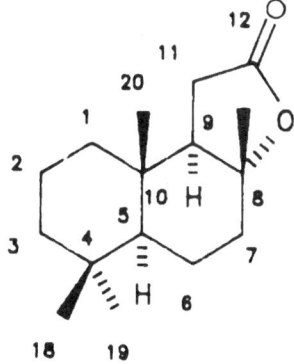

1156

87. Hexanorlabdane

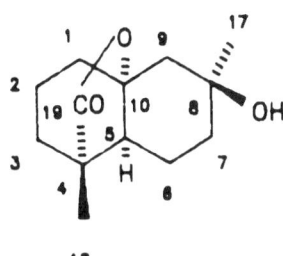

TABLE 2. *(contd.)*

Compound:	
No.	Substituents

28. Relhaniane

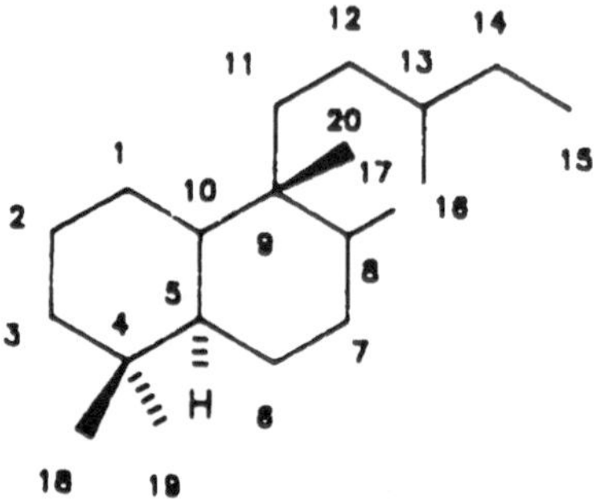

| 231 | 1(10);13(14)E;15(=O,OH);2(=O);8β(H) |
| 232 | 5(6);19(=O,OH);15(OH);10α(H);8α(H) |

233	5(6);13(16);14(15);19(=O,OH);8α(H)
234	5(6);13(14);19(=O,OH);16(=O);8α(H)
235	5(6);13(14);19(=O,OH);15(=O);8α(H)

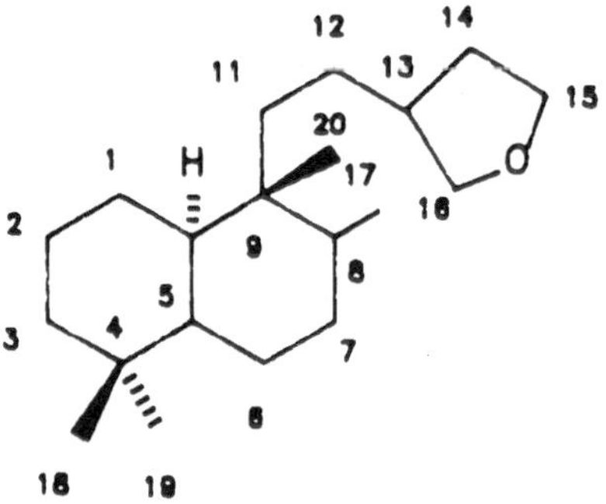

29. trans-Normal-Clerodane

(No proven structures of this type)

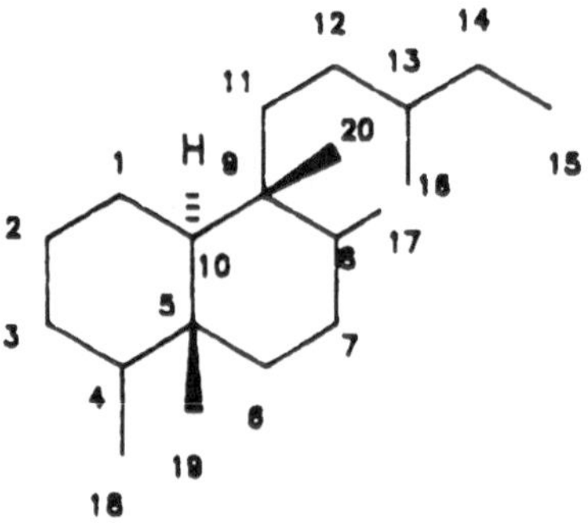

Table 2. Substitutional Patterns   87

TABLE 2. (*contd.*)

Compound:
No.   Substituents

30. Clerodane, 5,10-Seco-

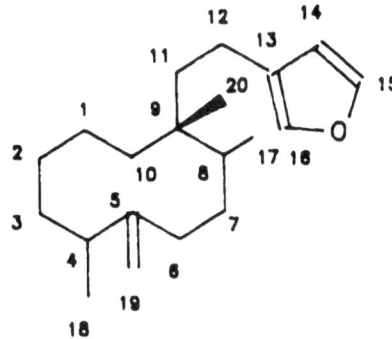

| 255 | 1(2);3(4);18(=O,OH);8β(H) [Revised to 9α-Me] |
| 256 | 1(2);3(4);18(=O,OMe);8β(H) [Revised to 9α-Me] |

31. <u>cis</u>-Normal-Clerodane

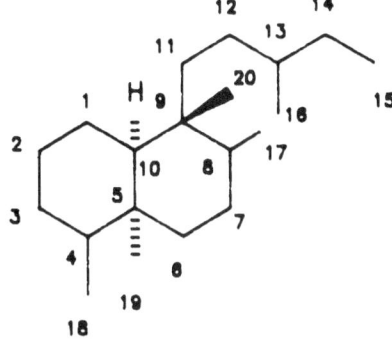

238	3(4);13(16);14(15);19(OH);15(-O-)16;8β(H)
253	3(4);13(14);15(=O);19(OH);8β(H)
254	3(4);13(14);16(=O);19(OH);8β(H)
257	3(4);15(=O,OH);8α(H)
258	3(4);13(14);15(=O);6β(OAng);15(-O-)16;8β(H)
259	3(4);13(14);15(=O);6β(OTig);15(-O-)16;8β(H)
260	13(14);15(=O);3β(-O-)4β;6β(OAng);15(-O-)16;8β(H)
261	13(14);15(=O);3α(-O-)4α;6β(OTig);15(-O-)16;8β(H)
262	3(4);13(14);6(=O);15(=O);15(-O-)16;8β(H)
263	3(4);13(14);15(=O);2α(OH);6β(OAng);15(-O-)16;8β(H)

TABLE 2. (*contd.*)

Compound: No.	Substituents
264	3(4);13(14);15(=O);2α(OH);6β(OTig);15(-O-)16;8β(H)
265	3(4);13(14);6(=O);15(=O);2β(OH);15(-O-)16;8β(H)
266	13(14);15(=O);3β(OH);4α(OH);6β(OAng);15(-O-)16;8β(H)
267	13(14);15(=O);3β(OH);4α(OH);6β(OTig);15(-O-)16;8β(H)
268	13(14);3(=O);15(=O);6β(OAng);15(-O-)16;8β(H)
269	13(14);3(=O);15(=O);6β(OTig);15(-O-)16;8β(H)
545	3(4);13(14);15(=O);6β(OH);15(-O-)16;8β(H)
546	3(4);13(14);15(=O);6β(OAc);15(-O-)16;8β(H)
551	13(14);15(=O);2α(OH);3α(-O-)4α;15(-O-)16;8β(H)
552	13(14);15(=O);6β(OAng);3α(-O-)4α;15(-O-)16;8β(H)
553	13(14);15(=O);3β(-O-)4β;6β(OAc);15(-O-)16;8β(H)
988	13(14);15(=O);3β(-O-)4β;6β(OTig);15(-O-)16;8β(H)
239	3(4);13(16);14(15);19(=O,OH);15(-O-)16;8β(H)
240	3(4);13(16);14(15);19(=O);15(-O-)16;8β(H)
241	3(4);13(16);14(15);19(OH);15(-O-)16;8β(H)
242	3(4);13(16);14(15);18(OH);19(OH);15(-O-)16;8β(H)
243	3(4);13(16);14(15);18(=O);19(=O);15(-O-)16;8β(H)
244	3(4);13(16);14(15);19(=O,OH);18(OAng);15(-O-)16;8β(H)
246	3(4);13(16);14(15);18(-O-)19;19(OH);15(-O-)16;8β(H)
247	18(-O-)19;3(-O-)4;19(OH);15(-O-)16;8β(H)
270	13(14);15(=O);3α(Cl);4β(OH);19(OMe);15(-O-)16; 18(-O-)19
556	13(14);15(=O);3β(-O-)4β;15(-O-)16
557	13(14);15(=O);3(-O-)4;7β(OAng);15(-O-)16
1025	3(4);14(15);2(=O);13β(OH);6β(OAc);8β(H)
1062	3(4);13(16);14(15);19(=O);15(-O-)16;8β(H);
1063	3(4);13(16);14(15);18(=O);19(=O);15(-O-)16;8β(H)
1064	3(4);13(16);14(15);19(OH);15(-O-)16;8β(H)
1065	3(4);13(16);14(15);18(OH);15(-O-)16;8β(H)
1066	3(4);13(16);14(15);18(=O);15(-O-)16;18(-O-)19;8β(H)
1067	3(4);13(16);14(15);19(OH);15(-O-)16;18(-O-)19;8β(H)

Table 2. Substitutional Patterns   89

TABLE 2. (*contd.*)

Compound: No.	Substituents
1068	13(16);14(15);19(OH);3α(-O-)4α;15(-O-)16; 18(-O-)19;8β(H)
1083	3(4);13(14);15(=O);6α(OH);18(OH);15(-O-)16;8α(H)
1084	3(4);13(14);15(=O);18(OH);19(OH);15(-O-)16;8α(H)
1085	3(4);13(14);15(=O);18(=O);15(-O-)16;18(-O-)19;8α(H)
1086	3(4);13(14);15(=O);19α(OH);15(-O-)16; 18(-O-)19;8α(H);19β(H);
1087	13(14);15(=O);18β(OH);19α(OH);3α(-O-)4α;15(-O-)16; 18(-O-)19;8α(H);18α(H);19β(H)
1088	13(14);15(=O);18(=O);19α(OH);3α(-O-)4α; 15(-O-)16;18(-O-)19;8α(H);19β(H)
1089	13(14);15(=O);19α(OH);3α(-O-)4α;15(-O-)16; 18(-O-)19;8α(H);19β(H)
1090	13(14);15(=O);3α(OH);4β(OH);19α(OH);15(-O-)16; 18(-O-)19;8α(H);19β(H)
1091	3(4);13(14);15(=O);19(=O,O-Arab);15(-O-)16;8α(H)
1092	3(4);13(14);15(=O);18(=O);2β(OH);15(-O-)16; 18(-O-)6α;8α(H)

90. Tricycloclerodane

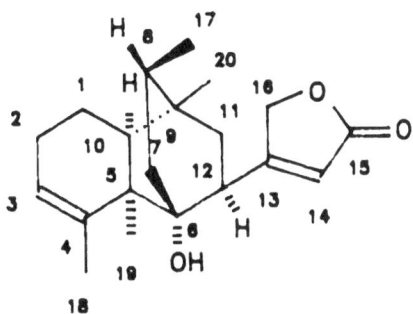

947

72. Chrysolane

271

TABLE 2. *(contd.)*

Compound:

No.   Substituents

III. Normal-Tricarbocyclics and Derivatives

32. Sandaracopimarane (Normal-Pimarane)

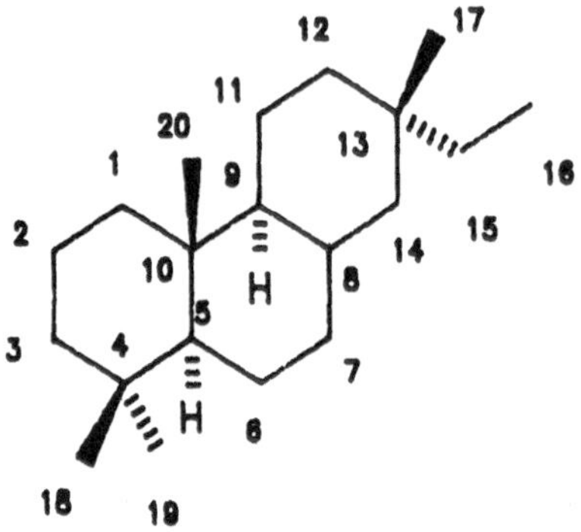

274	8(14);15(16)
927	8(14);15(16);7α(OH)
928	8(14);15(16);1β(OH);7α(OH)
929	8(14);15(16);7α(OH);1β(OAc)
930	8(14);15(16);1β(OH);7α(OAc)
931	8(14);15(16);1β(OH);11α(OH)
932	8(14);15(16);1β(OAc);11α(OAc)
933	8(14);15(16);1β(OH);11α(OH);7α(OAc)
934	8(14);15(16);7α(OH);1β(OAc);11α(OAc)
935	8(14);15(16);1β(OAc);7α(OAc);11α(OAc)
936	8(14);15(16);6(-O-)7;1β(OAc);11α(OAc)
937	8(14);15(16);7(=O);1β(OAc);11α(OAc)
275	7(8);15(16)
276	15(16);8β(OH)
277	15(16);6β(OH);8β(OH)
278	15(16);8β(OH);6β(OAc)
279	15(16);8β(OH);11α(OH)
280	15(16);8β(OH);11α(OSen)

Table 2. Substitutional Patterns 91

TABLE 2. *(contd.)*

Compound: No.	Substituents
281	15(16);8β(OH);11α(OTig)
282	15(16);8β(OH);11β(OAc)
283	15(16);11(=O);8β(OH)
284	15(16);8β(OH);12β(OAc)
285	15(16);8β(OH);18(OH)
286	15(16);8β(OH);20(OH)
287	15(16);20(OH);8β(-O-)20
288	15(16);8β(OH);11α(OH);12β(OH)
289	15(16);8β(OH);12β(OH);11α(OAc)
290	15(16);8β(OH);12β(OH);11α(OSen)
291	15(16);8β(OH);12β(OH);11α(OTig)
292	15(16);8β(OH);11α(OH);12β(OAc)
293	15(16);8β(OH);11α(OH);12β(OCoum)
294	15(16);8β(OH);11α(OAc);12β(OAc)
295	15(16);8β(OH);12α(OH);11β(OAc)
296	15(16);11(=O);8β(OH);12α(OH)
297	15(16);11(=O);8β(OH);12α(OAc)
298	15(16);6β(OH);8β(OH);11α(OH)
299	15(16);8β(OH);11α(OH);6β(OAc)
300	8(14);15(16);18(=O,OH)
301	7(8);15(16);18(=O,OH)
302	8(14);15(16);18(=O,OH);3β(OAc)
807	8(14);3β(OH);15(OH);16(OH);17(OH)
303	7(8);15(16);18(=O,OH);3β(OAc)
921	8(14);19(=O,OH);6β(OH);15(OH);16(OH)
304	7(8);3α(OH);15(OH);16(OH)
305	7(8);9α(OH);15(OH);16(OH)
922	8(14);15(OH);16(OH);6β(OGlu)
923	8(14);6α(OH);15(OH);16(OH);18(OH)

TABLE 2. (*contd.*)

Compound:
No.   Substituents

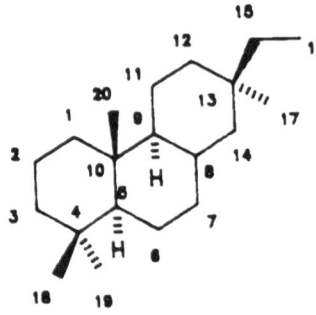

891   15(16);8β(OH);5α(H);5α(H)

306   7(8);3α(OH);15(OH);16(OH);19(OH);5α(H)   [Revised   to
             rosane]

307   7(8);3α(OH);9α(OH);15(OH);16(OH);5α(H)   [Revised   to
             rosane]

308   5(6);3(OH);15(OH);16(OH);19(OH)

34. Cassane

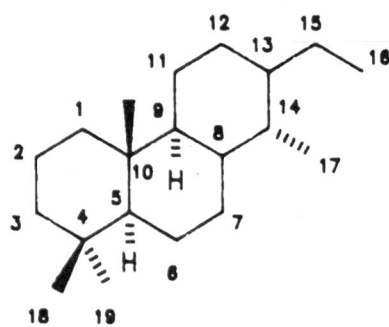

310   12(13);15(16);11(=O);2α(OH)

311   12(13);15(16);11(=O);1β(OH);2α(OH)

312   12(13);11(=O);1β(OH);2α(OH)

35. Rosane

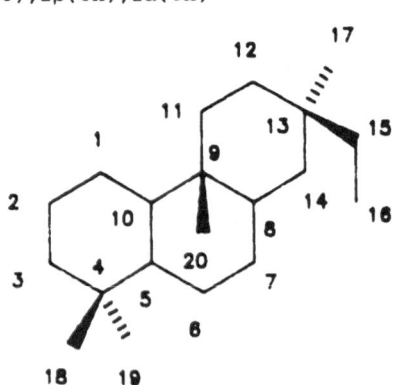

Table 2. Substitutional Patterns 93

TABLE 2. (*contd.*)

Compound:
No.   Substituents

306   5(6);3α(OH);15(OH);16(OH);19(OH);10α(H)

307   5(6);3α(OH);15(OH);16(OH);10α(H)

313   5(6);3β(OH);15(OH);16(OH);10α(H)

314   5(6);3β(OH);15(OH);16(OH);18(OH);10α(H)

315   5(6);3(=O);15(OH);16(OH);10α(H)

881   15(16);5β(OH);10α(H)

882   15(16);5β(OH);18(OH);10α(H)

883   15(16);5β(OH);18(OAc);10α(H)

884   15(16);3α(OH);5β(OH);10α(H)

885   15(16);5β(OH);18(OTig);10α(H)

886   15(16);3α(OH);5β(OH);18(OH);10α(H)

887   5(6);15(16);3α(OH);18(OH);10α(H)

888   5(6);15(16);18(OH);3α(OAc);10α(H)

889   5(6);15(16);3α(OH);18(OAc)

890   1(10);15(16);18(=O,OH)

1161   5(6);14(OH);15(OH);3β(OAc);18(OAc)

1162   5(6);14(OH);15(OH);18(OH);3β(OAc)

36. Sandaracopimarane, 19-Nor-13-Epi-

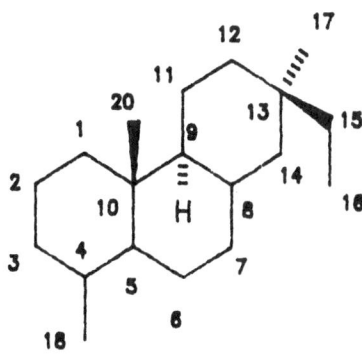

309   4(5);3(=O);15(OH);16(OH)

TABLE 2. *(contd.)*

Compound:
No.   Substituents

37. Normal-Abietane

316	7(8);13(14);2α(OAc)
317	7(8);13(14);3β(OH)
318	7(8);13(14);3(=O)
319	8(9);11(12);13(14);12(OH)
320	8(9);11(12);13(14);7(=O);11(OH);14(OH);12(OMe)
321	8(9);12(13);11(=O);14(=O);12(OH)
322	6(7);8(9);12(13);11(=O);14(=O);12(OH)
323	8(9);12(13);11(=O);14(=O);12(OH);7α(OAc)
324	8(9);12(13);7(=O);11(=O);14(=O);12(OH)
1072	8(14);6α(OH);13(OH);7α(OSen)
1073	8(14);7α(OH);13(OH);6α(OSen)
1074	8(14);7α(OH);13(OH)

1163   8(9);12(13);11(=O);14(=O);7α(OH);12(OH);20(OH)

Table 2. Substitutional Patterns 95

TABLE 2. (*contd.*)

Compound:
No.   Substituents

94. Stevisalane

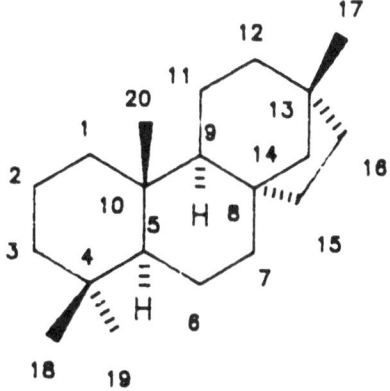

1026   14(15);2(=O);6β(OH);13β(OH);4β(H);8β(H)

IV. Normal-Tetracarbocyclics

73. Normal-Stachane

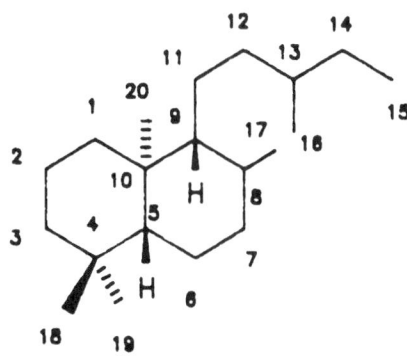

325   15(16);18(=O,OH)

V. ent-Bicarbocyclics and Derivatives

38. ent-Labdane

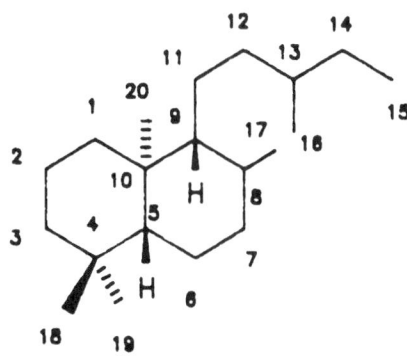

TABLE 2. *(contd.)*

Compound: No.	Substituents
326	8(17);13(14)E;15(OH)
848	8(17);13(14)E;2α(OH);15(OH)
327	8(17);13(14)E;15(Obern)
849	8(17);13(14)E;2(=O);15(OH)
328	8(17);13(14)E;15(OH);7β(OAc)
844	8(17);13(14)E;15(OH);18(OH);3α(OAng)
329	8(17);13(14)E;18(=O,OH);15(OH)
330	8(17);13(14)E;18(=O,OH);15(OAc)
331	8(17);13(14)E;15(=O);2α(OH)
332	8(17);13(14)Z;15(=O);2α(OH)
333	8(17);13(14)E;2(=O);15(=O)
334	8(17);13(14)Z;2(=O);15(=O)
335	8(17);13(14)E;15(=O,OH)
336	8(17);13(14)Z;15(=O,OH)
337	8(17);13(14)Z;15(=O,OH);3α(OH)
338	8(17);13(14)Z;15(=O,OH);3α(OH);7α(OH)
339	13(14)Z;15(=O,OH);17(=O);3α(OH);8β(H)
340	13(14)Z;15(=O,OH);17(=O);3α(OH);8α(H)
341	13(14)Z;15(=O,OH);3(=O);17(=O);8β(H)
342	13(14)Z;15(=O,OH);3(=O);17(=O);8α(H)
343	13(14)Z;15(=O,OH);17(=O);3α(OH);8α(OH)
344	13(14)Z;15(=O,OH);17(=O,OH);8β(H)
345	13(14)E;15(=O,OH);17(=O,OH);8β(H)
346	13(14)Z;15(=O,OH);3(=O);8(-O-)17
347	13(14)E;8β(-O-)17;15(Obern)
348	13(14)E;8(OH);15(OH)
349	13(14)E;15(=O);8(OH)
350	13(14)Z;15(=O);8(OH)
351	7(8);13(14)E;6β(OH);15(OH)
352	7(8);13(14)Z;6(=O);15(OH)

Table 2. Substitutional Patterns   97

TABLE 2. *(contd.)*

Compound: No.	Substituents
353	7(8);13(14)E;11(OH);3α(OAc);15(OAc)
354	7(8);13(14)E;15(=O,OH)
355	7(8);13(14)E;15(=O,OH);2β(OAng)
356	6(7);8(17);15(=O,OH)
357	8(17);2α(OH);15(OH)
358	13(S);8(17);15(=O,OH);2α(Oi-Val)
857	8(17);15(=O,OH);2α(OH)
858	8(17);15(=O,OH);2α(OAng)
859	8(17);15(=O,OH);2α(OTig)
860	8(17);15(=O,OH);2α(O2-Mebut)
861	8(17);15(=O,OH);2(=O)
862	15(=O,OH);2α(OAng);8β(-O-)17
863	15(=O,OH);2α(O2-Mebut);8β(-O-)17
359	8(17);2(=O);15(OH)
360	7(8);15(=O,OH);17(OMe)
361	7(8);15(=O,OH);2β(OAng)
362	7(8);13(14)Z;15(OH);16(OH);17(OH)
363	7(8);13(14)Z;15(OAc);16(OAc)
364	7(8);13(14)Z;15(OAc);16(OAc);17(OAc)
365	6(7);8(17);13(14)E;15(=O,OH);16(OH)
366	7(8);13(14)Z;15(=O,OH);16(OH)
367	7(8);13(14)Z;15(=O,OH);16(OH);17(OH)
368	7(8);13(14)Z;15(=O,OH);16(=O);12(OH)
369	7(8);13(14)Z;15(=O,OH);17(=O);16(OH)
370	7(8);13(14)Z;15(=O,OH);17(OH);16(OAc)
371	7(8);13(14)E;15(=O,OH);17(=O);16(OAc)
372	7(8);13(14)E;15(=O,OH);17(=O);6α(OH);16(OAc)
373	7(8);13(14)E;15(=O,OH);17(=O);6β(OH);16(OAc)
374	7(8);13(14)Z;15(=O,OH);6(=O);16(OAc)
375	7(8);13(14)Z;15(=O,OH);6(=O);16(=O)

TABLE 2. *(contd.)*

Compound: No.	Substituents
376	7(8);13(14)E;15(=O,OH);6(=O);16(OH)
377	7(8);13(14)E;15(=O,OH);6(=O);16(OAc)
378	7(8);13(14)E;15(=O,OH);6(=O);16(OMe)
379	6(7);13(14)E;15(=O,OH);8α(OH);16(OH)
380	6(7);13(14)E;15(=O,OH);17(=O);16(OAc);8α(H)
381	6(7);13(14)E;15(=O,OH);17(=O,OH);16(OAc);8α(H)
382	8(17);13(14)Z;7β(OH);15(OH);16(OH)
383	8(17);15(=O,OH);16(=O)
384	6(7);8(17);13(14);15(=O);15(-O-)16
385	8(17);13(14);15(=O);19(OH);15(-O-)16
386	7(8);13(14);15(=O);15(-O-)16
387	7(8);13(14);15(=O);12(OH);15(-O-)16
388	7(8);13(14);15(=O);16α(OH);15(-O-)16
389	7(8);13(14);15(=O);16β(OH);15(-O-)16
390	7(8);13(14);15(=O);12(OH);16(OH);15(-O-)16
391	7(8);13(14);15(=O);12(OH);16α(OH);15(-O-)16
392	7(8);13(14);15(=O);12(OH);16β(OH);15(-O-)16
393	7(8);13(14);6(=O);15(=O);15(-O-)16
394	7(8);13(14);6(=O);15(=O);16(OH);15(-O-)16
395	7(8);13(14);15(=O);17(=O);16α(OH);15(-O-)16
396	7(8);13(14);15(=O);17(=O);16β(OH);15(-O-)16
397	8(17);13(14);16(=O);7β(OH);15(-O-)16
843	8(17);13(14);18(=O,OH);16(=O);15(-O-)16
398	7(8);13(14);16(=O);15(-O-)16
399	7(8);13(14);16(=O);12(OH);15β(OH);15(-O-)16
400	7(8);13(14);16(=O);12(OH);15α(OH);15(-O-)16
401	8(17);13(16);14(15);7β(OH);15(-O-)16
402	8(17);13(16);14(15);7β(OAc);15(-O-)16
403	8(17);13(16);14(15);19(OH);15(-O-)16
404	8(17);13(16);14(15);6β(OH);19(OH);15(-O-)16

Table 2. Substitutional Patterns 99

TABLE 2. *(contd.)*

Compound: No.	Substituents
839	8(17);13(16);14(15);19(=O,OH);15(-O-)16
840	8(17);13(16);14(15);19(=O,OH);3α(OH);15(-O-)16
841	8(17);13(16);14(15);18(OH);19(OH);15(-O-)16
842	8(17);13(16);14(15);3α(OH);18(OH);15(-O-)16
974	8(17);13(16);14(15);18(OH);3α(OAng);15(-O-)16
975	8(17);13(16);14(15);3α(OH);18(OAng);15(-O-)16
830	8(17);13(16);14(15);18(=O,OH);15(-O-)16
831	8(17);13(16);14(15);18(=O,OH);3α(OH);15(-O-)16
832	8(17);13(16);14(15);18(=O,OH);3α(OAng);15(-O-)16
833	8(17);13(16);14(15);18(=O,OH);3α(O2-Mebut);15(-O-)16
834	8(17);13(16);14(15);18(=O,OH);3α(Oi-But);15(-O-)16
835	8(17);13(16);14(15);18(=O,OH);19(OH);15(-O-)16
836	13(16);14(15);18(=O,OH);8α(-O-)17;15(-O-)16
837	13(16);14(15);18(=O,OH);8β(-O-)17;15(-O-)16
838	13(16);14(15);18(=O,OH);17(=O);8β(H);15(-O-)16
405	1(2);8(17);13(16);14(15);18(=O,OH);15(-O-)16
406	8(17);13(16);14(15);2(=O);12(OH);15(-O-)16
407	8(17);13(16);14(15);2(=O);12(=O);15(-O-)16
408	8(17);13(16);14(15);18(OH);7β(OAc);15(-O-)16
409	8(17);13(16);14(15);7β(OAc);18(OAc);15(-O-)16
410	8(17);13(16);14(15);18(=O);7β(OH);15(-O-)16
411	8(17);13(16);14(15);18(=O);7β(OAc);15(-O-)16
412	8(17);13(16);14(15);2(=O);12(=O);3α(OH);15(-O-)16
413	8(17);13(16);14(15);2(=O);3α(OH);12(OH);15(-O-)16
414	7(8);13(16);14(15);17(OH);15(-O-)16
415	7(8);13(16);14(15);18(=O,OH);15(-O-)16
416	13(16);14(15);8(OH);3α(OAng);15(-O-)16
417	8(17);13(14);2(=O);12(=O);15α(OH);15(-O-)16
418	8(17);13(14);2(=O);12(=O);15β(OH);15(-O-)16
421	8(17);12(13);14(15);6β(OH);12(-O-)15

TABLE 2. (*contd.*)

Compound: No.	Substituents
422	8(17);12(13)Z;14(15)
423	8(17);12(13)Z;14(15);18(=O)
424	8(17);12(13)Z;14(15);18(=O,OH)
425	8(17);12(13)E;14(15);18(=O,OH)
426	8(17);12(13)Z;14(15);18(=O,OH);2α(OH)
427	8(17);12(13)Z;14(15);18(=O,OH);2(=O)
428	12(13)Z;14(15);6β(OH);7α(OH);8(OH)
429	12(13)Z;14(15);7α(OH);8(OH);6β(OAc)
808	12(13)Z;14(15);8β(OH);6β(OAng);7α(OAc)
430	8(17);12(13)E;14(15)
431	8(17);12(13)E;14(15);19(OH)
432	8(17);12(13)E;14(15);3α(OH);19(OH)
433	8(17);12(13)E;14(15);18(OH)
434	8(17);12(13)E;14(15);18(OH);19(OH)
435	8(17);12(13)E;14(15);18(=O,OH);1β(OAc)
436	12(13)E;14(15);3β(OH);8β(OH)
809	12(13)E;14(15);6β(OH);7α(OH);8β(OH)
810	12(13)E;14(15);7α(OH);8β(OH);6β(OAc)
811	12(13)E;14(15);8β(OH);6β(OAng);7α(OAc)
441	13(16);14(15);8α(OH)
442	8(17);14(15);13α(OH)
443	8(17);14(15);2β(OH);13α(OH)
444	8(17);14(15);2α(OH);13α(OH)
445	8(17);14(15);13α(OH);2β(OSucc)
446	8(17);14(15);2(=O);13α(OH)
447	7(8);14(15);13(OH)
448	7(8);14(15);3(OH);13(OH)
449	7(8);14(15);3(=O);13(OH)
450	14(15);8α(OH);13α(OH)
460	13(14)Z;15(=O,OH);6(=O);16(=O);8(-O-)12

Table 2. Substitutional Patterns 101

TABLE 2. (*contd.*)

Compound: No.	Substituents
461	13(S);1(OH);6β(OH);7β(-O-)8β;15(OSen);5β(H);9β(-O-)13
1012	7(8);13(14)E;2β(OH);15(OAc);18(OAc)
1013	7(8);13(14)E;2β(OH);15(OH);18(OAc)
1014	8(17);13(14)E;15(OH);3α(OAc);18(OTig)
1015	12(13)E;14(15);8β(OH);18(OH)
1016	8(9);14(15);7(=O);13β(OH);18(OH)
1017	14(15);7(=O);18(OH);9β(-O-)13β;8β(H)
1018	14(15);7(=O);18(=O);9β(-O-)13β;8β(H)
1019	14(15);18(=O,OH);7(=O);9β(-O-)13β;8β(H)
1037	8(17);13(16);14(15);18(OH);3α(OAng);15(-O-)16
1038	8(17);13(16);14(15);2β(OH);18(OH);3α(OAng);15(-O-)16
1039	8(17);13(14)Z;2β(OH);15(OH);16(OH);18(OH);3α(OAng)
1040	8(17);13(14)E;15(=O,OH)
1041	8(17);13(14)E;15(=O,OH);18(OH)
1042	7(8);13(14)E;15(=O,OH);18(=O);6β(OH)
1043	7(8);13(14)E;15(=O,OH);18(=O,OH);6β(O-i-Val)
1044	13(14)E;15(=O,OH);5β(-O-O-)8β
1045	7(8);13(14)E;15(=O,OH);6β(OH);18(OH)
1046	7(8);15(=O,OH);17(OH)
1047	7(8);15(=O,OH);17(OAc)
1048	7(8);15(=O,OH);17(=O)
1049	7(8);15(=O,OH);2α(OH);17(OH)
1050	7(8);15(=O,OH);2α(OH);17(OAc)
1051	7(8);15(=O,OH);17(=O);2α(OH)
1052	7(8);15(=O,OH);17(=O,OH)
1053	7(8);15(=O,OH);19(OH);17(OAc)
1164	8(17);13(14)E;15(=O,OH);18(=O,OH)
1173	8(17);19(=O,OH);15(OH)
1174	13(14)E;2β(OH);8β(OH);15(OH)
1175	7(8);15(=O,OH);3α(OH);17(OH)

TABLE 2. (*contd.*)

Compound: No.	Substituents
1176	7(8);15(=O,OH);3α(OH);17(OAc)
1177	7(8);15(=O,OH);17(=O);3α(OH)
1178	7(8);15(=O,OH);3α(OH);17(OMe)
1179	7(8);15(=O,OH);2(OH);17(OMe)
1180	7(8);13(14)Z;15(=O,OH);16(OH);17(OMe)
1181	8(17);13(14)Z;15(=O);18(OH);2β(OH);15(-O-)16
1182	8(17);13(14)Z;15(=O);18(OH);3α(OAng);15(-O-)16
1183	8(17);13(14)Z;15(=O);18(OGlu);3α(OAng);15(-O-)16
1184	8(17);13(14)Z;15(=O);2β(OH);18(OH);3α(OAng);15(-O-)16
1185	8(17);12(13)E;14(15);18(OAc);19(OAc)
1186	8(17);12(13)E;3α(OH);14(OH);15(OH)
1188	7(8);13(16);14(15);2β(OH);3α(OH);15(-O-)16
1189	8(17);13(14)E;15(=O,OH);3α(OH);18(OH)
1190	8(17);13(14)E;15(=O,OH);18(OH);3α(OAng)
1191	8(17);13(14)Z;15(=O,OH);2β(OH);16(OH);18(OH);3α(OAng)
1192	7(8);13(14)E;2β(OH);15(OH);3α(OAng)
1193	7(8);13(14)E;2β(OH);3β(OH);15(OH)
1194	6(7);8(17);13(14)E;15(=O,OH)
1195	6(7);8(17);13(14)E;15(=O,OH);18(OH)
1196	7(8);15(=O,OH);2β(OH);3β(OH)
1197	7(8);2β(OH);3β(OH);15(OH)

1165	7(8);13(14)E;15(OH);3α(OAng);2'β(OH);3'α(OH);4'α(OH)
1166	7(8);13(14)E;15(OH);3α(OAng);2'β(OH);3'α(OH);4'α(OAc)
1167	7(8);13(14)E;15(OH);3α(OAng);2'β(OH);3'α(OAc);4'α(OH)

Table 2. Substitutional Patterns   103

TABLE 2. *(contd.)*

Compound:
No.   Substituents

1168   7(8);13(14)Z;15(OH);16(OH);3α(OAng);2'β(OH);
             3'α(OH);4'α(OAc)

1169   7(8);13(16);14(OH);15(OH);3α(OAng);2'β(OH);
             3'α(OH);4'α(OAc)

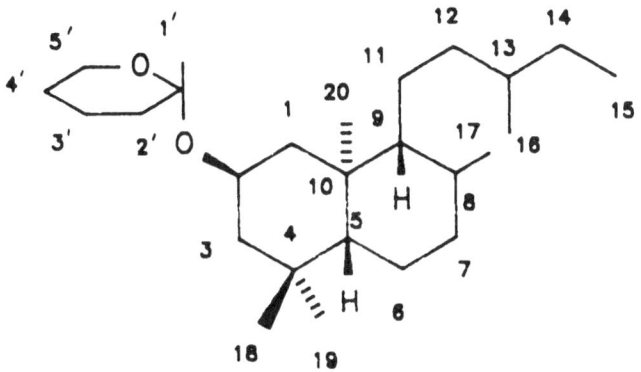

1170   7(8);13(14)E;15(OH);3α(OAng);2'β(OH);3'α(OH);4'β(OAc)

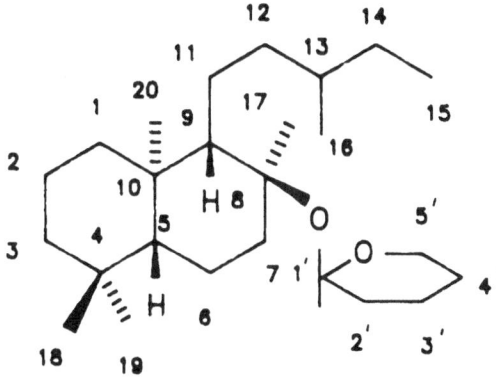

1171   7(8);13(14)E;15(OH);3α(OAng);2'β(OH);3'α(OH);4'β(OH)

1172   13(14)E;15(OH);2'α(OH);3'β(OH);4'α(OH)

TABLE 2. (*contd.*)

Compound:

No.   Substituents

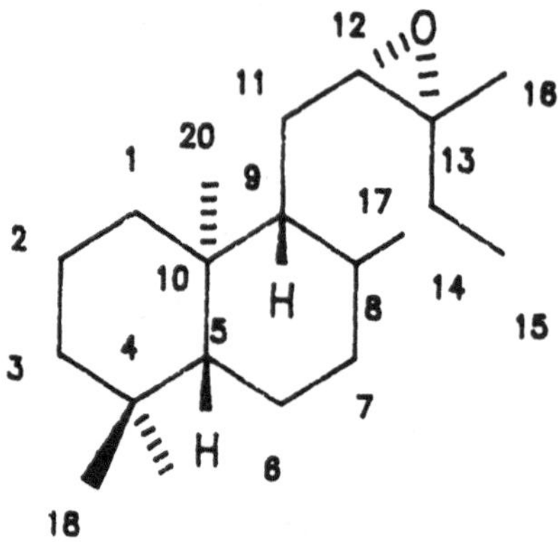

437   8(17);14(15);18(=O,OH)

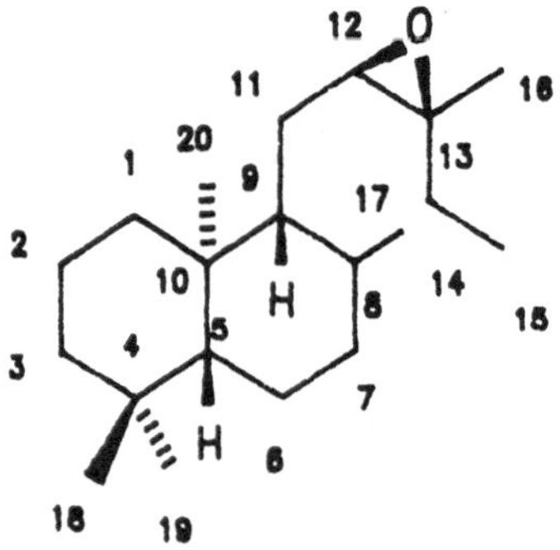

438   8(17);14(15);18(=O,OH)

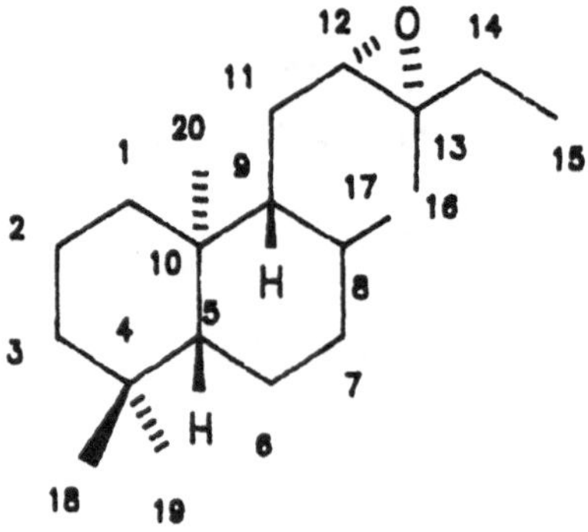

439   8(17);14(15);18(=O,OH)

Table 2. Substitutional Patterns   105

TABLE 2. *(contd.)*

Compound:
No.   Substituents

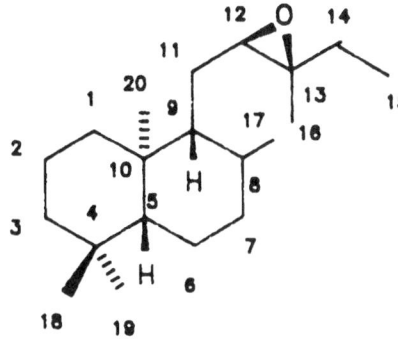

440   8(17);14(15);18(=O,OH)

39. ent-Labdane-19,6-Lactone

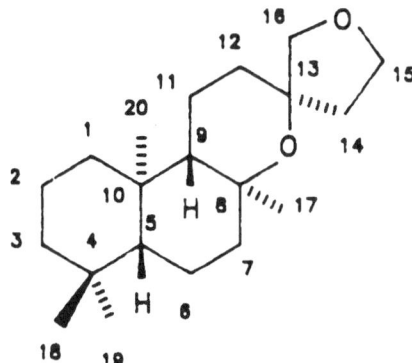

462   7(8);15(=O,OH)

40. ent-Labdane-8,13-Epoxide-15,16-Hemiacetal

463   14(OH);16β(OH)

TABLE 2. (*contd.*)

Compound:	
No.	Substituents
464	14(OH);16β(OAc)

97. ent-Labdane Dimer

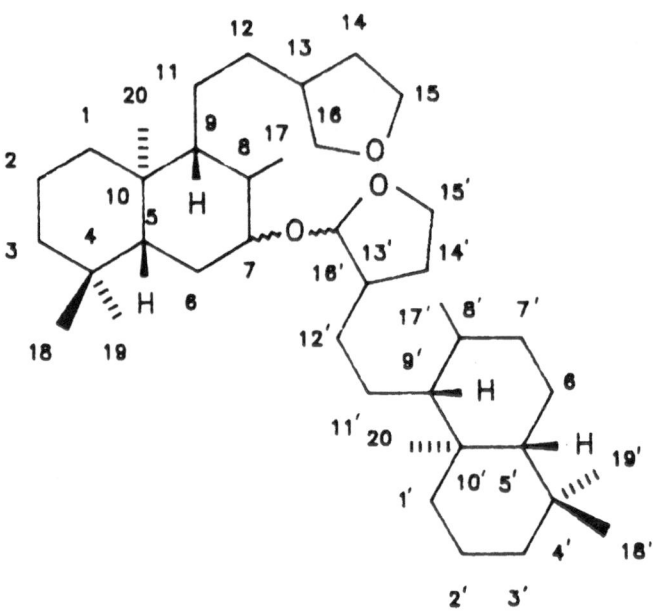

419	8(17);13(16);14(15);7α(H);8'(17');15'(=O);16'α(H)
420	8(17);13(16);14(15);7α(H);8'(17');15'(=O);16'β(H)

98. ent-Manoyl Oxide, 13-Epi-

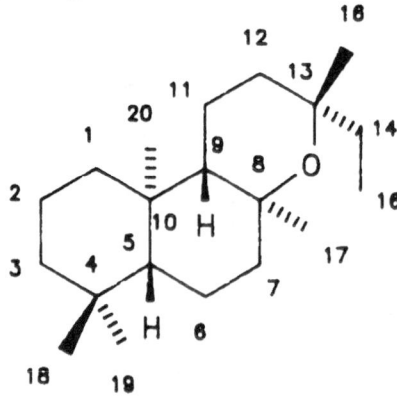

451	14(15)

Table 2. Substitutional Patterns   107

TABLE 2. *(contd.)*

Compound: No.	Substituents
452	14(15);3α(OH)
453	14(15);3(=O)
454	14(15);2(=O)
455	14(15);19(OH)
456	14(15);2(=O);19(OH)
457	14(-O-)15;16(OAc)
1198	14(15);3α(OAc)
1199	14(15);3α(OH);19(OCinn)
1200	14(15);3α(OH);19(OH)

99. ent-Manoyl Oxide

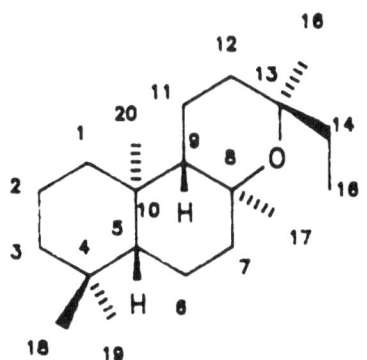

458	14(15)
459	14(15);3β(OH)

41. ent-Labdane, 17-Nor-

TABLE 2. *(contd.)*

Compound:

No.	Substituents
465	6(7);13(14)E;15(=O,OH);8β(OCHO);16(OAc)
466	6(7);13(14)E;15(=O,OH);8(=O);16(OAc)
467	13(14)Z;15(=O,OH);8(=O);3α(OH)
468	13(14)Z;15(=O,OH);3(=O);8(=O)
1201	13(14)Z;15(=O,OH);8(=O)
1202	13(14)Z;15(=O,OH);3(=O);8(=O)

42. ent-Labdane, 15-Nor-

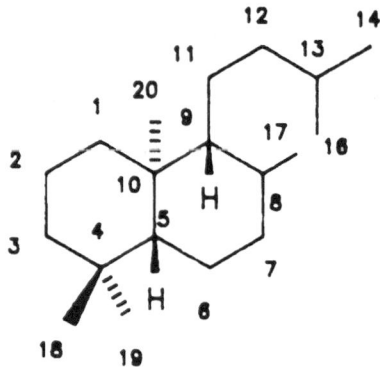

469	8(17);12(13)E;19(=O,OH);14(=O)

43. ent-Labdane, 14,15-Bis-Nor-

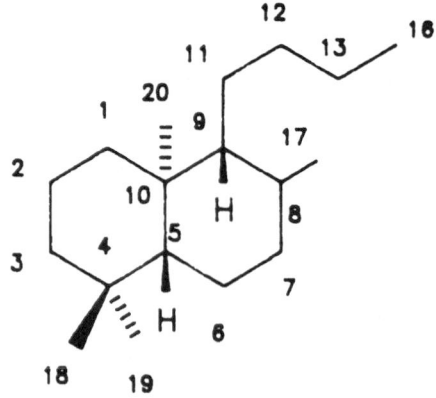

470	8(17);13(=O)

Table 2. Substitutional Patterns   109

TABLE 2. *(contd.)*

Compound:
No.   Substituents

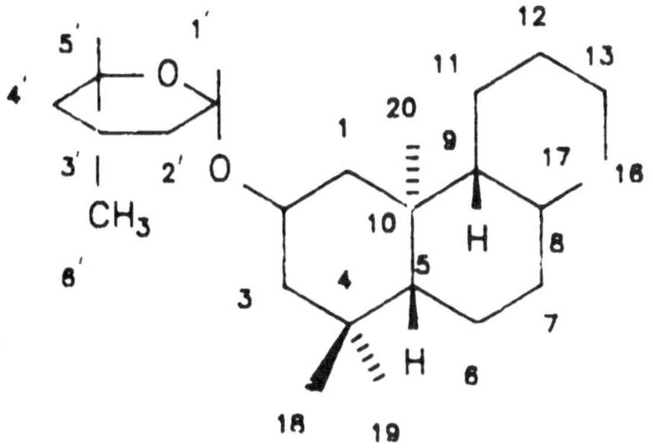

1203   7(8);13(=O);3α(OAng);2'β(OH);3'α(OH);4'α(OAc)

44. ent-Labdane, 13,14,15,16-Tetra-Nor-

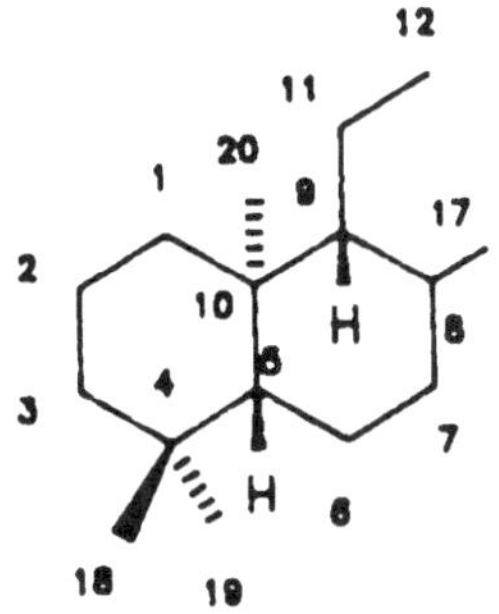

471   7(8);12(=O)

472   7(8);12(=O,OH);17(OH)

45. ent-Labdane, 7,8-Seco-

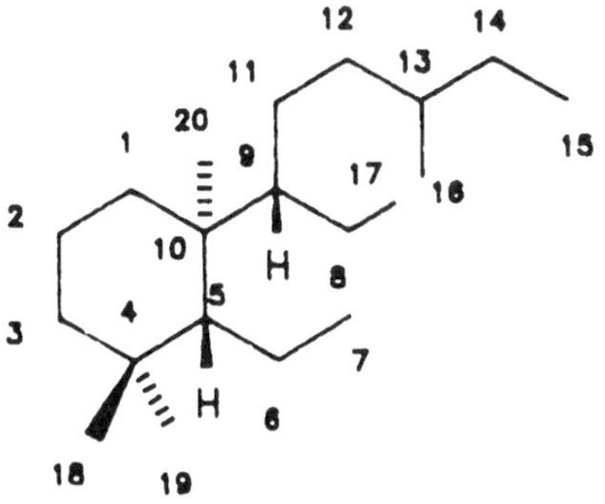

473   12(13)E;14(15);7(=O,OH);8(=O)

TABLE 2. *(contd.)*

Compound:

No.   Substituents

74. <u>ent</u>-Labdane, 18-Nor-

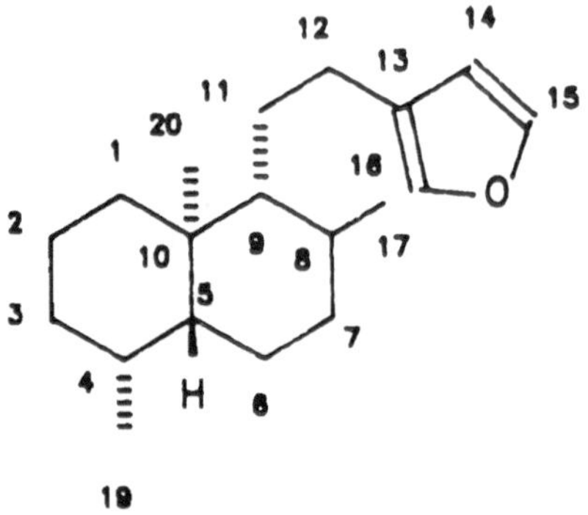

828	7(=O);2α(OH);3α(OH);19(OH);8α(H)
829	7(=O);12(=O);2α(OH);3α(OH);19(OH);8α(H)
1205	7(=O);3α(OH);19(OH);2α(OAc);8α(H)
1206	7(=O);2α(OH);3α(OH);19(OH);8α(H)
1207	7(=O);12(=O);3α(OH);19(OH);2α(OAc);8α(H)

95. <u>ent</u>-Friedolabdane

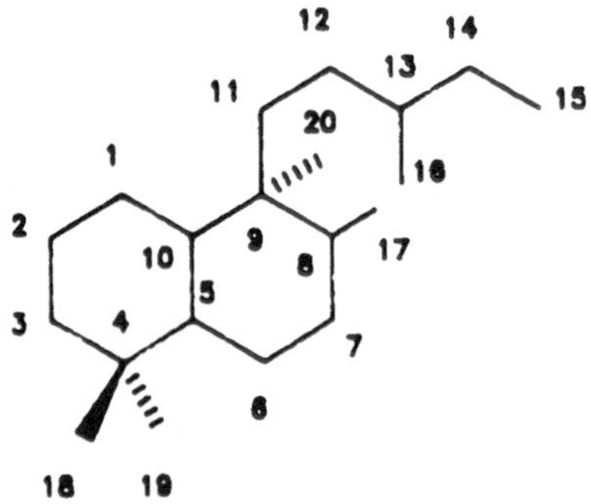

1059	5(6);13(14)E;15(=O,OH);18(OH);10β(H)
1208	5(6);13(14)E;15(=O,OH);18(OH);10β(H)
1209	1(10);13(14)E;2(=O);15(OH);5β(H)
1210	1(10);13(14)E;2α(OH);15(OH);5β(H)
1211	1(10);14(15) ;19(=O,OH);13(OH);5β(H)

Table 2. Substitutional Patterns   111

TABLE 2. (*contd.*)

Compound:
No.   Substituents

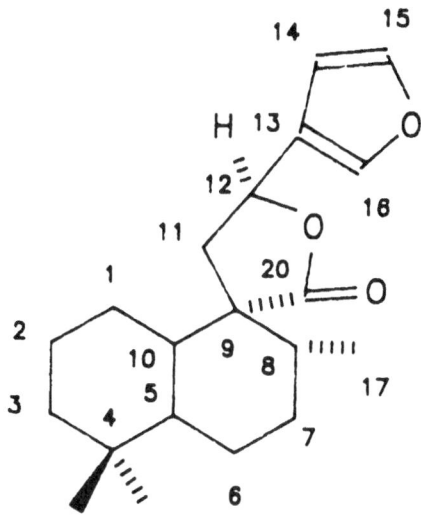

1212   3α(OH);5α(OH)

1213   5(6);3α(OH)

1214   3α(OH);5β(-O-)10β

102. <u>ent</u>-Friedolabdane, 3,4-Seco-

1215   3(=O)

1216   3(=O,OH)

TABLE 2. (*contd.*)

Compound:
No.   Substituents

1217   3(=O,OH);19(OH)

1218   3(=O,OMe);19(OH)

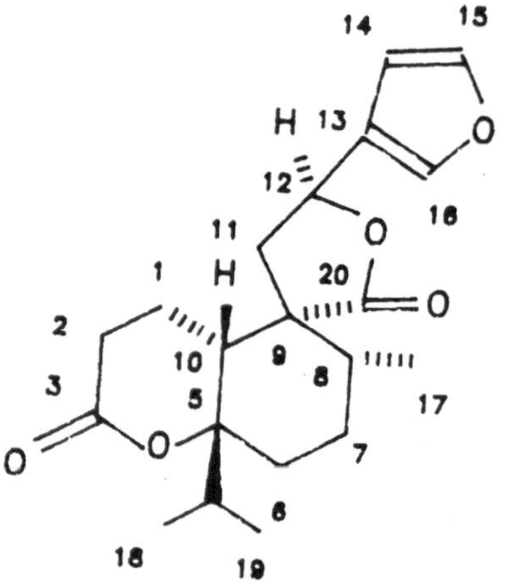

1219   4(18)

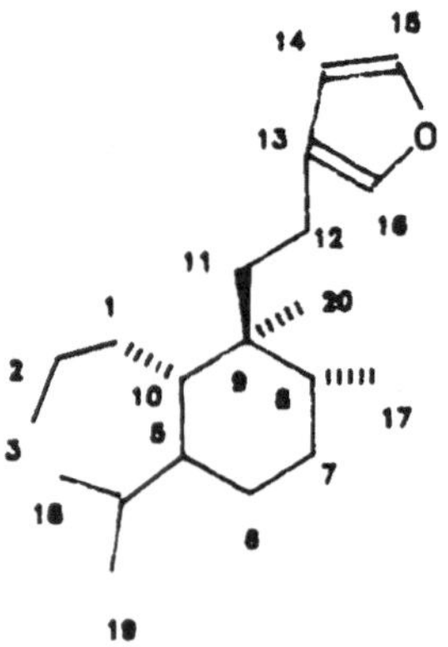

1220   4(5);3(=O,OH)

Table 2. Substitutional Patterns   113

TABLE 2. *(contd.)*

Compound:

No.   Substituents

103. ent-Friedolabdane, Rearranged 3,4-Seco-

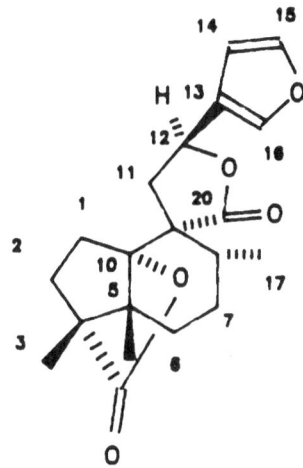

1221

46. trans-ent-Clerodane

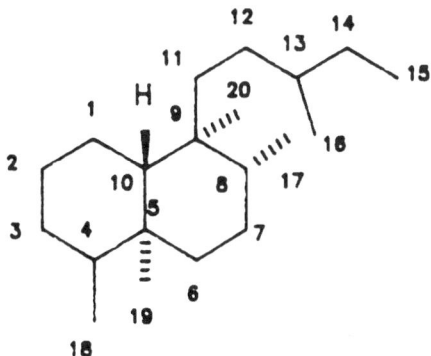

236	4(18);13(14)E;15(OH)
248	1(2);3(4);13(16);14(15);18(=O,OH);15(-O-)16
249	3(4);13(16);14(15);18(=O,OH);15(-O-)16

TABLE 2. *(contd.)*

Compound: No.	Substituents
250	1(2);3(4);13(16);14(15);18(=O,OH);15(-O-)16
251	1(2);3(4);13(16);14(15);18(=O,OMe);15(-O-)16
245	3(4);13(16);14(15);18(=O,OH);15(-O-)16
475	3(4);13(14)E;15(OH)
476	3(4);13(14)E;15(OH);17(OH)
477	13(14)E;3β(-O-)4β;2α(OH);15(OAc)
478	13(14)Z;3β(-O-)4β;2α(OH);16(OH);15(OAc)
1010	3(4);13(14)E;17(OH);15(OArac)
1011	3(4);13(14)E;17(OH);15(OBehen)
479	3(4);14(15);13(OH)
480	3(4);14(15);13(OH);6β(OAng)
481	3(4);15(OH);17(OH)
482	3(4);17(OH);15(OArac)
483	3(4);17(OH);15(OBehen)
484	3(4);15(OH);18(OSucc)
485	3α(-O-)4α;2α(OH);15(OH)
486	3(4);17(=O);15(OH)
487	3(4);18(=O);18(-O-)6α;15(OH);17(OAc)
488	3(4);6α(OH);15(OH);18(OH);17(OAc)
489	3(4);6α(OH);15(OH);18(OH);17(OPhe-Ac)
490	3(4);18(=O);6α(OH);15(OH);17(OPhe-Ac)
491	3(4);18(=O);6α(OH);15(OH);17(OAc)
492	3(4);15(=O);17(OH)
493	3(4);13(14)Z;15(=O,OH)
494	3(4);13(14)Z;15(=O,OH);2β(OAc)
495	3(4);13(14)E;15(=O,OMe)
496	3(4);13(14)E;15(=O,OMe);6β(OAc)
497	3(4);13(14)E;15(=O,OMe);6β(OAng)
498	3(4);13(14)E;15(=O,OH);6α(OAng)
499	3(4);13(14)E;15(=O,OH);6α(OTig)

Table 2. Substitutional Patterns   115

TABLE 2. *(contd.)*

Compound: No.	Substituents
500	3(4);13(14)E;15(=O,OH);7α(OAc)
962	13(14)Z;15(=O,OH);2α(OH);3α(OH);4β(OH)
501	3(4);13(14)E;15(=O,OH);6α(OH);18(OH);17(OAc)
502	3(4);13(14)E;15(=O,OH);6α(OH);18(OH);17(OPhe-Ac)
503	3(4);15(=O,OH)
504	3(4);15(=O,OH);18(OAc)
961	3(4);13(14)E;15(=O,OH);2(=O)
505	13(S);3(4);15(=O,OH);18(=O,OH)
506	15(=O,OH);2(=O);3β(-O-)4β
507	3(4);13(16);14(15);15(-O-)16
508	3(4);13(16);14(15);18(OMal);15(-O-)16
509	3(4);13(16);14(15);7α(OH);18(OH);15(-O-)16
846	3(4);13(16);14(15);18(OH);19(OH);15(-O-)16
510	3(4);13(16);14(15);19(OAc);18(OMal);15(-O-)16
967	3(4);13(16);14(15);17(=O);15(-O-)16
511	3(4);13(16);14(15);18(=O,OH);15(-O-)16
953	3(4);13(16);14(15);18(=O,OH);12α(O2-Mebut);15(-O-)16
512	1(2);3(4);13(16);14(15);18(=O,OH);15(-O-)16
911	1(2);3(4);13(16);14(15);18(=O,OH);19(OAc);15(-O-)16
513	3(4);13(16);14(15);18(=O,OH);2α(OH);15(-O-)16
514	7(8);13(16);14(15);3α(OH);18(Oi-Val);4β(H);15(-O-)16
515	7(8);13(16);14(15);3α(OH);18(O2-Mebut);4β(H);15(-O-)16
516	7(8);13(16);14(15);3α(OH);18(OAng);4β(H);15(-O-)16
518	3(4);13(16);14(15);18(=O,OH);19(OH);15(-O-)16
519	3(4);13(16);14(15);18(=O,OH);2β(OH);19(OH);15(-O-)16
900	3(4);13(16);14(15);18(=O,OH);19(OAc);15(-O-)16
901	3(4);13(16);14(15);18(=O,OH);19(OMe);15(-O-)16
902	3(4);13(16);14(15);18(=O,OH);19(OAng);15(-O-)16
903	3(4);13(16);14(15);18(=O,OH);19(Oi-Val);15(-O-)16
904	3(4);13(16);14(15);18(=O,OH);19(=O);15(-O-)16

TABLE 2. *(contd.)*

Compound: No.	Substituents
524	3(4);13(16);14(15);20(=O,OH);15(-O-)16
520	3(4);13(16);14(15);18(=O);15(-O-)16;18(-O-)19
950	3(4);13(16);14(15);18(=O);12α(OH);15(-O-)16;18(-O-)19
951	3(4);13(16);14(15);18(=O);7α(OH);12α(OH); 15(-O-)16;18(-O-)19
521	3(4);13(16);14(15);18(=O);1α(OH);15(-O-)16;18(-O-)19
522	3(4);15(=O,OH);18(=O);17(OPhe-Ac);6α(-O-)18
523	3(4);13(16);14(15);7(=O);18(=O);6α(-O-)18;15(-O-)16
534	3(4);13(14)Z;15(=O);17(OH)
535	3(4);13(14)E;15(=O);17(OH)
536	3(4);13(14)Z;17(=O,OH);15(=O)
537	3(4);13(14)E;17(=O,OH);15(=O)
865	13(14)E;3(4);15(=O,OH);16(OH)
962	13(14)E;3(4);15(=O,OH);6α(OAc)
538	3(4);13(14)Z;15(=O,OH);16(OH);18(OH)
539	3(4);13(14)Z;15(=O,OH);18(OH);16(OAc)
540	3(4);13(14)Z;15(=O,OH);18(=O);16(OH)
541	3(4);13(14)Z;15(=O,OH);18(=O);16(OAc)
542	3(4);15(=O,OH);16(=O)
543	3(4);13(14);15(=O);15(-O-)16
544	3(4);13(14);15(=O);2α(OH);15(-O-)16
864	3(4);13(14);15(=O);16(OH);15(-O-)16
550	3(4);13(14);15(=O);18(OH);15(-O-)16
558	3(4);13(14);18(=O,OH);15(=O);15(-O-)16
909	3(4);13(14);18(=O,OH);15(=O);16α(OAc);19(OAc); 15(-O-)16
910	3(4);13(14);18(=O,OH);15(=O);16β(OAc);19(OAc); 15(-O-)16
559	3(4);13(14);15(=O);18(=O);7α(OH);15(-O-)16;18(-O-)19
560	3(4);13(14);15(=O);18(=O);12(OH);15(-O-)16;18(-O-)19
561	3(4);7(=O);15(=O);18(=O);15(-O-)16;18(-O-)19

Table 2. Substitutional Patterns   117

TABLE 2. (*contd.*)

Compound: No.	Substituents
562	3(4);15(=O);18(=O);7α(OH);15(-O-)16;18(-O-)19
563	3(4);15(=O);18(=O);15(-O-)16;18(-O-)19
564	3(4);15(=O);15(-O-)16
565	3(4);15(=O);16(OH);15(-O-)16
963	3(4);15(=O);18(OAc);16(OMe);15(-O-)16
566	3(4);13(14);16(=O);20(=O);15(-O-)16
905	3(4);13(14);18(=O,OH);16(=O);15α(OH);19(OH);15(-O-)16
906	3(4);13(14);18(=O,OH);16(=O);15β(OH);19(OH);15(-O-)16
907	3(4);18(=O,OH);16(=O);19(OH);14α(-O-)15α;15(-O-)16
908	3(4);18(=O,OH);16(=O);19(OH);14β(-O-)15β;15(-O-)16
567	3(4);13(14);16(=O);2β(OH);15(-O-)16
568	3(4);13(14);2(=O);16(=O);15(-O-)16
569	3(4);15(OH);18(OMal);15(-O-)16
1027	3(4);13(14)Z;15(=O,OH);6α(O-i-But);7β(OAc)
1028	3(4);13(14)Z;15(=O,OH);6α(OAng);7β(OAc)
1029	3(4);13(14)Z;15(=O,OH);6α(OAc);7β(OAc)
1030	3(4);13(14)E;15(=O,OH);2(=O);6α(OH);7β(OH)
1031	3(4);15(=O,OH);6α(O-i-But);7β(OAng)
1033	3(4);13(16);14(15);18(=O,OH);19(OH);15(-O-)16;8β(H)
1034	3(4);13(16);14(15);18(=O,OH);2β(OH);19(OH); 15(-O-)16;8β(H)
1070	1(2);3(4);13(16);14(15);18(=O,OH);19(OH);15(-O-)16
1071	13(16);14(15);20(=O,OH);3α(-O-)4α;15(-O-)16;8β(H)
1075	13(14)Z;15(=O);3α(-O-)4α;8β(H)
1076	13(16);14(15);3α(-O-)4α;15(-O-)16;8β(H)
1077	13(14)E;15(=O);3α(-O-)4α;8β(H)
1078	13(14)Z;3(=O);15(=O);4β(H);8β(H)
1079	13(14)E;3(=O);15(=O);4β(H);8β(H)
1080	13(14);15(=O);3α(-O-)4α;15(-O-)16;8β(H)
1081	13(14);3(=O);15(=O);15(-O-)16;4β(H);8β(H)

TABLE 2. *(contd.)*

Compound: No.	Substituents
1082	13(14);15(=O);3α(OH);4β(OH);15(-O-)16;8β(H)
1222	3(4);13(14)Z;15(OH);16(OH);18(OH)
1223	3(4);13(14)E;18(=O,OH);6α(OH);15(OH)
1224	3(4);13(14)E;17(=O,OH);15(OH)
1225	3(4);13(14)Z;17(=O,OH);16(OH);15(OAc)
1227	3(4);13(14)E;15(OAc);18(OAc)
1228	13(14)Z;16(OH);3α(OAc);15(OAc);3β(-O-)4β
1229	3(4);15(OAc);18(OAc)
1230	3(4);15(OAc);16(OAc);18(OAc)
1231	3(4);13(16);14(15);18(OH);19(OMal);15(-O-)16
1232	3(4);13(16);14(15);18(OH);15(-O-)16
1233	3(4);13(16);14(15);18(OAc);19(OAc);15(-O-)16
1234	3(4);13(16);14(15);18(=O);15(-O-)16;18(-O-)6α
1235	3(4);13(16);14(15);18(=O);7α(OH);15(-O-)16;18(-O-)6α
1236	3(4);13(14)Z;15(=O);16(OMe);18(OAc);15(-O-)16
1237	3(4);13(14)Z;15(=O);18(=O);2(OH);15(-O-)16;18(-O-)19
1238	3(4);13(14)Z;15(=O);18(=O);2(OAc);15(-O-)16;18(-O-)19
1239	3(4);18(OAc);15(OMe);15(-O-)16
1240	3(4);13(14)Z;16(=O);18(OAc);15(-O-)16
1241	3(4);18(OH);15(OMe);15(-O-)16
1242	3(4);18(OH);19(OH);15(OMe);15(-O-)16
1243	3(4);18(OAc);19(OAc);15(OMe);15(-O-)16

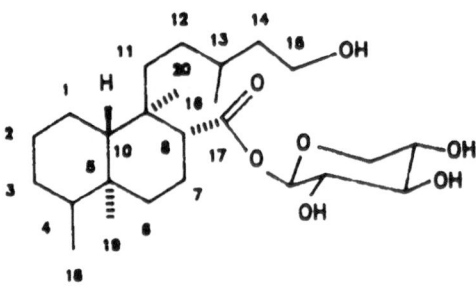

1226	3(4);13(14)E

Table 2. Substitutional Patterns   119

TABLE 2. *(contd.)*

Compound:
No.   Substituents

570   3(4);13(14)Z;15(=O,OH);20(OH)

525   3(4);8(17);7(=O);12α(H)

526   3(4);7(8);17(=O);7(OH);12α(H)

527   3(4);17(=O);8β(H);12β(H)

528   3(4);17(-O-)20;8β(H);12β(H);17β(H)

969   3(4);8(17);7(=O);12β(H)

970   3(4);7(=O);8β(H);12β(H)

971   1(=O);17(=O);6α(OAng);8α(H);12α(H)

972   3(4);7(8);17(=O);6β(OH);12β(H)

973   2(3);17(=O);8α(H);12α(H)

1069   3(4);17(=O);1α(OH);12β(H)

TABLE 2. *(contd.)*

Compound:
No.   Substituents

1257   3(4);17(=O);1α(OAc);8β(H);12β(H)

1258   3(4);17(=O);7α(OH);8β(H);12β(H)

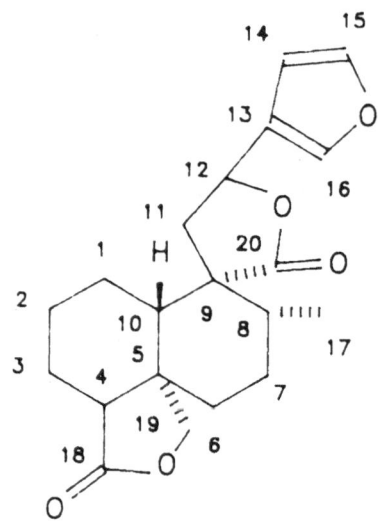

529   3(4);12β(H)

968   3(4);2α(OAng);12β(H)

530   3(4);7β(OAng);12β(H)

531   2α(OAng);3α(OSen);12β(H)

532   3(4);2α(OSen);12β(H)

533   3(4);2α(O2-Mebut);12β(H)

1244   3(4);3'(4');18(OMal);18'(OMal)

Table 2. Substitutional Patterns   121

TABLE 2. (*contd.*)

Compound:
No.   Substituents

1245   3(4);3'(4');18(OSucc);18'(OSucc)

47. Printziane

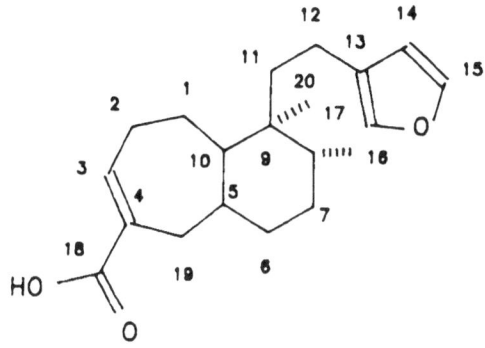

571   5(10)

572   5(6);10β(H)

913   5α(OH);10β(H)

914   5α(OMe);10β(H)

915   1(2);5α(OH);10β(H)

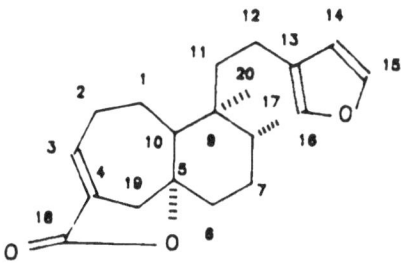

916   10β(H)

76. Conyscabrane

TABLE 2. *(contd.)*

Compound: No.	Substituents
917	5(6);18(=O,OH);1α(H);10β(H)
919	18(=O,OH);5α(OH);1α(H);10β(H)

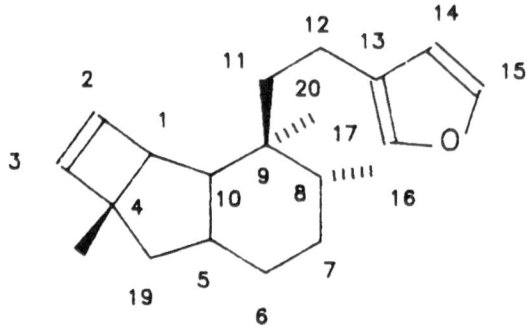

| 918 | 5(6);18(=O,OH);1β(H);10β(H) |
| 920 | 18(=O,OH);5α(OH);1β(H);10β(H) |

48. <u>ent</u>-Clerodane, 5,10-Seco-

573	1(2);3(4);13(16);14(15);18(=O,OH)
912	1(2);3(4);13(16);14(15);18(=O,OH);19(OAc)Z
985	1(2);3(4);13(14);18(=O,OH);16(=O)
986	1(2);3(4);13(14);18(=O,OH);15(=O)
987	1(2);3(4);18(=O,OH);16(=O);14(-O-)15;
952	1(2);3(4);13(16);14(15);18(=O,OH);12α(O2-Mebut)

Table 2. Substitutional Patterns   123

TABLE 2. *(contd.)*

Compound:

No.   Substituents

96. <u>trans</u>-<u>ent</u>-Clerodane, 17-Nor-

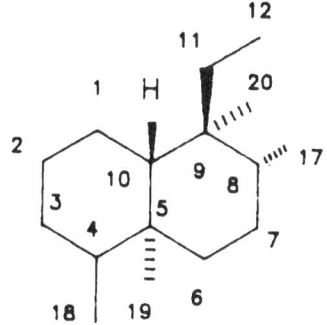

1061   3(4);13(16);14(15);7(=O);18(=O);12(OH);18(-O-)19

49. <u>trans</u>-<u>ent</u>-Clerodane, 13,14,15,16-Tetra-Nor-

574   3(4);12(OH);18(OH)

575   3(4);12(=O,OH);18(OH)

1032   3(4);12(=O,OH);6α(OH);7β(OH)

1093   1(2);3(4);12(=O,OH);18(=O,OH);19(OH);8β(H)

Undetermined

TABLE 2. *(contd.)*

Compound:	
No.	Substituents

237   8(S);3(4);13(14)E;15(OH)

50. <u>cis</u>-<u>ent</u>-Clerodane

578   3(4);13(16);14(15);15(-O-)16

579   3(4);13(16);14(15);18(OH);15(-O-)16

580   3(4);13(16);14(15);18(OAc);15(-O-)16

576   3(4);13(14);15(=O,OH);18(=O,OH)

577   3(4);13(4);15(=O,OH);18(=O,OMe)

581   3(4);13(16);14(15);6α(OH);18(OH);15(-O-)16

582   3(4);13(16);14(15);6α(OH);18(OAc);15(-O-)16

583   3(4);13(16);14(15);18(=O);15(-O-)16;6α(-O-)18

584   3(4);15(=O);18(=O);15(-O-)16;6α(-O-)18

585   3(4);15(=O);18(=O);2β(OH);15(-O-)16;6α(-O-)18

1055   3(4);13(14)E;15(=O,OH);18(OAc)

1056   3(4);13(14)E;15(=O,OH);18(OH)

1057   3(4);15(=O,OH);18(OH)

1058   3(4);15(=O,OH);19(OH)

1246   13(16);14(15);20(=O);3β(-O-)4β;20(-O-)12;
           15(-O-)16;12α(H)

Table 2. Substitutional Patterns   125

TABLE 2. (*contd.*)

Compound:
No.   Substituents

82. Solidagonane

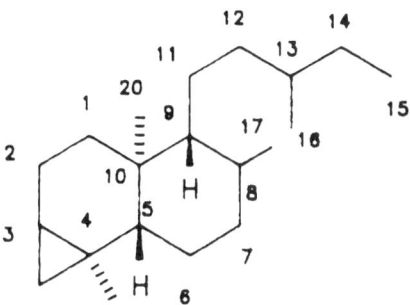

960

VI. ent-Tricarbocyclics and Derivatives

51. ent-Labdane, Cyclopropyl-

586   14(15);8α(OH);13α(OH)

77. Gutierreziane

845   17(=O);18(=O,OH)

TABLE 2. *(contd.)*

Compound:

No.   Substituents

52. ent-Pimarane

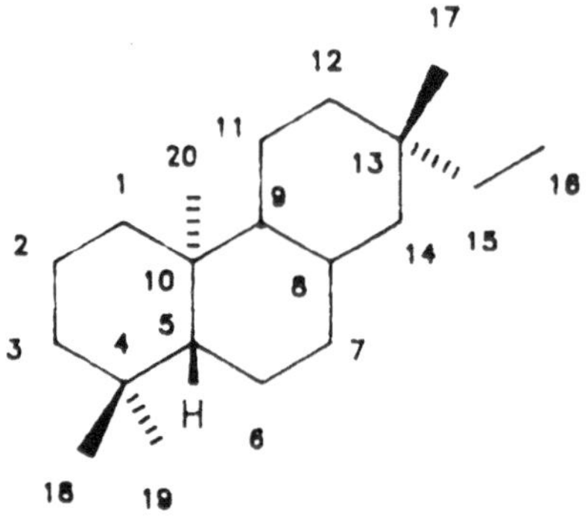

587	15(16);8β(OH);9β(H)
588	15(16);8β(OH);11α(OH);12β(OAc);9β(H)
589	8(14);2β(OH);15(OH);16(OH);19(OH);9β(H)
590	8(14);15(16);19(=O,OH);9β(H)
820	7(8);15(16);19(=O,OH);9β(H)
821	8(9);15(16);19(=O,OH);7(=O)
823	9(11);19(=O,OH);12β(OAc)
591	8(14);3α(OH);15β(OH);16(OH);9β(H)
592	8(14);15β(OH);16(OH);3α(OGlu);9β(H)
593	8(14);15β(OH);16(OH);6β(OGlu);9β(H)
594	8(14);6β(OH);15β(OH);16(OH);9β(H)
595	8(14);2β(OH);15β(OH);16(OH);19(OH);9β(H)
944	8(14);6β(OH);15(OH);16(OH);18(OH);9β(H)
1247	8(14);15(16);3α(OH);19(OH);9β(H)
1248	8(14);15(16);3α(OH);18(OH);9β(H)
1249	8(9);15(16);7(=O);19(OH)
1250	9(11);15(16);19(OH)

Table 2. Substitutional Patterns   127

TABLE 2. *(contd.)*

Compound:
No.   Substituents

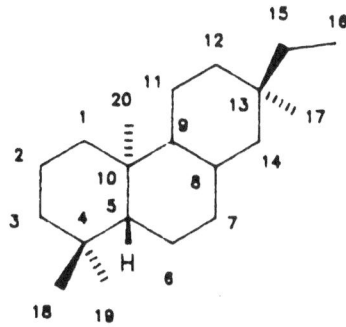

599	9(11);15(16);19(OH)
600	9(11);15(16);19(=O,OH)
1251	9(11);15(16);18(OH)
1252	9(11);15(16);18(OMal)
1253	9(11);15(16);19(OMal)
1254	9(11);15(16);3α(OH);19(OH)

53. ent-Pimarane-8,15-Tetrahydrofuran

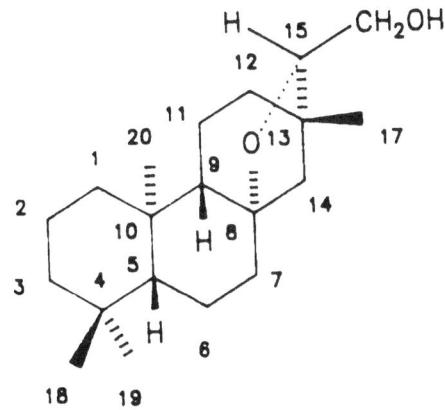

596	3α(OH);12β(OH);16(OH)
598	3(=O);12β(OH);16(OH)

TABLE 2. *(contd.)*

Compound:
No.   Substituents

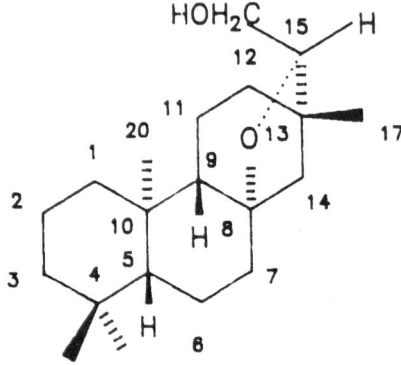

| 824 | 9(11);19(=O,OH);12β(OAc) |
| 597 | 3α(OH);12β(OH) |

54. Cleistanthane (Cleistanthol)

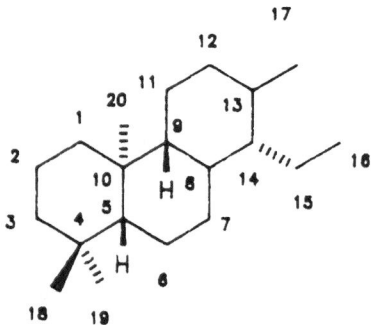

601	12(13);11(=O);15(-O-)16
602	12(13);11(=O);17(=O);15(-O-)16
603	12(13);11(=O);15(-O-)16;17(OAc)
604	12(13);11(=O);15(-O-)16;3α(OAng)
605	12(13);11(=O);14β(OH);15(-O-)16;17(OAc)
606	12(13);11(=O);15(-O-)16;3α(OAng);17(OAc)
607	12(13);11(=O);17(=O);8(-O-)9;15(-O-)16
608	12(13);11(=O);14β(OH);15β(OH);16(-O-)17;15α(H)

Table 2. Substitutional Patterns 129

TABLE 2. (*contd.*)

Compound:
No.   Substituents

85. Cleistanthane, Iso-

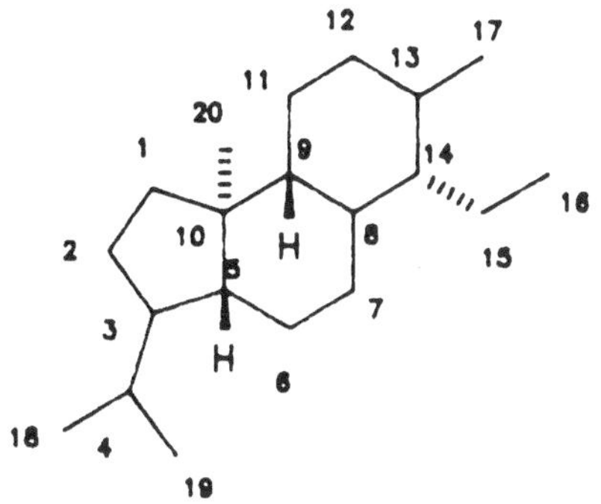

609   3(4);12(13);11(=O);15(-O-)16;17(OAc)

610   12(13);11(=O);3(-O-)4;15(-O-)16;17(OAc)

55. ent-Abietane

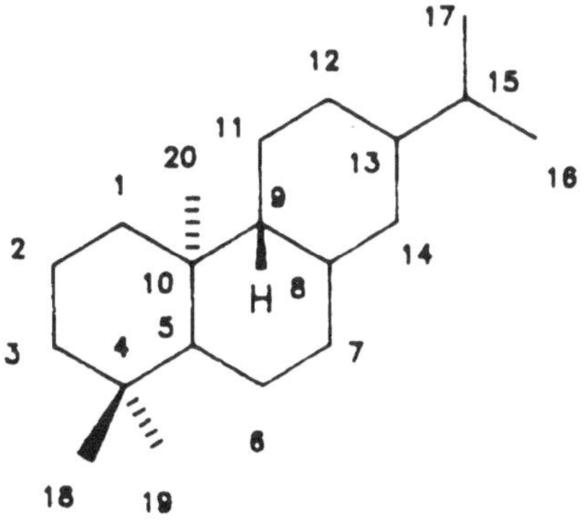

611   8(14);3(=O);13(OH);5β(H)

612   8(14);3β(OH);13(OH);5β(H)

613   8(14);13α(OH);2β(OAc);5β(H)

614   7(8);13(14);5β(H)

615   7(8);13(14);5(OAc)

616   7(8);13(14);19(=O,OH);5β(H)

1054   7(8);13(14);19(O-Succ);5β(H)

TABLE 2. *(contd.)*

Compound:
No.   Substituents

56. Acritoconfertane

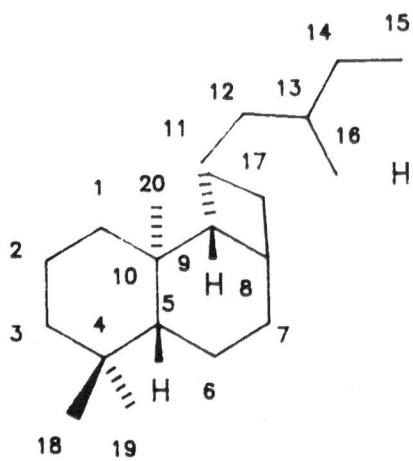

617   7(8);13(14)E;15(=O,OH);6(=O);17(OH);16(OAc)

57. Acritoconfertane-Acetal

1100   13(14)Z;15(=O,OH);6(=O)

Table 2. Substitutional Patterns   131

TABLE 2. *(contd.)*

Compound:
No.   Substituents

83. Erythroxane

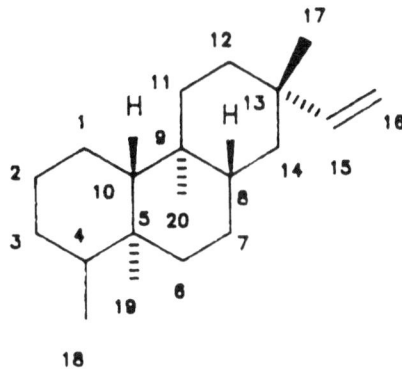

976   3(4);15(16);18(=O,OH)

VII. ent-Tetracarbocyclics and Derivatives

58. ent-Kaurane

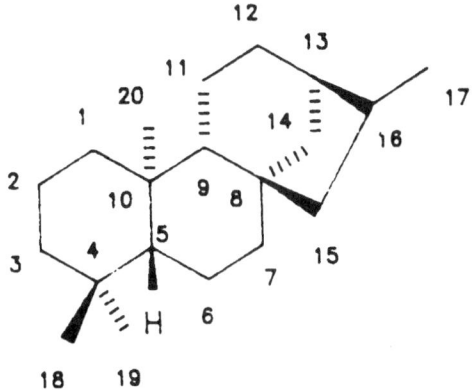

618	16(17)
619	16(17);19(OH)
997	16α(OH)
620	16(17);19(OAc)
621	16(17);19(OMeform)
622	16(17);19(O-p-OH-H-Cinn)

TABLE 2. (*contd.*)

Compound: No.	Substituents
623	16(17);3α(OH);19(OH)
624	16(17);19(=O)
625	16(17);19(=O,OH)
626	16(17);18(OAc)
926	16(17);18(=O,OH)
939	16(17);18(=O.OH);19(OAng)
627	16(17);19(=O,OGly)
628	16(17);19(=O,OH);3α(OH)
629	16(17);19(=O,OH);3α(OAc)
630	16(17);19(=O,OH);3α(OAng)
631	16(17);19(=O,OH);3α(OTig)
632	16(17);19(=O,OH);3α(OCinn)
633	16(17);19(=O,OH);3α(OSen)
634	16(17);19(=O,OH);3α(Oi-Val)
635	16(17);18(=O)
636	16(17);19(=O,OH);9β(OH);13β(OH);15α(OAng)
637	16(17);19(=O,OH);11β(OH);15α(OH)
638	18(OH);16α(-O-)17
1000	16(17);19(=O,OH);7β(OH)
639	16(17);19(=O,OMe);9β(OH)
640	16(17);19(=O,OH);11β(OAc)
641	16(17);19(=O,OH);12α(OH)
642	16(17);19(=O,OH);12α(OAc) [=1108]
796	16(17);19(=O,OH);12β(OAc)
643	16(17);19(=O,OH);18(OAng)
644	16(17);19(=O,OH);18(OSen)
645	16(17);19(=O,OH);18(Oi-Val)
646	16(17);19(=O,OH);15α(OH)
647	16(17);19(=O,OH);15β(OH)
648	16(17);19(=O,OH);15α(OAc)

Table 2. Substitutional Patterns   133

TABLE 2. (*contd.*)

Compound: No.	Substituents
649	16(17);19(=O,OH);15α(OAng)
1007	16(17);19(=O,OMe);9β(OH);15α(OAng)
650	16(17);19(=O,OH);15α(OTig)
651	16(17);19(=O,OH);15α(OSen)
652	16(17);19(=O,OH);15α(Oi-Val)
653	16(17);19(=O,OH);15α(O2-Mac)
654	16(17);19(=O,OH);15α(Oi-But)
655	16(17);19(=O,OH);15α(OCinn)
656	16(17);19(=O,OH);15α(OEpoxyang)
657	16(17);19(=O,OH);15α(OBenz)
818	16(17);19(=O,OH);15α(O2-Mebut-2,3-OH)
892	16(17);19(=O,OH);15α(O2-Mebut)
658	16(17);19(=O,OH);15β(OAng)
659	16(17);19(=O,OH);15β(OTig)
660	16(17);19(=O,OH);15β(OSen)
661	16(17);19(=O,OMe);15α(OH)
662	16(17);19(=O,OH);9β(OH);3α(OAng)
663	16(17);19(=O,OH);9β(OH);3α(OCinn)
664	16(17);19(=O,OH);9β(OH);15α(OH)
665	16(17);19(=O,OH);9β(OH);15α(OAng)
666	16(17);19(=O,OH);9β(OH);15α(OTig)
667	16(17);19(=O,OH);9β(OH);15α(OAc)
668	16(17);19(=O,OH);9β(OH);15α(OSen)
669	16(17);19(=O,OH);9β(OH);15α(Oi-Val)
670	16(17);19(=O,OH);9β(OH);15α(OCinn)
819	16(17);19(=O,OH);9β(OH);15α(O2-Mebut-3-OH)
671	16(17);19(=O,OH);11β(OH);15β(OH)
812	19(=O,OH);17(=O,OMe);16β(H)
672	16(17);19(=O,OH);11β(OH);15β(OAc)
673	16(17);19(=O,OH);11(=O);15β(OH)

TABLE 2. (*contd.*)

Compound: No.	Substituents
674	16(17);19(=O,OH);15(=O);11β(OH)
675	16(17);19(=O,OH);3β(OH);9β(OH);15α(OAng)
869	16(17);19(=O,OH);9β(OH);11β(OH);15α(OAng)
676	16(17);19(=O,OH);3β(OH);9β(OH);15α(OSen)
677	16(17);19(=O,OH);3β(OH);9β(OH);15α(OTig)
690	9(11);16(17);19(=O,OH)
691	9(11);16(17);19(=O,OH);2β(OH)
692	9(11);16(17);19(=O,OH);3β(OH)
693	9(11);16(17);19(=O,OH);7β(OH)
694	9(11);16(17);19(=O,OH);12β(OH)
998	9(11);16(17);19(=O,OH);12β(OEt)
695	9(11);16(17);19(=O,OH);12(=O)
696	9(11);16(17);19(=O,OH);15α(OH)
697	15(16);17(OH)
698	15(16);17(OH);19(OH)
699	15(16);17(OH);19(Op-OH-H-Cinn)
946	15(16);19(=O)
700	15(16);19(=O,OH)
701	15(16);19(=O,OH);17(OH)
702	15(16);19(=O,OH);17(=O)
703	15(16);17(=O,OH);19(=O,OH)
704	9(11);15(16);19(=O,OH);3α(Oi-Val)
705	9(11);15(16);19(=O,OH);3α(Oi-But)
706	9(11);15(16);19(=O,OH);3α(OTig)
707	15(16);19(=O,OH);9β(OH);3α(Oi-Val)
708	15(16);19(=O,OH);9β(OH);3α(Oi-But)
733	16α(-O-)17
709	16α(-O-)17;3α(OAng)
734	19(OH);16α(-O-)17
710	16α(-O-)17;19(OAng)
711	19(=O);16α(-O-)17;3α(OAng)

Table 2. Substitutional Patterns   135

TABLE 2. *(contd.)*

Compound: No.	Substituents
735	19(=O);16α(-O-)17
712	19(=O);16β(-O-)17
713	19(=O,OH);16α(-O-)17
714	19(=O,OH);16β(-O-)17
715	19(=O,OH);16(-O-)17;15α(OAng)
716	19(=O,OH);16(-O-)17;15α(OAc)
717	19(=O,OH);16(-O-)17;15β(OAng)
718	19(=O,OH);16(-O-)17;15β(OTig)
719	9(11);19(=O,OH);12β(OH);16α(-O-)17
720	16α(OH)
1001	16β(OH)
721	16α(OH);3β(OAc)
722	16α(OH);19(OAc)
723	19(=O,OH);16α(OH)
724	11(12);19(=O,OH);16α(OH)
725	16α(OH);17(OH)
726	16α(OH);17(OH);19(OH)
727	3(=O);16α(OH);17(OH)
851	16β(OH)
852	16β(OH);19(OH)
728	19(=O,OH);16α(OH);17(OH)
924	19(=O,OH);17(OH);16α(H)
925	17(=O,OH);19(=O,OH);16α(H)
822	19(=O,OH);16β(OH);17(OH)
729	17(=O);16β(H)
730	17(=O);19(OH);16β(H)
731	17(=O,OH);16β(H)
732	17(=O,OH);19(OH);16β(H)
736	17(=O,Me);19(OH);16β(H)
737	17(=O);19(=O);16β(H)
738	19(=O,OH);17(OH);16β(H)

TABLE 2. (*contd.*)

Compound:
No.   Substituents

739   9(11);19(=O,OH);16α(OH);17(OH)

989   17(=O,OH);19(=O);16β(H)

990   17(=O,OH);19(=O,OH);16β(H)

740   17(=O);19(Op-OH-H-Cinn)

741   19(=O,OH);17(Oi-But);16β(H)

742   19(=O,OH);17(ODodec);16β(H)

743   19(=O,OH);17(OTetradec);16β(H)

744   19(=O,OH);17(OHexadec);16β(H)

745   19(=O,OH);17(OOctadec);16β(H)

746   19(=O,OH);15α(-O-)16α;17(OH)

747   19(=O,OH);15(=O);11β(OH);16α(H)

748   19(=O,OH);15(=O);11β(OH);16α(H)

815   19(=O,OH);15(=O);11β(OH);16α(H)

942   9(11);16(17);19(=O)

943   9(11);16(17);19(=O,OH);15α(OCinn)

1255   19(=O);16α(OH)

1256   9(11);19(=O,OH);16α(OH)

1260   16(17);19(=O,OH);11β(OH);12β(OH);15β(OH)

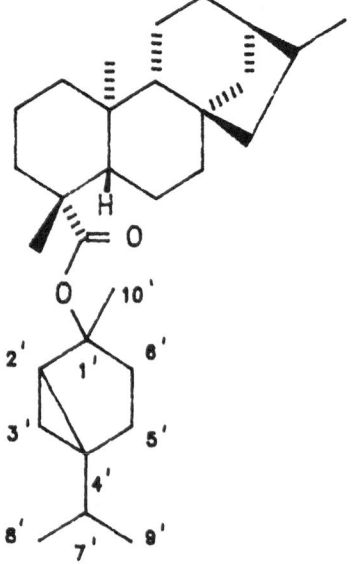

850

Table 2. Substitutional Patterns   137

TABLE 2. (*contd.*)

Compound:
No.   Substituents

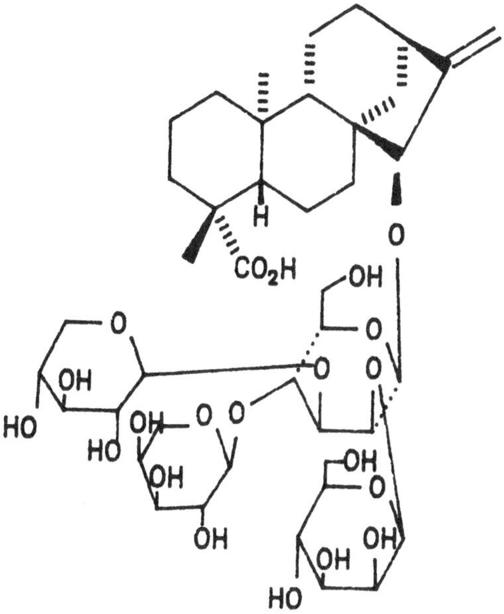

945

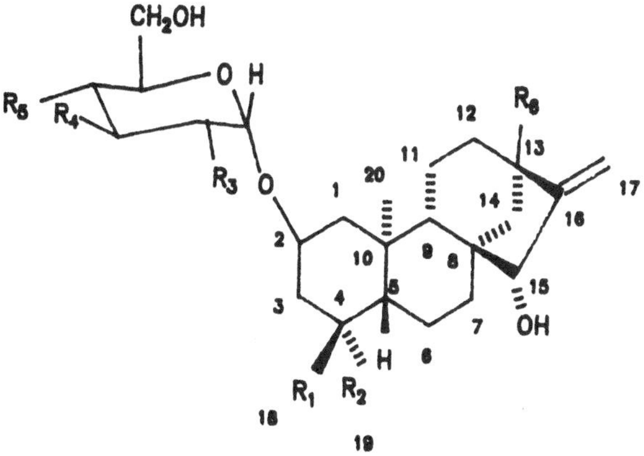

678   R1=R2=COOH;R3=OCOCH2CH(CH3)2;R4=R5=HO3SO;15α(OH)

679   R1=R2=COOH;R3=NHCOCH2CH(CH3)2;R4=CO(CH2)2C6H5;
           R5=OH;13(OH);15(OH)

680   R1=R2=COOH;R3=NHCOCH2CH(CH3)2;R4=CO(CH2)2C6H5;
           R5=OL-Rha;13(OH);15(OH)

681   R1=H;R2=COOH;R3=OCOCH2CH(CH3)2;R4=R5=HO3SO;15α(OH)

TABLE 2. (*contd.*)

Compound:
No.   Substituents

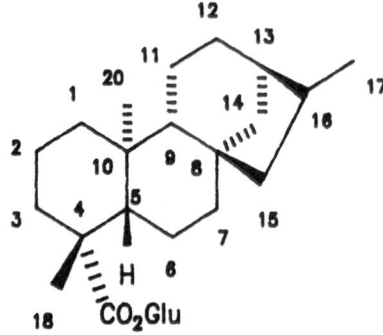

895   16(17);15β(OH)

896   16(17);11β(OH);15β(OH)

897   16(17);15(=O);11β(OH)

898   16α(OH);17(OH)

899   16(17);15β(OGlu)

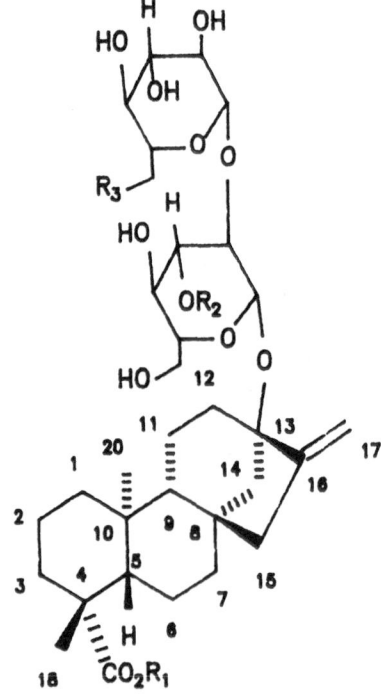

682   R1=R2=H;R3=OH

683   R1=βGlu;R2=H;R3=OH

Table 2. Substitutional Patterns   139

TABLE 2. (*contd.*)

Compound:
No.   Substituents

684   R1=H;R2=βGlu;R3=OH

685   R1=R2=βGlu;R3=OH

686   R1=βGlu;R2=R3=H

687   R1=R2=βGlu;R3=H

688   R1=βGlu[2]-[1]βGlu;R2=βGlu;R3=H

689   R1=βGlu[2]-[1]βGlu;R2=H;R3=OH

59. ent-Kaurane, 17-Nor-

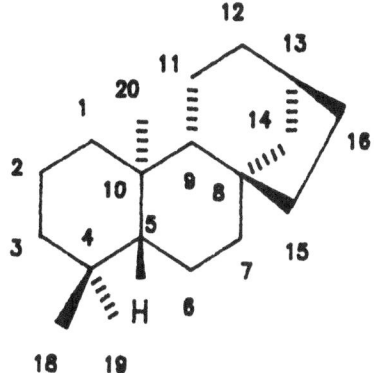

749   19(=O,OH);16(=O)

60. ent-Kaurane, 18-Nor-

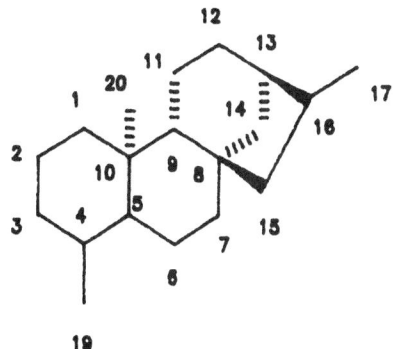

752   16(17);4β(H);5β(H)

753   16(17);4β(OH);5β(H)

754   4(5);6(7);16(17);3(=O);19(OH)

TABLE 2. *(contd.)*

Compound:

No.	Substituents
755	4(5);6(7);16(17);3(=O);19(=O)
756	4(5);6(7);16(17);3(=O);19(=O,OMe)
757	4(5);6(7);3(=O);16(-O-)17
758	16(17);19(=O,OH);2β(OH);15α(OH);4β(H);5β(H)
813	16(17);4β(-O-)19;5β(H)

61. ent-Kaurane, 19-Nor-

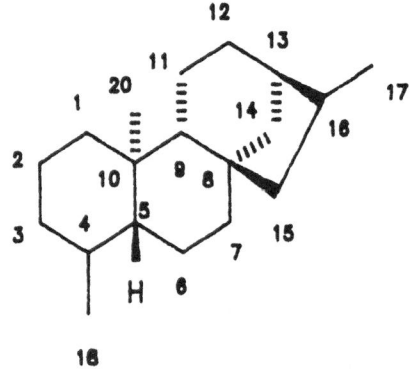

| 750 | 16(17);4α(H) |
| 751 | 16(17);4α(OH) |

62. ent-Kaurane, 9,10-Seco-

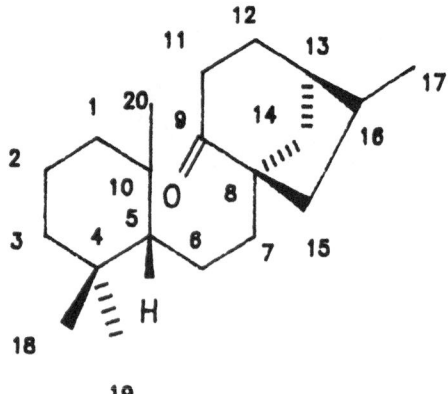

| 759 | 1(10);16(17);19(=O,OH) |

Table 2. Substitutional Patterns   141

TABLE 2. (*contd.*)

Compound:
No.   Substituents

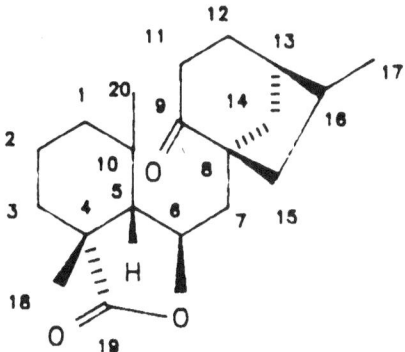

866	16(17);10β(OH);15α(OH)
867	16(17);10β(OH);15β(OH)
868	16(17);10β(OH);15β(OAc)

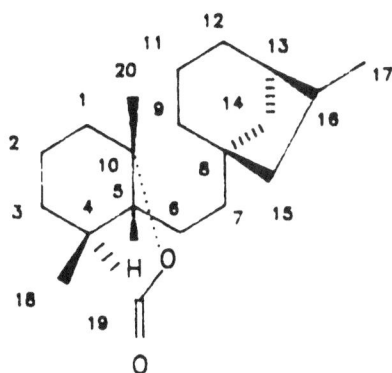

760	16(17);9(=O)
761	16(17);9α(OH)
762	16(17);9β(OH)
763	16(17);9(=O);15β(OH)

TABLE 2. (*contd.*)

Compound:
No.   Substituents

64. ent-Stachane (ent-Beyerane)

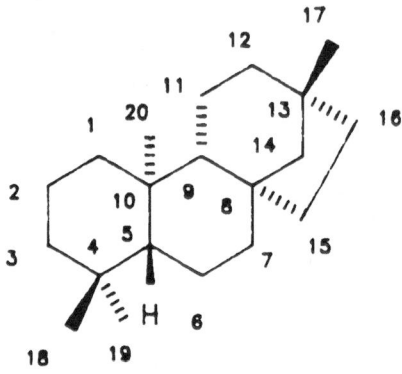

764	15(16);3α(OH)
765	15(16);19(OH)
766	15(16);18(OH)
767	15(16);3α(OH);19(OH)
768	15(16);3α(OH);12β(OH)
769	15(16);19(OMal)
770	15(16);3α(OH);17(OH)
817	15(16);12β(OH);19(OH)
771	15(16);17(OH);19(OH)
772	15(16);18(OH);19(OH)
774	15(16);19(=O,OH)
773	15(16);19(=O)
775	15(16);19(=O,OH);3α(OH)
776	15(16);19(=O,OH);3α(OTig)
777	9(11);15(16);19(=O,OH)
778	9(11);15(16);19(=O,OH);3α(OTig)
779	18(OH);15β(-O-)16β
780	19(OH);15β(-O-)16β

Table 2. Substitutional Patterns   143

TABLE 2. *(contd.)*

Compound:
No.   Substituents

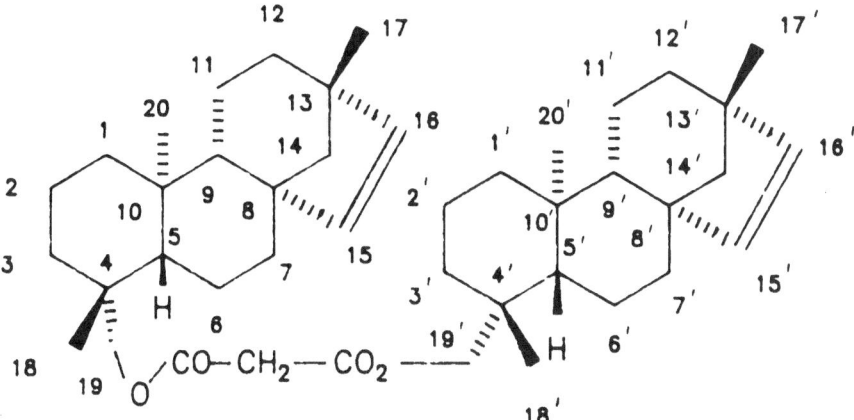

781

93. **<u>ent</u>-Stachane, 19-Nor-**

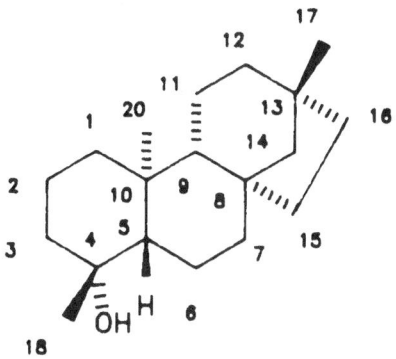

782   15(16)

65. Atisirane (Atisane)

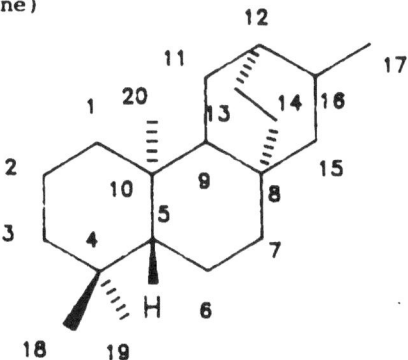

TABLE 2. (*contd.*)

Compound: No.	Substituents
783	16(17);19(=O,OH);9β(H)
853	16α(OH);9β(H)
854	16β(OH);9β(H)
816	13(14);3α(OH);16β(OH);9β(H)
784	16(17);19(=O,OH);7α(OH);9β(H)
785	16(17);19(=O,OH);11α(OAc);9β(H)
786	16(17);19(=O,OH);13α(OAng);9β(H)
787	16(17);19(=O,OH);13α(Oi-Val);9β(H)
788	16(17);19(=O,OH);13α(Oi-But);9β(H)
789	9(11);16(17);19(=O,OH);7α(OAc)
790	9(11);16(17);19(=O,OH);14β(OAc)
814	16(17);19(=O,OH);13(OH);9β(H)

66. Trachylobane

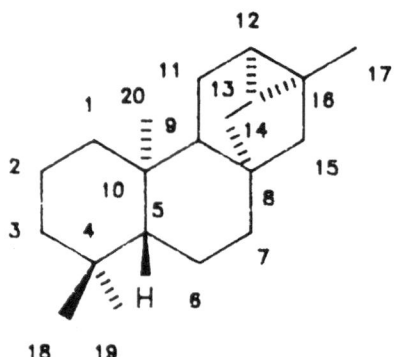

855	19(=O);9βH
791	19(=O,OH);9β(H)
825	19(=O,OH);15α(OAc);9β(H)
826	19(=O,OH);15α(OAng);9β(H)
827	19(=O,OH);15α(Oi-Val);9β(H)
893	19(=O,OH);15α(Oi-But);9β(H)
792	19(=O,OH);7α(OH);9β(H)
793	19(=O,OH);11(=O);9β(H)

Table 2. Substitutional Patterns   145

TABLE 2. *(contd.)*

Compound: No.	Substituents
794	9(11);19(=O,OH)
1204	19(=O,OMe);9β(H)
1259	19(=O,OH);7β(OH);9β(H)

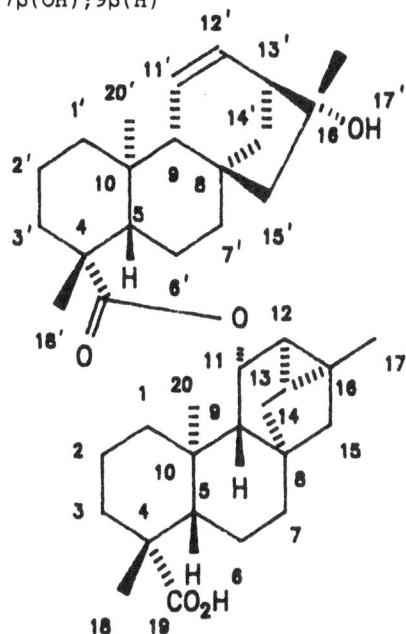

795

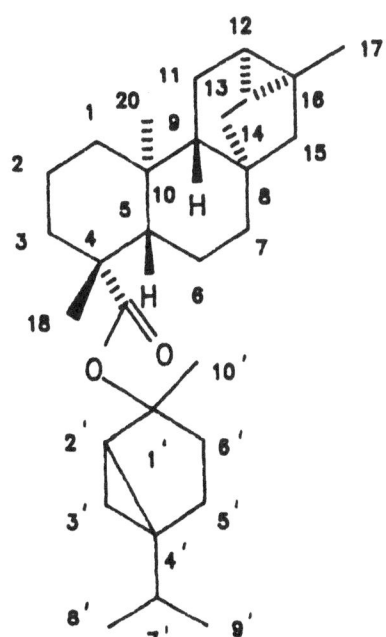

856

TABLE 2. *(contd.)*

Compound:
No.   Substituents

67. Helifulvane

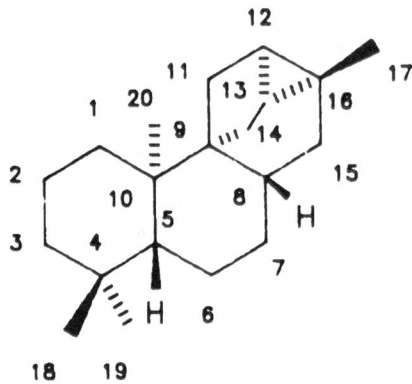

796	19(OH)
797	19(=O,OH)
798	19(=O,OH);11α(OH)
799	19(=O,OH);11α(OAc)

68. Tetrachyrane

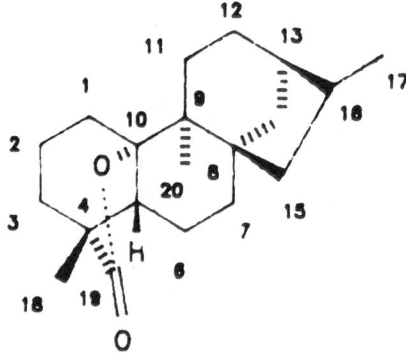

800	16(17)
847	16(17);15(=O)
805	7α(OH);16α(OH)
806	6α(OH);16α(OH)

Table 2. Substitutional Patterns   147

TABLE 2. *(contd.)*

Compound:
No.   Substituents

78. Villanovane

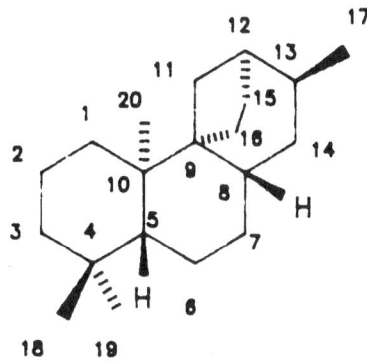

875	13α(OH);17(OH);19(Oi-Val)
876	13α(OH);17(OAc);19(Oi-Val)
877	13α(OH);17(OH);19(OMeVal)
878	13α(OH);17(OAc);19(OMeVal)
879	13α(OH);17(OAc);3α(Oi-Val)
880	13α(OH);17(OAc);3α(OMeVal)
894	13α(OH);15α(OH)

69. Lycoctonine

TABLE 2. (*contd.*)
___

Compound:
No.   Substituents

801

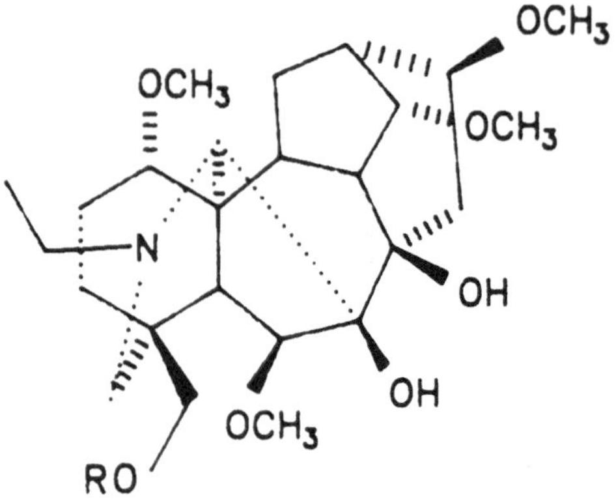

802   R=H
803   R=Anth

VIII. Undetermined Stereochemistry

79. Inulaefane

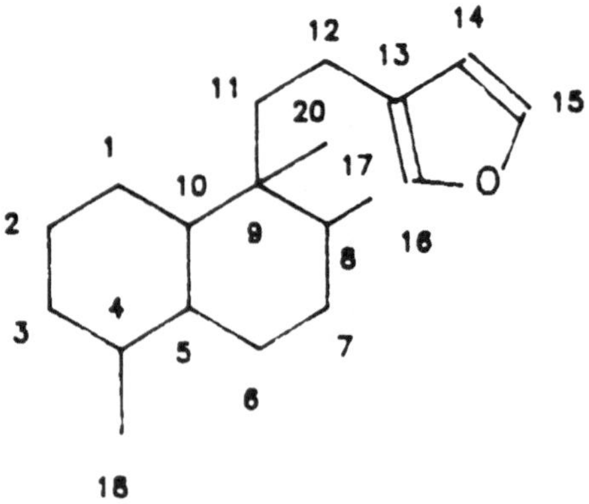

804   2(OH);3(OH);19(OH)

Table 2. Substitutional Patterns 149

TABLE 2. (*contd.*)

Compound:
No.   Substituents

84. Hebeclinane (9,10-Seco-Labdane)

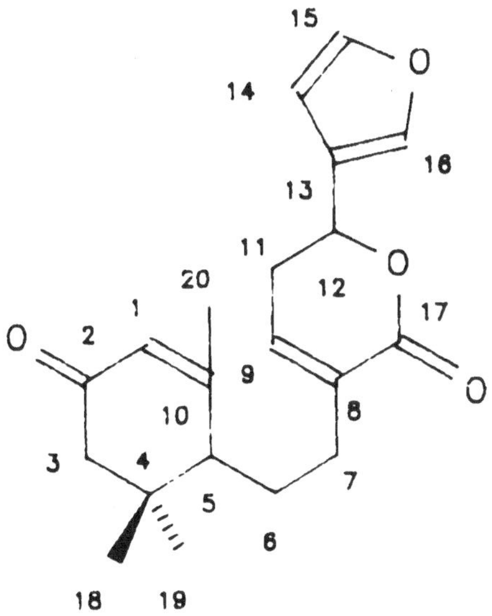

940

999

TABLE 3. Diterpene compound numbers, the species from which these compounds were reported and references (an asterisk indicates unpublished result)

Compound: No.	Species	Reference
1.	Helichrysum mimetes	351
	Mikania goyazensis	46
2.	Pegolettia senegalensis	104
3.	Bejaranoa semistriata	44
	Disynaphia halimifolia	*
	Kingianthus paradoxus	216
	Stevia myriadenia	191
4.	Bejaranoa semistriata	44
	Trichogoniopsis morii	192
5.	Mikania sessilifolia	48
6.	Wyethia helenioides	202
7.	Stylotrichium rotundifolium	148
8.	Mikania officinalis	48
9.	Lasiolaena morii	106
10.	Lasiolaena morii	106
11.	Grazielia intermedia	189
	Lasiolaena morii	
	Lasiolaena santosii	99
12.	Grazielia intermedia	189
13.	Mikania goyazensis	46
14.	Disynaphia halimifolia	67
15.	Lasiolaena morii	106
16.	Zexmenia gnaphaloides	125
17.	Mikania sessilifolia	48
18.	Mikania sessilifolia	48
19.	Koanophyllon admantium	42
20.	Koanophyllon conglobatum	42
21.	Lasiolaena santosii	99
22.	Zinnia tenuiflora	215
	Zinnia verticillata	215
23.	Geigeria burkei subsp. burkei var. burkei	172
	Geigeria burkei subsp. fruticulosa	172
	Helichrysum mundii	171

Table 3. Diterpene Compound Numbers   151

TABLE 3. (*contd.*)

Compound: No.	Species	Reference
	Helichrysum nanum	*
	Helichrysum subfacultatum	171
	Inula cuspidata	140
	Mikania officinalis	48
	Stevia myriadenia	191
24.	Geigeria burkei subsp. burkei var. burkei	172
	Geigeria burkei subsp. diffusa	172
25.	Geigeria burkei subsp. burkei var. burkei	172
26.	Geigeria burkei subsp. burkei var. elata	172
27.	Geigeria burkei subsp. burkei var. elata	172
	Geigeria burkei subsp. burkei var. intermedia	172
28.	Geigeria burkei subsp. burkei var. elata	172
	Geigeria burkei subsp. burkei var. intermedia	172
	Geigeria burkei subsp. burkei var. zeyheri	172
29.	Geigeria burkei subsp. burkei var. burkei	172
	Geigeria burkei subsp. burkei var. elata	172
	Geigeria burkei subsp. burkei var. intermedia	172
30.	Geigeria burkei subsp. burkei var. burkei	172
31.	Geigeria burkei subsp. burkei var. burkei	172
32.	Helichrysum Krebsianum	171
	Helichrysum oreophilum	351
	Helichrysum oxyphyllum	171
33.	Geigeria burkei subsp. diffusa	172
34.	Geigeria burkei subsp. diffusa	172
35.	Geigeria burkei subsp. burkei var. burkei	172
	Geigeria burkei subsp. diffusa	172
36.	Geigeria burkei subsp. burkei var. burkei	172
37.	Geigeria burkei subsp. fruticulosa	172

TABLE 3. (*contd.*)

Compound: No.	Species	Reference
38.	Geigeria burkei subsp. burkei var. zeyheri	172
39.	Geigeria burkei subsp. burkei var. burkei	172
	Geigeria burkei subsp. burkei var. zeyheri	172
40.	Acanthospermum australe	96
41.	Acanthospermum australe	96
42.	Melampodium longipilum	434
43.	Melampodium longipilum	434
44.	Melampodium diffusum	435
45.	Melampodium diffusum	435
46.	Melampodium longipilum	434
47.	Melampodium longipilum	434
48.	Acanthospermum australe	96
49.	Ichthyothere ulei	107
50.	Ichthyothere ulei	107
51.	Ichthyothere ulei	107
52.	Ichthyothere ulei	107
53.	Ichthyothere ulei	107
54.	Ichthyothere ulei	107
55.	Montanoa leucantha subsp. arborescens	382
	Montanoa leucantha subsp. leucantha	382
	Montanoa mollissima	382
	Montanoa tomentosa subsp. microcephala	382
	Montanoa tomentosa subsp. tomentosa	356
	Montanoa tomentosa subsp. xanthiifolia	382
56.	Montanoa tomentosa subsp. xanthiifolia	459
57.	Montanoa tomentosa subsp. tomentosa	356
58.	Mikania sessilifolia	48

Table 3. Diterpene Compound Numbers　　153

TABLE 3. (*contd.*)

Compound: No.	Species	Reference
59.	Baccharis trinervis	117
	Centaurea cineraria	*
	Conyza canadensis	*
	Conyza ulmifolia	*
	Felicia erigeroides	*
	Heteromma simplicifolium	*
	Nidorella auriculata subsp. polycephala	*
	Osteospermum barberia	151
	Plagiocheilus prostratus	133
60.	Centipeda orbicularis	128,133
	Plagiocheilus prostratus	133
61.	Conyza podocephala	154
62.	Dimerostemma asperatum	214
	Dimerostemma brasilianum	141
63.	Dimerostemma asperatum	214
64.	Dimerostemma brasilianum	141
65.	Helichrysum calliconum	39
66.	Helichrysum calliconum	39
67.	Helichrysum calliconum	39
68.	Achyrocline vargasiana	*
	Gnaphalium microcephalum	*
	Helichrysum Krebsianum	*
	Helichrysum odoratissimum	298
	Helichrysum nudifolium	39
	Senecio pseudoorientalis	*
	Senecio trichopterygius	*
69.	Helichrysum acutatum	40
70.	Greenmaniella resinosa	*
	Helichrysum acutatum	
	Helichrysum Krebsianum	171
	Helichrysun odoratissimum	*
71.	Helichrysum acutatum	40
72.	Helichrysum acutatum	40
73.	Helichrysum subfalcatum	171
74.	Liatris elegans	335
75.	Liatris elegans	335
76.	Hemizonia lutescens	93
	Zexmenia gnaphaloides	*

TABLE 3. (*contd.*)

Compound: No.	Species	Reference
77.	Acanthospermum hispidum	398
78.	Brickellia eupatoriedes	59
79.	Baccharis scoparia Oxylobus arbutifolius Oxylobus glanduliferus	57 87 6
80.	Gymnosperma glutinosa	393
81.	Oxylobus glanduliferus	6
82.	Carterothamnus anomalochaeta	185
83.	Carterothamnus anomalochaeta	185
84.	Carterothamnus anomalochaeta	185
85.	Hemizonia lutescens	93
86.	Madia sativa	100
87.	Madia sativa	100
88.	Hemizonia congesta Hemizonia lutescens	93 93
89.	Hemizonia congesta Hemizonia lutescens Madia sativa	93 93 100
90.	Hemizonia lutescens	93
91.	Hemizonia lutescens	93
92.	Hemizonia lutescens	93
93.	Hemizonia lutescens	93
94.	Carterothamnus anomalochaeta	185
95.	Gutierrezia sphaerocephala	272
96.	Gutierrezia sphaerocephala	272
97.	Brickellia diffusa	50
98.	Brickellia eupatoriedes	59
99.	Brickellia eupatoriedes	59
100.	Pleurocoronis pluriseta	147
101.	Brickellia argyrolepis	147
102.	Brickellia argyrolepis	147
103.	Brickellia eupatoriedes	59

Table 3. Diterpene Compound Numbers   155

TABLE 3. (*contd.*)

Compound: No.	Species	Reference
104.	Brickellia eupatoriedes	59
105.	Brickellia eupatoriedes	59
106.	Brickellia corymbosa Brickellia eupatoriedes Brickellia squarrosa	158 59 158
107.	Brickellia squarrosa	158
108.	Brickellia corymbosa Brickellia squarrosa	158 158
109.	Brickellia squarrosa	158
110.	Baccharis tucumanensis	477
111.	Senecio erosus	209
112.	Trichogoniopsis morii	192
113.	Solidago sempervirens	430
114.	Chrysothamnus nauseusus	70
115.	Chrysothamnus nauseusus	70
116.	Chrysothamnus nauseusus	70
117.	Chrysothamnus nauseusus	70
118.	Stevia jaliscensis	206
119.	Brickellia vernicosa Chrysothamnus nauseusus	29 70
120.	Chrysothamnus nauseusus Grindelia discoidea	70 471
121.	Macowania glandulosa Stevia salicifolia	163 170
122.	Brickellia argyrolepis	147
123.	Brickellia argyrolepis	147
124.	Baccharis salicifolia Brickellia veronicaefolia	503 222
125.	Brickellia veronicaefolia	222
126.	Brickellia eupatoriedes Brickellia paniculata Brickellia sp.	59 287 59,286
127.	Brickellia eupatoriedes	59
128.	Brickellia eupatoriedes	59

TABLE 3. (*contd.*)

Compound: No.	Species	Reference
129.	Brickellia veronicaefolia	158
	Brickellia veronicaefolia var. typica	158
	Chrysothamnus nauseusus	70
130.	Gutierrezia lucida	188
131.	Gutierrezia lucida	188
132.	Denekia capensis	180
	Helichrysum confertum	180
	Helichrysum sutherlandia	171
	Palafoxia rosea	66
133.	Palafoxia rosea	66
	Palafoxia texana var. ambigua	*
134.	Palafoxia rosea	66
	Palafoxia texana var. ambigua	*
135.	Palafoxia rosea	66
	Palafoxia texana var. ambigua	*
136.	Palafoxia rosea	66
137.	Chromolaena collina	175
	Fleischmannia deborabellae	68
	Helichrysum confertum	180
138.	Chromolaena collina	175
139.	Fleischmannia pycnocephaloides	68
	Geigeria burkei Harv. subsp. burkei var. intermedia	172
	Geigeria burkei Harv. subsp. burkei var. zeyheri	172
	Inula crithmoides	
	Koanophyllon conglobatum	*
	Podachaenium eminens	121
	Silphium connatum	*
140.	Hofmeisteria fasciculata	135
	Stevia andina	*
141.	Nidorella auriculata DC. subsp. polycephala	72
	Stevia andina	*
142.	Nidorella auriculata DC. subsp. polycephala	72
	Stevia andina	*
143.	Stevia monardaefolia	433
144.	Koanophyllon conglobatum	135
145.	Koanophyllon conglobatum	135
	Stevia berlandiera	170

Table 3. Diterpene Compound Numbers   157

TABLE 3. (*contd.*)

Compound: No.	Species	Reference
146.	Koanophyllon conglobatum	135
147.	Koanophyllon conglobatum	135
148.	Koanophyllon conglobatum	135
149.	Aristeguietia buddleaefolia	134
150.	Aristeguietia buddleaefolia	134
151.	Aristeguietia buddleaefolia	134
152.	Bishovia boliviensis	186
153.	Stevia monardaefolia	433
154.	Stevia salicifolia	415
155.	Smallanthus fruticosus	211
	Helichrysum coriaceum	*
	Helichrysum nudifolium var. nudifolium	*
	Helichrysum pinifolium	171
	Helichrysum vernum	171
156.	Koanophyllon conglobatum	135
157.	Koanophyllon conglobatum	135
158.	Solidago canadensis	16,17,73
	Solidago gigantea	
	Solidago sempervirens	249
159.	Gutierrezia dracunculoides	80
160.	Gutierrezia mandonii	188
161.	Gutierrezia mandonii	188
162.	Gutierrezia mandonii	188
163.	Gutierrezia dracunculoides	80
164.	Gutierrezia lucida	188
165.	Chrysothamnus paniculatus	475
	Grindelia boliviana	348
	Grindelia camporum	49,473,475
	Grindelia chiloensis	294
	Grindelia paludosa	49
	Grindelia pulchella	294
	Grindelia squarrosa	181
	Grindelia stricta	49
166.	Chrysothamnus paniculatus	475
	Grindelia camporum	473,475
	Grindelia chiloensis	294
	Grindelia pulchella	475
	Grindelia squarrosa	473,475

TABLE 3. *(contd.)*

Compound: No.	Species	Reference
	Heterotheca subaxillaris	*
167.	Chrysothamnus paniculatus	475
	Grindelia aphanactis	348
	Grindelia acutifolia	*
	Grindelia boliviana	348
	Grindelia camporum	49,473
	Grindelia chiloensis	294,348
	Grindelia nauseosus	440
	Grindelia paludosa	49
	Grindelia perennis	348
	Grindelia pulchella	294
	Grindelia robusta	440
	Grindelia squarrosa	68,220,473
	Grindelia stricta	49
	Haplopappus venetus	74
	Heterotheca subaxillaris	*
	Isocoma coronopifolia	253
	Solidago petradoria	*
168.	Grindelia pulchella	293
169.	Grindelia stricta	49
170.	Grindelia camporum	49
	Grindelia humilis	441
	Grindelia paludosa	49
	Grindelia squarrosa	473
	Grindelia stricta	49
171.	Grindelia camporum	49
	Grindelia stricta	49
172.	Chrysothamnus paniculatus	475
	Grindelia camporum	473,475
	Grindelia humilis	441
	Grindelia stricta	475
173.	Grindelia aphanactis	348
	Grindelia boliviana	348
	Grindelia chiloensis	294
	Grindelia perennis	348
	Grindelia pulchella	294
	Grindelia squarrosa	220,473
	Grindelia stricta	49
	Haplopappus tenuisecta	74
	Heterotheca subaxillaris	*
	Isocoma coronopifolia	253
174.	Chrysothamnus paniculatus	475
	Grindelia camporum	473,475
	Grindelia chiloensis	294
	Grindelia perennis	348
	Grindelia pulchella	294
	Grindelia squarrosa	473
	Grindelia stricta	49
	Solidago petradoria	*

Table 3. Diterpene Compound Numbers   159

TABLE 3. (*contd.*)

Compound: No.	Species	Reference
175.	Chrysothamnus paniculatus	475
	Grindelia camporum	473,475
	Grindelia squarrosa	473
	Grindelia stricta	49
176.	Chrysothamnus paniculatus	475
	Grindelia camporum	475
	Grindelia perennis	348
	Grindelia squarrosa	473
	Grindelia stricta	49
177.	Chrysothamnus paniculatus	475
	Grindelia camporum	475
	Grindelia squarrosa	473
	Grindelia stricta	49
178.	Grindelia perennis	348
	Grindelia stricta	49
179.	Chrysothamnus paniculatus	475
	Grindelia camporum	475
	Grindelia squarrosa	473
180.	Chrysothamnus nauseusus	440
	Grindelia camporum	49
	Grindelia chiloensis	294
	Grindelia paludosa	49
	Grindelia pulchella	294
	Grindelia stricta	49
	Isocoma coronopifolia	253
181.	Grindelia camporum	49
	Grindelia paludosa	49
	Grindelia stricta	49
182.	Grindelia stricta	49
183.	Grindelia stricta	49
184.	Grindelia stricta	49
185.	Chrysothamnus nauseusus	440
186.	Grindelia stricta	49
187.	Grindelia stricta	49
188.	Grindelia camporum	49
	Grindelia chiloensis	294
	Grindelia paludosa	49
	Grindelia pulchella	294
	Grindelia stricta	49
189.	Chrysothamnus paniculatus	341,475
	Grindelia acutifolia	
	Grindelia aphanactis	*
	Grindelia camporum	49,473,475
	Grindelia paludosa	49
	Grindelia robusta	380

TABLE 3. (*contd.*)

Compound: No.	Species	Reference
	Grindelia squarrosa	473
	Grindelia stricta	49,475
	Heterotheca subaxillaris	*
190.	Grindelia stricta	49
191.	Chrysothamnus paniculatus	475
	Grindelia camporum	475
	Grindelia chiloensis	475
	Grindelia pulchella	475
	Grindelia stricta	49
192.	Grindelia camporum	473
	Grindelia pulchella	294
	Grindelia robusta	381
	Grindelia squarrosa	473
193.	Solidago canadensis	17,73
194.	Solidago canadensis	73
195.	Eupatorium jhanii	289
	Stevia rebaudiana	461
196.	Eupatorium jhanii	289
197.	Eupatorium jhanii	289
198.	Aristeguietia buddleaefolia	134
	Eupatorium jhanii	289
199.	Eupatorium jhanii	289
200.	Palafoxia rosea	66,250
201.	Palafoxia rosea	250
202.	Baccharis tola	450
203.	Lindheimera texana	337
	Silphium asteriscus	91
	Silphium compositum	*
	Silphium integrifolium	91
	Silphium perfoliatum	91
	Silphium terebinthinaceum	91
204.	Echinacea purpurea	88
	Lindheimera connata	
	Lindheimera texana	337
	Silphium asteriscus	91
	Silphium compositum	*
	SIlphium connatum	*
	Silphium integrifolium	91
	Silphium perfoliatum	91
	Silphium terebinthinaceum	91
	Trichogoniopsis morii	
205.	Silphium perfoliatum	91

Table 3. Diterpene Compound Numbers 161

TABLE 3. (*contd.*)

Compound: No.	Species	Reference
206.	Silphium asteriscus	91
	Silphium integrifolium	91
	Silphium perfoliatum	91
	Silphium terebinthinaceum	91
207.	Silphium asteriscus	91
	Silphium integrifolium	91
	Silphium perfoliatum	91
	Silphium terebinthinaceum	91
208.	Silphium perfoliatum	91
209.	Silphium perfoliatum	91
210.	Carterothamnus anomalochaeta	185
211.	Carterothamnus anomalochaeta	185
212.	Silphium perfoliatum	91
	Silphium terebinthinaceum	91
213.	Silphium perfoliatum	91
214.	Nidorella hottentotica	155
215.	Brickellia diffusa	50
216.	Brickellia diffusa	50
217.	Brickellia diffusa	50
218.	Brickellia diffusa	50
219.	Koanophyllon conglobatum	135
220.	Brickellia veronicaefolia	158
221.	Grindelia paludosa	49
	Grindelia stricta	49
222.	Grindelia stricta	49
223.	Chrysothamnus paniculatus	341
	Grindelia acutifolia	
	Grindelia stricta	49
224.	Grindelia stricta	49
225.	Nidorella hottentotica	155
226.	Nidorella hottentotica	155
227.	Nidorella hottentotica	155
228.	Nidorella hottentotica	155
229.	Nidorella hottentotica	155
231.	Relhania acerosa	90

TABLE 3. *(contd.)*

Compound: No.	Species	Reference
232.	Koanophyllon conglobatum	42
233.	Koanophyllon conglobatum	42
234.	Koanophyllon conglobatum	42
235.	Koanophyllon conglobatum	42
236.	Bedfordia salicinia	122
237.	Bedfordia salicinia	122
238.	Nidorella agria	72
239.	Solidago gigantea var. serotina Solidago serotina	15,401 14
240.	Solidago serotina	14
241.	Solidago serotina	14
242.	Solidago serotina	14
243.	Solidago serotina	14
244.	Solidago gigantea var. serotina Solidago serotina	15,401 14
245.	Koanophyllon admantium	42
246.	Solidago serotina	14
247.	Solidago serotina	14
248.	Centipeda orbicularis	128
249.	Centipeda orbicularis	128
250.	Koanophyllon admantium	42
251.	Nidorella resedifolia	72
253.	Nidorella agria	72
254.	Nidorella agria	72
255.	Conyza stricta Grangea maderaspatana Koanophyllon admantium	376 42
256.	Nidorella resedifolia	42
257.	Macowania glandulosa	163
258.	Solidago altissima Solidago virgaurea	143 291

Table 3. Diterpene Compound Numbers 163

TABLE 3. (*contd.*)

Compound: No.	Species	Reference
259.	Solidago altissima	143
	Solidago virgaurea	291
260.	Solidago altissima	143
	Solidago rugosa	73
	Solidago virgaurea	291
261.	Solidago altissima	143
	Solidago virgaurea	291
262.	Solidago altissima	143
	Solidago virgaurea	291
263.	Solidago virgaurea	291
264.	Solidago virgaurea	291
265.	Solidago virgaurea	291
266.	Solidago virgaurea	291
267.	Solidago virgaurea	291
268.	Solidago virgaurea	291
269.	Solidago virgaurea	291
270.	Gutierrezia dracunculoides	235
271.	Chrysothamnus paniculatus	474
272.	Othonna sedifolia	*
	Palafoxia rosea	66
	Senecio variabilis	*
	Stevia berlandiera	170
273.	Palafoxia rosea	66
	Stevia salicifolia	170
274.	Senecio subrubriflorus	169
275.	Senecio subrubriflorus	169
276.	Chrysanthemoides monolifera subsp. canescens	79
	Garuleum bipinnatum	156
	Garuleum pinnatifidum	156
	Garuleum sonchifolium	
	Osteospermum auriculatum	151
	Osteospermum barberiae	151
	Osteospermum ciliatum	151
	Osteospermum corymbosum	157
	Osteospermum fruticosum	156
	Osteospermum jucundum	151
	Osteospermum junceum	156
	Osteospermum muricatum subsp. muricatum	151
	Osteospermum rotundifolium	157
	Senecio subrubriflorus	169

TABLE 3. (*contd.*)

Compound: No.	Species	Reference
277.	Garuleum bipinnatum	156
	Garuleum pinnatifidum	156
	Osteospermum auriculatum	151
278.	Garuleum bipinnatum	151
	Garuleum pinnatifidum	151
	Osteospermum auriculatum	151
279.	Chrysanthemoides monolifera	79
	Chrysanthemoides monolifera subsp. canescens	79
	Garuleum bipinnatum	156
	Garuleum pinnatifidum	156
280.	Senecio subrubriflorus	79
281.	Senecio subrubriflorus	79
282.	Senecio subrubriflorus	79
283.	Senecio subrubriflorus	169
284.	Garuleum bipinnatum	156
	Garuleum pinnatifidum	156
	Senecio subrubriflorus	79
285.	Osteospermum fruticosum	156
286.	Osteospermum barberiae	151
287.	Osteospermum barberiae	151
288.	Chrysanthemoides monolifera subsp. canescens	79
	Senecio subrubriflorus	169
289.	Chrysanthemoides monolifera subsp. canescens	79
	Garuleum bipinnatum	156
	Garuleum pinnatifidum	156
	Osteospermum auriculatum	151
	Osteospermum ciliatum	151
	Osteospermum corymbosum	157
	Osteospermum incanum	*
	Osteospermum jucundum	151
	Osteospermum junceum	156
	Osteospermum muricatum subsp. muricatum	151
	Osteospermum oppositifolium	156
	Osteospermum polygaloides	*
	Osteospermum rotundifolium	157
	Osteospermum scariosum subsp. scariosum	151
	Osteospermum thodei	151
290.	Senecio subrubriflorus	169
291.	Senecio subrubriflorus	169

Table 3. Diterpene Compound Numbers   165

TABLE 3. (*contd.*)

Compound: No.	Species	Reference
292.	Chrysanthemoides incana	79
	Chrysanthemoides monolifera	79
	Garuleum bipinnatum	156
	Garuleum pinnatifidum	156
	Garuleum sonchifolium	
	Osteospermum corymbosum	157
	Osteospermum junceum	156
	Osteospermum polygalioides	157
	Osteospermum rotundifolium	157
	Senecio sandersonii	173
	Senecio subrubriflorus	*
293.	Garuleum bipinnatum	156
	Garuleum pinnatifidum	156
294.	Chrysanthemoides monolifera subsp. canescens	79
	Helichrysum diosmifolium	*
	Osteospermum subulatum	157
295.	Chrysanthemoides monolifera subsp. canescens	79
	Osteospermum auriculatum	151
	Osteospermum ciliatum	151
	Osteospermum jucundum	151
	Osteospermum thodei	151
	Senecio sandersonii	173
296.	Garuleum bipinnatum	156
	Garuleum pinnatifidum	156
297.	Garuleum pinnatifidum	156
298.	Garuleum bipinnatum	156
299.	Garuleum bipinnatum	156
300.	Achillea filipendulina	25
	Dimorphotheca pluvialis	118
	Mikania pyramidata	131
301.	Dimorphotheca pluvialis	118
302.	Dimorphotheca pluvialis	118
303.	Dimorphotheca pluvialis	118
304.	Palafoxia rosea	66
305.	Palafoxia rosea	66
306.	Palafoxia rosea	66
307.	Palafoxia rosea	66
308.	Croptilon divaricatum	252
309.	Croptilon divaricatum	252
	Palafoxia rosea	66

TABLE 3. (*contd.*)

Compound: No.	Species	Reference
310.	Osteospermum muricatum subsp. muricatum	151
311.	Osteospermum muricatum subsp. muricatum	151
312.	Osteospermum muricatum subsp. muricatum	151
313.	Palafoxia rosea	66,168
314.	Palafoxia rosea	66,168,251
315.	Palafoxia rosea	66,168
316.	Solidago missouriensis	10
317.	Solidago missouriensis	10
318.	Solidago missouriensis	10
319.	Inula royleana	34
320.	Inula royleana	34
321.	Inula royleana	34,254
322.	Inula royleana	34,254
323.	Inula royleana	34,254
324.	Inula royleana	34
325.	Dimorphotheca pluvialis	118
326.	Zexmenia gnaphaloides	125
327.	Relhania acerosa	90
328.	Austroeupatorium chaparense	146
329.	Chrysothamnus viscidiflorus	370
330.	Chrysothamnus viscidiflorus Grindelia aegialitis	370 414
331.	Baccharis sternbergiana	136
332.	Baccharis sternbergiana	136
333.	Baccharis sternbergiana	136
334.	Baccharis sternbergiana	136
335.	Morithamnus crassus Relhania acerosa	102 90
336.	Morithamnus crassus	102

Table 3. Diterpene Compound Numbers   167

TABLE 3. (*contd.*)

Compound: No.	Species	Reference
337.	Ayapana amygdalina	113
338.	Ayapana amygdalina	113
339.	Ayapana amygdalina	113
340.	Ayapana amygdalina	113
341	Ayapana amygdalina	113
342.	Ayapana amygdalina	113
343.	Ayapana amygdalina	113
344.	Ayapana amygdalina	113
345.	Ayapana amygdalina	113
346.	Ayapana amygdalina	113
347.	Relhania acerosa	90
348.	Acritopappus longifolius Helichrysum confertum	85 180
349.	Acritopappus longifolius	85
350.	Acritopappus longifolius	85
351.	Solidago canadensis	73
352.	Solidago canadensis	73
353.	Lasiolaena santosii	99
354.	Morithamnus crassus	102
355.	Ageratina dendroides Pleurocoronis pluriseta	77 62
356.	Hartwrightia floridana	95
357.	Baccharis sternbergiana	136
358.	Eupatorium petiolare	223
359.	Baccharis sternbergiana	136
360.	Haplopappus pectinatus	348
361.	Pleurocoronis pluriseta	62
362.	Achyrocline alata	43
363.	Baccharis intermixta	178
364.	Baccharis intermixta Lasiolaena santosii	178 99

TABLE 3. (*contd.*)

Compound: No.	Species	Reference
365.	Acritopappus confertus	184
366.	Acritopappus confertus	179,184
367.	Acritopappus confertus	179
368.	Acritopappus confertus	184
369.	Acritopappus confertus	179
370.	Acritopappus confertus	179
371.	Acritopappus confertus	184
372.	Acritopappus confertus	184
373.	Acritopappus confertus	184
374.	Acritopappus confertus	184
375.	Acritopappus confertus	184
376.	Acritopappus confertus	184
377.	Acritopappus confertus	184
378.	Acritopappus confertus	184
379.	Acritopappus confertus	184
380.	Acritopappus confertus	184
381.	Acritopappus confertus	184
382.	Acritopappus morii	179
383.	Acritopappus morii	179
384.	Acritopappus longifolius	85
385.	Xanthocephalum linearifolium	110
386.	Acritopappus longifolius	85
387.	Acritopappus confertus	184
388.	Ageratum fastigiatum	53,126
389.	Ageratum fastigiatum	53,126
390.	Ageratum fastigiatum	126
391.	Ageratum fastigiatum	53
392.	Ageratum fastigiatum	53
393.	Acritopappus confertus	184
394.	Acritopappus confertus	184

Table 3. Diterpene Compound Numbers   169

TABLE 3. (*contd.*)

Compound: No.	Species	Reference
395.	Acritopappus longifolius	85
396.	Acritopappus longifolius	85
397.	Acritopappus morii	179
398.	Ageratum fastigiatum	126
399.	Acritopappus hagei	179
400.	Acritopappus hagei	179
401.	Acritopappus morii	179
402.	Austroeupatorium chaparense	146
403.	Xanthocephalum linearifolium	110
404.	Psiadia altissima	224
405.	Conyza stricta	460
406.	Austroeupatorium inulaefolium	177
407.	Austroeupatorium inulaefolium	177
408.	Austroeupatorium chaparense	146
409.	Austroeupatorium chaparense	146
410.	Austroeupatorium chaparense	146
411.	Austroeupatorium chaparense	146
412.	Planaltoa lychnophoroides	203
413.	Planaltoa lychnophoroides	203
414.	Acritopappus morii	179
415.	Desmanthodium fruticosum Stevia lucida Lag. var. Bipontini	* 448
416.	Xanthocephalum linearifolium	110
417.	Austroeupatorium inulaefolium	177
418.	Austroeupatorium inulaefolium	177
419.	Acritopappus morii	179
420.	Acritopappus morii	179
421.	Stevia myriadenia	191
422.	Ichthyothere connata Viguiera bishopii	201 92

TABLE 3. (*contd.*)

Compound: No.	Species	Reference
423.	Helianthus decapetalus	98
424.	Helianthus angustifolius	98,409
	Helianthus decapetalus	98
	Helianthus maximiliani	280
	Helianthus occidentalis	
	Helianthus tuberosus	98
	Ichthyothere connata	201
	Ichthyothere latifolia	201
	Ichthyothere terminalis	107
	Mikania banisteriae	228
	Viguiera stenoloba	
425.	Helianthus angustifolius	98
	Helianthus maximiliani	*
	Helianthus occidentalis	
	Helianthus tuberosus	98
	Mikania alvimii	47
	Mikania pyramidata	131
426.	Ichthyothere terminalis	107
427.	Ichthyothere terminalis	107
428.	Austroeupatorium inulaefolium	177
	Stevia andina	350
	Stevia aristata	*
	Stevia origanoides	221
	Stevia rebaudiana	461
429.	Stevia berlandiera	170
	Stevia origanoides	221
	Stevia rebaudiana	461
	Stevia salicifolia	221
430.	Helichrysum confertum	180
	Helichrysum suterlandii	171
	Ichthyothere connata	201
431.	Mikania alvimii	47
432.	Mikania alvimii	47
433.	Mikania alvimii	47
434.	Mikania alvimii	47
435.	Mikania pyramidata	131
436.	Grazielia dimorpholepsis	189
437.	Mikania alvimii	47
438.	Mikania alvimii	47
439.	Mikania alvimii	47
	Mikania pyramidata	131
440.	Mikania alvimii	47

Table 3. Diterpene Compound Numbers   171

TABLE 3. (*contd.*)

Compound: No.	Species	Reference
	Mikania pyramidata	131
441.	Helichrysum sp.	171
442.	Aspilia parvifolia	213
	Baccharis oxydonta	116
	Denekia capensis	180
	Viguiera bishopii	92
443.	Baccharis oxydonta	116
444.	Baccharis oxydonta	116
445.	Baccharis oxydonta	116
446.	Baccharis oxydonta	116
447.	Helichrysum albirosulatum	171
	Helichrysum pinifolium	171
	Helichrysum vernum	171
448.	Helichrysum albirosulatum	171
449.	Helichrysum albirosulatum	171
450.	Gnaphalium gaudichaudianum	
	Gnaphalium undulatum	210
451.	Coespeletia moritziana	145
	Espeletia hartwegiana	*
	Espeletiopsis guacharaca	145
	Olearia paniculata	426
	Othonna sedifolia	*
	Solidago missouriensis	10,12
452.	Coespeletia moritziana	145
453.	Solidago missouriensis	10,12
454.	Printzia laxa	165
455.	Baccharis tola	451
456.	Printzia laxa	165
457.	Schkuhria multiflora	101
458.	Libanothamnus spectabilis	174
	Libanothamnus wurdackii	174
459.	Coespeletia lutescens	144
	Coespeletia moritziana	145
	Libanothamnus occultus	174
	Libanothamnus spectabilis	174
460.	Acritopappus confertus	184
461.	Erigeron philadelphicus	487

TABLE 3. (*contd.*)

Compound: No.	Species	Reference
462.	Hartwrightia floridana	95
463.	Schkuhria multiflora	101
464.	Schkuhria multiflora	101
465.	Acritopappus confertus	184
466.	Acritopappus confertus	184
467.	Ayapana amygdalina	113
468.	Ayapana amygdalina	113
469.	Mikania alvimii	47
470.	Morithamnus crassus	102
471.	Acritopappus confertus	184
472.	Acritopappus confertus	184
473.	Athrixia elata	150
475.	Melampodium divaricatum Solidago elongata	343 18
476.	Symphyopappus reticulatus	190
477.	Goyazianthus tetrastichus	60
478.	Goyazianthus tetrastichus	60
479.	Solidago elongata	18
480.	Solidago elongata	18
481.	Symphyopappus reticulatus	190
482.	Symphyopappus reticulatus	190
483.	Symphyopappus reticulatus	190
484.	Baccharis alaternoides	117
485.	Symphyopappus compressus	190
486.	Symphyopappus reticulatus	190
487.	Gochnatia paniculata	51
488.	Gochnatia paniculata	51
489.	Gochnatia paniculata	51
490.	Gochnatia paniculata	51
491.	Gochnatia paniculata	51

Table 3. Diterpene Compound Numbers 173

TABLE 3. (*contd.*)

Compound: No.	Species	Reference
492.	Symphyopappus reticulatus	190
493.	Fleischmannia sinclairii Stevia myriadenia	217 191
494.	Stevia myriadenia	191
495.	Solidago elongata	18
496.	Solidago elongata	18
497.	Solidago elongata	18
498.	Solidago altissima	412
499.	Solidago altissima	412
500.	Solidago altissima	143,364,365,412
501.	Gochnatia paniculata	51
502.	Gochnatia paniculata	51
503.	Fleischmannia sinclairii Hartwrightia floridana Neomiranda angularis	217 95 *
504.	Liatris scariosa	54
505.	Haplopappus ciliatus	35,36
506.	Hartwrightia floridana	95
507.	Baccharis hutchisonii Solidago arguta	57 7,386
508.	Baccharis incanum Baccharis tricuneata var. lineata	284 200
509.	Baccharis tricuneata (L. f.)Pers. var. tricuneata	490
510.	Baccharis articulata Baccharis rhetinoides	283 503
511.	Baccharis macraei Brickellia annulosa Grangea maderaspatana Heteropappus altaicus Koanophyllon adamanticum Printzia laxa Solidago juncea	 158 423 181 42 165 318
512.	Grangea maderaspatana	423
513.	Conyza japonica	423
514.	Hinterhubera imbricata	82

TABLE 3. *(contd.)*

Compound: No.	Species	Reference
515.	Hinterhubera imbricata	82
516.	Hinterhubera imbricata	82
518.	Baccharis calvescens	178
	Baccharis chilco	57
	Baccharis crispa	
	Baccharis macraei	
	Baccharis salicifolia	*
	Baccharis sarothroides	
	Baccharis tricuneata var. lineata	200
	Conyza ivaefolia	76
	Conyza scabrida	83
	Heteropappus althaicus	181
519.	Olearia muelleri	352
520.	Baccharis tricuneata var. lineata	200
	Conyza ivaefolia	76
	Conyza scabrida	83
	Heteropappus altaicus	181
	Printzia laxa	165
521.	Baccharis conferta	295
522.	Gochnatia paniculata	51
523.	Pulicaria gnaphaloides	443
524.	Solidago canadensis	73
	Solidago juncea	318
	Solidago rugosa	73
525.	Conyza podocephala	154
526.	Conyza podocephala	154
527.	Baccharis articulata	283
	Baccharis gilliesii	503
	Baccharis magellanica	503
	Baccharis rhetinoides	503
	Baccharis tricuneata var. tricuneata	490
	Liatris spicata	
528.	Baccharis tricuneata var. tricuneata	283
529.	Baccharis chilco	57
	Baccharis gilliesii	503
	Baccharis subdentata	116
	Baccharis tricuneata var. tricuneata	489,490
530.	Baccharis tricuneata var. tricuneata	489,490
531.	Baccharis cassinaefolia	178

Table 3. Diterpene Compound Numbers   175

TABLE 3. (*contd.*)

Compound: No.	Species	Reference
532.	Baccharis cassinaefolia	178
533.	Baccharis cassinaefolia	178
534.	Symphyopappus reticulatus	190
535.	Symphyopappus reticulatus	190
536.	Symphyopappus reticulatus	190
537.	Synphyopappus reticulatus	190
538.	Acritopappus confertus Acritopappus hagei	179 179
539.	Acritopappus hagei Morithamnus crassus	179 102
540.	Acritopappus hagei	179
541.	Acritopappus hagei	179
542.	Acritopappus teixeirae	179
543.	Acritopappus longifolius Eupatorium turbinatum Solidago altissima	52 350 143,413
544.	Symphyopappus compressus	190
545.	Solidago altissima Solidago elongata Solidago rugosa	401,413 18,402 73
546.	Solidago elongata	18,402
550.	Acritopappus hagei	179
551.	Symphyopappus compressus	190
552.	Solidago elongata Solidago rugosa	402 73
553.	Solidago elongata	18,402
556.	Solidago Shortii	8,11
557.	Solidago Shortii	8,11
558.	Printzia laxa	165
559.	Baccharis trimera	334
560.	Olearia heterocarpa	426
561.	Baccharis genistelloides Baccharis microcephala	111 57
562.	Baccharis trimera	334

TABLE 3. (*contd.*)

Compound: No.	Species	Reference
563.	Baccharis trimera	334
564.	Bahianthus viscidus	84
565.	Acritopappus teixeirae Baccharis truncata Bahianthus viscidus	179 116 84
566.	Solidago rugosa	73
567.	Acritopappus hagei	179
568.	Acritopappus hagei	179
569.	Baccharis alaternoides Baccharis polyphylla	117 117
570.	Stevia polycephala	8
571.	Printzia laxa	165
572.	Printzia laxa	165
573.	Conyza japonica Conyza scabrida Conyza stricta	423 83 470
574.	Acritopappus hagei	179
575.	Acritopappus hagei	179
576.	Haplopappus angustifolius Haplopappus foliosus	339,462 339,462
577.	Haplopappus angustifolius	462
578.	Solidago arguta	7,386
579.	Solidago arguta	7,386
580.	Solidago arguta	7,386
581.	Solidago arguta	7,386
582.	Solidago arguta	7,386
583.	Solidago arguta Solidago itatiayensis	7,263,386 483
584.	Solidago itatiayensis	483
585.	Solidago itatiayensis	483
586.	Gnaphalium undulatum	210
587.	Garuleum sonchifolium	78
588.	Garuleum sonchifolium	78

Table 3. Diterpene Compound Numbers   177

TABLE 3. (*contd.*)

Compound: No.	Species	Reference
589.	Sigesbeckia pubescens	397
590.	Achillea filipendula	25
	Aspilia ovalifolia	*
	Helianthus strumosus	327
	Mikania alvimii	47
	Mikania pyramidata	131
591.	Palafoxia arida	320
	Sigesbeckia orientalis	28
592.	Sigesbeckia orientalis	358
	Sigesbeckia pubescens	358
593.	Sigesbeckia pubescens	358
594.	Sigesbeckia pubescens	358
595.	Sigesbeckia pubescens	358
596.	Liatris laevigata	325
597.	Liatris laevigata	325
598.	Liatris laevigata	325
599.	Othonna floribunda	112
600.	Othonna cylindrica	112
	Othonna floribunda	112
601.	Brickellia eupatoriedes	59
602.	Brickellia eupatoriedes	59
603.	Brickellia eupatoriedes	59
604.	Brickellia eupatoriedes	59
605.	Brickellia eupatoriedes	59
606.	Brickellia eupatoriedes	59
607.	Brickellia eupatoriedes	59
608.	Brickellia eupatoriedes	59
609.	Brickellia eupatoriedes	59
610.	Brickellia eupatoriedes	59
611.	Solidago missouriensis	12
612.	Solidago missouriensis	12
613.	Solidago missouriensis	12
614.	Helichrysum chionosphaerum	45

TABLE 3. (*contd.*)

Compound:		
No.	Species	Reference
615.	Solidago nemoralis	73
616.	Mikania banisteriae	228
	Solidago rugosa	73
617.	Acritopappus confertus	184
618.	Arctotis revoluta	119
	Aspilia foliacea	*
	Athrixia pinifolia	164
	Baccharis minutiflora	117
	Balsamorhiza sagittata	130
	Caramboa pithieri	*
	Coespeletia marcana	145
	Espeletia garciae	
	Espeletia uribei	144
	Espeletiopsis guacharaca	145
	Espeletiopsis purpurascens	145
	Espeletiopsis tachirensis	145
	Gnaphalium undulatum	210
	Helichrysum davyi	*
	Libanothamnus granatesianus	174
	Libanothamnus tamanus	174
	Libanothamnus wurdackii	174
	Lychnophora sellowii	204
	Montanoa pteropoda	123
	Montanoa tomentosa	*
	Ruilopezia jahnii	144
	Ruilopezia lindenii	144
	Ruilopezia marcescens	*
	Ruilopezia palanoides	*
	Ruilopezia ruizii	*
	Tamania chardonii	*
619.	Abrotanella nivigena	13
	Arctotis revoluta	119
	Aspilia foliacea	*
	Aspilia jolyana	*
	Aspilia parvifolia	213
	Austroeupatorium inulaefolium	177
	Baccharis minutiflora	117
	Baccharis quitensis	*
	Cacalia bulbifera	257,373
	Campovassouria bupleurifolia	*
	Coespeletia lutescens	144
	Coespeletia marcana	145
	Coespeletia moritziana	145
	Critonia daleoides	217
	Disynaphia multicrenulata	*
	Espeletia garciae	
	Espeletia grandiflora	*
	Espeletia schultzii	132
	Espeletia wedellii	144
	Espeletiopsis garciae	145
	Espeletiopsis glandulosa	145
	Espeletiopsis guacharaca	145
	Espeletiopsis purpurascens	145
	Espeletiopsis tachirensis	145
	Goyazianthus tetrastichus	60

Table 3. Diterpene Compound Numbers   179

TABLE 3. (*contd.*)

Compound:

No.	Species	Reference
	Helianthus giganteus	98
	Helichrysum davyi	*
	Helichrysum dendroideum	373
	Helichrysum heterolasium	39
	Helichrysum miconiifolium	171
	Helichrysum pilosellum	351
	Helichrysum vernum	*
	Ichthyothere rufa	201
	Ichthyothere terminalis	201
	Libanothamnus granatesianus	174
	Libanothamnus neriifolia	144
	Mikania arrojadoi	142
	Mikania belemii	48
	Mikania luetzelburgii	48
	Mikania sessilifolia	48
	Montanoa pteropoda	123
	Oyedaea lanceolata	166
	Oyedaea rusbyi	*
	Oyedaea verbesinoides	*
	Perymenium ecuadoricum	162
	Perymenium featherstoni	*
	Perymenium serratum	*
	Polymnia fruticosum	161
	Polymnia pyramidalis	161
	Ruilopezia bromeloides	*
	Ruilopezia figerasii	*
	Ruilopezia jahnii	144
	Ruilopezia paltonioides	*
	Ruilopezia lindenii	144
	Smallanthus fruticosus	211
	Smallanthus uvedalia	211
	Steiractinia mollis	115
	Steiractinia sodiroi	124
	Tamania chardonii	*
	Viguiera grammatoglossa	*
	Wedelia grandiflora	120
	Wedelia hookeriana	193
620.	Campovassouria bupleurifolia	146
	Espeletiopsis guacharaca	145
	Helichrysum miconiifolium	171
621.	Baccharis minutiflora	117
622.	Baccharis quitensis	117
	Polymnia fruticosa	116
623.	Helichrysum dendroideum	374
624.	Arctotis revoluta	119
	Aspilia jolyana	*
	Aspilia ovalifolia	*
	Aspilia parvifolia	213
	Aspilia pluriseta subsp. pluriseta	*
	Aster bakeranus	478
	Austroeupatorium inulaefolium	139
	Baccharis minutiflora	117
	Baccharis ramosissima	116
	Baccharis truncata	116

TABLE 3. (*contd.*)

Compound: No.	Species	Reference
	Cacalia bulbifera	257
	Campovassouria bupleurifolia	146
	Coespeletia lutescens	144
	Coespeletia marcana	145
	Coespeletia moritziana	145
	Disynaphia multicrenulata	*
	Espeletia garciae	
	Espeletia grandiflora	
	Espeletia hartwegiana	*
	Espeletia neriifolia	
	Espeletia schultzii	132
	Espeletia uribei	144
	Espeletia wedellii	144
	Espeletiopsis guacharaca	145
	Gnaphalium undulatum	219
	Goyazianthus tetrastichus	60
	Grazielia intermedia	*
	Helianthus annuus	446
	Helichrysum davyi	*
	Helichrysum heterolasium	39
	Helichrysum miconiifolium	171
	Helichrysum pallidum	171
	Helichrysum pilosellum	351
	Ichthyothere terminalis	201
	Ichthyothere ulei	201
	Libanothamnus neriifolia	144
	Mikania arrojadoi	142
	Mikania belemii	48
	Mikania luetzelburgii	48
	Mikania micrantha	*
	Mikania sessilifolia	48
	Montanoa tomentosa subsp. tomentosa	227
	Montanoa pteropoda	123
	Oyedaea buphthalmoides	
	Oyedaea lanceolata	166
	Oyedaea rusbyi	*
	Perymenium ecuadoricum	162
	Perymenium klattianum	*
	Ruilopezia figueirasii	*
	Ruilopezia lindenii	144
	Ruilopezia marcescens	*
	Ruilopezia paltonioides	*
	Smallanthus fruticosus	211
	Smallanthus glabratus	108
	Smallanthus uvedalia	211
	Solidago multiradiata var. multiradiata	*
	Solidago rugosa	73
	Steiractinia mollis	115
	Viguiera grammatoglossa	*
	Wedelia grandiflora	120
	Zaluzania angusta	*
	Zaluzania subcordata	*
625.	Abrotanella nivigena	13
	Arctotheca prostata	*
	Arctotis revoluta	119

Table 3. Diterpene Compound Numbers   181

TABLE 3. (*contd.*)

Compound:		
No.	Species	Reference
	Aspilia floribunda	
	Aspilia foliacea	*
	Aspilia jolyana	*
	Aspilia mossambicensis	*
	Aspilia ovalifolia	*
	Aspilia parvifolia	213
	Aspilia pluriseta	*
	Aspilia pluriseta subsp. pluriseta	94
	Aspilia riddellii	*
	Aster bakeranus	478
	Aster pleiocephalus	*
	Athrixia elata	150
	Austroeupatorium inulaefolium	177
	Baccharis concinnia	116
	Baccharis elliptica	*
	Baccharis intermixta	178
	Baccharis minutiflora	117
	Baccharis polyphylla	178
	Baccharis pylicoides	57
	Baccharis quintensis	116
	Baccharis ramosissima	116
	Baccharis salzmannii	178
	Baccharis truncata	116
	Cacalia bulbifera	257
	Campovassouria bupleurifolia	146
	Caramboa pithieri	*
	Caramboa trugillensis	*
	Chromalaena cryptantha	*
	Coespeletia lutescens	144
	Coespeletia marcana	145
	Coespeletia moritziana	145
	Critonia daleoides	217
	Critonia hebebotrya	*
	Disynaphia multicrenulata	137
	Enhydra fluctuans	422
	Espeletia figueirasii	480
	Espeletia floccosa	480
	Espeletia garciae	
	Espeletia grandiflora	428
	Espeletia moritziana	480
	Espeletia schultzii	132
	Espeletia uribei	144
	Espeletiopsis garciae	145
	Espeletiopsis glandulosa	145
	Espeletiopsis guacharaca	145
	Espeletiopsis purpurascens	145
	Espeletiopsis tachirensis	145
	Eupatorium petiolare	223
	Gnaphalium oligandrum	210
	Gnaphalium undulatum	210
	Goyazianthus tetrastichus	60
	Grazielia dimorpholepsis	189
	Grazielia intermedia	*
	Grazielia serrata	189
	Helianthopsis bishopii	198
	Helianthopsis microphylla	198
	Helianthopsis utcubambensis	198

TABLE 3. (*contd.*)

Compound:

No.	Species	Reference
	Helianthus angustifolius	98
	Helianthus annuus	264,358,491
	Helianthus debilis subsp. cucumerifolius	332
	Helianthus debilis subsp. debilis	411
	Helianthus decapetalus	31,98
	Helianthus occidentalis var. dowellianus	105
	Helianthus giganteus	31,98
	Helianthus heterophyllus	321
	Helianthus hirsutus	31
	Helianthus niveus subsp. canescens	410
	Helianthus nuttallii	31
	Helianthus occidentalis	332
	Helianthus petiolaris	327
	Helianthus radula	326
	Helianthus rigidus	
	Helianthus simulans	332
	Helianthus tuberosus	98
	Helichrysum argentissimum	171
	Helichrysum argyrolepis	*
	Helichrysum aureum var. monocephalum	
	Helichrysum bellum	167
	Helichrysum chionosphaerum	45
	Helichrysum confertum	
	Helichrysum cooperi	
	Helichrysum davyi	*
	Helichrysum fulvum	208
	Helichrysum heterolasium	39
	Helichrysum miconiifolium	171
	Helichrysum pallidum	171
	Helichrysum pilosellum	351
	Helichrysum platypterum	167
	Helichrysum refluxum	86
	Helichrysum subulifolium	351
	Helichrysum vernum	171
	Heterocondylus grandis	*
	Ichthyothere connata	201
	Ichthyothere cunabi	
	Ichthyothere latifolia	201
	Ichthyothere rufa	201
	Ichthyothere terminalis	201
	Ichthyothere ulei	201
	Iostephane heterophylla	*
	Lagascea rigida	89
	Lasianthaea podocephala	124
	Libanothamnus granatesianus	174
	Libanothamnus neriifolia	144
	Libanothamnus occultus	174
	Libanothamnus spectabilis	174
	Libanothamnus tamanus	174
	Libanothamnus wurdackii	174
	Loxothysanus sinuatus	129
	Melampodium paniculatum	*
	Melampodium perfoliatum	*
	Mikania arrojadoi	142
	Mikania belemii	48

Table 3. Diterpene Compound Numbers 183

TABLE 3. (*contd.*)

Compound: No.	Species	Reference
	Mikania cordata	131
	Mikania hirsutissima var. hirsutissima	396
	Mikania luetzelburgii	48
	Mikania micrantha	*
	Mikania monagasensis	383,384
	Mikania schenkii	*
	Mikania sessilifolia	48
	Montanoa atriplicifolia	64
	Montanoa frutescens	*
	Montanoa tomentosa subsp. xanthiifolia	227
	Montanoa pteropoda	123
	Othonna cylindrica	112
	Othonna floribunda	112
	Othonna sedifolia	*
	Oyedaea boliviana	166
	Oyedaea buphthalmoides	*
	Oyedaea lanceolata	166
	Oyedaea rusbyi	*
	Oyedaea verbesinoides	*
	Perymeniopsis ovalifolia	103
	Perymenium ecuadoricum	162
	Perymenium discolor	
	Perymenium featherstoni	*
	Perymenium klattianum	103
	Perymenium serratum	*
	Polymnia fruticosum	161
	Polymnia maculata var. maculata	371
	Polymnia pyramidalis	161
	Robinsonia thurifera	339
	Ruilopezia bromeloides	*
	Ruilopezia jahnii	144
	Ruilopezia lindenii	144
	Ruilopezia paltonioides	*
	Ruilopezia ruizii	*
	Simsia dombeyana	*
	Simsia holwayi	*
	Smallanthus fruticosus	211
	Smallanthus glabratus	108
	Smallanthus reparius	109
	Smallanthus siegesbeckia	*
	Smallanthus uvedalia	114,211
	Solidago juncea	
	Solidago missouriensis	369
	Solidago multiradiata	*
	Solidago rigida	11,369
	Solidago rugosa	73
	Steiractinia mollis	115
	Steiractinia sodiroi	124
	Stevia andina	*
	Stevia monardaefolia	433
	Stevia myriadenia	191
	Stevia setifera	69
	Tetrachyron manicata	*
	Tetrachyron orizabaensis var. websteri	411
	Tithonia longiradiata	*
	Verbesina angustifolia	160

TABLE 3. (*contd.*)

Compound: No.	Species	Reference
	Verbesina oncophora	160
	Vernonia divaricata	*
	Vernonia venosissima	*
	Viguiera bishopii	92
	Viguiera cordata	*
	Viguiera cordifolia	199
	Viguiera dentata	275
	Viguiera excelsa	242
	Viguiera grammatoglossa	199
	Viguiera hypargyrea	5
	Viguiera incana	*
	Viguiera insignis	245
	Viguiera linearifolia	*
	Viguiera linearis	454
	Viguiera maculata	241
	Viguiera oaxacana	242
	Viguiera pazensis	*
	Viguiera porteri	331
	Viguiera potosina	276
	Viguiera procumbens	*
	Viguiera trichophylla	*
	Wedelia buphthalmiflora	455
	Wedelia calycina	328
	Wedelia glauca	405
	Wedelia grandiflora	120
	Wedelia helianthoides	120
	Wedelia hispida	328
	Wedelia hookeriana	193
	Wedelia scaberrima	476
	Wedelia speciosa	182
	Wedelia trilobata	120,213
	Wedelia villosa	*
	Zaluzania angusta	*
	Zaluzania cinarescens	*
	Zaluzania discoidea	*
	Zaluzania subcordata	*
	Zaluzania triloba	*
	Zexmenia hispida	124
626.	Mikania banisteriae	228
627.	Mikania sessilifolia	48
	Ruilopezia jahnii	144
628.	Bedfordia salicinia	122
	Helichrysum dendroideum	374
629.	Helichrysum aureum var. monocephalum	180
	Helichrysum cooperi	180
	Helichrysum heterolasium	39
	Mikania arrojadoi	142
630.	Smallanthus uvedalia	114
	Wedelia trilobata	120,213
631.	Wedelia trilobata	114,120
	Wulffia maculata	*

Table 3. Diterpene Compound Numbers   185

TABLE 3. (*contd.*)

Compound: No.	Species	Reference
632.	Wedelia trilobata	114,120
	Wulffia maculata	*
633.	Smallanthus uvedalia	114
	Wedelia grandiflora	120
	Wedelia trilobata	120
634.	Smallanthus uvedalia	114
635.	Mikania banisteriae	228
636.	Aspilia pluriseta subsp. pluriseta	94
637.	Sciadocephala schultze-rhonhofiae	217
638.	Mikania banisteriae	228
639.	Viguiera stenoloba	237,243
640.	Helichrysum aureum var. monocephalum	180
	Helichrysum cooperi	180
	Helichrysum davyi	*
	Helichrysum heterolasium	39
	Helichrysum pallidum	171
	Helichrysum pilosellum	351
	Helichrysum ruderale	171
641.	Aster tanacetifolius	*
	Ruilopezia figueirasii	*
	Smallanthus uvedalia	114
	Stevia eupatoria	416
642.	Helianthus decapetalus	98
	Helianthus radula	326
	Helianthus simulans	332
	Helianthus tuberosus	98
	Mikania arrojadoi	142
643.	Smallanthus fruticosus	211
	Smallanthus reparius	109
	Smallanthus siegesbeckia	*
	Smallanthus uvedalia	114
644.	Smallanthus fruticosus	211
	Smallanthus glabratus	108
	Smallanthus uvedalia	114
645.	Smallanthus uvedalia	114
646.	Austroeupatorium inulaefolium	406
	Baccharis concinna	116
	Baccharis intermixta	178
	Baccharis polyphylla	178
	Coespeletia lutescens	144
	Coespeletia marcana	145
	Disynaphia multicrenulata	137
	Enhydra fluctuans	421
	Espeletia grandiflora	*

TABLE 3. *(contd.)*

Compound: No.	Species	Reference
	Espeletia humbertii	479
	Espeletia littlei	479
	Espeletia schultzii	219
	Espeletia tenor	394
	Espeletia uribei	144
	Espeletia weddellii	394
	Espeletiopsis glandulosa	145
	Espeletiopsis guacharaca	145
	Espeletiopsis tachirensis	145
	Gnaphalium undulatum	210
	Helianthus annuus	387
	Helianthus debilis subsp. cucumerifolius	332
	Helianthus grosseserratus	332
	Helianthus niveus subsp. canescens	410
	Helianthus nuttallii subsp. nuttallii	367
	Helianthus occidentalis	332
	Helianthus petiolaris	327
	Helianthus simulans	332
	Ichthyothere terminalis	*
	Mikania arrojadoi	142
	Mikania luetzelburgii	48
	Oyedaea verbesinoides	*
	Ruilopezia jahnii	144
	Ruilopezia lindenii	144
	Ruilopezia ruizii	*
	Viguiera dentata	275
	Viguiera maculata	241
	Viguiera porteri	331
	Viguiera potosina	276
	Wedelia calycina	328
	Wedelia scaberrima	476
647.	Mikania arrojadoi	142
648.	Arctotis revoluta	119
	Baccharis truncata	116
	Coespeletia lutescens	144
	Coespeletia marcana	145
	Coespeletia moritziana	145
	Espeletia curialensis	*
	Espeletia hartwegiana	*
	Espeletia humbertii	479
	Espeletia schultzii	132
	Espeletia uribei	144
	Espeletiopsis garciae	145
	Espeletiopsis glandulosa	145
	Espeletiopsis guacharaca	145
	Espeletiopsis purpurascens	145
	Espeletiopsis tachirensis	145
	Helianthus annuus	264
	Helichrysum aureum var. monocephalum	180
	Helichrysum heterolasium	39
	Helichrysum miconiifolium	171
	Helichrysum pallidum	171
	Libanothamnus granatesianus	174

Table 3. Diterpene Compound Numbers 187

TABLE 3. (*contd.*)

Compound: No.	Species	Reference
	Libanothamnus neriifolia	144
	Libanothamnus occultus	174
	Libanothamnus spectabilis	174
	Libanothamnus tamanus	174
	Libanothamnus wurdackii	174
	Mikania arrojadoi	142
	Ruilopezia figuerasii	*
	Ruilopezia jahnii	144
	Ruilopezia lindenii	144
	Ruilopezia marcescens	*
	Tamania chardonii	*
	Tetrachyron manicata	*
	Wedelia hookeriana	193
649.	Aster tanacetifolius	*
	Austroeupatorium inulaefolium	139
	Baccharis concinna	116
	Baccharis polyphylla	178
	Caramboa pithieri	*
	Disynaphia multicrenulata	*
	Enhydra fluctuans	285
	Goyazianthus tetrastichus	60
	Helianthus angustifolius	98
	Helianthus annuus	264
	Helianthus debilis subsp. cucumerifolius	332
	Helianthus debilis subsp. debilis	411
	Helianthus giganteus	98
	Helianthus occidentalis	332
	Helianthus radula	326
	Helianthus simulans	332
	Ichthyothere terminalis	201
	Ichthyothere ulei	107
	Montanoa pteropoda	123
	Montanoa tomentosa subsp. xanthiifolia	351
	Oyedaea buphthalmoides	
	Oyedaea rusbyi	*
	Perymenium ecuadoricum	162
	Perymenium featherstoni	*
	Perymenium ovalifolia	*
	Polymnia fruticosum	161
	Polymnia pyramidalis	161
	Smallanthus fruticosus	211
	Smallanthus reparius	109
	Smallanthus siegesbeckia	*
	Smallanthus uvedalia	211
	Steiractinia mollis	115
	Stevia monardaefolia	433
	Tamania chardonii	*
	Tetrachyron orizabaensis var. websteri	411
	Viguiera dentata	275
	Viguiera pazensis	205
	Viguiera quinqueradiata	240
	Wedelia buphthalmiflora	455
	Wedelia calycina	328
	Wedelia grandiflora	120

TABLE 3. (*contd.*)

Compound: No.	Species	Reference
	Wedelia helianthoides	120
	Wedelia hookeriana	193
	Wedelia pinetorum	
	Zexmenia pinetorum	
650.	Aspilia jolyana	*
	Aspilia ovalifolia	*
	Aspilia parvifolia	213
	Aspilia pluriseta subsp. pluriseta	94
	Aster pleiocephalus	*
	Baccharis salzmannii	178
	Baccharis truncata	116
	Grazielia dimorpholepsis	189
	Grazielia intermedia	189
	Grazielia serrata	189
	Helianthus occidentalis subsp. dowellianus	105
	Helianthus tuberosus	98
	Ichthyothere connata	201
	Ichthyothere latifolia	201
	Ichthyothere rufa	201
	Ichthyothere terminalis	107,201
	Ichthyothere ulei	107
	Lagascea rigida	89
	Oyedaea verbesinoides	*
	Stevia bupthalmiflora	
	Stevia myriadenia	191
	Viguiera pazensis	205
	Viguiera quinqueradiata	240
	Wedelia buphthalmiflora	455
	Wedelia helianthoides	120
	Wedelia hookeriana	193
	Wedelia scaberrima	476
	Wedelia villosa	*
651.	Aspilia jolyana	*
	Baccharis elliptica	*
	Caramboa pithieri	*
	Coespeletia moritziana	145
	Espeletiopsis garciae	145
	Espeletiopsis glandulosa	145
	Espeletiopsis purpurascens	145
	Espeletiopsis tachirensis	145
	Ichthyothere connata	201
	Ichthyothere latifolia	201
	Ichthyothere rufa	201
	Ichthyothere terminalis	107,201
	Ichthyothere ulei	107
	Ruilopezia bromeloides	*
	Ruilopezia lindenii	144
	Ruilopezia marcescens	*
	Ruilopezia paltonioides	*
	Ruilopezia ruizii	*
	Stevia bupthalmiflora	
	Stevia myriadenia	191
	Tamania chardonii	*
	Wedelia grandiflora	120
	Wedelia helianthoides	*

Table 3. Diterpene Compound Numbers 189

TABLE 3. (*contd.*)

Compound: No.	Species	Reference
	Wedelia hookeriana	193
652.	Aspilia pluriseta subsp. pluriseta	94
	Austroeupatorium inulaefolium	139
	Campovassouria bupleurifolia	146
	Coespeletia moritziana	145
	Disynaphia multicrenulata	*
	Enhydra fluctuans	421
	Espeletiopsis garciae	145
	Espeletiopsis glandulosa	145
	Espeletiopsis guacharaca	145
	Grazielia intermedia	*
	Helianthus annuus	264
	Ichthyothere terminalis	107
	Libanothamnus granatesianus	174
	Mikania sessilifolia	48
	Montanoa frutescens	
	Montanoa pteropoda	123
	Ruilopezia jahnii	144
	Stevia bupthalmiflora	
	Tamania chardonii	*
	Wedelia buphthalmiflora	455
653.	Ichthyothere connata	201
	Ichthyothere latifolia	201
	Ichthyothere rufa	201
	Ichthyothere terminalis	201
	Mikania arrojadoi	142
	Mikania luetzelburgii	48
	Viguiera pazensis	205
654.	Aspilia pluriseta subsp. pluriseta	94
	Espeletiopsis purpurascens	145
	Loxothysanus sinuatus	129
	Mikania cordata	131
	Wedelia buphthalmiflora	455
655.	Helianthus debilis subsp. curcumerifolius	
	Mikania arrojadoi	142
	Mikania belemii	48
	Mikania luetzelburgii	48
	Mikania oblongifolia	482
	Mikania sessilifolia	48
	Montanoa pteropoda	123
	Montanoa tomentosa subsp. xanthiifolia	227
	Perymeniopsis ovalifolia	103
	Wedelia glauca	405
	Wedelia trilobata	120
656.	Viguiera porteri	331
	Wedelia calycina	328
	Wedelia grandiflora	120
657.	Mikania cordata	131

TABLE 3. (*contd.*)

Compound: No.	Species	Reference
658.	Aspilia foliacea	*
	Aspilia parvifolia	213
	Solidago rugosa	73
659.	Aspilia parvifolia	213
	Aspilia pluriseta subsp. pluriseta	*
	Baccharis ramoisissima	*
	Solidago rugosa	73
660.	Aspilia parvifolia	*
	Baccharis ramoisissima	*
	Solidago rugosa	73
661.	Baccharis concinna	117
	Baccharis polyphylla	178
662.	Wedelia trilobata	213
663.	Wedelia trilobata	213
664.	Grazielia intermedia	*
	Ichthyothere terminalis	107
	Ichthyothere ulei	107
	Wedelia calycina	328
665.	Aspilia floribunda	
	Aspilia foliacea	*
	Campovassouria bupleurifolia	146
	Goyazianthus tetrastichus	60
	Helianthus occidentalis	332
	Oyedaea boliviana	166
	Oyedaea lanceolata	166
	Oyedaea rusbyi	*
	Steiractinia mollis	115
	Viguiera porteri	331
	Viguiera trichophylla	*
	Wedelia grandiflora	*
	Wedelia helianthoides	120
	Wedelia hookeriana	193
	Wedelia pinetorum	*
	Wedelia trilobata	120
666.	Ichthyothere terminalis	*
	Ichthyothere ulei	107
	Oyedaea boliviana	166
	Viguiera hypargyrea	5
667.	Wedelia hookeriana	193
668.	Espeletiopsis tachirensis	145
	Ichthyothere terminalis	107
	Ichthyothere ulei	107
	Ruilopezia lindenii	144
	Ruilopezia marcescens	*
	Ruilopezia ruizii	*
	Wedelia helianthoides	
	Wedelia hookeriana	193

Table 3. Diterpene Compound Numbers 191

TABLE 3. (*contd.*)

Compound: No.	Species	Reference
669.	Campovassouria bupleurifolia	146
	Espeletiopsis guacharaca	145
	Oyedaea rusbyii	*
670.	Mikania sessilifolia	48
671.	Adenostemma caffrum	127
	Adenostemma lavenia	230
	Amphiachyris dracunculoides	*
	Eupatorium album	127
	Grazielia serrata	189
672.	Adenostemma lavenia	230
673.	Adenostemma caffrum	127
	Eupatorium album	127
674.	Adenostemma caffrum	127
	Adenostemma lavenia	230
	Grazielia serrata	189
675.	Ichthyothere terminalis	107
676.	Ichthyothere terminalis	107
677.	Ichthyothere terminalis	107
678.	Atractylis gummifera	239
	Xanthium strumarium	232
	Xanthium strumarium var. strumarium	238
679.	Wedelia asperrima	408
680.	Wedelia asperrima	408
681.	Atractylis gummifera	239
	Wedelia glauca	456
682.	Stevia rebaudiana	362
683.	Stevia rebaudiana	361,362
684.	Stevia rebaudiana	362
685.	Stevia rebaudiana	361,362
686.	Stevia rebaudiana	361
687.	Stevia rebaudiana	361
688.	Stevia rebaudiana	447
689.	Stevia rebaudiana	447
690.	Athrixia elata	150
	Coespeletia lutescens	144
	Coespeletia marcana	145

TABLE 3. (*contd.*)

Compound: No.	Species	Reference
	Coespeletia moritziana	145
	Doronicum hungaricum	*
	Espeletia hartwegiana	*
	Espeletia littlei	
	Espeletia neriifolia	
	Espeletia timotensis	
	Espeletia schultzii	132,218
	Espeletia uribei	144
	Espeletia wedellii	144
	Espeletiopsis garciae	145
	Espeletiopsis glandulosa	145
	Espeletiopsis guacharaca	145
	Espeletiopsis purpurascens	145
	Espeletiopsis tachirensis	145
	Eupatorium deltoideum	
	Gnaphalium oligandrum	210
	Goyazianthus tetrastichus	60
	Helianthus giganteus	98
	Helianthus grosseserratus	31
	Helianthus heterophyllus	321
	Helianthus maximiliani	
	Helichrysum aureum var. monocephalum	
	Helichrysum pilosellum	351
	Helichrysum platypterum	171
	Helichrysum refluxum	*
	Helichrysum ruderale	171
	Ichthyothere terminalis	107
	Lagascea rigida	89
	Lasianthaea podocephala	124
	Libanothamnus neriifolia	144
	Libanothamnus occultus	174
	Libanothamnus spectabilis	174
	Libanothamnus tamanus	174
	Libanothamnus wurdackii	174
	Montanoa pteropoda	123
	Montanoa tomentosa subsp. xanthiifolia	227
	Othonna cylindrica	*
	Oyedaea boliviana	166
	Oyedaea buphthalmoides	
	Oyedaea lanceolata	166
	Oyedaea rusbyi	*
	Perymenium klattianum	103
	Polymnia maculata var. maculata	371
	Riencourtia oblongifolia	*
	Ruilopezia bromeloides	*
	Ruilopezia figueirasii	*
	Ruilopezia jahnii	*
	Ruilopezia lindenii	*
	Ruilopezia marcescens	*
	Ruilopezia margarita	481
	Ruilopezia paltonioides	*
	Ruilopezia ruizii	*
	Smallanthus fruticosus	211
	Steiractinia mollis	115
	Steiractinia sodiroi	124
	Tetrachyron manicata	*

Table 3. Diterpene Compound Numbers   193

TABLE 3. *(contd.)*

Compound: No.	Species	Reference
	Tithonia longiradiata	*
	Verbesina angustifolia	160
	Verbesina oncophora	160
	Viguiera bishopii	92
	Viguiera dentata	92
	Viguiera excelsa	242
	Viguiera grammatoglossa	
	Viguiera linearis	454
	Viguiera insignis	245
	Viguiera oaxacana	242
	Viguiera procumbens	92
	Viguiera stenoloba var. chihuahuensis	
	Wedelia buphthalmiflora	455
	Wedelia calycina	328
	Wedelia grandiflora	120
	Wedelia helianthoides	120
	Wedelia hispida	328
	Wedelia hookeriana	193
	Wedelia trilobata	120
	Zaluzania discoidea	*
	Zaluzania subcordata	*
	Zaluzania triloba	*
	Zexmenia hispida	124
691.	Helianthus maximiliani	
	Ichthyothere terminalis	107
692.	Ichthyothere terminalis	107
693.	Ichthyothere terminalis	107
694.	Coespeletia lutescens	144
	Coespeletia marcana	145
	Espeletia uribei	144
	Espeletiopsis guacharaca	145
	Espeletiopsis purpurascens	145
	Espeletiopsis tachirensis	145
	Helianthus heterophyllus	321
	Ichthyothere terminalis	*
	Montanoa pteropoda	123
	Perymenium klattianum	103
	Ruilopezia lindenii	*
	Ruilopezia marcescens	*
	Ruilopezia ruizii	*
	Tetrachyron manicata	*
	Viguiera excelsea	242
	Viguiera latibracteata	
	Viguiera linearis	454
	Wedelia calycina	328
	Wedelia hookeriana	193
695.	Ambrosia hispida	322
	Coespeletia lutescens	144
	Coespeletia marcana	145
	Espeletiopsis glandulosa	145
	Espeletiopsis guacharaca	145
	Espeletiopsis purpurascens	145
	Espeletiopsis tachirensis	145

TABLE 3. (*contd.*)

Compound:		
No.	Species	Reference

	Helianthus heterophyllus	321
	Ichthyothere terminalis	*
	Ruilopezia jahnii	*
	Ruilopezia lindenii	*
	Ruilopezia marcescens	*
	Ruilopezia ruizii	*
	Viguiera excelsa	242
	Viguiera hypargyrea	5
696.	Coespeletia lutescens	144
	Espeletiopsis tachirensis	145
	Ichthyothere terminalis	*
	Mikania arrojadoi	*
	Smallanthus fruticosus	211
697.	Baccharis minutiflora	117
698.	Baccharis minutiflora	117
	Helichrysum dendroideum	374
699.	Baccharis quitensis	116
700.	Espeletia figueirasii	480
	Espeletia floccosa	480
	Espeletia moritziana	480
	Helianthus occidentalis var. dowellianus	105
701.	Austroeupatorium inulaefolium	406
	Coespeletia lutescens	144
	Coespeletia marcana	145
	Coespeletia moritziana	145
	Espeletia uribei	144
	Espeletiopsis guacharaca	145
	Espeletiopsis purpurascens	145
	Espeletiopsis tachirensis	145
	Helianthus annuus	387
	Helianthus debilis Nutt. subsp. cucumerifolius	332
	Helianthus grosseserratus	332
	Helianthus nuttallii subsp. nuttallii	367
	Helianthus occidentalis	332
	Helianthus petiolaris	327
	Helianthus radula	326
	Mikania luetzelburgii	48
	Othonna sedifolia	*
	Ruilopezia jahnii	144
	Ruilopezia lindenii	144
	Ruilopezia marcescens	*
	Ruilopezia paltonioides	*
	Ruilopezia ruizii	*
	Viguiera latibracteata	
	Viguiera porteri	331
	Wedelia helianthoides	*
702.	Coespeletia moritziana	145
	Espeletiopsis guacharaca	145
	Helianthus debilis subsp.	332

Table 3. Diterpene Compound Numbers   195

Table 3. (*contd.*)

Compound: No.	Species	Reference
	cucumerifolius	
	Helianthus occidentalis	332
	Helianthus radula	326
	Helianthus simulans	332
	Mikania arrojadoi	142
	Ruilopezia paltonioides	*
	Ruilopezia ruizii	*
703.	Helianthus debilis subsp.	332
	cucumerifolius	
	Helianthus occidentalis	332
	Helianthus occidentalis var.	105
	dowellianus	
	Viguiera latibracteata	
704.	Polymnia canadensis	187
705.	Polymnia canadensis	187
706.	Polymnia canadensis	187
707.	Polymnia canadensis	187
708.	Polymnia canadensis	187
709.	Smallanthus uvedalia	114
710.	Smallanthus uvedalia	117
711.	Smallanthus uvedalia	117
712.	Mikania sessilifolia	48
713.	Baccharis elliptica	*
	Baccharis minutiflora	117
	Disynaphia multicrenulata	137
	Goyazianthus tetrastichus	60
	Mikania arrojadoi	142
	Mikania sessilifolia	48
	Othonna sedifolia	*
	Wedelia helianthoides	*
714.	Baccharis elliptica	*
	Helianthus occidentalis var.	105
	dowellianus	
	Mikania arrojadoi	142
	Mikania luetzelburgii	48
	Mikania sessilifolia	48
715.	Perymenium ecuadoricum	163
	Smallanthus fruticosus	211
716.	Espeletia uribei	144
717.	Aspilia parvifolia	213
	Aspilia pluriseta	
718.	Aspilia parvifolia	213

TABLE 3. (*contd.*)

Compound: No.	Species	Reference
719.	Ichthyothere terminalis	107
720.	Arctotis revoluta	119
	Aspilia parvifolia	213
	Austroeupatorium inulaefolium	177
	Baccharis minutiflora	117
	Cacalia bulbifera	257
	Coespeletia marcana	145
	Enhydra fluctuans	422
	Espeletia humbertii	479
	Espeletia littlei	
	Espeletia schultzii	132,219
	Espeletiopsis guacharaca	145
	Espeletiopsis purpurascens	145
	Espeletiopsis tachirensis	145
	Helianthus annuus	446
	Helianthus argophyllus	493
	Helianthus occidentalis	332
	Helianthus salicifolius	324
	Helichrysum pallidum	171
	Helichrysum refluxum	*
	Perymenium ecuadoricum	162
	Perymenium serratum	*
	Oyedaea buphthalmoides	
	Robinsonia gayana	*
	Ruilopezia jahnii	144
	Ruilopezia lindenii	144
	Ruilopezia margarita	481
	Steiractinia mollis	115
	Steiractinia sodiroi	124
	Verbesina oncophora	160
	Viguiera bishopii	92
	Wedelia grandiflora	120
	Zaluzania cinarescens	*
721.	Ichthyothere terminalis	107
722.	Helichrysum davyi	*
	Helichrysum pallidum	171
	Ichthyothere terminalis	107
723.	Coespeletia marcana	145
	Helianthus debilis subsp. cucumerifolius	332
	Perymenium discolor	
	Ruilopezia margarita	481
	Steiractinia mollis	*
	Viguiera bishopii	92
	Viguiera linearis	
	Viguiera stenoloba	
	Zaluzania angustifolia	
724.	Helianthus angustifolius	409
	Helianthus argophyllus	493
	Helianthus radula	326
	Helianthus rigidus	324
	Helianthus salicifolius	324
725.	Morithamnus crassus	102

Table 3. Diterpene Compound Numbers   197

TABLE 3. (*contd.*)

Compound: No.	Species	Reference
	Peteravenia malvaefolia	*
726.	Trichogoniopsis morii	192
727.	Hymenopappus newberryi	207
728.	Helianthus occidentalis	332
	Helianthus petiolaris	327
	Helianthus radula	326
	Helianthus simulans	332
	Helichrysum diosmifolium	366
	Sigesbeckia pubescens	285
	Steiractinia mollis	*
	Viguiera excelsa	242
	Viguiera latibracteata	
	Wedelia calycina	328
729.	Baccharis minutiflora	117
730.	Baccharis minutiflora	117
731.	Baccharis minutiflora	117
732.	Baccharis minutiflora	117
733.	Baccharis minutiflora	117
734.	Baccharis minutiflora	117
735.	Baccharis minutiflora	117
736.	Baccharis minutiflora	117
737.	Baccharis minutiflora	117
738.	Baccharis minutiflora	117
739.	Baccharis minutiflora	117
	Ichthyothere terminalis	*
740.	Baccharis quitensis	116
741.	Sigesbeckia pubescens	397
742.	Sigesbeckia pubescens	397
743.	Sigesbeckia pubescens	397
744.	Sigesbeckia pubescens	397
745.	Sigesbeckia pubescens	397
746.	Helianthus debilis subsp. cucumerifolius	332
	Viguiera latibracteata	
	Wedelia calycina	328
747.	Adenostemma caffrum	127

TABLE 3. (*contd.*)

Compound: No.	Species	Reference
748.	Adenostemma lavenia	230,336
749.	Helianthus radula	326
750.	Coespeletia marcana	145
	Coespeletia moritziana	145
	Espeletia hartwegiana	*
	Espeletiopsis guacharaca	145
	Espeletiopsis purpurascens	145
	Espeletiopsis tachirensis	145
	Libanothamnus granatesianus	174
	Libanothamnus tamanus	174
	Libanothamnus wurdackii	174
	Ruilopezia figueirasii	*
	Ruilopezia marcescens	*
	Ruilopezia paltonioides	*
	Ruilopezia ruizii	*
	Tamania chardonii	*
751.	Coespeletia marcana	145
	Coespeletia moritziana	145
	Espeletiopsis glandulosa	145
	Espeletiopsis guacharaca	145
	Espeletiopsis purpurascens	145
	Espeletiopsis tachirensis	145
	Ruilopezia lindenii	144
	Ruilopezia paltonioides	*
752.	Coespeletia marcana	145
	Coespeletia moritziana	145
	Espeletia hartwegiana	*
	Espeletiopsis guacharaca	145
	Espeletiopsis purpurascens	145
	Libanothamnus granatesianus	174
	Libanothamnus tamanus	174
	Libanothamnus wurdackii	174
	Ruilopezia figueirasii	*
	Ruilopezia marcescens	*
	Ruilopezia paltonioides	*
	Ruilopezia ruizii	*
	Tamania chardonii	*
753.	Coespeletia lutescens	144
	Coespeletia marcana	145
	Coespeletia moritziana	145
	Espeletiopsis purpurascens	145
	Espeletiopsis tachirensis	145
	Ruilopezia jahnii	144
	Ruilopezia lindenii	144
	Tamania chardonii	*
754.	Athrixia phylicoides	164
755.	Athrixia arachnoidea	150
	Athrixia elata	164
	Athrixia phylicoides	164
756.	Athrixia phylicoides	164

Table 3. Diterpene Compound Numbers   199

TABLE 3. (*contd.*)

Compound:
No.	Species	Reference
757.	Athrixia phylicoides	164
758.	Atractylis gummifera	239
759.	Ichthyothere terminalis	107
760.	Ageratum fastigiatum	126
	Ichthyothere terminalis	107
	Wedelia trilobata	213
761.	Ichthyothere terminalis	107
762.	Ichthyothere terminalis	107
763.	Gnaphalium undulatum	213
764.	Helichrysum dendroideum	373
	Viguiera insignis	245
765.	Baccharis tola	450
	Dimorphotheca pseudoaurantiaca	199
	Helichrysum dendroideum	373
	Nidorella anomala	153
	Viguiera grammatoglossa	199
	Wedelia hookeriana	193
	Viguiera pazensis	*
766.	Baccharis tola	450
767.	Helichrysum dendroideum	374
	Nidorella anomala	153
	Viguiera insignis	245
768.	Viguiera insignis	245
769.	Nidorella anomala	153
770.	Helichrysum dendroideum	374
771.	Baccharis tola	451
	Helichrysum dendroideum	374
772.	Baccharis tola	451
773.	Montanoa tomentosa subsp. xanthiifolia	351
	Viguiera grammatoglossa	199,425
774.	Dimorphotheca aurantiaca	199
	Dimorphotheca pseudoaurantiaca	199
	Helianthopsis bishopii	198
	Helianthopsis microphylla	198
	Helianthopsis utcubambensis	198
	Helianthus annuus	264
	Helichrysum fulvum	208
	Helichrysum refluxum	86
	Montanoa tomentosa subsp. xanthiifolia	351

TABLE 3. (*contd.*)

Compound: No.	Species	Reference
	Nidorella anomala	153
	Perymenium klattianum	103
	Senecio asperulus	*
	Stevia aristata	*
	Stevia myriadenia	191
	Viguiera bishopii	92
	Viguiera cordifolia	199
	Viguiera grammatoglossa	199
	Viguiera insignis	245
	Viguiera pazensis	*
	Wedelia hookeriana	193
775.	Nidorella anomala	153
	Peteravenia malvaefolia	*
	Peteravenia schultzii	*
776.	Dimorphotheca aurantiaca	199
	Dimorphotheca pseudoaurantiaca	199
777.	Viguiera bishopii	92
778.	Dimorphotheca aurantiaca	199
779.	Baccharis tola	451
780.	Baccharis tola	450
781.	Nidorella anomala	153
782.	Baccharis tola	451
783.	Helichrysum chionosphaerum	45
784.	Helianthus occidentalis	331
785.	Garuleum sonchifolium	78
	Helianthus decapetalus	31,98
786.	Helianthus decapetalus	98
787.	Helianthus decapetalus	98
788.	Helianthus decapetalus	98
789.	Garuleum sonchifolium	78
790.	Garuleum sonchifolium	78
791.	Helianthopsis bishopii	198
	Helianthopsis microphylla	198
	Helianthopsis utcubambensis	198
	Helianthus annuus	31,392,491
	Helianthus debilis subsp. cucumerifolius	31
	Helianthus giganteus	31
	Helianthus hirsutus	31
	Helianthus occidentalis var. dowellianus	105

Table 3. Diterpene Compound Numbers   201

TABLE 3. (*contd.*)

Compound: No.	Species	Reference
	Helianthus rigidus	31
	Helianthus tomentosus	31
	Helichrysum fulvum	208
	Helichrysum refluxum	86
	Viguiera bishopii	92
	Viguiera dentata	275
	Viguiera lanceolata	92
	Viguiera pazensis	92
	Viguiera procumbens	92
	Viguiera trichopylla	*
792.	Helianthus annuus	387
	Helianthus argophyllus	493
	Helianthus californicus	282
	Helianthus ciliaris	37
	Helianthus grosseserratus	332
	Helianthus niveus subsp. canescens	410
	Helianthus nuttallii subsp. nuttallii	367
	Helianthus occidentalis	331
	Helianthus petiolaris	327
	Helianthus rigidus	332
	Helianthus salicifolius	324
793.	Helianthus radula	326
794.	Viguiera bishopii	92
	Viguiera trichophylla	*
795.	Helianthus radula	326
796	Helichrysum chionosphaerum	45
	Mikania arrojadoi	*
797.	Helichrysum chionosphaerum	45
798.	Helichrysum fulvum	208
799.	Helichrysum fulvum	208
800.	Helianthus debilis subsp. debilis	411
	Montanoa tomentosa subsp. xanthiifolia	
	Tetrachyron orizabaensis var. websteri	411
	Viguiera latibracteata	
801.	Inula royleana	255
802.	Inula royleana	255
803.	Inula royleana	255
804.	Eupatorium inulaefolium	23
805.	Eupatorium album	323
806.	Eupatorium album	323

TABLE 3. (*contd.*)

Compound: No.	Species	Reference
807.	Piqueria trinervia	354
808.	Stevia origanoides	221
809.	Stevia andina	*
	Stevia origanoides	221
810.	Stevia origanoides	221
811.	Stevia origanoides	221
812.	Eupatorium petaloideum	336
813.	Mikania banisteriae	228
814.	Eupatorium petaloideum	336
816.	Viguiera insignis	244
817.	Viguiera insignis	244
818.	Wedelia calycina	328
819.	Wedelia calycina	328
820.	Helianthus strumosus	327
	Mikania triangularis	360
821.	Helianthus strumosus	327
822.	Helianthus petiolaris	327
823.	Helianthus hirsutus	327
824.	Helianthus hirsutus	327
825.	Helianthus annuus	264
826.	Helianthus annuus	264
	Viguiera pazensis	205
827.	Helianthus annuus	264
	Viguiera pazensis	205
828.	Austroeupatorium inulaefolium	406
829.	Austroeupatorium inulaefolium	406
830.	Critonia daleoides	217
	Critonia hebebotrya	*
	Gutierrezia sarothrae	197
831.	Gutierrezia sarothrae	197
832.	Gutierrezia sarothrae	197
833.	Gutierrezia sarothrae	197
834.	Gutierrezia sarothrae	197

Table 3. Diterpene Compound Numbers 203

TABLE 3. (*contd.*)

Compound: No.	Species	Reference
835.	Gutierrezia sarothrae	197
836.	Gutierrezia sarothrae	197
837.	Gutierrezia sarothrae	197
838.	Gutierrezia sarothrae	197
839.	Gutierrezia sarothrae	197
840.	Gutierrezia sarothrae	197
841.	Gutierrezia sarothrae	197
842.	Gutierrezia grandis Gutierrezia sarothrae	271 197
843.	Gutierrezia sarothrae	197
844.	Gutierrezia sarothrae	197
845.	Gutierrezia sarothrae	197
846.	Baccharis incanum Baccharis kingii Conyza scabrida	 196 83
847.	Viguiera maculata	241
848.	Ophryosporus chilca	152
849.	Ophryosporus chilca	152
850.	Helianthus annuus	446
851.	Helianthus annuus	446
852.	Helianthus annuus	446
853.	Helianthus annuus	446
854.	Helianthus annuus	446
855.	Helianthus annuus	446
856.	Helianthus annuus	446
857.	Fleischmannia viscidipes	81
858.	Fleischmannia viscidipes	81
859.	Fleischmannia viscidipes	81
860.	Fleischmannia viscidipes	81
861.	Fleischmannia viscidipes Oxylobus canescens	81 *
862.	Fleischmannia viscidipes	81

TABLE 3. (*contd.*)

Compound: No.	Species	Reference
863.	Fleischmannia viscidipes	81
864.	Acritopappus longifolius	52
865.	Acritopappus longifolius	52
866.	Wedelia regis	75
867.	Wedelia regis	75
868.	Wedelia regis	75
869.	Aspilia floribunda	75
870.	Aristeguieta pseudoarborea	194
871.	Aristeguietia pseudoarborea	194
872.	Aristeguietia pseudoarborea	194
873.	Aristeguietia pseudoarborea	194
874.	Aristeguietia pseudoarborea	194
875.	Villanova titicaensis	195
876.	Villanova titicaensis	195
877.	Villanova titicaensis	195
878.	Villanova titicaensis	195
879.	Villanova titicaensis	195
880.	Villanova titicaensis	195
881.	Trichogonia salvaiaefolia Trichogonia villosa	183 183
882.	Trichogonia salvaiaefolia	183
883.	Trichogonia villosa	183
884.	Trichogonia salvaiaefolia Trichogonia villosa	183 183
885.	Trichogonia salvaiaefolia	183
886.	Trichogonia salvaiaefolia	183
887.	Trichogonia salviaefolia Trichogonia villosa	* 183
888.	Trichogonia villosa	183
889.	Trichogonia villosa	183
890.	Trichogonia salvaiaefolia Trichogonia villosa	183 183

Table 3. Diterpene Compound Numbers   205

TABLE 3. (*contd.*)

Compound: No.	Species	Reference
891.	Trichogonia villosa	183
892.	Aspilia mossambicensis	*
	Viguiera pazensis	205
893.	Viguiera pazensis	205
894.	Viguiera pazensis	205
895.	Stevia ovata	355,464
	Stevia paniculata	501
896.	Stevia ovata	355
	Stevia paniculata	501
897.	Stevia ovata	355
	Stevia paniculata	501
898.	Stevia ovata	355
	Stevia paniculata	501
899.	Stevia ovata	355
	Stevia paniculata	501
900.	Conyza scabrida	83
901.	Conyza scabrida	83
902.	Conyza scabrida	83
903.	Conyza scabrida	83
904.	Baccharis salicifolia	492
	Conyza scabrida	83
905.	Conyza scabrida	83
906.	Conyza scabrida	83
907.	Conyza scabrida	83
908.	Conyza scabrida	83
909.	Conyza scabrida	83
910.	Conyza scabrida	83
911.	Conyza scabrida	83
912.	Conyza scabrida	83
913.	Conyza scabrida	83
914.	Conyza scabrida	83
915.	Conyza scabrida	83
916.	Conyza scabrida	83

TABLE 3. (*contd.*)

Compound: No.	Species	Reference
917.	Conyza scabrida	83
918.	Conyza scabrida	83
919.	Conyza scabrida	83
920.	Conyza scabrida Pulicaria angustifolia	83 463
921.	Sigesbeckia pubescens	285
922.	Sigesbeckia pubescens	285
923.	Sigesbeckia pubescens	285
924.	Sigesbeckia pubescens	285
925.	Sigesbeckia pubescens	285
926.	Lasianthaea podocephala Melampodium perfoliatum Mikania banisteriae	124 159 228
927.	Zexmenia phyllocephala	124
928.	Zexmenia phyllocephala	124
929.	Zexmenia phyllocephala	124
930.	Zexmenia phyllocephala	124
931.	Zexmenia phyllocephala	124
932.	Zexmenia phyllocephala	124
933.	Zexmenia phyllocephala	124
934.	Zexmenia phyllocephala	124
935.	Zexmenia phyllocephala	124
936.	Zexmenia phyllocephala	124
937.	Zexmenia phyllocephala	124
938.	Eupatorium villosum	379
939.	Melampodium divaricatum Melampodium paludosum Melampodium perfoliatum Smallanthus glabratus	* * 159 108
940.	Eupatorium jhanii	285
941.	Eupatorium jhanii	285
942.	Montanoa tomentosa subsp. xanthiifolia Perymenium klattianum	227 103

Table 3. Diterpene Compound Numbers   207

TABLE 3. (*contd.*)

Compound: No.	Species	Reference
943.	Montanoa tomentosa subsp. xanthiifolia	227
944.	Sigesbeckia pubescens	225
945.	Dornoicum macrophyllum Doronicum microphyllum	3 285
946.	Espeletia wedellii	394
947.	Solidago altissima	500
948.	Balsamorhiza sagittata	130
949.	Balsamorhiza sagittata	130
950.	Heteropappus altaicus	181
951.	Heteropappus altaicus	181
952.	Heteropappus altaicus	181
953.	Heteropappus altaicus	181
954.	Heteropappus altaicus	181
955.	Grindelia squarrosa	473
956.	Grindelia squarrosa	473
957.	Helianthopsis bishopii	198
958.	Macowania glandulosa Oxylobus arbutifolius	* 87
959.	Oxylobus adscendens	87
960.	Solidago altissima	143
961.	Solidago altissima	143
962.	Baccharis tucumanensis	442
963.	Baccharis grandicapitulata	57
964.	Baccharis scoparia	57
965.	Baccharis eggersii	57
966.	Baccharis eggersii	57
967.	Baccharis hutchisonii	57
968.	Baccharis nitida	57
969.	Baccharis scoparia	57
970.	Baccharis scoparia	57

TABLE 3. (*contd.*)

Compound: No.	Species	Reference
971.	Aster alpinus	97
972.	Aster alpinus	97
973.	Tridax peruviensis	97
974.	Gutierrezia gilliesii Gutierrezia grandis Gutierrezia sobrigii	348 271 347
975.	Gutierrezia grandis	271
976.	Helichrysum refluxum	86
977.	Rudbeckia fulgida var. fulgida	329
978.	Smallanthus glabratus	108
979.	Smallanthus glabratus	108
980.	Smallanthus glabratus	108
981.	Smallanthus glabratus	108
982.	Smallanthus glabratus	108
983.	Smallanthus glabratus	108
984.	Ageratina tristis	58
985.	Pulicaria angustifolia	463
986.	Pulicaria angustifolia	463
987.	Pulicaria angustifolia	463
988.	Solidago altissima Solidago virgaurea	401 378
989.	Viguiera potosina	276
990.	Viguiera potosina	276
991.	Brickellia vernicosa Eupatorium maretiana	2 2
992.	Brickellia vernicosa	2
993.	Brickellia vernicosa	2
994.	Brickellia vernicosa	2
995.	Brickellia vernicosa	2
996.	Denekia capensis Helichrysum confertum Viguiera dentata	180 180 275
997.	Disynaphia multicrenulata	137

Table 3. Diterpene Compound Numbers 209

TABLE 3. (*contd.*)

Compound: No.	Species	Reference
	Viguiera dentata	275
998.	Stevia eupatoria	416
	Viguiera excelsa	242
999.	Hebeclinium macrophyllum	77
1000.	Solidago missouriensis	369
1001.	Solidago missouriensis	369
1002.	Mikania congesta	330
1003.	Mikania congesta	330
1004.	Mikania congesta	330
1005.	Mikania congesta	330
1006.	Melampodium leucanthum	436
1007.	Viguiera stenoloba	243
1008.	Eupatorium inulaefolium	23
1010.	Symphyopappus reticulatus	190
1011.	Symphyopappus reticulatus	190
1012.	Piptothrix jaliscensis	269
1013.	Piptothrix jaliscensis	269
1014.	Piptothrix sinaloae	269
1015.	Austrobrickellia patens	349
1016.	Austrobrickellia patens	349
1017.	Austrobrickellia patens	349
1018.	Austrobrickellia patens	349
1019.	Austrobrickellia patens	349
1020.	Stevia lemmonia	170
1021.	Stevia berlandiera	170
1022.	Stevia berlandiera	170
1023.	Stevia andina	*
	Stevia berlandiera	170
102?.	Stevia salicifolia	170
1025.	Stevia salicifolia	170
1026.	Stevia salicifolia	170

TABLE 3. (*contd.*)

Compound: No.	Species	Reference
1027.	Chromolaena laevigata	390
1028.	Chromolaena laevigata	390
1029.	Chromolaena laevigata	390
1030.	Chromolaena laevigata	390
1031.	Chromolaena laevigata	390
1032.	Chromolaena laevigata	390
1033.	Baccharis salicifolia Baccharis sarothroides Baccharis vaccinoides	22 22 22
1034.	Baccharis sarothroides	22
1035.	Grindelia boliviana	348
1036.	Grindelia boliviana	348
1037.	Gutierrezia gilliesii Gutierrezia solbrigii	348 347
1038.	Gutierrezia gilliesii Gutierrezia solbrigii	348 347
1039.	Gutierrezia gilliesii	348
1040.	Gutierrezia spathulata	348
1041.	Gutierrezia spathulata	348
1042.	Gutierrezia spathulata	348
1043.	Gutierrezia spathulata	348
1044.	Gutierrezia spathulata	348
1045.	Haplopappus glutinosus	348
1046.	Haplopappus pectinatus	348
1047.	Haplopappus pectinatus	348
1048.	Haplopappus pectinatus	348
1049.	Haplopappus pectinatus	348
1050.	Haplopappus pectinatus	348
1051.	Haplopappus pectinatus	348
1052.	Haplopappus pectinatus	348
1053.	Haplopappus pectinatus	348
1054.	Solidago petradoria	348

Table 3. Diterpene Compound Numbers   211

TABLE 3. (*contd.*)

Compound: No.	Species	Reference
1055.	Haplopappus paucidentatus	348
1056.	Haplopappus paucidentatus	348
1057.	Haplopappus paucidentatus	348
1058.	Haplopappus paucidentatus	348
1059.	Haplopappus paucidentatus	348
1060.	Grindelia boliviana	348
1061.	Baccharis scoparia	57
1062.	Solidago gigantea var. serotina	317
1063.	Solidago gigantea var. serotina	317
1064.	Solidago gigantea var. serotina	317
1065.	Solidago gigantea var. serotina	317
1066.	Solidago gigantea var. serotina	317
1067.	Solidago gigantea var. serotina	317
1068.	Solidago gigantea var. serotina	317
1069.	Liatris spicata	29
1070.	Baccharis salicifolia	492
1071.	Solidago juncea	318
1072.	Solidago juncea	318
1073.	Solidago juncea	318
1074.	Solidago juncea	318
1075.	Solidago serotina	385
1076.	Solidago serotina	385
1077.	Solidago gigantea var. serotina	385
1078.	Solidago gigantea var. serotina	385
1079.	Solidago gigantea var. serotina	385
1080.	Solidago gigantea var. serotina	385
1081.	Solidago gigantea var. serotina	385
1082.	Solidago gigantea var. serotina	385
1083.	Gutierrezia texana	274
1083.	Gutierrezia texana	274

TABLE 3. (*contd.*)

Compound: No.	Species	Reference
1084.	Gutierrezia texana	274
1085.	Gutierrezia texana	274
1086.	Gutierrezia texana	274
1087.	Gutierrezia texana	274
1088.	Gutierrezia texana	274
1089.	Gutierrezia texana	274
1090.	Gutierrezia texana	274
1091.	Gutierrezia texana	274
1092.	Gutierrezia texana	274
1093.	Baccharis salicifolia	492
1094.	Montanoa tomentosa	432
1095.	Montanoa tomentosa	432
1096.	Montanoa tomentosa	432
1097.	Montanoa tomentosa subsp. tomentosa	431
1098.	Montanoa tomentosa subsp. tomentosa	431
1099.	Montanoa tomentosa subsp. tomentosa	417
1100.	Acritopappus confertus	184
1101.	Helichrysum acutatum	40
1102.	Denekia capensis	180
	Helichrysum confertum	180
1103.	Helichrysum confertum	180
1104.	Helichrysum confertum	180
1105.	Helichrysum albirosulatum	180
1106.	Helichrysum albirosulatum	180
1107.	Helichrysum albirosulatum	180
1108.	Helianthus decapetalus	31
	Helianthus decapetalus var. multiflorus	31
	Helianthus hirsutus	31
	Helianthus rigidus	31
1109.	Cronquistianthus bishopii	61

Table 3. Diterpene Compound Numbers   213

TABLE 3. (*contd.*)

Compound: No.	Species	Reference
1110.	Cronquistianthus bishopii	61
1111.	Solidago drummondii	65
1112.	Mikania periplocifolia	296
1113.	Mikania periplocifolia	296
1114.	Mikania periplocifolia	296
1115.	Mikania periplocifolia	296
1116.	Bejaranoa balansae	453
1117.	Bejaranoa balansae	453
1118.	Chiliotrichium rosarinifolium	346
1119.	Solidago drummondii	65
1120.	Aspilia parvifolia	182
1121.	Vernonia glabra var. glabra	55
1122.	Heterochrysum oreophilum	351
1123.	Heterochrysum oreophilum	351
1124.	Acanthospermum australe	138
1125.	Solidago flexicaulis Solidago racemosa Solidago drummondii	65 65 65
1126.	Solidago drummondii	65
1127.	Solidago drummondii	65
1128.	Solidago drummondii	65
1129	Solidago drummondii	65
1130.	Gutierrezia solbrigii	347
1131.	Gutierrezia solbrigii	347
1132.	Gutierrezia solbrigii	347
1133.	Nordophyllum lanatum	346
1134.	Nordophyllum lanatum	346
1135.	Gutierrezia solbrigii	347
1136.	Gutierrezia solbrigii	347
1137.	Solidago drummondii	65

TABLE 3. *(contd.)*

Compound: No.	Species	Reference
1138.	Helichrysum nudifolium var. nudifolium	351
1139.	Oxylobus glanduliferus	6
1140.	Grindelia discoidea	472
1141.	Grindelia discoidea	472
1142.	Grindelia discoidea	472
1143.	Eupatorium salvia	339
1144.	Eupatorium salvia	339
1145.	Grindelia discoidea	471
1146.	Chrysothamnus nauseosus Grindelia discoidea	471 471
1147.	Helichrysum confertum	180
1148.	Gutierrezia sphaerocephala	272
1149.	Gutierrezia sphaerocephala	272
1150.	Gutierrezia sphaerocephala	272
1151.	Helichrysum nudifolium var. nudifolium	351
1152.	Helichrysum nudifolium var. nudifolium	351
1153.	Isocoma coronopifolia	253
1154.	Isocoma coronopifolia	253
1155.	Lindheimera texana	337
1156.	Lindheimera texana	337
1157.	Stevia rebaudiana	418
1158.	Stevia rebaudiana	418
1159.	Stevia rebaudiana	418
1160.	Stevia rebaudiana	418
1161.	Palafoxia texana	290
1162.	Palafoxia texana	290
1163.	Eupatorium turbinatum	350
1164.	Grindelia aegialitis	414
1165.	Baccharis pingraea	503

Table 3. Diterpene Compound Numbers   215

TABLE 3. (*contd.*)

Compound: No.	Species	Reference
	Baccharis salicifolia	503
1166.	Baccharis pingraea	503
	Baccharis salicifolia	503
1167.	Baccharis pingraea	503
	Baccharis salicifolia	503
1168.	Baccharis pingraea	503
1169.	Baccharis pingraea	503
1170.	Baccharis salicifolia	503
1171.	Baccharis pingraea	503
1172.	Baccharis salicifolia	503
1173.	Baccharis trinervis	27
1174.	Gutierrezia solbrigii	347
1175.	Haplopappus pectinatus	348
1176.	Haplopappus pectinatus	348
1177.	Haplopappus pectinatus	348
1178.	Haplopappus pectinatus	348
1179.	Haplopappus pectinatus	348
1180.	Acritopappus morii	179
1181.	Gutierrezia solbrigii	347
1182.	Gutierrezia solbrigii	347
1183.	Gutierrezia solbrigii	347
1184.	Gutierrezia solbrigii	347
1185.	Mikania alvimii	47
1186.	Palafoxia rosea	26
1187.	Grindelia boliviana	348
1188.	Baccharis salicifolia	503
1189.	Gutierrezia solbrigii	347
1190.	Gutierrezia solbrigii	347
1191.	Gutierrezia solbrigii	347
1192.	Baccharis salicifolia	503
1193.	Baccharis salicifolia	503

TABLE 3. (*contd.*)

Compound: No.	Species	Reference
1194..	Gutierrezia spathulata	348
1195.	Gutierrezia spathulata	348
1196.	Baccharis salicifolia	503
1197.	Baccharis salicifolia	503
1198.	Palafoxia rosea	26
1199.	Palafoxia rosea	26
1200.	Palafoxia rosea	26
1201.	Ayapana amygdalina	113
1202.	Ayapana amygdalina	113
1203.	Baccharis pingraea	503
1204.	Helichrysum argolepis	*
1205.	Austroeupatorium inulaefolium	139
1206.	Austroeupatorium inulaefolium	139
1207.	Austroeupatorium inulaefolium	139
1208.	Haplopappus paucidentatus	348
1209.	Baccharis polifera	503
1210.	Baccharis polifera	503
1211.	Eupatorium turbinatum	350
1212.	Chiliotrichium rosmarinifolium	346
1213.	Chiliotrichium rosmarinifolium	346
1214.	Chiliotrichium rosmarinifolium	346
1215.	Chiliotrichium rosmarinifolium	346
1216.	Chiliotrichium rosmarinifolium Nardophyllum lanatum	346 346
1217.	Nardophyllum lanatum	346
1218.	Nardophyllum lanatum	346
1219.	Nardophyllum lanatum	346
1221.	Chiliotrichium rosmarinifolium	346
1222.	Baccharis incarum	449
1223.	Pulicaria salviifolia	403

Table 3. Diterpene Compound Numbers   217

TABLE 3. (contd.)

Compound: No.	Species	Reference
1224.	Baccharis salicifolia	503
1225.	Baccharis salicifolia	503
1226.	Baccharis salicifolia	503
1227.	Baccharis incarum	284
1228.	Goyazianthus tetrastichus	60
1229.	Baccharis rhomboidalis	452
1230.	Baccharis rhomboidalis	452
1231.	Baccharis rhetinodes	503
1232.	Baccharis incarum	449
1233.	Baccharis incarum	449
1234.	Pulicaria salviifolia	404
1235.	Pulicaria salviifolia	404
1236.	Baccharis incarum	284
1237.	Baccharis articulata	466
1238.	Baccharis articulata	466
1239.	Baccharis grandicapitulata	57
1240.	Baccharis incarum	284
1241.	Baccharis rhomboidalis	452
1242.	Baccharis rhomboidalis	452
1243.	Baccharis rhomboidalis	452
1244.	Baccharis magellanica	503
1245.	Baccharis magellanica	503
1246.	Chiliotrichium rosmarinifolium	346
1247.	Gochnatia glutinosa	278
1248.	Gochnatia glutinosa	278
1249.	Senecio hypochoerideus	71
1250.	Mikania triangularis	360
1251.	Helichrysum setosum	351
1252.	Helichrysum setosum	351

TABLE 3. (*contd.*)

Compound: No.	Species	Reference
1253.	Helichrysum setosum	351
1254.	Helichrysum setosum	351
1255.	Ruilopezia margarita [=Espeletia margarita]	481
1256.	Ruilopezia margarita	481
1257.	Baccharis incarum	284
1258.	Baccharis incarum	284
1259.	Helianthus grosseserratus Helianthus laciniatus Helianthus rigidus	* * *
1260.	Eupatorium petaloideum	336

Figure 1. Structures of Diterpenes   219

I. Linear or Unicarbocyclic

1. Geranylnerol

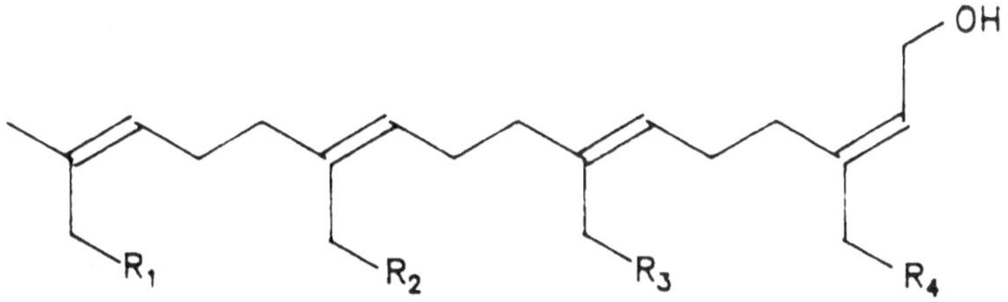

1	$R_1 = R_2 = R_3 = R_4 = H$
2	$R_1 = R_3 = R_4 = H$, $R_2 = OH$
3	$R_1 = R_2 = R_3 = H$, $R_4 = OH$
4	$R_1 = R_4 = OH$, $R_2 = R_3 = H$
984	$R_1 = R_2 = H$, $R_3 = OAc$, $R_4 = OH$
1109	$R_1 = R_2 = OH$, $R_3 = R_4 = H$
1110	$R_1 = R_2 = OH$, $R_3 = H$, $R_4 = OH$

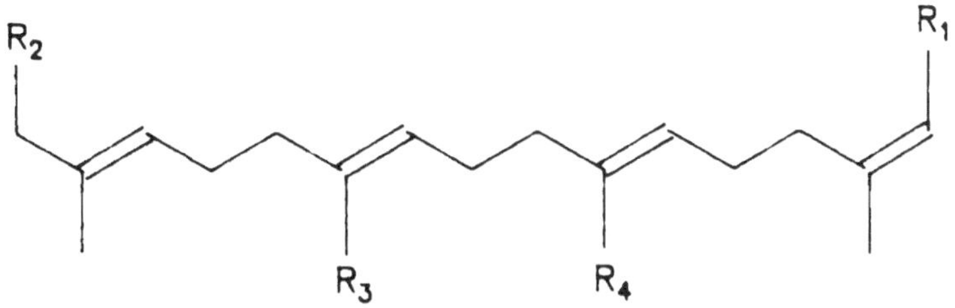

5	$R_1 = CHO$, $R_2 = H$, $R_3 = R_4 = Me$
6	$R_1 = CH_2OH$, $R_2 = OH$, $R_3 = CO_2H$, $R_4 = Me$
1002	$R_1 = CH_2OH$, $R_2 = H$, $R_3 = Me$, $R_4 = CO_2H$
1003	$R_1 = CH_2OAc$, $R_2 = H$, $R_3 = Me$, $R_4 = CO_2H$

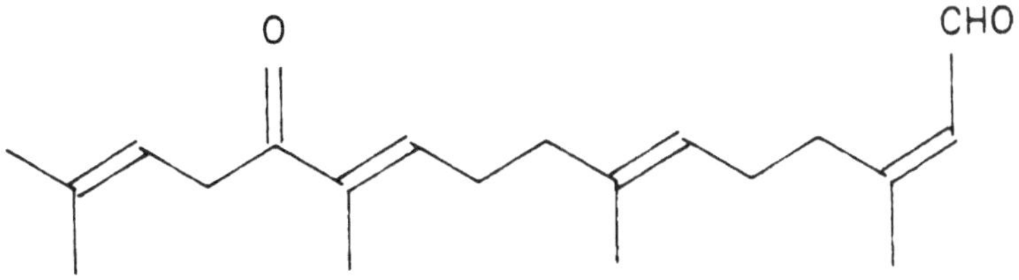

1111

FIGURE 1. Structures of diterpenes reported from the Compositae.

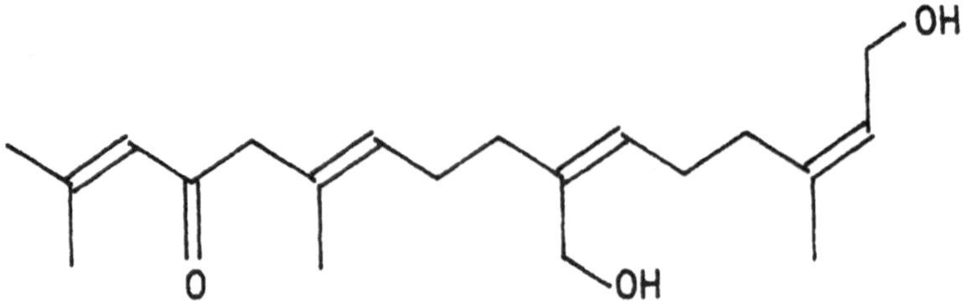

7

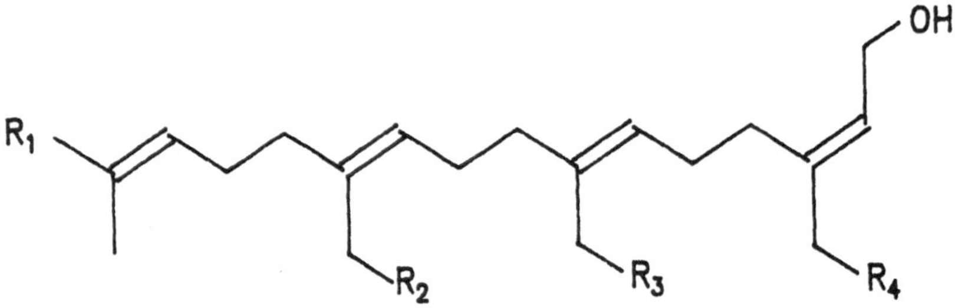

8 R_1 = CHO, R_2 = H, R_3 = OH, R_4 = OH

957 R_1 = CH$_2$OH, R_2 = OH, R_3 = OH, R_4 = H

1112 R_1 = CH$_2$OH, R_2 = R_3 = OAc, R_4 = OH

1113 R_1 = CH$_2$OH, R_2 = R_3 = R_4 = OAc

1114 R_1 = CH$_2$OH, R_2 = OAc, R_3 = H, R_4 = OH

1115 R_1 = CH$_2$OH, R_2 = H, R_3 = OAc, R_4 = OH

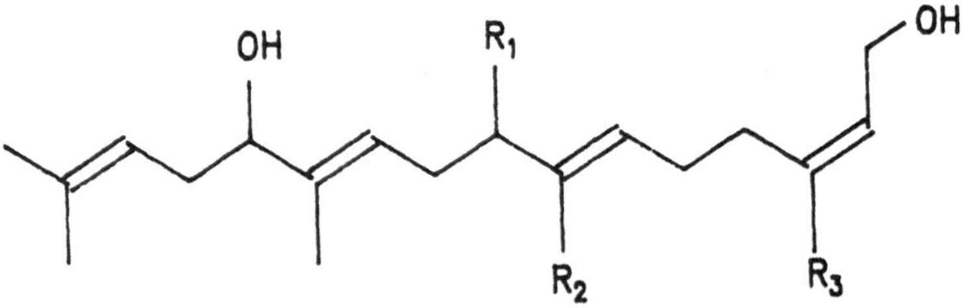

15 R_1 = H, R_2 = CH$_2$OAc, R_3 = CH$_2$OH

948 R_1 = OH, R_2 = CH$_2$OH, R_3 = Me

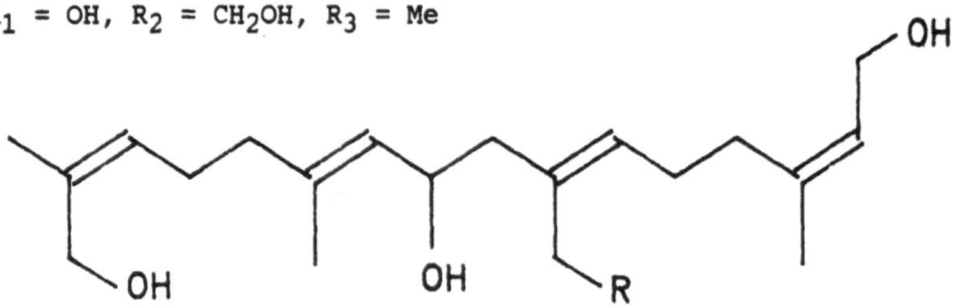

1116 R = H

1117 R = OH

FIGURE 1. (*contd.*)

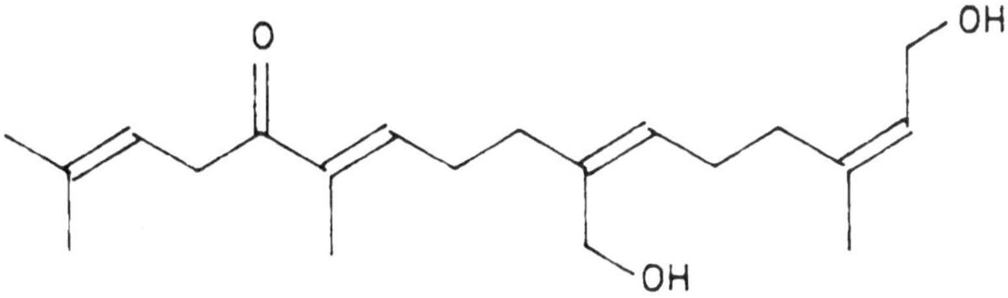

Figure 1. Structures of Diterpenes 221

978

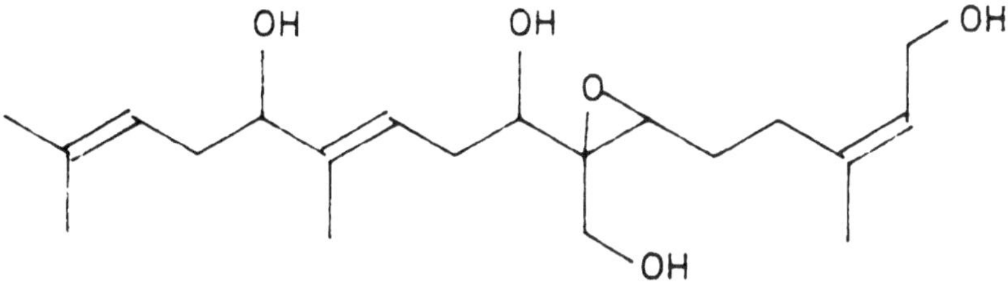

949

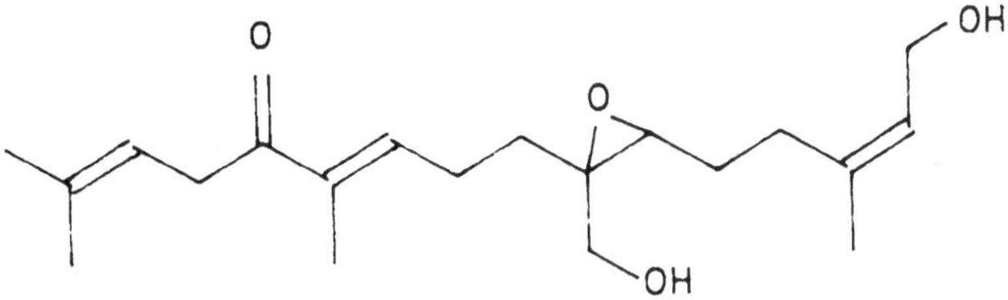

979

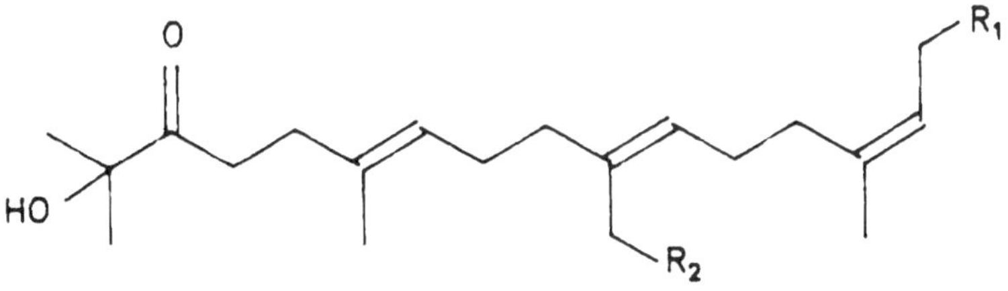

1004 R$_1$ = R$_2$ = OH
1005 R$_1$ = R$_2$ = OAc

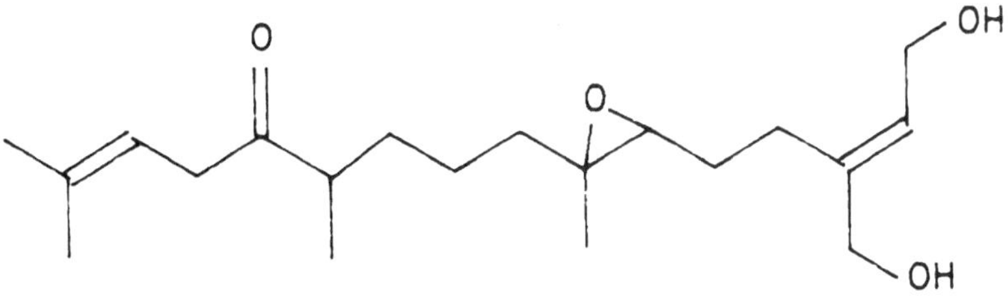

1095

FIGURE 1. (*contd.*)

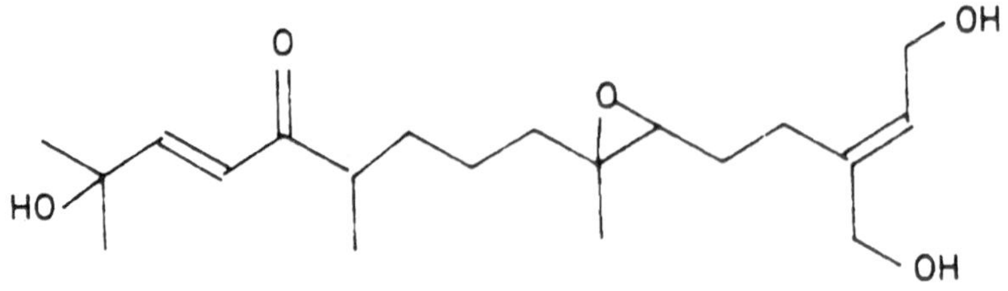

1096

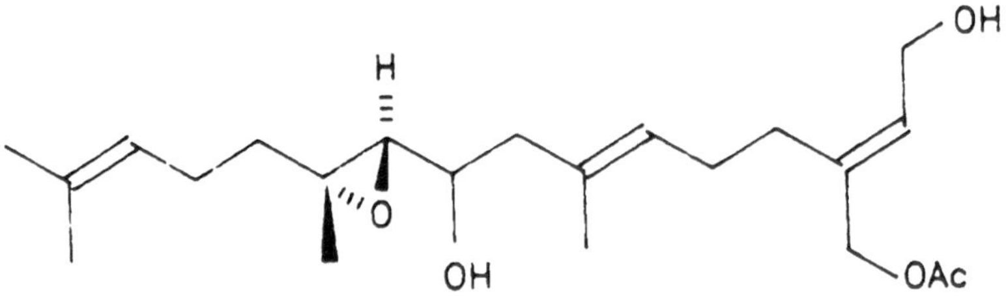

1020

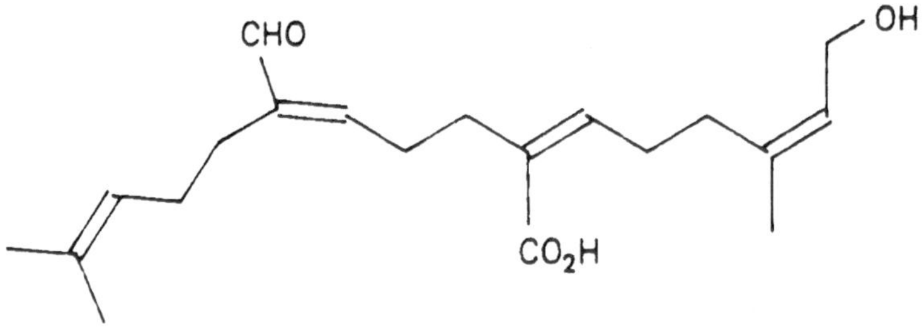

14

2. **Geranylgeraniol**

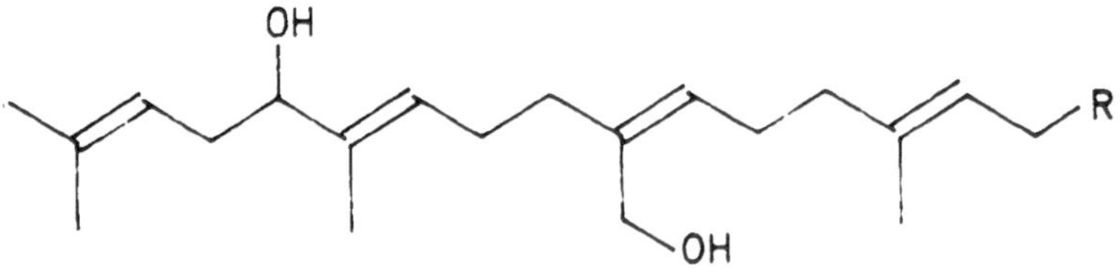

11　　R = OH

12　　R = OAc

FIGURE 1. (*contd.*)

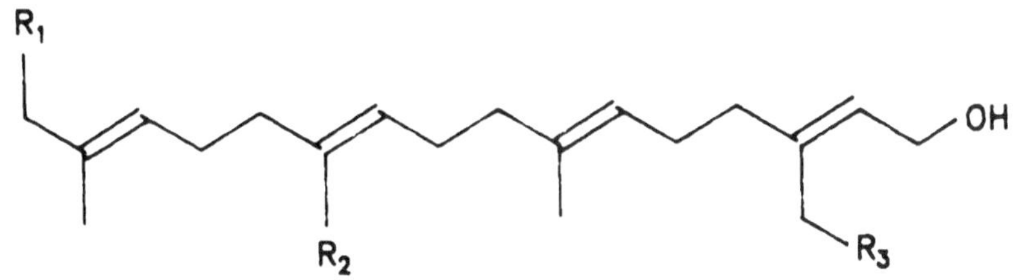

Figure 1. Structures of Diterpenes 223

13   R_1 = OH, R_2 = Me, R_3 = H

954 R_1 = H, R_2 = CO_2H, R_3 = H (10,11E)

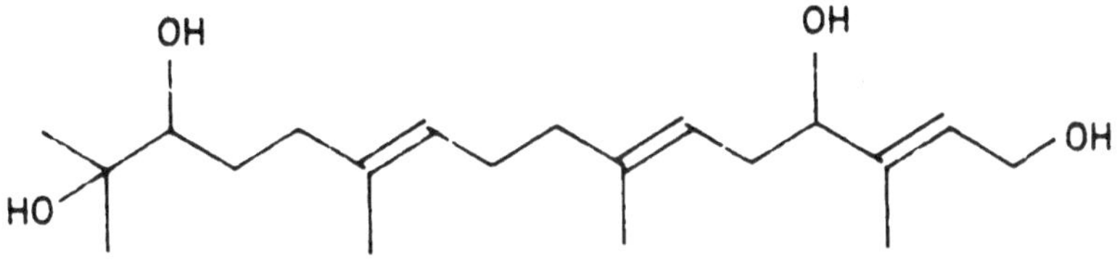

16

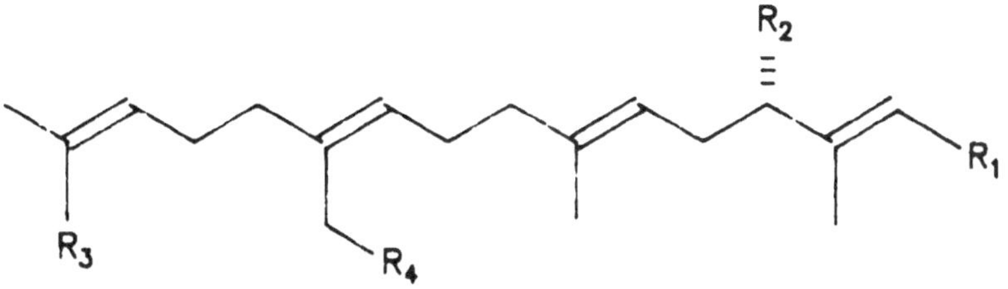

17   R_1 = CHO, R_2 = H, R_3 = Me, R_4 = H

18   R_1 = CHO, R_2 = OH, R_3 = Me, R_4 = H

19   R_1 = CH_2OAc, R_2 = H, R_3 = CO_2H, R_4 = H

1118 R_1 = CH_2OH, R_2 = H, R_3 = CH_2OH, R_4 = OAc

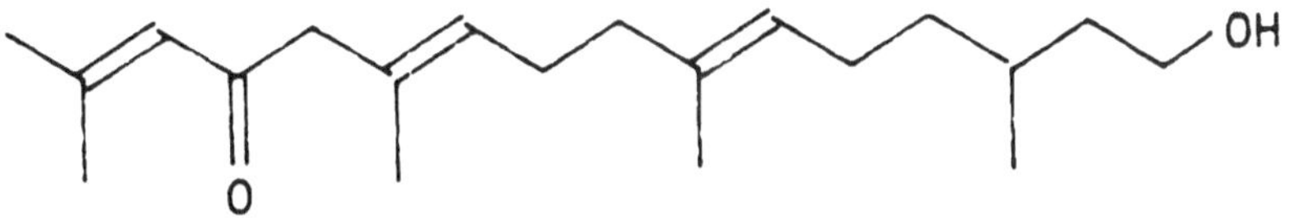

20

FIGURE 1. (*contd.*)

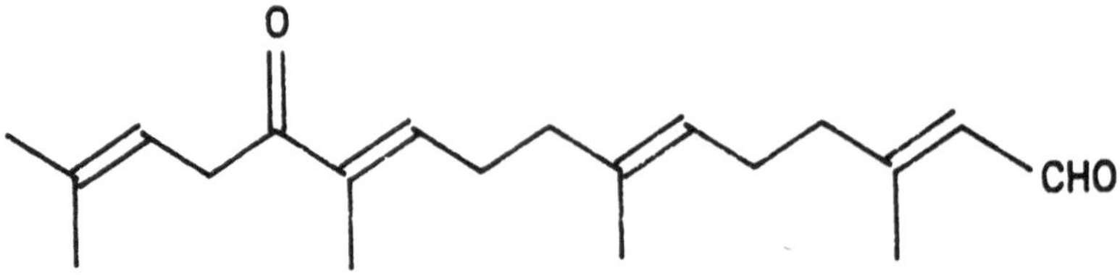

1119

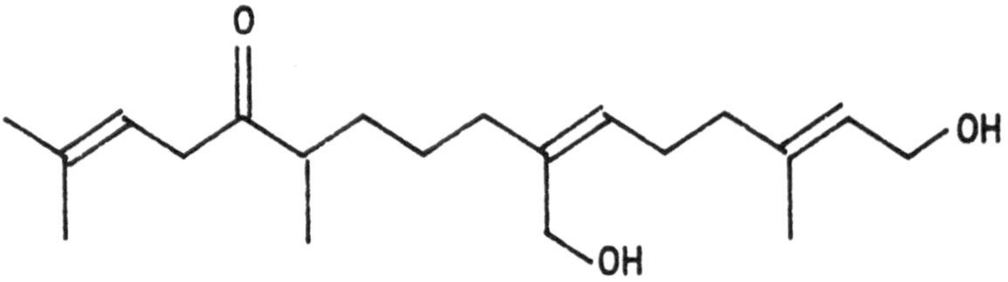

21

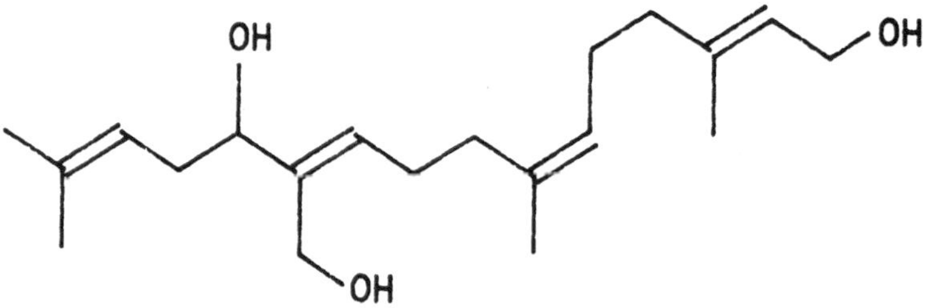

22

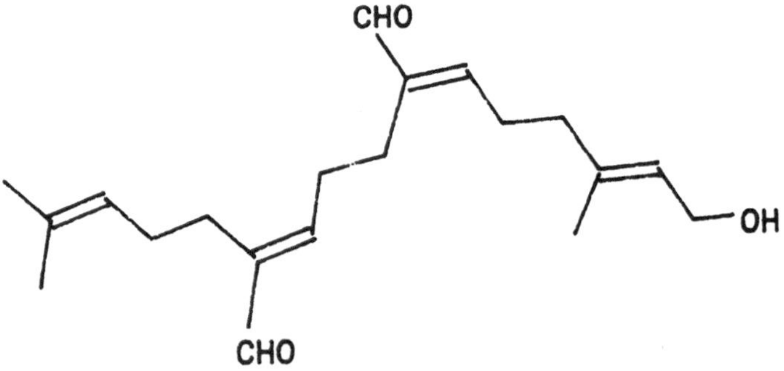

1120

3. Geranyllinalool

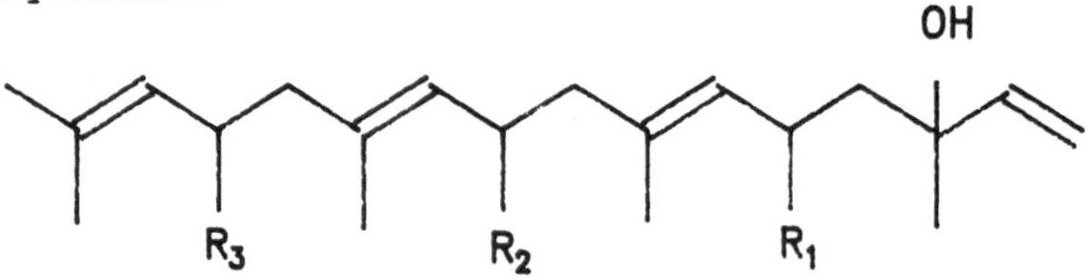

23    $R_1 = R_2 = R_3 = H$

FIGURE 1. (*contd.*)

Figure 1. Structures of Diterpenes   225

24 $R_1 = R_2 = H$, $R_3 = OH$

25 $R_1 = R_2 = H$, $R_3 = OAc$

26 $R_1 = H$, $R_2 = OH$, $R_3 = H$

27 $R_1 = H$, $R_2 = OAc$, $R_3 = H$

28 $R_1 = OH$, $R_2 = R_3 = H$

29 $R_1 = OH$, $R_2 = OAc$, $R_3 = H$

30 $R_1 = R_2 = OAc$, $R_3 = H$

31 $R_1 = OH$, $R_2 = H$, $R_3 = OAc$

1121 $R_1 = R_2 = H$, $R_3 = OAc$

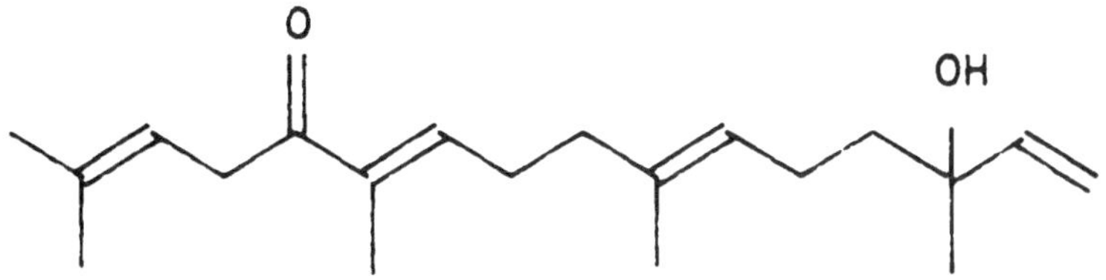

32

1122 (10,11-dihydro-homolog of **32**)

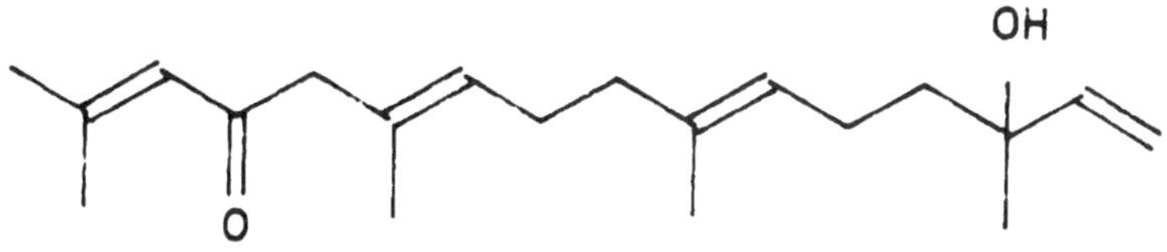

33

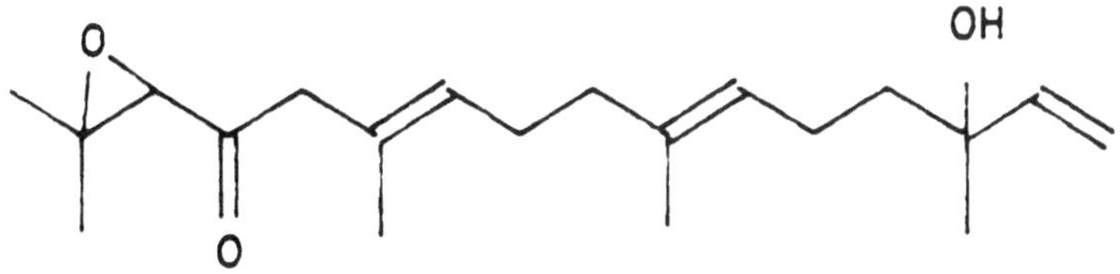

34

FIGURE 1. (*contd.*)

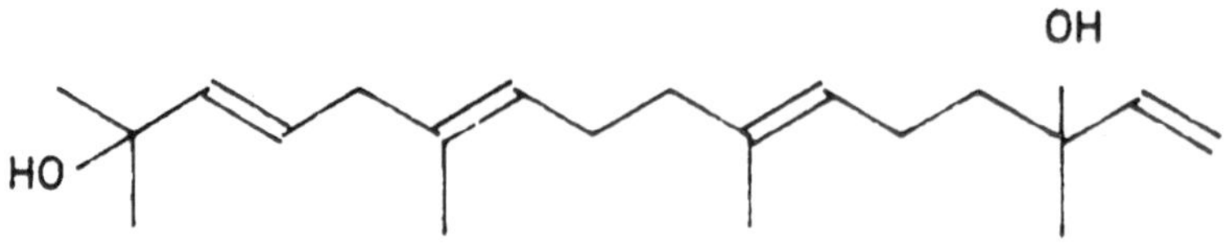

35

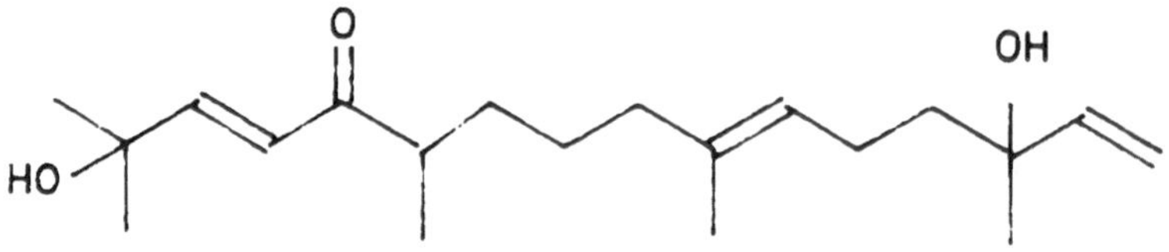

1123

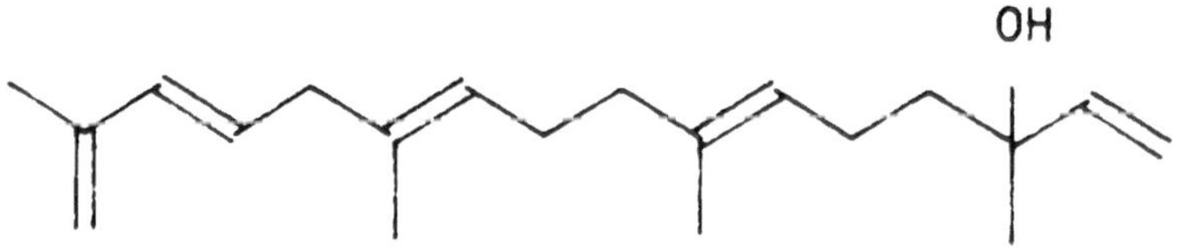

36

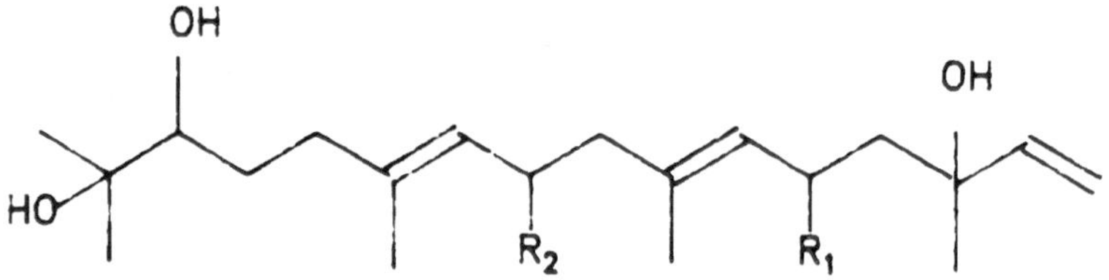

37 R_1 = R_2 = H

38 R_1 = OH, R_2 = H

39 R_1 = H, R_2 = OAc

4. Phytol

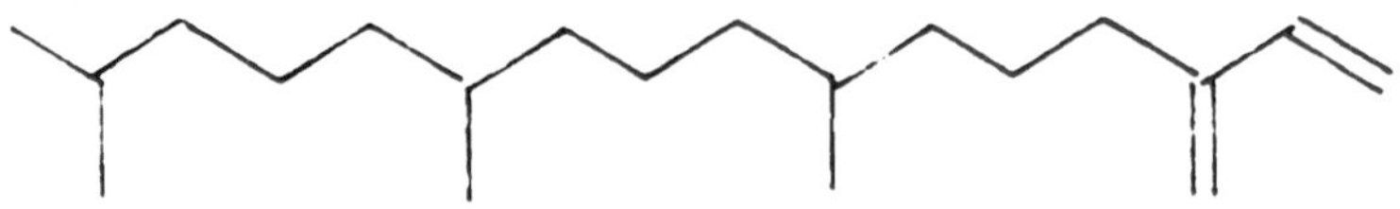

59

FIGURE 1. (*contd.*)

Figure 1. Structures of Diterpenes   227

80. 14-Methylgeranylnerol

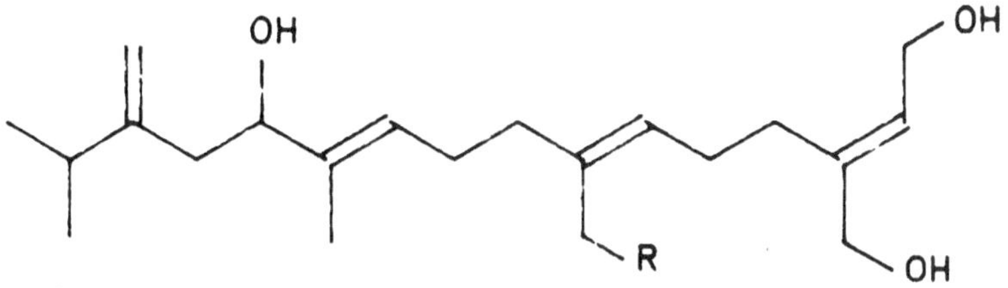

9   R = OH
10 R = OAc

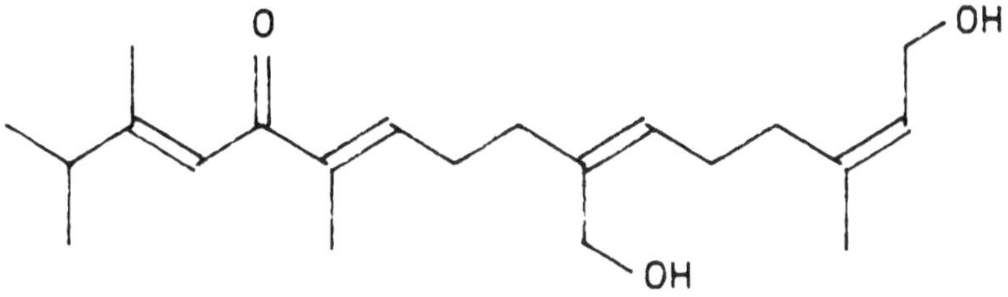

980

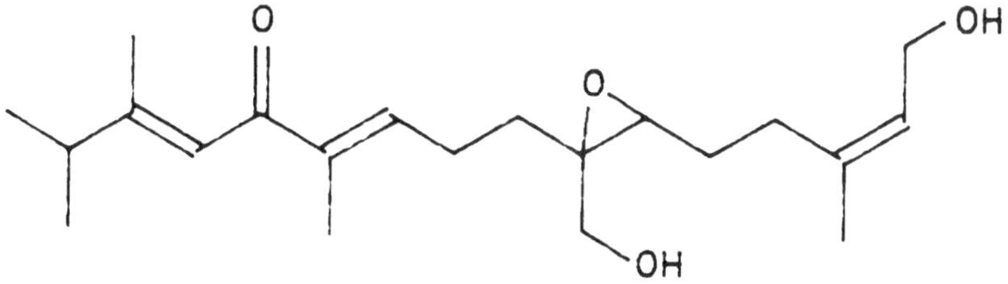

981

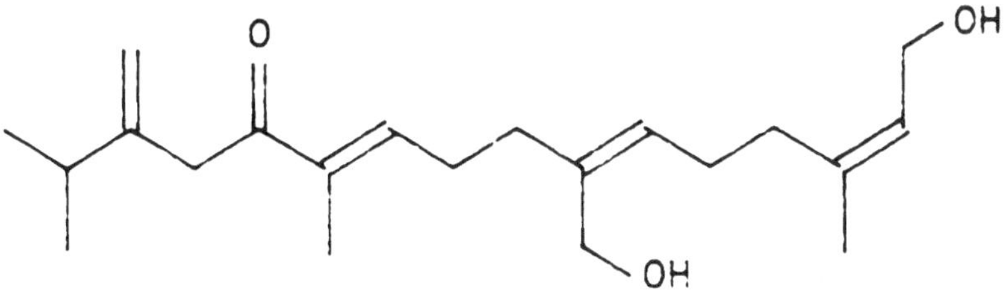

982

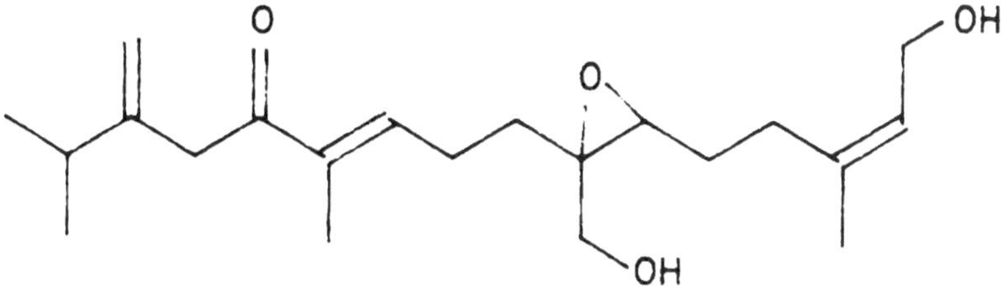

983

FIGURE 1. (*contd.*)

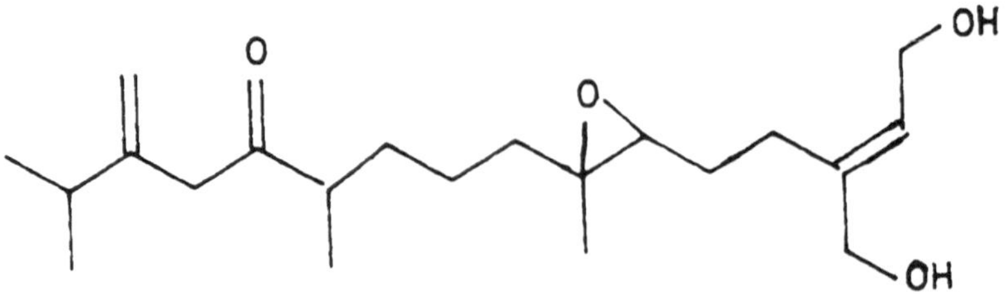

1094

5. Acanthoaustralane

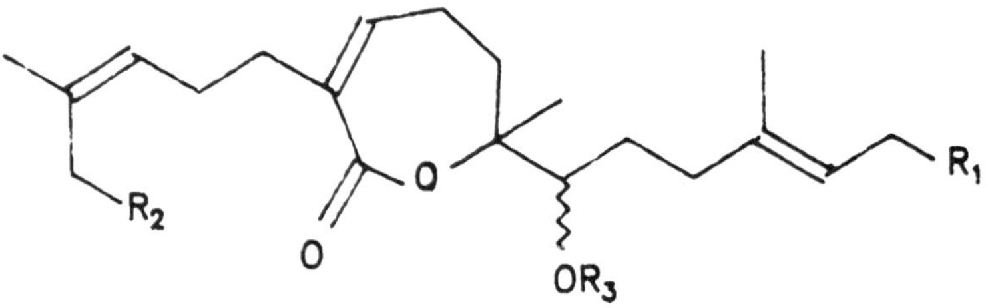

40 R_1 = OH, R_2 = R_3 = H

41 R_1 = OAc, R_2 = R_3 = H

42 R_1 = R_2 = OH, R_3 = H

43 R_1 = OH, R_2 = OAc, R_3 = H

1124 R_1 = OAc, R_2 = H, R_3 = Ac

6. Acanthoaustralane-6,11-Epoxide

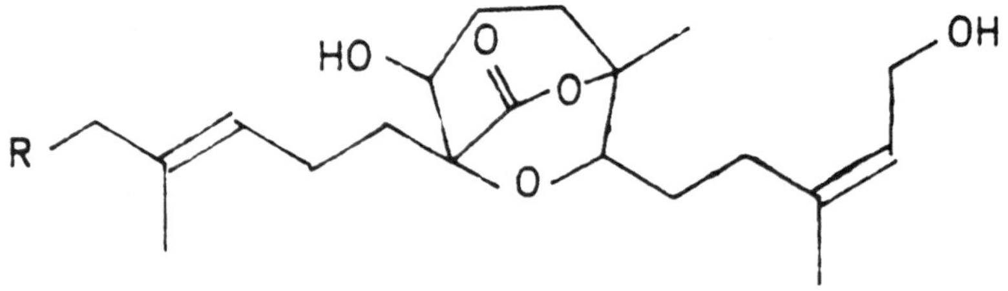

44   R = OAc

45   R = OH

FIGURE 1. (*contd.*)

Figure 1. Structures of Diterpenes 229

7. Isoacanthoaustralane

46   R_1 = R_2 = OH

47   R_1 = OH, R_2 = OAc

48   R_1 = OAc, R_2 = H

8. Geranylgeraniol-18,8-Lactone

49   R = H

50   R = OAc

51   R = OOH

52   R = OH

FIGURE 1. (*contd.*)

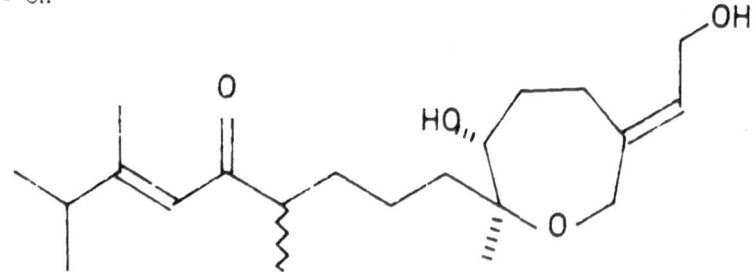

53   R = OOH

54   R = OH

9. Oxepane

55

56   R = OAc

1098 R = OH

57

FIGURE 1. (*contd.*)

Figure 1. Structures of Diterpenes 231

1097

1099

86. Melcanthane

1006

10. Mikanofurane

58

FIGURE 1. (*contd.*)

11. Centipedane

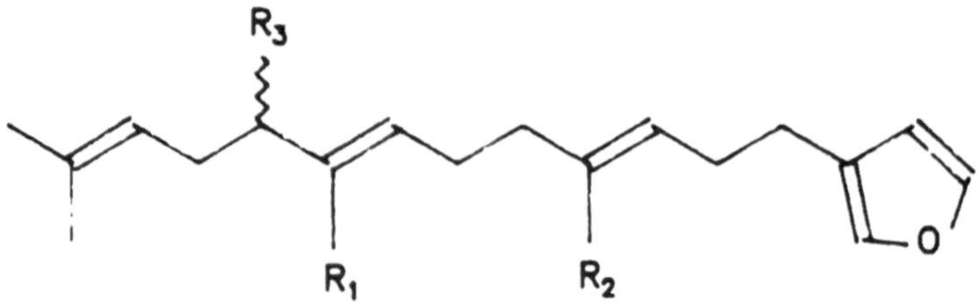

60    $R_1 = CO_2H$, $R_2 = Me$, $R_3 = H$

61    $R_1 = R_2 = CH_2OH$, $R_3 = H$

1126 $R_1 = R_2 = Me$, $R_3 = OH$

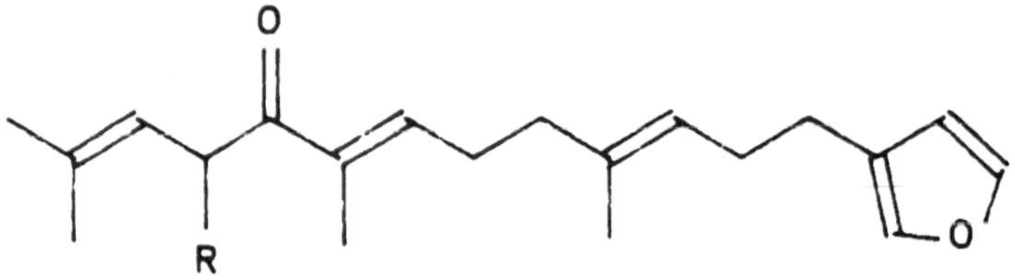

1125 R = H

1127 R = OH

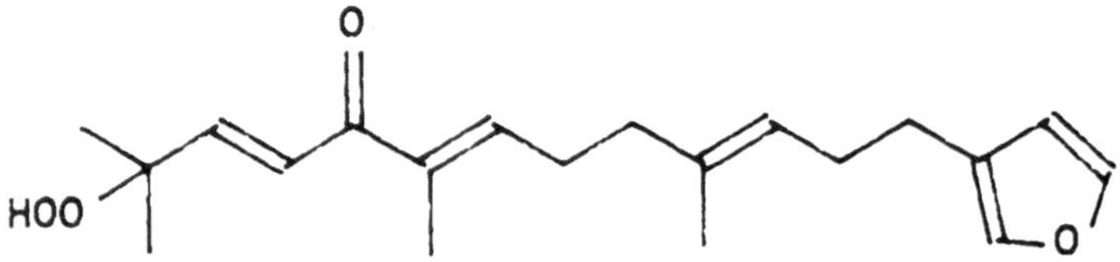

1128

12. Geranylgeraniol-1,20-Lactone

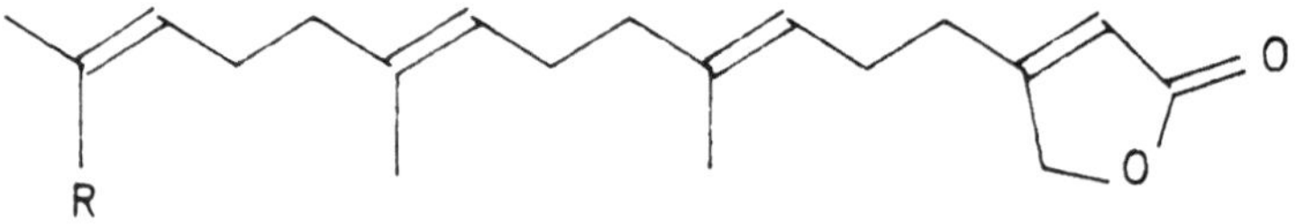

62    $R = CO_2H$

63    $R = CO_2Me$

FIGURE 1. (contd.)

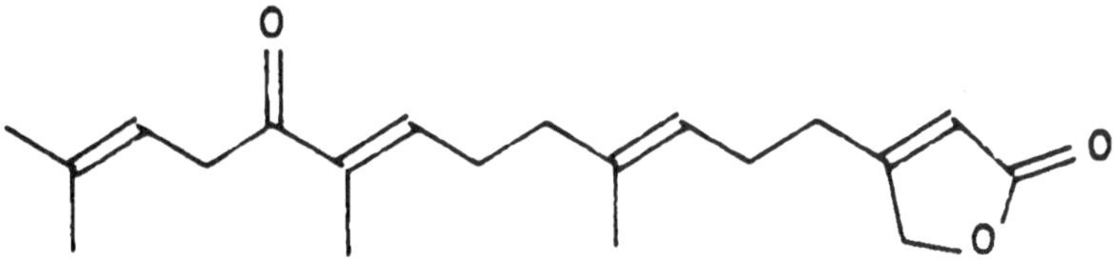

Figure 1. Structures of Diterpenes   233

1129

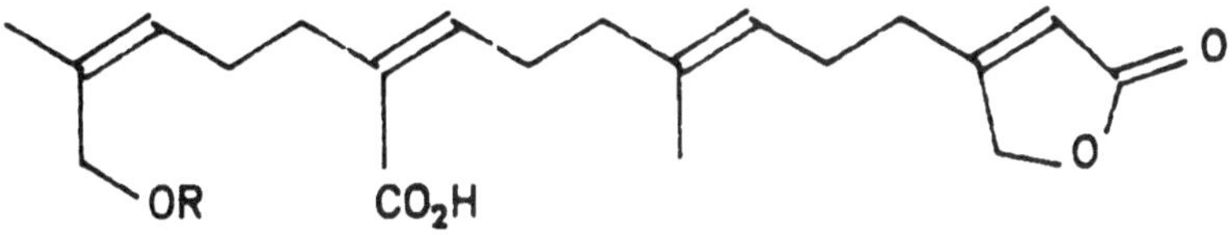

1130 R = H

1131 R = H (10,11E)

1132 R = Ac

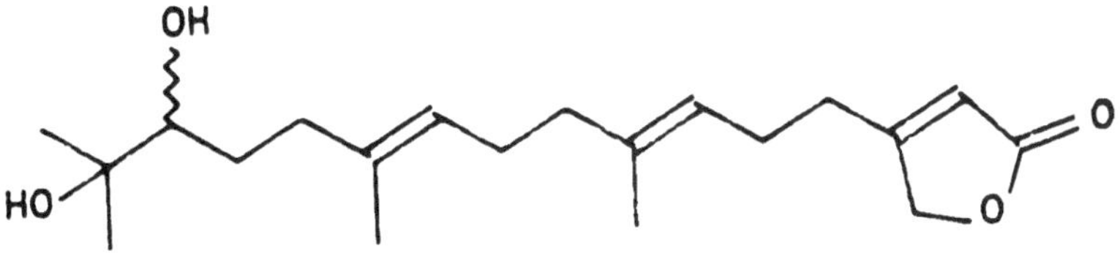

64

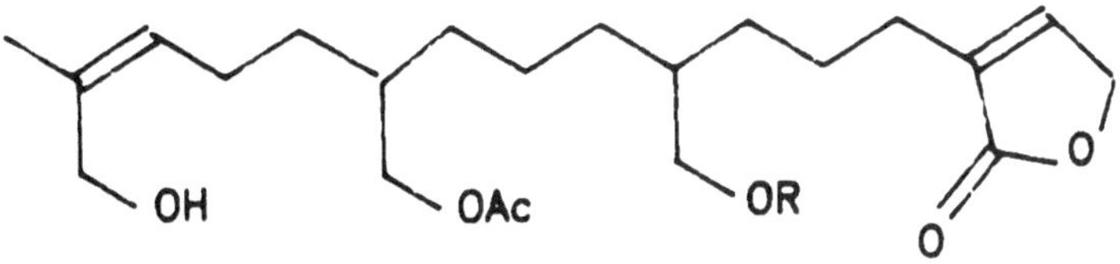

1133 R = H

1134 R = Ac

FIGURE 1. (*contd.*)

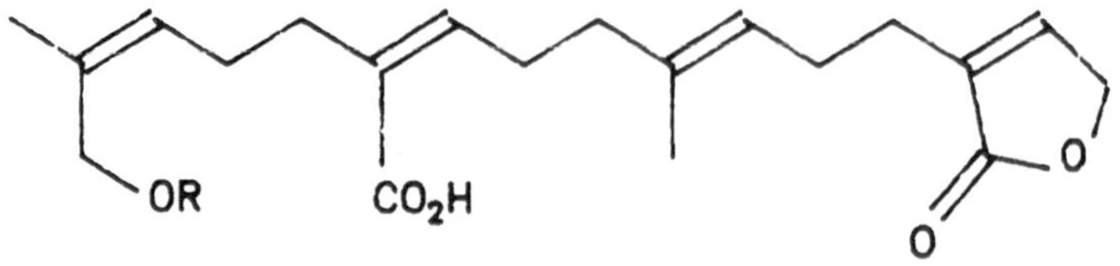

1135 R = H

1136 R = Ac

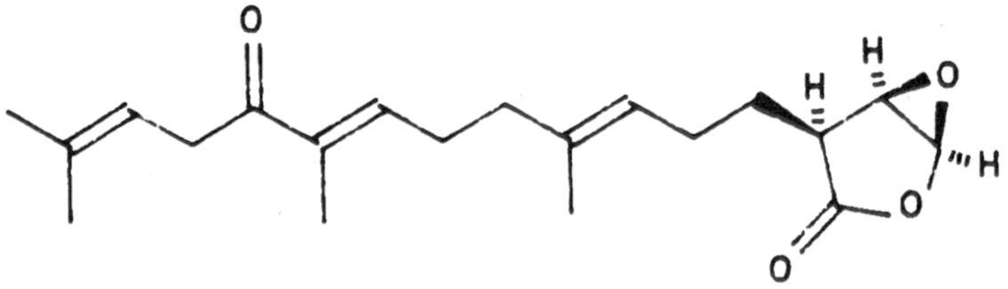

1137

13. Geranylterpinene

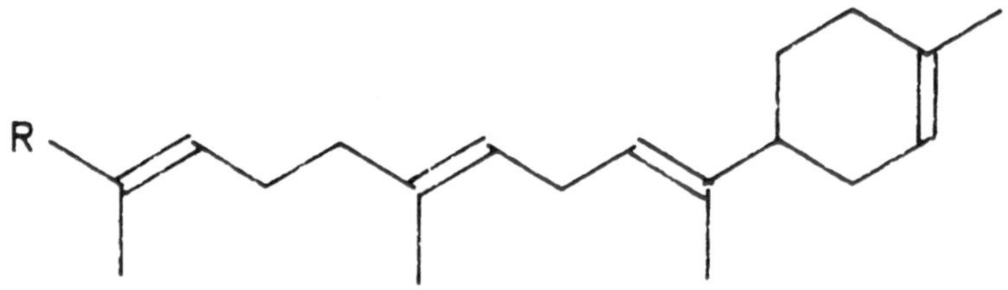

65   R = CH$_2$OH

66   R = CHO

67   R = CO$_2$H

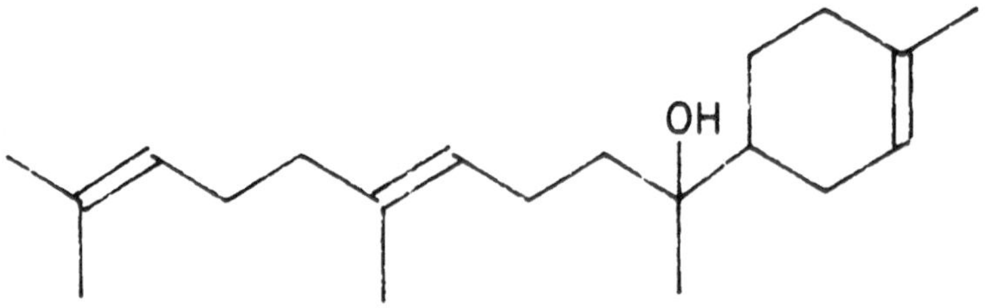

1138

FIGURE 1. (*contd.*)

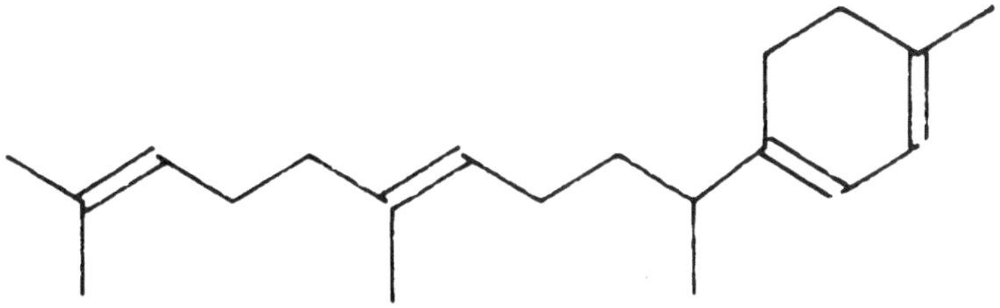

Figure 1. Structures of Diterpenes   235

68

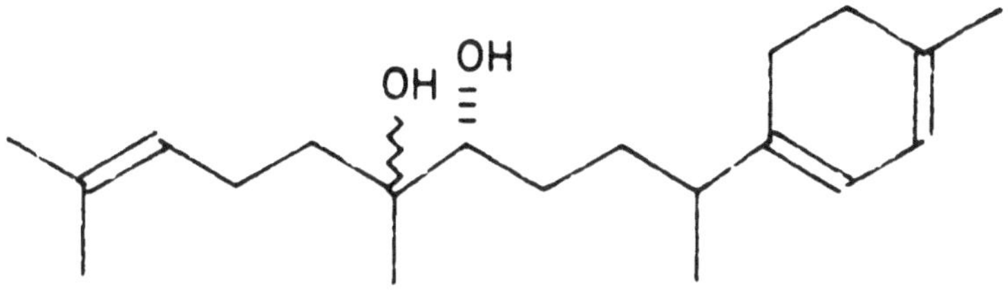

69

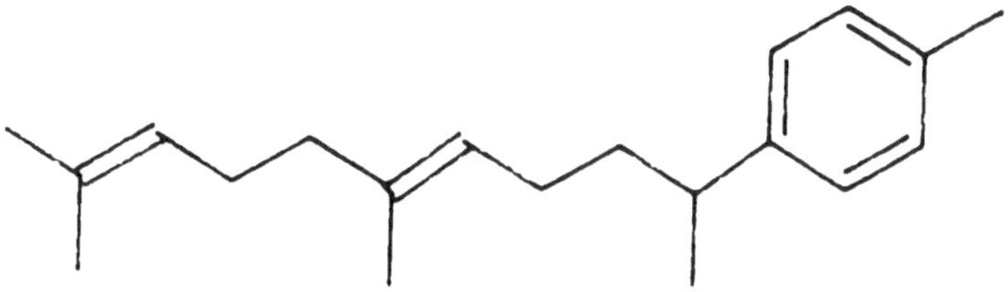

70

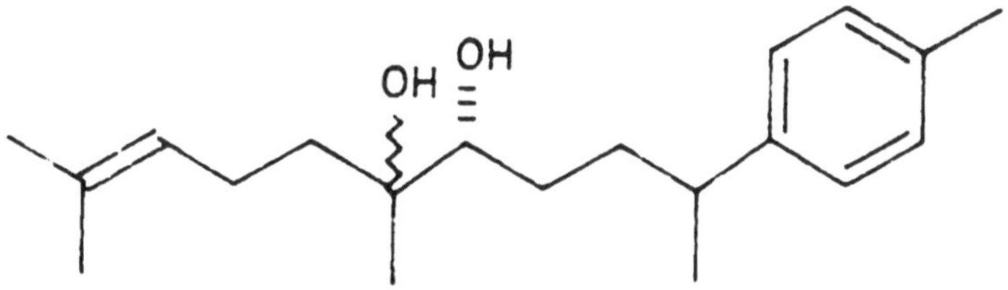

71

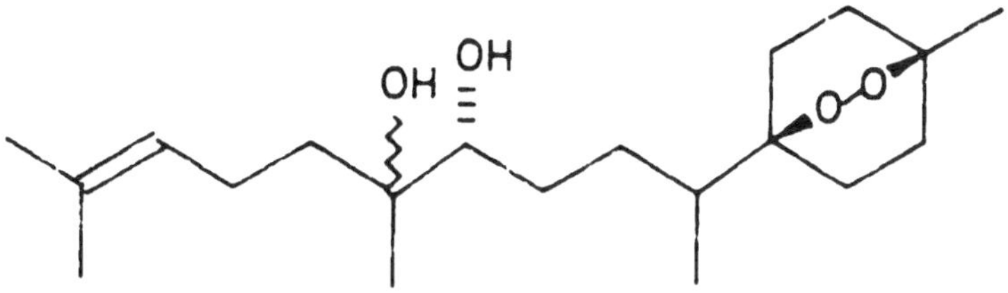

72

FIGURE 1. (*contd.*)

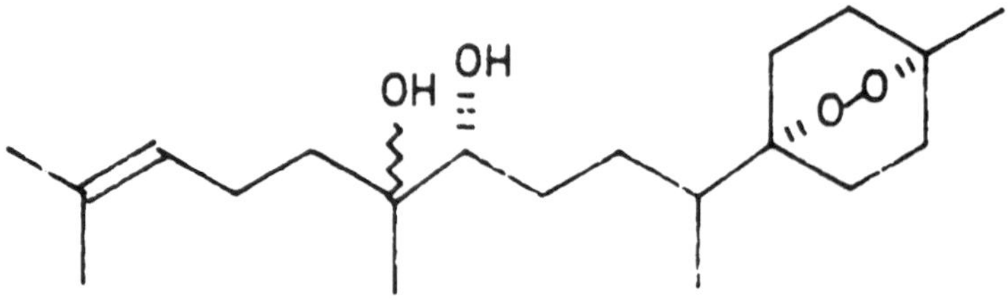

1101

14. Isocembrene

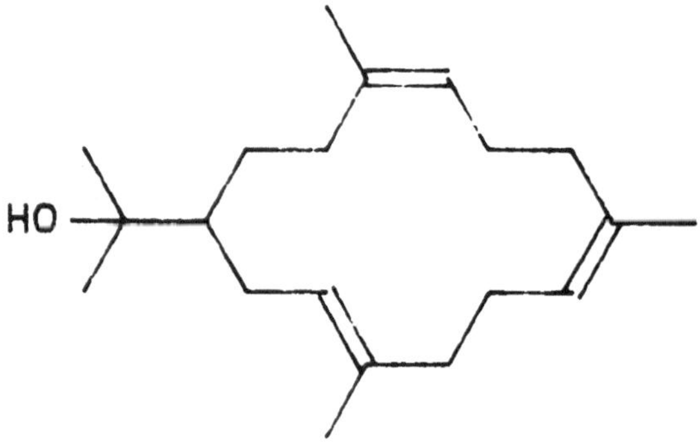

73

15. Geranylnerol-11,14-α-Epoxide

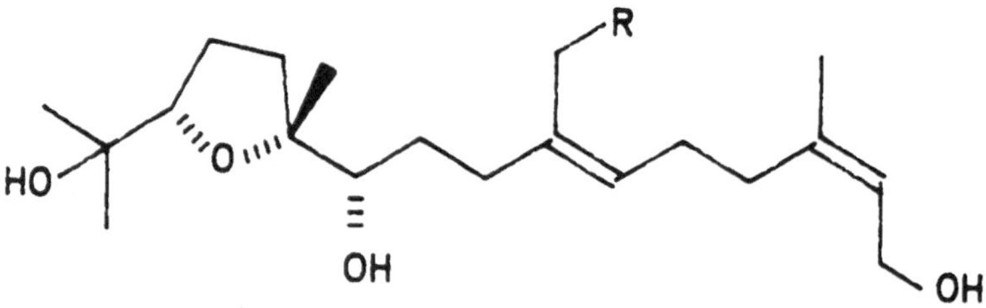

74   R = OH

75   R = OAc

FIGURE 1. (*contd.*)

Figure 1. Structures of Diterpenes   237

II. Normal-Bicarbocyclics and Derivatives

16. Normal-Labdane

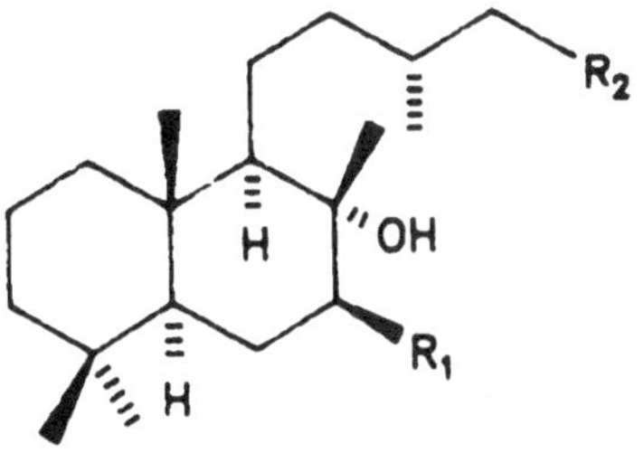

79   R_1 = H, R_2 = CH$_2$OH

80   R_1 = OH, R_2 = CH$_2$OH

958   R_1 = H, R_2 = CO$_2$H (=121?)

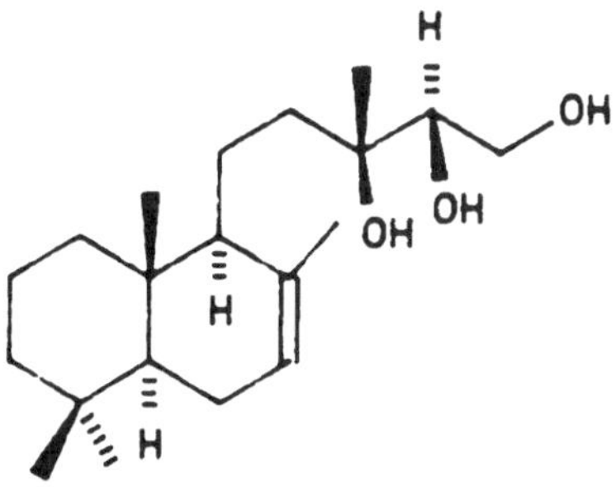

86

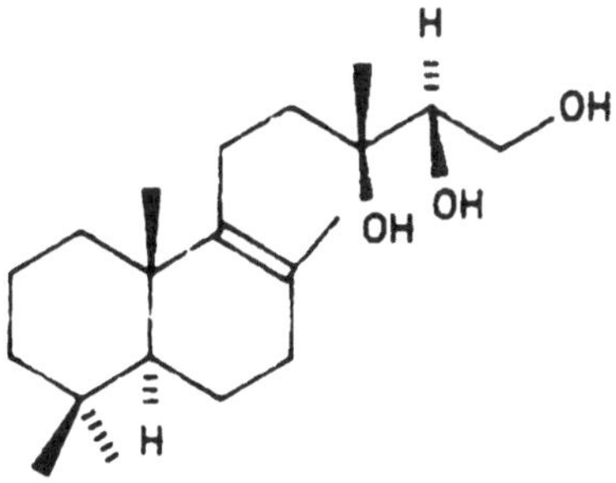

87

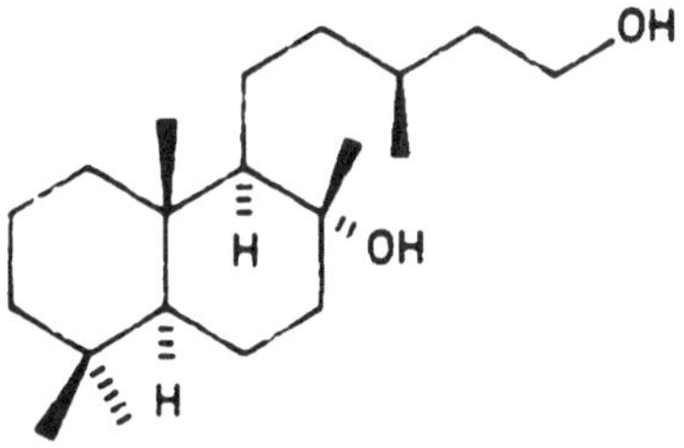

81

FIGURE 1. (contd.)

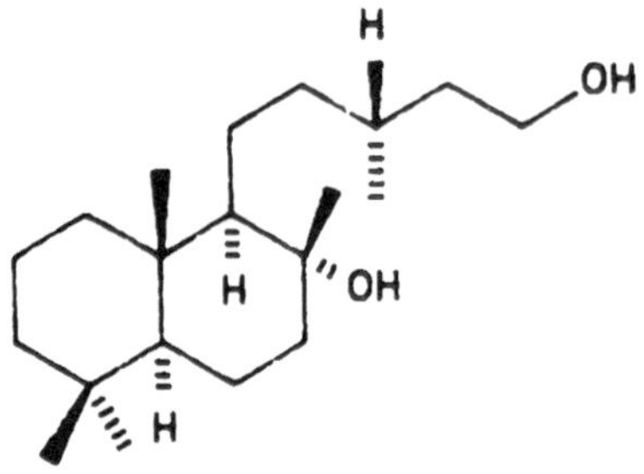

1139

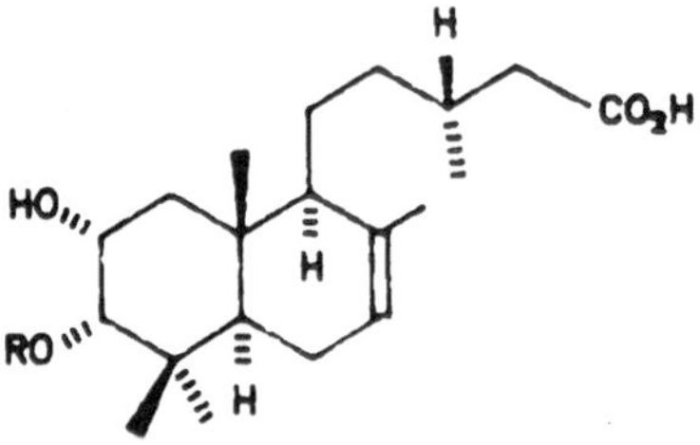

124   R = H

125   R = Mebut-2-OH

126   R = Ang

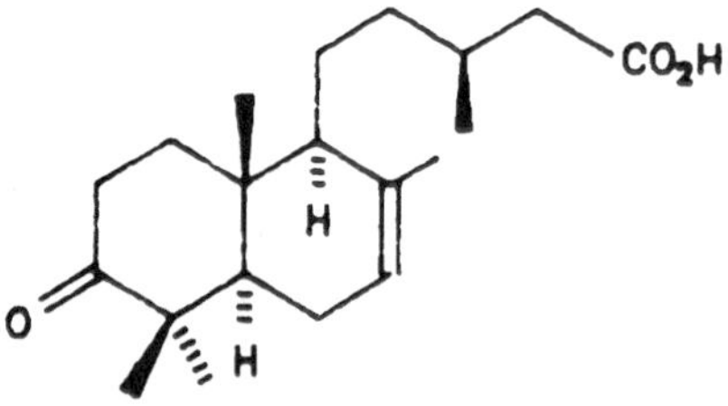

129

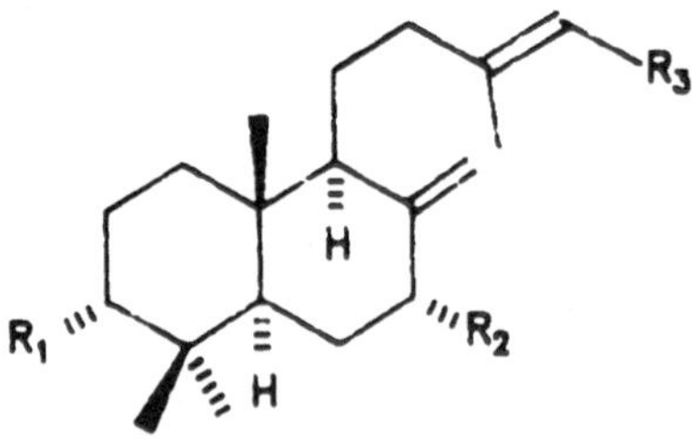

76 $R_1 = R_2 = H$, $R_3 = CH_2OH$

77 $R_1 = OH$, $R_2 = O\beta$-Gal, $R_3 = CH_2OH$

111 $R_1 = R_2 = H$, $R_3 = CO_2H$

FIGURE 1. (*contd.*)

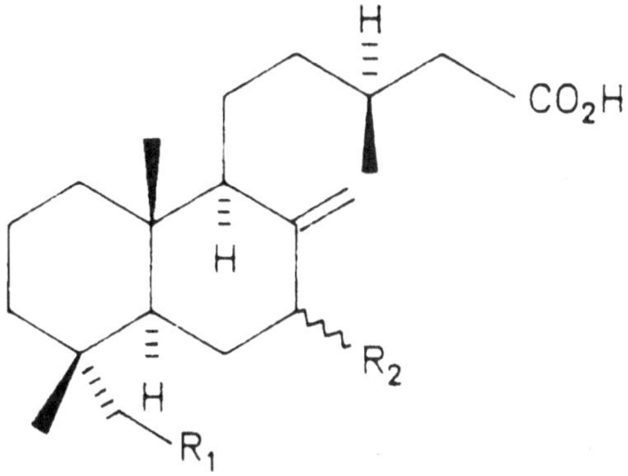

Figure 1. Structures of Diterpenes   239

1140 R_1 = OH, R_2 = α-OH

1141 R_1 = OAc, R_2 = α-OH

1142 R_1 = OH, R_2 = β-OH

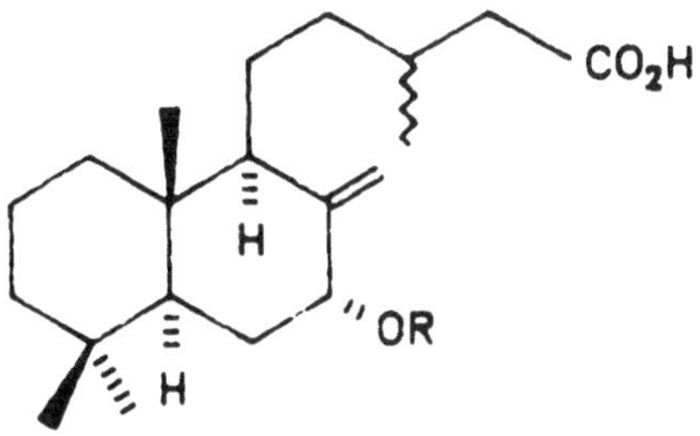

1143 R = H

1144 R = Ac

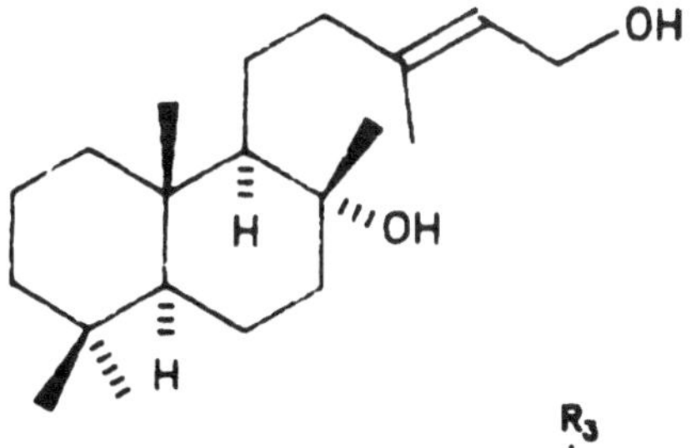

1103

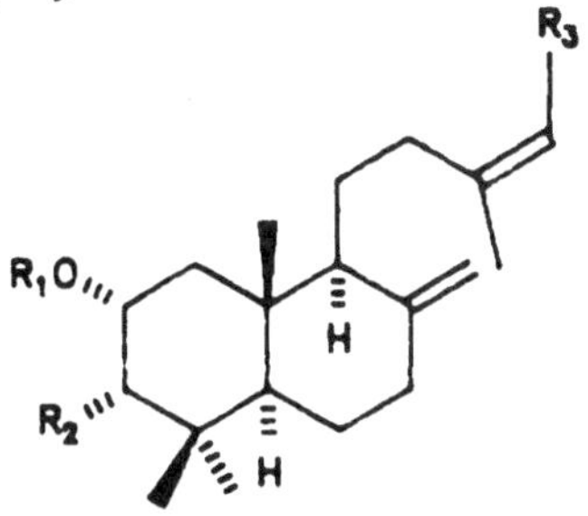

78 R_1 = H, R_2 = H, R_3 = CH$_2$OH

FIGURE 1. (*contd.*)

108 R_1 = H, R_2 = OAng, R_3 = CO_2H

109 R_1 = Ang, R_2 = OH, R_3 = CO_2H

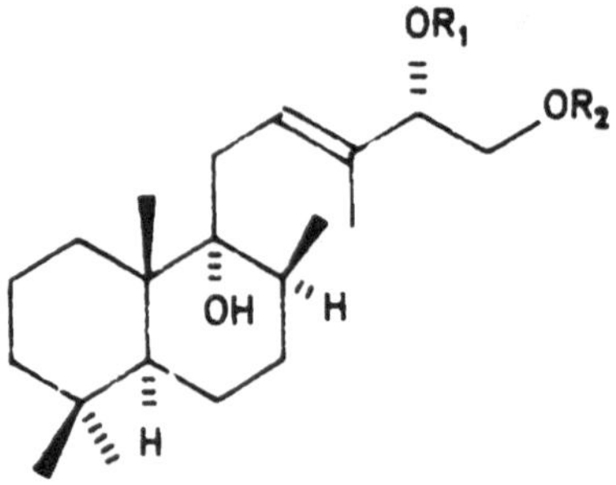

82 R_1 = R_2 = H

83 R_1 = H, R_2 = Ac (Acetylation product)

84 R_1 = R_2 = Ac (Acetylation product)

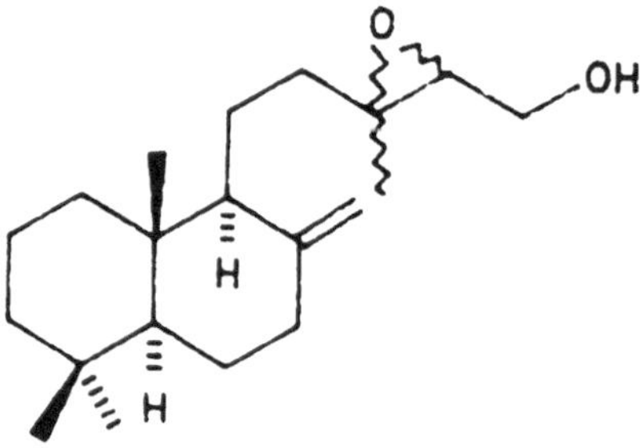

85

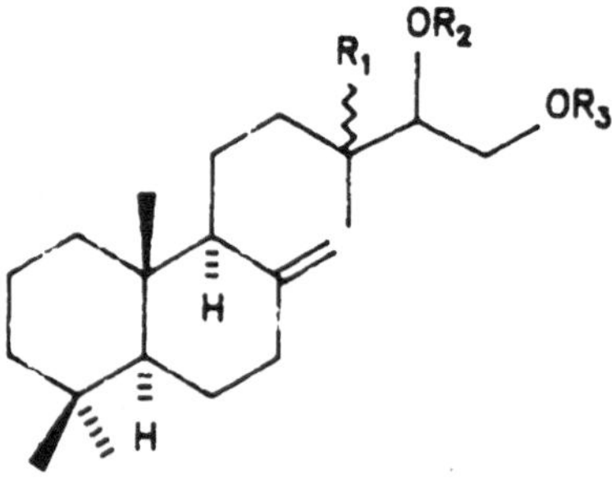

88 R_1 = R_2 = R_3 = H

89 R_1 = OH, R_2 = R_3 = H

90 R_1 = R_2 = H, R_3 = Ac

91 R_1 = OH, R_2 = H, R_3 = Ac

92 R_1 = H, R_2 = Ac, R_3 = H

93 R_1 = OH, R_2 = Pro, R_3 = Ac

FIGURE 1. (*contd.*)

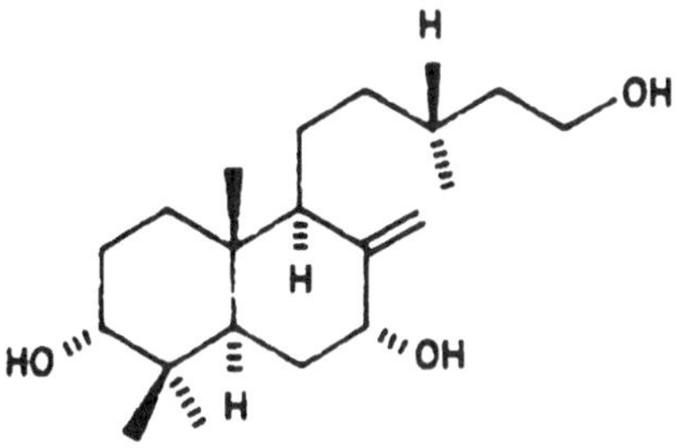

Figure 1. Structures of Diterpenes   241

995

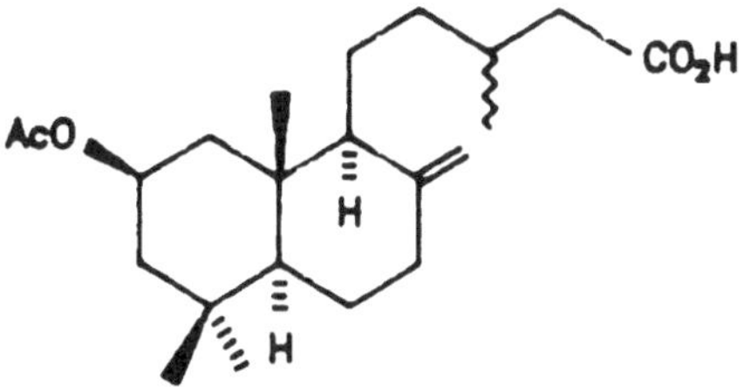

959

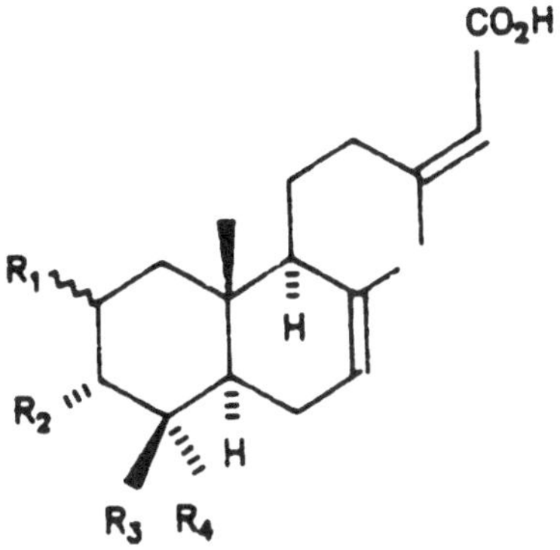

100 R_1 = OAng, R_2 = H, R_3 = R_4 = Me

116 R_1 = H, R_2 = OH, R_3 = R_4 = Me

101 R_1 = R_2 = H, R_3 = CH$_2$OAng, R_4 = Me

102 R_1 = R_2 = H, R_3 = CH$_2$OTig, R_4 = Me

103 R_1 = α-OH, R_2 = OAng, R_3 = R_4 = Me

104 R_1 = α-OH, R_2 = O-trans-Cinn, R_3 = R_4 = Me

105 R_1 = α-OH, R_2 = O-cis-Cinn, R_3 = R_4 = Me

106 R_1 = α-OH, R_2 = OAng, R_3 = R_4 = Me

FIGURE 1. (*contd.*)

107 R_1 = α-OAng, R_2 = OH, R_3 = R_4 = Me

110 R_1 = R_2 = H, R_3 = Me, R_4 = CO_2H

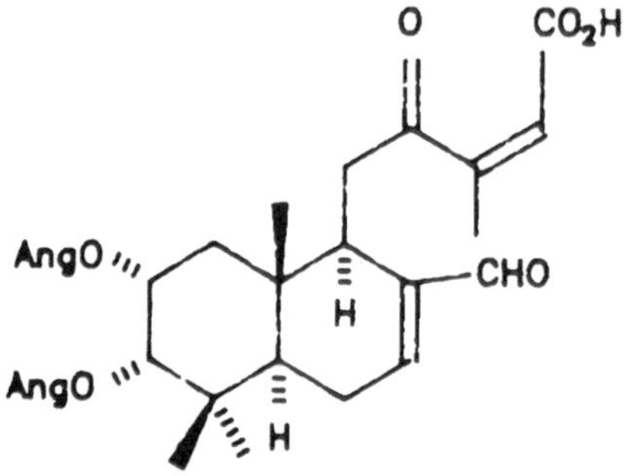

97

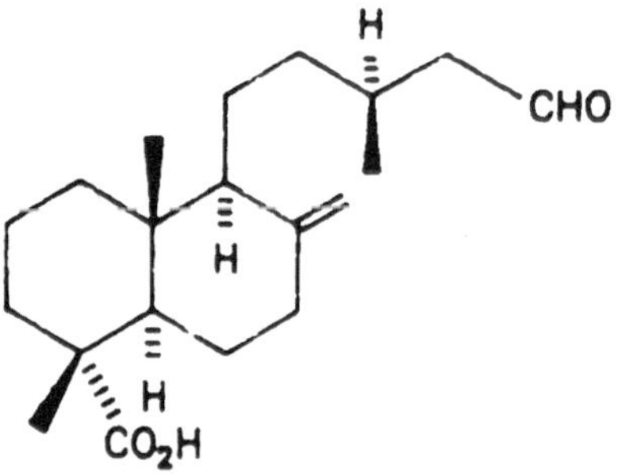

977

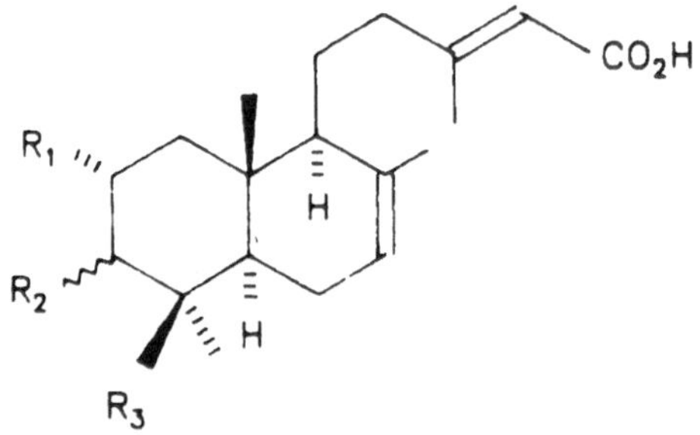

112 R_1 = OTig, R_2 = H, R_3 = Me

113 R_1 = H, R_2 = β-OAc, R_3 = Me

114 R_1 = H, R_2 = α-OH, R_3 = Me

115 R_1 = H, R_2 = β-OH, R_3 = Me

130 R_1 = R_2 = H, R_3 = CO_2H

FIGURE 1. (*contd.*)

Figure 1. Structures of Diterpenes 243

117

98 R_1 = OAng or OMebut-2-OH, R_2 = OAng or OMebut-2-OH,

       R_3 = O-trans-Cinn

99 R_1 = OAng or OMebut-2-OH, R_2 = OAng or OMebut-2-OH,

       R_3 = O-cis-Cinn

1024 R_1 = R_2 = R_3 = H

118 R_1 = R_2 = H, R_3 = CO_2H, R_4 = Me

119 R_1 = H, R_2 = α-OH, R_3 = CO_2H, R_4 = Me (=**991**?)

120 R_1 = H, R_2 = β-OH, R_3 = CO_2H, R_4 = Me

FIGURE 1. (*contd.*)

122 $R_1 = R_2 = H$, $R_3 = CO_2H$, $R_4 = CH_2OAng$

123 $R_1 = R_2 = H$, $R_3 = CO_2H$, $R_4 = CH_2OTig$

127 $R_1 = OH$, $R_2 = \alpha$-O-trans-Cinn, $R_3 = CO_2H$, $R_4 = Me$

128 $R_1 = OH$, $R_2 = \alpha$-O-cis-Cinn, $R_3 = CO_2H$, $R_4 = Me$

131 $R_1 = R_2 = H$, $R_3 = R_4 = CO_2H$

870 $R_1 = OSen$, $R_2 = \beta$-OAc, $R_3 = CO_2Me$, $R_4 = Me$

871 $R_1 = OH$, $R_2 = H$, $R_3 = CH_2OH$, $R_4 = Me$

872 $R_1 = OH$, $R_2 = H$, $R_3 = CH_2OCoum$, $R_4 = Me$

873 $R_1 = OH$, $R_2 = H$, $R_3 = CH_2O3$-Me-Coum, $R_4 = Me$

874 $R_1 = OH$, $R_2 = H$, $R_3 = CH_2O$-p-OH-Cinn, $R_4 = Me$

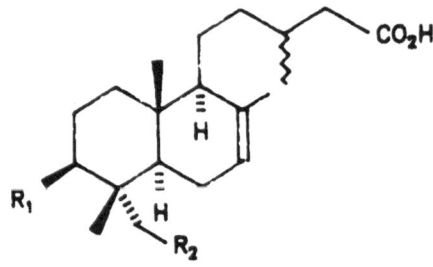

1145 $R_1 = H$, $R_2 = OH$

1146 $R_1 = OH$, $R_2 = H$ (=120)

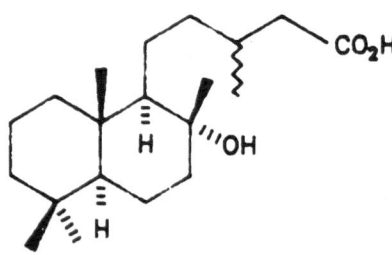

121   (=958?)

132 $R_1 = R_2 = H$, $R_3 = R_4 = Me$

FIGURE 1. (*contd.*)

Figure 1. Structures of Diterpenes   245

133 $R_1 = \alpha$-OH, $R_2 = H$, $R_3 = R_4 = Me$

134 $R_1 = \beta$-OH, $R_2 = H$, $R_3 = R_4 = Me$

135 $R_1 = \beta$-OAc, $R_2 = H$, $R_3 = R_4 = Me$

137 $R_1 = R_2 = H$, $R_3 = CO_2H$, $R_4 = Me$

138 $R_1 = H$, $R_2 = \alpha$-OAc, $R_3 = CO_2H$, $R_4 = Me$

1104 $R_1 = R_2 = H$, $R_3 = Me$, $R_4 = CO_2Me$

1147

991 R = CO_2H (=119?)

992 R = CO_2Me

993 R = CH_2OH

994 R = CH_2O2-Mebut

136

FIGURE 1. (*contd.*)

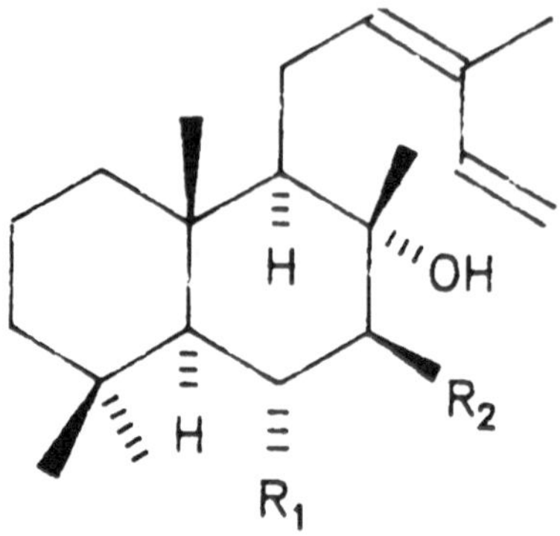

139 $R_1 = R_2 = H$

140 $R_1 = H$, $R_2 = OH$

1021 $R_1 = H$, $R_2 = OAc$

1023 $R_1 = R_2 = OH$

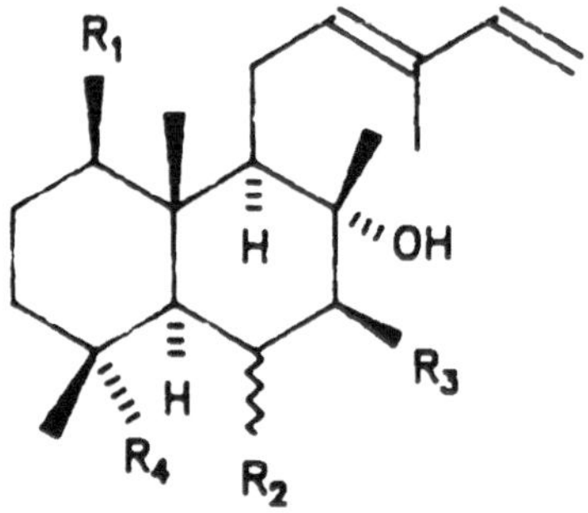

141 $R_1 = R_2 = H$, $R_3 = OH$, $R_4 = Me$

142 $R_1 = H$, $R_2 = \alpha\text{-OH}$, $R_3 = OH$, $R_4 = Me$

143 $R_1 = H$, $R_2 = \alpha\text{-OAng}$, $R_3 = OH$, $R_4 = Me$

144 $R_1 = H$, $R_2 = \beta\text{-OH}$, $R_3 = H$, $R_4 = Me$

145 $R_1 = R_2 = H$, $R_3 = OAc$, $R_4 = Me$

146 $R_1 = H$, $R_2 = \beta\text{-OH}$, $R_3 = OH$, $R_4 = Me$

147 $R_1 = H$, $R_2 = \beta\text{-OH}$, $R_3 = H$, $R_4 = CH_2OH$

148 $R_1 = H$, $R_2 = \beta\text{-OH}$, $R_3 = OAc$, $R_4 = Me$

149 $R_1 = OAc$, $R_2 = R_3 = H$, $R_4 = CH_2OArac$

150 $R_1 = OAc$, $R_2 = R_3 = H$, $R_4 = CH_2OCoum$

151 $R_1 = OAc$, $R_2 = R_3 = H$, $R_4 = CH_2O\text{-}cis\text{-}Coum$

1022 $R_1 = H$, $R_2 = \alpha\text{-OH}$, $R_3 = OAc$, $R_4 = Me$

FIGURE 1. (*contd.*)

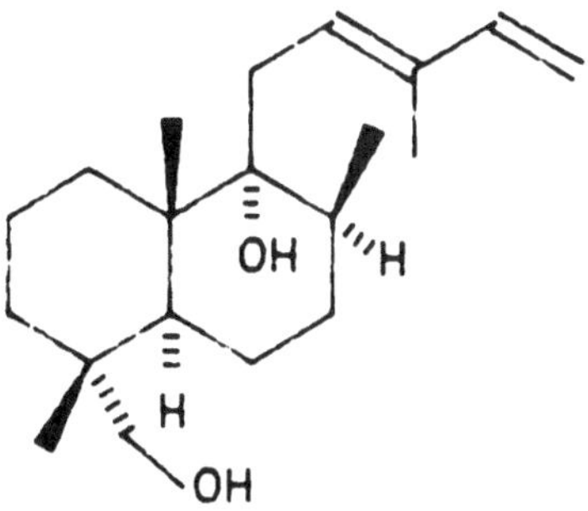

Figure 1. Structures of Diterpenes 247

152

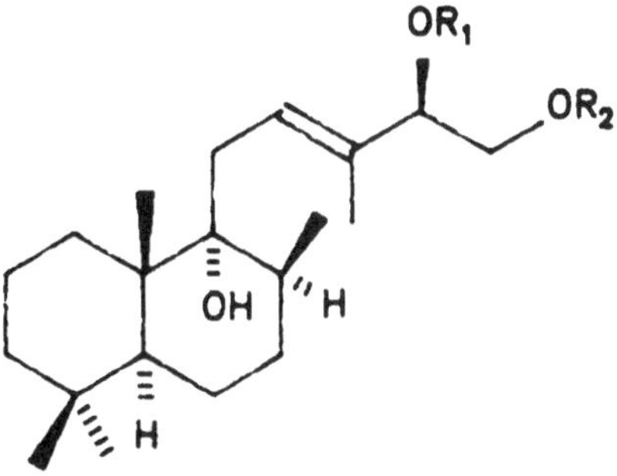

94 $R_1 = R_2 = H$

95 $R_1 = H$, $R_2 = Ac$ (Acetylation product)

96 $R_1 = R_2 = Ac$ (Acetylation product)

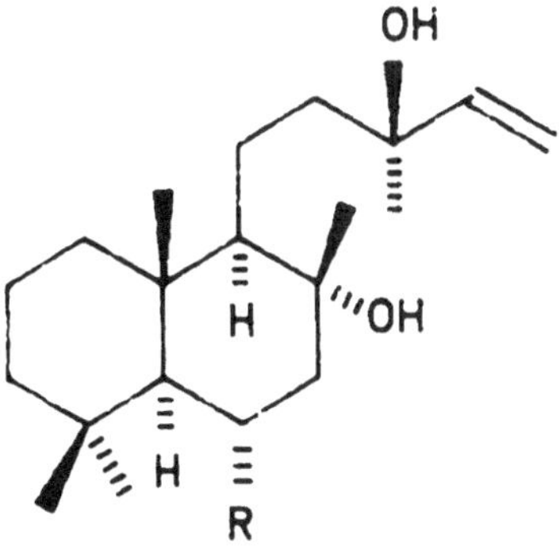

153 $R = \alpha$-OAng

1102 $R = H$

FIGURE 1. (*contd.*)

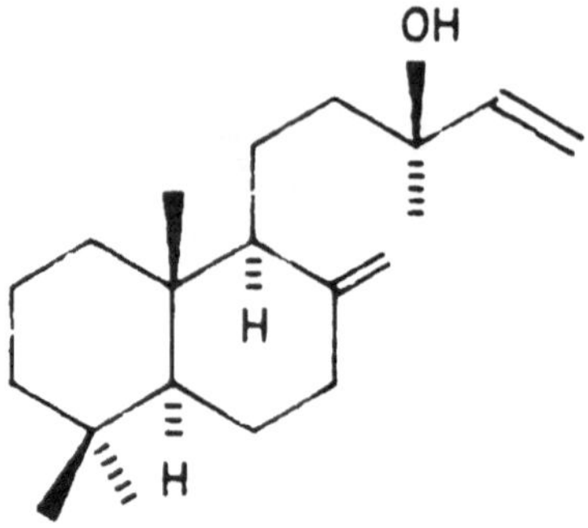

996

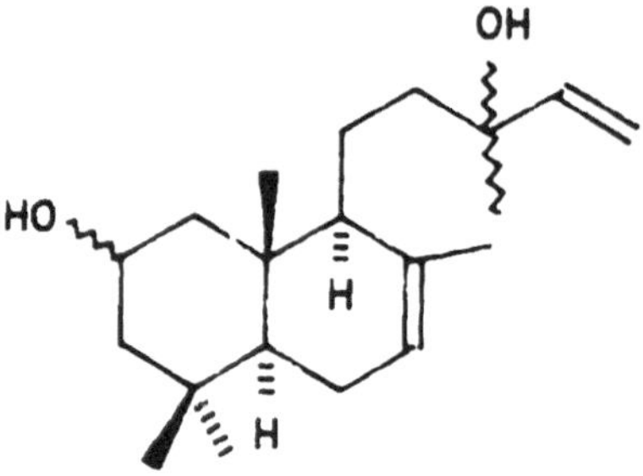

154

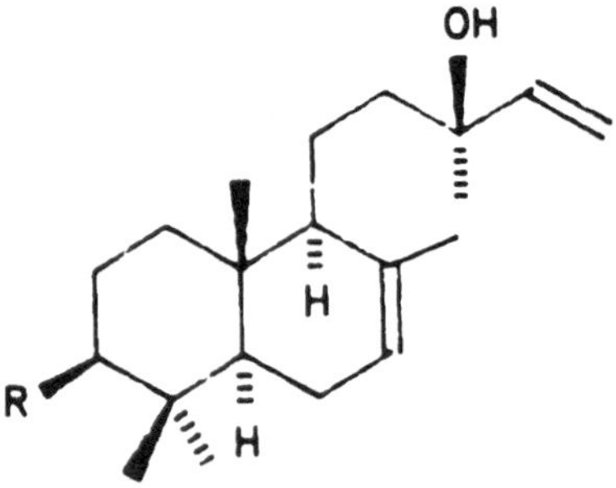

1105 R = H

1106 R = OH

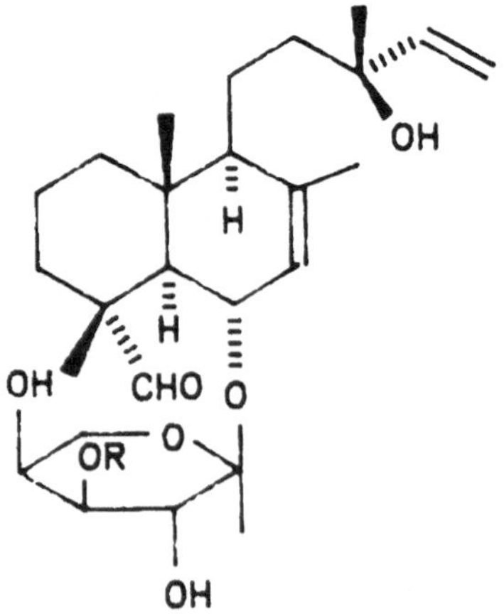

95   R = Tig

96   R = H

FIGURE 1. (*contd.*)

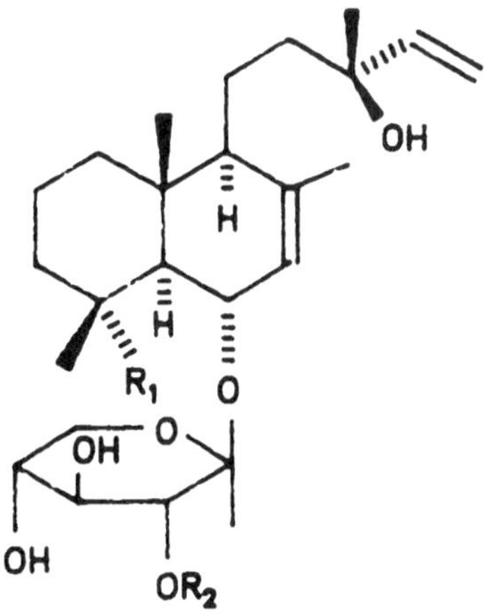

Figure 1. Structures of Diterpenes   249

1148 R_1 = CHO, R_2 = Ac
1149 R_1 = CO_2Me, R_2 = Ac
1150 R_1 = CO_2Me, R_2 = H

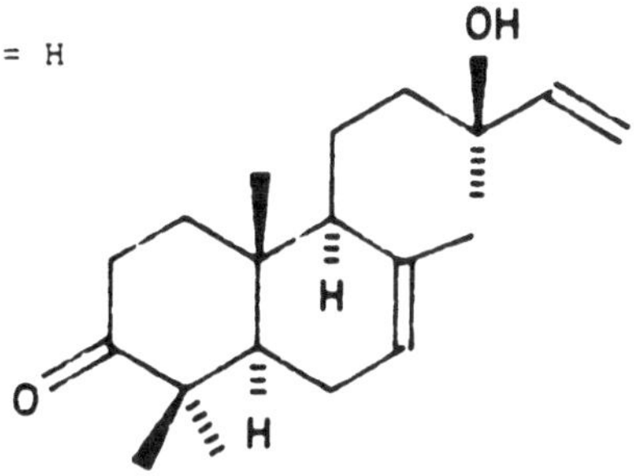

1107

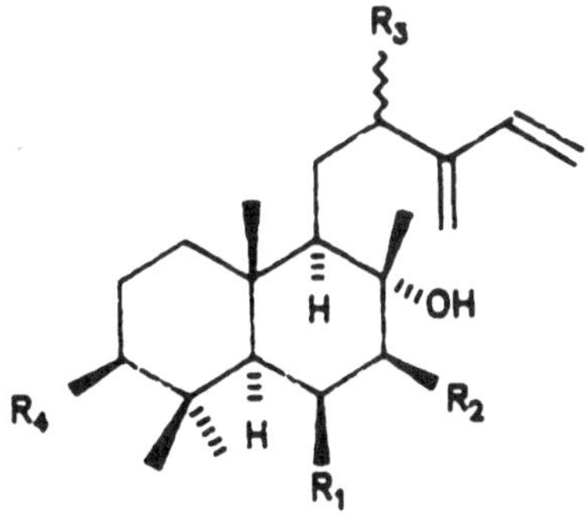

155   R_1 = R_2 = R_3 = R_4 = H
156   R_1 = OH, R_2 = OAc, R_3 = β-O-OH, R_4 = H
157   R_1 = OH, R_2 = OAc, R_3 = α-O-OH, R_4 = H
1151 R_1 = R_2 = R_3 = H, R_4 = OAc

FIGURE 1. (contd.)

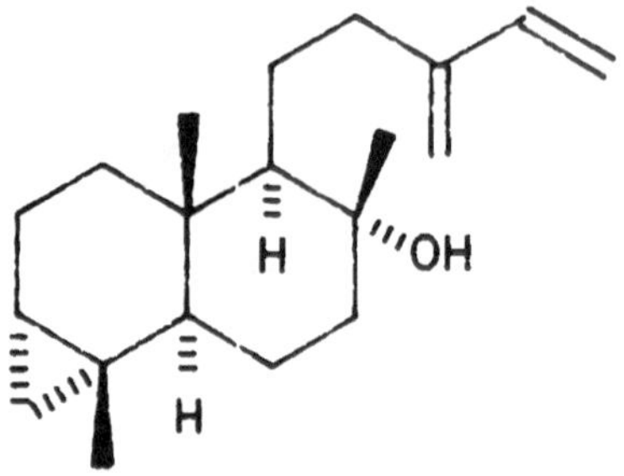

1152

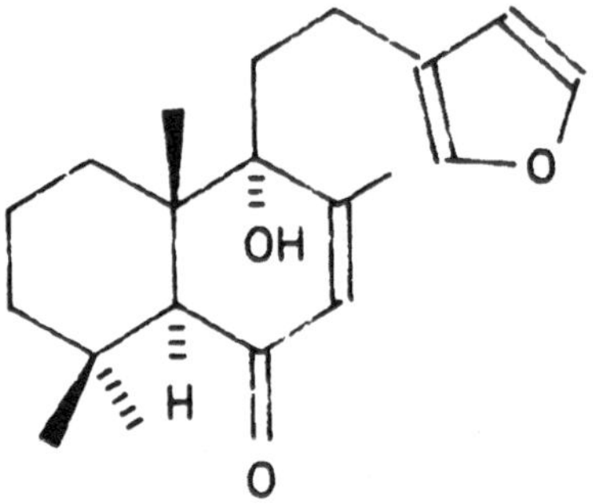

158

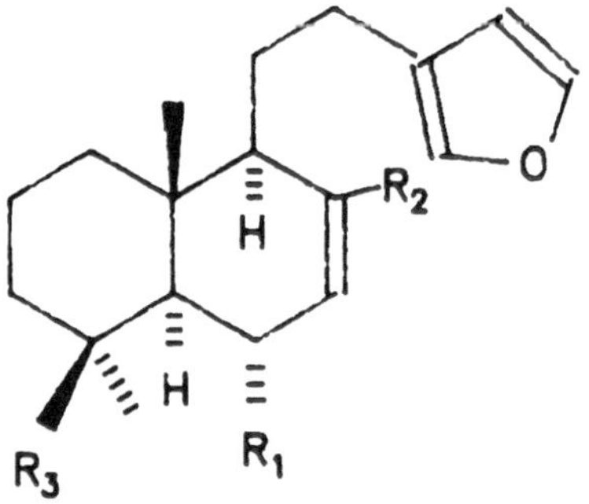

159   R_1 = H, R_2 = CH_2OH, R_3 = CO_2H

160   R_1 = O2-Mebut, R_2 = CO_2H, R_3 = Me

161   R_1 = Oi-But, R_2 = CO_2H, R_3 = Me

162   R_1 = Oi-Val, R_2 = CO_2H, R_3 = Me

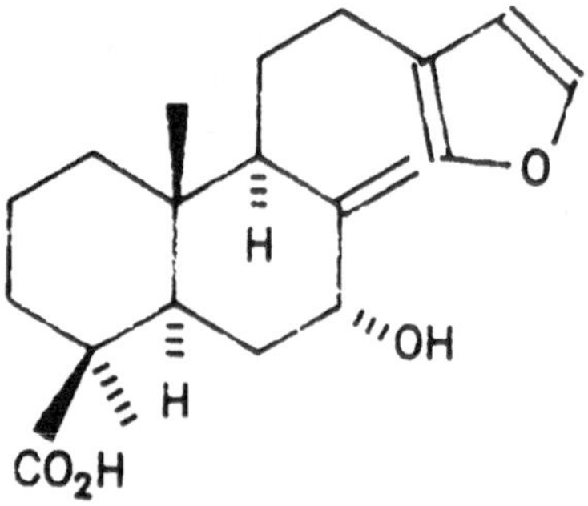

163

FIGURE 1. (*contd.*)

Figure 1. Structures of Diterpenes 251

17. Grindelane

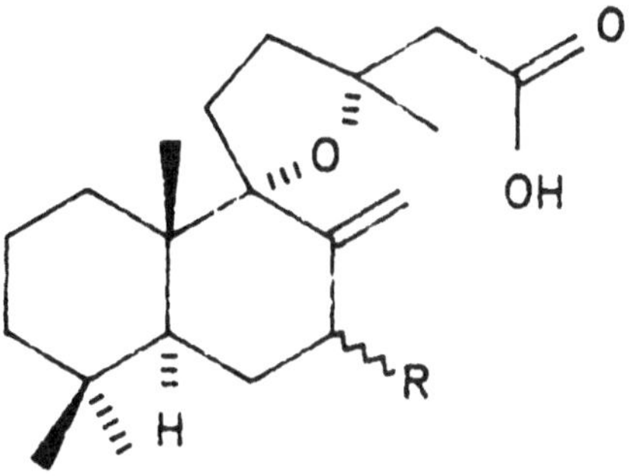

165 R = α-OH

1036 R = β-OH

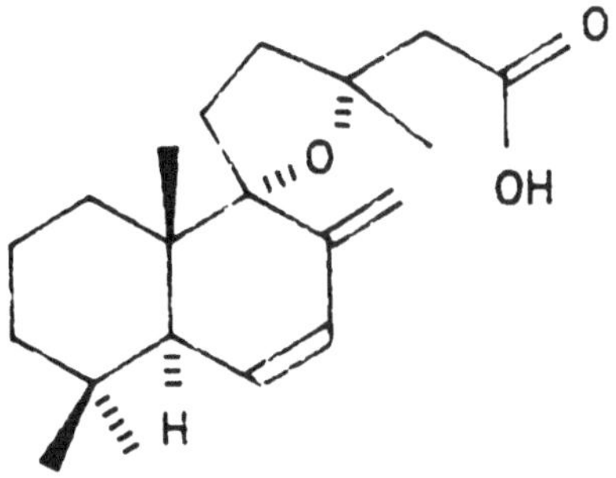

166

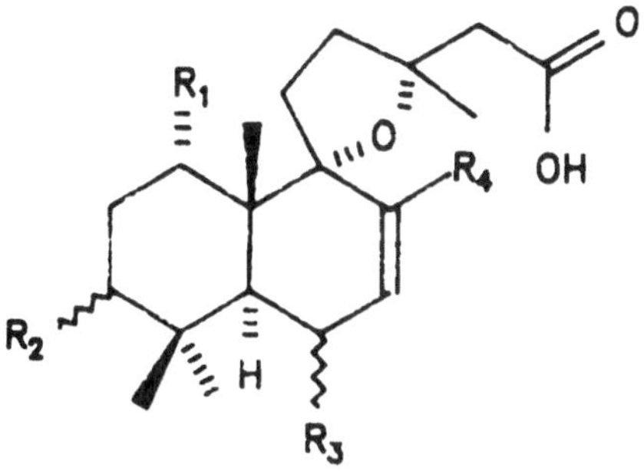

167 R$_1$ = R$_2$ = R$_3$ = H, R$_4$ = Me

168 R$_1$ = OH, R$_2$ = R$_3$ = H, R$_4$ = Me

169 R$_1$ = H, R$_2$ = β-OH, R$_3$ = H, R$_4$ = Me

170 R$_1$ = R$_2$ = H, R$_3$ = α-OH, R$_4$ = Me

171 R$_1$ = R$_2$ = H, R$_3$ = α-OCHO, R$_4$ = Me

172 R$_1$ = R$_2$ = H, R$_3$ = β-OH, R$_4$ = Me

173 R$_1$ = R$_2$ = R$_3$ = H, R$_4$ = CH$_2$OH

FIGURE 1. (*contd.*)

174 $R_1 = R_2 = R_3 = H$, $R_4 = CH_2OAc$

175 $R_1 = R_2 = R_3 = H$, $R_4 = CH_2OPro$

176 $R_1 = R_2 = R_3 = H$, $R_4 = CH_2Oi-But$

177 $R_1 = R_2 = R_3 = H$, $R_4 = CH_2Oi-Val$

178 $R_1 = R_2 = R_3 = H$, $R_4 = CH_2O2-Mebut$

179 $R_1 = R_2 = R_3 = H$, $R_4 = CH_2OMe$

1035 $R_1 = H$, $R_2 = \alpha-OH$, $R_3 = H$, $R_4 = Me$

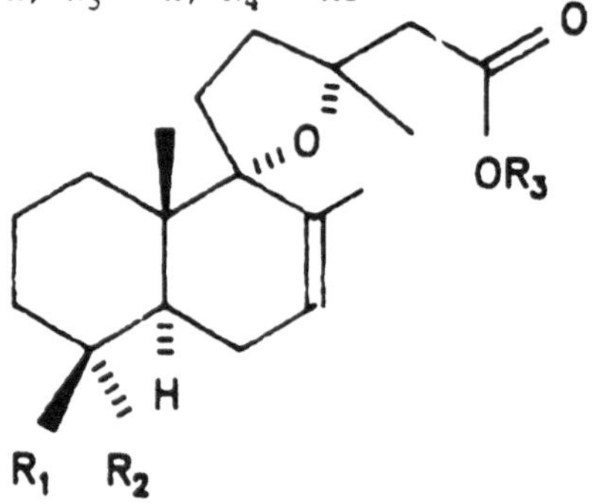

180 $R_1 = Me$, $R_2 = CH_2OH$, $R_3 = H$

181 $R_1 = Me$, $R_2 = CH_2OAc$, $R_3 = H$

182 $R_1 = Me$, $R_2 = CH_2Oi-Val$, $R_3 = H$

183 $R_1 = Me$, $R_2 = CH_2Oi-But$, $R_3 = H$

184 $R_1 = Me$, $R_2 = CH_2O2-Mebut$, $R_3 = H$

185 $R_1 = Me$, $R_2 = CH_2OSucc$, $R_3 = H$

186 $R_1 = Me$, $R_2 = CHO$, $R_3 = H$

187 $R_1 = Me$, $R_2 = CO_2H$, $R_3 = H$

188 $R_1 = CH_2OH$, $R_2 = Me$, $R_3 = H$

1153 $R_1 = Me$, $R_2 = CO_2H$, $R_3 = Me$

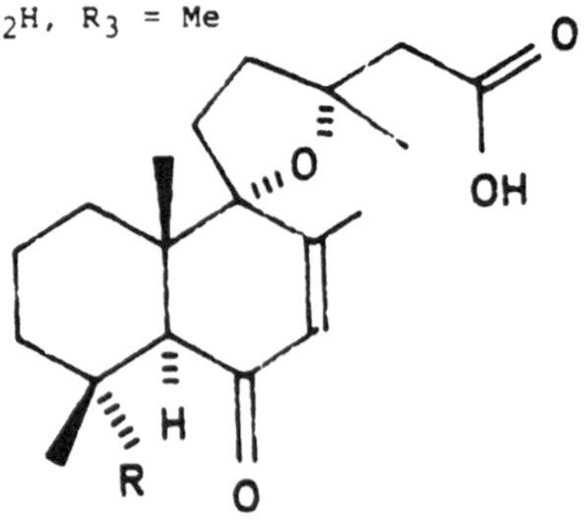

189 $R = Me$

190 $R = CH_2OAc$

FIGURE 1. (*contd.*)

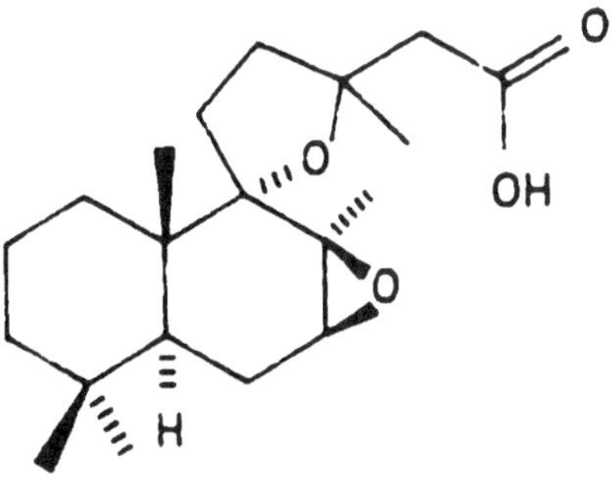

Figure 1. Structures of Diterpenes 253

191

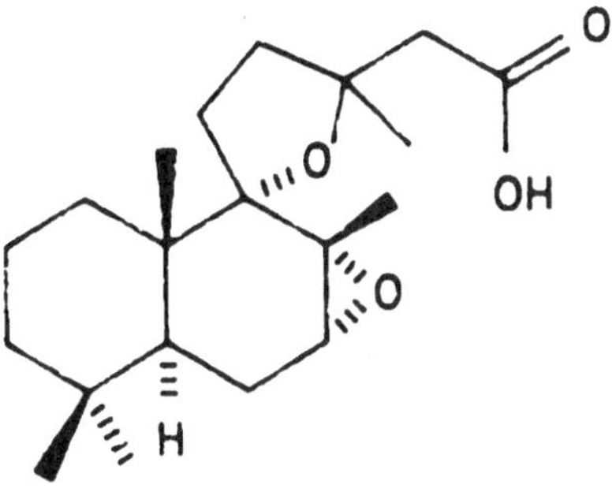

192

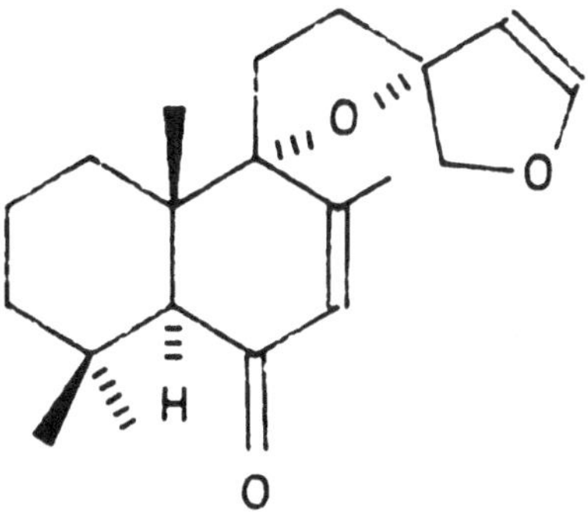

193

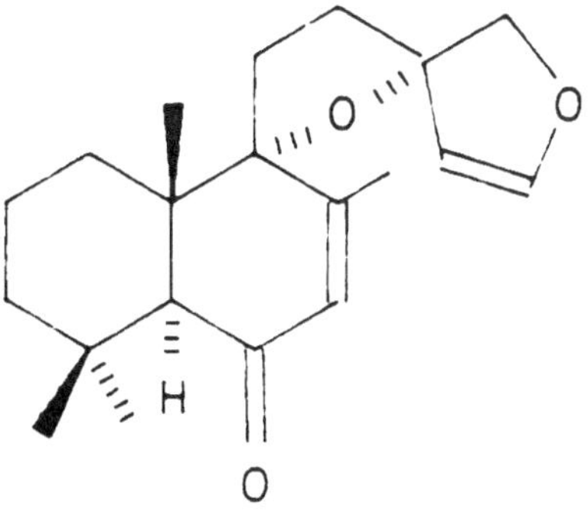

194

FIGURE 1. (contd.)

955

1060

956

1187

FIGURE 1. (*contd.*)

Figure 1. Structures of Diterpenes   255

18. Normal-Labdane-8,13-Epoxide

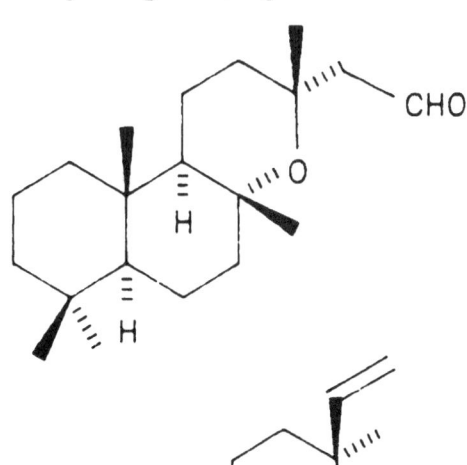

272 $R_1 = R_2 = H$, $R_3 = R_4 = Me$

195 $R_1 = R_2 = H$, $R_3 = Me$, $R_4 = CH_2OH$

196 $R_1 = R_2 = H$, $R_3 = Me$, $R_4 = CH_2OAc$

197 $R_1 = OH$, $R_2 = H$, $R_3 = Me$, $R_4 = CH_2OH$

198 $R_1 = OH$, $R_2 = H$, $R_3 = Me$, $R_4 = CH_2OAc$

199 $R_1 = OAc$, $R_2 = H$, $R_3 = Me$, $R_4 = CH_2OAc$

200 $R_1 = H$, $R_2 = OAc$, $R_3 = R_4 = Me$

201 $R_1 = H$, $R_2 = OH$, $R_3 = CH_2OBenz$, $R_4 = Me$

964

273 R = Me

202 R = CH_2OH

FIGURE 1. (*contd.*)

19. Normal-Labdane-8,12-Epoxide

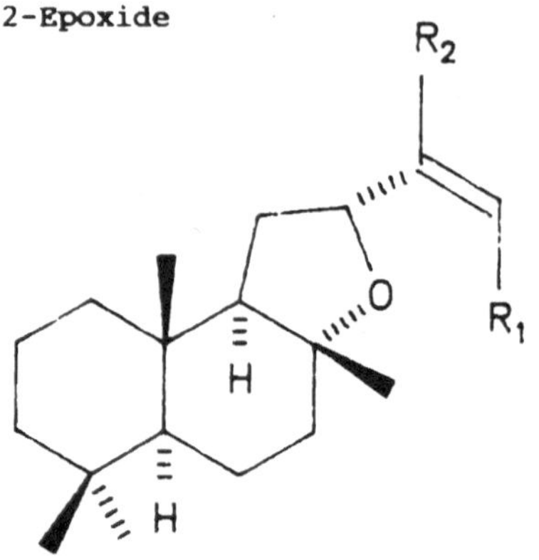

203 $R_1 = R_2 = CH_2OH$

204 $R_1 = CH_2OH$, $R_2 = CH_2OAc$

205 $R_1 = R_2 = CH_2OAc$

206 $R_1 = CH_2OH$, $R_2 = CHO$

207 $R_1 = R_2 = CHO$

210 $R1 = CH_2OH$, $R_2 = Me$

211 $R_1 = CO_2H$, $R_2 = Me$

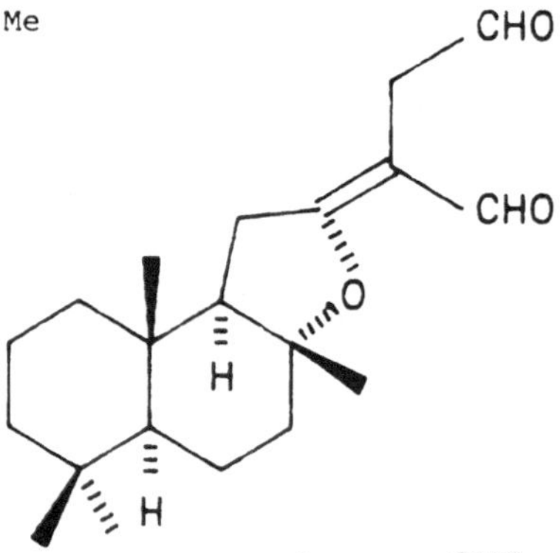

212

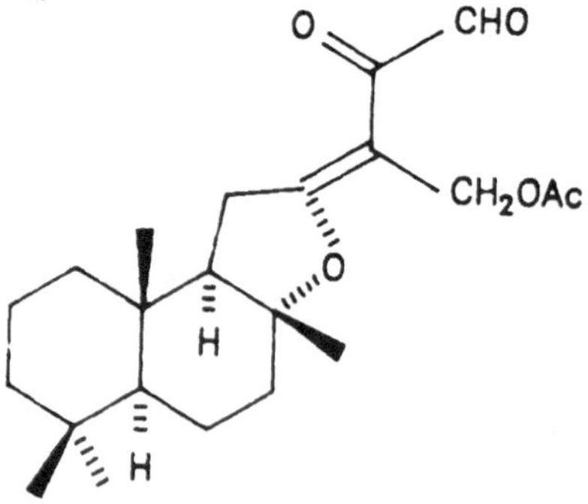

213

FIGURE 1. (*contd.*)

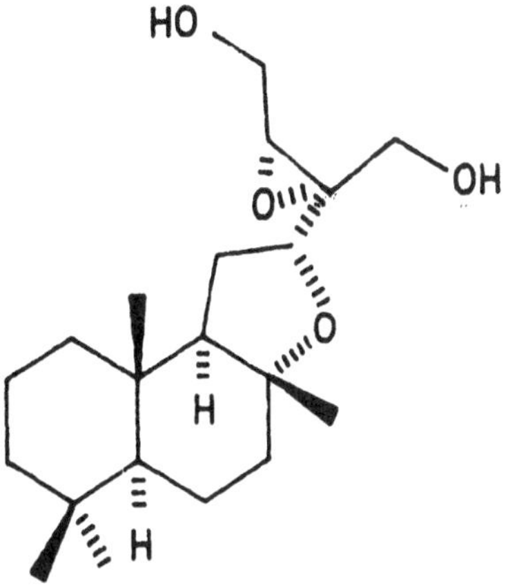

Figure 1. Structures of Diterpenes 257

208

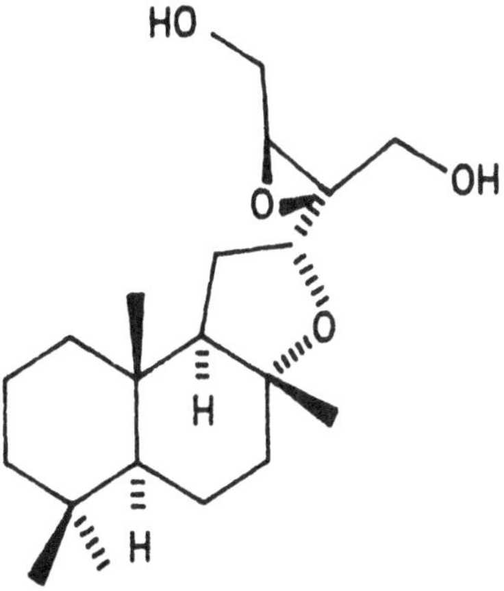

209

20. Normal-Labdane-15,16-Lactone

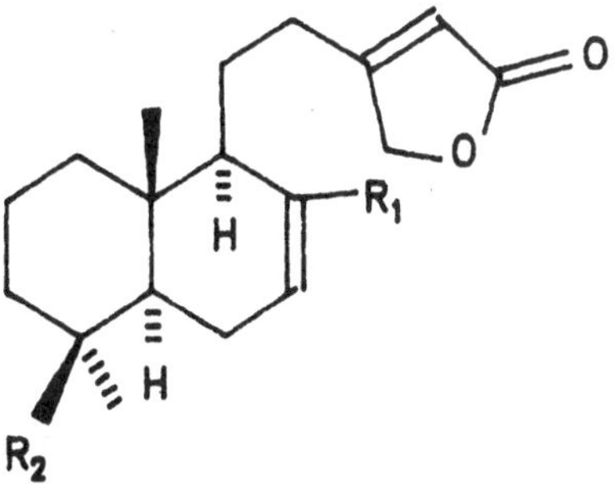

164 R_1 = Me, R_2 = CO_2H

965 R_1 = CH_2OH, R_2 = Me

966 R_1 = CHO, R_2 = Me

FIGURE 1. (*contd.*)

21. Normal-Labdane-20,12-Lactone

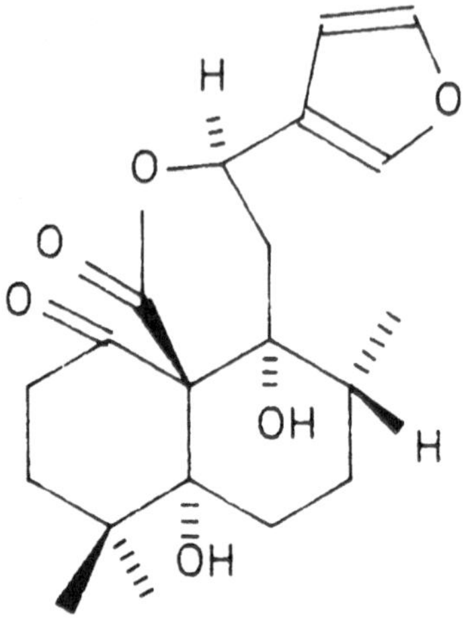

214

22. Normal-Labdane Spiroketal

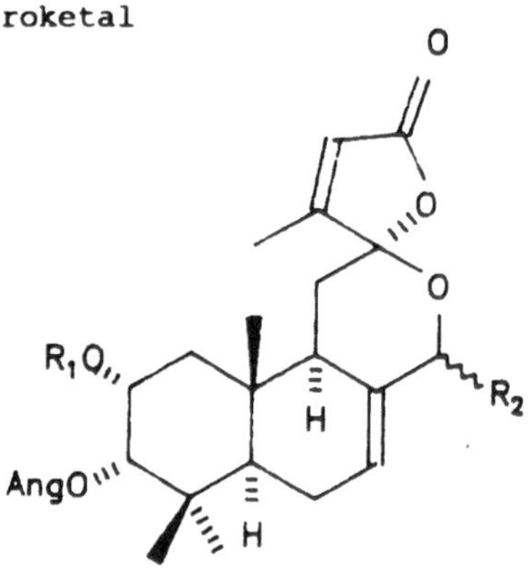

215   R_1 = Ang, R_2 = H

216   R_1 = Ang, R_2 = α-CH

217   R_1 = Ang, R_2 = β-OH

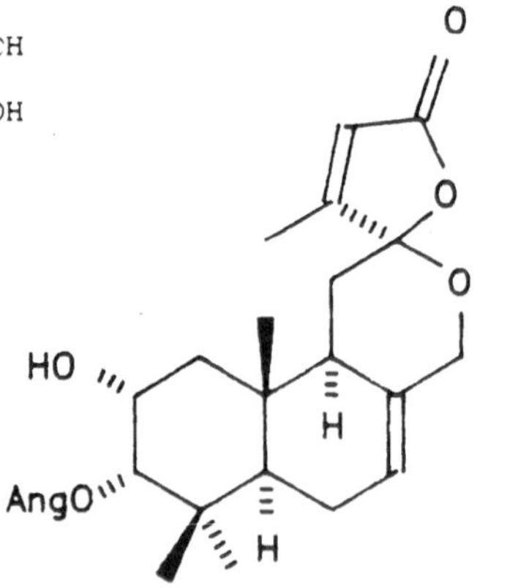

218

FIGURE 1. (*contd.*)

Figure 1. Structures of Diterpenes   259

88. Normal-Labdane-15,12-Lactone

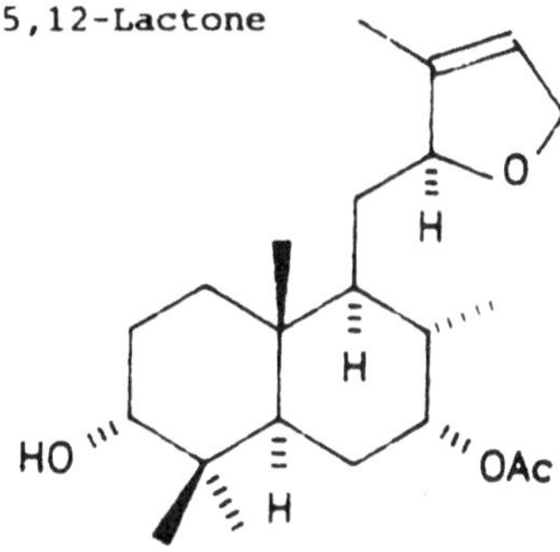

938

23. Normal-Labdane, 7,8-Seco-

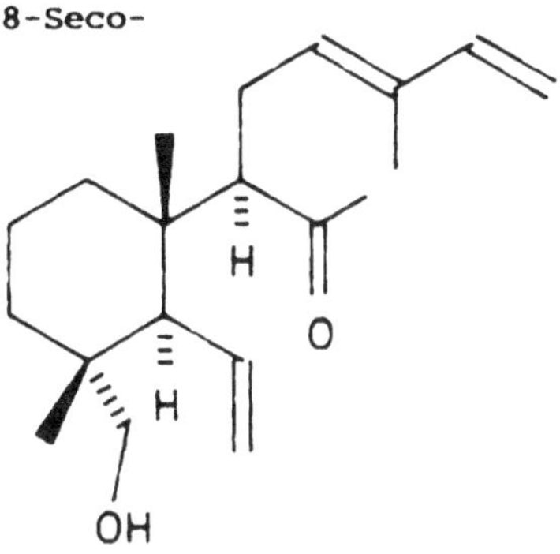

219

Normal-Labdane, 3,4-Seco-

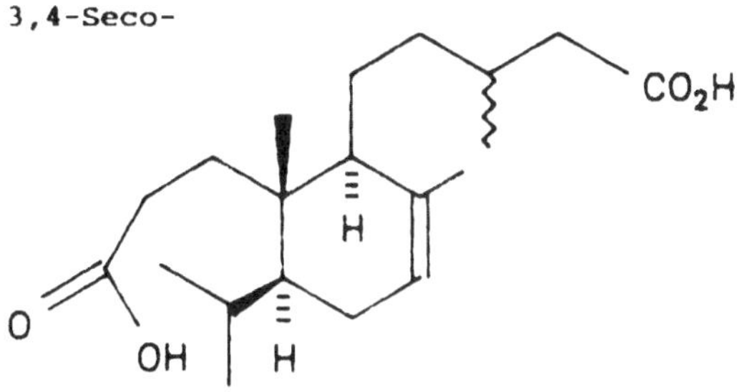

220

25. Grindelane, 8,9-Seco-

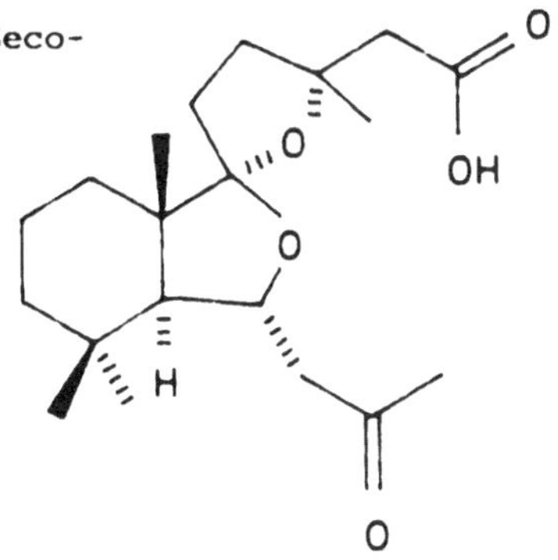

223

FIGURE 1. (contd.)

89. Grindelane, 19-Nor-

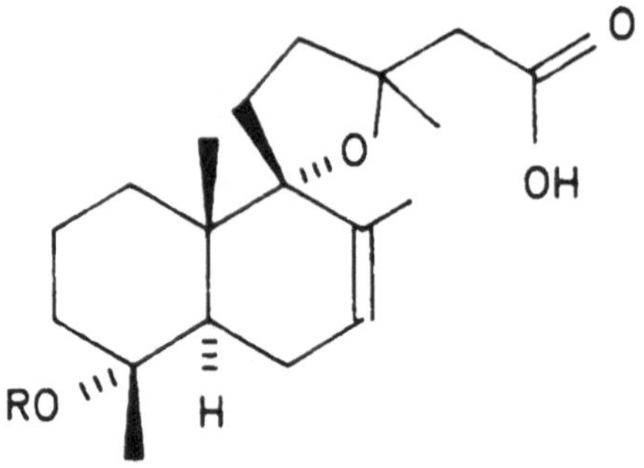

221 R = H
222 R = CHO

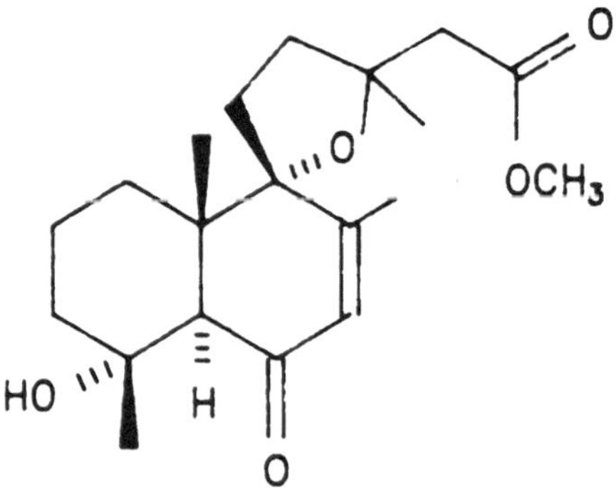

1154

26. Grindelane, 8,17-Bis-Nor-8,9-Seco-

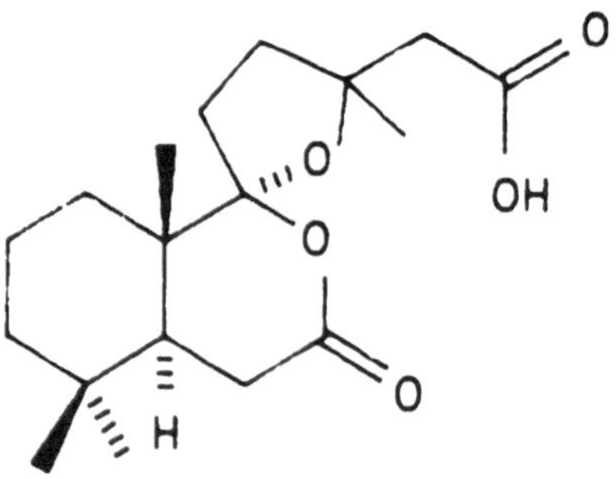

224

FIGURE 1. (*contd.*)

Figure 1. Structures of Diterpenes 261

27. Normal-Labdane-20,12-Lactone, 5,10-Seco-

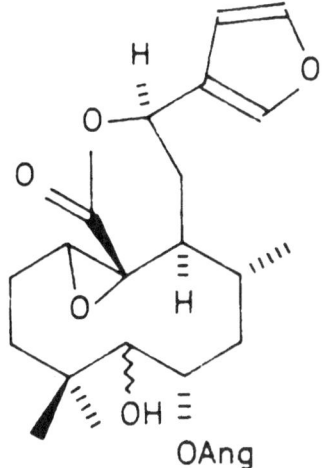

230 (Reaction product)

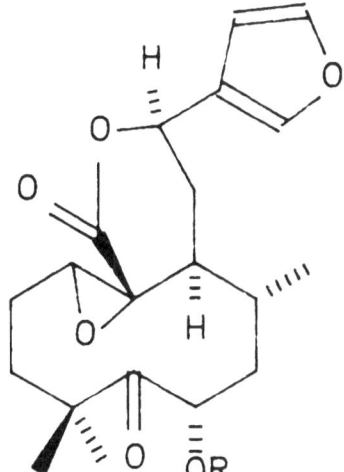

225 R = H

226 R = Ang

227 R = i-But

228 R = 2-Mebut

229 R = i-Val

FIGURE 1. (*contd.*)

100. Normal-Labdane, 14,15-Bisnor-

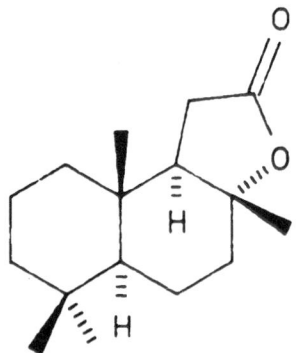

1155 $R_1 = R_2 = H$

1157 $R_1 = R_2 = OH$

1158 $R_1 = OAc,\ R_2 = OH$

1159 $R_1 = OH,\ R_2 = OAc$

1160 $R_1 = H,\ R_2 = OH$

101. Normal-Labdane, 13,14,15,16-Tetranor-

1156

87. Hexanorlabdane

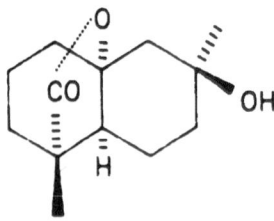

941

FIGURE 1. (*contd.*)

Figure 1. Structures of Diterpenes   263

28. Relhaniane

231

232

233

234

FIGURE 1. (*contd.*)

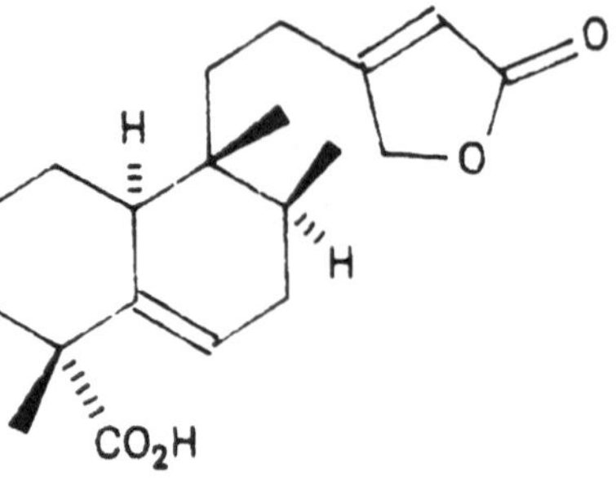

235

29. <u>trans</u>- Normal-Clerodane

   (No proven structures of this skeletal type)

30. Clerodane, 5,10-Seco-

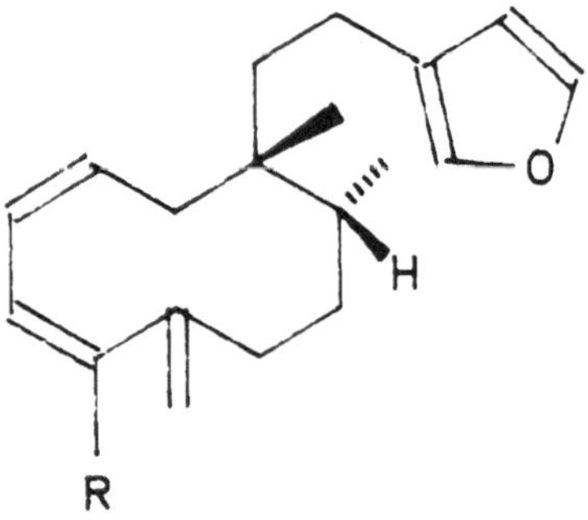

255 R = CO_2H (Revised to 9α-Me; = 573)

256 R = CO_2Me (Revised to 9α-Me; = 573 methyl derivative)

31. <u>cis</u>-Normal-Clerodane

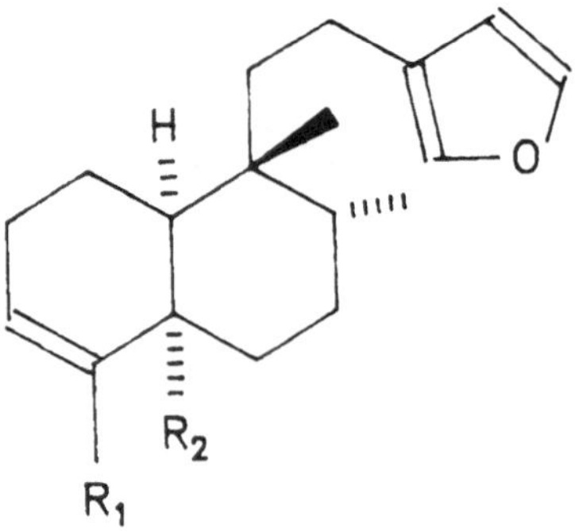

238 R_1 = Me, R_2 = CH_2OH (= **241**)

FIGURE 1. (*contd.*)

Figure 1. Structures of Diterpenes   265

239 R_1 = Me, R_2 = CO_2H

240 R_1 = Me, R_2 = CHO (= **1062**)

241 R_1 = Me, R_2 = CH_2OH (= **1064**)

242 R_1 = R_2 = CH_2OH

243 R_1 = R_2 = CHO (= **1063**)

244 R_1 = CH_2OAng, R_2 = CO_2H

1065 R_1 = CH_2OH, R_2 = Me

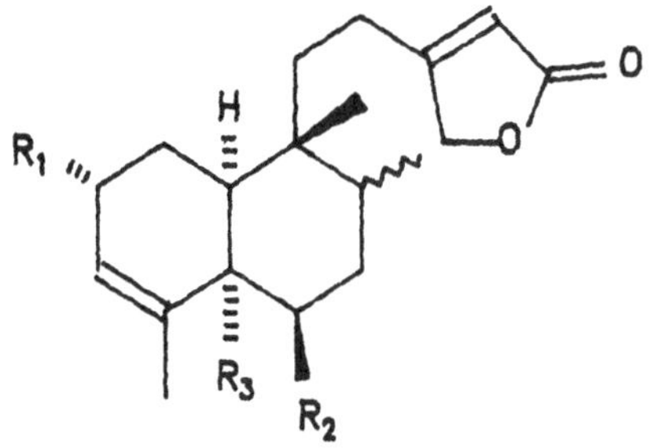

253 R_1 = R_2 = H, R_3 = CH_2OH

258 R_1 = H, R_2 = OAng, R_3 = Me

259 R_1 = H, R_2 = OTig, R_3 = Me

263 R_1 = OH, R_2 = OAng, R_3 = Me

264 R_1 = OH, R_2 = OTig, R_3 = Me

545 R_1 = H, R_2 = OH, R_3 = Me

546 R_1 = H, R_2 = OAc, R_3 = Me

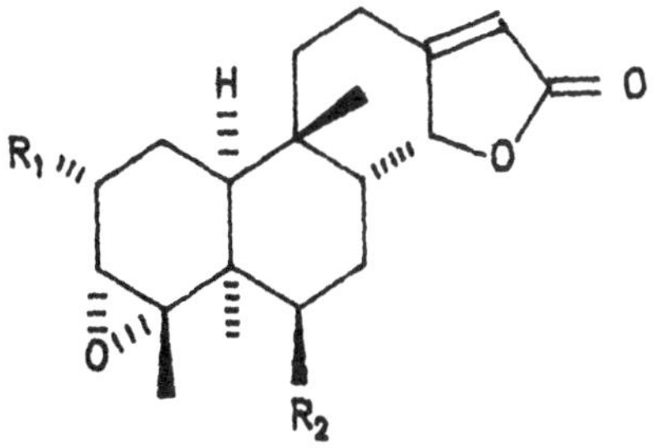

551 R_1 = OH, R_2 = H

552 R_1 = H, R_2 = OAng

FIGURE 1. (*contd.*)

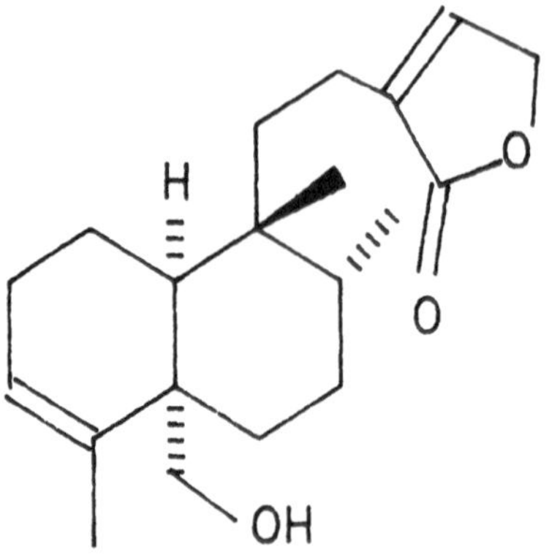

254

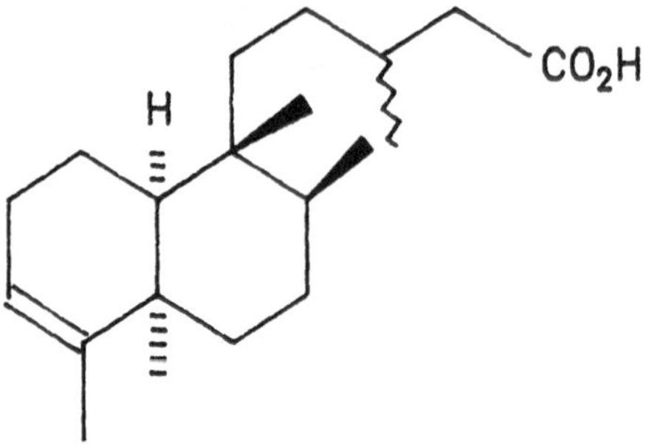

257

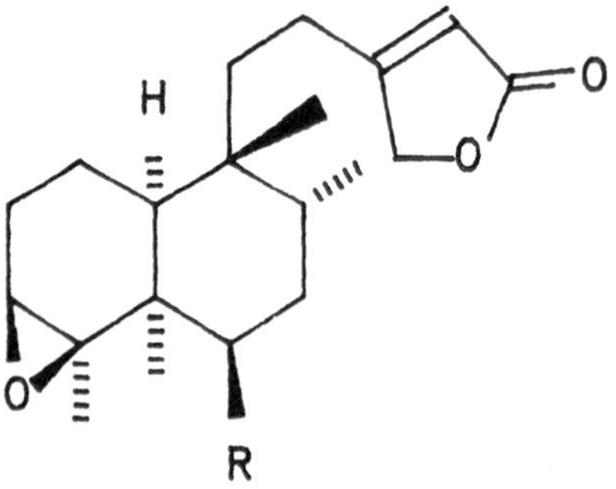

260   R = β-OAng

553   R = β-OAc

988   R = β-OTig

556   R = H

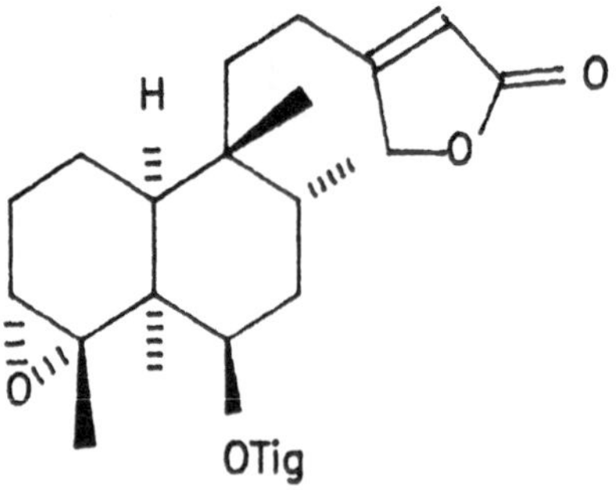

261

FIGURE 1. (*contd.*)

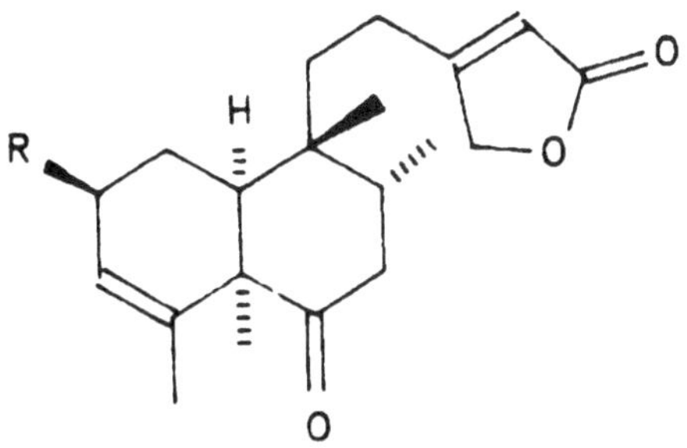

Figure 1. Structures of Diterpenes   267

262 R = H

265 R = OH

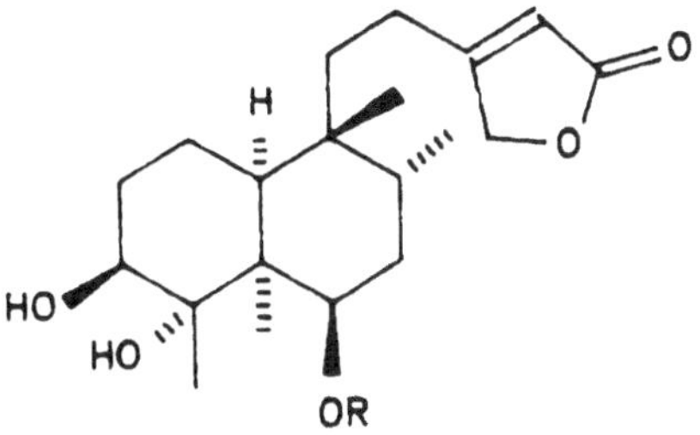

266 R = Ang

267 R = Tig

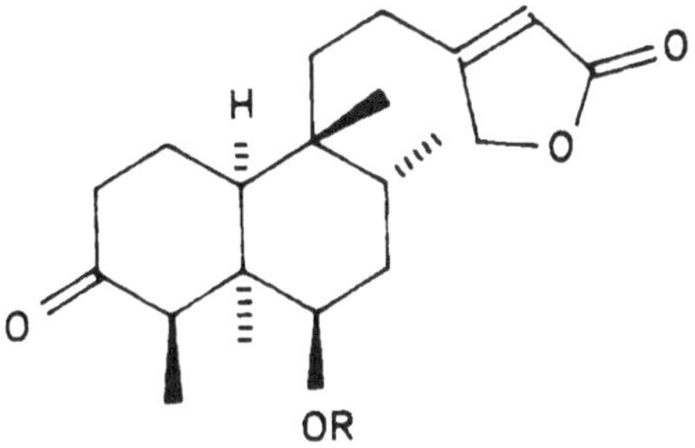

268 R = Ang

269 R = Tig

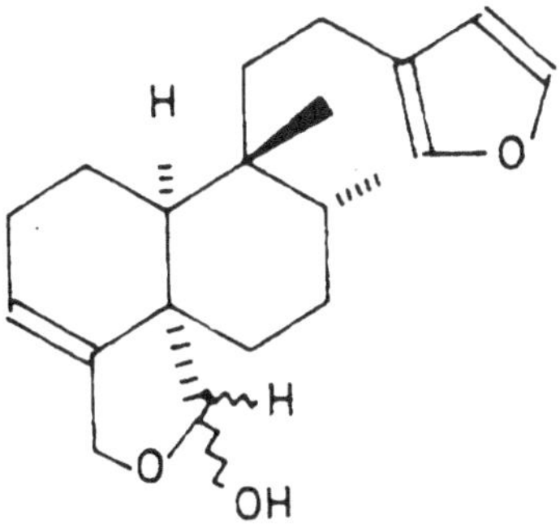

246

FIGURE 1. (*contd.*)

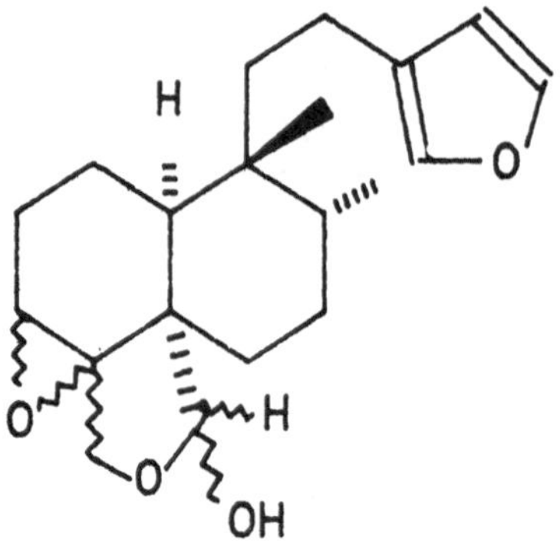

2. Compound Names, Structures and Sources

247

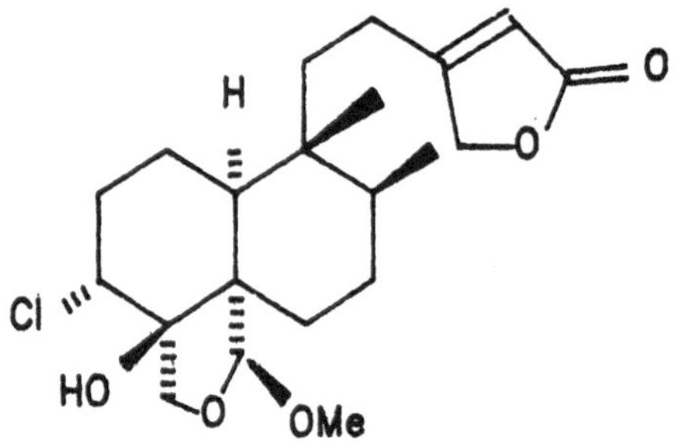

270

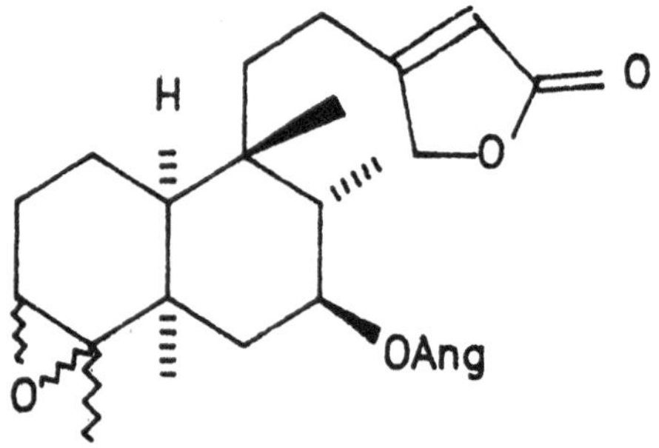

557

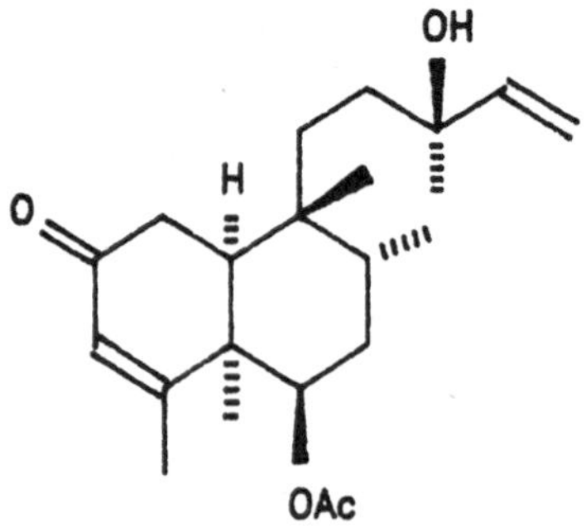

1025

FIGURE 1. (*contd.*)

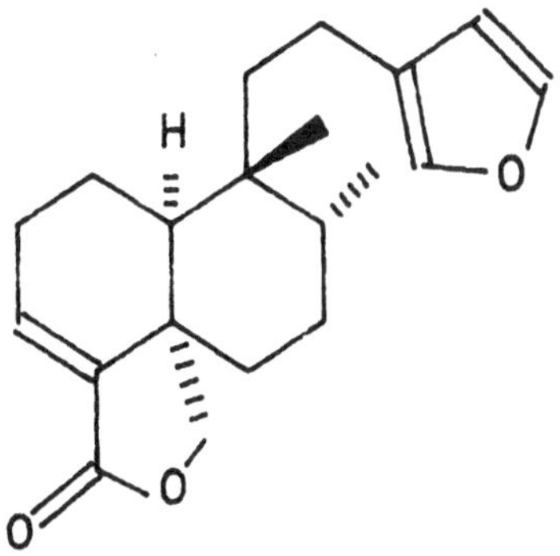

Figure 1. Structures of Diterpenes   269

1066

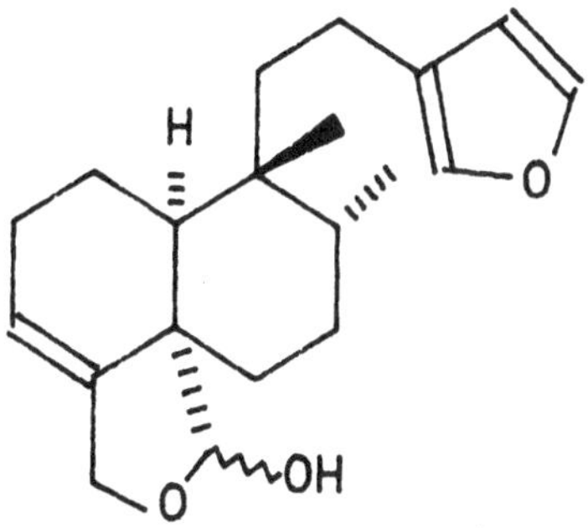

1067

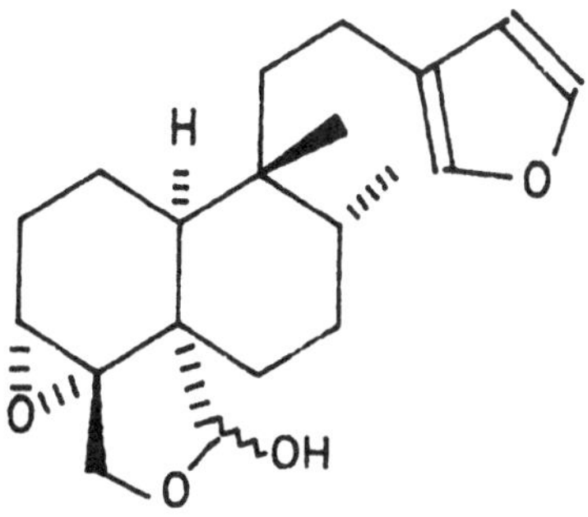

1068 (= 247?)

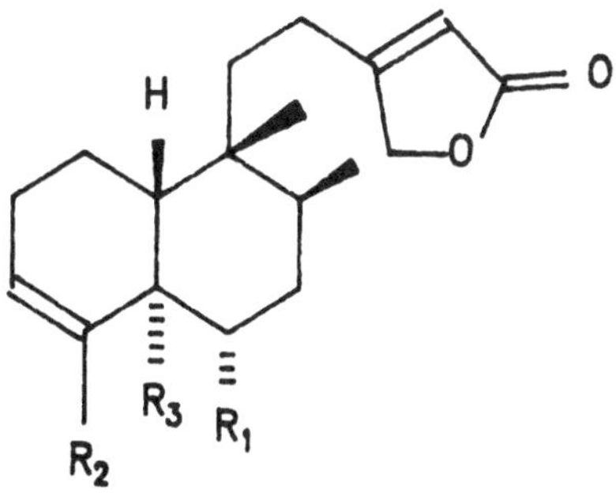

1083 R_1 = OH, R_2 = CH_2OH, R_3 = Me

1084 R_1 = H, R_2 = R_3 = CH_2OH

1091 R_1 = H, R_2 = Me, R_3 = CO_2Arab

FIGURE 1. (*contd.*)

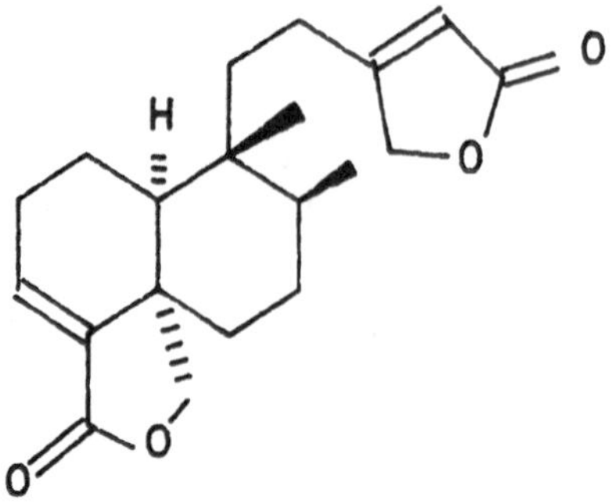

2. Compound Names, Structures and Sources

1085

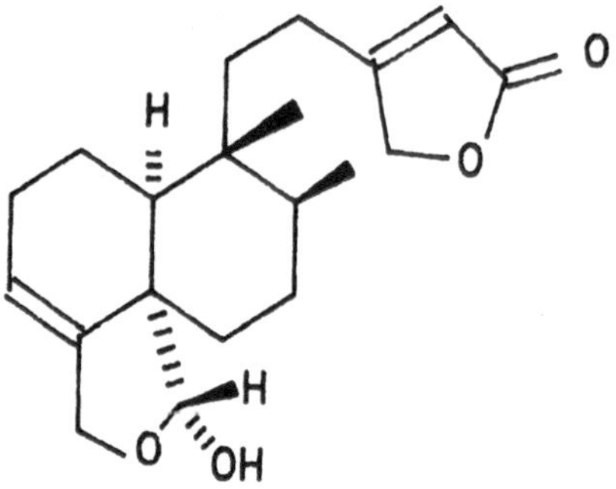

1086

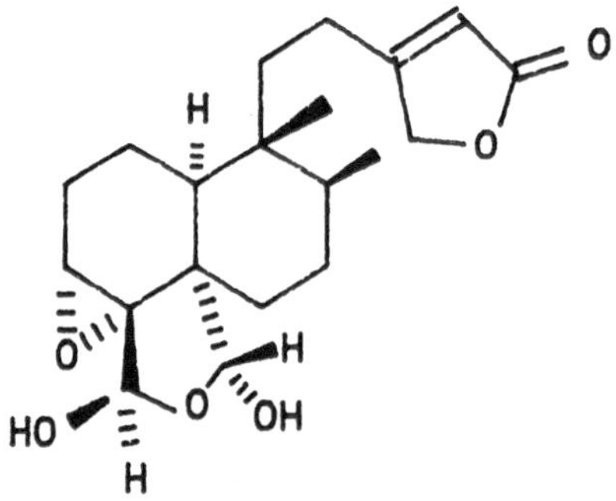

1087

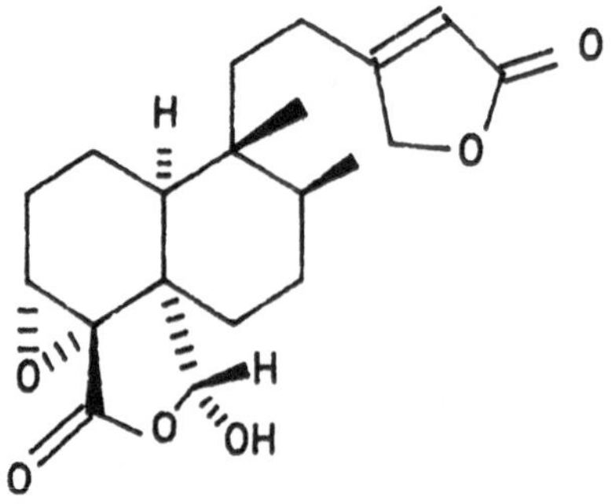

1088

FIGURE 1. (*contd.*)

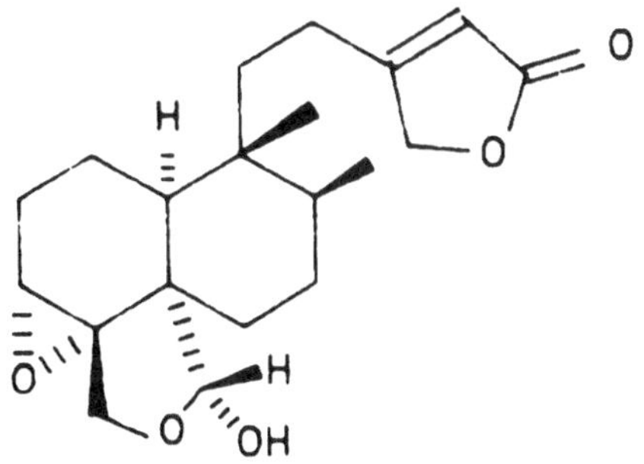

Figure 1. Structures of Diterpenes 271

1089

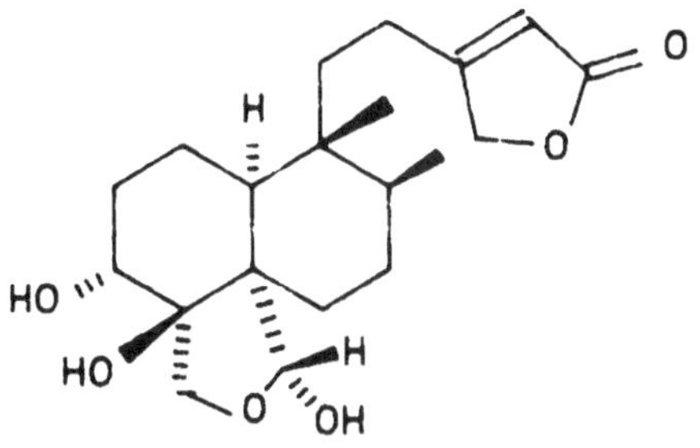

1090

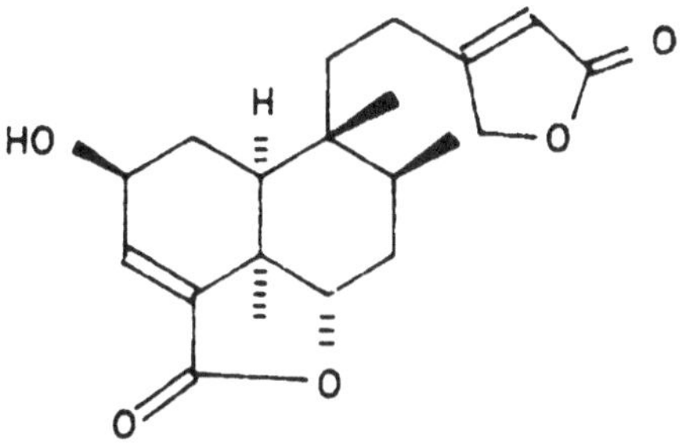

1092

90. Tricycloclerodane

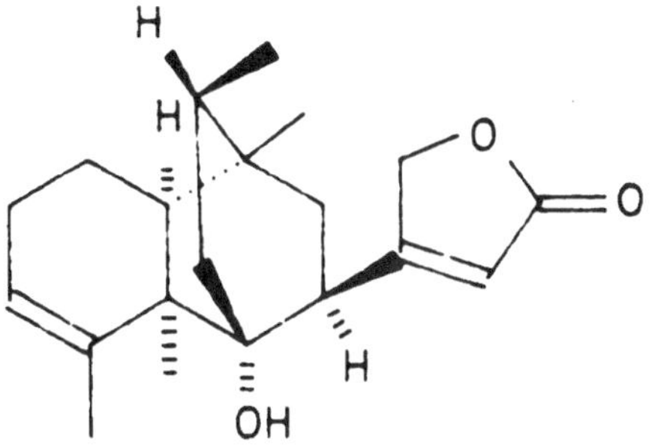

947

FIGURE 1. (contd.)

72. Chrysolane

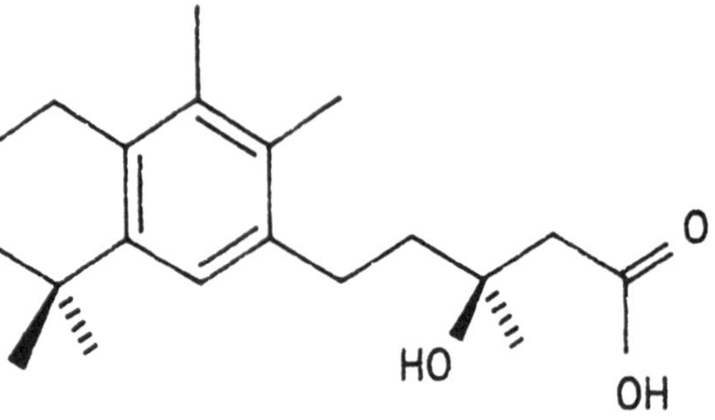

271

III. Normal Tricarbocyclics and Derivatives

32. Sandaracopimaranes (Normal-Pimaranes)

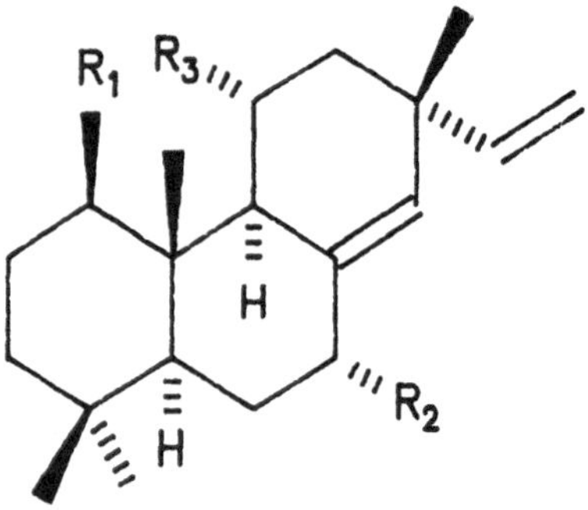

274   $R_1 = R_2 = R_3 = H$

927   $R_1 = H$, $R_2 = OH$, $R_3 = H$

928   $R_1 = R_2 = OH$, $R_3 = H$

929   $R_1 = OAc$, $R_2 = OH$, $R_3 = H$

930   $R_1 = OH$, $R_2 = OAc$, $R_3 = H$

931   $R_1 = OH$, $R_2 = H$, $R_3 = OH$

932   $R_1 = OAc$, $R_2 = H$, $R_3 = OAc$

933   $R_1 = OH$, $R_2 = OAc$, $R_3 = OH$

934   $R_1 = OAc$, $R_2 = OH$, $R_3 = OAc$

935   $R_1 = R_2 = R_3 = OAc$

FIGURE 1. (*contd.*)

Figure 1. Structures of Diterpenes   273

936

937

275 R_1 = H, R_2 = Me

301 R_1 = H, R_2 = CO$_2$H

303 R_1 = OAc, R_2 = CO$_2$H

276 R_1 = R_2 = R_3 = H, R_4 = Me

FIGURE 1. (*contd.*)

277 R_1 = OH, R_2 = R_3 = H, R_4 = Me

278 R_1 = OAc, R_2 = R_3 = H, R_4 = Me

279 R_1 = H, R_2 = OH, R_3 = H, R_4 = Me

280 R_1 = H, R_2 = OSen, R_3 = H, R_4 = Me

281 R_1 = H, R_2 = OTig, R_3 = H, R_4 = Me

282 R_1 = H, R_2 = H, R_3 = OAc, R_4 = Me

284 R_1 = R_2 = H, R_3 = OAc, R_4 = Me

285 R_1 = R_2 = R_3 = H, R_4 = CH_2OH

288 R_1 = H, R_2 = R_3 = OH, R_4 = Me

289 R_1 = H, R_2 = OAc, R_3 = OH, R_4 = Me

290 R_1 = H, R_2 = OSen, R_3 = OH, R_4 = Me

291 R_1 = H, R_2 = OTig, R_3 = OH, R_4 = Me

292 R_1 = H, R_2 = OH, R_3 = OAc, R_4 = Me

293 R_1 = H, R_2 = OH, R_3 = OCoum, R_4 = Me

294 R_1 = H, R_2 = R_3 = OAc, R_4 = Me

298 R_1 = R_2 = OH, R_3 = H, R_4 = Me

299 R_1 = OAc, R_2 = OH, R_3 = H, R_4 = Me

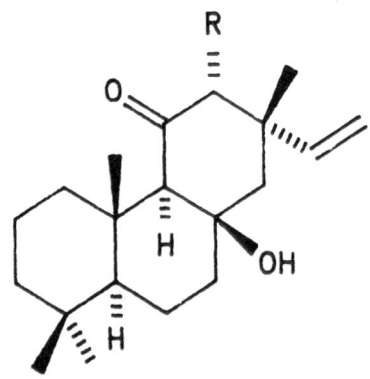

283 R = H

296 R = OH

297 R = OAc

FIGURE 1. (*contd.*)

Figure 1. Structures of Diterpenes 275

286

287

295

300 R = H

302 R = OAc

FIGURE 1. (*contd.*)

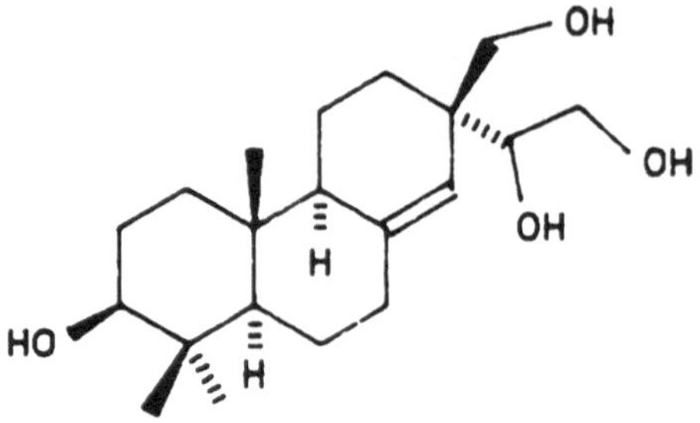

807

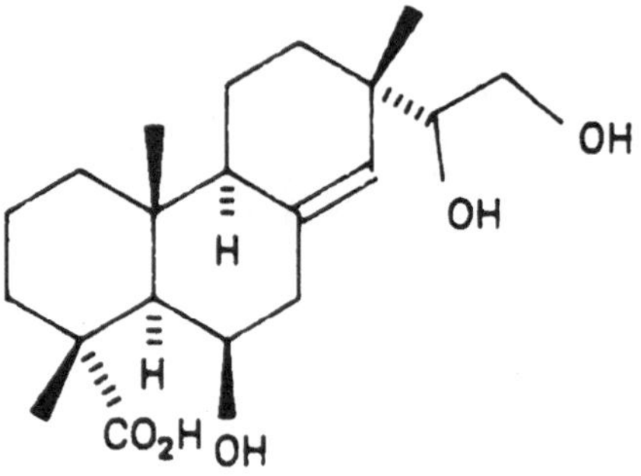

921

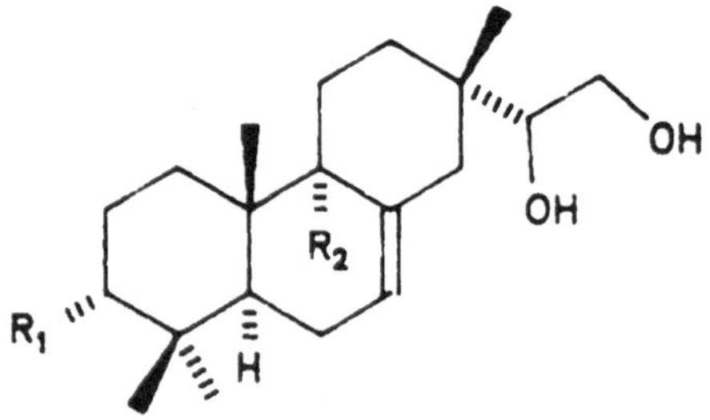

304   R_1 = OH, R_2 = H

305   R_1 = H, R_2 = OH

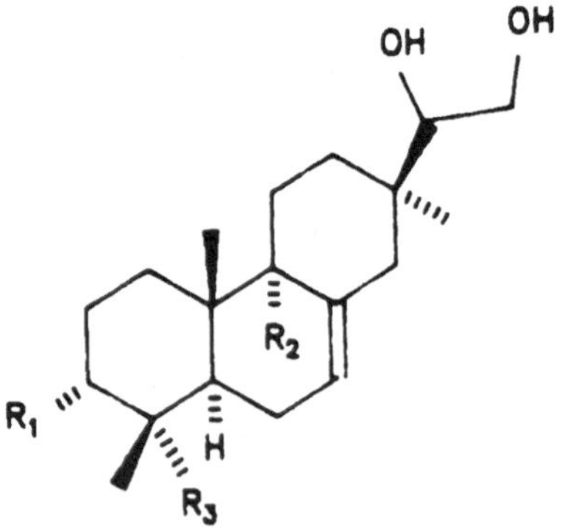

306   R_1 = OH, R_2 = H, R_3 = CH_2OH (Revised to rosane)

307   R_1 = R_2 = OH, R_3 = Me (Revised to rosane)

FIGURE 1. (contd.)

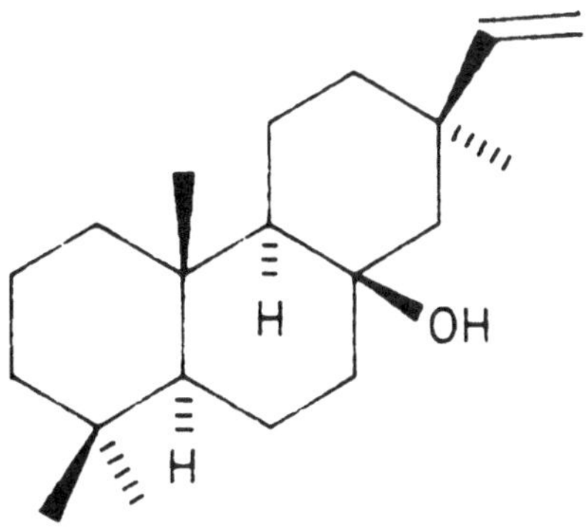

Figure 1. Structures of Diterpenes   277

891

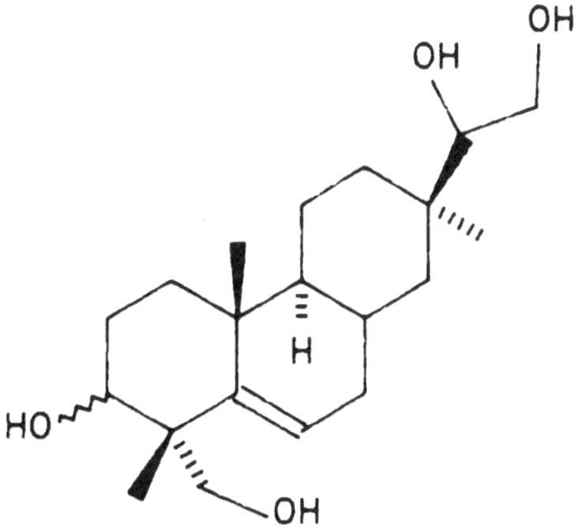

308

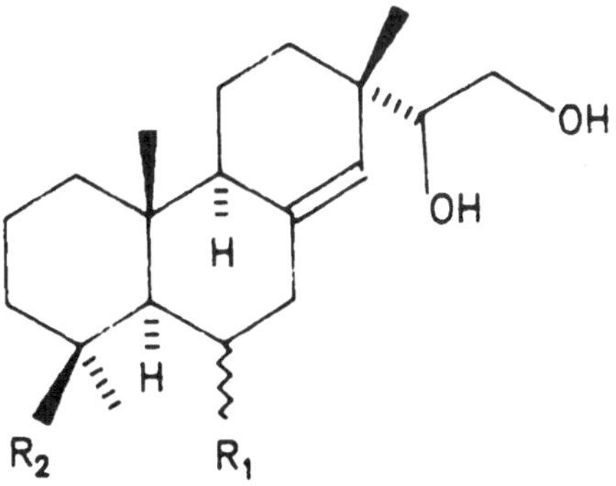

922 R_1 = β-OGlu, R_2 = Me (Probably <u>ent</u>-pimarane; = **593**?)

923 R_1 = α-OH, R_2 = CH_2OH (Probably <u>ent</u>-pimarane; = **944**?)

FIGURE 1. (*contd.*)

34. Cassane

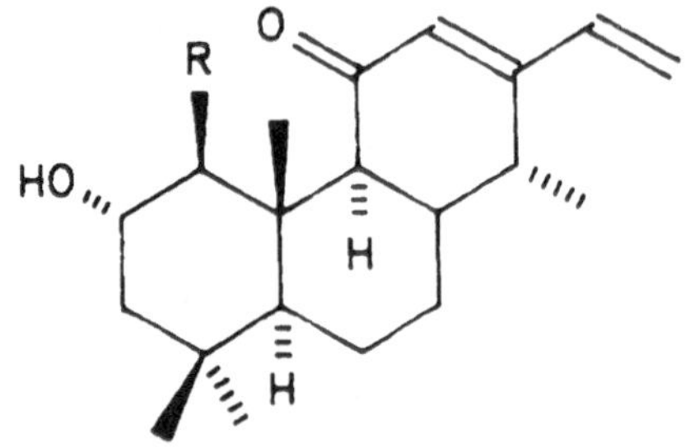

310   R = H

311   R = OH

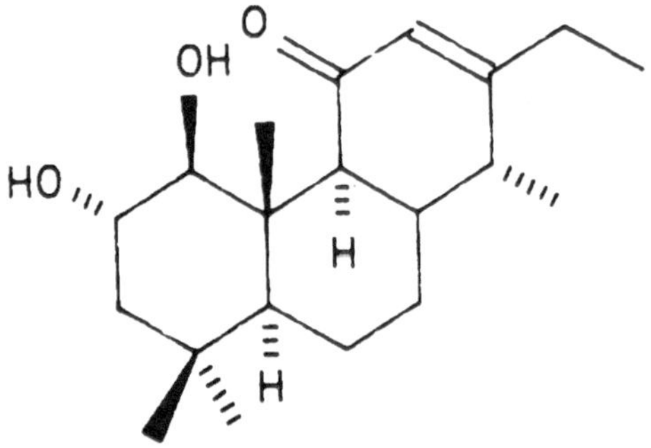

312

35. Rosane

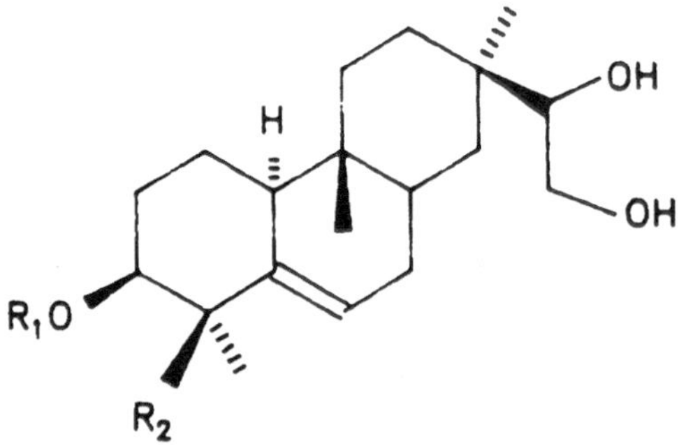

313   R$_1$ = H, R$_2$ = Me

314   R$_1$ = H, R$_2$ = CH$_2$OH

1161 R$_1$ = Ac, R$_2$ = CH$_2$OAc

1162 R$_1$ = Ac, R$_2$ = CH$_2$OH

FIGURE 1. (*contd.*)

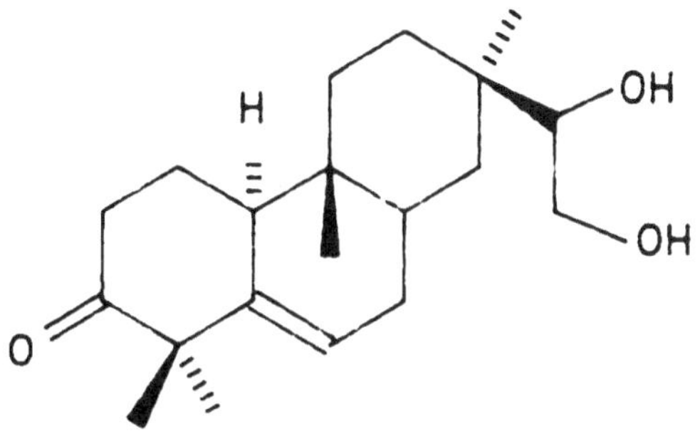

Figure 1. Structures of Diterpenes 279

315

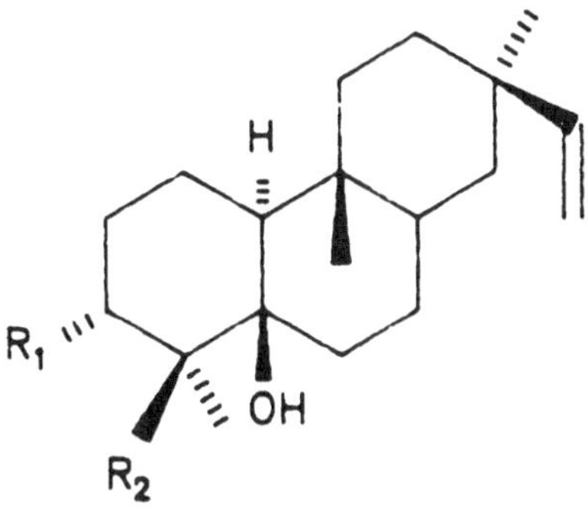

881   R_1 = H, R_2 = Me

882   R_1 = H, R_2 = CH$_2$OH

883   R_1 = H, R_2 = CH$_2$OAc

884   R_1 = R_2 = Me

885   R_1 = H, R_2 = CH$_2$OTig

886   R_1 = OH, R_2 = CH$_2$OH

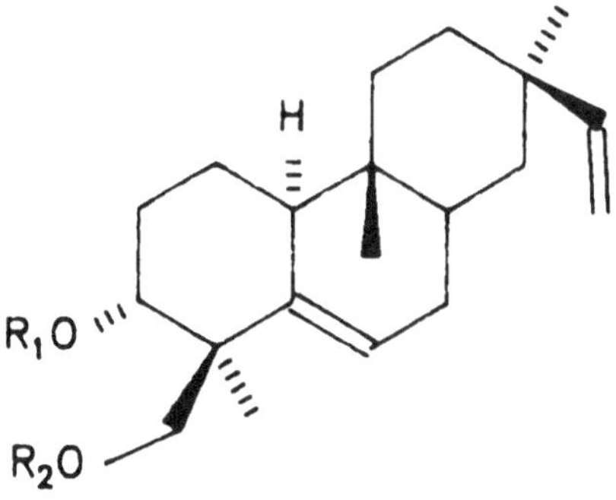

887   R_1 = R_2 = H

888   R_1 = Ac, R_2 = H

889   R_1 = H, R_2 = Ac

FIGURE 1. (contd.)

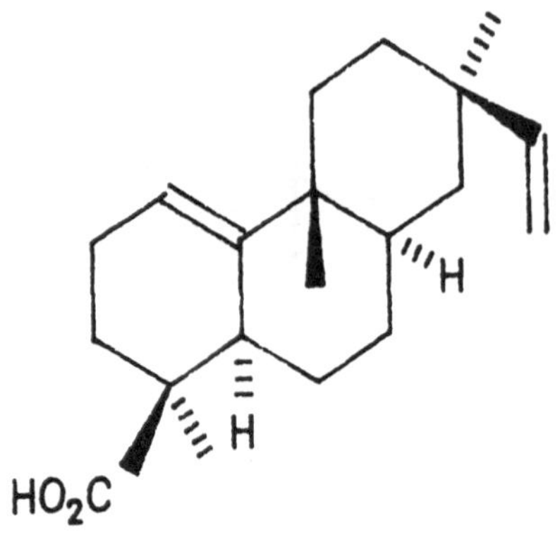

890

36. Sandaracopimarane, 19-Nor-13-Epi

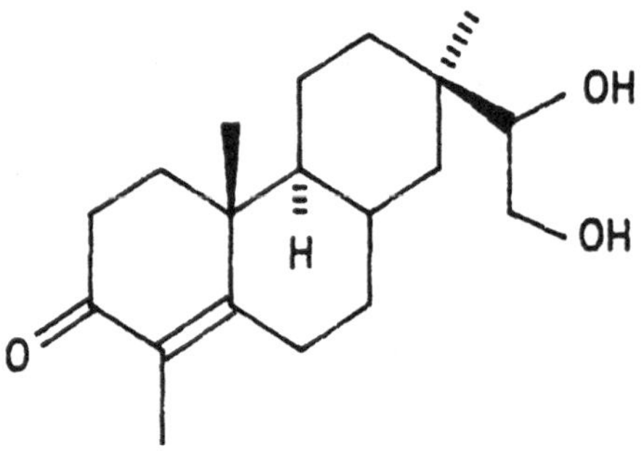

309

37. Normal-Abietane

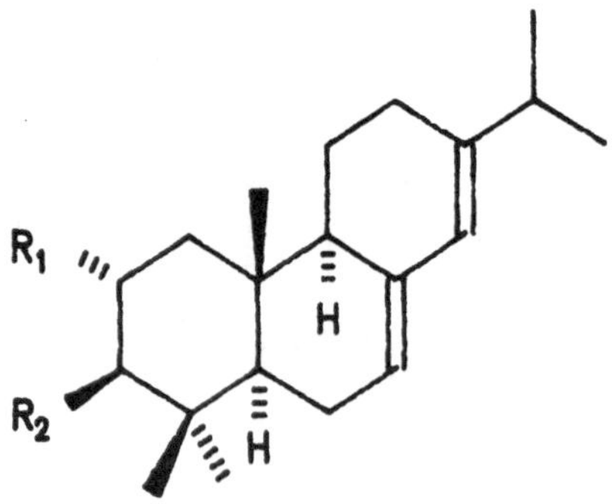

316 R_1 = OAc, R_2 = H (**316-318** probable artifacts; dehydration products ?)

317 R_1 = H, R_2 = OH

FIGURE 1. (*contd.*)

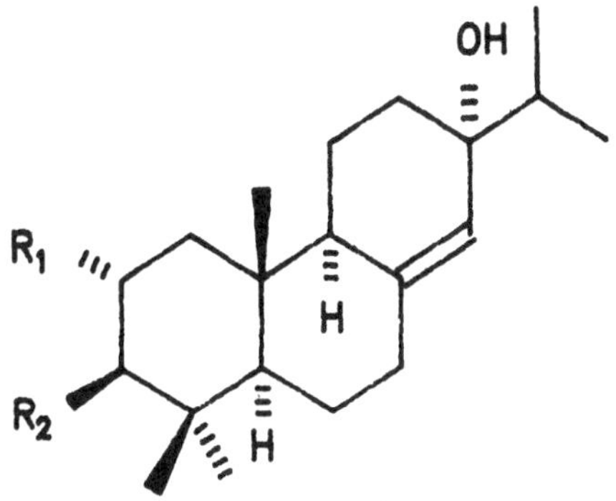

Figure 1. Structures of Diterpenes   281

316 R_1 = OAc, R_2 = H (Possible dehydration precursor)

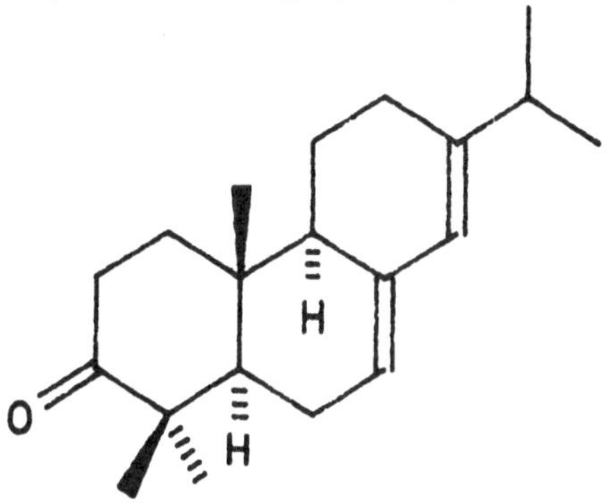

318

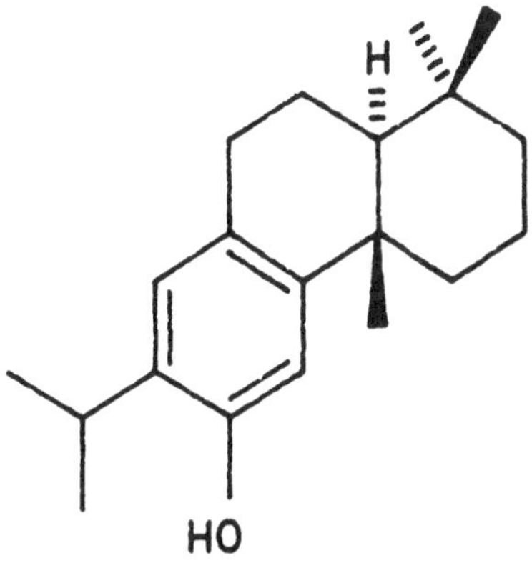

319

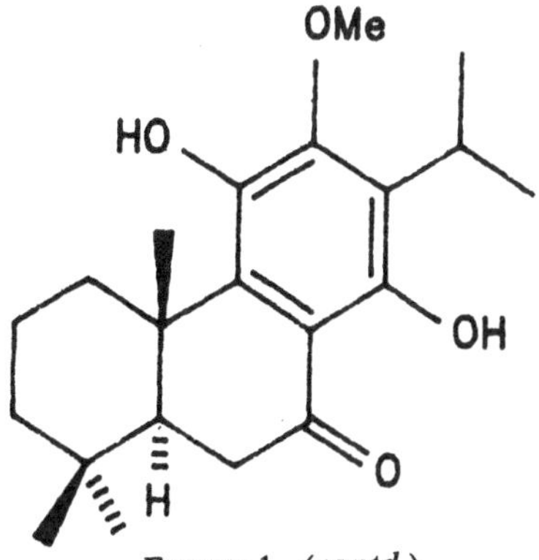

320

FIGURE 1. (contd.)

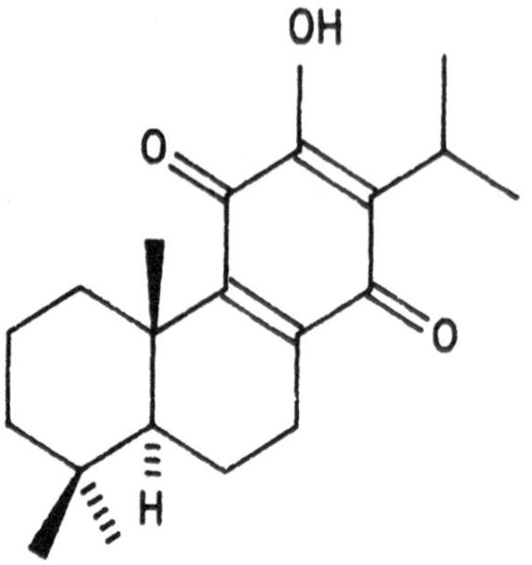

321

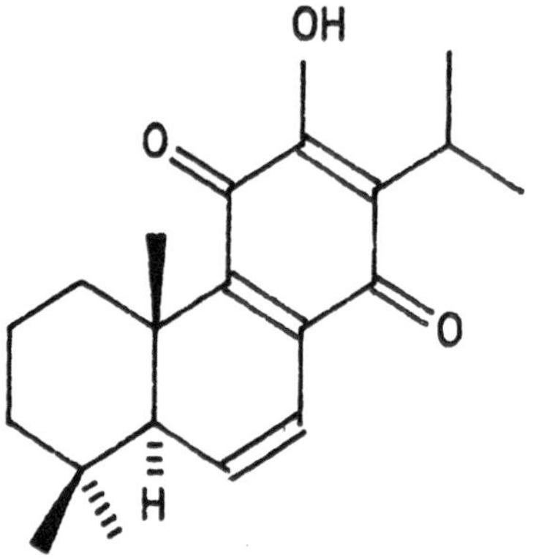

322

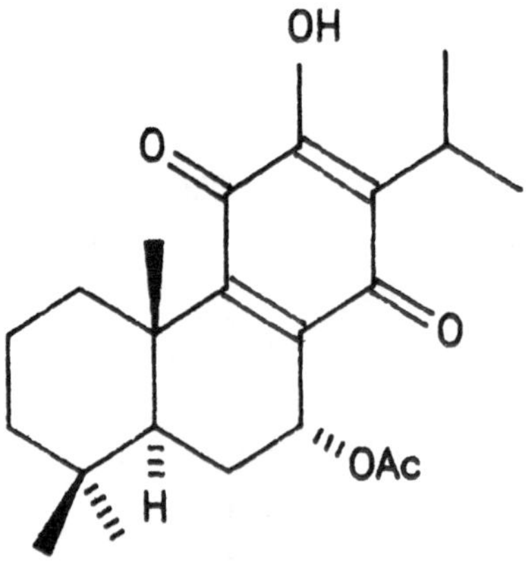

323

FIGURE 1. (*contd.*)

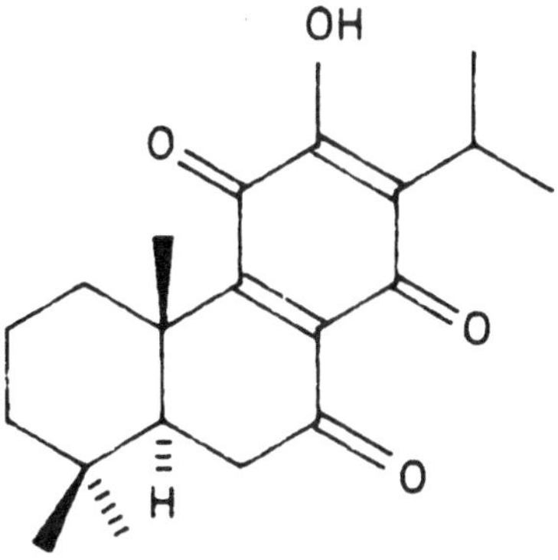

Figure 1. Structures of Diterpenes   283

324

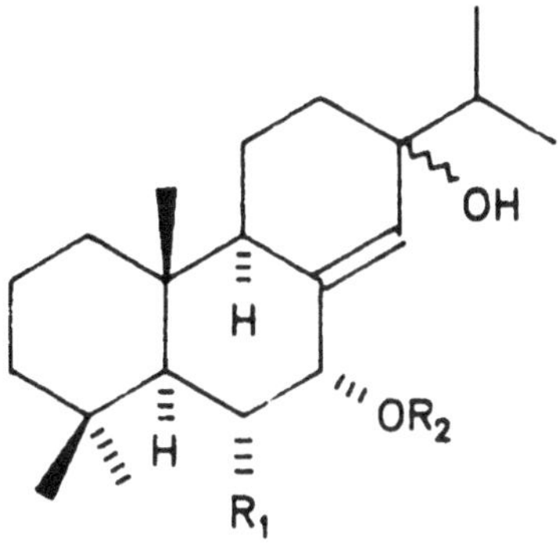

1072 R_1 = OH, R_2 = Sen

1073 R_1 = OSen, R_2 = H

1074 R_1 = R_2 = H

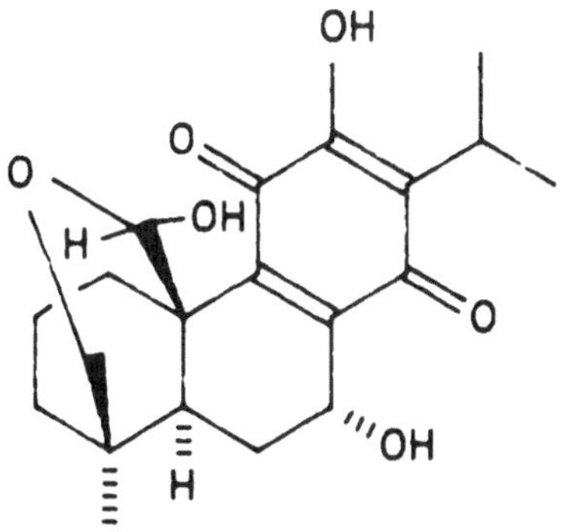

1163

FIGURE 1. (*contd.*)

94. Stevisalane

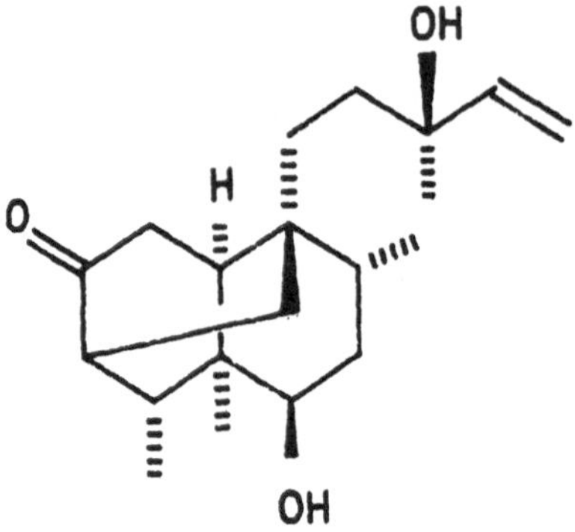

1026

IV. Normal-Tetracarbocyclics

73. Normal-Stachane

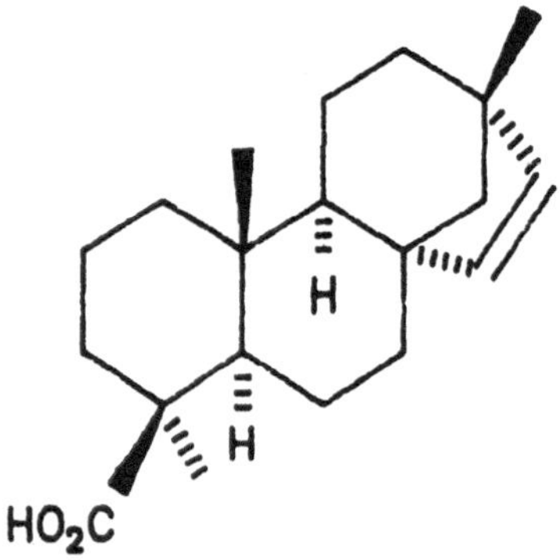

325 (Probable ent-stachane; = 774?)

V. ent-Bicarbocyclics and Derivatives

38. ent-Labdane

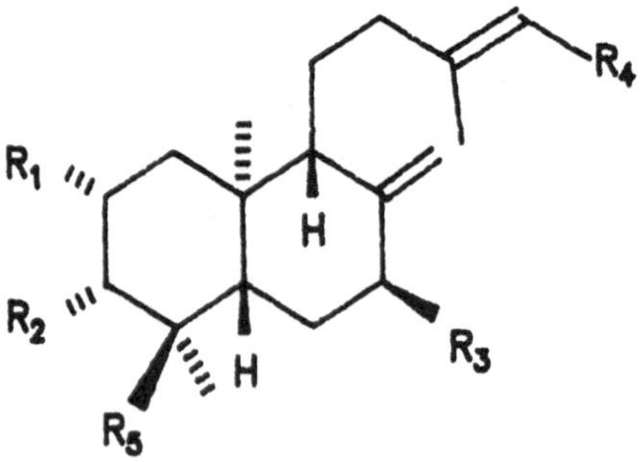

326 $R_1 = R_2 = R_3 = H$, $R_4 = CH_2OH$, $R_5 = Me$

327 $R_1 = R_2 = R_3 = H$, $R_4 = CH_2Obern$, $R_5 = Me$

FIGURE 1. (*contd.*)

Figure 1. Structures of Diterpenes   285

848 R_1 = OH, R_2 = R_3 = H, R_4 = CH_2OH, R_5 = Me

328 R_1 = R_2 = H, R_3 = OAc, R_4 = CH_2OH, R_5 = Me

844 R_1 = H, R_2 = OAng, R_3 = H, R_4 = R_5 = CH_2OH

329 R_1 = R_2 = R_3 = H, R_4 = CH_2OH, R_5 = CO_2H

330 R_1 = R_2 = R_3 = H, R_4 = CH_2OAc, R_5 = CO_2H

331 R_1 = OH, R_2 = R_3 = H, R_4 = CHO, R_5 = Me

335 R_1 = R_2 = R_3 = H, R_4 = CO_2H, R_5 = Me (= **1040**)

1014 R_1 = H, R_2 = OAc, R_3 = H, R_4 = CH_2OH, R_5 = CH_2OTig

1041 R_1 = R_2 = R_3 = H, R_4 = CO_2H, R_5 = CH_2OH

1164 R_1 = R_2 = R_3 = H, R_4 = R_5 = CO_2H

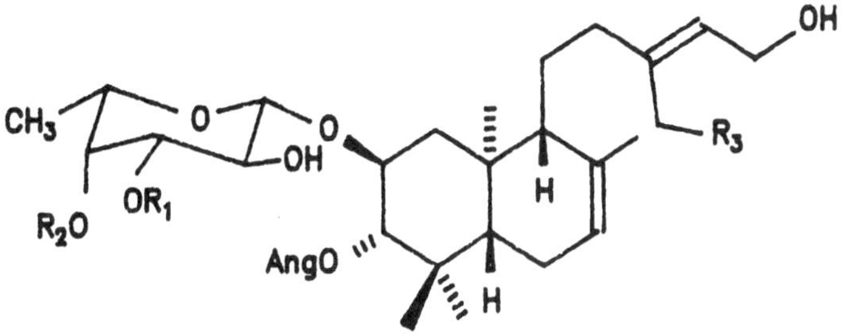

1165 R_1 = R_2 = R_3 = H

1166 R_1 = H, R_2 = Ac, R_3 = H

1167 R_1 = Ac, R_2 = R_3 = H

1168 R_1 = H, R_2 = Ac, R_3 = OH

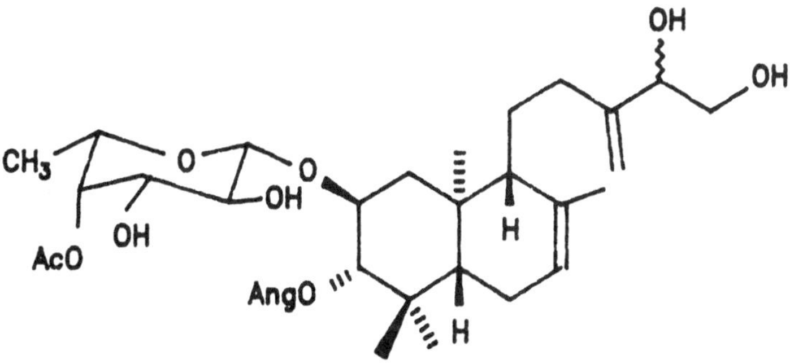

1169

FIGURE 1. (contd.)

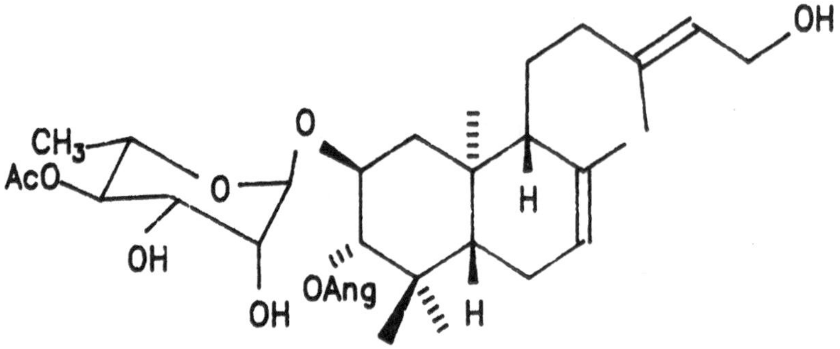

1170

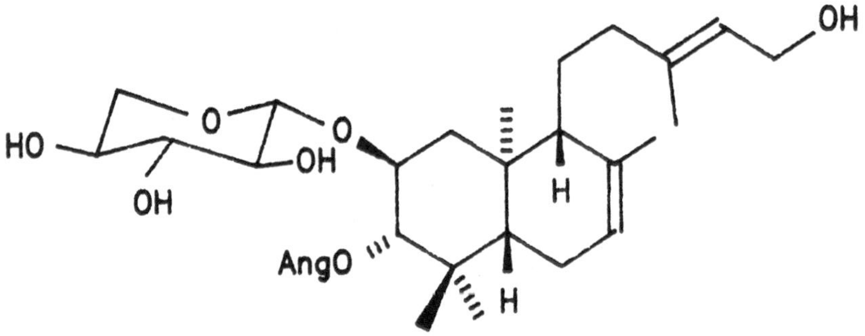

1171

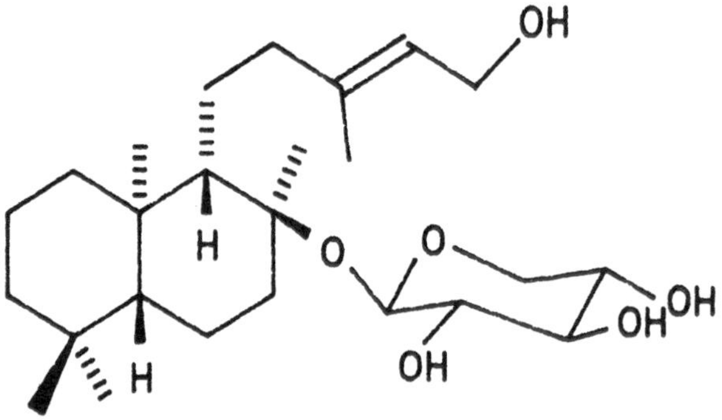

1172

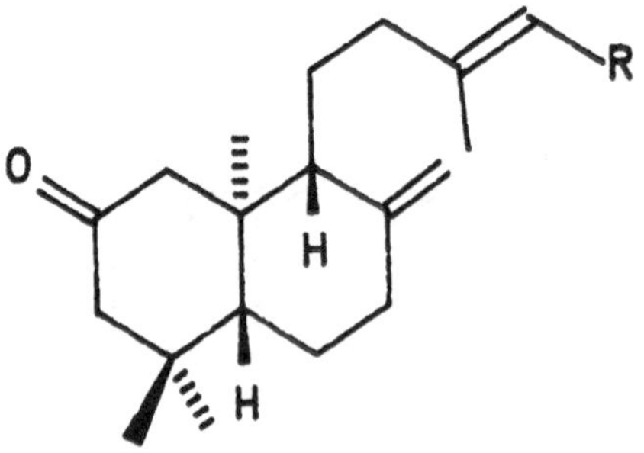

849 R = CH$_2$OH

333 R = CHO

FIGURE 1. (*contd.*)

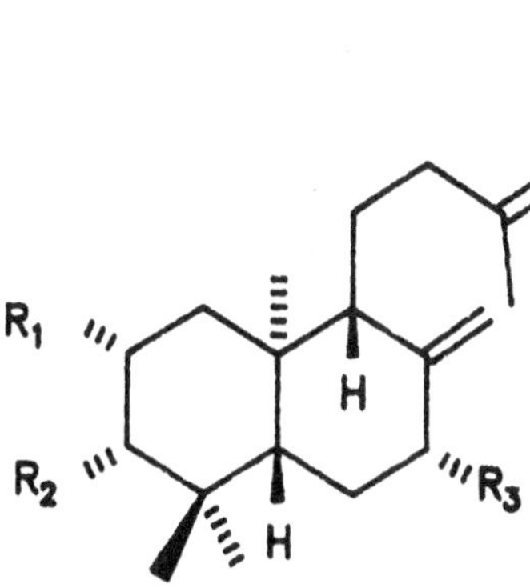

Figure 1. Structures of Diterpenes 287

332 R_1 = OH, R_2 = R_3 = H, R_4 = CHO

336 R_1 = R_2 = R_3 = H, R_4 = CO_2H

337 R_1 = H, R_2 = OH, R_3 = H, R_4 = CO_2H

338 R_1 = H, R_2 = R_3 = OH, R_4 = CO_2H

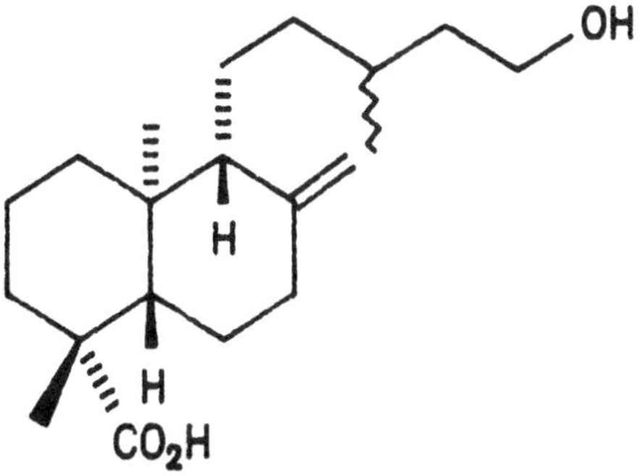

1173

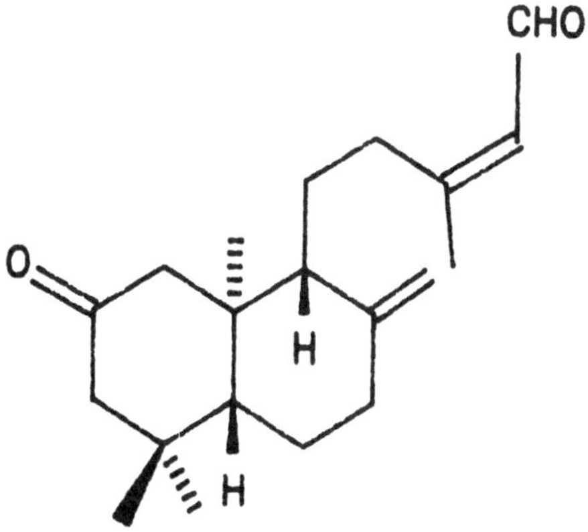

334

FIGURE 1. (contd.)

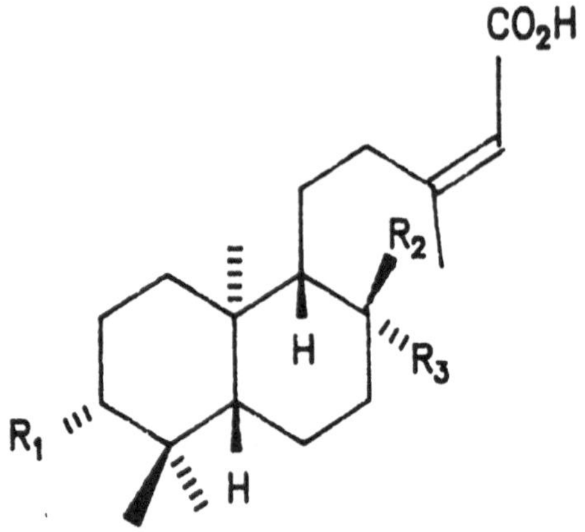

339 R_1 = OH, R_2 = H, R_3 = CHO

340 R_1 = OH, R_2 = CHO, R_3 = H

343 R_1 = OH, R_2 = CHO, R_3 = OH

344 R_1 = H, R_2 = H, R_3 = CO_2H

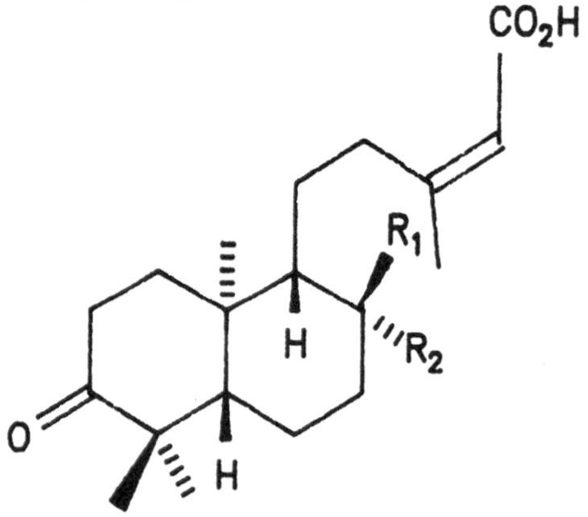

341 R_1 = H, R_2 = CHO

342 R_1 = CHO, R_2 = H

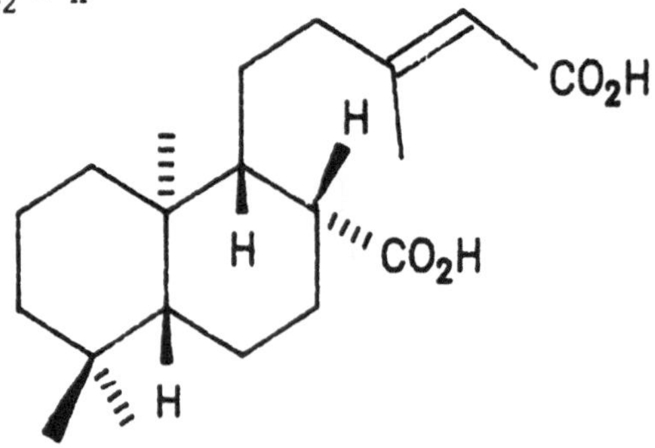

345

FIGURE 1. (*contd.*)

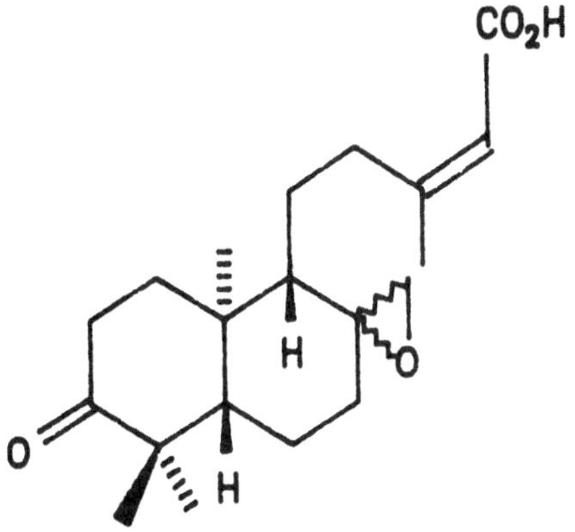

Figure 1. Structures of Diterpenes 289

346

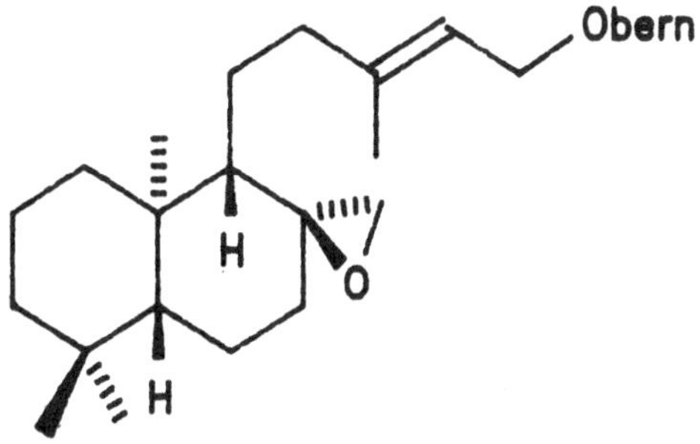

347

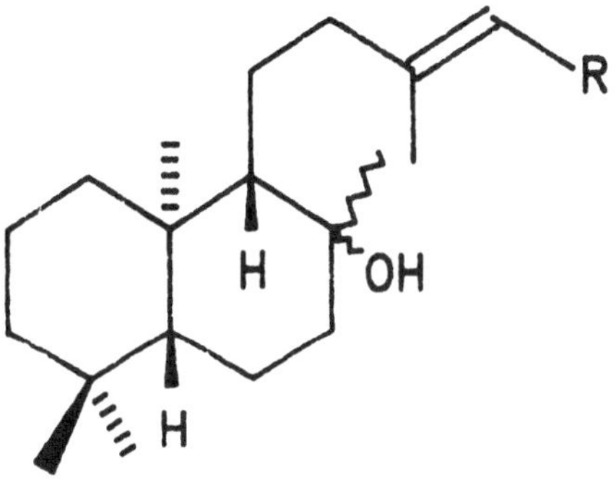

348 R = CH$_2$OH

349 R = CHO

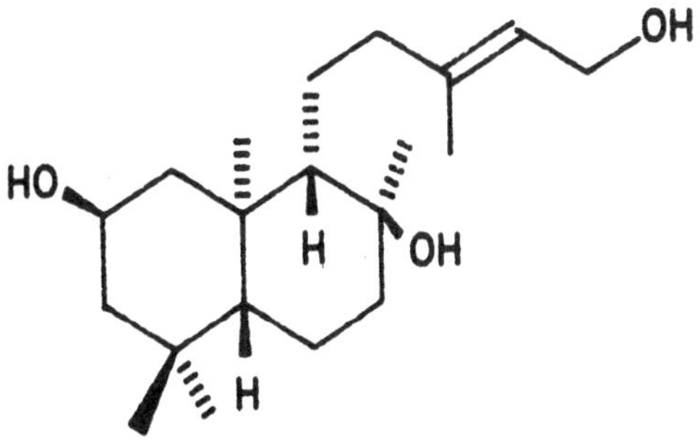

1174

FIGURE 1. (contd.)

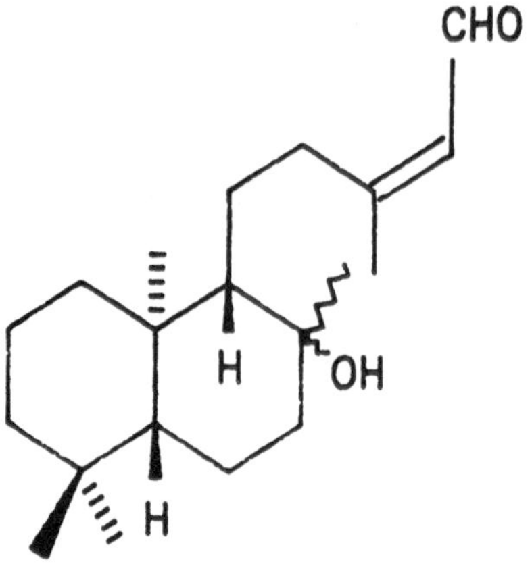

350

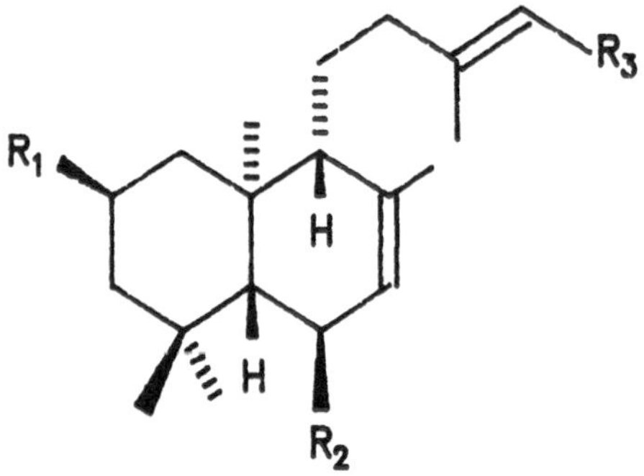

351   R_1 = H, R_2 = OH, R_3 = CH$_2$OH

355   R_1 = OAng, R_2 = H, R_3 = CO$_2$H (13,14 Z or E)

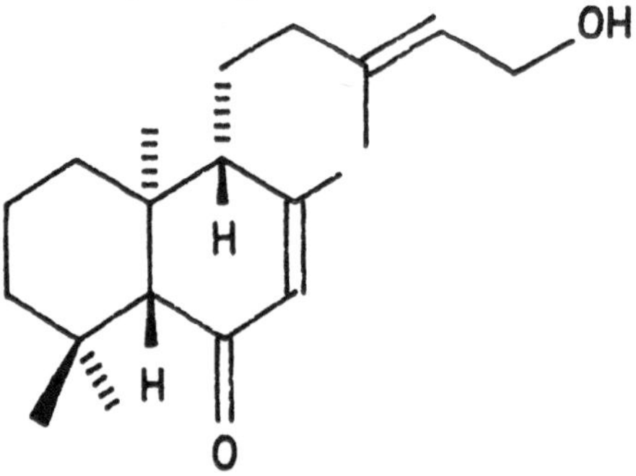

352

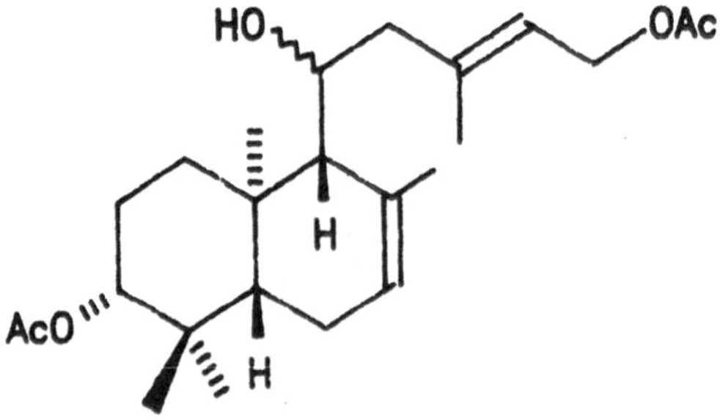

353

FIGURE 1. (*contd.*)

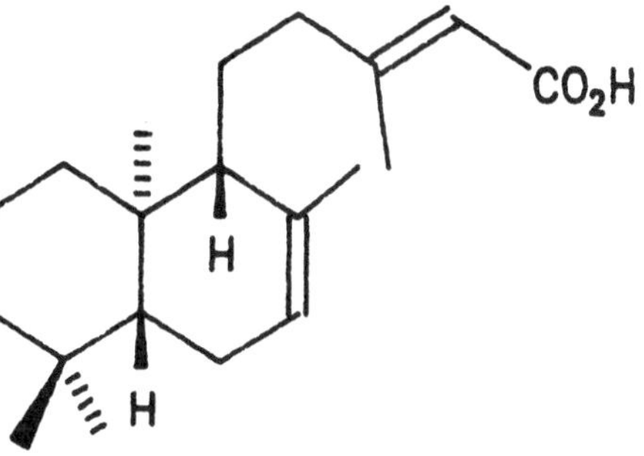

Figure 1. Structures of Diterpenes   291

354

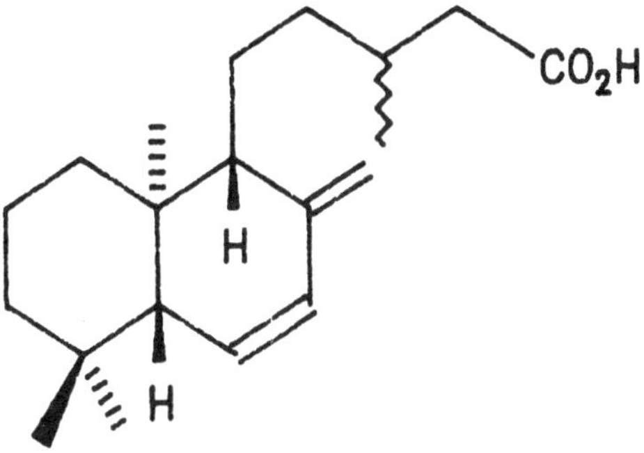

356

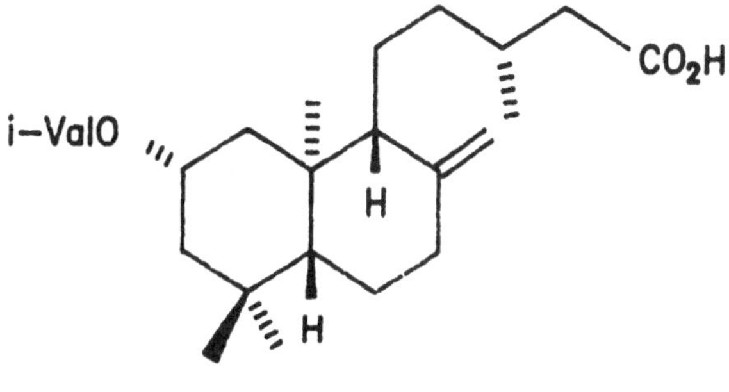

358

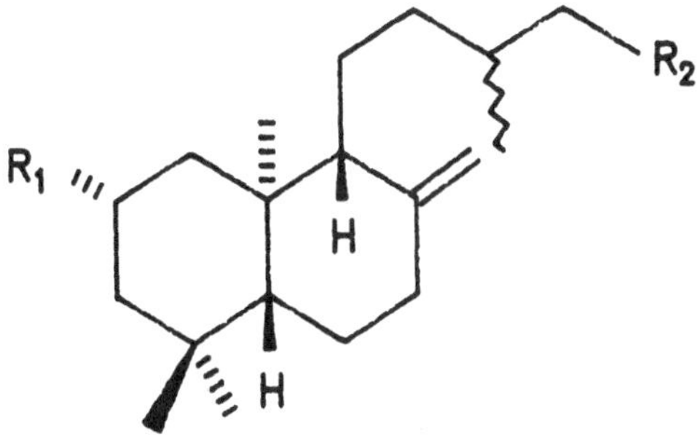

357 R_1 = OH, R_2 = CH$_2$OH

857 R_1 = OH, R_2 = CO$_2$H

FIGURE 1. (*contd.*)

858 R_1 = OAng, R_2 = CO_2H

859 R_1 = OTig, R_2 = CO_2H

860 R_1 = O2-Mebut, R_2 = CO_2H

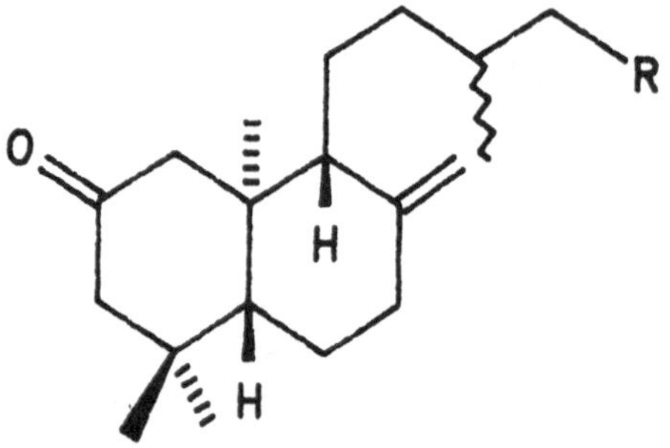

359 R = CH_2OH

861 R = CO_2H

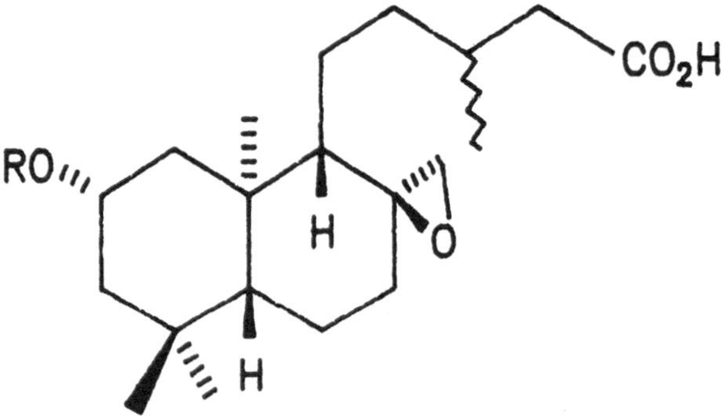

862 R = OAng

863 R = O2-Mebut

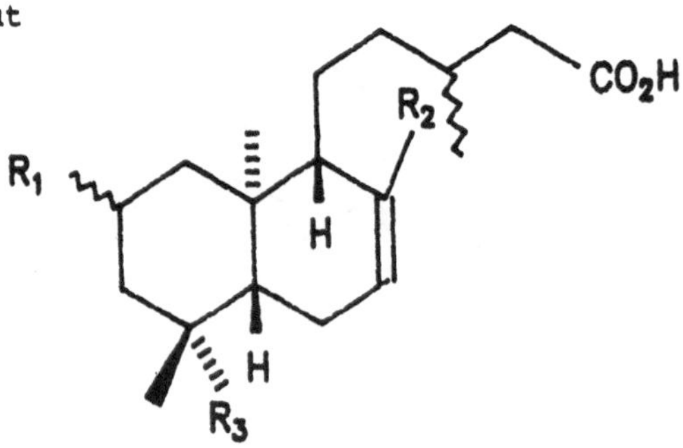

360 R_1 = H, R_2 = CH_2OMe, R_3 = Me

361 R_1 = β-OAng, R_2 = R_3 = Me

FIGURE 1. (*contd.*)

Figure 1. Structures of Diterpenes   293

1046 R$_1$ = H, R$_2$ = CH$_2$OH, R$_3$ = Me

1047 R$_1$ = H, R$_2$ = CH$_2$OAc, R$_3$ = Me

1048 R$_1$ = H, R$_2$ = CHO, R$_3$ = Me

1049 R$_1$ = α-OH, R$_2$ = CH$_2$OH, R$_3$ = Me

1050 R$_1$ = α-OH, R$_2$ = CH$_2$OAc, R$_3$ = Me

1051 R$_1$ = α-OH, R$_2$ = CHO, R$_3$ = Me

1052 R$_1$ = H, R$_2$ = CO$_2$H, R$_3$ = Me

1053 R$_1$ = H, R$_2$ = CH$_2$OAc, R$_3$ = CH$_2$OH

1179 R$_1$ = OH, R$_2$ = CH$_2$OMe, R$_3$ = Me

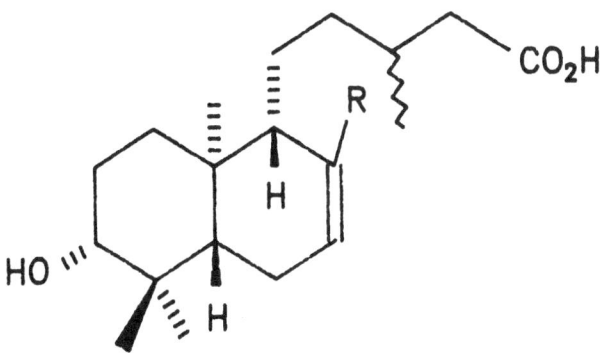

1175 R = CH$_2$OH

1176 R = CH$_2$OAc

1177 R = CHO

1178 R = CH$_2$OMe

362 R$_1$ = R$_2$ = R$_3$ = CH$_2$OH

363 R$_1$ = R$_2$ = CH$_2$OAc, R$_3$ = Me

FIGURE 1. (*contd.*)

364 $R_1 = R_2 = R_3 = CH_2OAc$

366 $R_1 = CO_2H$, $R_2 = CH_2OH$, $R_3 = Me$

367 $R_1 = CO_2H$, $R_2 = R_3 = CH_2OH$

369 $R_1 = CO_2H$, $R_2 = CH_2OH$, $R_3 = CHO$

370 $R_1 = CO_2H$, $R_2 = CH_2OAc$, $R_3 = CH_2OH$

1180 $R_1 = CO_2H$, $R_2 = CH_2OH$, $R_3 = CH_2OMe$

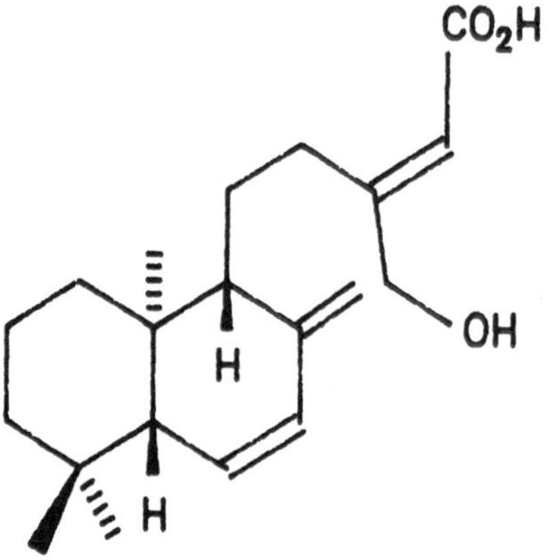

365

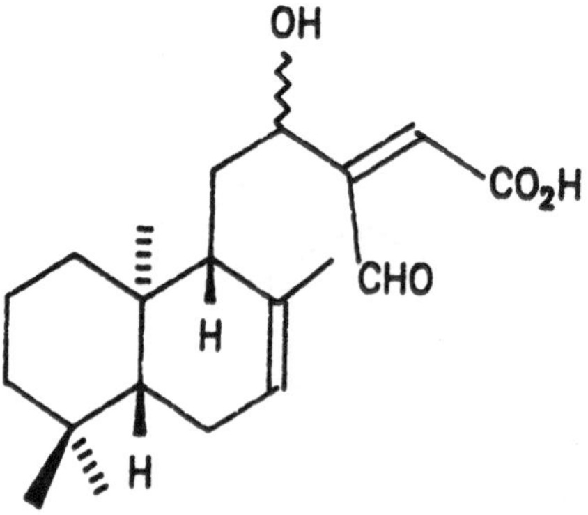

368

FIGURE 1. (*contd.*)

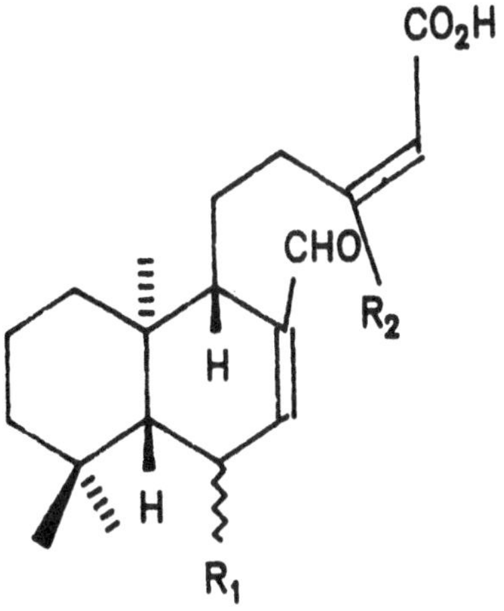

Figure 1. Structures of Diterpenes   295

371   R_1 = H, R_2 = CH$_2$OAc

372   R_1 = α-OH, R_2 = CH$_2$OAc

373   R_1 = β-OH, R_2 = CH$_2$OAc

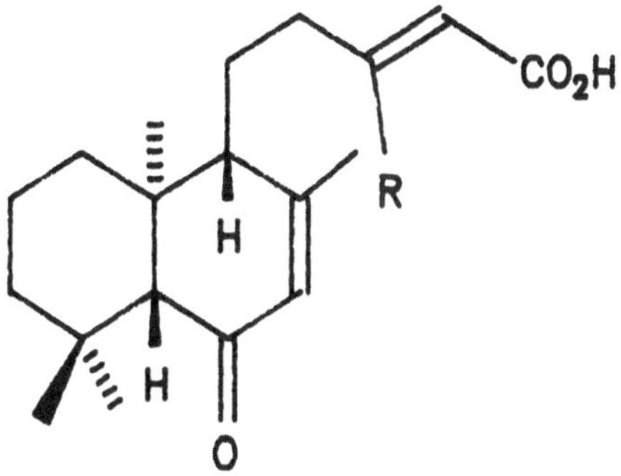

374   R = CH$_2$OAc

375   R = CHO

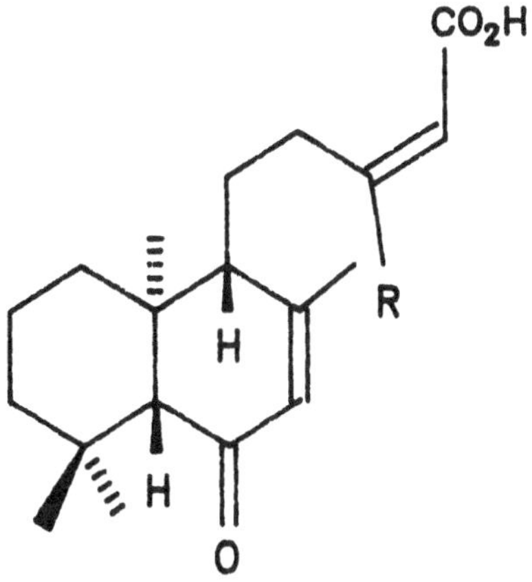

376   R = CH$_2$OH

377   R = CH$_2$OAc

378   R = CH$_2$OMe

FIGURE 1. (contd.)

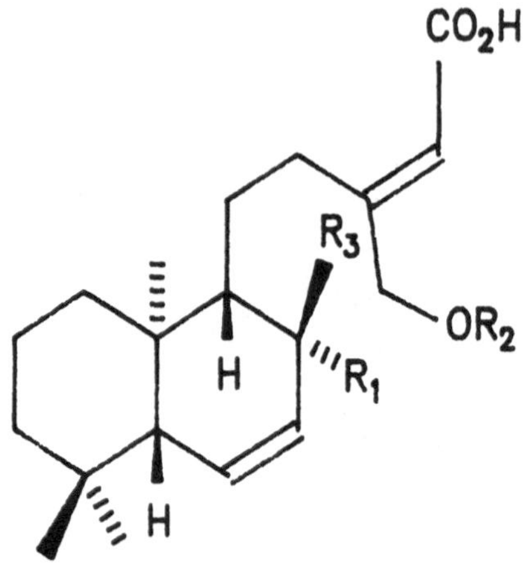

379   R_1 = OH, R_2 = H, R_3 = Me

380   R_1 = H, R_2 = Ac, R_3 = CHO

381   R_1 = H, R_2 = Ac, R_3 = CO_2H

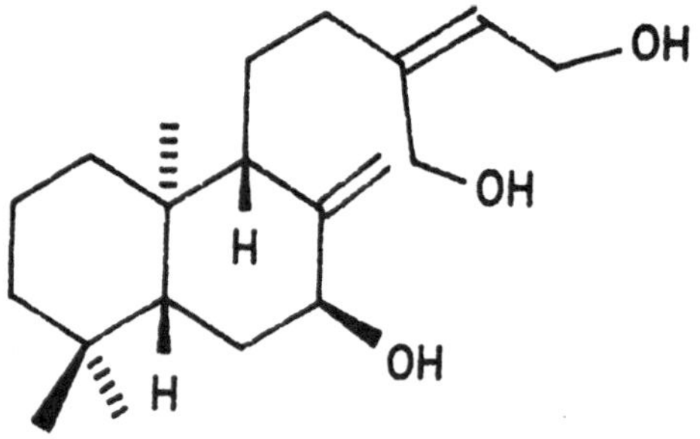

382

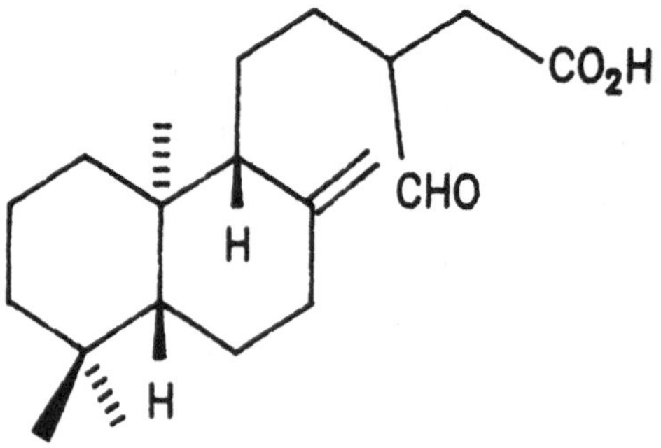

383

FIGURE 1. (contd.)

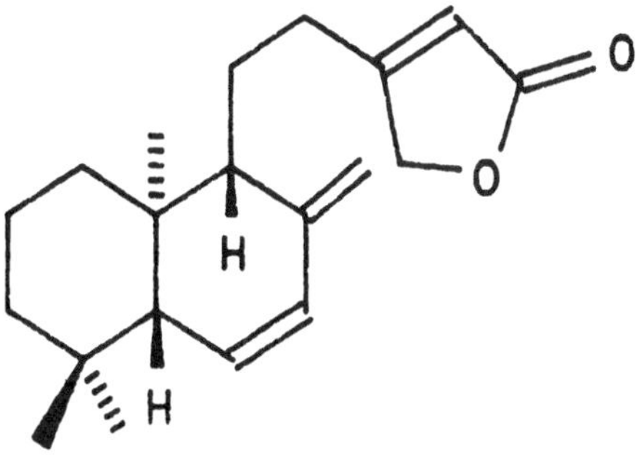

Figure 1. Structures of Diterpenes   297

384

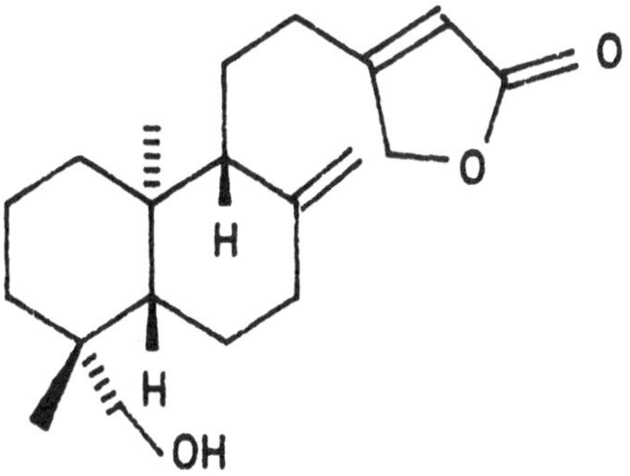

385

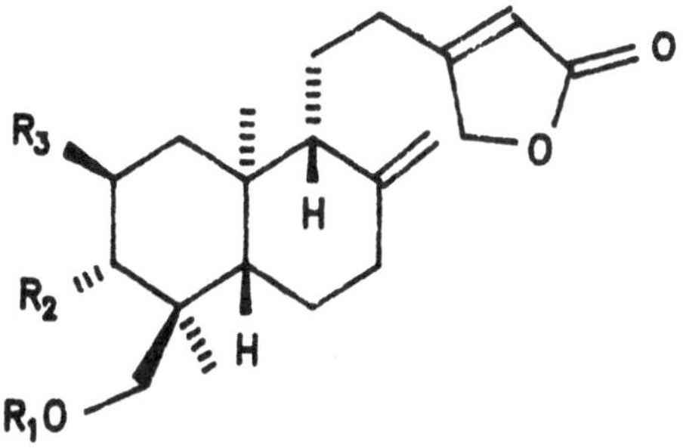

1181 R_1 = R_2 = H, R_3 = OH

1182 R_1 = H, R_2 = OAng, R_3 = H

1183 R_1 = Glu, R_2 = OAng, R_3 = H

1184 R_1 = H, R_2 = OAng, R_3 = OH

FIGURE 1. (*contd.*)

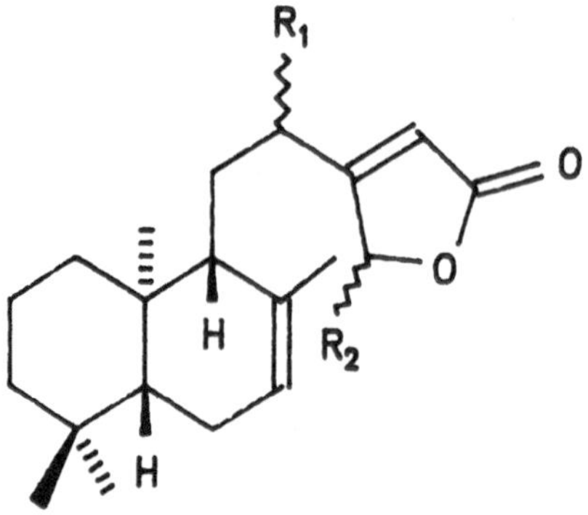

386 R$_1$ = R$_2$ = H

387 R$_1$ = OH, R$_2$ = H

388 R$_1$ = H, R$_2$ = α-OH

389 R$_1$ = H, R$_2$ = β-OH

390 R$_1$ = R$_2$ = OH

391 R$_1$ = OH, R$_2$ = α-OH

392 R$_1$ = OH, R$_2$ = β-OH

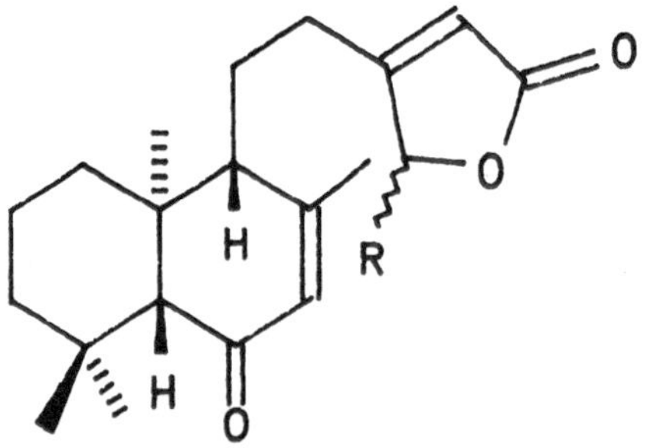

393 R = H

394 R = OH

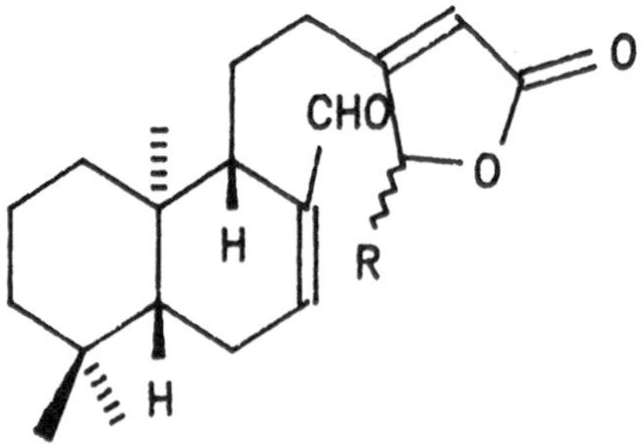

395 R = α-OH

396 R = β-OH

FIGURE 1. (*contd.*)

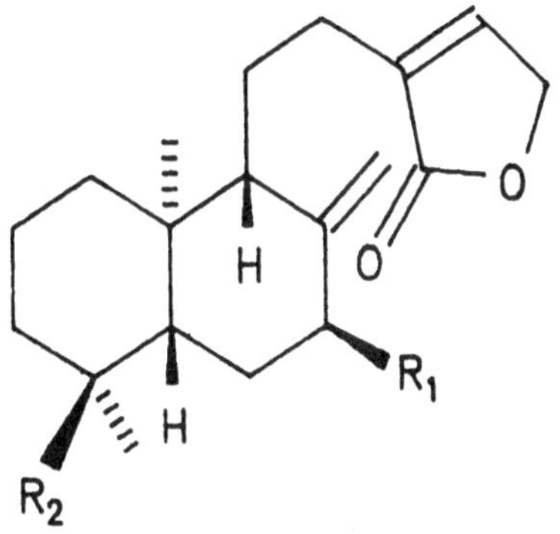

Figure 1. Structures of Diterpenes   299

397 R$_1$ = OH, R$_2$ = Me

843 R$_1$ = H, R$_2$ = CO$_2$H

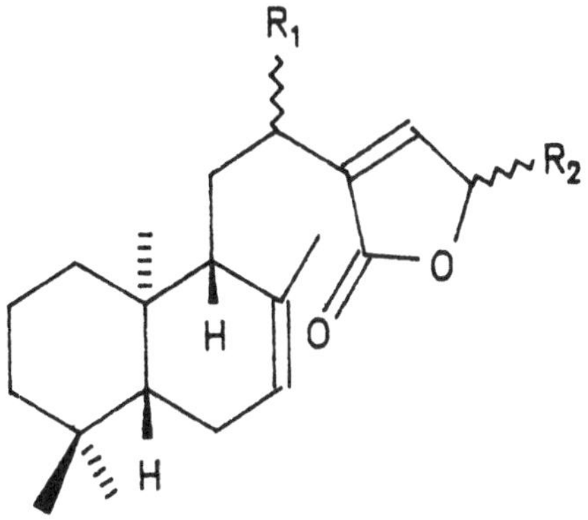

398 R$_1$ = R$_2$ = H

399 R$_1$ = OH, R$_2$ = β-OH

400 R$_1$ = OH, R$_2$ = α-OH

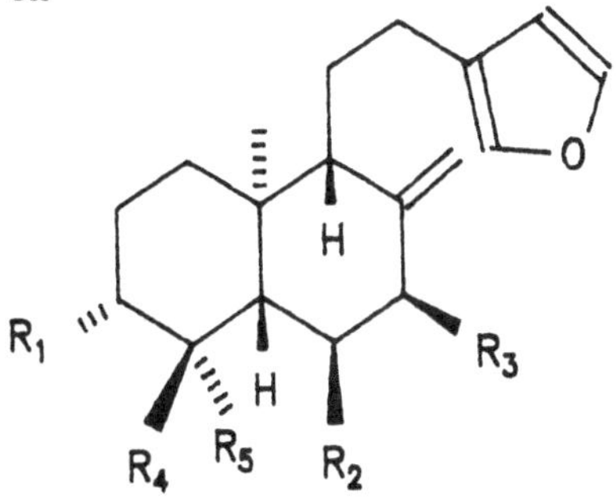

401 R$_1$ = R$_2$ = H, R$_3$ = OH, R$_4$ = R$_5$ = Me

402 R$_1$ = R$_2$ = H, R$_3$ = OAc, R$_4$ = R$_5$ = Me

403 R$_1$ = R$_2$ = R$_3$ = H, R$_4$ = Me, R$_5$ = CH$_2$OH

FIGURE 1. (*contd.*)

404 $R_1 = H$, $R_2 = OH$, $R_3 = H$, $R_4 = Me$, $R_5 = CH_2OH$

839 $R_1 = R_2 = R_3 = H$, $R_4 = Me$, $R_5 = CO_2H$

840 $R_1 = OH$, $R_2 = R_3 = H$, $R_4 = Me$, $R_5 = CO_2H$

841 $R_1 = R_2 = R_3 = H$, $R_4 = R_5 = CH_2OH$

842 $R_1 = OH$, $R_2 = R_3 = H$, $R_4 = CH_2OH$, $R_5 = Me$

974 $R_1 = OAng$, $R_2 = R_3 = H$, $R_4 = CH_2OH$, $R_5 = Me$

975 $R_1 = OH$, $R_2 = R_3 = H$, $R_4 = CH_2OAng$, $R_5 = Me$

830 $R_1 = R_2 = R_3 = H$, $R_4 = CO_2H$, $R_5 = Me$

831 $R_1 = OH$, $R_2 = R_3 = H$, $R_4 = CO_2H$, $R_5 = Me$

832 $R_1 = OAng$, $R_2 = R_3 = H$, $R_4 = CO_2H$, $R_5 = Me$

833 $R_1 = O2\text{-}Mebut$, $R_2 = R_3 = H$, $R_4 = CO_2H$, $R_5 = Me$

834 $R_1 = Oi\text{-}But$, $R_2 = R_3 = H$, $R_4 = CO_2H$, $R_5 = Me$

835 $R_1 = R_2 = R_3 = H$, $R_4 = CO_2H$, $R_5 = CH_2OH$

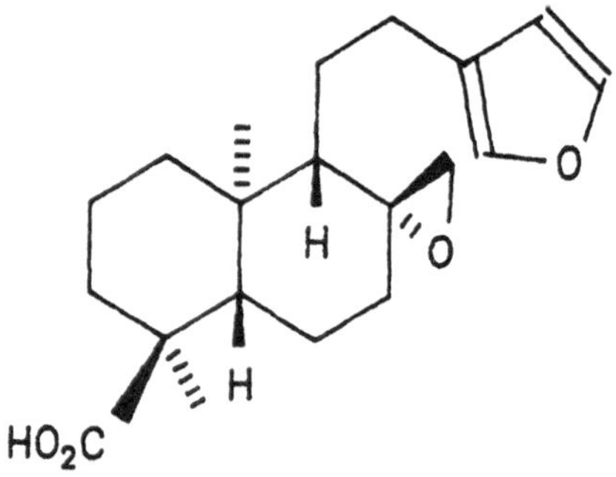

836

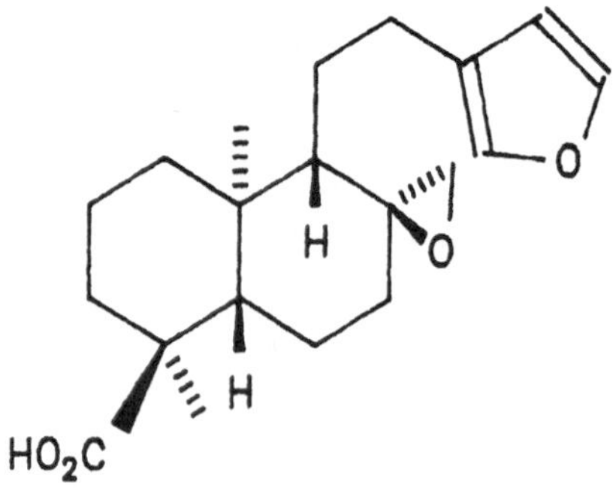

837

FIGURE 1. (*contd.*)

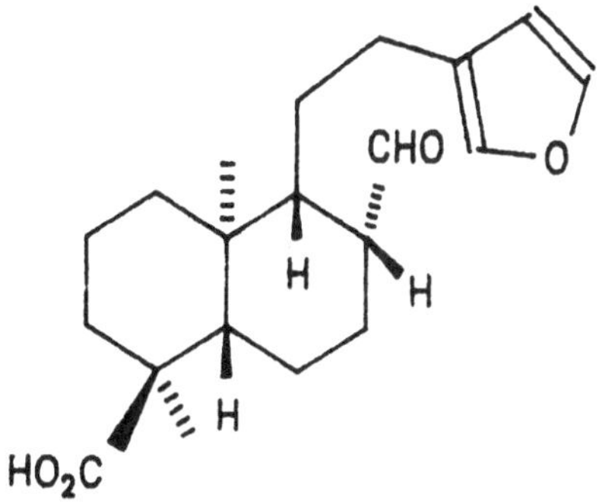

Figure 1. Structures of Diterpenes   301

838

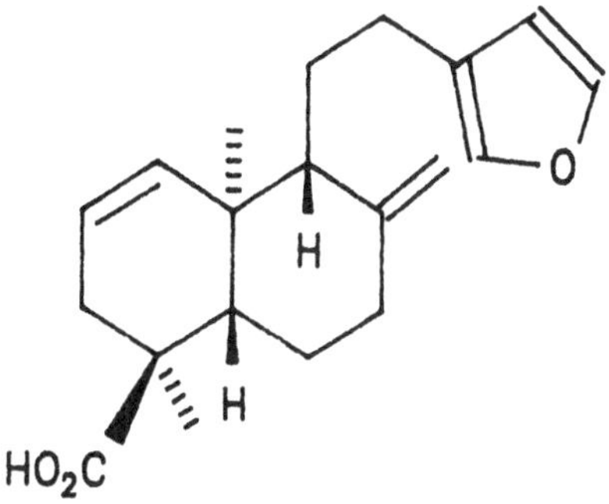

405

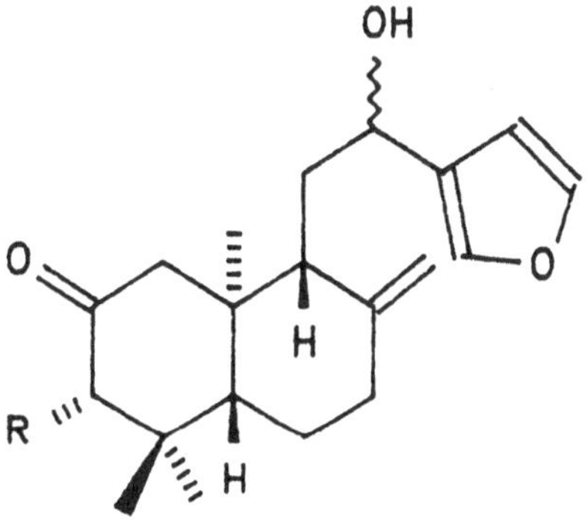

406 R = H

413 R = OH

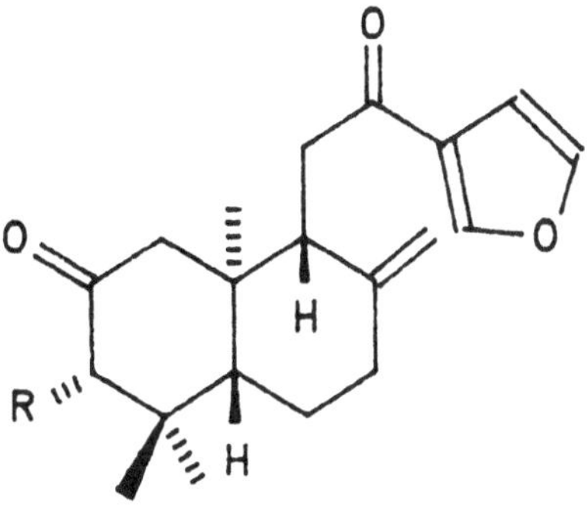

407 R = H

412 R = OH

FIGURE 1. (contd.)

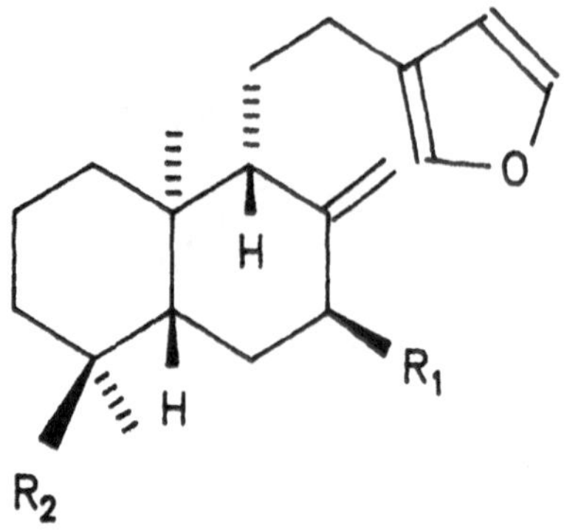

408 R$_1$ = OAc, R$_2$ = CH$_2$OH

409 R$_1$ = OAc, R$_2$ = CH$_2$OAc

410 R$_1$ = OH, R$_2$ = CHO (Saponification product)

411 R$_1$ = OAc, R$_2$ = CHO

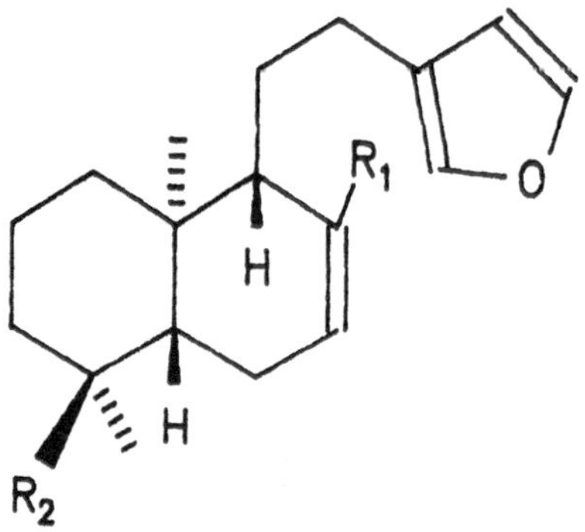

414 R$_1$ = CH$_2$OH, R$_2$ = Me

415 R$_1$ = Me, R$_2$ = CO$_2$H

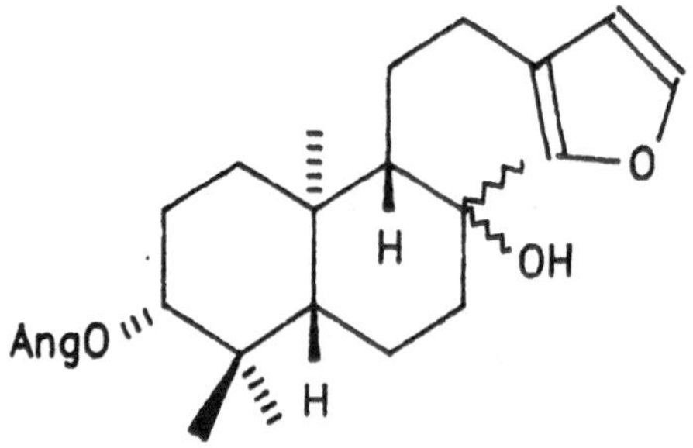

416

FIGURE 1. (*contd.*)

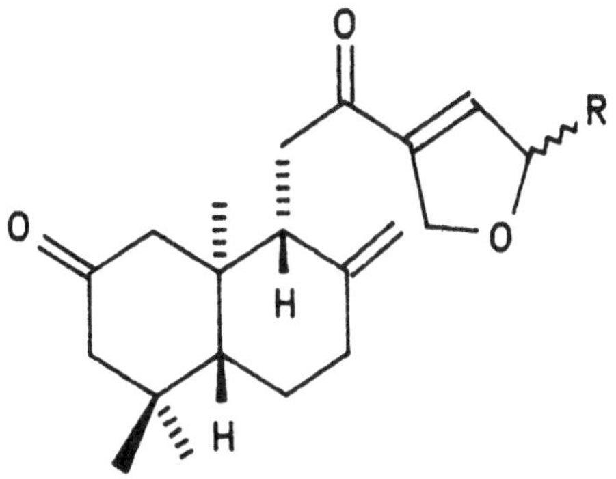

Figure 1. Structures of Diterpenes 303

417 R = α-OH

418 R = β-OH

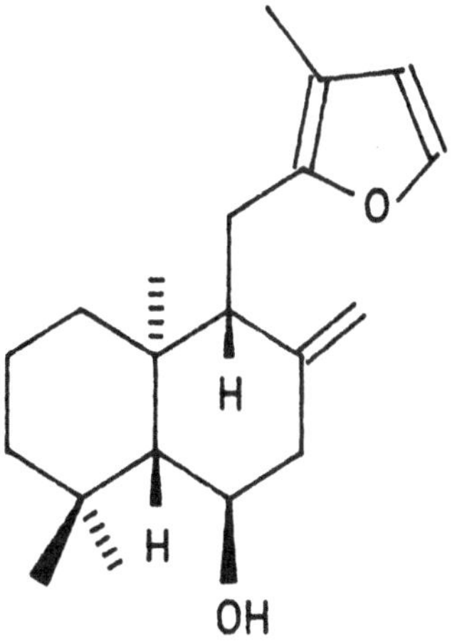

421

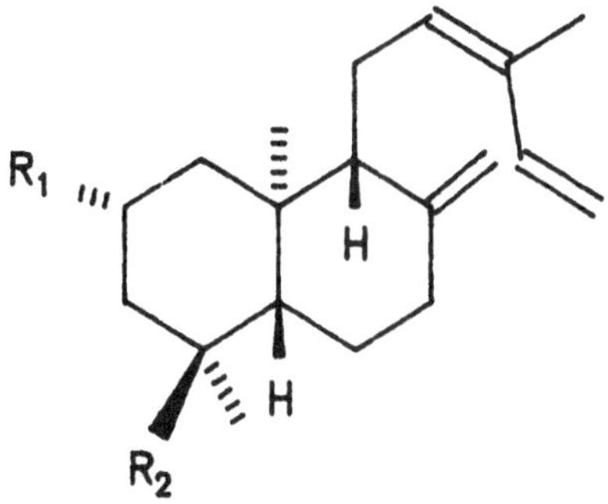

422 R_1 = H, R_2 = Me

423 R_1 = H, R_2 = CHO

424 R_1 = H, R_2 = CO_2H

426 R_1 = OH, R_2 = CO_2H

FIGURE 1. (*contd.*)

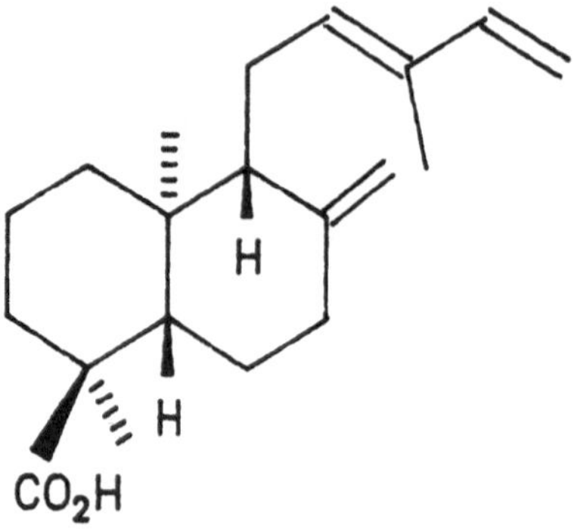

425

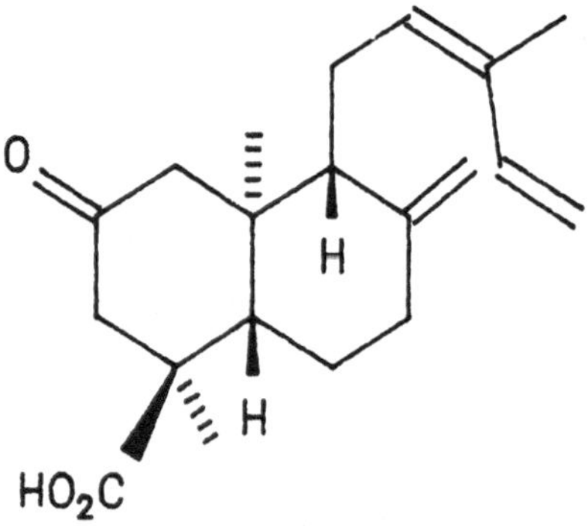

427

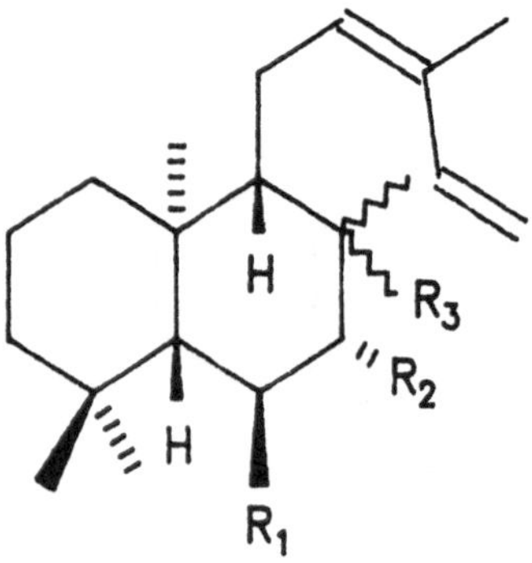

428 $R_1 = R_2 = R_3 = OH$

429 $R_1 = OAc$, $R_2 = R_3 = OH$

808 $R_1 = OAng$, $R_2 = OAc$, $R_3 = \beta\text{-OH}$

FIGURE 1. (*contd.*)

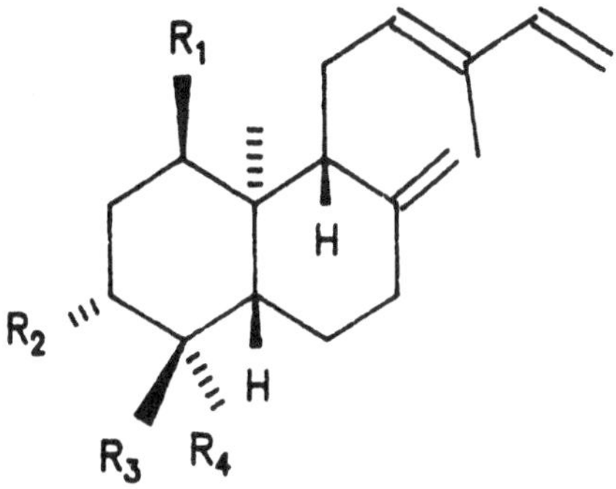

Figure 1. Structures of Diterpenes 305

430 $R_1 = R_2 = H$, $R_3 = R_4 = Me$

431 $R_1 = R_2 = H$, $R_3 = Me$, $R_4 = CH_2OH$

432 $R_1 = H$, $R_2 = OH$, $R_3 = Me$, $R_4 = CH_2OH$

433 $R_1 = R_2 = H$, $R_3 = CH_2OH$, $R_4 = Me$

434 $R_1 = R_2 = H$, $R_3 = R_4 = CH_2OH$

435 $R_1 = OAc$, $R_2 = H$, $R_3 = CO_2H$, $R_4 = Me$

1185 $R_1 = R_2 = H$, $R_3 = R_4 = CH_2OAc$

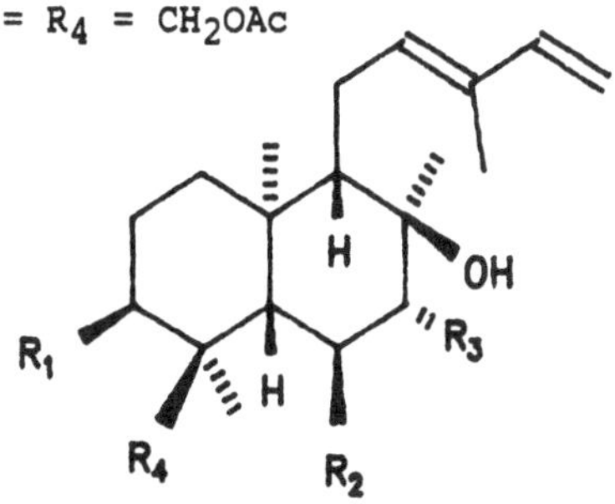

436 $R_1 = OH$, $R_2 = R_3 = H$, $R_4 = Me$

809 $R_1 = H$, $R_2 = R_3 = OH$, $R_4 = Me$

810 $R_1 = H$, $R_2 = OAc$, $R_3 = OH$, $R_4 = Me$

811 $R_1 = H$, $R_2 = OAng$, $R_3 = OAc$, $R_4 = Me$

1015 $R_1 = R_2 = R_3 = H$, $R_4 = CH_2OH$

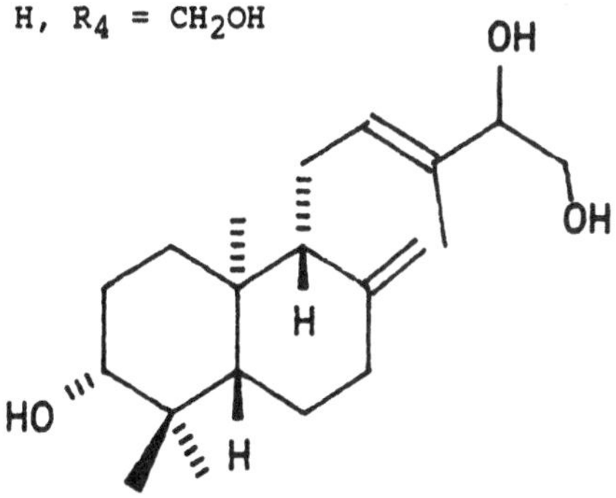

1186

FIGURE 1. (*contd.*)

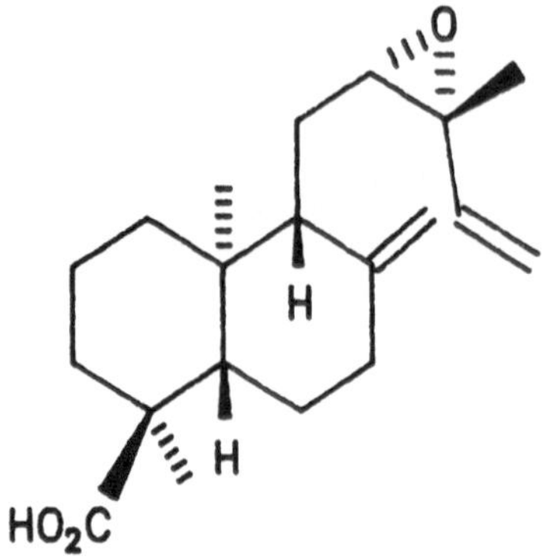

437

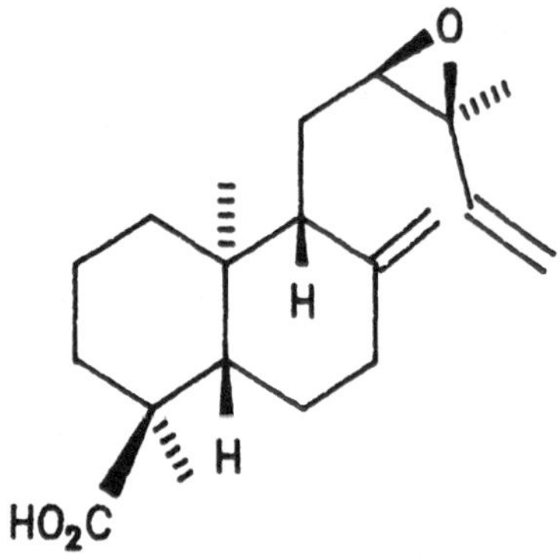

438

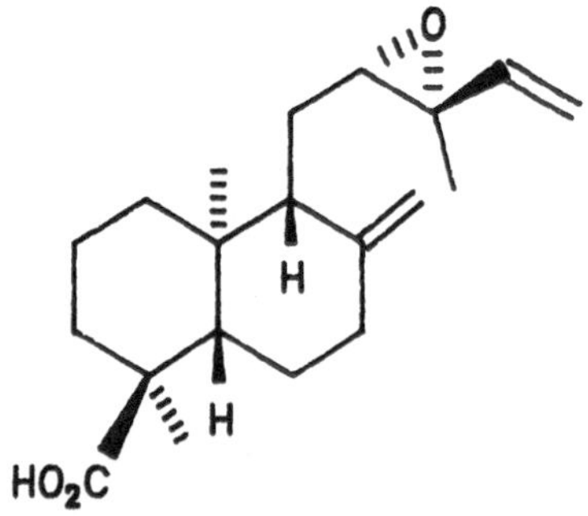

439

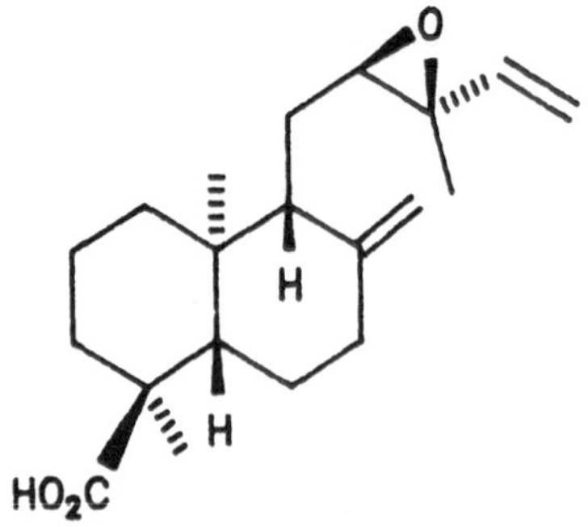

440

FIGURE 1. (contd.)

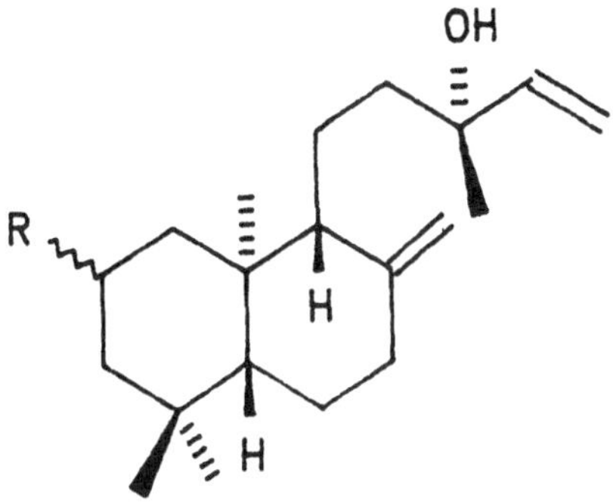

Figure 1. Structures of Diterpenes   307

442 R = H

443 R = β-OH

444 R = α-OH

445 R = β-OSucc

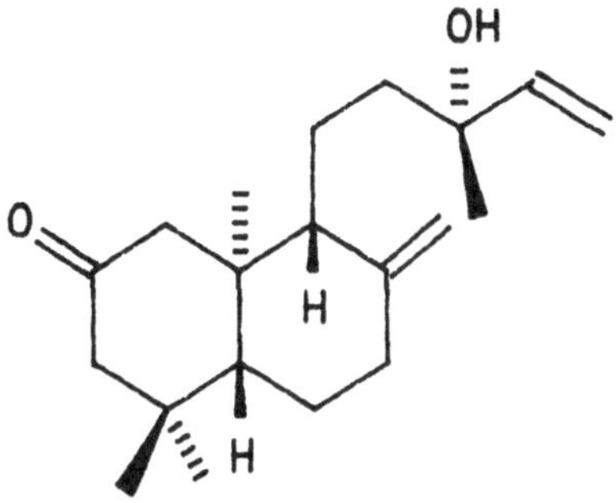

446

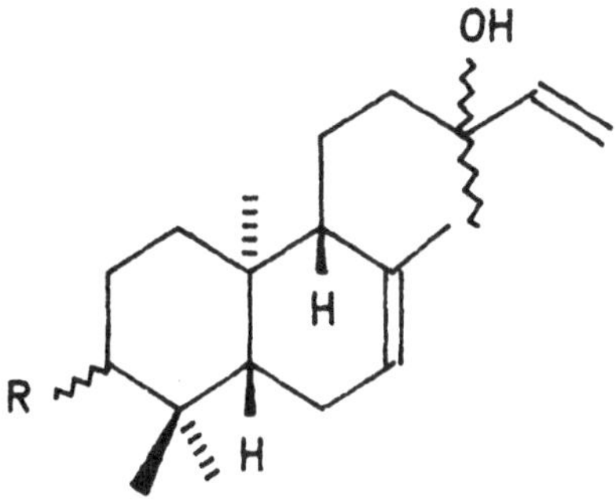

447 R = H

448 R = OH (Revised to normal-labdane; = 1106?)

FIGURE 1. (contd.)

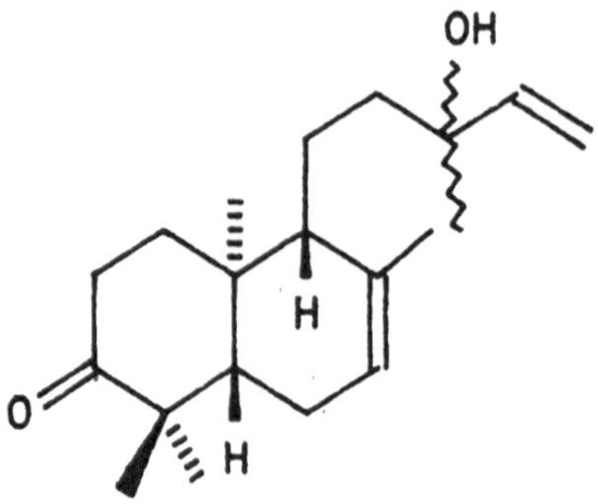

449   (Revised to normal-labdane; = 1107?)

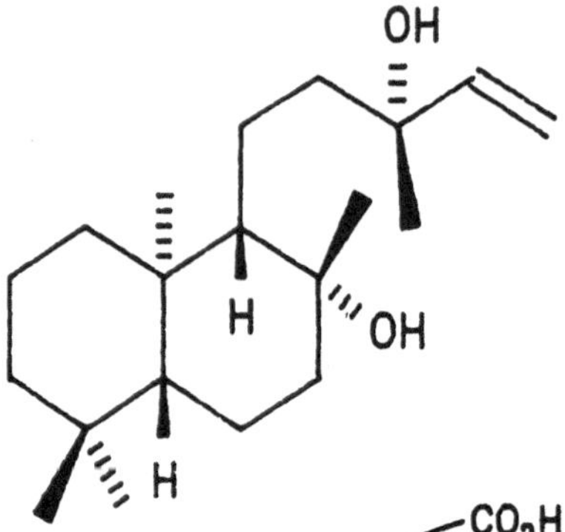

450

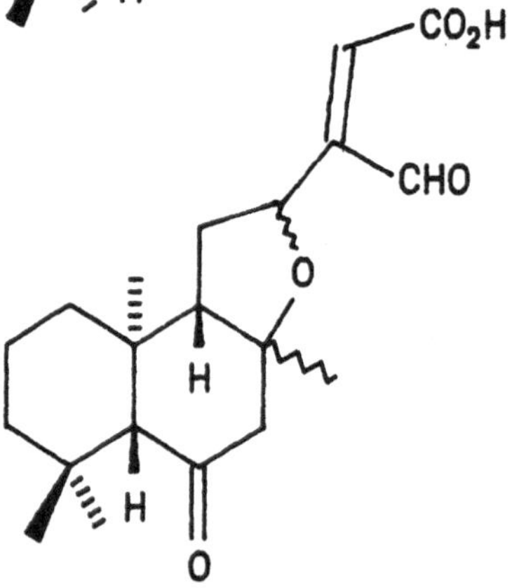

460

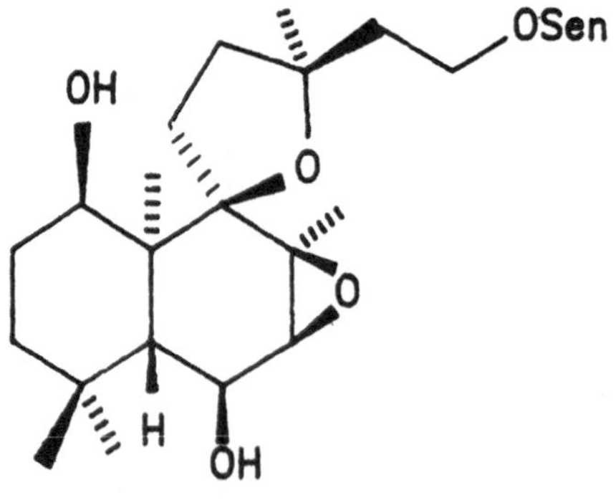

461

FIGURE 1. (contd.)

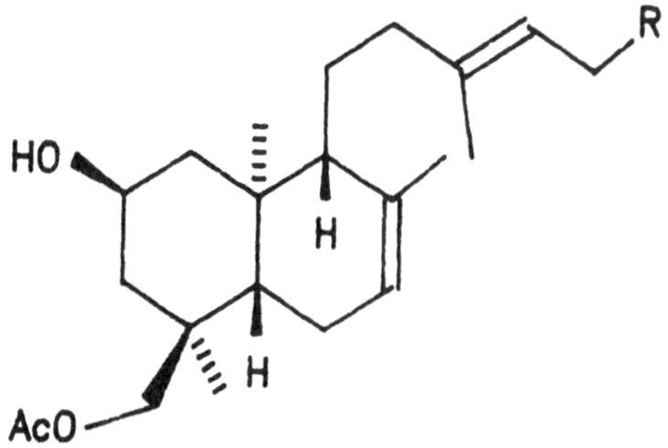

Figure 1. Structures of Diterpenes 309

1012 R = OAc

1013 R = OH

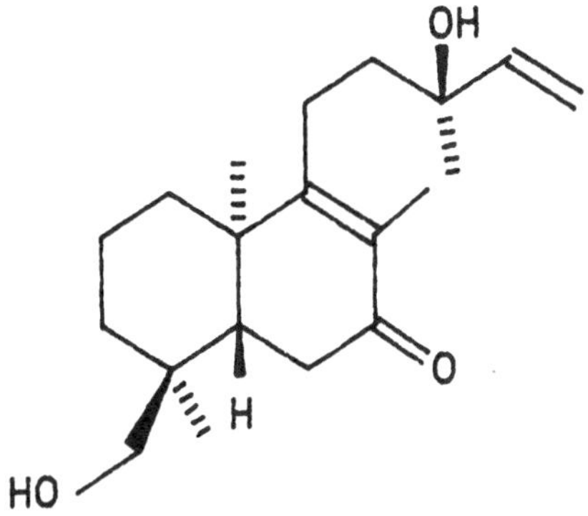

1016

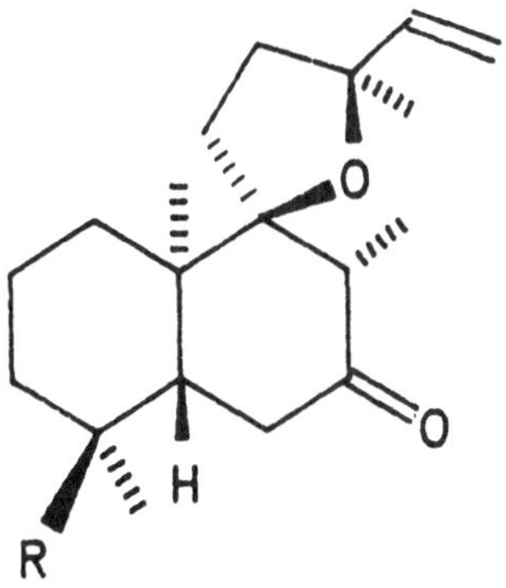

1017 R = CH$_2$OH

1018 R = CHO

1019 R = CO$_2$H

FIGURE 1. (*contd.*)

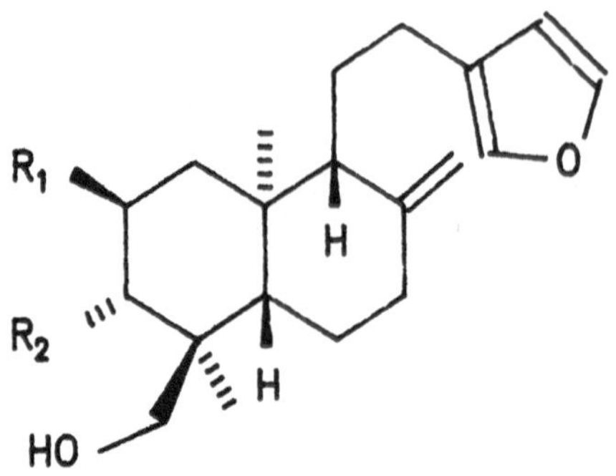

1037 R₁ = H, R₂ = OAng

1038 R₁ = OH, R₂ = OAng

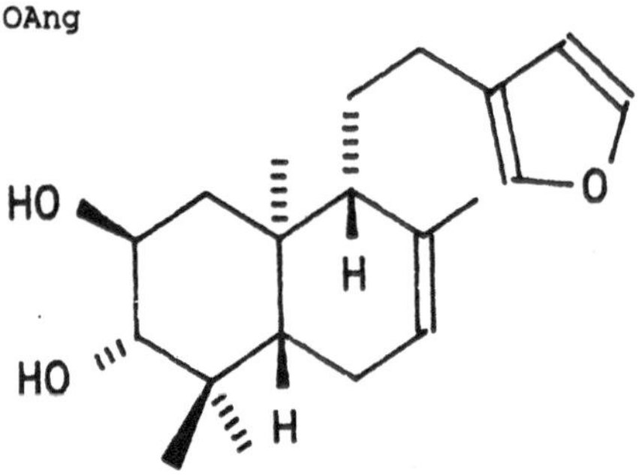

1188

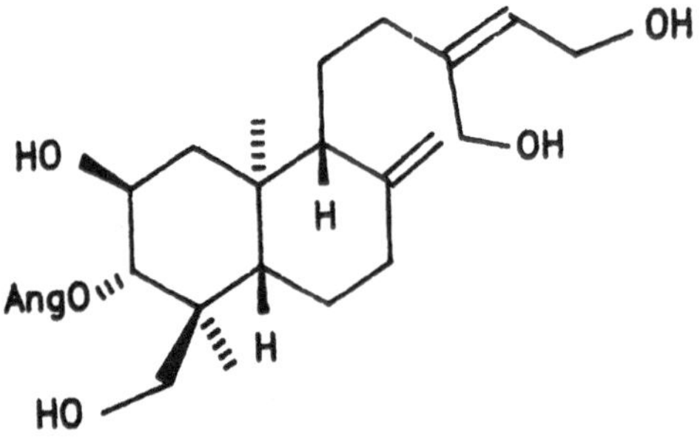

1039

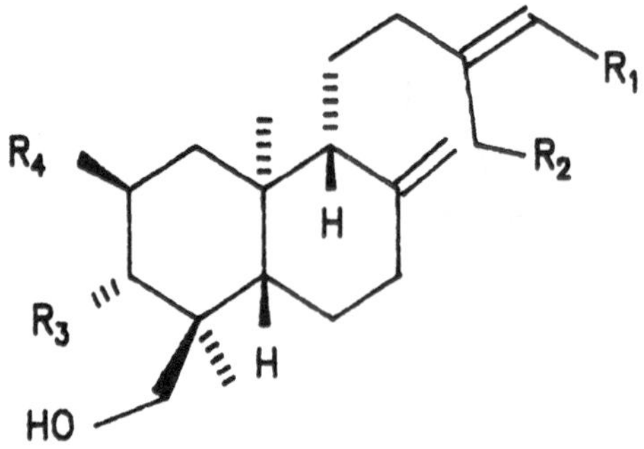

1189 R₁ = CO₂H, R₂ = H, R₃ = OH, R₄ = H

FIGURE 1. (contd.)

Figure 1. Structures of Diterpenes   311

1190 R_1 = CO_2H, R_2 = H, R_3 = OAng, R_4 = H

1191 R_1 = CH_2OH, R_2 = OH, R_3 = OAng, R_4 = OH

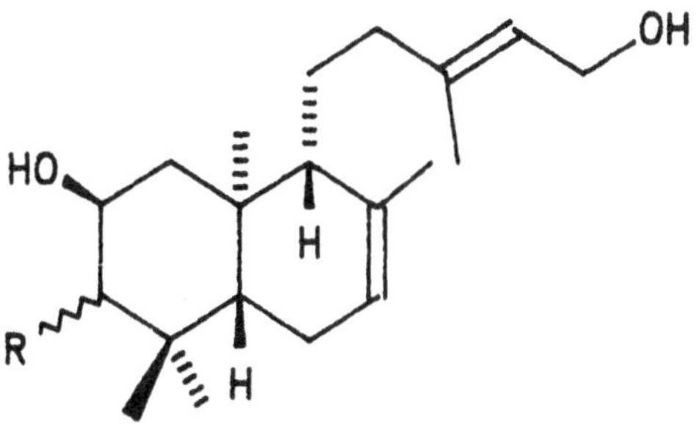

1192 R = α-OAng

1193 R = β-OH

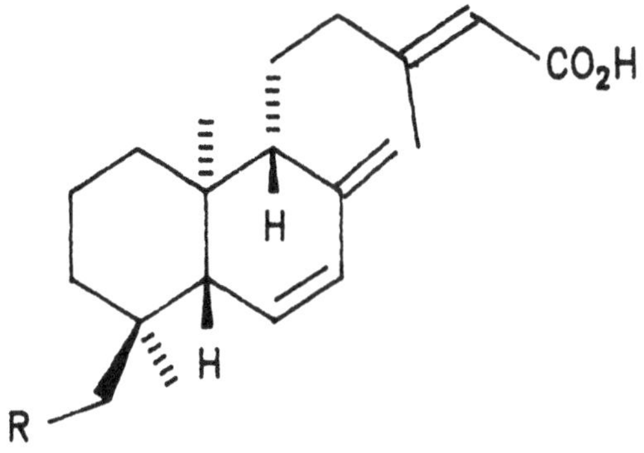

1194 R = H

1195 R = OH

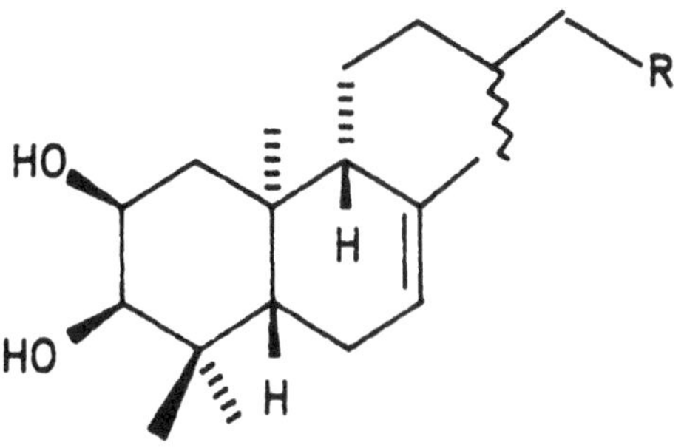

1196 R = CO_2H

1197 R = CH_2OH

FIGURE 1. (*contd.*)

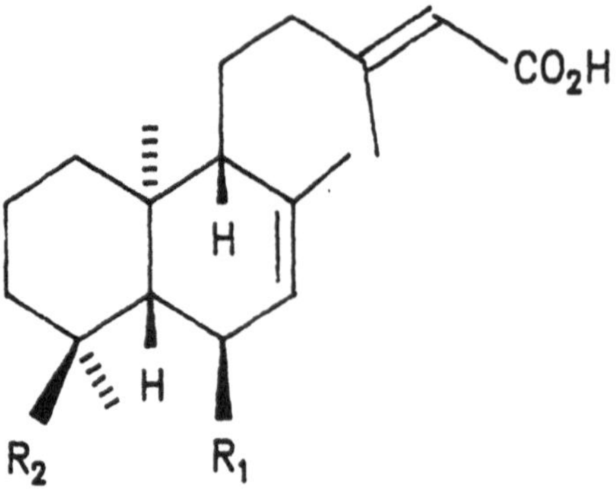

1042 R_1 = OH, R_2 = CHO

1043 R_1 = O-i-Val, R_2 = CO_2H

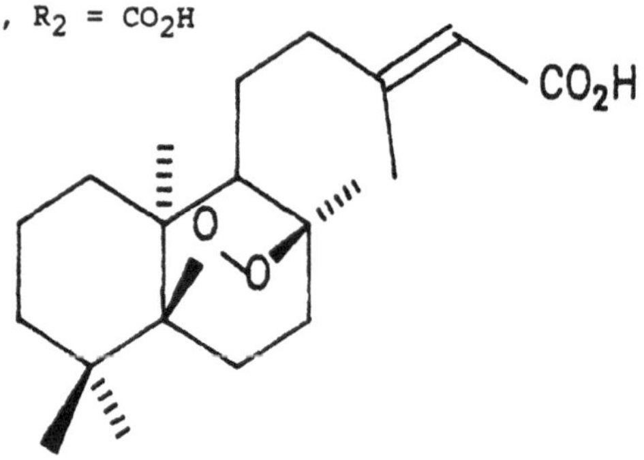

1044

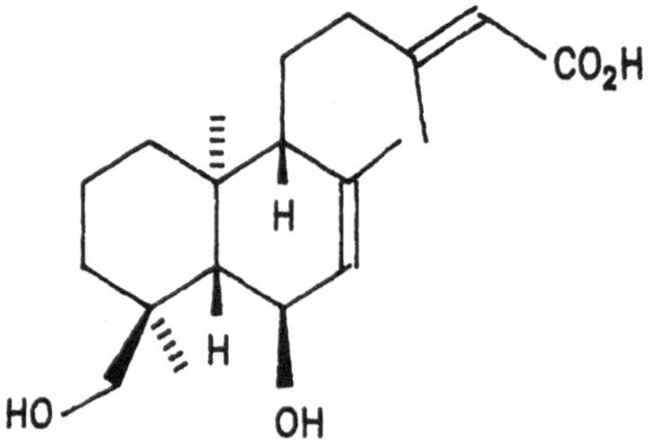

1045

39. ent-Labdane-19,6-Lactone

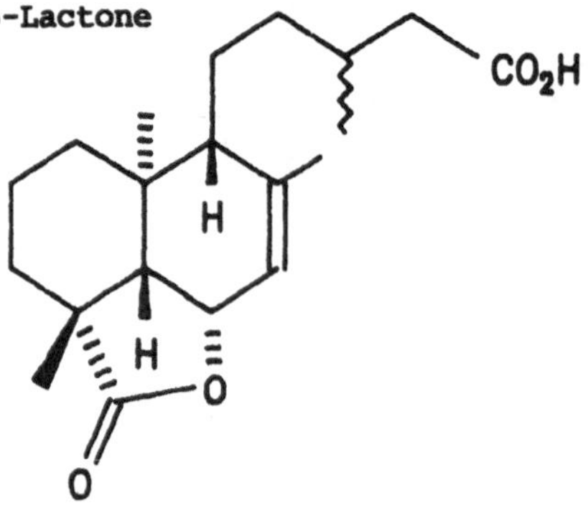

462

FIGURE 1. (*contd.*)

Figure 1. Structures of Diterpenes   313

40. ent-Labdane-8,13-Epoxide-15,16-Hemiacetal

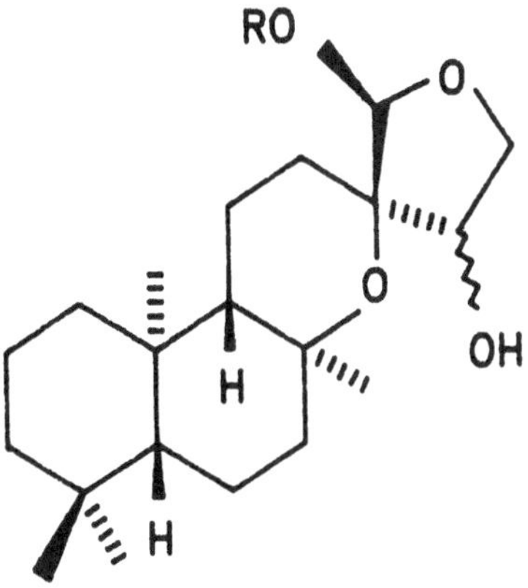

463 R = H

464 R = Ac

97. ent-Labdane Dimer

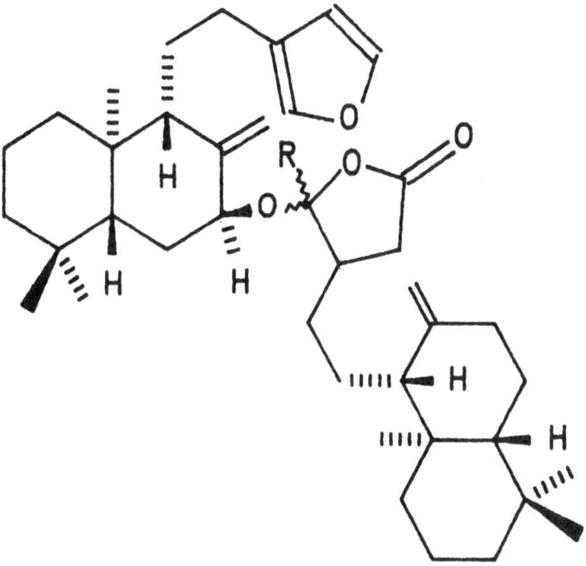

419 R = α-H

420 R = β-H

FIGURE 1. (*contd.*)

98. <u>ent</u>-Manoyl Oxide, 13-Epi-

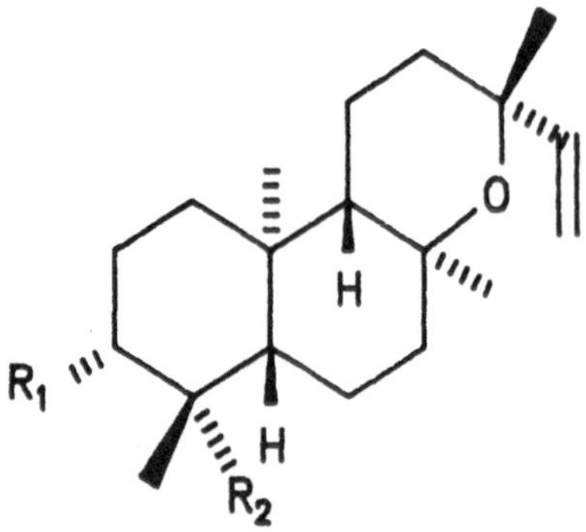

451 R$_1$ = H, R$_2$ = Me

452 R$_1$ = OH, R$_2$ = Me

455 R$_1$ = H, R$_2$ = CH$_2$OH

1198 R$_1$ = OAc, R$_2$ = Me

1199 R$_1$ = OH, R$_2$ = CH$_2$OCinn

1200 R$_1$ = OH, R$_2$ = CH$_2$OH

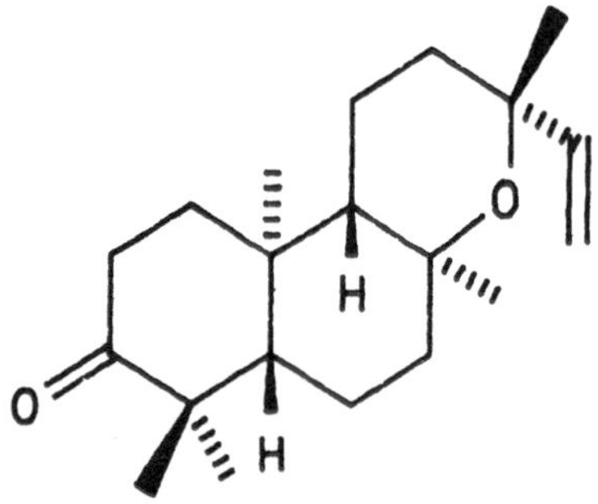

453

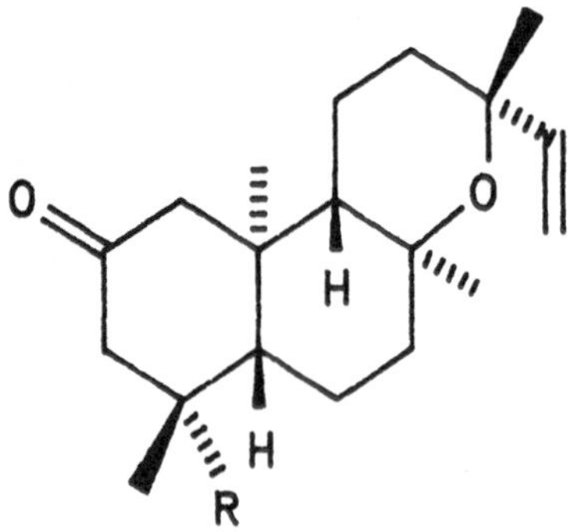

454 R = Me

456 R = CH$_2$OH

FIGURE 1. (*contd.*)

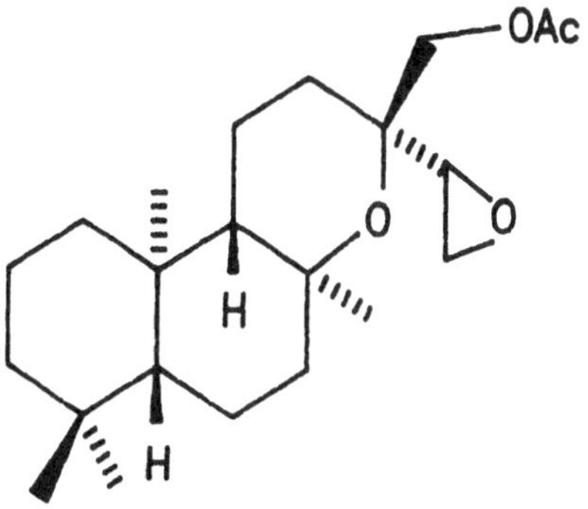

Figure 1. Structures of Diterpenes 315

457

99. ent-Manoyl Oxide

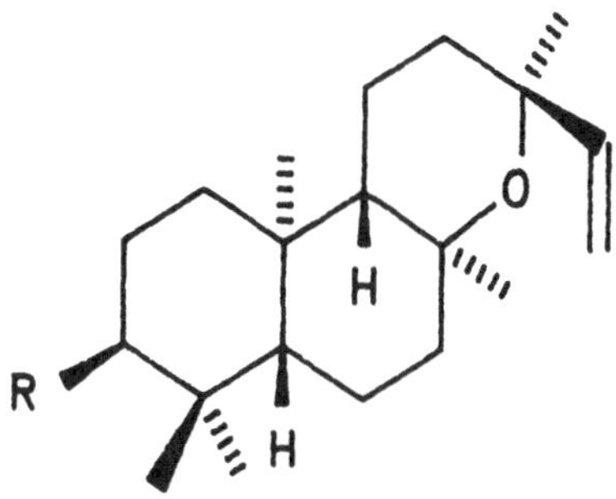

458 R = H

459 R = OH

41. ent-Labdane, 17-Nor-

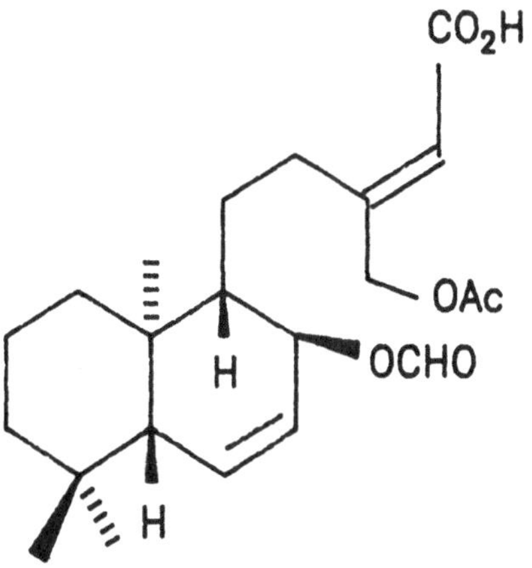

465

FIGURE 1. (*contd.*)

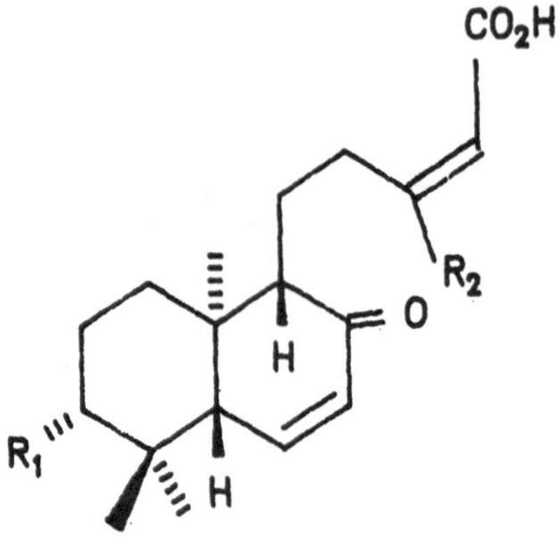

466 R_1 = H, R_2 = CH_2OAc

467 R_1 = OH, R_2 = Me

1201 R_1 = OH, R_2 = Me (6,7-dihydro)

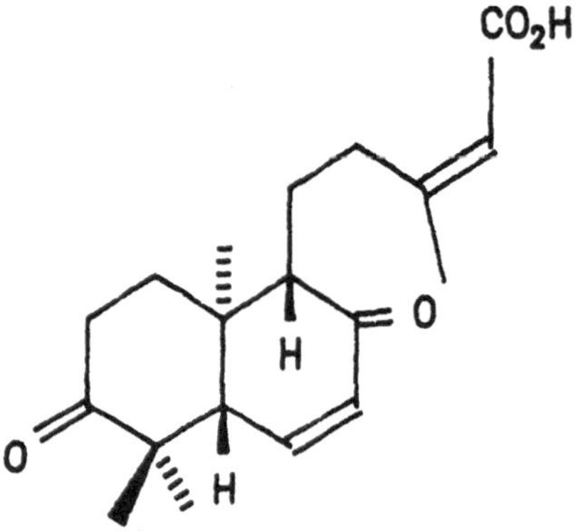

468

1202 (6,7-dihydro)

42. ent-Labdane, 15-Nor-

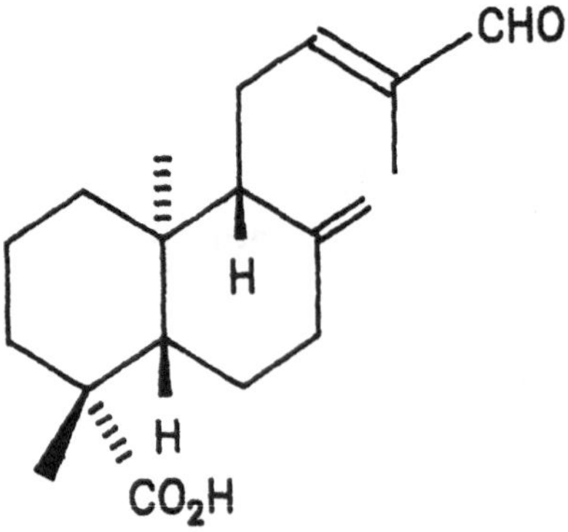

469

FIGURE 1. (*contd.*)

Figure 1. Structures of Diterpenes 317

43. <u>ent</u>-Labdane, 14,15-Bis-Nor-

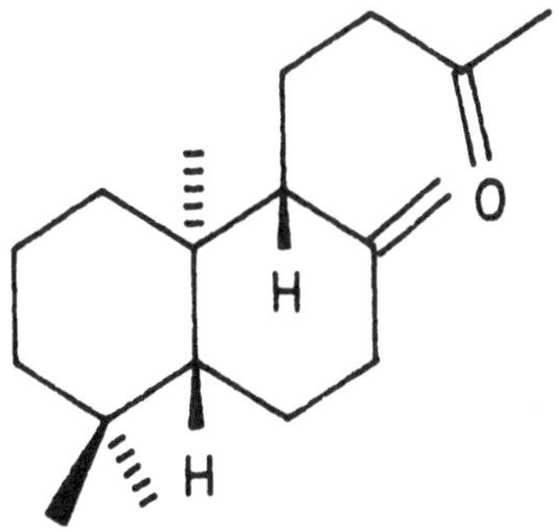

470

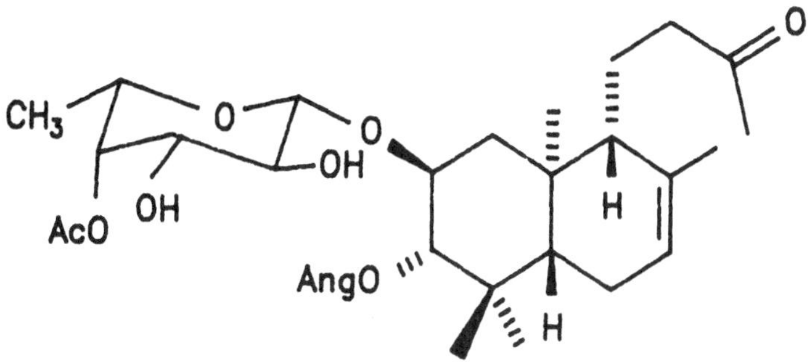

1203

44. <u>ent</u>-Labdane, 13,14,15,16-Tetra-Nor-

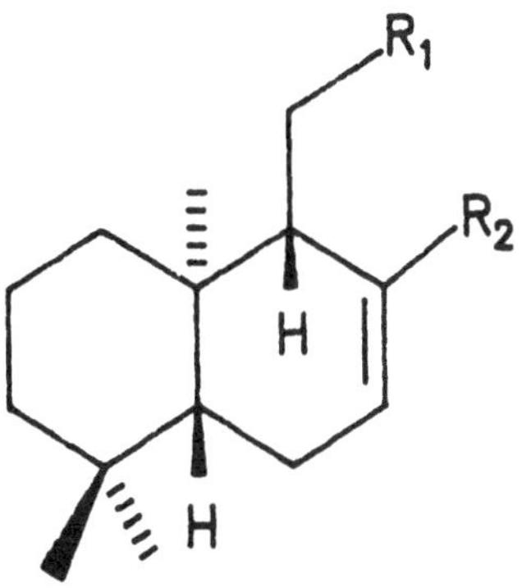

471 R_1 = CHO, R_2 = Me

472 R_1 = CO_2H, R_2 = CH_2OH

FIGURE 1. (*contd.*)

45. ent-Labdane, 7,8-Seco-

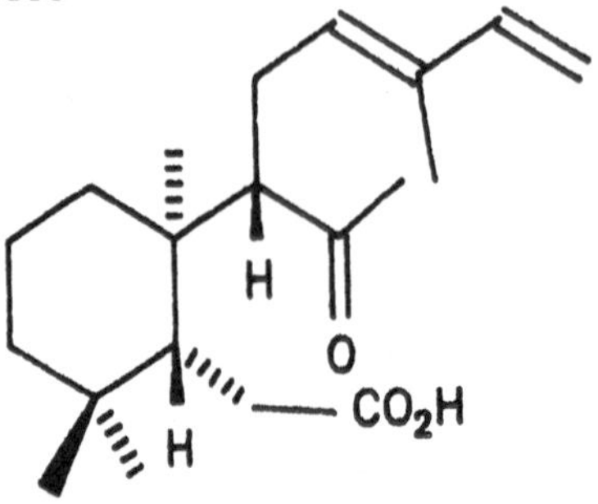

473

74. ent-Labdane, 18-Nor-

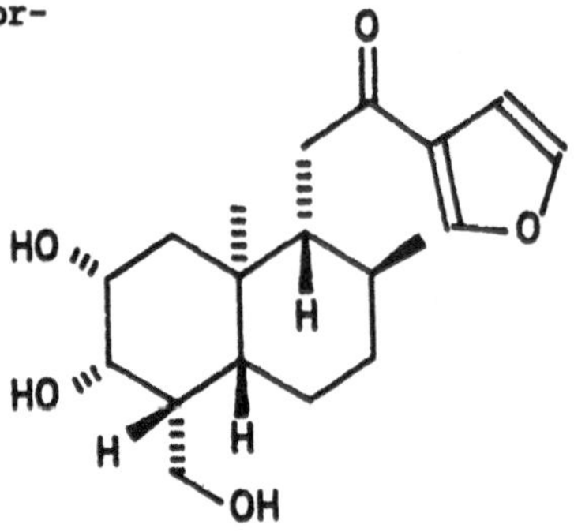

828   (Structure revised; = 1206)

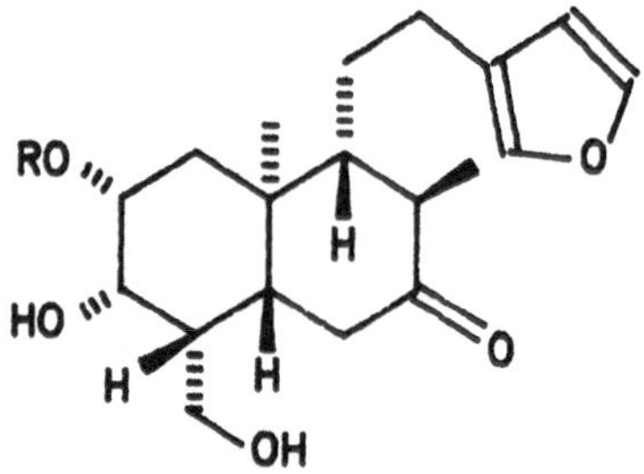

1205 R = Ac

1206 R = H

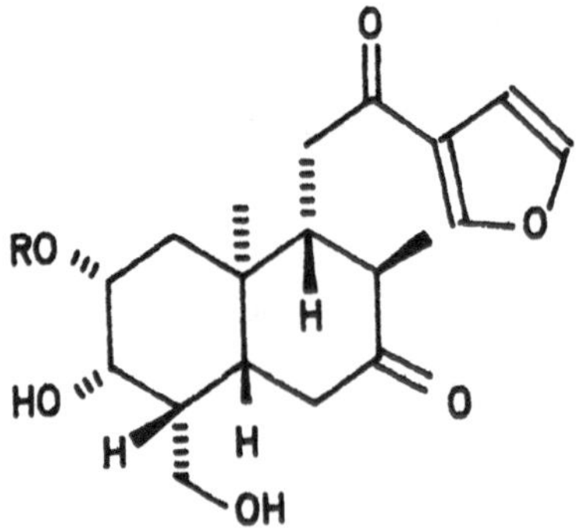

829 R = H

1207 R = Ac

FIGURE 1. (contd.)

Figure 1. Structures of Diterpenes   319

95. <u>ent</u>-Friedolabdane

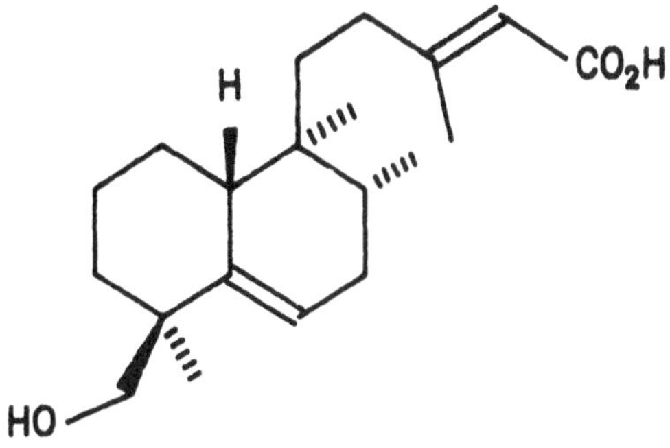

1059

1208 (Illustrated in report as 13,14-dihydro-homolog of **1059**,
   but text indicates that **1208** is identical to **1059**.)

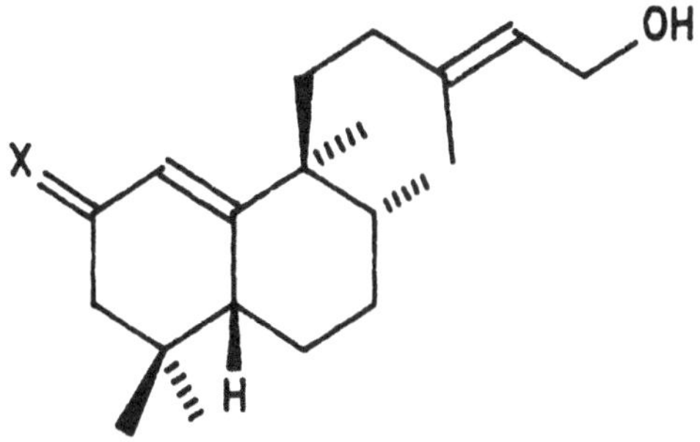

1209 X = O

1210 X = α-OH,H

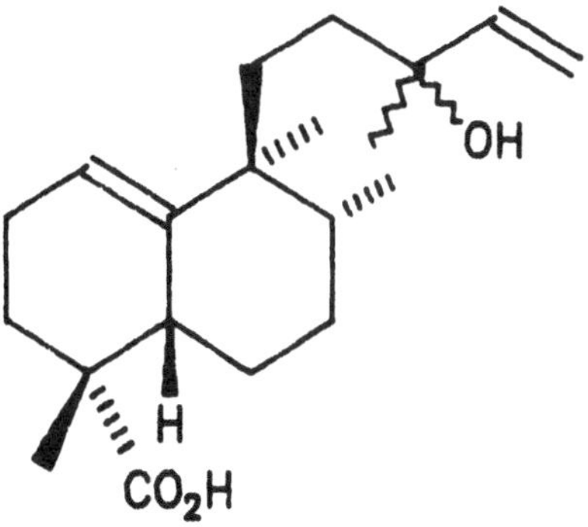

1211

FIGURE 1. (*contd.*)

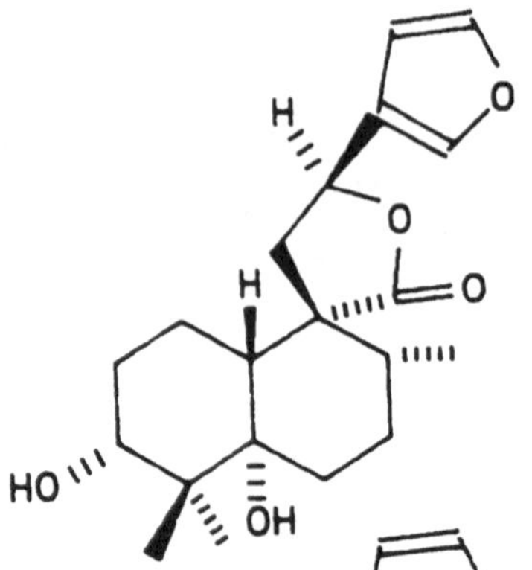

1212

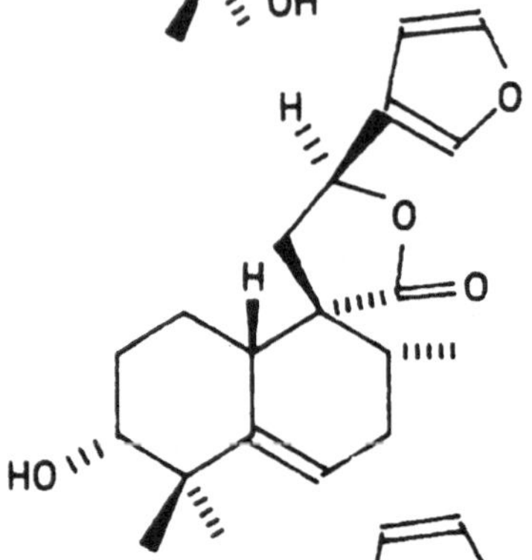

1213

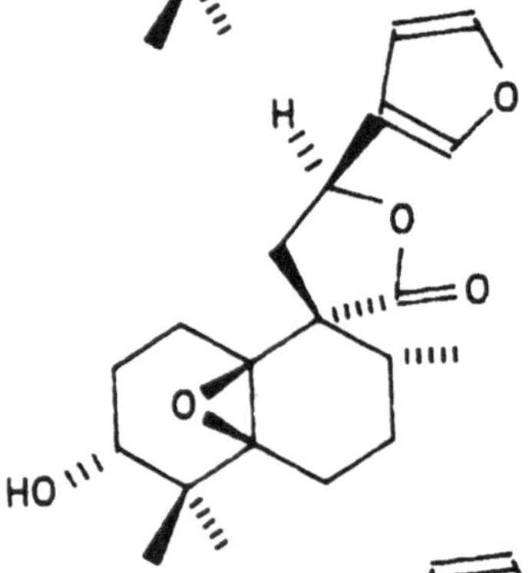

1214

102. ent-Friedolabdane, 3,4-seco-

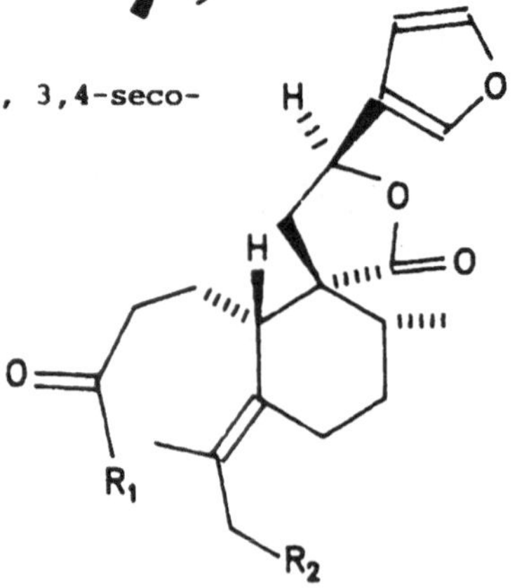

1215 R$_1$ = R$_2$ = H

FIGURE 1. (contd.)

Figure 1. Structures of Diterpenes   321

1216 R$_1$ = OH, R$_2$ = H

1217 R$_1$ = R$_2$ = OH

1218 R$_1$ = OMe, R$_2$ = OH

1219

1220

103. <u>ent</u>-Friedolabdane, rearranged-

1221

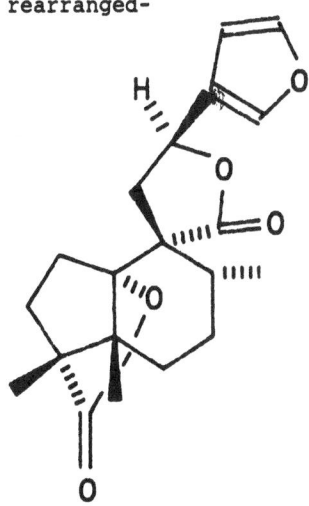

FIGURE 1. (*contd.*)

46. <u>trans</u>-<u>ent</u>-Clerodane

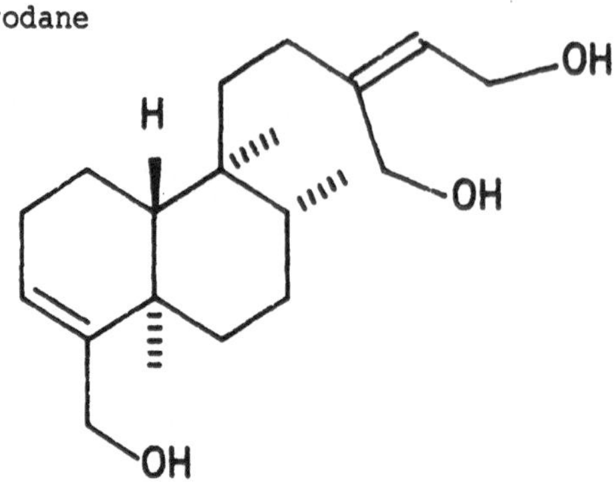

1222

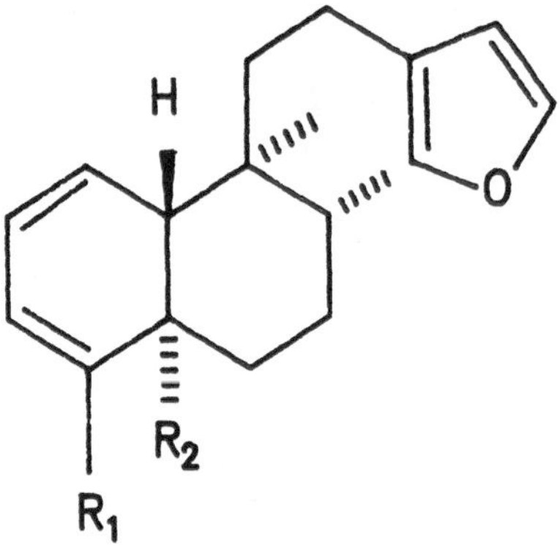

248 R_1 = CO_2H, R_2 = Me (= 512 = 250?)

251 R_1 = CO_2Me, R_2 = Me

911 R_1 = CO_2H, R_2 = CH_2OAc

1070 R_1 = CO_2H, R_2 = CH_2OH

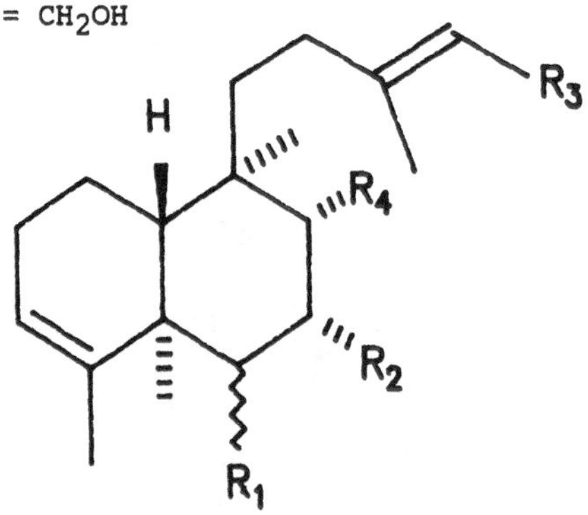

475 R_1 = R_2 = H, R_3 = CH_2OH, R_4 = Me

476 R_1 = R_2 = H, R_3 = R_4 = CH_2OH

1010 R_1 = R_2 = H, R_3 = CH_2OArac, R_4 = CH_2OH

1011 R_1 = R_2 = H, R_3 = $CH_2OBehen$, R_4 = CH_2OH

FIGURE 1. (*contd.*)

Figure 1. Structures of Diterpenes 323

495 $R_1 = R_2 = H$, $R_3 = CO_2Me$, $R_4 = Me$

496 $R_1 = \beta\text{-OAc}$, $R_2 = H$, $R_3 = CO_2Me$, $R_4 = Me$

497 $R_1 = \beta\text{-OAng}$, $R_2 = H$, $R_3 = CO_2Me$, $R_4 = Me$

498 $R_1 = \alpha\text{-OAng}$, $R_2 = H$, $R_3 = CO_2H$, $R_4 = Me$

499 $R_1 = \alpha\text{-OTig}$, $R_2 = H$, $R_3 = CO_2H$, $R_4 = Me$

500 $R_1 = H$, $R_2 = OAc$, $R_3 = CO_2H$, $R_4 = Me$

535 $R_1 = R_2 = H$, $R_3 = CHO$, $R_4 = CH_2OH$

537 $R_1 = R_2 = H$, $R_3 = CHO$, $R_4 = CO_2H$

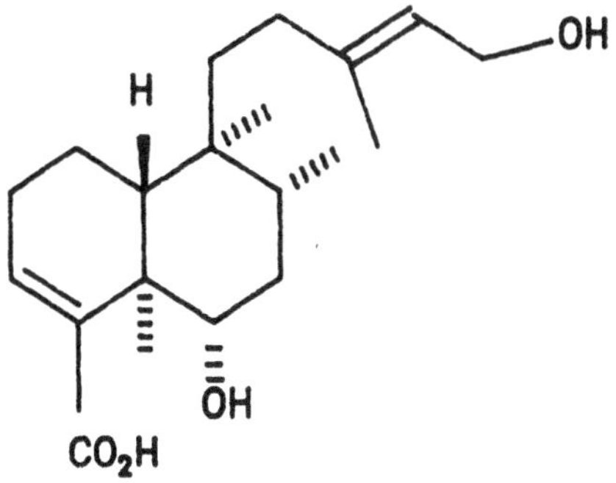

1223

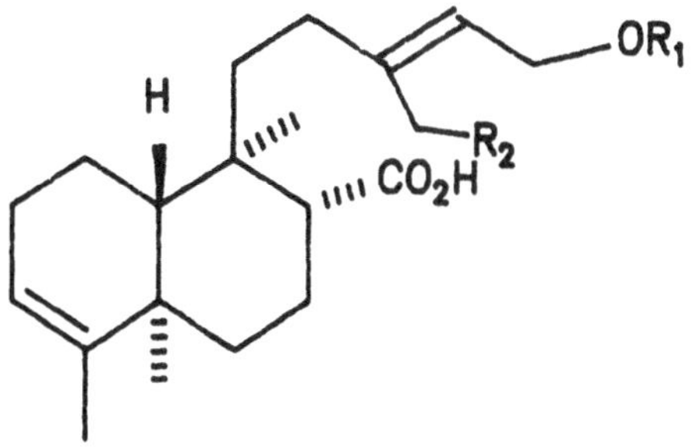

1224 $R_1 = R_2 = H$

1225 $R_1 = Ac$, $R_2 = OH$

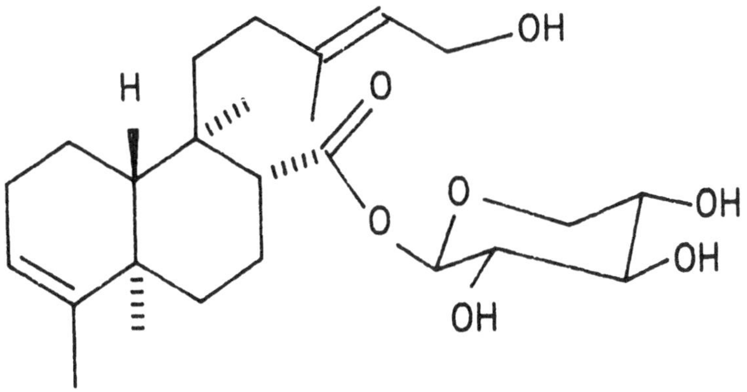

1226

FIGURE 1. (*contd.*)

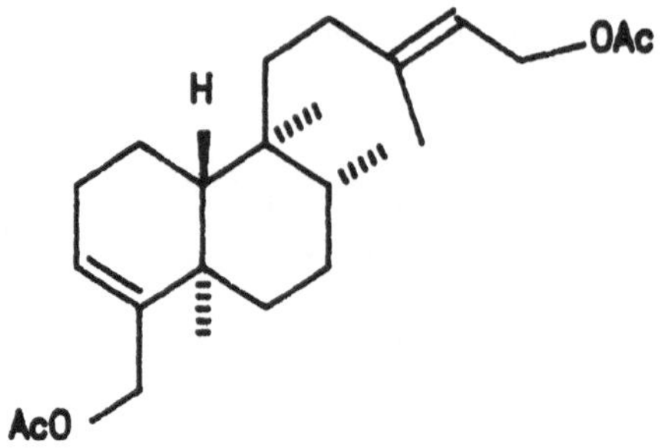

1227

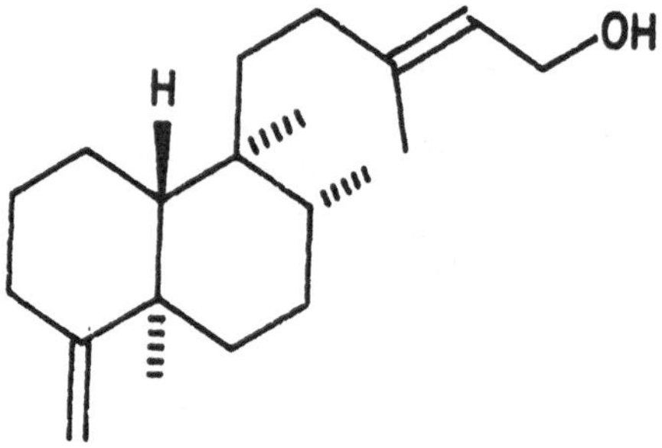

236

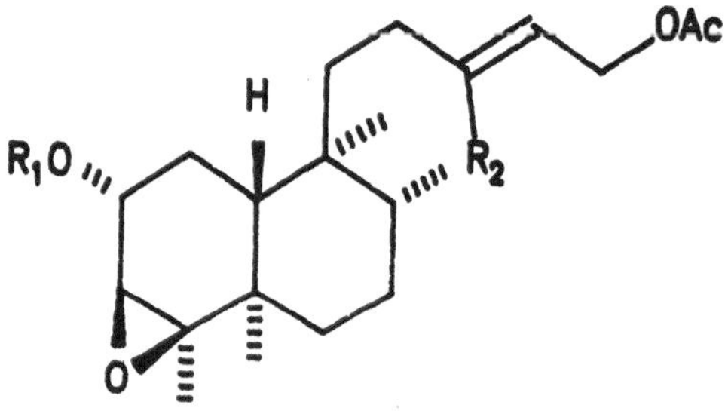

477 R$_1$ = H, R$_2$ = Me

478 R$_1$ = H, R$_2$ = CH$_2$OH

1228 R$_1$ = Ac, R$_2$ = CH$_2$OH

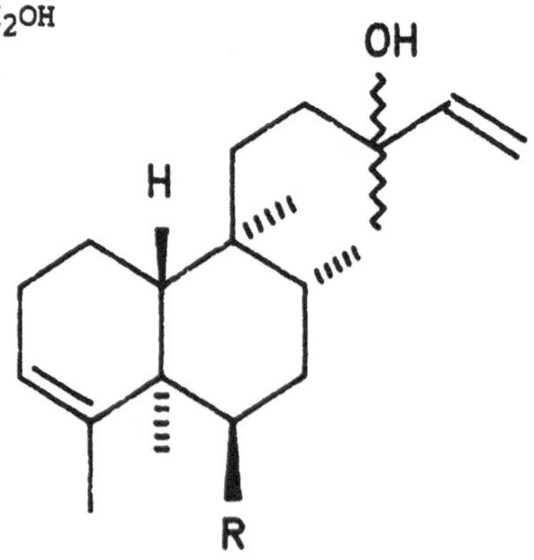

479 R = H

480 R = OAng

FIGURE 1. (*contd.*)

Figure 1. Structures of Diterpenes

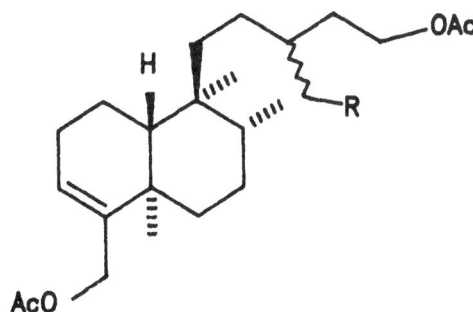

481 R_1 = H, R_2 = R_3 = CH_2OH, R_4 = Me

482 R_1 = H, R_2 = CH_2OArac, R_3 = CH_2OH, R_4 = Me

483 R_1 = H, R_2 = $CH_2OBehen$, R_3 = CH_2OH, R_4 = Me

484 R_1 = H, R_2 =CH_2OH, R_3 = Me, R_4 = CH_2OSucc

486 R_1 = H, R_2 = CH_2OH, R_3 = CHO, R_4 = Me

488 R_1 = OH, R_2 = CH_2OH, R_3 = CH_2OAc, R_4 = CH_2OH

489 R_1 = OH, R_2 = CH_2OH, R_3 = $CH_2OPhe-Ac$, R_4 = CH_2OH

490 R_1 = OH, R_2 = CH_2OH, R_3 = $CH_2OPhe-Ac$, R_4 = CHO

491 R_1 = OH, R_2 = CH_2OH, R_3 = CH_2OAc, R_4 = CHO

492 R_1 = H, R_2 = CHO, R_3 = CH_2OH, R_4 = Me

503 R_1 = H, R_2 = CO_2H, R_3 = R_4 = Me

504 R_1 = H, R_2 = CO_2H, R_3 = Me, R_4 = CH_2OAc

1229 R = H

1230 R = OAc

FIGURE 1. (*contd.*)

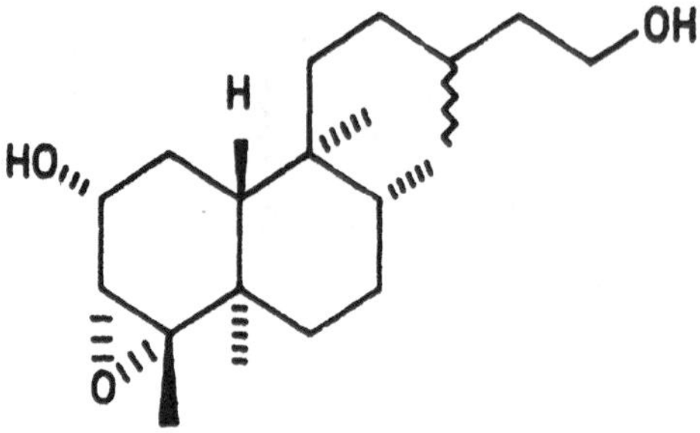

485

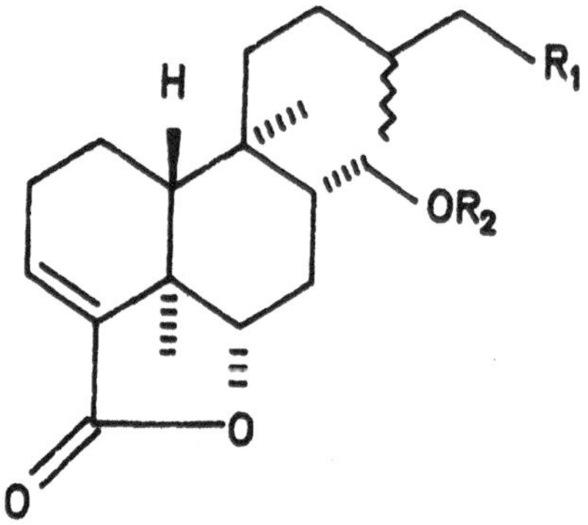

487 R$_1$ = CH$_2$OH, R$_2$ = Ac

522 R$_1$ = CO$_2$H, R$_2$ = Phe-Ac

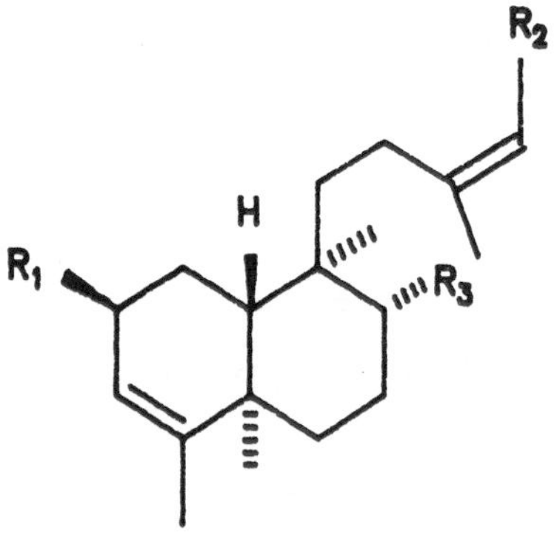

493 R$_1$ = H, R$_2$ = CO$_2$H, R$_3$ = Me

494 R$_1$ = OAc, R$_2$ = CO$_2$H, R$_3$ = Me

534 R$_1$ = H, R$_2$ = CHO, R$_3$ = CH$_2$OH

536 R$_1$ = H, R$_2$ = CHO, R$_3$ = CO$_2$H

FIGURE 1. (*contd.*)

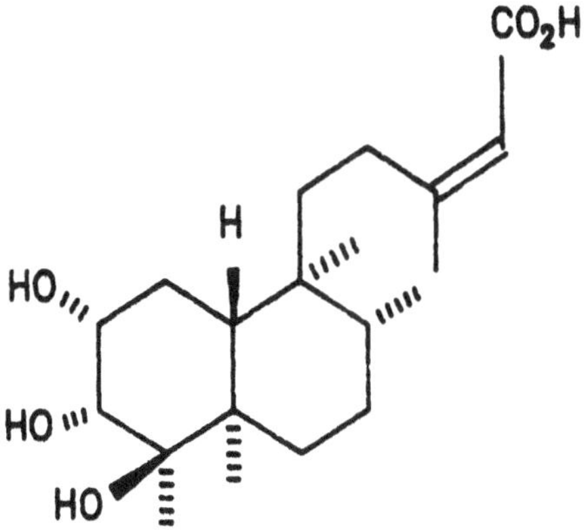

Figure 1. Structures of Diterpenes   327

962

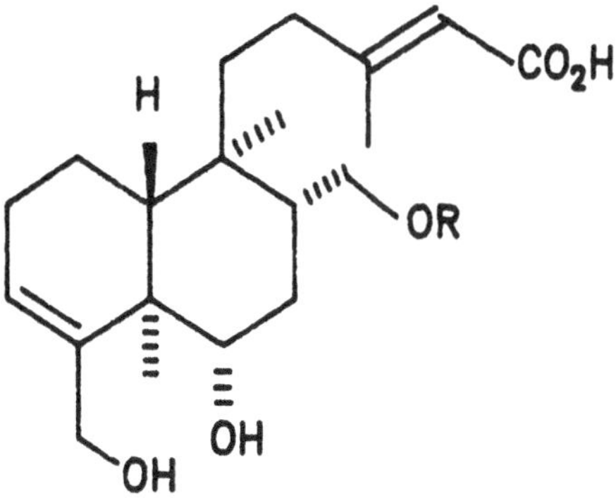

501 R = Ac

502 R = Phe-Ac

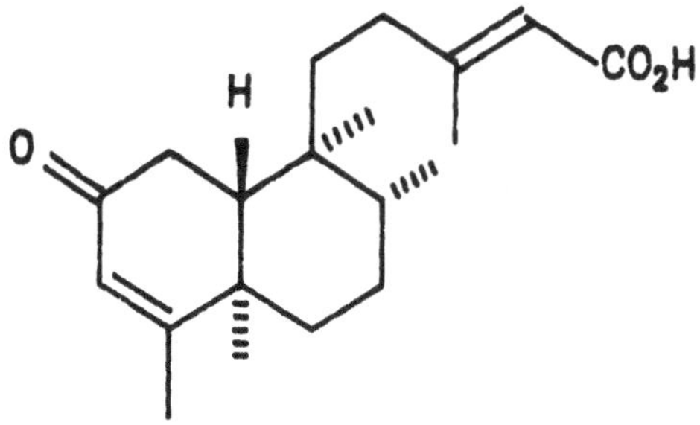

961

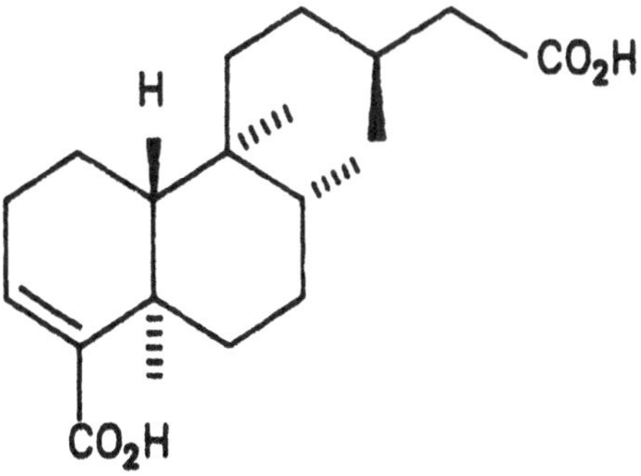

505

FIGURE 1. (*contd.*)

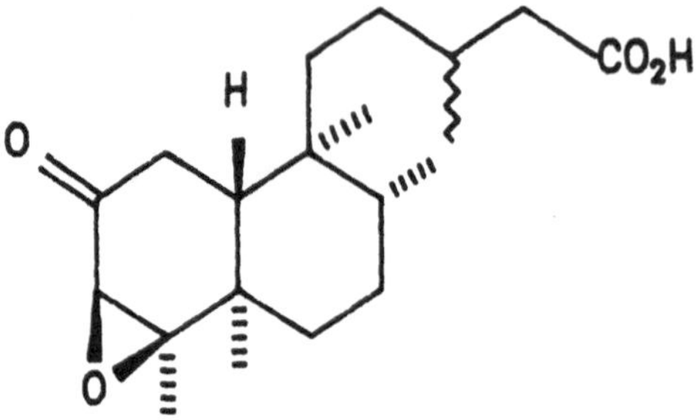

506

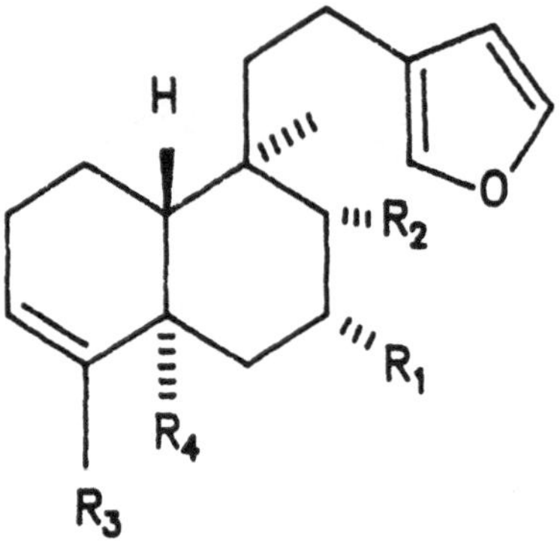

507 R_1 = H, R_2 = R_3 = R_4 = Me

508 R_1 = H, R_2 = Me, R_3 = CH_2OMal, R_4 = Me

509 R_1 = OH, R_2 = Me, R_3 = CH_2OH, R_4 = Me

510 R_1 = H, R_2 = Me, R_3 = CH_2OMal, R_4 = CH_2OAc

846 R_1 = H, R_2 = Me, R_3 = R_4 = CH_2OH

511 R_1 = H, R_2 = Me, R_3 = CO_2H, R_4 = Me (= **245**? = **249**?)

967 R_1 = H, R_2 = CHO, R_3 = R_4 = Me

518 R_1 = H, R_2 = Me, R_3 = CO_2H, R_4 = CH_2OH (=**1033**)

900 R_1 = H, R_2 = Me, R_3 = CO_2H, R_4 = CH_2OAc

901 R_1 = H, R_2 = Me, R_3 = CO_2H, R_4 = CH_2OMe

902 R_1 = H, R_2 = Me, R_3 = CO_2H, R_4 = CH_2OAng

903 R_1 = H, R_2 = Me, R_3 = CO_2H, R_4 = CH_2Oi-Val

904 R_1 = H, R_2 = Me, R_3 = CO_2H, R_4 = CHO

1231 R_1 = H, R_2 = Me, R_3 = CH_2OH, R_4 = CH_2OMal

1232 R_1 = H, R_2 = Me, R_3 = CH_2OH, R_4 = Me

1233 R_1 = H, R_2 = Me, R_3 = R_4 = CH_2OAc

FIGURE 1. (*contd.*)

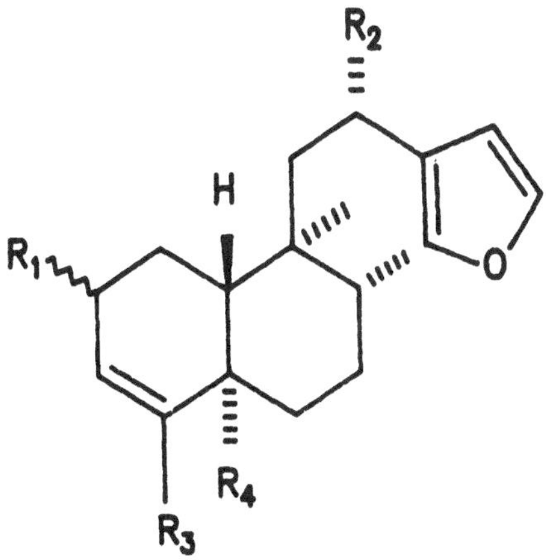

Figure 1. Structures of Diterpenes 329

953 R_1 = H, R_2 = O2-Mebut, R_3 = CO_2H, R_4 = Me

513 R_1 = α-OH, R_2 = H, R_3 = CO_2H, R_4 = Me

519 R_1 = β-OH, R_2 = H, R_3 = CO_2H, R_4 = CH_2OH

1034 R_1 = β-OH, R_2 = H, R_3 = CO_2H, R_4 = CH_2OH (=**519**)

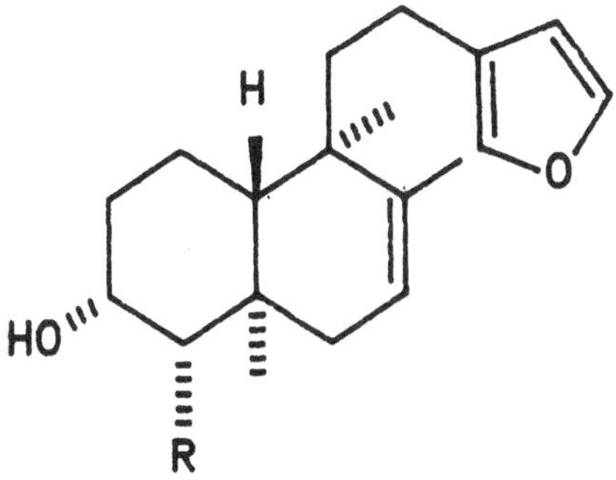

514 R = CH_2Oi-Val

515 R = CH_2O2-Mebut

516 R = CH_2OAng

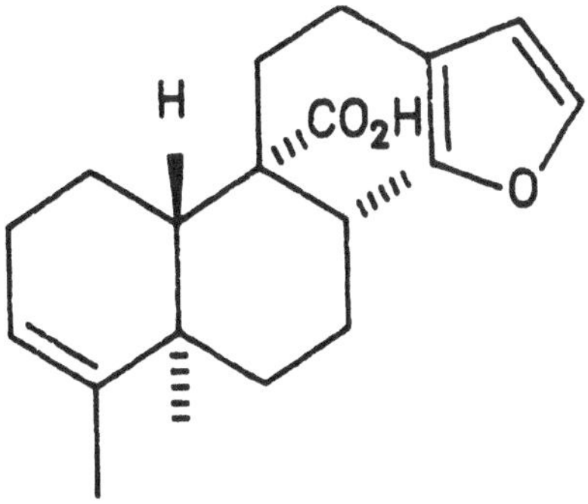

524

FIGURE 1. (*contd.*)

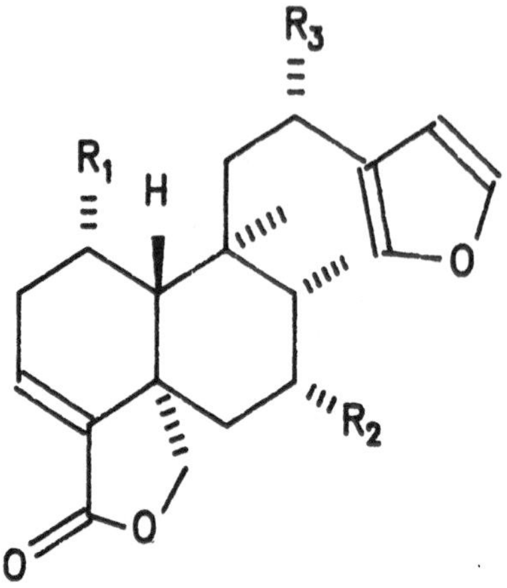

520    $R_1 = R_2 = R_3 = H$

521    $R_1 = OH$, $R_2 = R_3 = H$

950    $R_1 = R_2 = H$, $R_3 = OH$

951    $R_1 = H$, $R_2 = R_3 = OH$

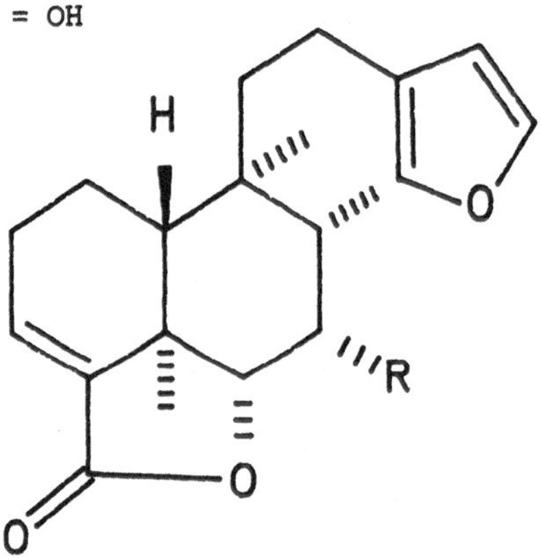

1234   R = H

1235   R = OH

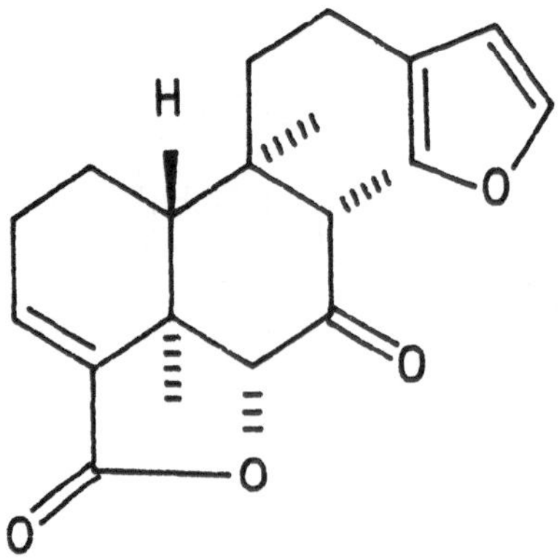

523

FIGURE 1. (*contd.*)

Figure 1. Structures of Diterpenes   331

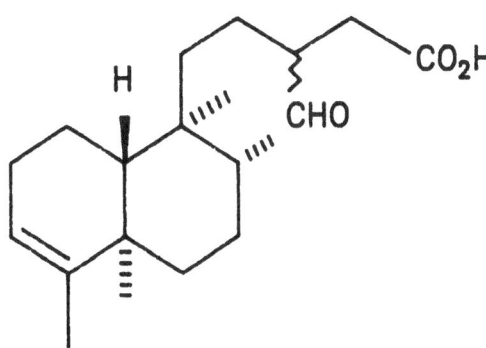

865

538 R_1 = H, R_2 = CH_2OH

539 R_1 = Ac, R_2 = CH_2OH

540 R_1 = H, R_2 = CHO

541 R_1 = Ac, R_2 = CHO

542

FIGURE 1. (*contd.*)

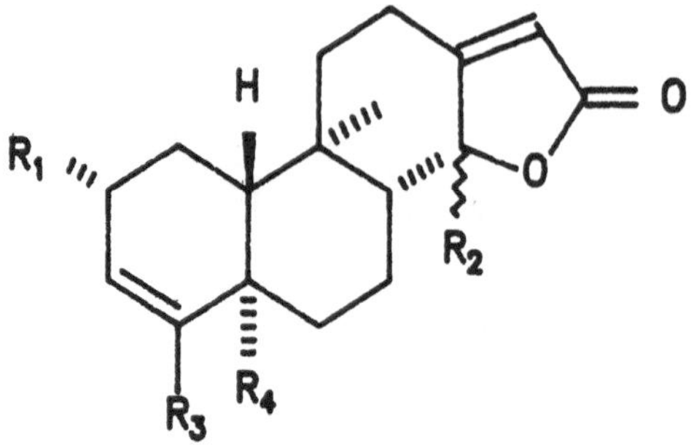

543   $R_1 = R_2 = H$, $R_3 = R_4 = Me$

544   $R_1 = OH$, $R_2 = H$, $R_3 = R_4 = Me$

864   $R_1 = H$, $R_2 = OH$, $R_3 = R_4 = Me$

550   $R_1 = R_2 = H$, $R_3 = CH_2OH$, $R_4 = Me$

558   $R_1 = R_2 = H$, $R_3 = CO_2H$, $R_4 = Me$

909   $R_1 = H$, $R_2 = \alpha\text{-}OH$, $R_3 = CO_2H$, $R_4 = CH_2OH$

910   $R_1 = H$, $R_2 = \beta\text{-}OH$, $R_3 = CO_2H$, $R_4 = CH_2OH$

1236 $R_1 = H$, $R_2 = OMe$, $R_3 = CH_2OAc$, $R_4 = Me$

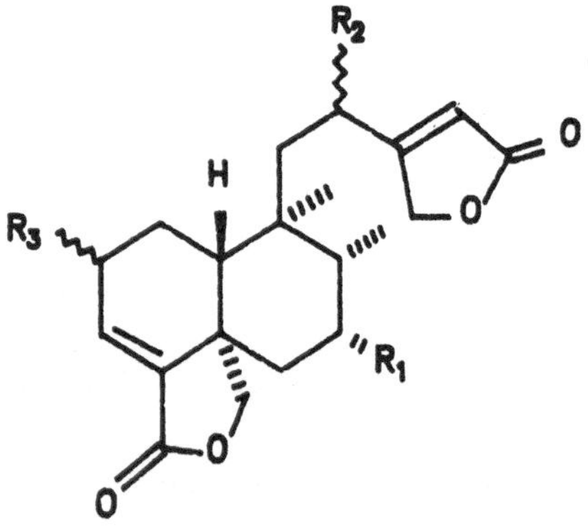

559   $R_1 = OH$, $R_2 = H$, $R_3 = H$

560   $R_1 = H$, $R_2 = OH$, $R_3 = H$

1237 $R_1 = R_2 = H$, $R_3 = OH$

1238 $R_1 = R_2 = H$, $R_3 = OAc$

FIGURE 1. (*contd.*)

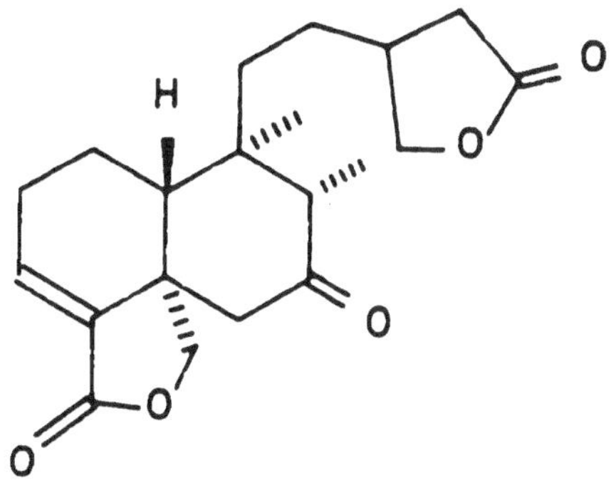

Figure 1. Structures of Diterpenes 333

561

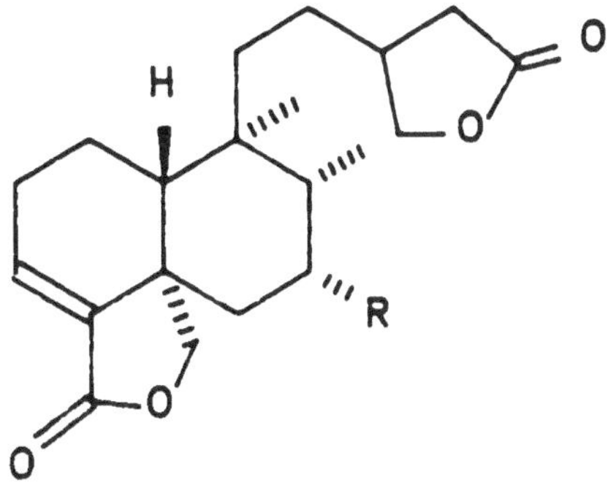

562 R = OH

563 R = H

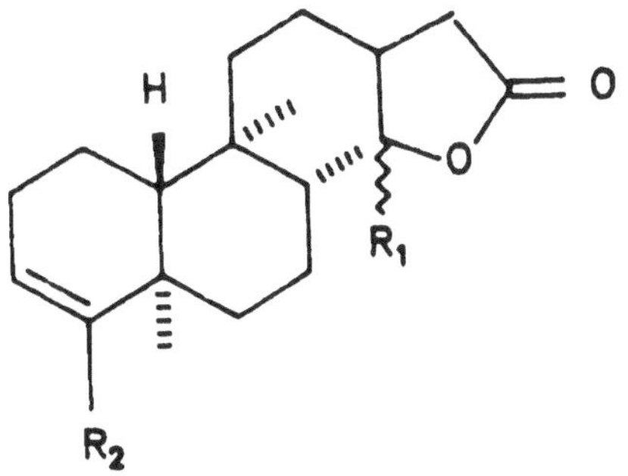

564 R$_1$ = H, R$_2$ = Me

565 R$_1$ = OH, R$_2$ = Me

963 R$_1$ = OMe, R$_2$ = CH$_2$OAc

FIGURE 1. (*contd.*)

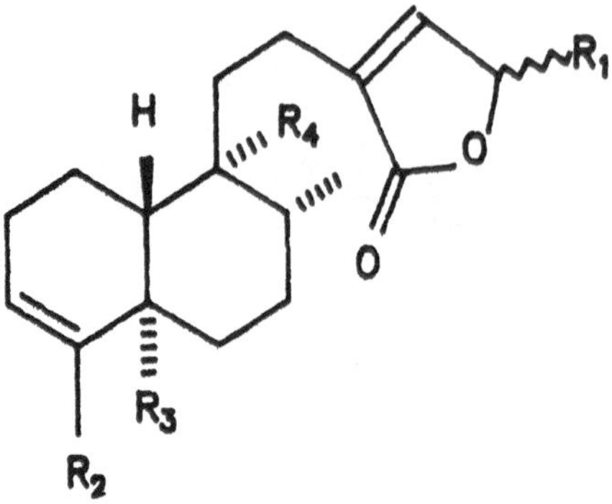

566 R_1 = H, R_2 = R_3 = Me, R_4 = CHO

905 R_1 = α-OH, R_2 = CO_2H, R_3 = CH_2OH, R_4 = Me

906 R_1 = β-OH, R_2 = CO_2H, R_3 = CH_2OH, R_4 = Me

1240 R_1 = H, R_2 = CH_2OAc, R_3 = R_4 = Me

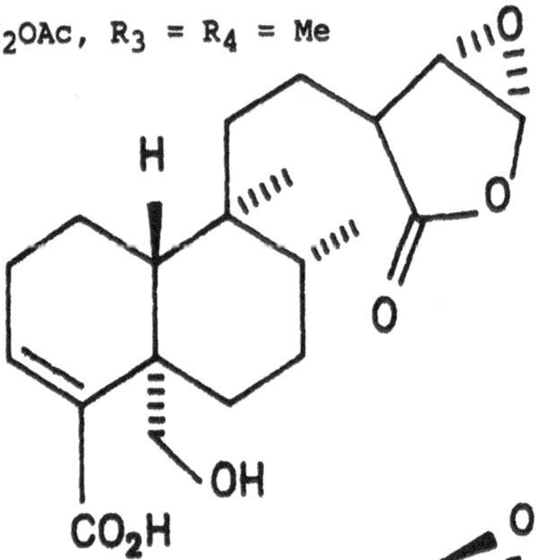

907

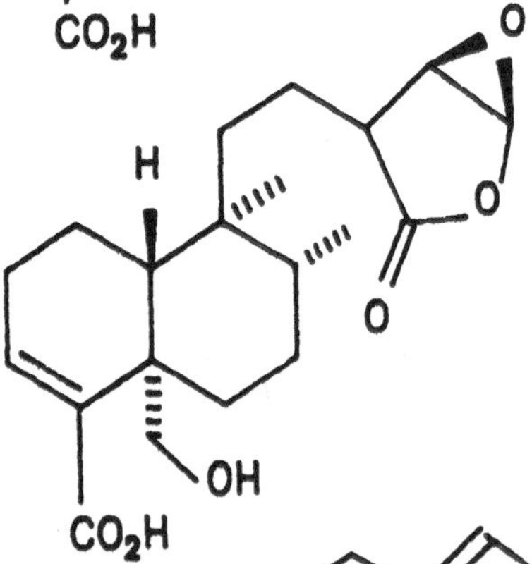

908

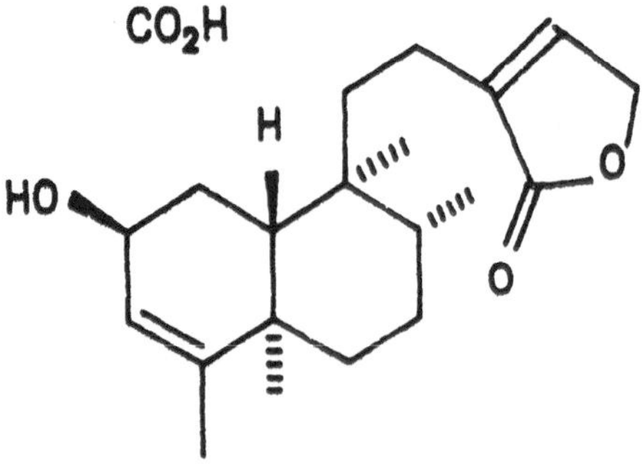

567

FIGURE 1. (contd.)

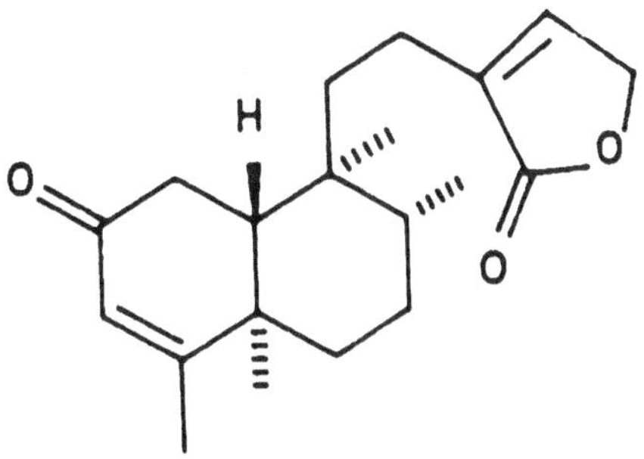

Figure 1. Structures of Diterpenes   335

568

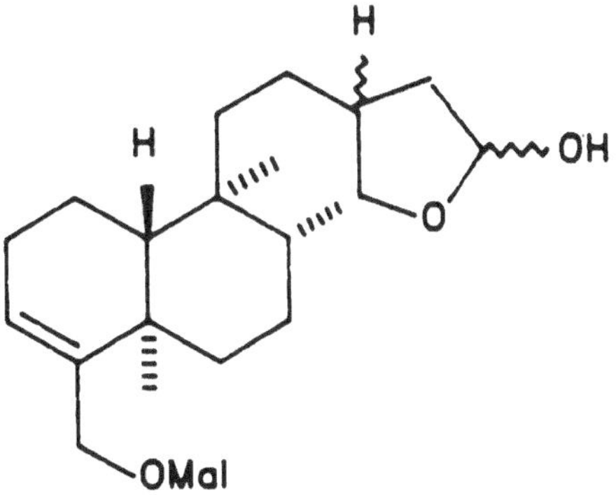

569

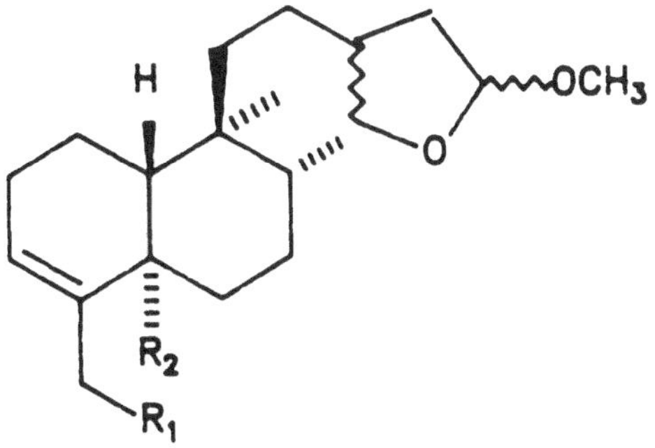

1239 R$_1$ = OAc, R$_2$ = Me

1241 R$_1$ = OH, R$_2$ = Me

1242 R$_1$ = OH, R$_2$ = CH$_2$OH

1243 R$_1$ = OAc, R$_2$ = CH$_2$OAc

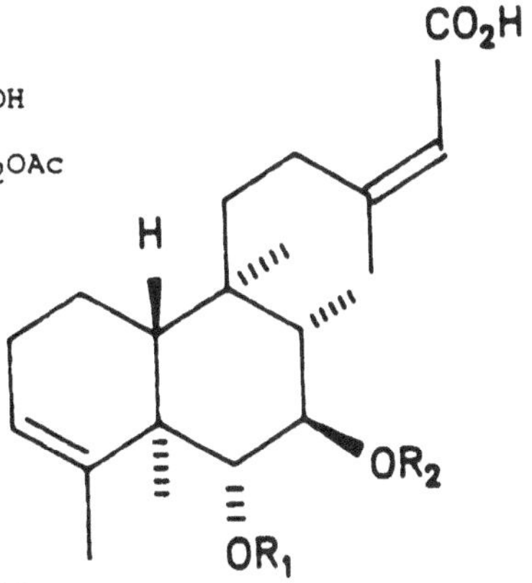

1027 R$_1$ = i-But, R$_2$ = Ac

FIGURE 1. (*contd.*)

1028 R_1 = Ang, R_2 = Ac

1029 R_1 = R_2 = Ac

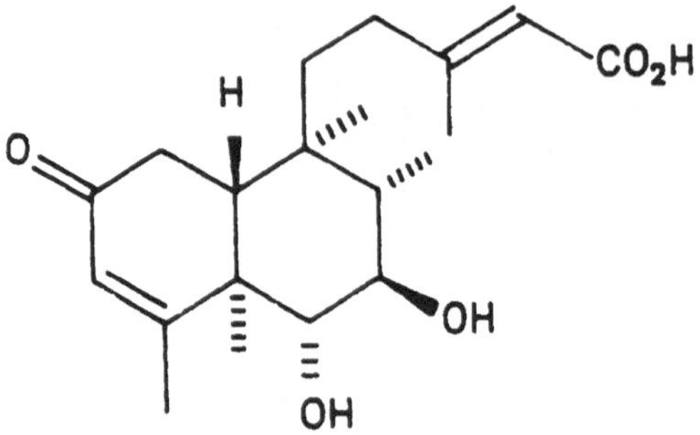

1030

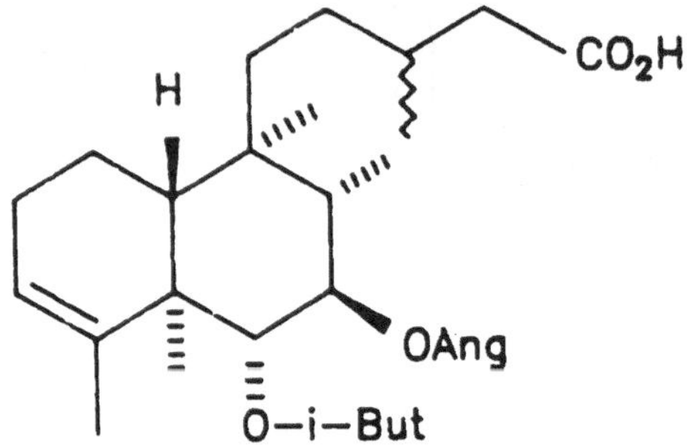

1031

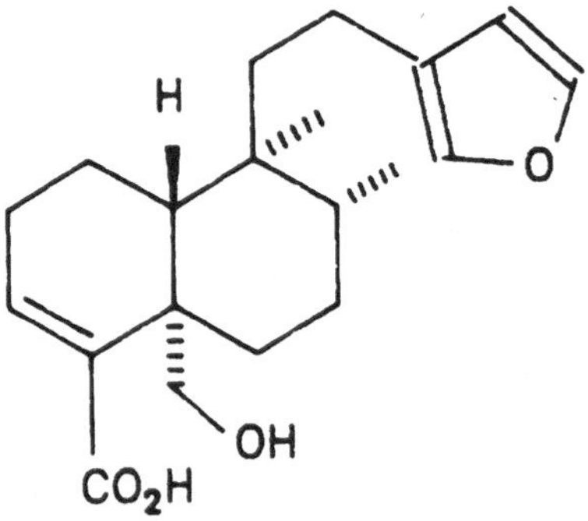

1033

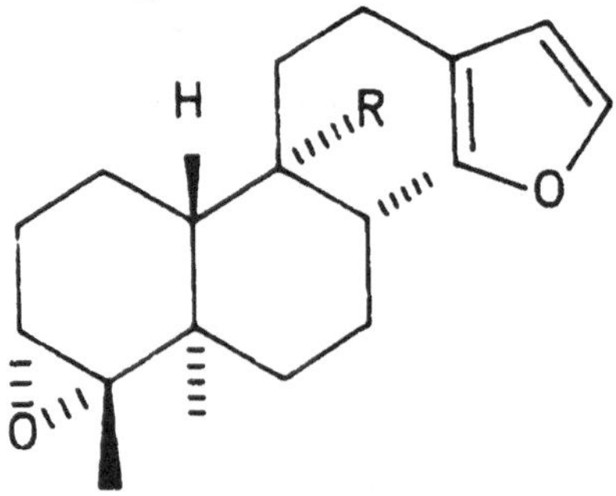

1071 R = CO_2H

1076 R = Me

FIGURE 1. (*contd.*)

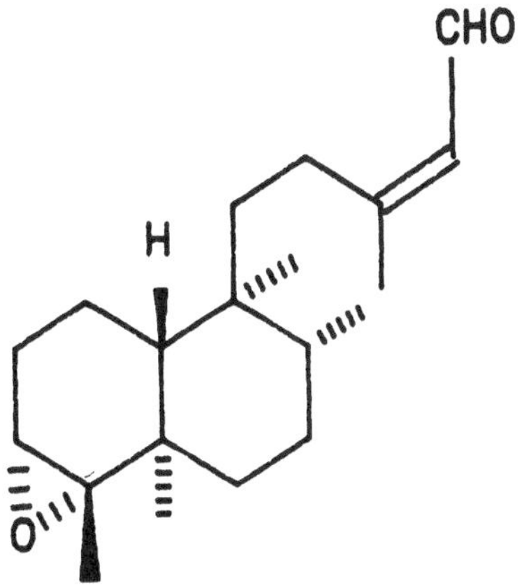

Figure 1. Structures of Diterpenes   337

1075

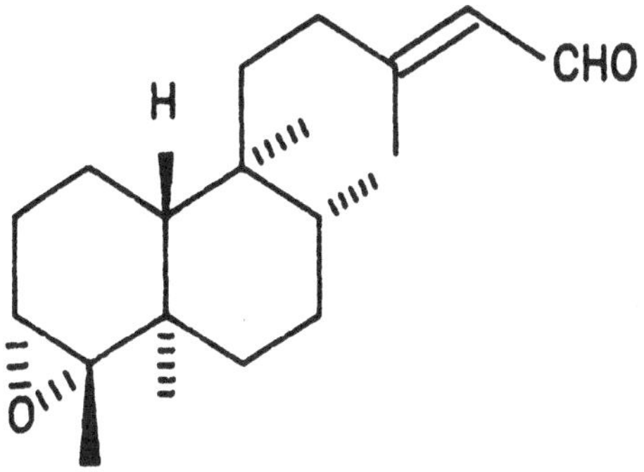

1077

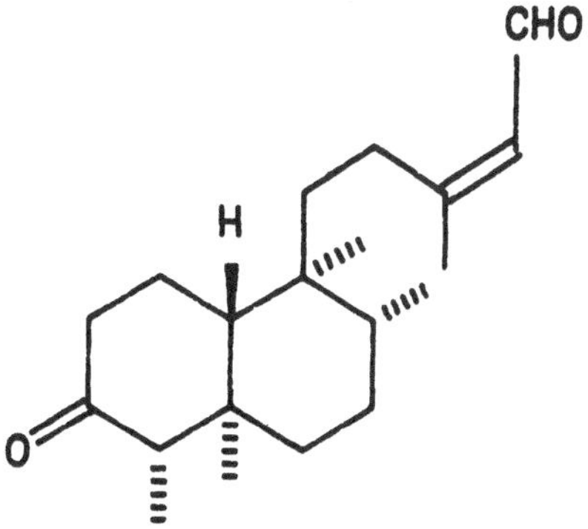

1078

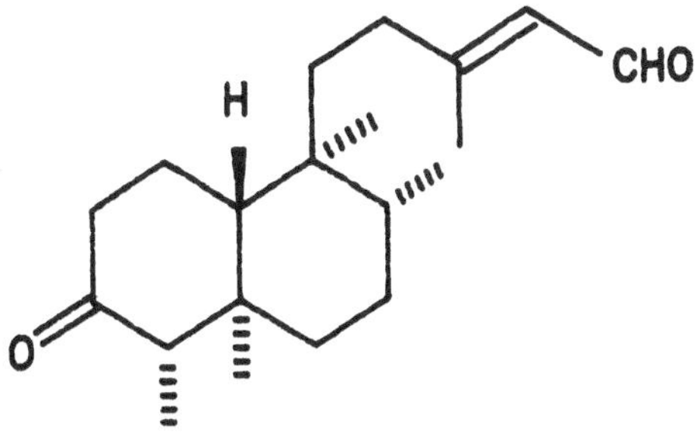

1079

FIGURE 1. (*contd.*)

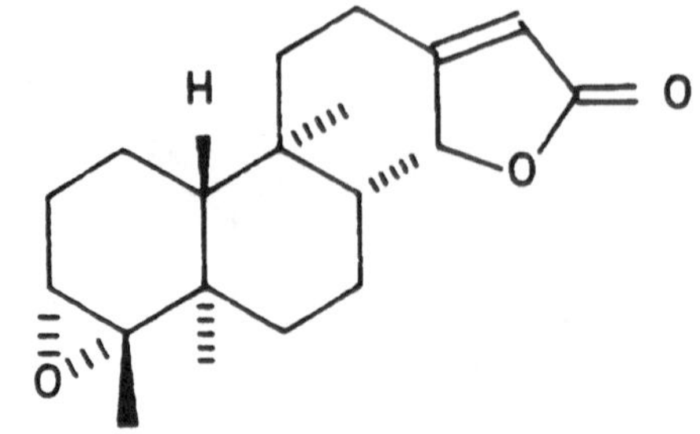

1080

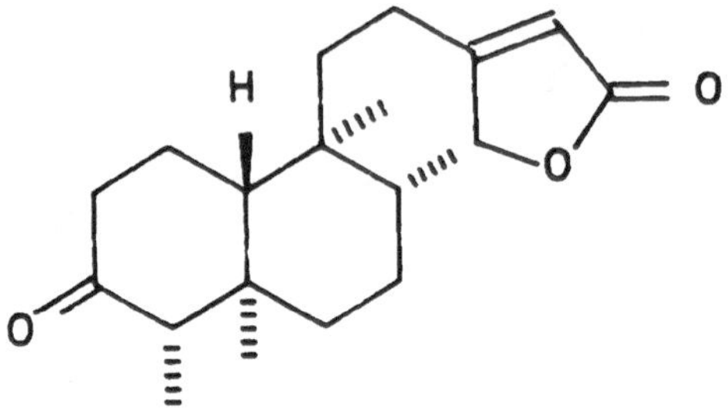

1081

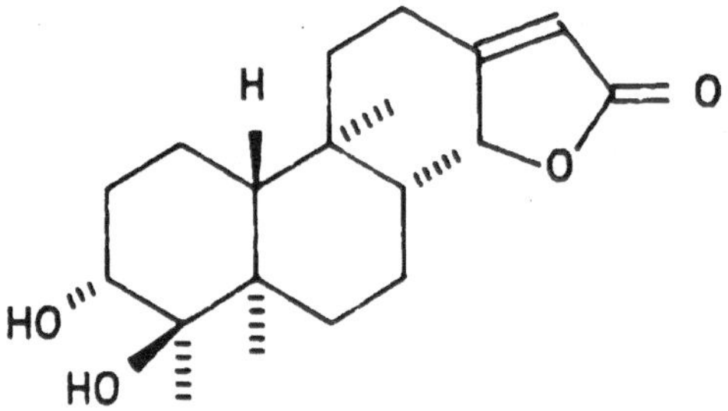

1082

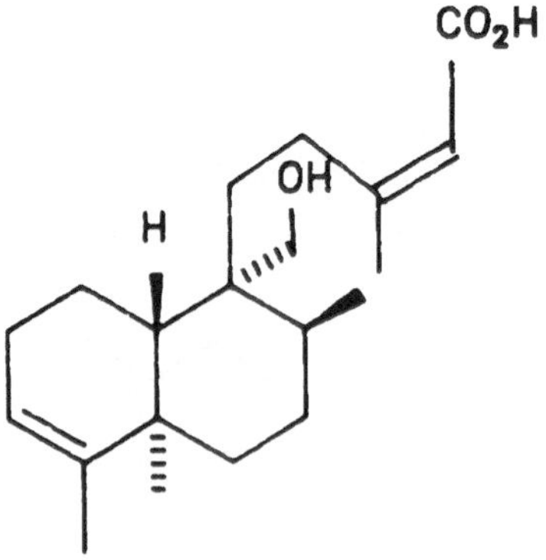

570

FIGURE 1. (*contd.*)

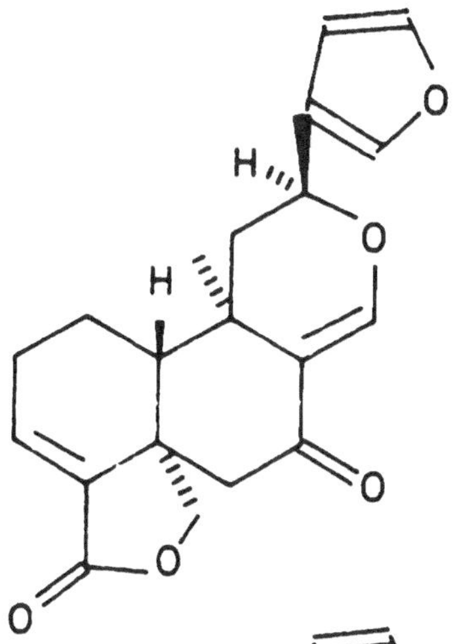

Figure 1. Structures of Diterpenes 339

525

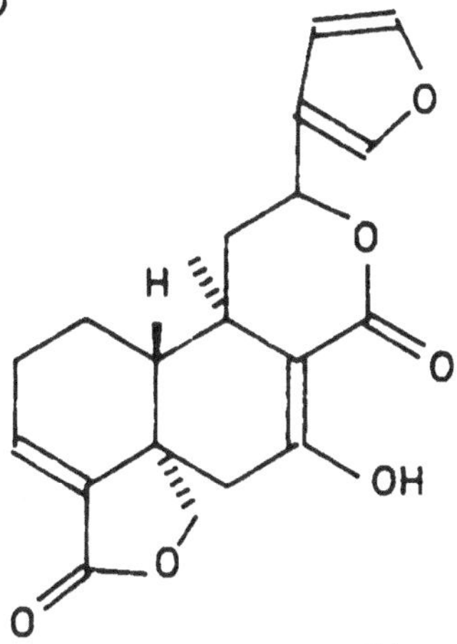

526

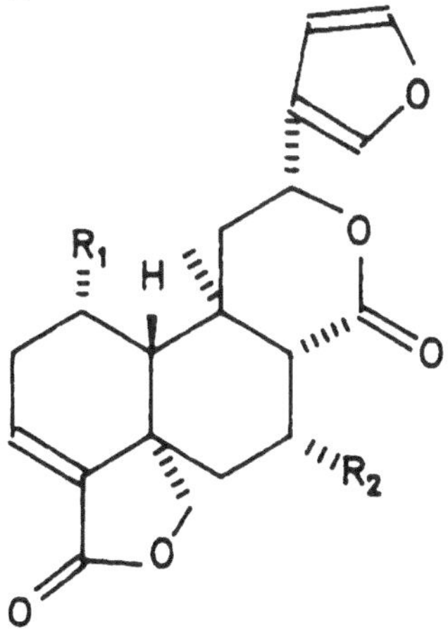

527 R$_1$ = R$_2$ = H

1069 R$_1$ = OH, R$_2$ = H

FIGURE 1. (*contd.*)

1257 R$_1$ = OAc, R$_2$ = H

1258 R$_1$ = H, R$_2$ = OH

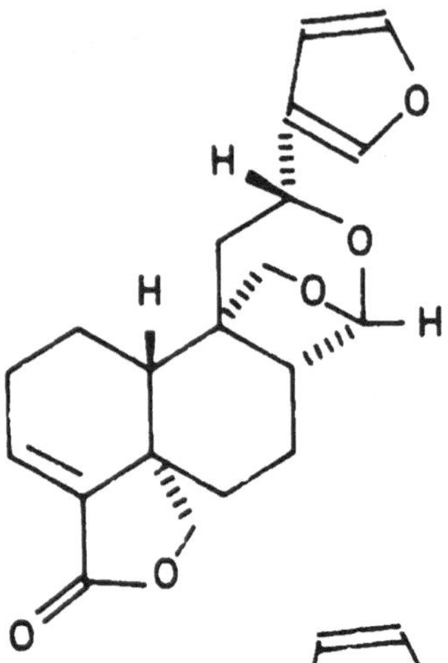

528

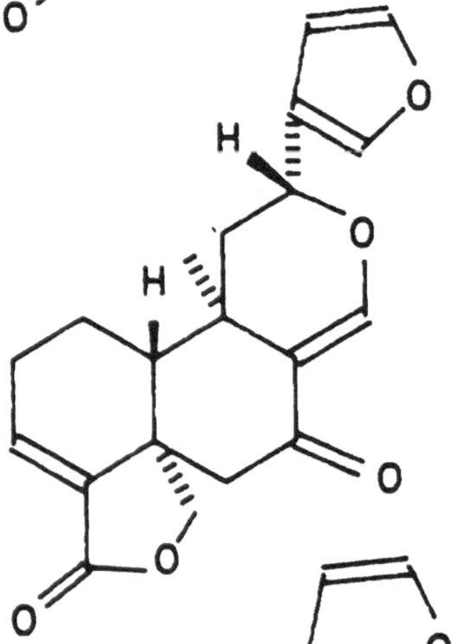

969

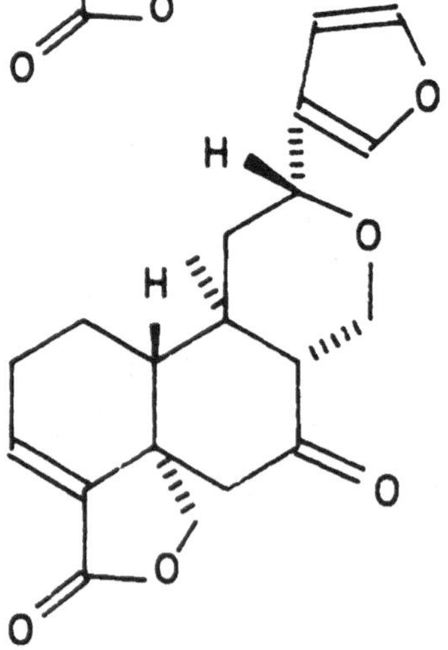

970

FIGURE 1. (*contd.*)

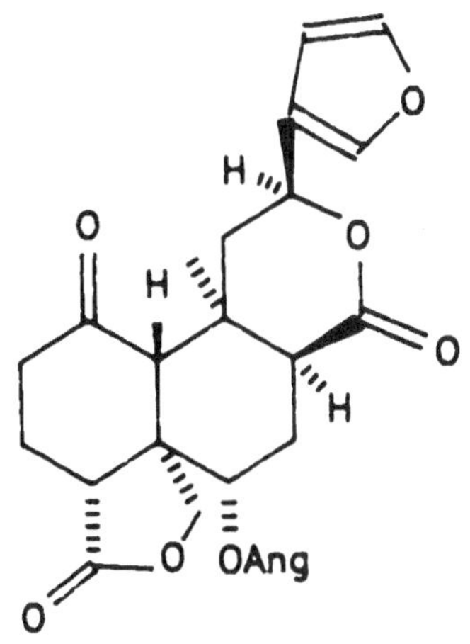

Figure 1. Structures of Diterpenes   341

971

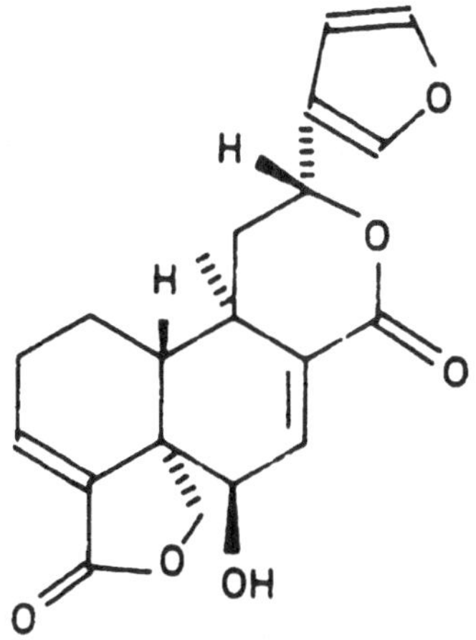

972

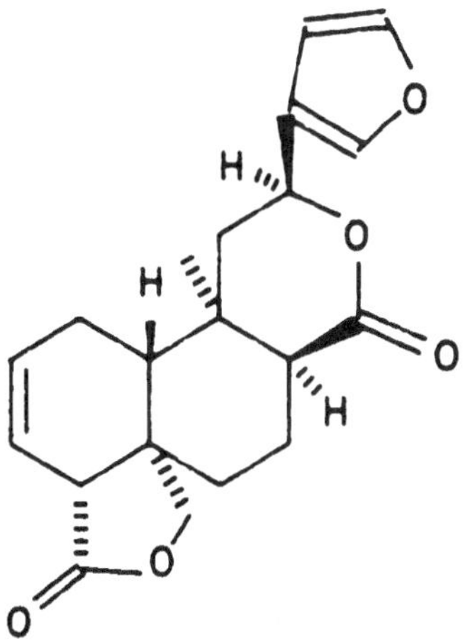

973

FIGURE 1. (*contd.*)

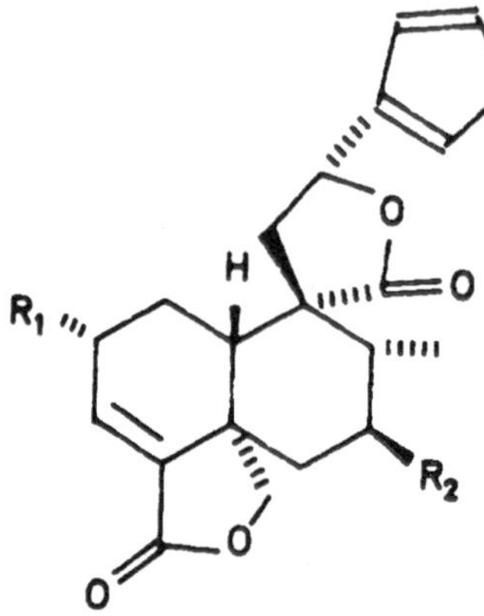

529   R_1 = R_2 = H

530   R_1 = H, R_2 = OAng

968   R_1 = OAng, R_2 = H

532   R_1 = OSen, R_2 = H

533   R_1 = O2-Mebut, R_2 = H

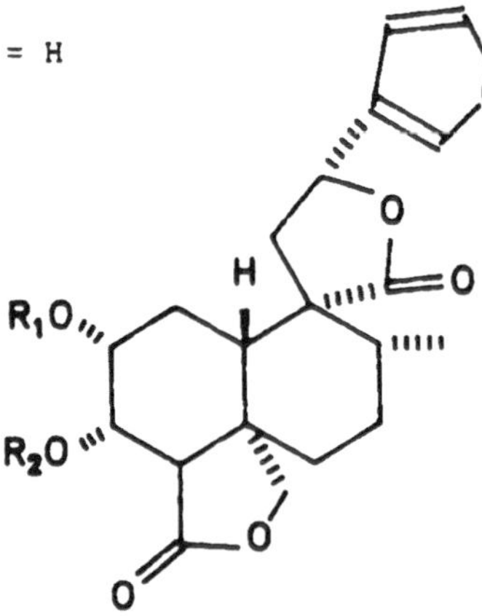

531   R_1 = Ang, R_2 = Sen

trans-ent-Clerodane Dimer

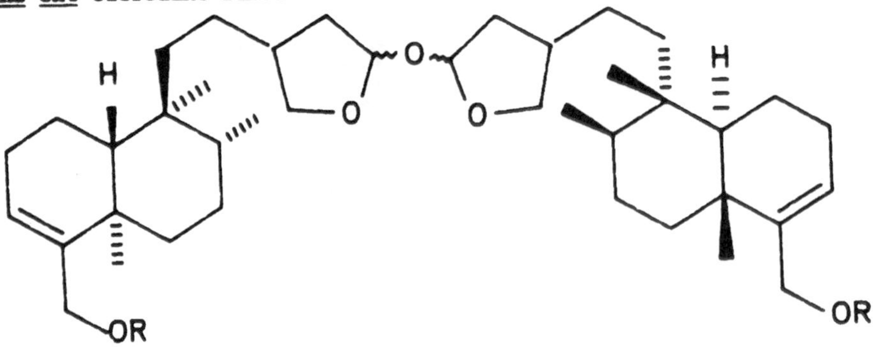

1244 R = Mal

1245 R = Succ

FIGURE 1. (contd.)

Figure 1. Structures of Diterpenes   343

47. Printziane

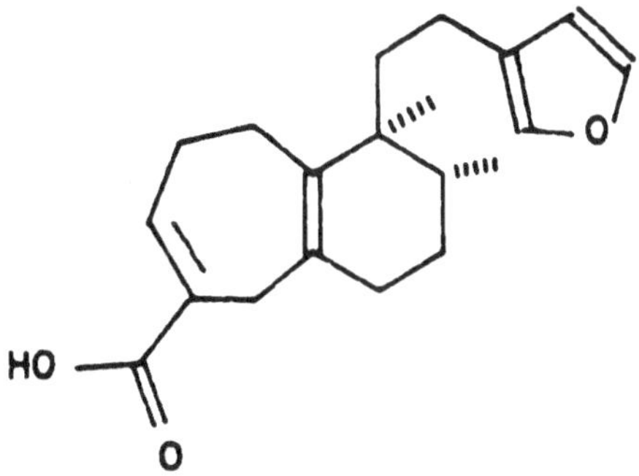

571

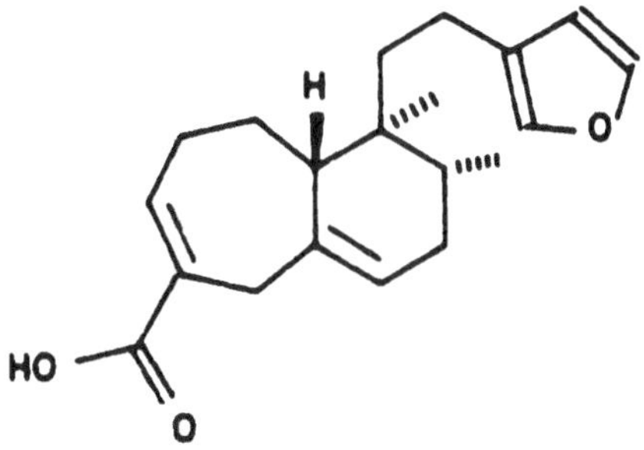

572

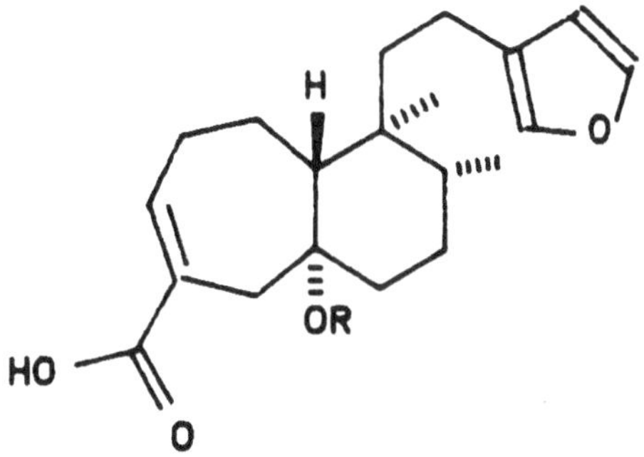

913   R = H
914   R = Me

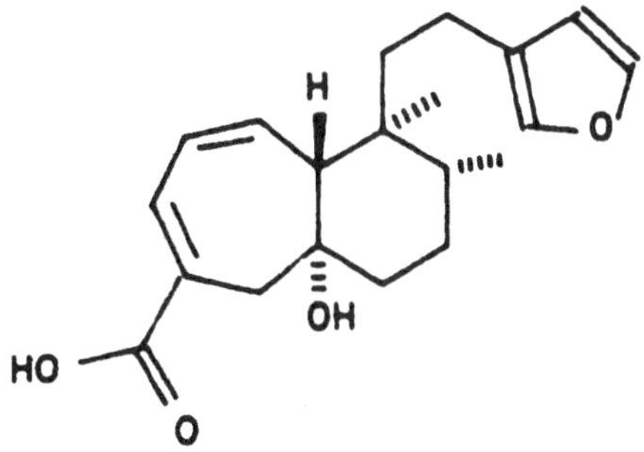

915

FIGURE 1. (*contd.*)

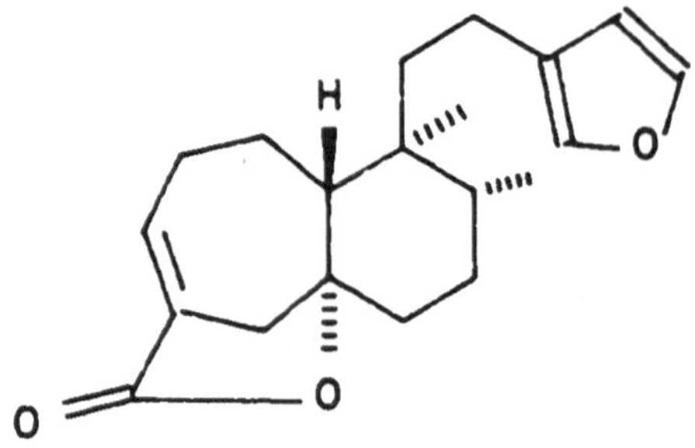

916

76. Conyscabrane

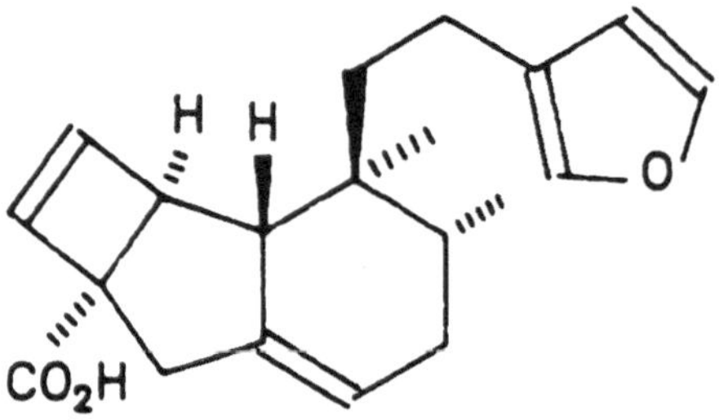

917

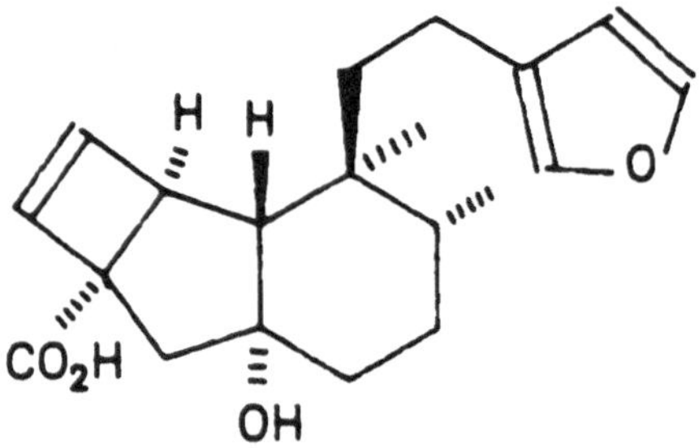

919

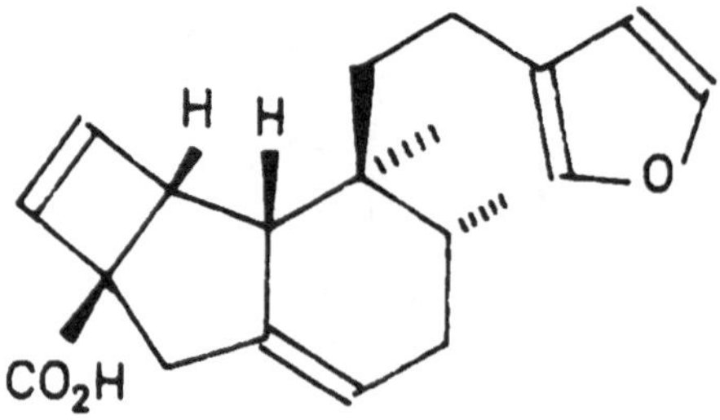

918

FIGURE 1. (*contd.*)

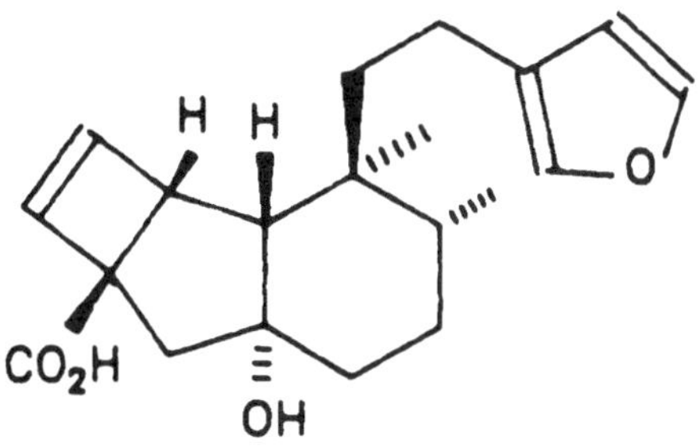

Figure 1. Structures of Diterpenes 345

920

48. ent-Clerodane, 5,10-Seco-

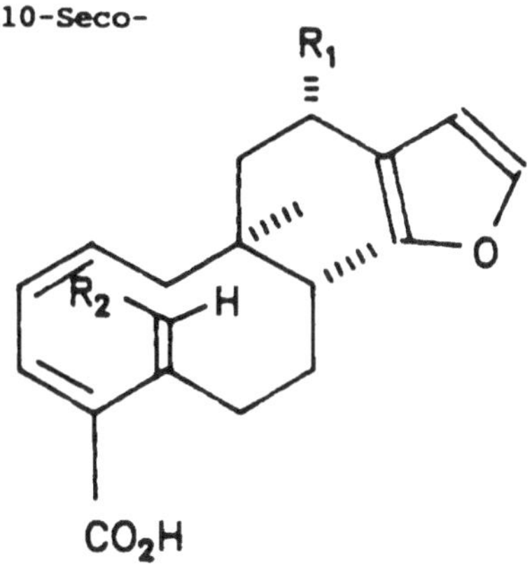

573   R$_1$ = R$_2$ = H

912   R$_1$ = H, R$_2$ = OAc

952   R$_1$ = O2-Mebut, R$_2$ = H

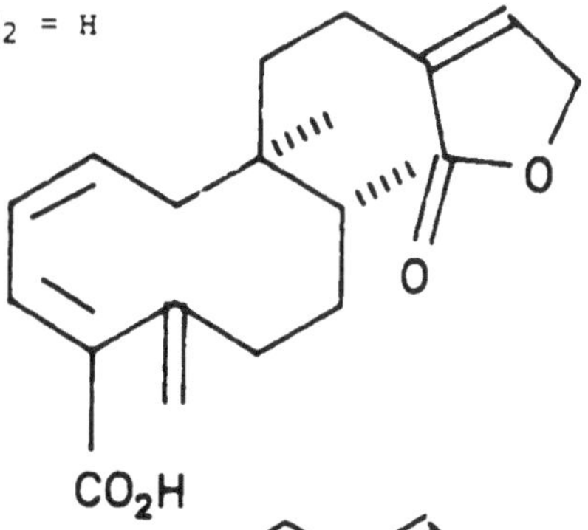

985

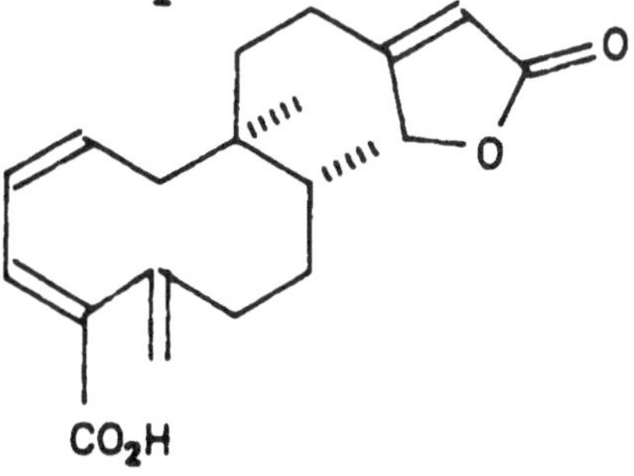

986

FIGURE 1. (*contd.*)

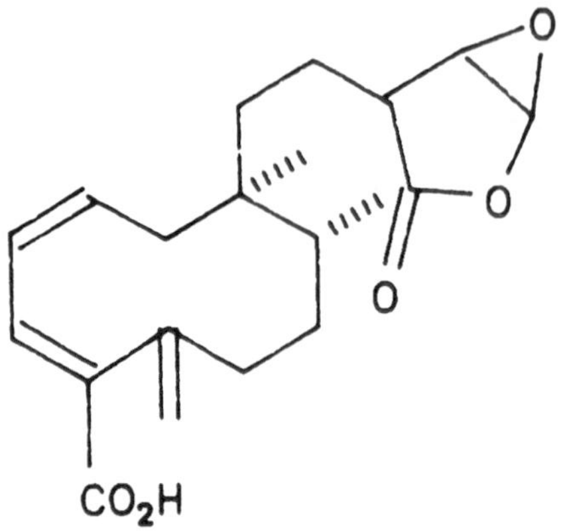

987

96. <u>trans</u>-<u>ent</u>-Clerodane, 17-Nor-

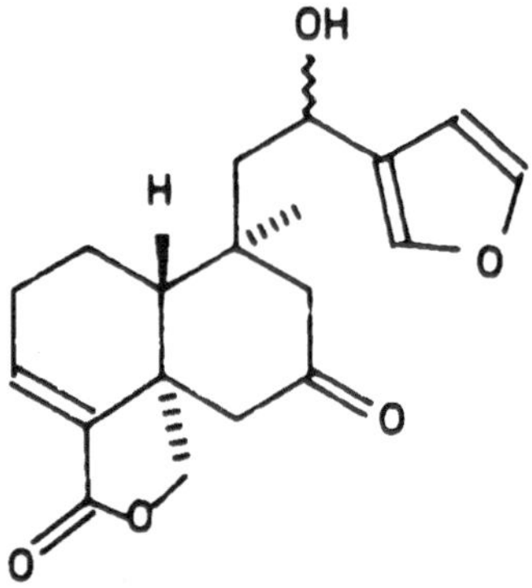

1061

49. <u>trans</u>-<u>ent</u>-Clerodane, 13,14,15,16-Tetra-Nor-

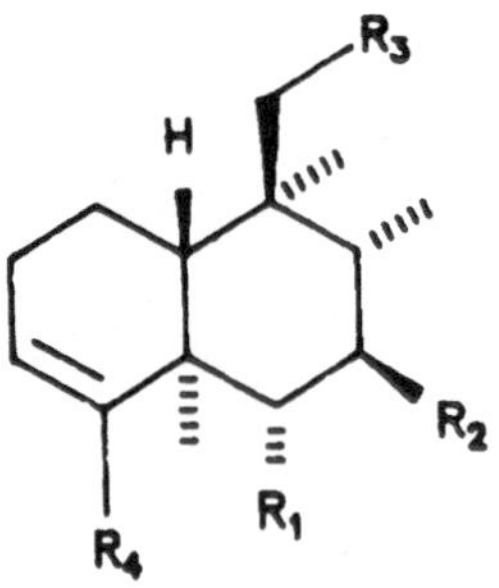

574 $R_1 = R_2 = H$, $R_3 = R_4 = CH_2OH$

575 $R_1 = R_2 = H$, $R_3 = CO_2H$, $R_4 = CH_2OH$

1032 $R_1 = R_2 = OH$, $R_3 = CO_2H$, $R_4 = Me$

FIGURE 1. (*contd.*)

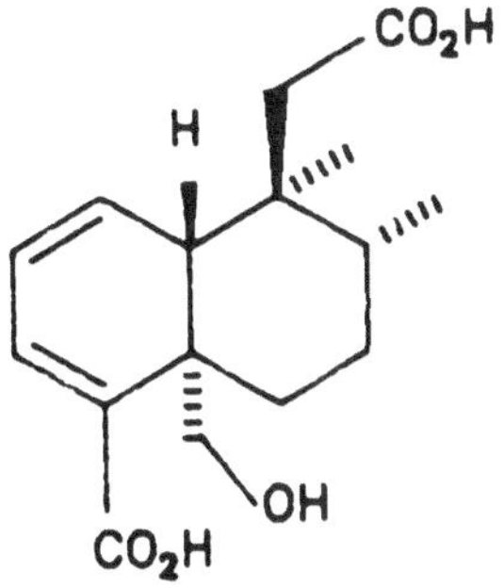

Figure 1. Structures of Diterpenes 347

1093

Undetermined

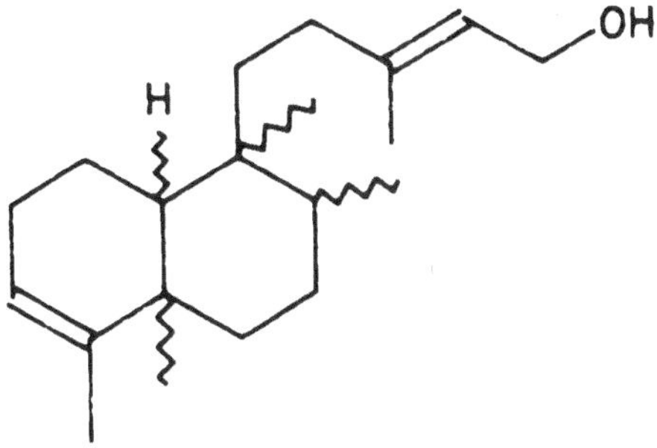

237

50. cis-ent-Clerodane

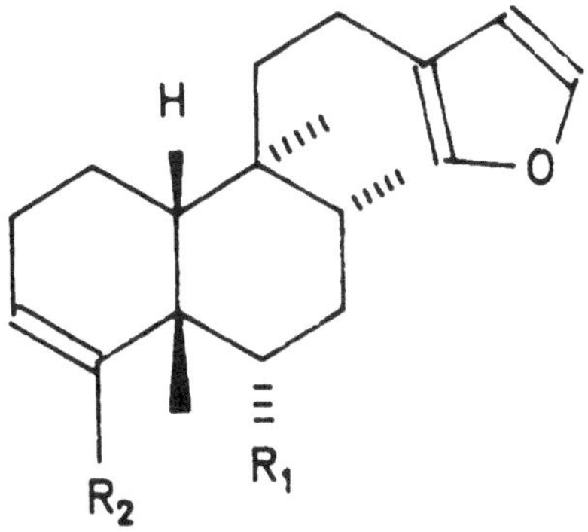

578 R_1 = H, R_2 = Me

579 R_1 = H, R_2 = CH_2OH

580 R_1 = H, R_2 = CH_2OAc

581 R_1 = OH, R_2 = CH_2OH

582 R_1 = OH, R_2 = CH_2OAc

FIGURE 1. (*contd.*)

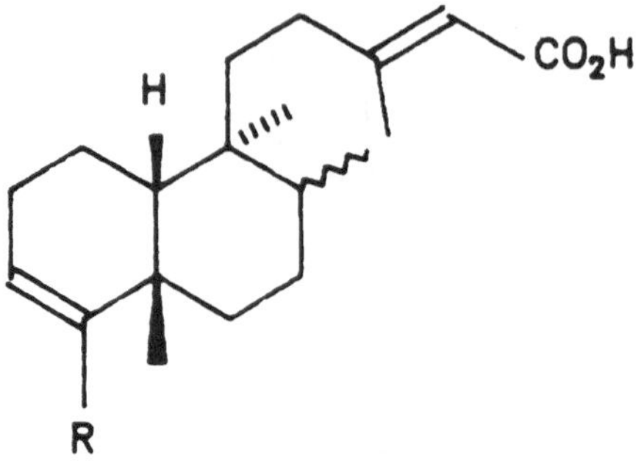

576 R = CO$_2$H

577 R = CO$_2$Me

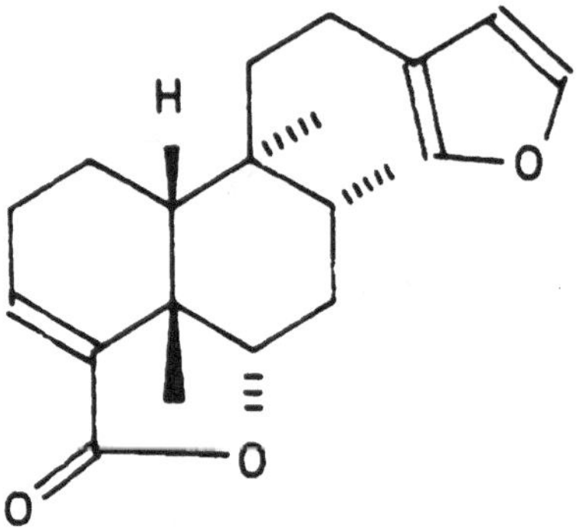

583

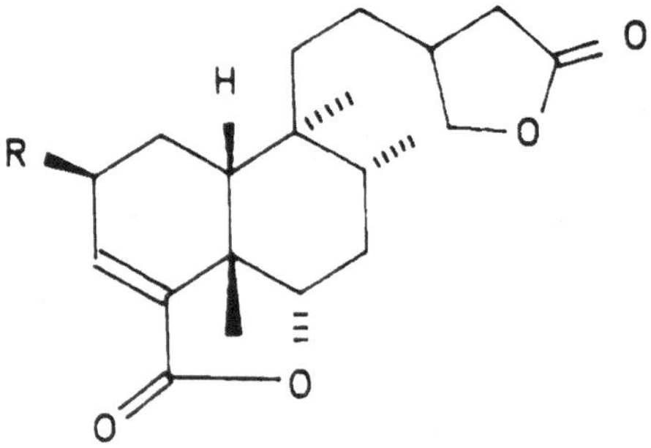

584 R = H

585 R = OH

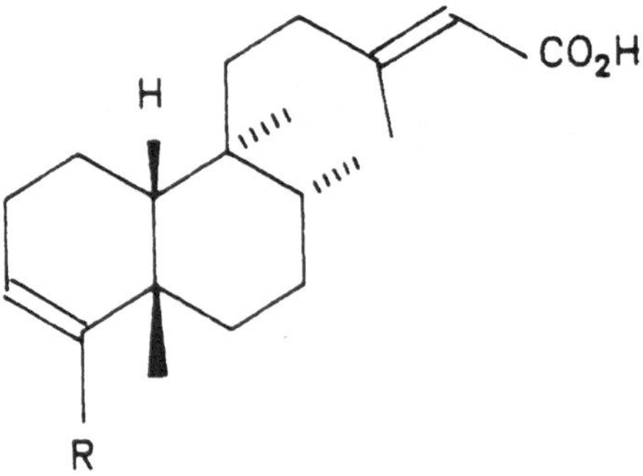

1055 R = CH$_2$OAc

1056 R = CH$_2$OH

FIGURE 1. (*contd.*)

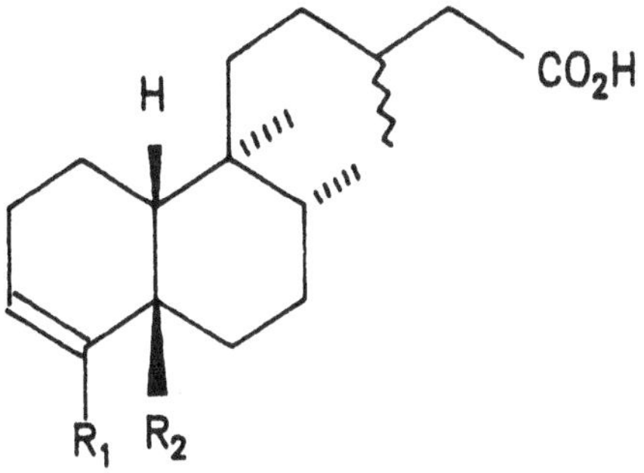

Figure 1. Structures of Diterpenes   349

1057 R_1 = CH_2OH, R_2 = Me

1058 R_1 = Me, R_2 = CH_2OH

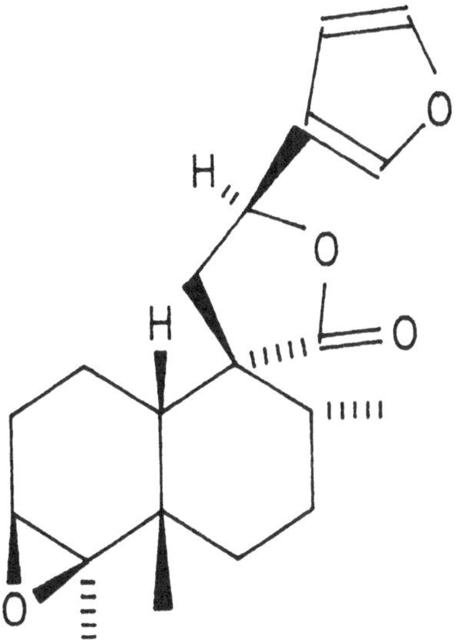

1246

82. Solidagonane

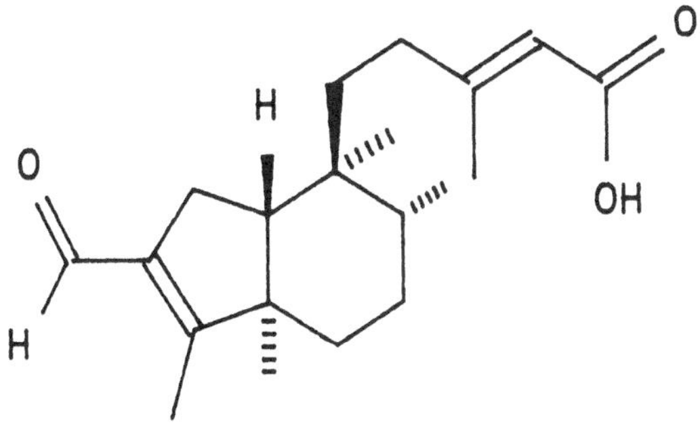

960

FIGURE 1. (*contd.*)

VI. <u>ent</u>-Tricarbocyclics and Derivatives

51. <u>ent</u>-Labdane, Cyclopropyl-

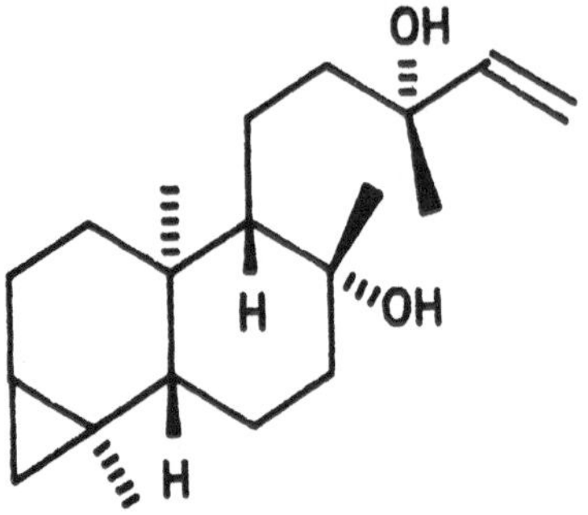

586

77. Gutierreziane

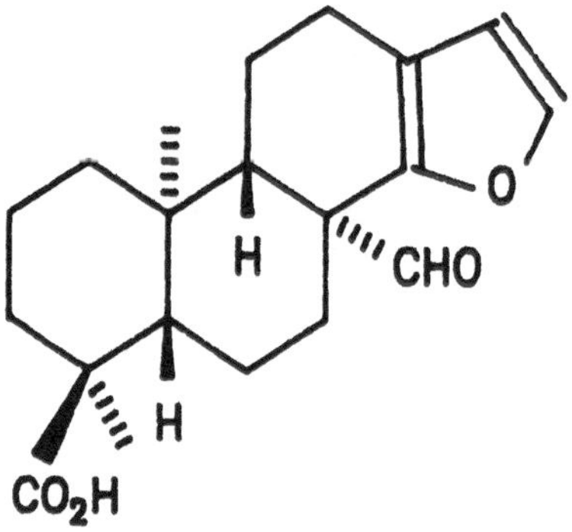

845

52. <u>ent</u>-Pimarane

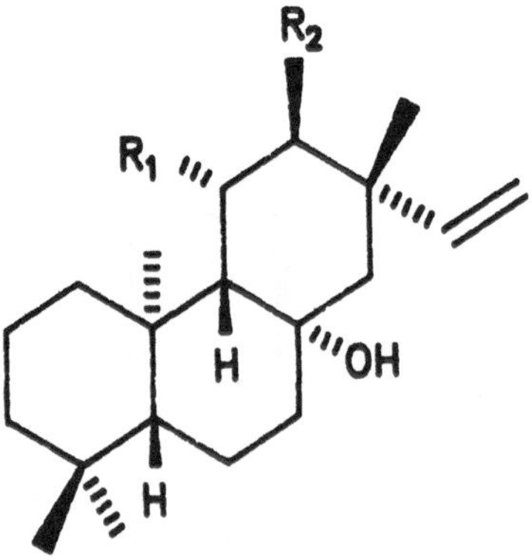

587 $R_1 = R_2 = H$

588 $R_1 = OH, R_2 = OAc$

FIGURE 1. (*contd.*)

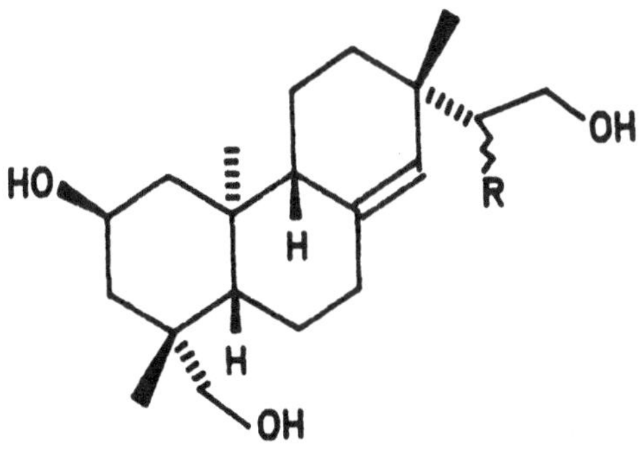

Figure 1. Structures of Diterpenes

351

589 R = OH

595 R = β-OH

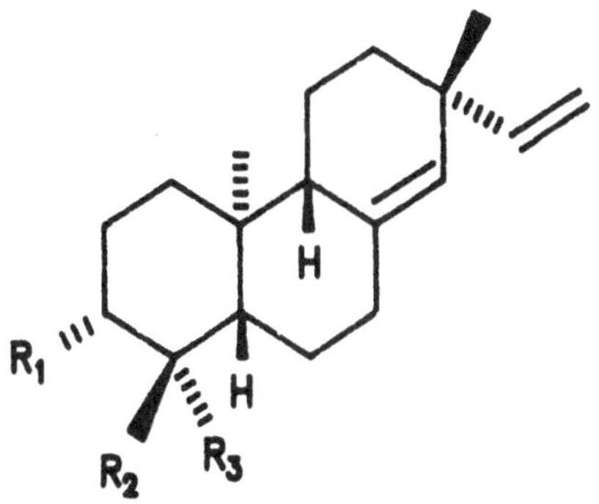

590 R_1 = H, R_2 = Me, R_3 = CO_2H

1247 R_1 = OH, R_2 = Me, R_3 = CH_2OH

1248 R_1 = OH, R_2 = CH_2OH, R_3 = Me

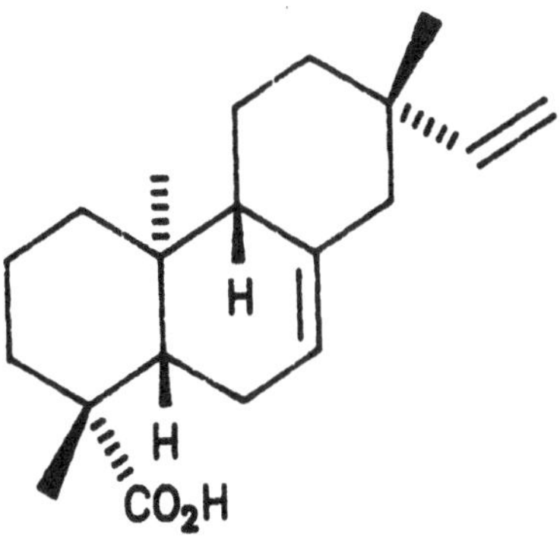

820

FIGURE 1. (*contd.*)

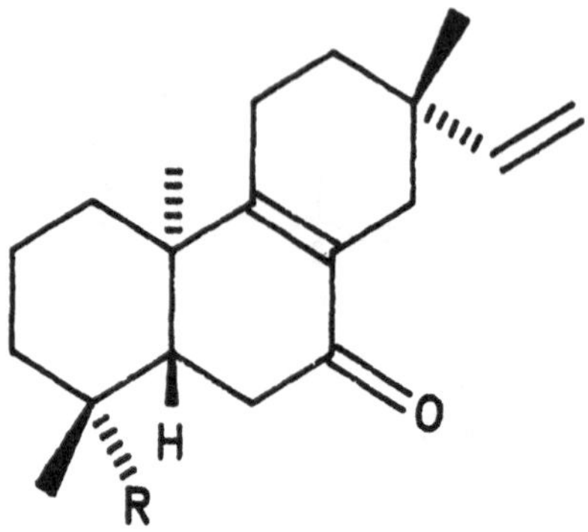

821   R = CO$_2$H

1249 R = CH$_2$OH

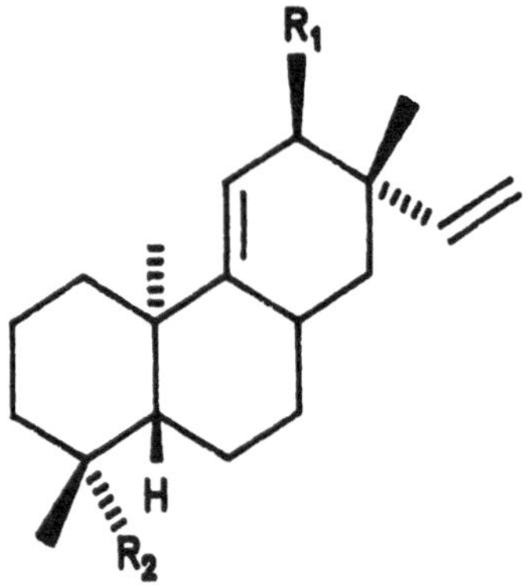

823   R$_1$ = OAc, R$_2$ = CO$_2$H

1250 R$_1$ = H, R$_2$ = CH$_2$OH

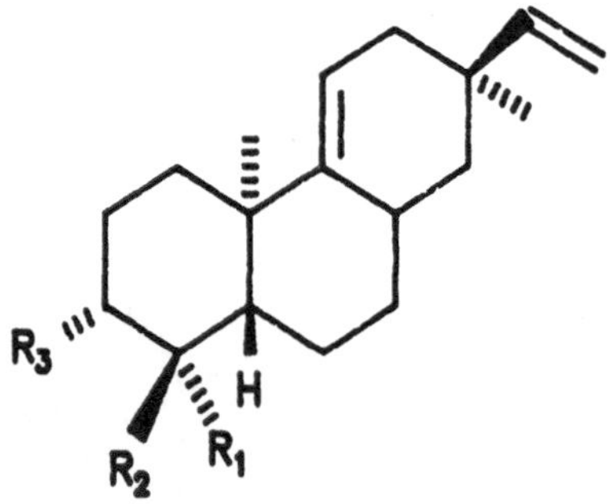

599   R$_1$ = CH$_2$OH, R$_2$ = Me, R$_3$ = H

600   R$_1$ = CO$_2$H, R$_2$ = Me, R$_3$ = H

1251 R$_1$ = Me, R$_2$ = CH$_2$OH, R$_3$ = H

1252 R$_1$ = Me, R$_2$ = CH$_2$OMal, R$_3$ = H

1253 R$_1$ = CH$_2$OMal, R$_2$ = Me, R$_3$ = H

1254 R$_1$ = CH$_2$OH, R$_2$ = Me, R$_3$ = OH

FIGURE 1. (*contd.*)

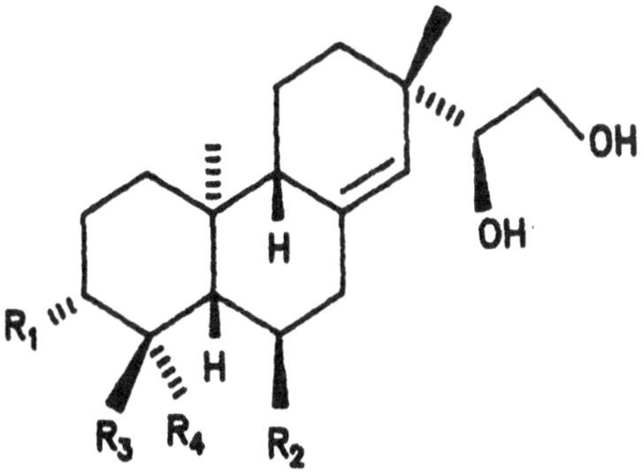

Figure 1. Structures of Diterpenes 353

591 R$_1$ = OH, R$_2$ = H, R$_3$ = R$_4$ = Me

592 R$_1$ = OGlu, R$_2$ = H, R$_3$ = R$_4$ = Me

593 R$_1$ = H, R$_2$ = OGlu, R$_3$ = R$_4$ = Me

594 R$_1$ = H, R$_2$ = OH, R$_3$ = R$_4$ = Me

944 R$_1$ = H, R$_2$ = OH, R$_3$ = CH$_2$OH, R$_4$ = Me

53. <u>ent</u>-Pimarane-8,15-Tetrahydrofuran

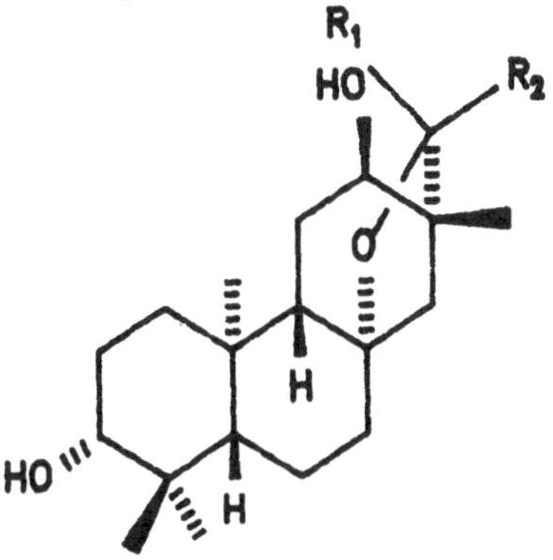

596 R$_1$ = H, R$_2$ = CH$_2$OH

597 R$_1$ = CH$_2$OH, R$_2$ = H

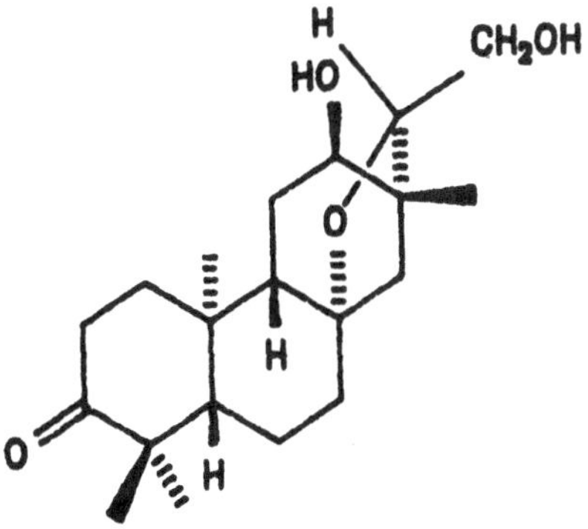

598

FIGURE 1. (*contd.*)

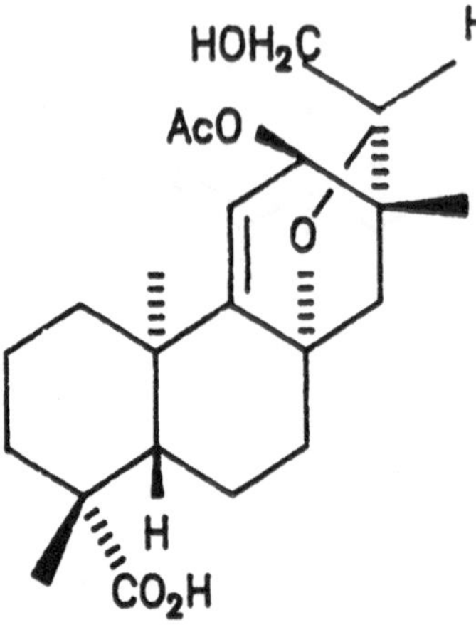

824

54. Cleistanthane

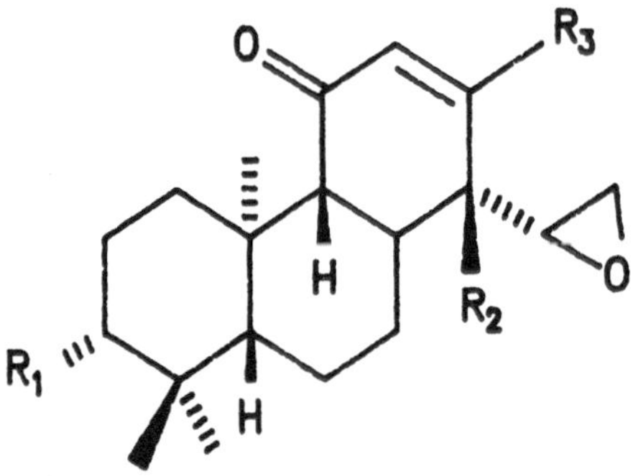

601 R_1 = H, R_2 = H, R_3 = Me

602 R_1 = H, R_2 = H, R_3 = CHO

603 R_1 = H, R_2 = H, R_3 = CH$_2$OAc

604 R_1 = OAng, R_2 = H, R_3 = Me

605 R_1 = H, R_2 = OH, R_3 = CH$_2$OAc

606 R_1 = OAng, R_2 = H, R_3 = CH$_2$OAc

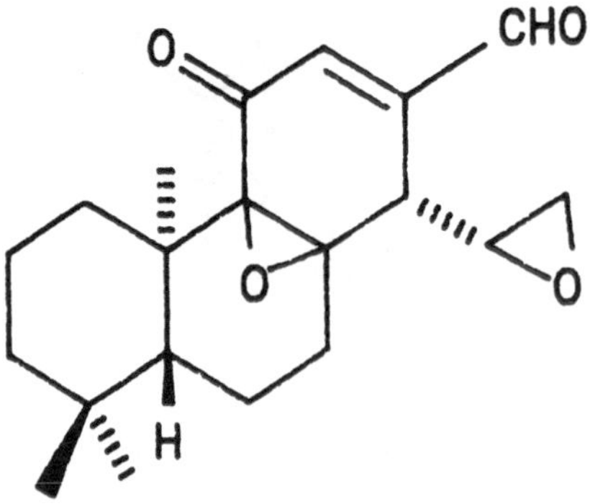

607

FIGURE 1. (*contd.*)

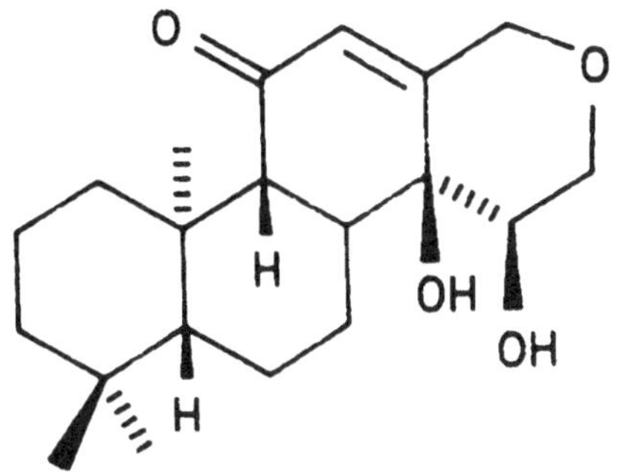

Figure 1. Structures of Diterpenes   355

608

85. Cleistanthane, Iso-

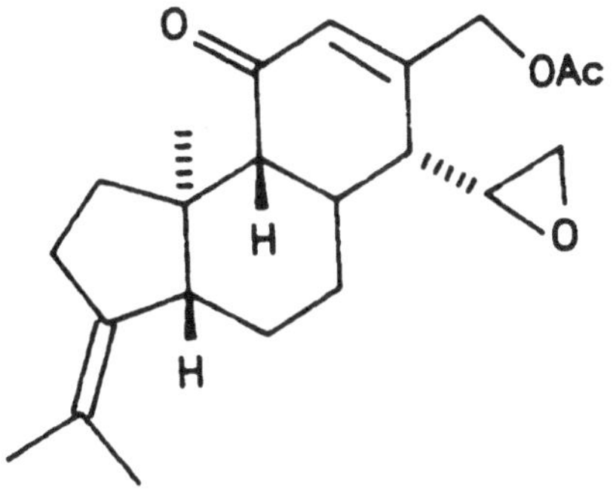

609

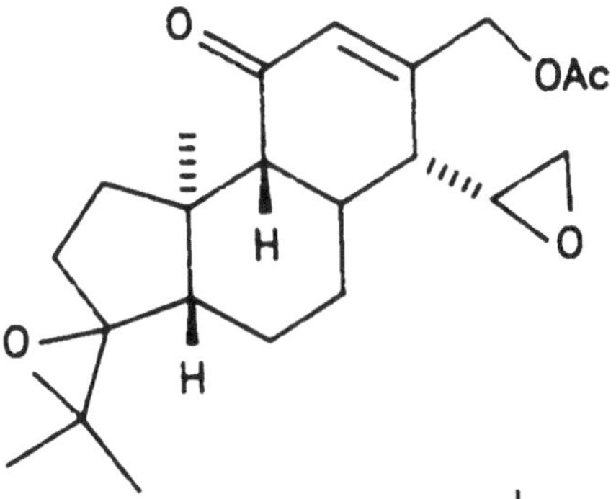

610

55. <u>ent</u>-Abietane

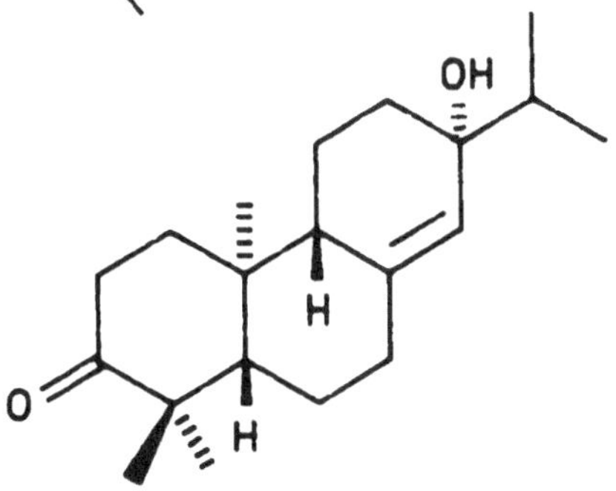

611

FIGURE 1. (*contd.*)

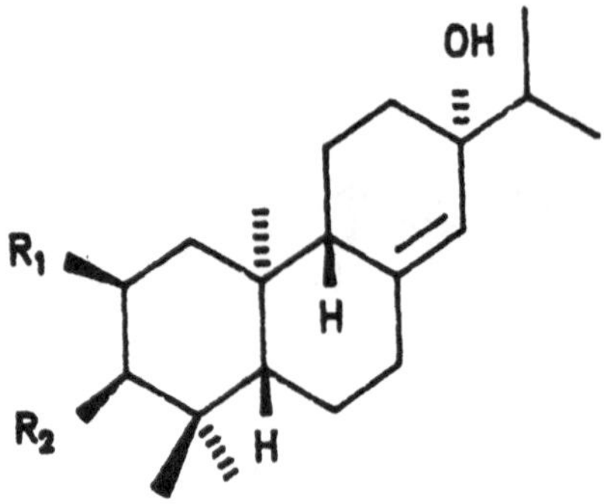

612 R$_1$ = H, R$_2$ = OH

613 R$_1$ = OAc, R$_2$ = H

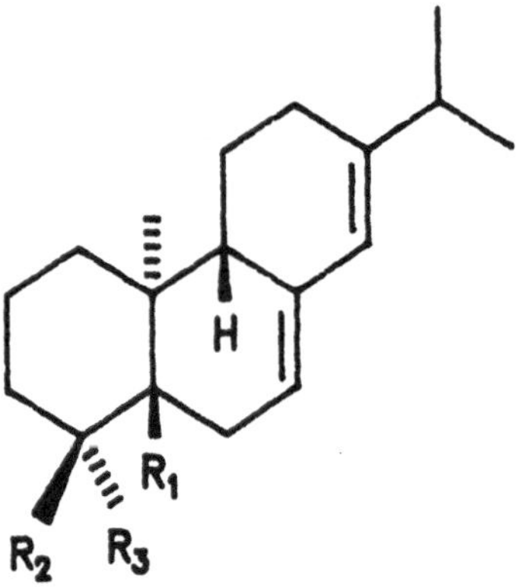

614 R$_1$ = H, R$_2$ = R$_3$ = Me

615 R$_1$ = OAc, R$_2$ = R$_3$ = Me

616 R$_1$ = H, R$_2$ = Me, R$_3$ = CO$_2$H

1054 R$_1$ = H, R$_2$ = Me, R$_3$ = CH$_2$OSucc

56. Acritoconfertane

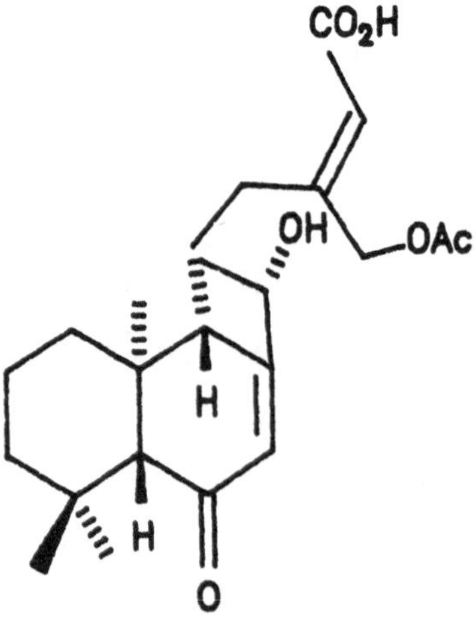

617

FIGURE 1. (contd.)

Figure 1. Structures of Diterpenes   357

57. Acritoconfertane-Acetal

1100

83. Erythroxane

976

VII. <u>ent</u>-Tetracarbocyclics and Derivatives

58. <u>ent</u>-Kaurane

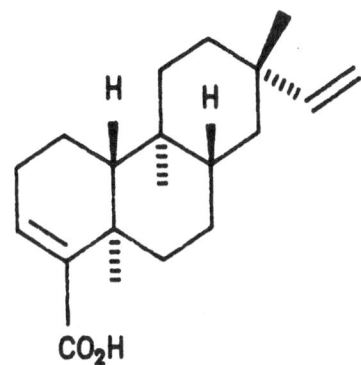

618 $R_1 = R_2 = H$, $R_3 = R_4 = Me$

619 $R_1 = R_2 = H$, $R_3 = Me$, $R_4 = CH_2OH$

620 $R_1 = R_2 = H$, $R_3 = Me$, $R_4 = CH_2OAc$

621 $R_1 = R_2 = H$, $R_3 = Me$, $R_4 = CH_2OMeform$

FIGURE 1. (*contd.*)

622 $R_1 = R_2 = H$, $R_3 = Me$, $R_4 = CH_2O\text{-}p\text{-}OH\text{-}H\text{-}Cinn$

623 $R_1 = OH$, $R_2 = H$, $R_3 = Me$, $R_4 = CH_2OH$

624 $R_1 = R_2 = H$, $R_3 = Me$, $R_4 = CHO$

625 $R_1 = R_2 = H$, $R_3 = Me$, $R_4 = CO_2H$

626 $R_1 = R_2 = H$, $R_3 = CH_2OAc$, $R_4 = Me$

635 $R_1 = R_2 = H$, $R_3 = CHO$, $R_4 = Me$

926 $R_1 = R_2 = H$, $R_3 = CO_2H$, $R_4 = Me$

939 $R_1 = R_2 = H$, $R_3 = CO_2H$, $R_4 = CH_2OAng$

627 $R_1 = R_2 = H$, $R_3 = Me$, $R_4 = CO_2Gly$

628 $R_1 = OH$, $R_2 = H$, $R_3 = Me$, $R_4 = CO_2H$

629 $R_1 = OAc$, $R_2 = H$, $R_3 = Me$, $R_4 = CO_2H$

630 $R_1 = OAng$, $R_2 = H$, $R_3 = Me$, $R_4 = CO_2H$

631 $R_1 = OTig$, $R_2 = H$, $R_3 = Me$, $R_4 = CO_2H$

632 $R_1 = OCinn$, $R_2 = H$, $R_3 = Me$, $R_4 = CO_2H$

633 $R_1 = OSen$, $R_2 = H$, $R_3 = Me$, $R_4 = CO_2H$

634 $R_1 = Oi\text{-}Val$, $R_2 = H$, $R_3 = Me$, $R_4 = CO_2H$

643 $R_1 = R_2 = H$, $R_3 = CH_2OAng$, $R_4 = CO_2H$

644 $R_1 = R_2 = H$, $R_3 = CH_2OSen$, $R_4 = CO_2H$

645 $R_1 = R_2 = H$, $R_3 = CH_2Oi\text{-}Val$, $R_4 = CO_2H$

1000 $R_1 = H$, $R_2 = OH$, $R_3 = Me$, $R_4 = CO_2H$

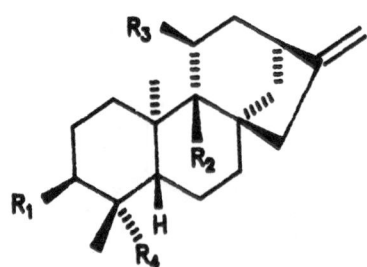

639 $R_1 = H$, $R_2 = OH$, $R_3 = H$, $R_4 = CO_2Me$

640 $R_1 = R_2 = H$, $R_3 = OAc$, $R_4 = CO_2H$

FIGURE 1. (*contd.*)

Figure 1. Structures of Diterpenes 359

641 $R_1 = R_2 = H, R_3 = \alpha\text{-OH}, R_4 = H$

642 $R_1 = R_2 = H, R_3 = \alpha\text{-OAc}, R_4 = H$

796 $R_1 = R_2 = H, R_3 = \beta\text{-OAc}, R_4 = H$

646 $R_1 = R_2 = R_3 = H, R_4 = \alpha\text{-OH}$

647 $R_1 = R_2 = R_3 = H, R_4 = \beta\text{-OH}$

648 $R_1 = R_2 = R_3 = H, R_4 = \alpha\text{-OAc}$

649 $R_1 = R_2 = R_3 = H, R_4 = \alpha\text{-OAng}$

650 $R_1 = R_2 = R_3 = H, R_4 = \alpha\text{-OTig}$

651 $R_1 = R_2 = R_3 = H, R_4 = \alpha\text{-OSen}$

652 $R_1 = R_2 = R_3 = H, R_4 = \alpha\text{-Oi-Val}$

653 $R_1 = R_2 = R_3 = H, R_4 = \alpha\text{-O2-Mac}$

654 $R_1 = R_2 = R_3 = H, R_4 = \alpha\text{-Oi-But}$

655 $R_1 = R_2 = R_3 = H, R_4 = \alpha\text{-OCinn}$

656 $R_1 = R_2 = R_3 = H, R_4 = \alpha\text{-OEpoxyang}$

657 $R_1 = R_2 = R_3 = H, R_4 = \alpha\text{-OBenz}$

818 $R_1 = R_2 = R_3 = H, R_4 = \alpha\text{-O2-Mebut-2,3-OH}$

892 $R_1 = R_2 = R_3 = H, R_4 = \alpha\text{-O2Mebut}$

658 $R_1 = R_2 = R_3 = H, R_4 = \beta\text{-OAng}$

659 $R_1 = R_2 = R_3 = H, R_4 = \beta\text{-OTig}$

660 $R_1 = R_2 = R_3 = H, R_4 = \beta\text{-OSen}$

662 $R_1 = \alpha\text{-OAng}, R_2 = OH, R_3 = R_4 = H$

FIGURE 1. (*contd.*)

663 R_1 = α-OCinn, R_2 = OH, R_3 = R_4 = H

664 R_1 = H, R_2 = OH, R_3 = H, R_4 = α-OH

665 R_1 = H, R_2 = OH, R_3 = H, R_4 = α-OAng

666 R_1 = H, R_2 = OH, R_3 = H, R_4 = α-OTig

667 R_1 = H, R_2 = OH, R_3 = H, R_4 = α-OAc

668 R_1 = H, R_2 = OH, R_3 = H, R_4 = α-OSen

669 R_1 = H, R_2 = OH, R_3 = H, R_4 = α-Oi-Val

670 R_1 = H, R_2 = OH, R_3 = H, R_4 = α-OCinn

819 R_1 = H, R_2 = OH, R_3 = H, R_4 = α-O2-Mebut-3-OH

675 R_1 = β-OH, R_2 = OH, R_3 = H, R_4 = α-OAng

676 R_1 = β-OH, R_2 = OH, R_3 = H, R_4 = α-OSen

677 R_1 = β-OH, R_2 = OH, R_3 = H, R_4 = α-OTig

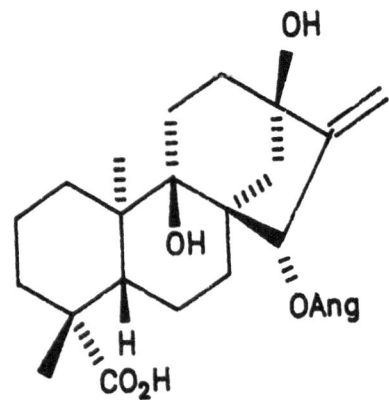

636

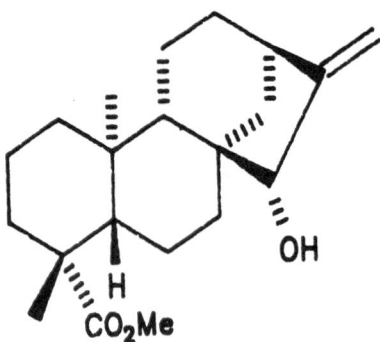

661

FIGURE 1. (*contd.*)

Figure 1. Structures of Diterpenes   361

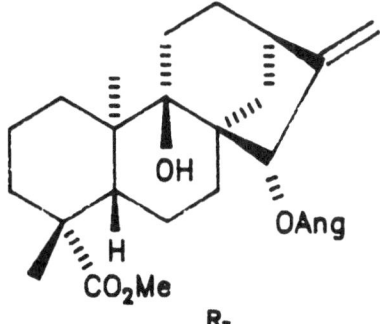

1007

671 $R_1 = H$, $R_2 = \beta\text{-OH}$, $R_3 = H$, $R_4 = \beta\text{-OH}$

672 $R_1 = H$, $R_2 = \beta\text{-OH}$, $R_3 = H$, $R_4 = \beta\text{-OAc}$

637 $R_1 = H$, $R_2 = \beta\text{-OH}$, $R_3 = H$, $R_4 = \alpha\text{-OH}$

869 $R_1 = OH$, $R_2 = \beta\text{-OH}$, $R_3 = H$, $R_4 = \alpha\text{-OAng}$

1260 $R_1 = H$, $R_2 = \beta\text{-OH}$, $R_3 = \beta\text{-OH}$, $R_4 = \beta\text{-OH}$

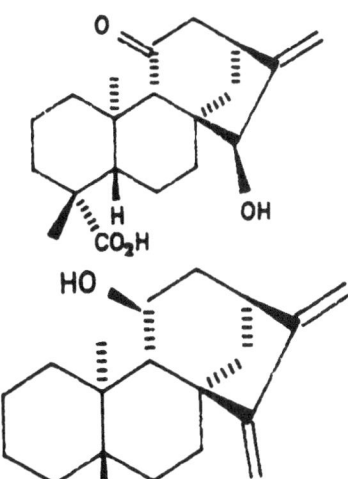

673 (Revised to 674)

674

FIGURE 1. (*contd.*)

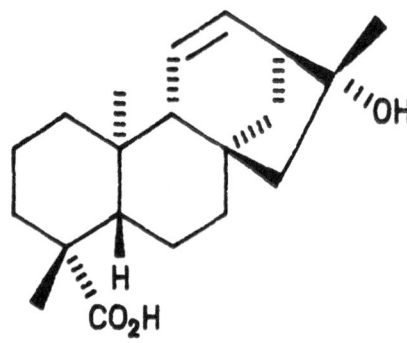

997 R_1 = H, R_2 = R_3 = Me (= 720)

721 R_1 = OAc, R_2 = R_3 = Me

722 R_1 = H, R_2 = Me, R_3 = CH_2OAc

1255 R_1 = H, R_2 = Me, R_3 = CHO

723 R_1 = H, R_2 = Me, R_3 = CO_2H

725 R_1 = H, R_2 = CH_2OH, R_3 = Me

726 R_1 = H, R_2 = R_3 = CH_2OH

728 R_1 = H, R_2 = CH_2OH, R_3 = CO_2H

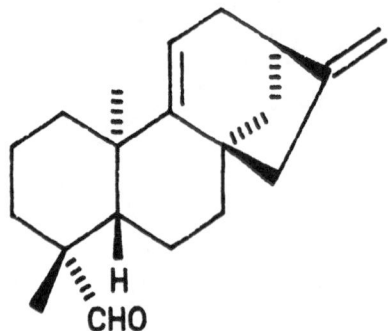

724

942

FIGURE 1. (*contd.*)

Figure 1. Structures of Diterpenes 363

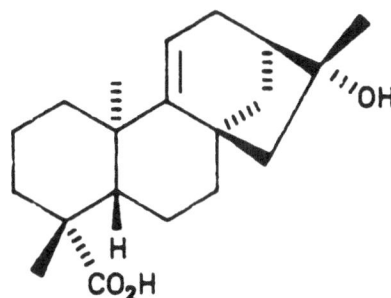

690 $R_1 = R_2 = R_3 = R_4 = R_5 = H$

691 $R_1 = OH, R_2 = R_3 = R_4 = R_5 = H$

692 $R_1 = H, R_2 = OH, R_3 = R_4 = R_5 = H$

693 $R_1 = R_2 = H, R_3 = OH, R_4 = R_5 = H$

694 $R_1 = R_2 = R_3 = H, R_4 = OH, R_5 = H$

998 $R_1 = R_2 = R_3 = H, R_4 = OEt, R_5 = H$

696 $R_1 = R_2 = R_3 = R_4 = H, R_5 = OH$

943 $R_1 = R_2 = R_3 = R_4 = H, R_5 = OCinn$

1256

695

FIGURE 1. (*contd.*)

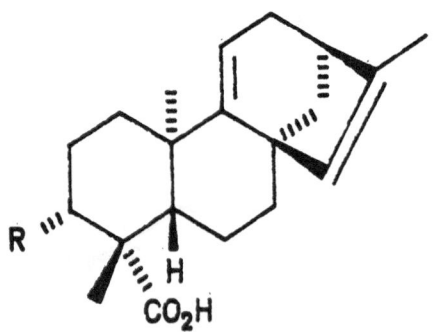

697 $R_1 = R_2 = H$, $R_3 = CH_2OH$, $R_4 = Me$

698 $R_1 = R_2 = H$, $R_3 = R_4 = CH_2OH$

699 $R_1 = R_2 = H$, $R_3 = CH_2OH$, $R_4 = CH_2Op-OH-H-Cinn$

946 $R_1 = R_2 = H$, $R_3 = Me$, $R_4 = CHO$

700 $R_1 = R_2 = H$, $R_3 = Me$, $R_4 = CO_2H$

701 $R_1 = R_2 = H$, $R_3 = CH_2OH$, $R_4 = CO_2H$

702 $R_1 = R_2 = H$, $R_3 = CHO$, $R_4 = CO_2H$

703 $R_1 = R_2 = H$, $R_3 = R_4 = CO_2H$

707 $R_1 = Oi-Val$, $R_2 = OH$, $R_3 = Me$, $R_4 = CO_2H$

708 $R_1 = Oi-But$, $R_2 = OH$, $R_3 = Me$, $R_4 = CO_2H$

704 $R = Oi-Val$

705 $R = Oi-But$

706 $R = OTig$

FIGURE 1. *(contd.)*

Figure 1. Structures of Diterpenes 365

733	R_1 = H, R_2 = R_3 = Me
709	R_1 = OAng, R_2 = R_3 = Me
638	R_1 = H, R_2 = CH$_2$OH, R_3 = Me
734	R_1 = H, R_2 = Me, R_3 = CH$_2$OH
710	R_1 = H, R_2 = Me, R_3 = CH$_2$OAng
711	R_1 = OAng, R_2 = Me, R_3 = CHO
735	R_1 = H, R_2 = Me, R_3 = CHO
713	R_1 = H, R_2 = Me, R_3 = CO$_2$H

712	R = CHO
714	R = CO$_2$H

715	R = α-OAng

FIGURE 1. (contd.)

716 R = α-OAc

717 R = β-OAng

718 R = β-OTig

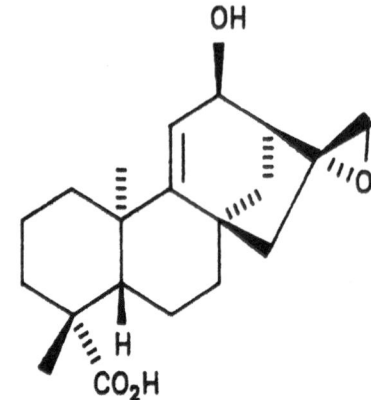

719

727

1001 $R_1 = R_2 = Me$ (= **851**)

852 $R_1 = Me$, $R_2 = CH_2OH$

822 $R_1 = CH_2OH$, $R_2 = CO_2H$

FIGURE 1. (*contd.*)

Figure 1. Structures of Diterpenes   367

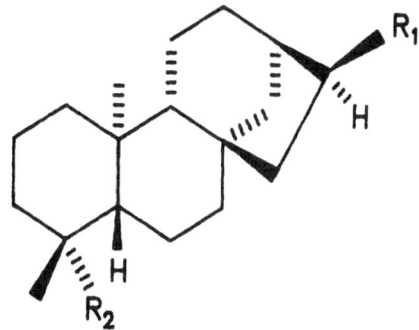

924 R_1 = CH_2OH, R_2 = CO_2H

925 R_1 = R_2 = CO_2H

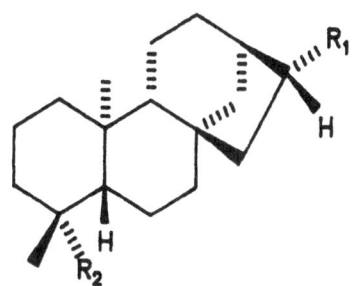

729 R_1 = CHO, R_2 = Me

730 R_1 = CHO, R_2 = CH_2OH

731 R_1 = CO_2H, R_2 = Me

732 R_1 = CO_2H, R_2 = CH_2OH

736 R_1 = COMe, R_2 = CH_2OH

737 R_1 = R_2 = CHO

738 R_1 = CH_2OH, R_2 = CO_2H

989 R_1 = CO_2H, R_2 = CHO

990 R_1 = R_2 = CO_2H

740 R_1 = CHO, R_2 = CH_2Op-OH-H-Cinn

741 R_1 = CH_2Oi-But, R_2 = CO_2H

742 R_1 = CH_2ODodec, R_2 = CO_2H

743 R_1 = CH_2OTetradec, R_2 = CO_2H

744 R_1 = CH_2OHexadec, R_2 = CO_2H

FIGURE 1. (*contd.*)

745 R_1 = CH$_2$OOctadec, R_2 = CO$_2$H

812 R_1 = CO$_2$Me, R_2 = CO$_2$H

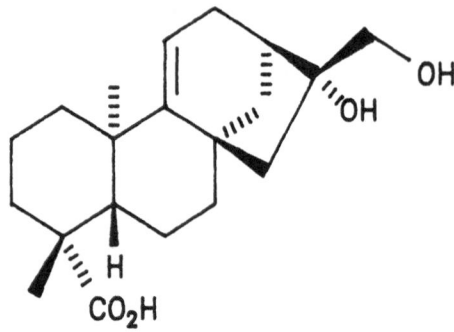

739

746

747 R = β-OH (= **815**)

748 R = β-OH

FIGURE 1. (*contd.*)

Figure 1. Structures of Diterpenes 369

850

945

678 $R_1 = R_2 = CO_2H$, $R_3 = OCOCH_2CH(CH_3)_2$, $R_4 = R_5 = HO_3SO$, $R_6 = H$

FIGURE 1. (*contd.*)

679 $R_1 = R_2 = CO_2H$, $R_3 = NHCOCH_2CH(CH_3)_2$, $R_4 = CO(CH_2)_2C_6H_5$,
$R_5 = OH$, $R_6 = OH$

680 $R_1 = R_2 = CO_2H$, $R_3 = NHCOCH_2CH(CH_3)_2$, $R_4 = CO(CH_2)_2C_6H_5$,
$R_5 = OL-Rha$, $R_6 = OH$

681 $R_1 = H$, $R_2 = CO_2H$, $R_3 = OCOCH_2CH(CH_3)_2$, $R_4 = R_5 = HO_3SO$,
$R_6 = H$

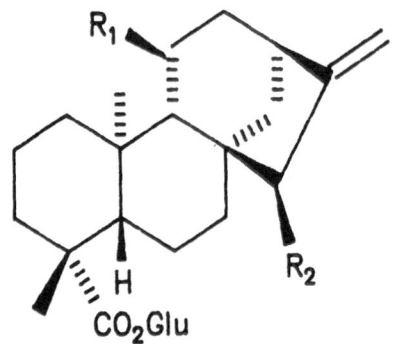

895 $R_1 = H$, $R_2 = OH$
896 $R_1 = R_2 = OH$

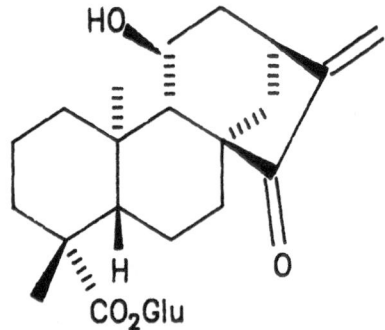

897

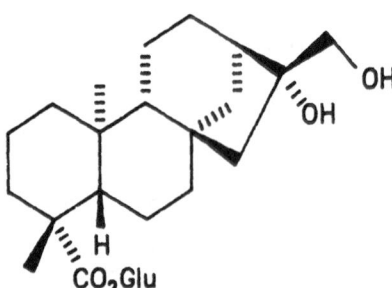

898

FIGURE 1. (*contd.*)

Figure 1. Structures of Diterpenes   371

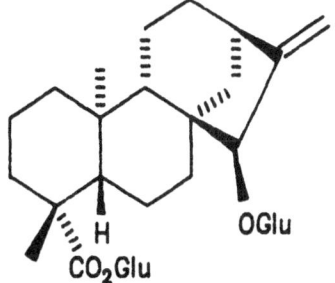

899

682 $R_1 = R_2 = H$, $R_3 = OH$

683 $R_1 = Glu$, $R_2 = H$, $R_3 = OH$

684 $R_1 = H$, $R_2 = Glu$, $R_3 = OH$

685 $R_1 = R_2 = Glu$, $R_3 = OH$

686 $R_1 = Glu$, $R_2 = R_3 = H$

687 $R_1 = R_2 = Glu$, $R_3 = H$

688 $R_1 = \beta Glu[2]-[1]Glu$, $R_2 = Glu$, $R_3 = H$

689 $R_1 = \beta Glu[2]-[1]Glu$, $R_2 = H$, $R_3 = OH$

FIGURE 1. (*contd.*)

59. ent-Kaurane, 17-Nor-

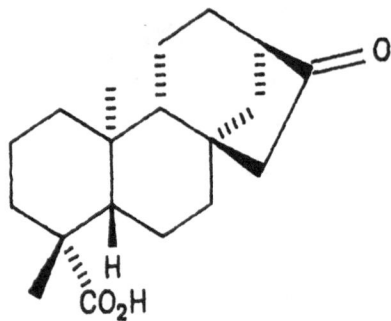

749

60. ent-Kaurane, 18-Nor-

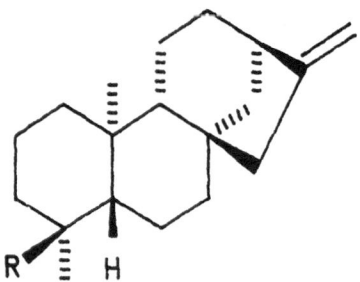

752 R = H

753 R = OH

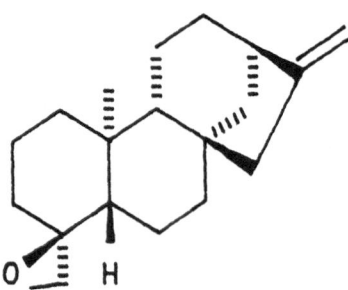

813

FIGURE 1. (*contd.*)

Figure 1. Structures of Diterpenes   373

754 R = CH$_2$OH

755 R = CHO

756 R = CO$_2$Me

757

758

61. ent-Kaurane, 19-Nor-

750 R = H

FIGURE 1. (*contd.*)

751   R = OH

62. ent-Kaurane, 9,10-Seco-

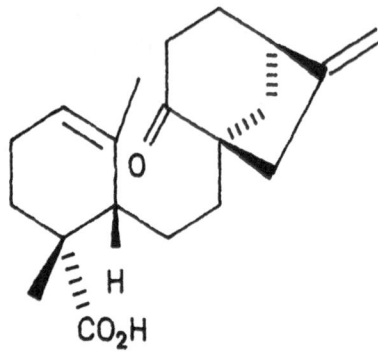

759

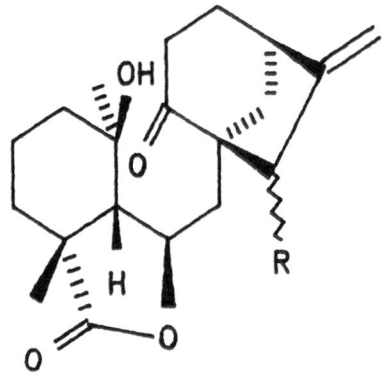

866   R = α-OH

867   R = β-OH

868   R = β-OAc

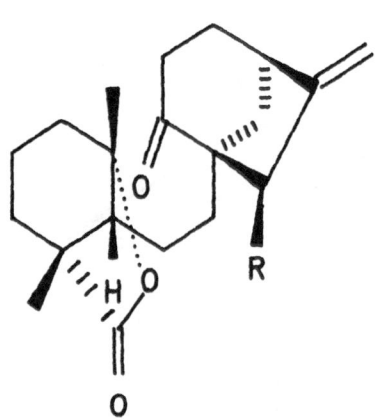

760   R = H

763   R = OH

FIGURE 1. *(contd.)*

Figure 1. Structures of Diterpenes   375

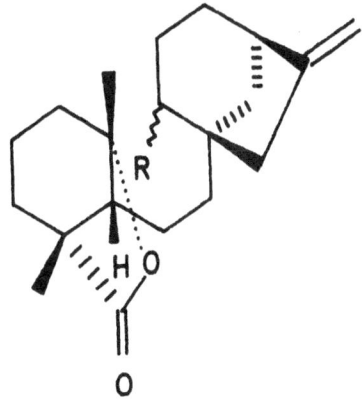

761 R = α-OH

762 R = β-OH

64. ent-Stachane (ent-Beyerane)

764 R$_1$ = OH, R$_2$ = H, R$_3$ = R$_4$ = Me

765 R$_1$ = R$_2$ = H, R$_3$ = Me, R$_4$ = CH$_2$OH

766 R$_1$ = R$_2$ = H, R$_3$ = CH$_2$OH, R$_4$ = Me

767 R$_1$ = OH, R$_2$ = H, R$_3$ = Me, R$_4$ = CH$_2$OH

768 R$_1$ = R$_2$ = OH, R$_3$ = R$_4$ = Me

769 R$_1$ = R$_2$ = H, R$_3$ = Me, R$_4$ = CH$_2$OMal

773 R$_1$ = R$_2$ = H, R$_3$ = Me, R$_4$ = CHO

774 R$_1$ = R$_2$ = H, R$_3$ = Me, R$_4$ = CO$_2$H

775 R$_1$ = OH, R$_2$ = H, R$_3$ = Me, R$_4$ = CO$_2$H

776 R$_1$ = OTig, R$_2$ = H, R$_3$ = Me, R$_4$ = CO$_2$H

FIGURE 1. (*contd.*)

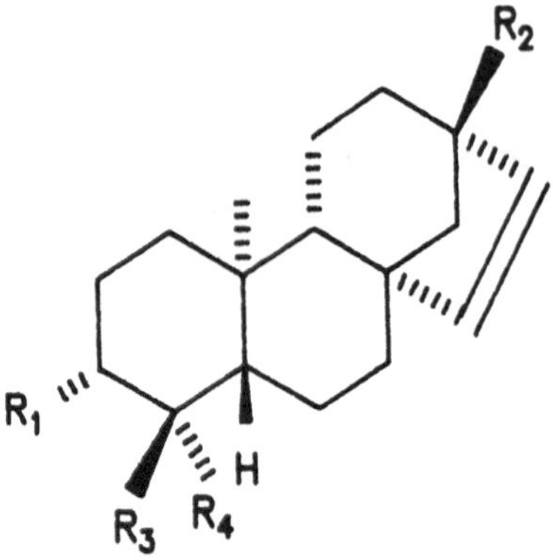

770   R_1 = OH, R_2 = CH_2OH, R_3 = R_4 = Me

771   R_1 = H, R_2 = CH_2OH, R_3 = Me, R_4 = CH_2OH

772   R_1 = H, R_2 = Me, R_3 = R_4 = CH_2OH

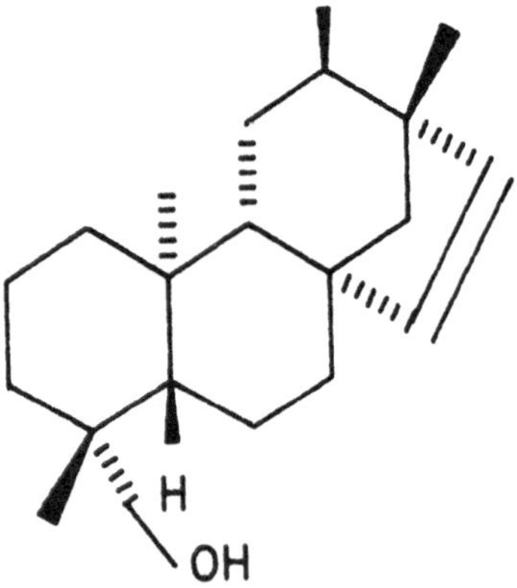

817

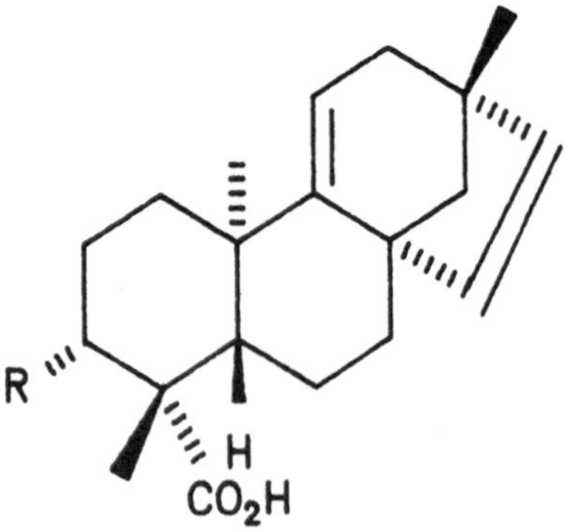

777   R = H

778   R = OTig

FIGURE 1. (contd.)

Figure 1. Structures of Diterpenes 377

779

780

781

FIGURE 1. (*contd.*)

93. ent-Stachane, 19-Nor-

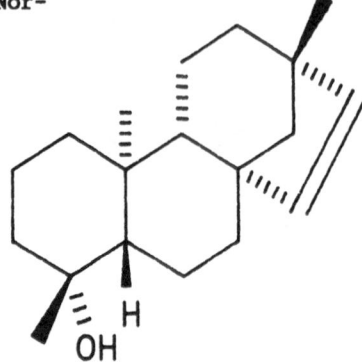

782

65. Atisirane (Atisane)

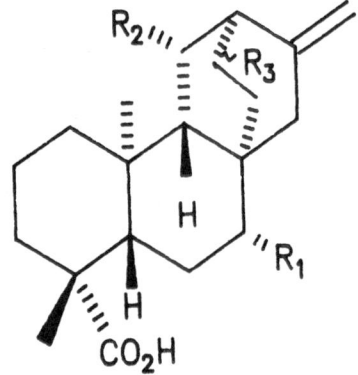

783   $R_1 = R_2 = R_3 = H$

784   $R_1 = OH, R_2 = R_3 = H$

785   $R_1 = H, R_2 = OAc, R_3 = H$

786   $R_1 = R_2 = H, R_3 = OAng$

787   $R_1 = R_2 = H, R_3 = Oi\text{-}Val$

788   $R_1 = R_2 = H, R_3 = Oi\text{-}But$

814   $R_1 = R_2 = H, R_3 = OH$

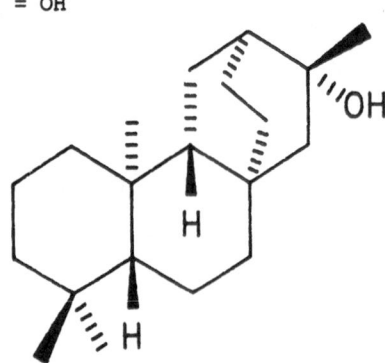

853

FIGURE 1. (*contd.*)

Figure 1. Structures of Diterpenes 379

854

816

789 $R_1 = OAc$, $R_2 = H$

790 $R_1 = H$, $R_2 = \beta$-OAc

66. Trachylobane

FIGURE 1. (*contd.*)

855 $R_1 = R_2 = H$, $R_3 = CHO$

791 $R_1 = R_2 = H$, $R_3 = CO_2H$

825 $R_1 = H$, $R_2 = OAc$, $R_3 = CO_2H$

826 $R_1 = H$, $R_2 = OAng$, $R_3 = CO_2H$

827 $R_1 = H$, $R_2 = Oi\text{-}Val$, $R_3 = CO_2H$

893 $R_1 = H$, $R_2 = Oi\text{-}But$, $R_3 = CO_2H$

792 $R_1 = \alpha\text{-}OH$, $R_2 = H$, $R_3 = CO_2H$

1204 $R_1 = R_2 = H$, $R_3 = CO_2Me$

1259 $R_1 = \beta\text{-}OH$, $R_2 = H$, $R_3 = CO_2H$

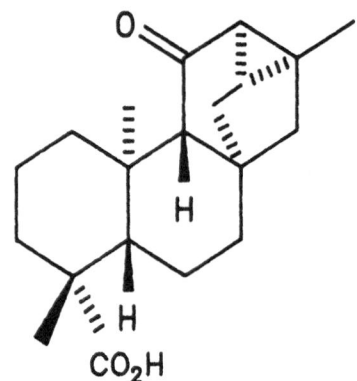

793

794

FIGURE 1. (*contd.*)

Figure 1. Structures of Diterpenes   381

795

856

67. Helifulvane

FIGURE 1. (*contd.*)

796 R_1 = H, R_2 = CH$_2$OH

797 R_1 = H, R_2 = CO$_2$H

798 R_1 = OH, R_2 = CO$_2$H

799 R_1 = OAc, R_2 = CO$_2$H

68. Tetrachyrane

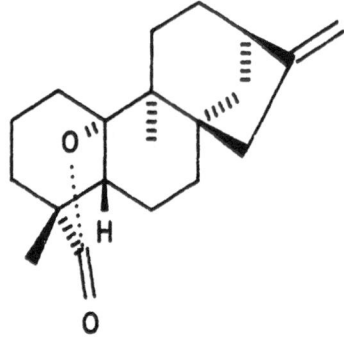

800

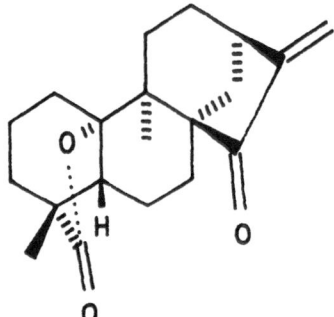

847

805 R_1 = H, R_2 = OH

806 R_1 = OH, R_2 = H

FIGURE 1. (*contd.*)

Figure 1. Structures of Diterpenes   383

78. Villanovane

875 $R_1 = R_2 = H$, $R_3 = CH_2OH$, $R_4 = CH_2Oi\text{-}Val$

876 $R_1 = R_2 = H$, $R_3 = CH_2OAc$, $R_4 = CH_2Oi\text{-}Val$

877 $R_1 = R_2 = H$, $R_3 = CH_2OH$, $R_4 = CH_2OMeVal$

878 $R_1 = R_2 = H$, $R_3 = OAc$, $R_4 = CH_2OMeVal$

879 $R_1 = Oi\text{-}Val$, $R_2 = H$, $R_3 = CH_2OAc$, $R_4 = Me$

880 $R_1 = OMeVal$, $R_2 = H$, $R_3 = CH_2OAc$, $R_4 = Me$

894 $R_1 = H$, $R_2 = OH$, $R_3 = R_4 = Me$

69. Lycoctonine

801

FIGURE 1. (*contd.*)

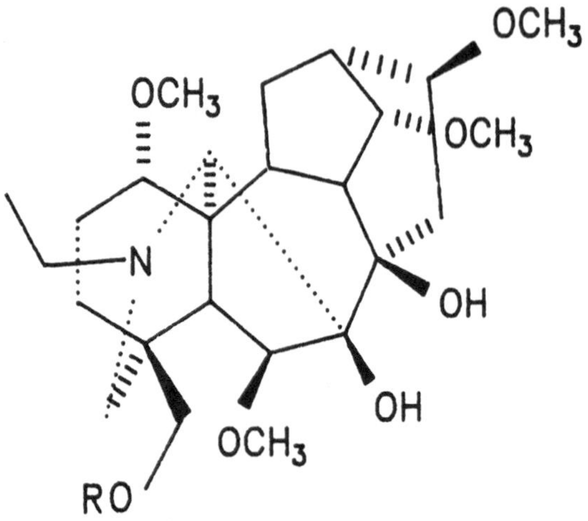

802 R = H

803 R = Anth

VIII. Undetermined Stereochemistry

79. Inulaefane

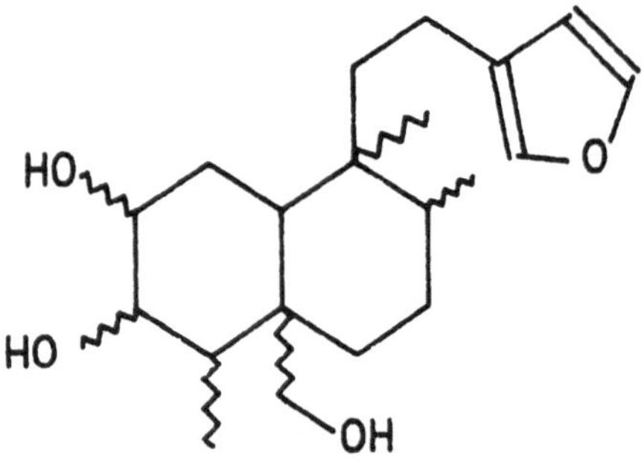

804

84. Hebeclinane (9,10-Seco-Labdane)

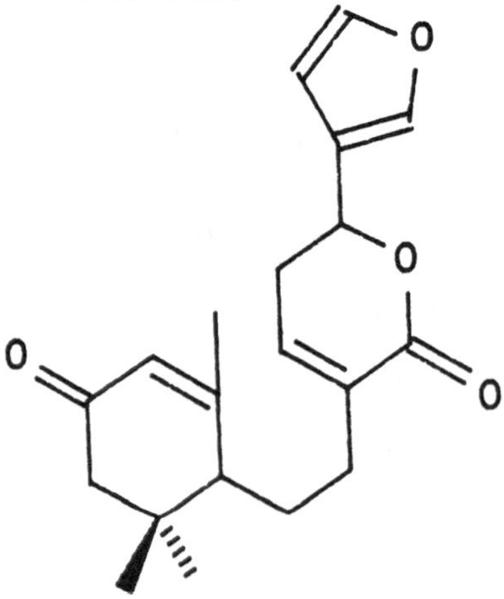

940 (= 999)

FIGURE 1. (contd.)

CHAPTER 3

Biogenesis

"...apart from the carbon skeleton of abietic acid with an irregular sequence of 4 isoprene units, the carbon skeletons of all diterpenes are derived from a regular isoprene tetramer consisting of 4 isoprene units connected head-to-tail... This structural relationship may be termed the "phytol rule", in analogy to the farnesol rule of the sesquiterpenes."

Ruzicka (*444*, p. 358)

III.1. *ent*- and Normal-Labdanes

Cyclization of the linear diterpene precursor usually proceeds along two alternative parallel routes (Figure 2 [p. 392]), one leading to the bicyclic "normal" labdane (normal-labdane) absolute stereochemistry and the other resulting in its antipodal, enantiomeric (*ent*-labdane) absolute stereochemistry. As a consequence of these two parallel biosynthetic routes, diterpenes lack the homogeneous absolute stereochemistry of the triterpenes.

C-5, C-9, and C-10 substituents are oppositely oriented in the normal- and *ent*-labdane cyclization products. This distinction between normal- and *ent*-labdanes suggests that the linear precursor adopts one of two possible conformations at the onset of cyclization. The formation of the normal- and antipodal *ent*-AB ring junctions and the *trans*-anti- configuration of these junctions are governed by the relative orientation of the cyclizing double bonds within the chairlike conformation of the folded C20 linear precursor (Figure 2). These alternative cyclizations may be visualized as arising through two different coilings of the open-chain geranylgeranyl pyrophosphate (GGPP) precursor on an enzyme surface followed by (1) initiation of cyclization by proton addition to C-14 of the 14,15-double bond and (2) antiparallel cationic 1,2-additions resulting from the migration of electrons from the two internal 6,7- and 10,11-double bonds. Following cyclization to the *trans*-fused decalin ring system, the C-4 axial methyl group and H-5 are oriented in the same plane and arranged *trans* to one another, a relationship that also exists between the C-10 methyl and H-9. Likewise, the newly formed C-4-C-5 and C-9-C-10 bonds are in the same plane and are *trans* about the C-5-C-10 axis (Figure 2).

The hypothetical C-8 carbonium ion resulting from this initial cyclization is stabilized by several alternative events: (1) proton-loss yielding either the C-7-C-8 or C-8-C-17 double bond (e.g., Figure 2), (2) neutralization via hydration or lactone or ether formation and (3) rearrangement. Much of the diterpene structural variability (Figure 3 [pp. 393–415]) is a consequence of the variety of rearrangement products resulting from this last mode of stabilization.

In the following discussion, diterpene biogenesis is described in terms of a hypothetical reaction sequence leading from an appropriate labdane precursor to a product representative of one of the 103 skeletal types reported from the Compositae. These skeletal types are arranged in Figure 3 according to their putative biogenetic relationship.

Regardless of the extensive variability in leaf-resin chemistries found in the Compositae, all plants of this family presumably biosynthesize gibberellin diterpenes via a pathway outlined in Figure 4 (p. 416). As gibberellin biosynthesis (1) appears to be anatomically and functionally distinct from that of leaf-resin diterpenoids and (2) is regularly reviewed in the literature, it will not be discussed here.

III.1.1. Seco- and Nor-Labdane Skeletons

Ring cleavage and decarboxylation of labdanes have generated several normal- (Figure 3, Skeleton Nos. 23, 24, 25, 26, 27, 87 and 89) and *ent-* (Figure 3, Nos. 41, 42, 43, 44, 45 and 75) seco- and nor-labdane skeletons.

ent-Labdanes of *Ayapana amygdalina* and their C-17 nor-homologs (Figure 5 (p. 417); *113*) fit well into a biogenetic scheme originating from the proposed precursor **337**. 8,17-epoxidation and subsequent hydration to produce **343** facilitates loss of C-17 to give compounds with the C-17-nor labdane skeleton (i.e. **467** and **468**). An alternative route leads to compounds (**339–342, 344, 345**) with a C-17 aldehyde or carboxyl function.

Chrysothamnus paniculatus yielded a skeleton resembling the seco- (cleaved) product of a common grindelane precursor, **172** (Figure 6 (p. 418); *341*). This 8,9-seco-grindelane skeleton (Figure 3, No. 25) distinguishes **223**, a compound variously named chrysothame and strictanonic acid. Another skeleton of seco- origin reported from *Nidorella hottentotica* displays a ten-membered ring produced by the cleavage of the C-5-C-10 bond. The proposed relationship of representatives of this skeleton and structurally related, co-occurring normal-labdanes is depicted in Figure 7 (p. 419) (*155*).

III.1.2. Rearranged Labdanes

Rearranged labdane skeletons arise via a series of associated hydride and methyl shifts, usually involving concerted *trans*-migrations of the participating substituents.

III.1.2.1. Normal-Labdane Series

Among the normal-labdanes, the simplest rearrangement product, the relhaniane skeleton (Figure 3, No. 28), results from the concerted migration of C-9α-H to

C-8 and C-10β-methyl to C-9 (Figure 8 Scheme 2 [p. 420]). C-1 proton loss and C-1-C-10 double bond formation complete the biogenesis proposed for the *Relhania acerosa* compound, relhania acid, **231** (Figure 8, Scheme 1 (p. 420); *90*). Alternatively, migration of H-5 to C-10 and deprotonation of C-6 are the final steps of the proposed biogenesis for the *Koanophyllon conglobatum* compounds koanophyllic acids A-D (**232–235**; Scheme 2, Figure 8). As the absolute stereochemistry of the koanophyllic acid series is not established, their normal-labdane derivation is unproven and only follows the authors' tentative assignment (*42*).

The novel chrysolane skeleton (Figure 3, No. 72) probably results from the rearrangement of a precursor resembling 6β-hydroxy-grindelic acid (**172**). In this scheme, migration of the C-10 methyl sets in motion a complicated series of events terminating in the aromatic chrysolane skeleton reported from *Chrysothamnus paniculatus* (Figure 9 [p. 421]; *341*).

Examples of the clerodane skeleton, the most common rearrangement product, can be further subdivided into *cis-* and *trans-*fused decalin forms (Figure 10 [p. 422]). In the series of methyl and hydride shifts which make up the concerted "friedo" or "backbone" rearrangement process culminating in the clerodane skeleton, the migrating groups display a *trans-*diaxial relationship. For normal-labdanes, the expected product of a concerted series of *trans-*migrations (C-9α-H to C-8, C-10β-methyl to C-9, C-5α-H to C-10, and C-4β-methyl to C-5) would be the *trans-*clerodane (Figure 3, No. 29; Figure 10). There is as yet no proof that this normal-labdane-based pathway operates within the Compositae. However, the *cis-*clerodane skeleton (Figure 3, No. 31; Figure 10) which differs from the *trans-*version in the orientation of the C-5 methyl is well represented. Biogenetically, the existence of the *cis-*form is explained by a sequence of concerted *trans-*migrations (C-9α-H to C-8, C-10β-methyl to C-9 and C-5α-H to C-10) culminating in the formation of a carbonium ion at C-5. The AB *cis-*fused ring system would then result from the migration of the C-4α-methyl to C-5, followed by deprotonation at C-3.

III.1.2.2. ent-Labdane Series

Steps proposed for the biogenesis of rearranged *ent-*labdanes appear to correspond to the mechanisms indicated for the normal-labdanes, differing only in the opposite orientivity of the migrating groups.

The simplest rearranged *ent-*labdane is the *ent-*friedolabdane skeleton (Figure 3, No. 95). A proposed migration sequence for this structure begins with a hydride shift from C-9(β-H) to C-8, followed by a methyl shift from C-10(α-methyl) to C-9 and hydride shift from C-5(β-H) to C-10 and terminates in C-6 deprotonation and C-5-C-6 double bond formation (e.g., **1059**).

In contrast to the normal-clerodanes, both *cis-* and *trans-*fused *ent-*clerodanes are well documented (Figure 3, Nos. 50 and 46, respectively). Mechanistically, the proposed steps of *ent-*clerodane biogenesis differ from those of the normal-series only in the opposite orientation of the corresponding migrating hydride or methyl group (Figure 10).

III.1.3. Seco-, Nor- and Rearranged Clerodane Skeletons

The next level of skeletal complexity is expressed by seco-, rearranged and nor-derivatives of both normal- and *ent*-clerodanes. Such skeletons (e.g., Figure 3, No. 41) tend to have relatively narrow distributions, often being reported from a single source.

Rearrangement of an *ent*-clerodane precursor results in a series of skeletal types some of which are shared by two genera, the Astereae member, *Conyza* (*83*), and the Inuleae taxon, *Printzia* (*165*). These include the printzianes (Figure 3, No. 47; Figure 11 [p. 423], **571, 572, 915**), conyscabranes (Figure 3, No. 76; Figure 11, **917–920**) and 5,10-seco-clerodanes (Figure 3, No. 48; Figure 11, **573, 912**). Another rearranged *ent*-clerodane skeleton, solidagonane, (Figure 3, No. 47; Figure 11 [p. 423], **571, 572, 915**), conyscabranes (Figure 3, solidagonal acid (**960**), which was probably derived from a kolavenic acid (**495**)-like precursor (Figure 12 [p. 424]).

III.2. Tricyclic Diterpenes

Instead of rearrangement to clerodanes and related bicyclic skeletons, the hypothetical product of labdane formation, labdadienyl pyrophosphate, can cyclize to produce either tricyclic pimara-8(9),15-diene or its 8(14)-isomer. The simplest mechanism for this process is as follows: attack at C-13 by electrons of the exocyclic double bond (S_N' reaction) is followed by migration of the double bond from C-13 and elimination of the pyrophosphate group from the allylic position. The resulting intermediate tricyclic cation at C-8 is stabilized by proton elimination at C-9 or C-14 (e.g., Figure 13 [p. 425] **274**). The resulting skeletons are referred to as sandaracopimaranes or normal-pimaranes (Figure 3, No. 32) if derived from the normal-labdane precursor or *ent*-pimaranes (Figure 3, No. 52) if formed from an *ent*-labdane.

Tricyclic compounds with the epimeric arrangement of methyl and vinyl substituents at C-13 often occur as pairs, e.g., rimuene and *ent*-rimuene (Figure 14A [p. 426]), suggesting that electron addition from the exocyclic methylene can occur at either face of the double bond depending on the stereoselectivity of the enzyme involved. (The two examples, rimuene and *ent*-rimuene, also illustrate that the prefix *ent*- does not always refer to compounds of *ent*-labdane origin.) Frequent reports of such C-13 epimeric pairs suggest that enzyme stereospecificity is low.

Instead of stabilization of the pimarane intermediate by proton elimination at C-9 or C-14, several further rearrangement processes can occur. For example, the cassane (Figure 3, No. 34) and abietane (Figure 3, No. 37) skeletons can result from alternative methyl migrations (Figure 13). The rosane skeleton (Figure 3, No. 35; Figure 14B [p. 426], **881–889**) can also be generated by methyl migration and stabilized by a variety of routes as demonstrated by the chemistry of *Trichogonia* (Figure 14B, *183*).

The abietane skeleton that Ruzicka excluded from the "phytol rule" (*444*) can be derived in a straightforward manner from either a normal- or *ent*-pimarane precursor (Figures 13 and 15 [p. 427], respectively). Representatives of both normal- and *ent*-abietanes (Figure 3, Nos. 37 and 55, respectively) have been reported from the Compositae.

In addition to the *ent*-abietane skeleton, other tricyclic *ent*-pimarane derived ring skeletons have been reported (Figure 15): Cleistanthanes and isocleistanthanes were detected in *Brickellia eupatoriedes* (*59*) and erythroxanes were identified in *Helichrysum refluxum* (*450*). The absolute configuration (5R, 10R) of the *Brickellia* cleistanthanes is identical to that reported for the cleistanthanes of the Euphorbiaceae, but antipodal to those (5S, 10S) of the Velloziaceae (*Vellozia*) (Figure 16 [p. 428]; *427*).

Acritopappus confertus (Eupatorieae) produces a pair of unusual tricarbocyclic compounds, **617** and **618**, that do no result from the cyclization processes described above. A biogenetic scheme for these constituents (Figure 17 [p. 429]) invokes an aldol condensation reaction to link C-11 and C-17 (*184*).

III.3. Tetracyclic Diterpenes

The hypothetical tricyclic pimarenyl cation can either undergo further cyclization to yield a tetracyclic cation which eliminates a proton to give an *ent*-stachane (*ent*-beyerane; Figure 18, Route I [p. 430]) or form alternative protonated-cyclopropane-containing intermediates (Figure 18, Routes II and III). Through either simple proton elimination or more complicated processes, a variety of tetra- and pentacyclic skeletons are produced via these cyclopropanoid intermediates. With one reported exception, normal-stachane (Figure 3, No. 73), all these polycyclics are members of the *ent*-series (Figure 18): *ent*-stachane, *ent*-atisirane, *ent*-kaurane, *ent*-trachylobane, *ent*-helifulvane, *ent*-tetrachyrane and *ent*-villanovane (Figure 3, Nos. 64, 65, 58, 66, 67, 68, and 78, respectively).

III.4. Biosynthetic Evidence

Assuming that similar enzymes and enzyme systems are involved in the biosynthesis of both gibberellins and resin diterpenes, the relative wealth of biosynthetic data concerning the *ent*-kaurane-derived gibberellins may provide some insight into resin diterpene biogenesis. Three aspects of gibberellin biosynthesis seem pertinent to understanding resin diterpene distribution in the Compositae. The first is the existence of two discrete and separable enzymatic processes, activities A and B. Activity A cyclizes geranylgeranyl pyrophosphate (GGPP) to give the labdane intermediate and B completes the cyclization process to yield *ent*-kaurene and/or other polycyclic structures. The second is the role of plastid compartmentation in the segregation of diterpene synthesis from other terpenoid biosynthetic processes. Further separation of biosynthetic processes is suggested by evidence that AB cyclization activity is a product of soluble enzymes (*313*) from which the diterpene is passed to membrane-bound mixed function oxygenases (*497*). The third is the role of GGPP as a key branch-point metabolite in polyisoprenoid metabolism. Each aspect will be discussed in detail below.

III.4.1. Diterpene Biosynthesis: AB Activity

The presence of two separable enzymatic activities associated with polycyclic diterpene (i.e., *ent*-kaurene) biosynthesis suggests that inactivation of the first (A inactivation) could result in the total absence of cyclic diterpenes and inactivation of the second (B inactivation) would lead to production of only labdanes and related bicyclic structures. Inactivation of A activity would prevent the cyclization of GGPP to the bicyclic intermediate. For the bicyclic product of A activity, absence of B activity would eliminate its role as precursor to *ent*-kaurane. If the enzymes of resin diterpene biosynthesis resemble those of gibberellin biosynthesis, the A and B activities serve as a model regulatory system accounting for many observed resin chemistry differences between plant groups. The largely tetracyclic diterpenoid chemistry of the tribe Heliantheae is radically different from the predominantly labdane and clerodane chemistries of the Astereae and Eupatorieae. Another large tribe, the Vernonieae yielded no significant reports of diterpenes despite intensive chemical analysis. Such widespread patters could be explained by a regulatory system based on either A or B inactivation.

In castor bean cell-free extracts, enzyme activity could be further differentiated into discrete (1) A activity, (2) AB activity which produced a range of polycyclic diterpenes including *ent*-sandaracopimara-8(14),16-diene, *ent*-beyerene, *ent*-trachylobane and *ent*-kaurene, and (3) discrete B activities that each correspond to distinct catalytic sites for the production of individual end products (e.g. sandaracopimara-8(14),16-diene) (*497*). The activation/inactivation of different B activities could account for major differences observed in the resin chemistries of large taxa. For example, a segment of the tribe Calenduleae is characterized by sandaracopimarane-type diterpenes.

III.4.2. Soluble Cyclization and Membrane-bound Oxgenation Enzymes

Kaurene synthetase (a term used collectively for A and B activities) appears to be a soluble stromal enzyme weakly associated with plastid membranes (*437*). Evidence from gibberellin biosynthesis suggests that subsequent to cyclization, the lipophilic *ent*-kaurene intermediate proceeds through a series of oxidation reactions catalyzed by a group of membrane-bound microsomal mixed function oxidases. Further oxidation of the more-polar product is evidently carried out by soluble oxidase enzymes to yield various gibberellins. The membrane-bound enzymes belong to the class of oxidases that require reduced pyridine nucleotide and O_2 as co-substrates and are cytochrome P-450 dependent. Members of this enzyme class may be more generally involved with diterpene oxygenation (*497*).

These results suggest a clear distinction between the biogenetic steps associated with the cyclization processes (AB activity) and subsequent oxidative modifications. Although the extent to which gibberellin biosynthesis can be generalized to other diterpenes is unknown, the organization of its biosynthesis provides a useful framework for interpreting resin diterpene structural variability.

III.4.3. GGPP and Diterpene Distribution

GGPP is a key branch point metabolite of polyisoprenoid metabolism. Apart from the cyclic diterpenes, it is the precursor of the tetraterpenes (carotenes, xanthophylls), and acyclic diterpenes (phytol). Cyclic diterpenes compete with compounds of established importance for either GGPP or its precursors (e.g., farnesyl pyrophosphate). The "spotty" nature of resin diterpene distribution may in part be explained by the obvious demand for carotenoids and phytyls and the relatively unestablished requirement for these nongibberellin diterpenes.

III.4.4. Compartmentation and Diterpene/Sesquiterpene Expression

GGPP biosynthesis and its conversion to acyclic, macrocyclic, bicyclic and polycyclic diterpenes may be localized within specific plastid compartments (*498*). This proposal fits into a broader scheme in which key isoprenoid metabolic branch points leading to mono-, sesqui-, and diterpenes are selectively activated only within certain membrane-bound sites. Compartmentation would then help to regulate the proximity of chain-length specific prenyl transferases to the appropriate C_{10}, C_{15}, and C_{20} cyclases, and modifying enzymes. Strict metabolic regulation would then be possible within a given compartment (*234*).

This model can be applied to the terpene-producing trichomes that typify Compositae species. Trichomes are clearly the primary sites of physical compartmentation of terpenoid biosynthesis. "Multiple compartmentalized biosynthetic sites" probably occur within the secretory cells of the trichomes (*234*). Thus, expression of a well-developed sesquiterpene and/or diterpene chemistry in a plant may only be the result of the activation of specific, preexisting "biosynthetic sites."

This view of compartmentation as an important regulatory factor in terpenoid biosynthesis (*498*) may help explain the extraordinary diversity of sesqui- and diterpene chemistries found even within groups of closely related species. A survey of glandular terpenoid chemistries reveals that some species are predominantly sesquiterpene producers, others are diterpene producers, and still others produce copious amounts of both. Such diversity even within clusters of related species suggests that the expression of the regulatory process controlling the activation/inactivation of compartmentalized sesqui- and diterpene biosynthesis may play a role in the speciation process.

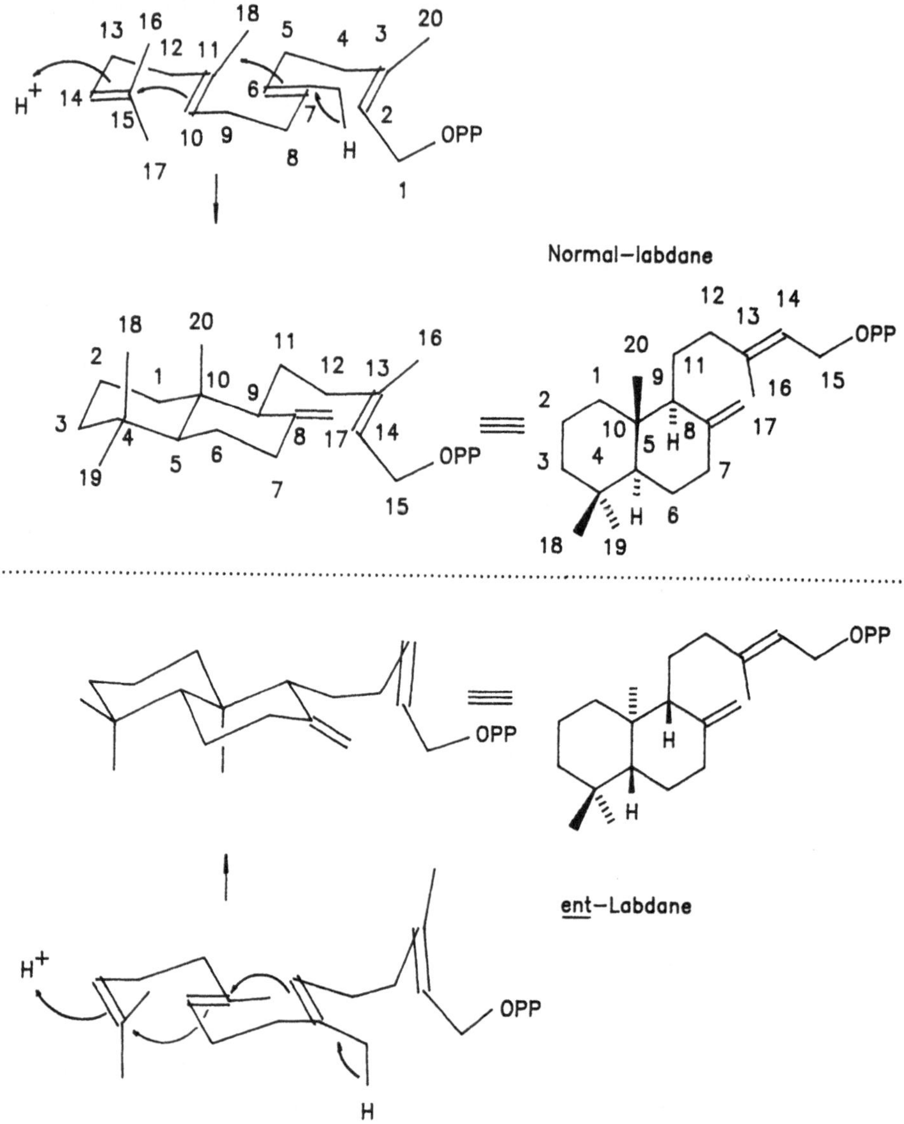

FIGURE 2. Alternative cyclizations of geranylgeranyl pyrophosphate to produce normal- and *ent*-labdane products.

Figure 3. Linear and Cyclic Diterpene Skeletal Types   393

I. Linear or Unicarbocyclic

1. Geranylnerol

2. Geranylgeraniol

3. Gerranyllinalool

4. Phytol-derived

80. 14-Methylgeranylnerol

5. Acanthoaustralane

6. Acanthoaustralane-6,11-epoxide

7. Isoacanthoaustralane

8. Geranylgeraniol-18,8-lactone

9. Oxepane

86. Melcanthane

10. Mikanofurane

11. Centipedane

12. Geranylgeraniol-1,20-lactone

13. Geranylterpinene

14. Isocembrene

15. Geranylnerol-11,14-α-epoxide

II. Normal-Bicarbocyclics and Derivatives

A. Bicyclic (regular)

16. Normal-labdane

17. Grindelane

18. Normal-labdane, 8,13-epoxide-

19. Normal-labdane, 8,12-epoxide-

20. Normal-labdane-15,16-lactone

FIGURE 3. Linear diterpene skeletal types and cyclic diterpene skeletal types derived from normal and *ent*-labdane precursors.

21. Normal-labdane-20,12-lactone

22. Normal-labdane spiroketal

88. Normal-labdane-15,12-lactone

B. Bicyclic (seco- or nor-)

23. Normal-labdane, 7,8-seco-

24. Normal-labdane, 3,4-seco-

25. Grindelane, 8,9-seco-

89. Grindelane, 19-nor-

26. Grindelane, 8,17-bis-nor-8,9-seco-

27. Normal-labdane-20,12-lactone, 5,10-seco-

87. Hexanorlabdane

100. Normal labdane, 14,15-bisnor-

101. Normal labdane, 13,14,15,16-tetranor-

C. Bicyclic (irregular)

28. Relhaniane

29. <u>trans</u>-Normal-clerodane

30. Clerodane, 5,10-seco-

31. <u>cis</u>-Normal-clerodane

72. Chrysolane

90. Tricycloclerodane

III. Normal-Tricarbocyclics and Derivatives

32. Sandaracopimarane (Normal-pimarane)

33. Sandaracopimarane-20,8-hemiacetal

34. Cassane

35. Rosane

36. Sandaracopimarane, 19-nor-13-epi-

FIGURE 3. (*contd.*)

Figure 3. Linear and Cyclic Diterpene Skeletal Types   395

37. Normal-abietane

94. Stevisalane

IV. Normal Tetracarbocyclics

73. Normal-stachane

V. ent-Bicarbocyclics and Derivatives

A. Bicyclic (regular)

38. ent-Labdane

39. ent-Labdane-19,6-lactone

40. ent-Labdane-8,13-epoxide-15,16-hemiacetal

97. ent-Labdane dimer

98. ent-Manoyl oxide, 13-epi-

99. ent-Manoyl oxide

B. Bicyclic (seco- or nor-)

41. ent-Labdane, 17-nor-

42. ent-Labdane, 15-nor-

43. ent-Labdane, 14,15-bis-nor-

44. ent-Labdane, 13,14,15,16-tetra-nor-

45. ent-Labdane, 7,8-seco-

74. ent-Labdane, 18-nor-

C. Bicyclic (irregular)

95. Friedolabdane

46. trans-ent-Clerodane

47. Printziane

FIGURE 3. (contd.)

76. Conyscabrane

48. <u>ent</u>-Clerodane, 5,10-seco-

96. <u>trans</u>-<u>ent</u>-clerodane, 17-nor-

49. <u>trans</u>-<u>ent</u>-Clerodane, 13,14,15,16-tetra-nor-

50. <u>cis</u>-<u>ent</u>-Clerodane

82. Solidagonane

102. Friedolabdane, 3,4-seco-<u>ent</u>-

103. Friedolabdane, Rearranged 2,3-seco-<u>ent</u>-
VI. <u>ent</u>-Tricarbocyclics

51. <u>ent</u>-Labdane, cyclopropyl-

77. Gutierreziane

52. <u>ent</u>-Pimarane

53. <u>ent</u>-Pimarane-8,15-tetrahydrofuran

54. Cleistanthane (Cleistanthol)

85. Cleistanthane, iso-

55. <u>ent</u>-Abietane

56. Acritoconfertane

57. Acritoconfertane-acetal

83. Erythroxane

VII. <u>ent</u>-Tetracarbocyclics and Derivatives

58. <u>ent</u>-kaurane

59. <u>ent</u>-Kaurane, 17-nor-

60. <u>ent</u>-Kaurane, 18-nor-

61. <u>ent</u>-Kaurane, 19-nor-

62. <u>ent</u>-Kaurane, 9,10-seco-

63. <u>ent</u>-Kaurane-19,10-lactone, 9,10-seco-

FIGURE 3. (*contd.*)

Figure 3. Linear and Cyclic Diterpene Skeletal Types   397

81. <u>ent</u>-Kaurane-19,6-lactone, 9,10-seco-

64. <u>ent</u>-Stachane (<u>ent</u>-Beyerane)

93. <u>ent</u>-Stachane, 19-nor-

65. Atisirane (Atisane)

66. Trachylobane

67. Helifulvane

68. Tetrachyrane

78. Villanovane

69. Lycoctonine-diterpene

VIII. Undetermined Stereochemistry

70. Pimarane, 18-nor-

79. Inulaefane

84. Hebeclinane (9,10-seco-labdane)

FIGURE 3. (*contd.*)

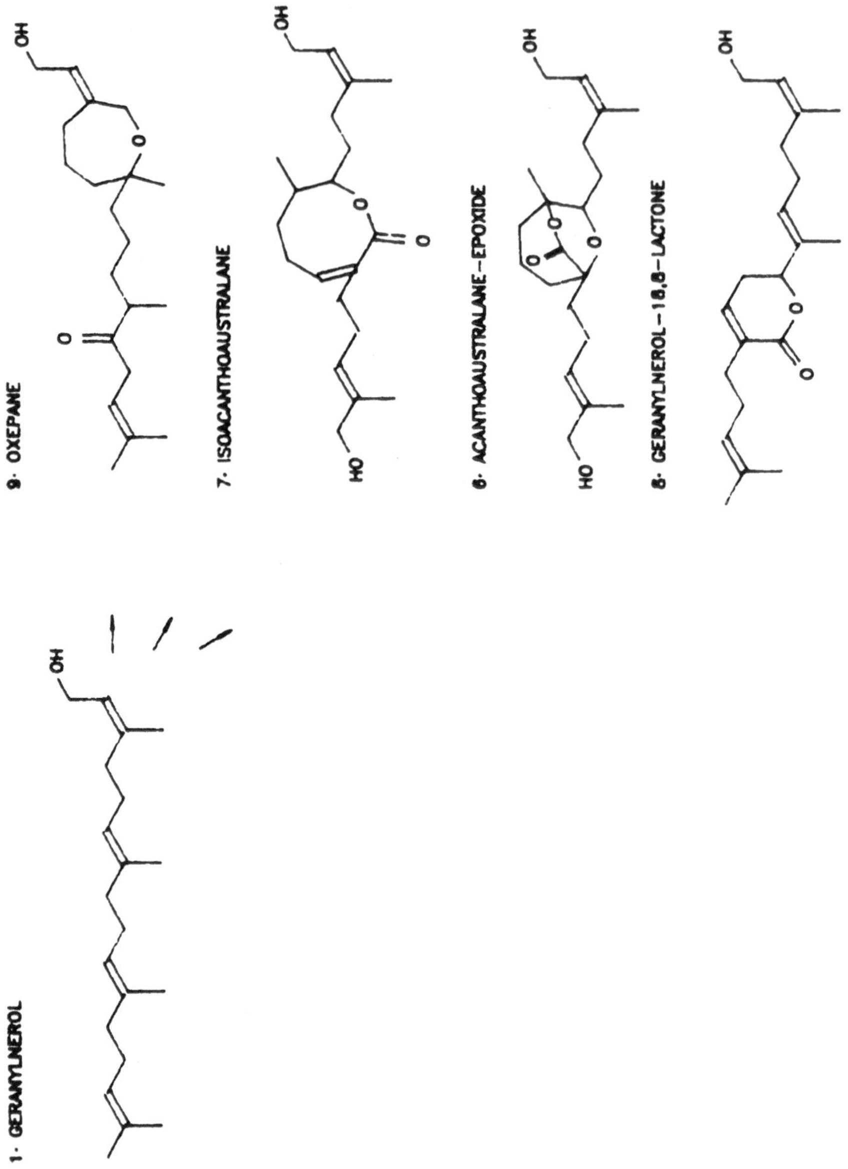

3. Biogenesis

9 · OXEPANE

7 · ISOACANTHOAUSTRALANE

6 · ACANTHOAUSTRALANE - EPOXIDE

8 · GERANYLNEROL - 1,8,8 - LACTONE

1 · GERANYLNEROL

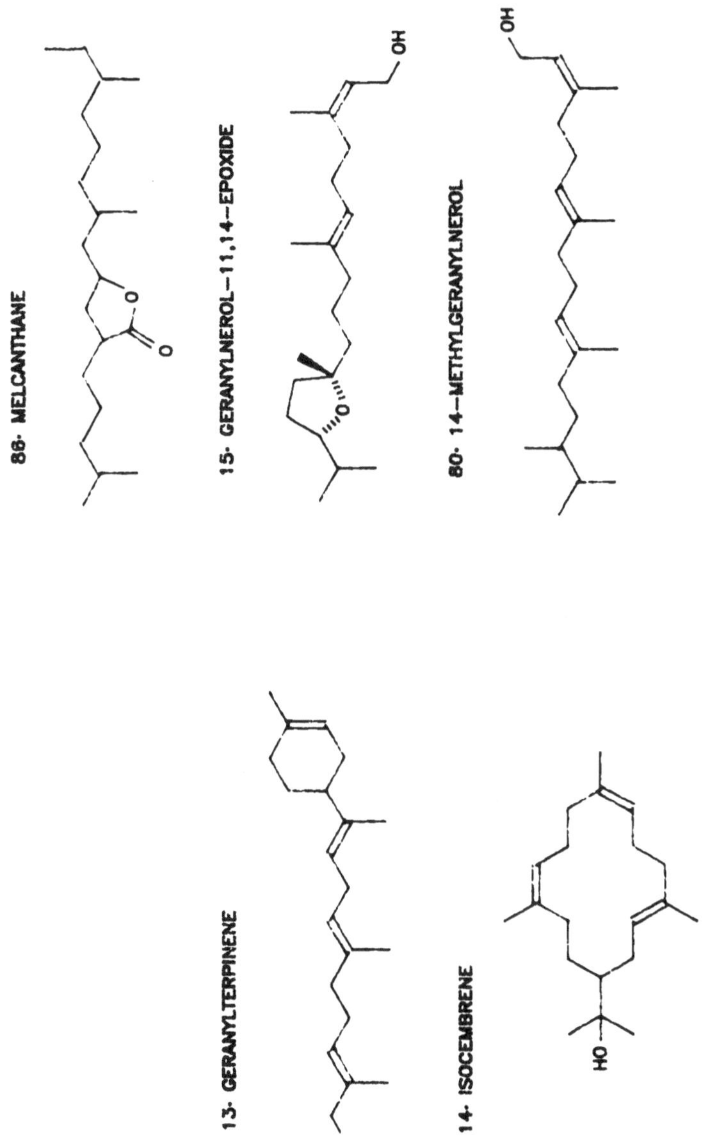

Figure 3. Linear and Cyclic Diterpene Skeletal Types 399

86· MELCANTHANE

15· GERANYLNEROL—11,14—EPOXIDE

80· 14—METHYLGERANYLNEROL

13· GERANYLTERPINENE

14· ISOCEMBRENE

FIGURE 3. (*contd.*)

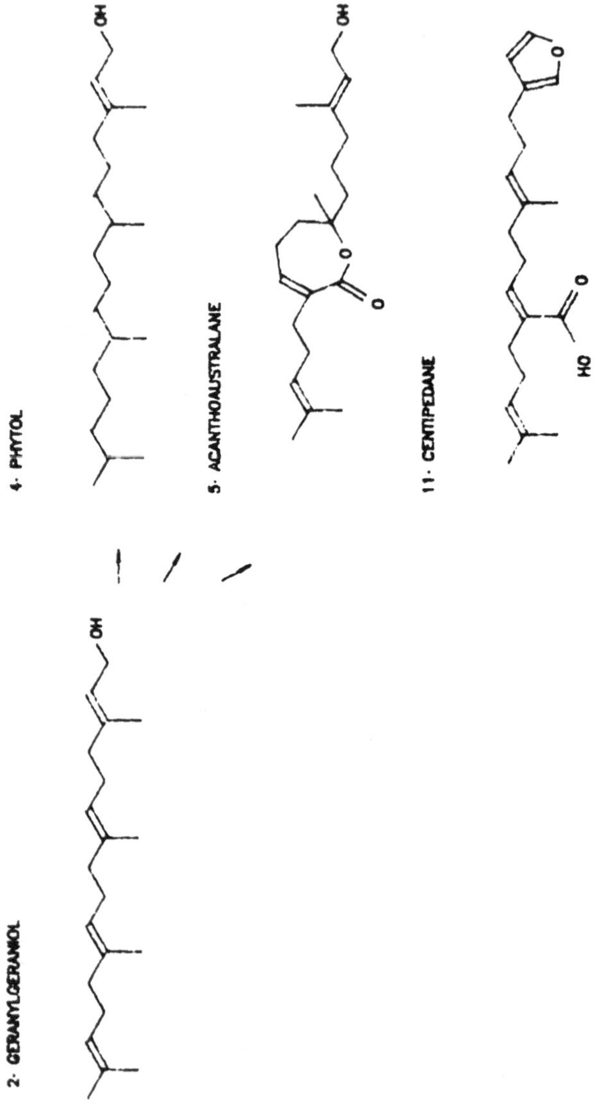
2· GERANYLGERANIOL
OH
4· PHYTOL
OH
5· ACANTHOAUSTRALANE
OH
O
O
11· CENTIPEDANE
O
O
HO

Figure 3. Linear and Cyclic Diterpene Skeletal Types   401

10· MIKANOFURANE

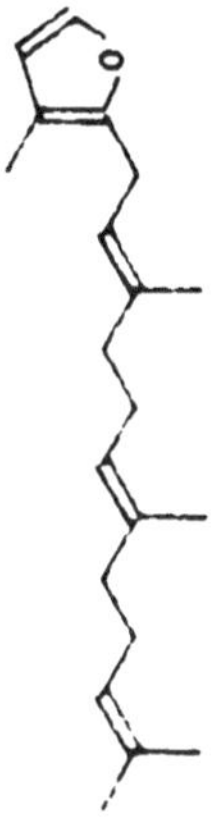

12· GERANYLGERANIOL—1,20—LACTONE

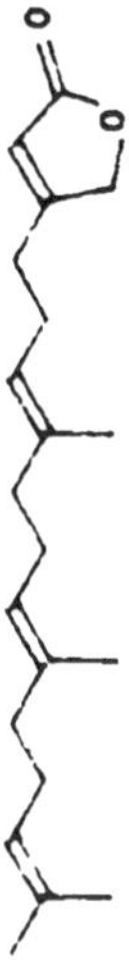

3· GERANYLLINALOOL

FIGURE 3. (contd.)

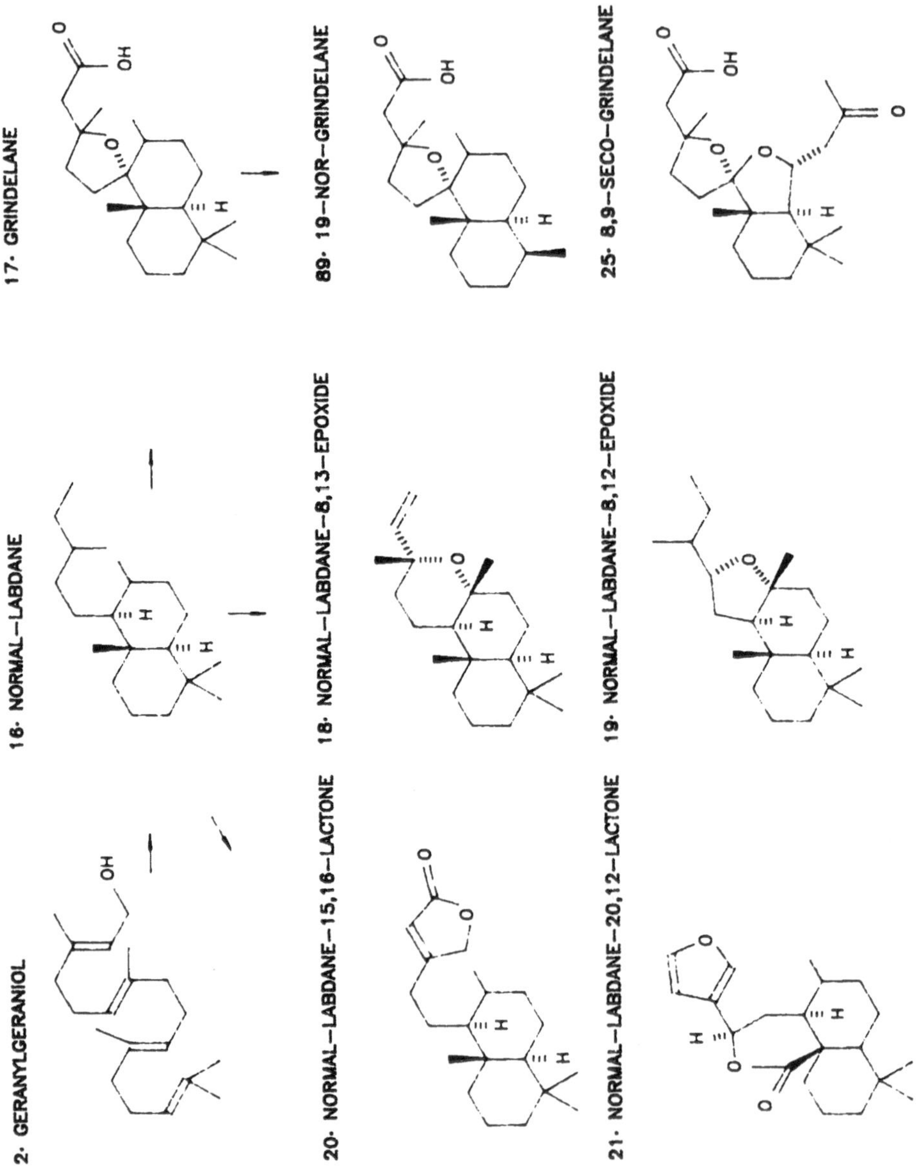

17· GRINDELANE

89· 19-NOR-GRINDELANE

25· 8,9-SECO-GRINDELANE

16· NORMAL-LABDANE

18· NORMAL-LABDANE-8,13-EPOXIDE

19· NORMAL-LABDANE-8,12-EPOXIDE

2· GERANYLGERANIOL

20· NORMAL-LABDANE-15,16-LACTONE

21· NORMAL-LABDANE-20,12-LACTONE

Figure 3. Linear and Cyclic Diterpene Skeletal Types   403

88· NORMAL—LABDANE—15,12—LACTONE

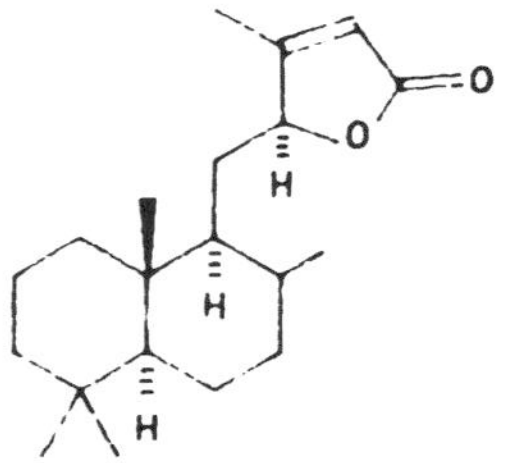

22· NORMAL—LABDANE SPIROKETAL

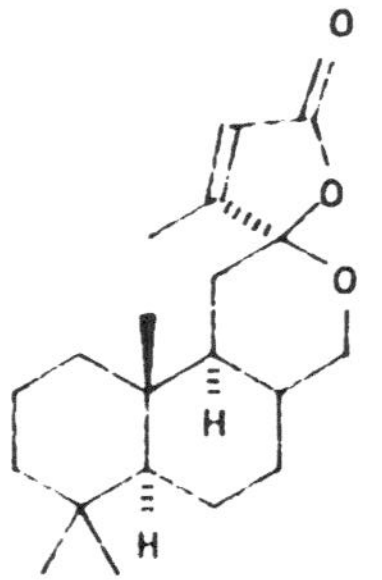

26· 8,17—BIS—NOR—8,9—SECO—
GRINDELANE

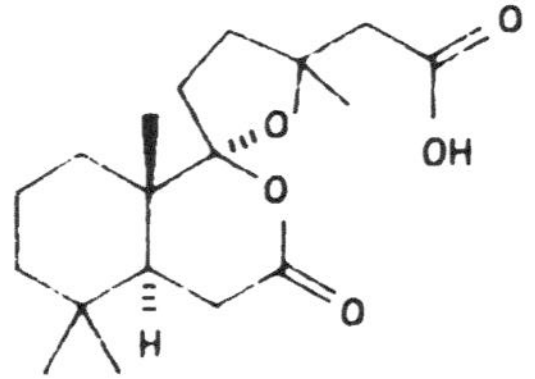

27· 5,10—SECO—NORMAL—
LABDANE—20,12—LACTONE

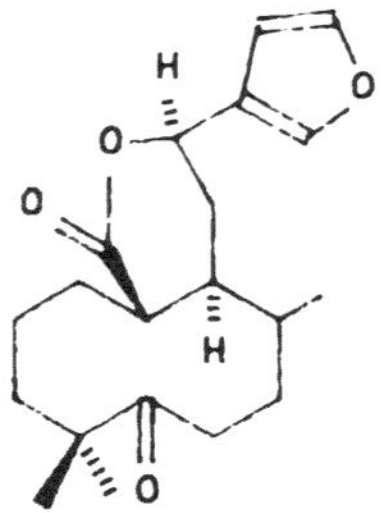

Figure 3. (contd.)

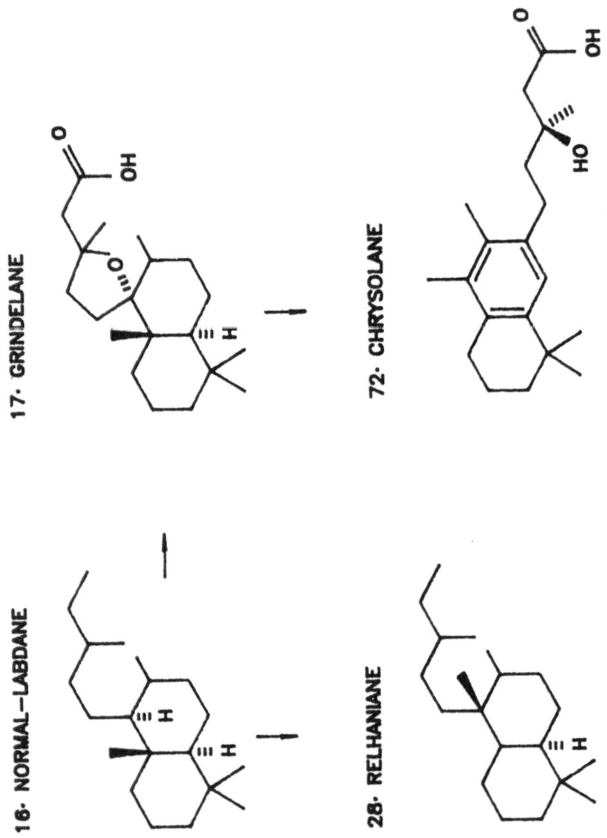

17· GRINDELANE

72· CHRYSOLANE

16· NORMAL—LABDANE

28· RELHANIANE

Figure 3. Linear and Cyclic Diterpene Skeletal Types 405

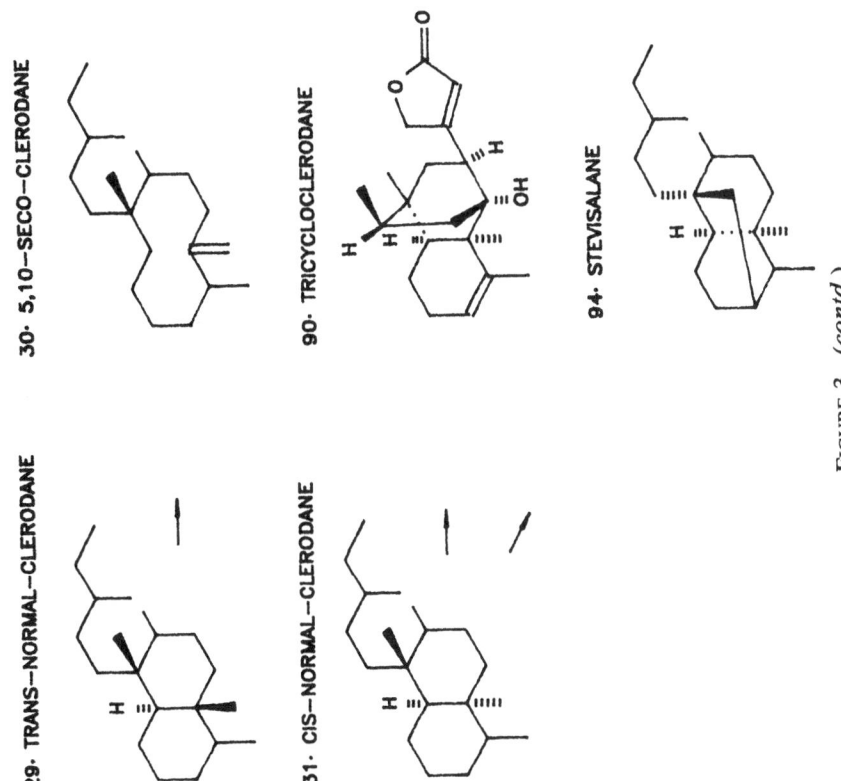

29· TRANS—NORMAL—CLERODANE

30· 5,10—SECO—CLERODANE

31· CIS—NORMAL—CLERODANE

90· TRICYCLOCLERODANE

94· STEVISALANE

FIGURE 3. (*contd.*)

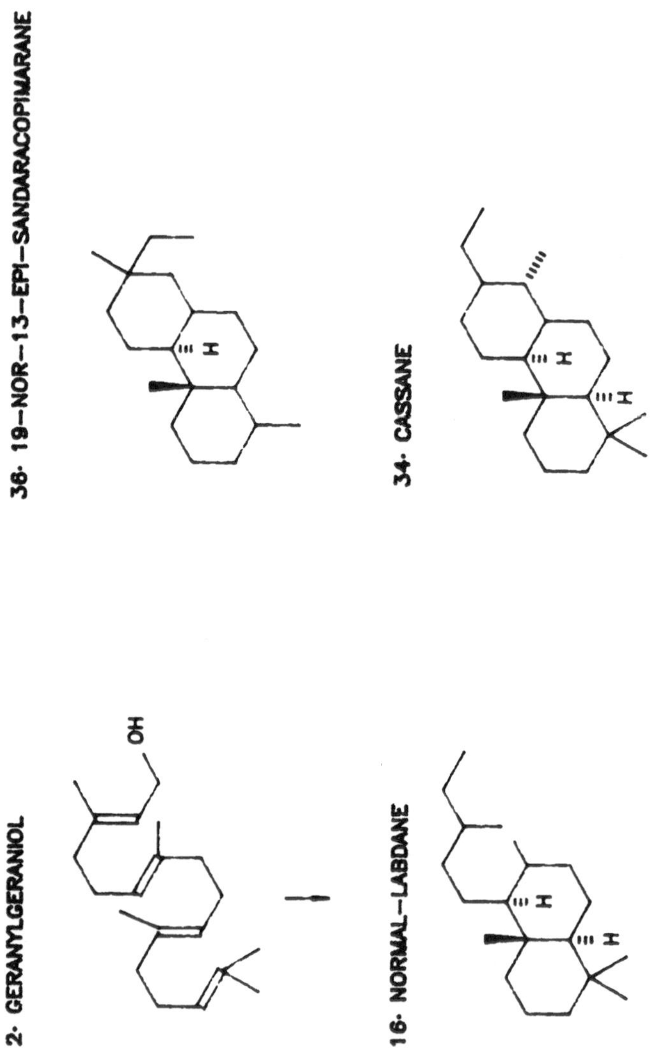
2· GERANYLGERANIOL
OH
16· NORMAL–LABDANE
H
36· 19–NOR–13–EPI–SANDARACOPIMARANE
34· CASSANE

Figure 3. Linear and Cyclic Diterpene Skeletal Types   407

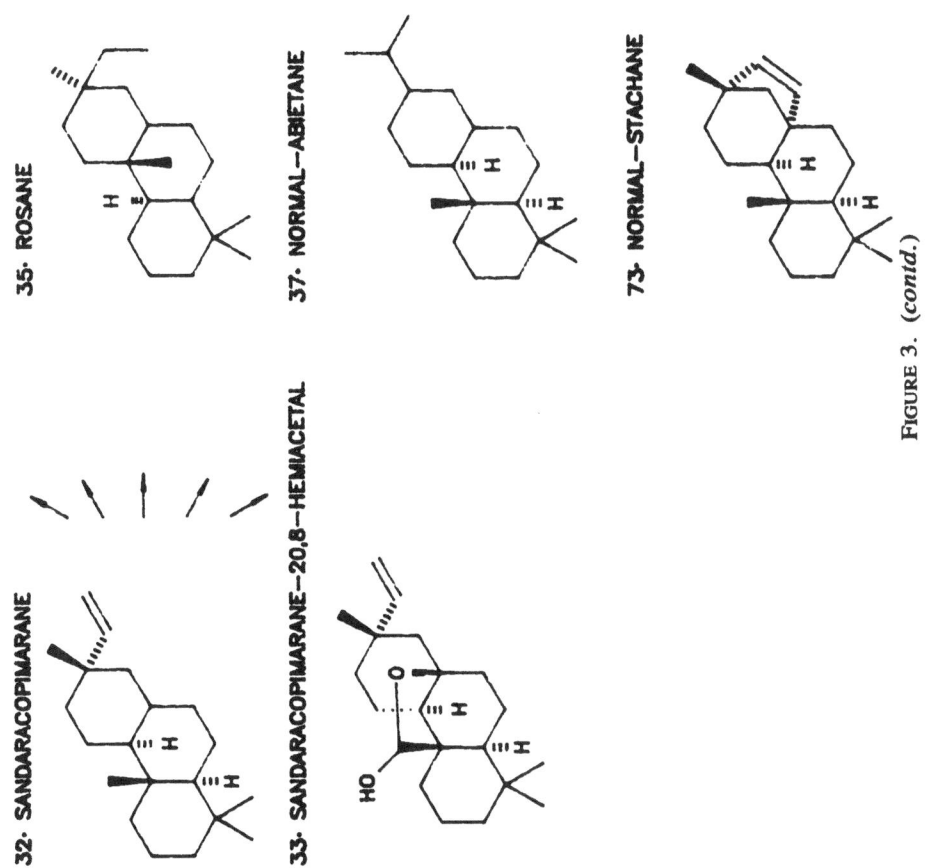

32· SANDARACOPIMARANE

33· SANDARACOPIMARANE—20,8—HEMIACETAL

35· ROSANE

37· NORMAL—ABIETANE

73· NORMAL—STACHANE

FIGURE 3. (contd.)

42· 15—NOR—ENT—LABDANE

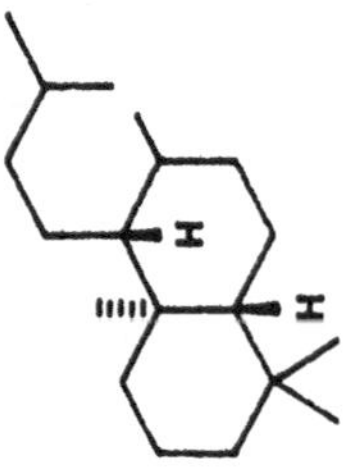

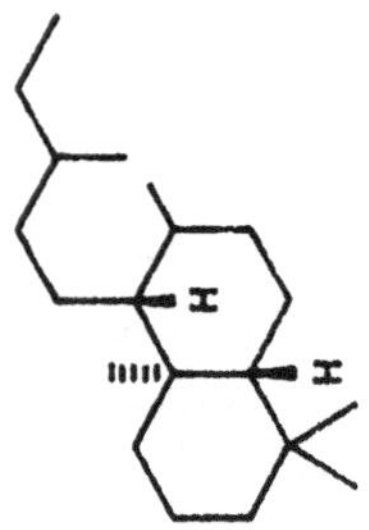

41· 17—NOR—ENT—LABDANE

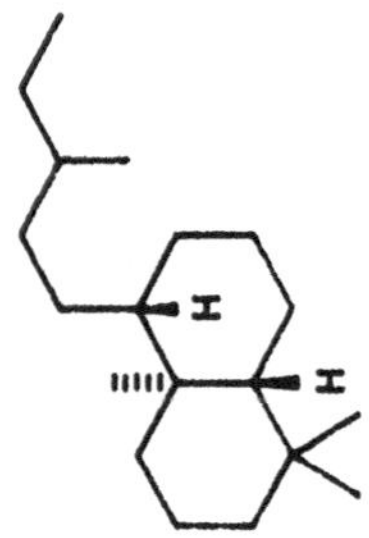

74· 18—NOR—ENT—LABDANE

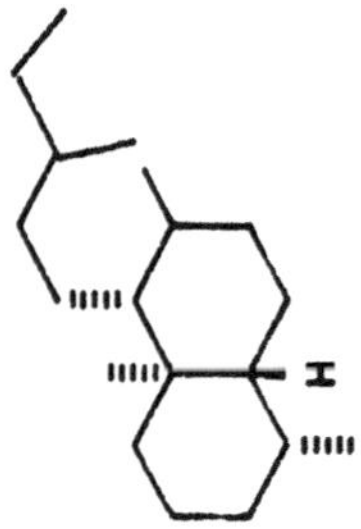

2· GERANYLGERANIOL

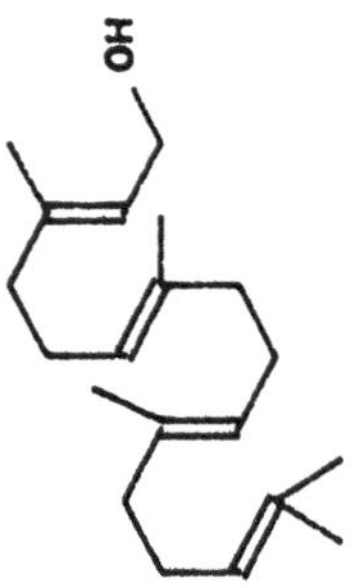

38· ENT—LABDANE

Figure 3. Linear and Cyclic Diterpene Skeletal Types   409

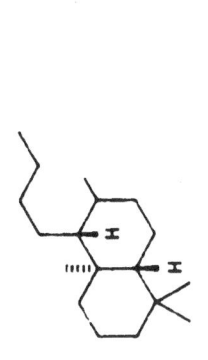

43· 14,15—BIS—NOR—ENT—LABDANE

44· 13,14,15,16—TETRA—NOR—ENT—LABDANE

45· 7,8—SECO—ENT—LABDANE

FIGURE 3. (contd.)

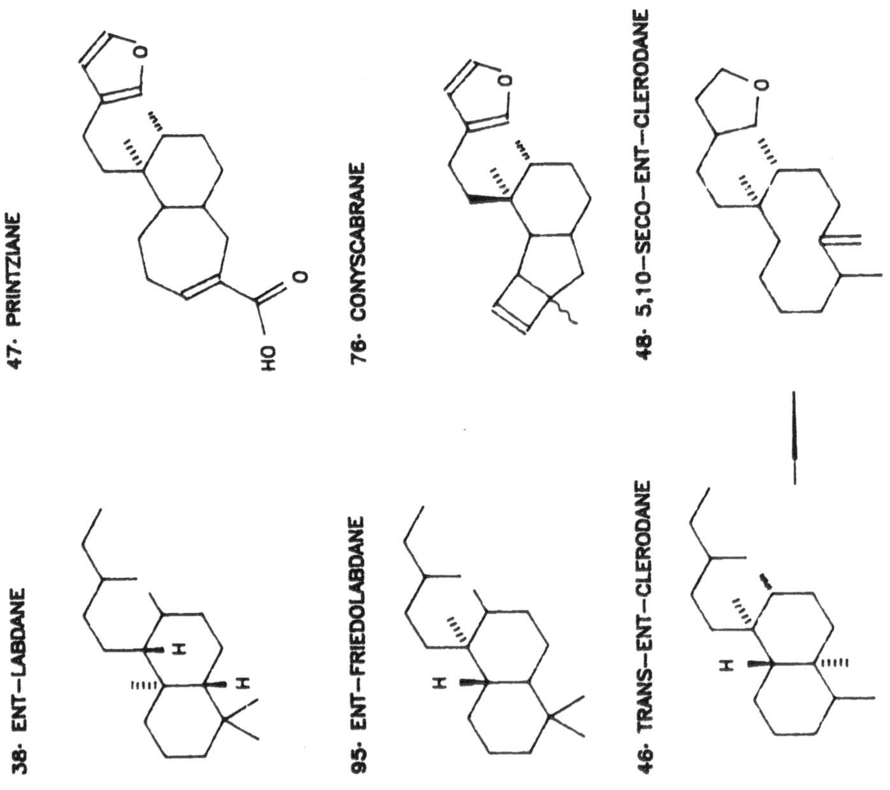
38· ENT—LABDANE
H
H
95· ENT—FRIEDOLABDANE
H
46· TRANS—ENT—CLERODANE
H
47· PRINTZIANE
O
HO
76· CONYSCABRANE
O
48· 5,10—SECO—ENT—CLERODANE
O

Figure 3. Linear and Cyclic Diterpene Skeletal Types   411

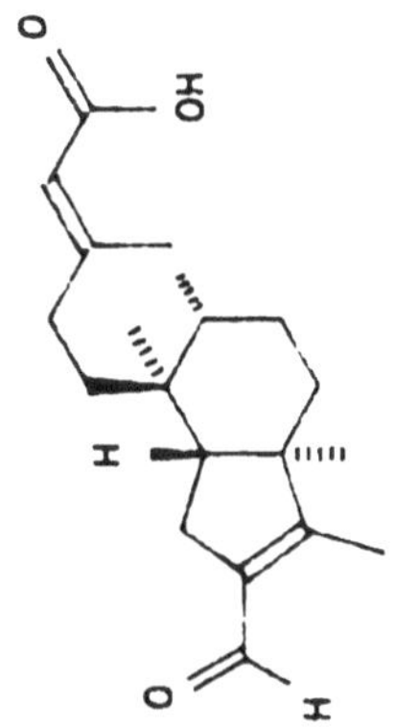

82· SOLIDAGONANE

50· CIS–ENT–CLERODANE

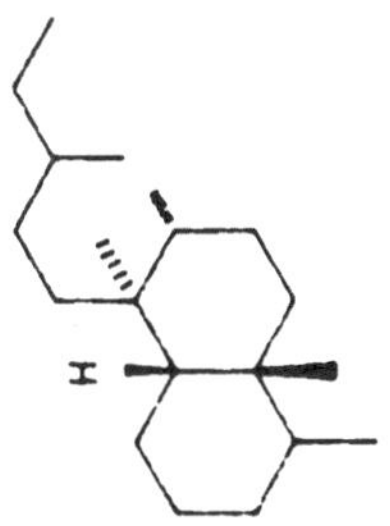

96· 17–NOR–TRANS–ENT–CLERODANE

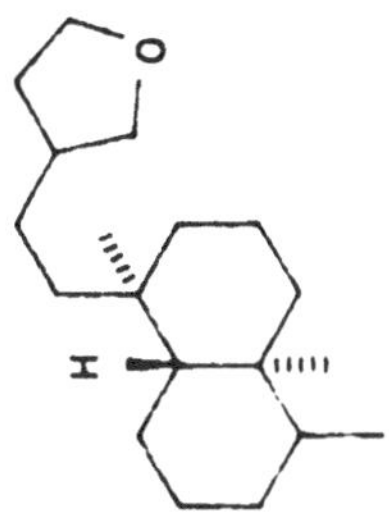

49· 13,14,15,16–TETRA–NOR–TRANS–ENT–CLERODANE

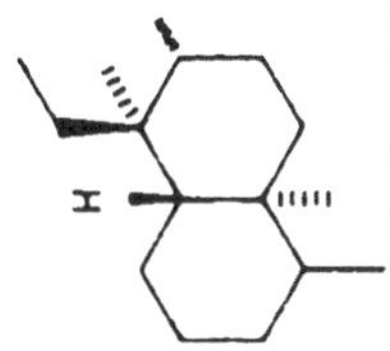

FIGURE 3. (contd.)

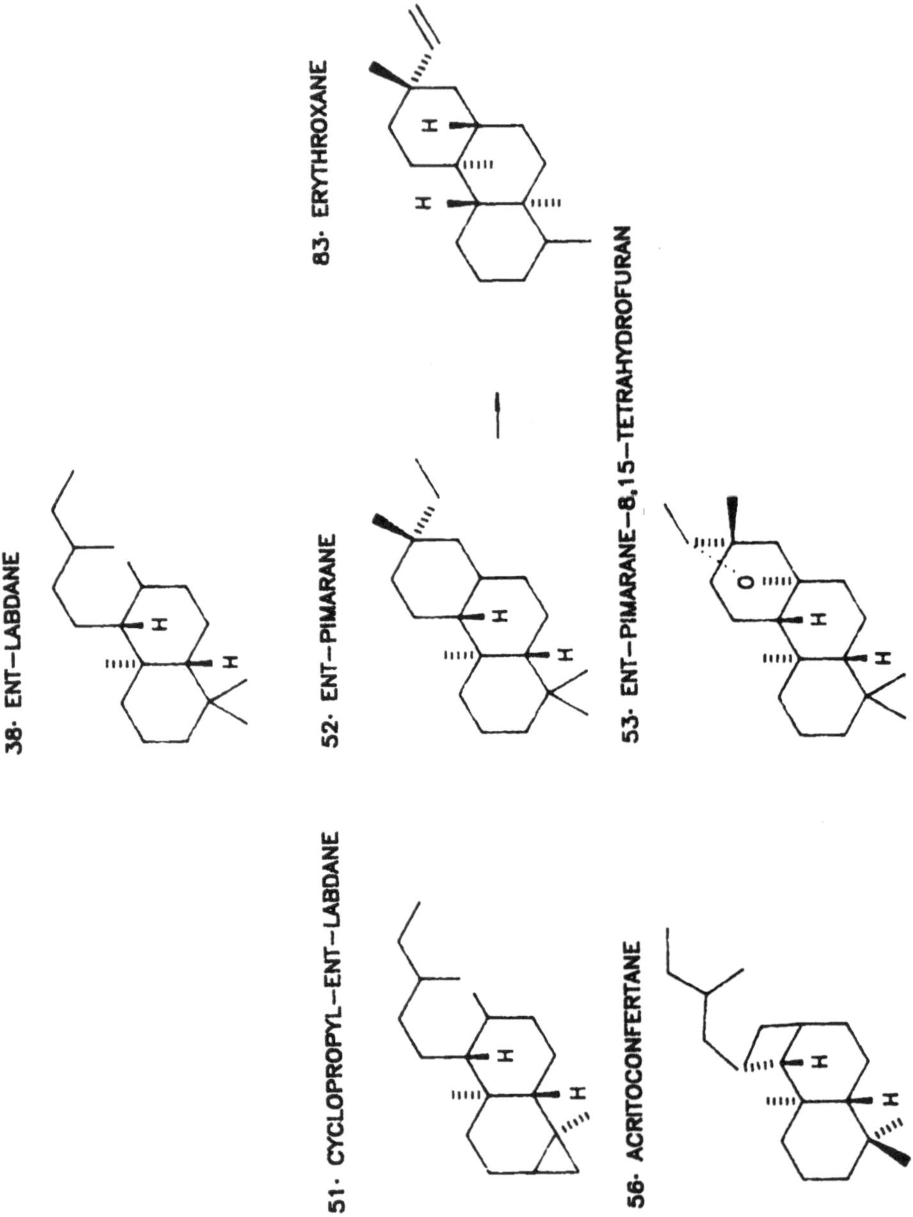

83· ERYTHROXANE

38· ENT—LABDANE

52· ENT—PIMARANE

53· ENT—PIMARANE—8,15—TETRAHYDROFURAN

51· CYCLOPROPYL—ENT—LABDANE

56· ACRITOCONFERTANE

Figure 3. Linear and Cyclic Diterpene Skeletal Types 413

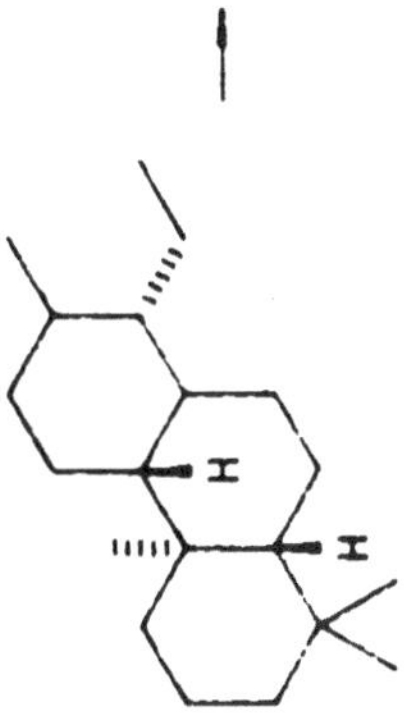

85· ISOCLEISTANTHANE

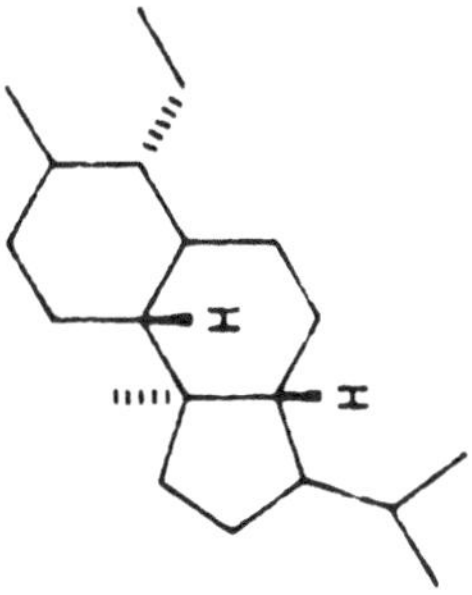

54· CLEISTANTHANE

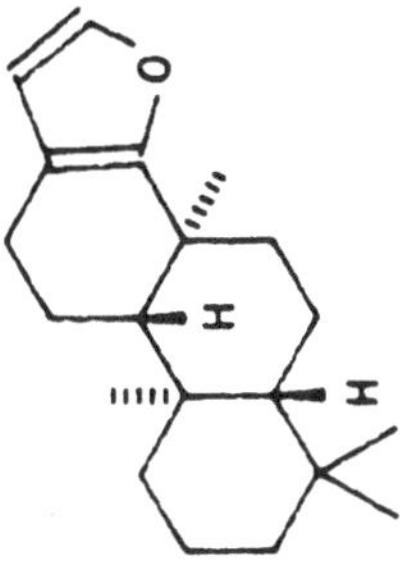

55· ENT–ABIETANE

57· ACRITOCONFERTANE–ACETAL

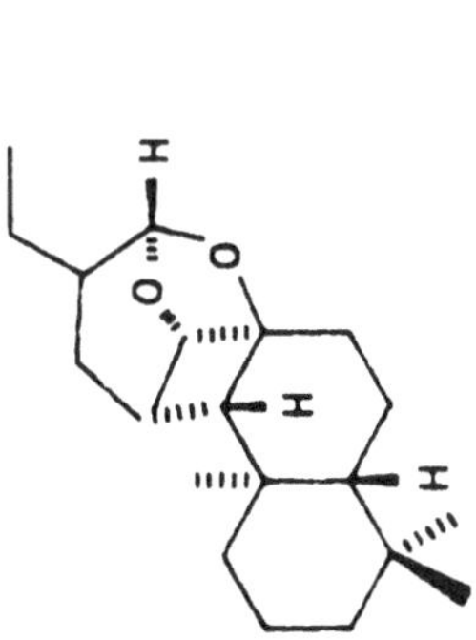

77· GUTIERREZIANE

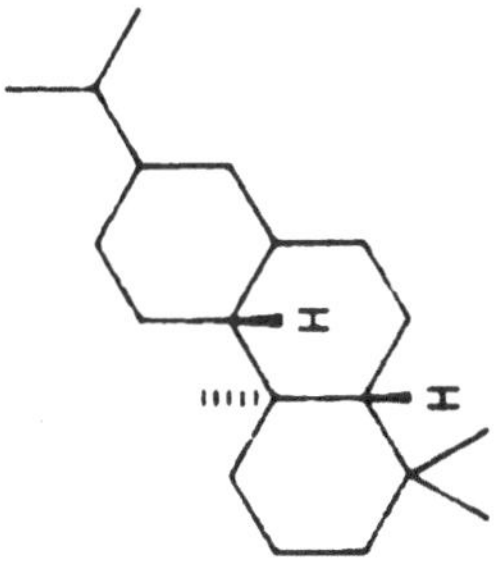

FIGURE 3. (*contd.*)

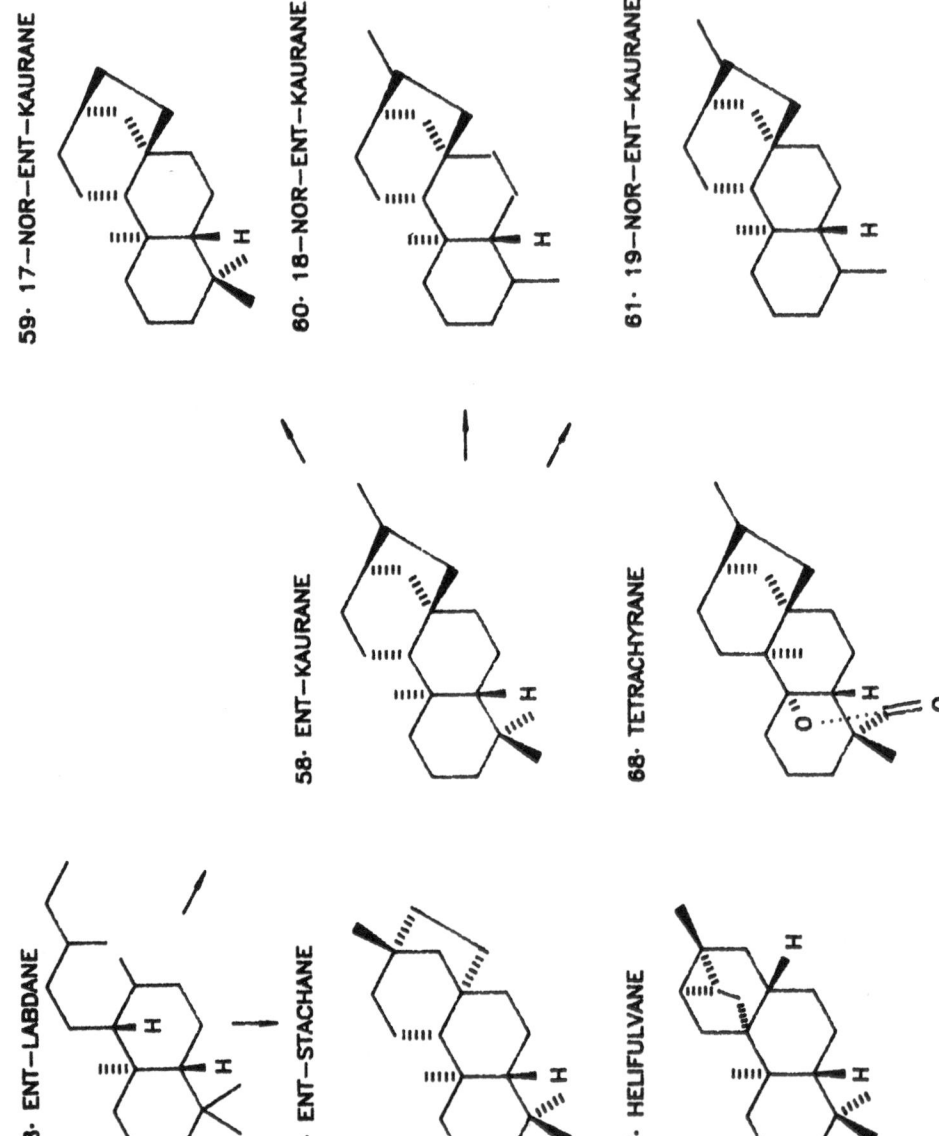

59· 17—NOR—ENT—KAURANE

60· 18—NOR—ENT—KAURANE

61· 19—NOR—ENT—KAURANE

58· ENT—KAURANE

68· TETRACHYRANE

38· ENT—LABDANE

64· ENT—STACHANE

67· HELIFULVANE

Figure 3. Linear and Cyclic Diterpene Skeletal Types   415

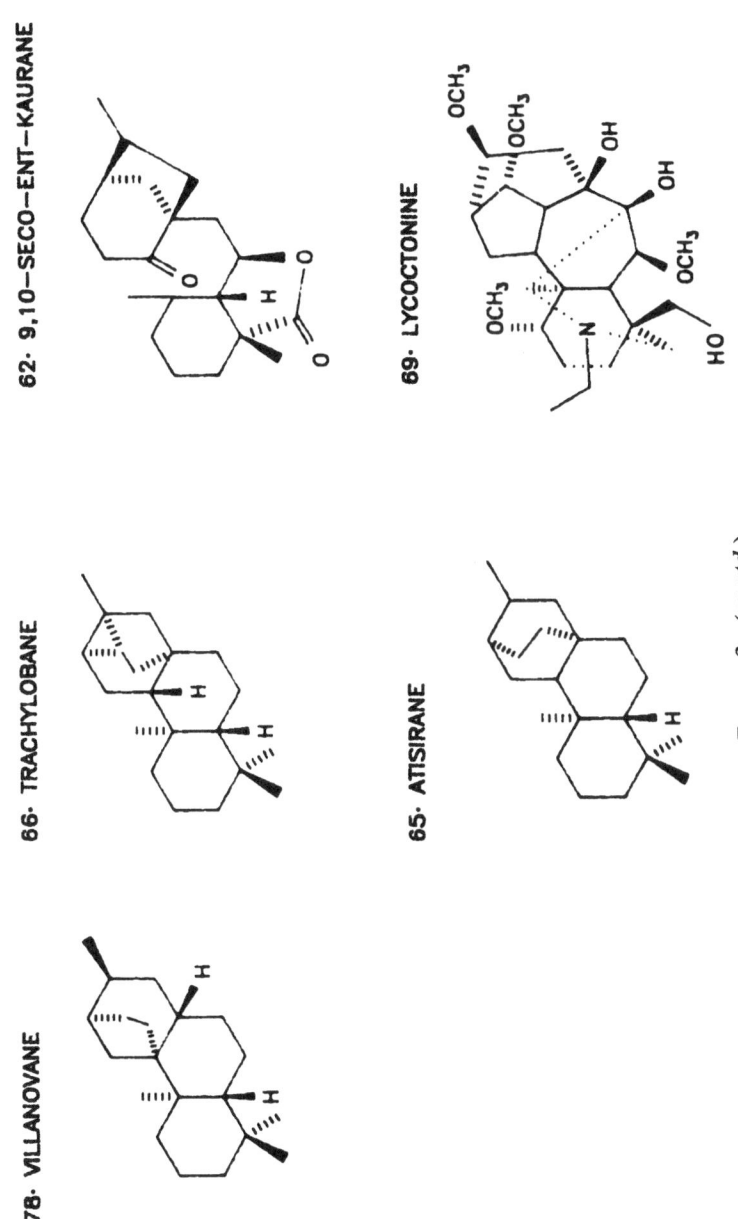

62· 9,10-SECO-ENT-KAURANE

69· LYCOCTONINE

86· TRACHYLOBANE

65· ATISIRANE

78· VILLANOVANE

FIGURE 3. (*contd.*)

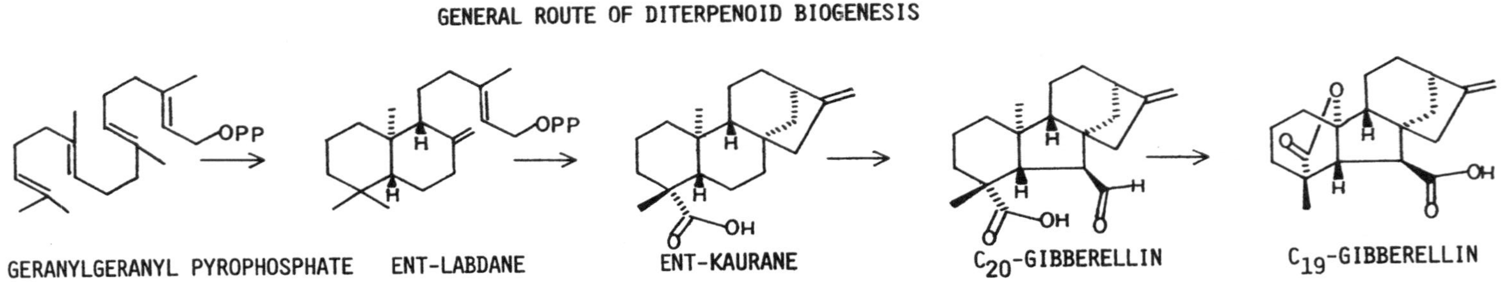

FIGURE 4. Generalized route of "universal" diterpene biogenesis culminating in C_{19}-gibberellin.

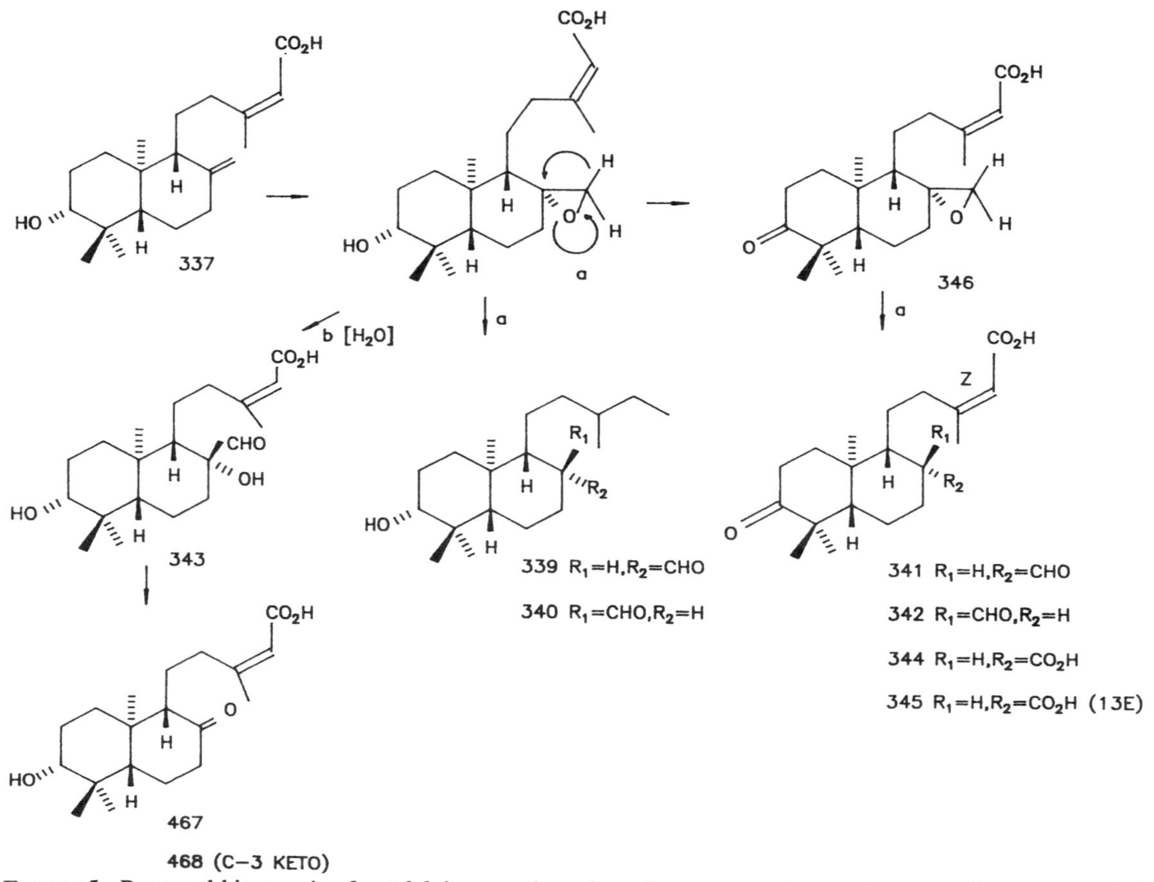

Figure 5. Proposed Biogenesis of *ent*-labdanes and nor homologs   417

FIGURE 5. Proposed biogenesis of *ent*-labdanes and nor-homologs reported from *Ayapana* (Eupatorieae, *113*).

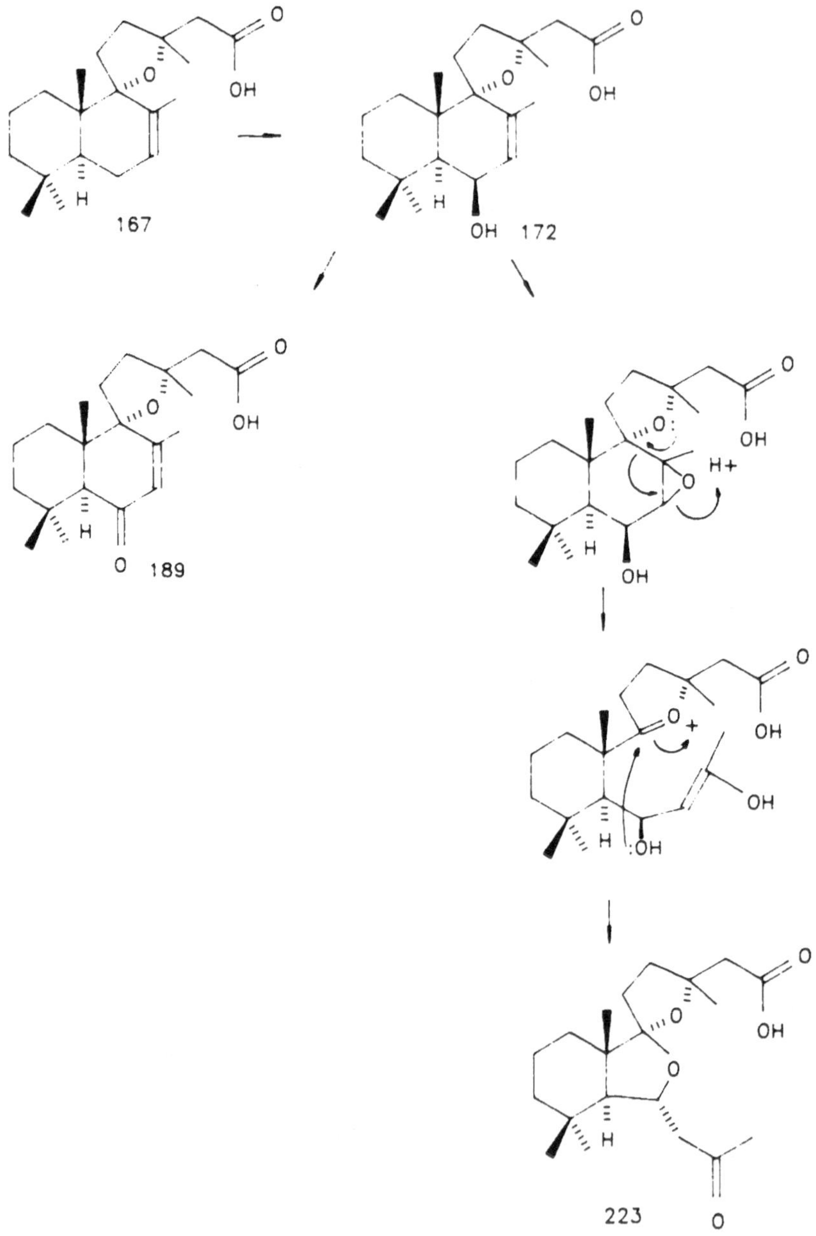

FIGURE 6. Proposed route of biogenesis of the 8,9-seco-grindelane chrysothame (223) from grindelic acid (167) in *Chrysothamnus paniculatus* (Astereae) (*341*).

Figure 7. Proposed Route of Biogenesis   419

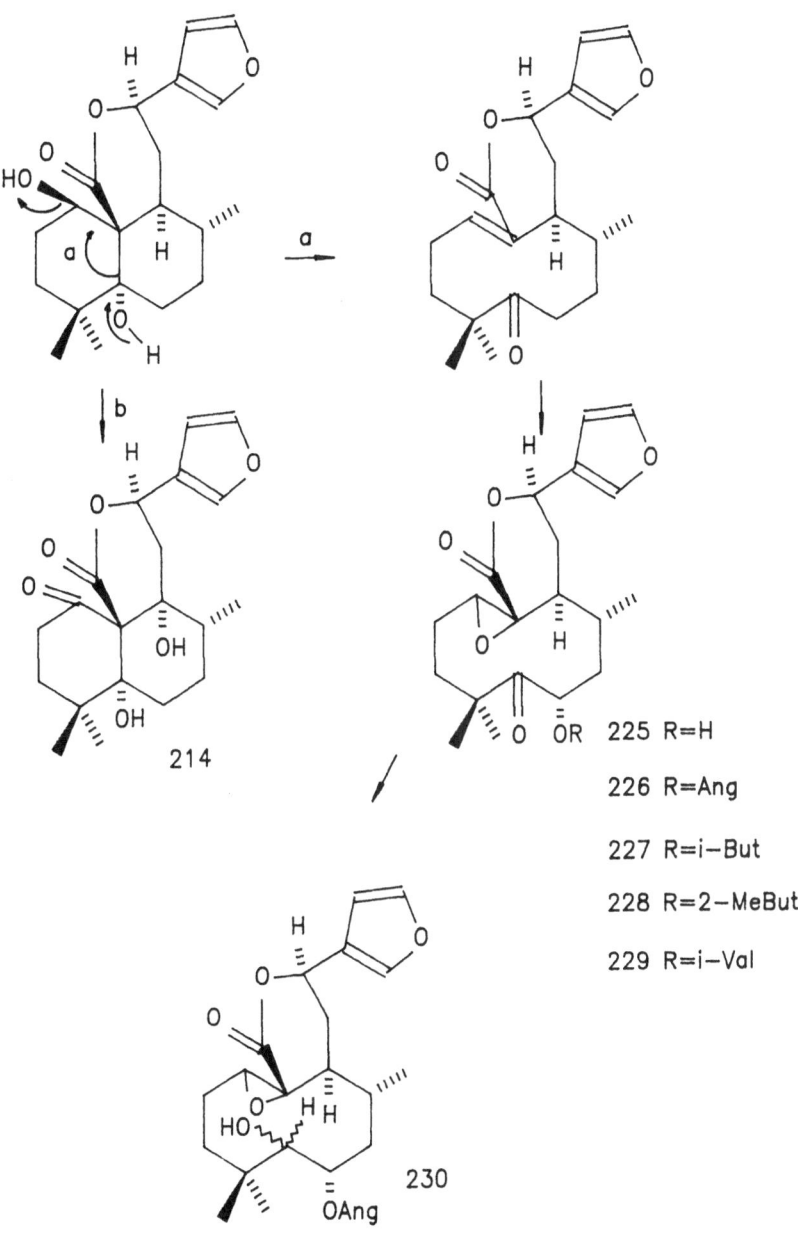

225 R=H

226 R=Ang

227 R=i-But

228 R=2-MeBut

229 R=i-Val

214

230

FIGURE 7. Proposed route of biogenesis of the 5,10-seco-normal-labdanes, 225–230, from a common precursor (similar to **214**) occurring in *Nidorella hottentotica* (*155*).

1· Relhania acerosa

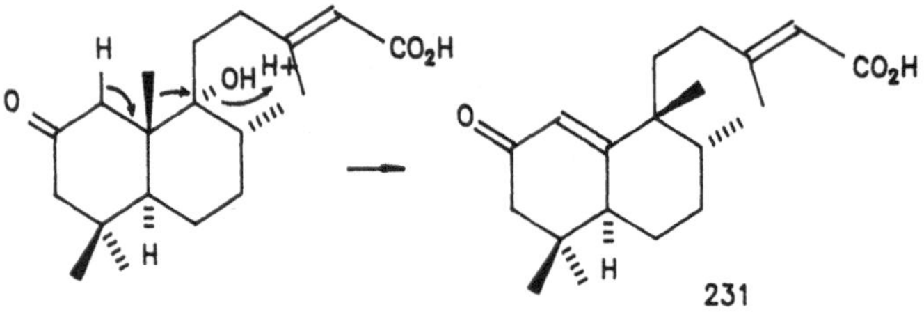

2· Koanophyllon conglobatum

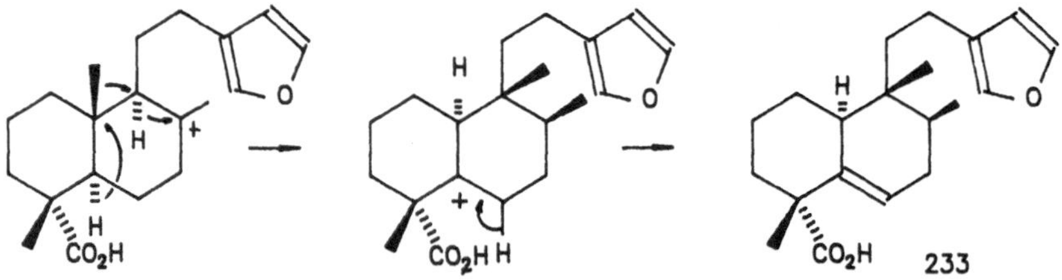

FIGURE 8. Diterpene skeleton products of the rearrangement of normal-labdane precursors: alternative routes for the biogenesis of diterpenes with the relhaniane skeleton, scheme 1 for compounds reported from *Relhania acerosa* (*90*) and scheme 2 for compounds from *Koanophyllon conglobatum* (*42*).

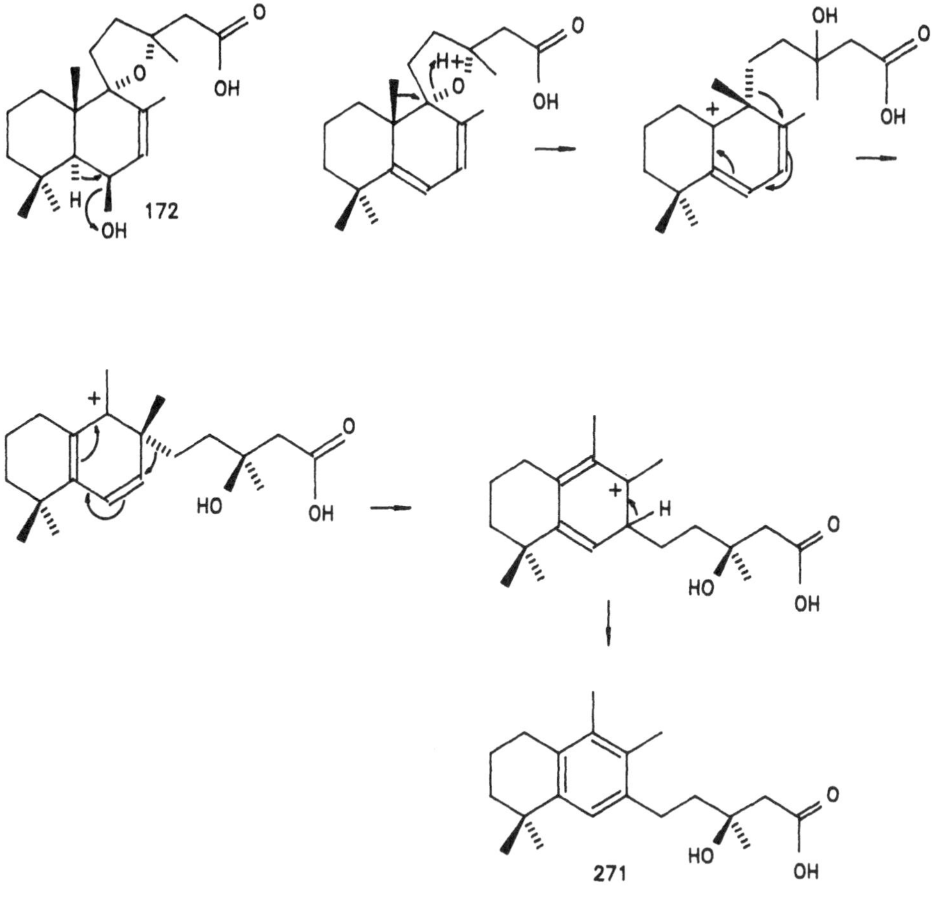

Figure 9. Proposed Biogenesis of the Chrysolane Compound   421

FIGURE 9. Proposed biogenesis of the chrysolane compound, 271, from the 6β-hydroxy-grindelic acid precursor, 172 in *Chrysothamnus paniculatus* (*341*).

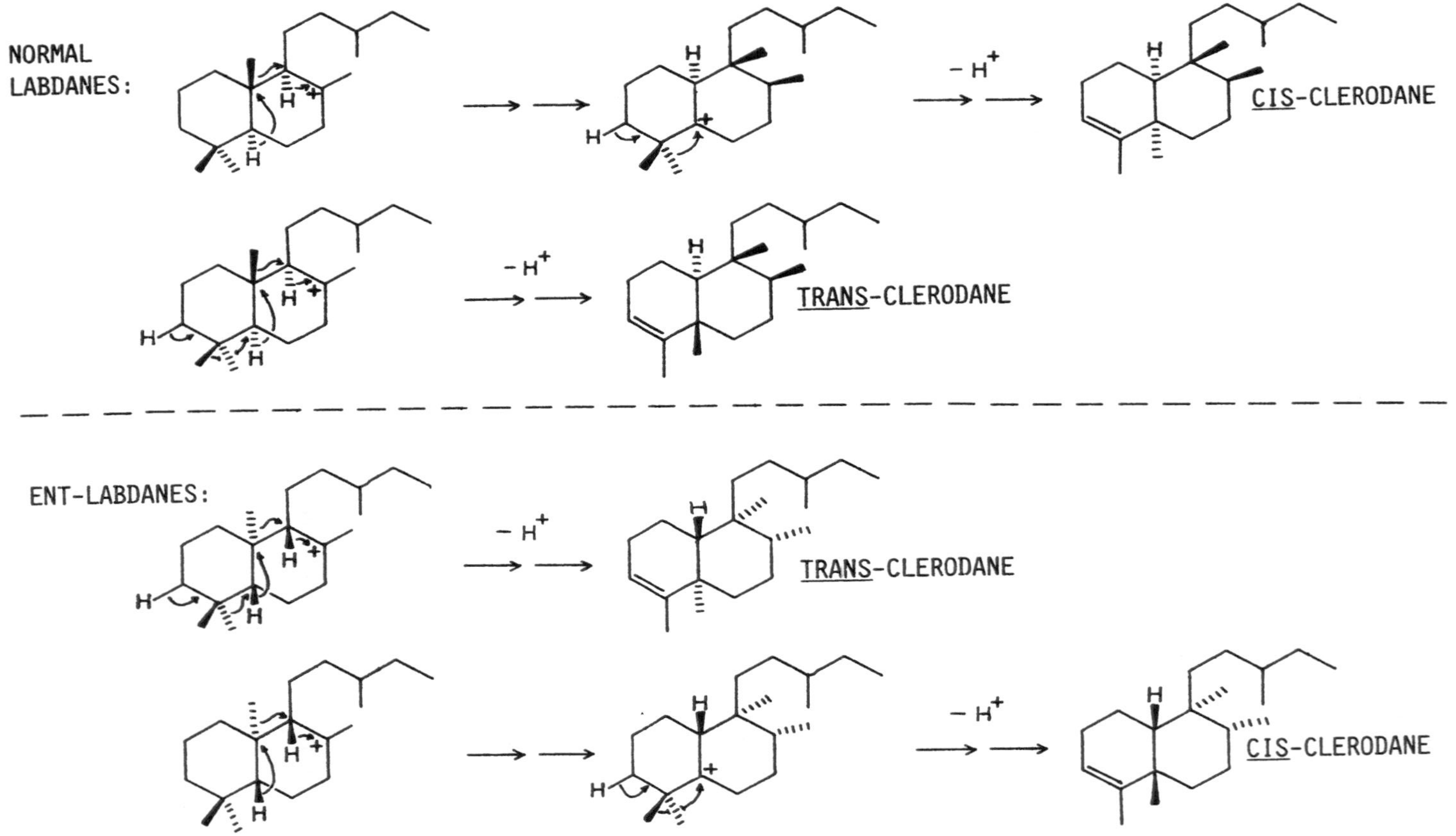

3. Biogenesis

FIGURE 10. Proposed biogenetic routes for the four clerodane (kolavane) sketetal types.

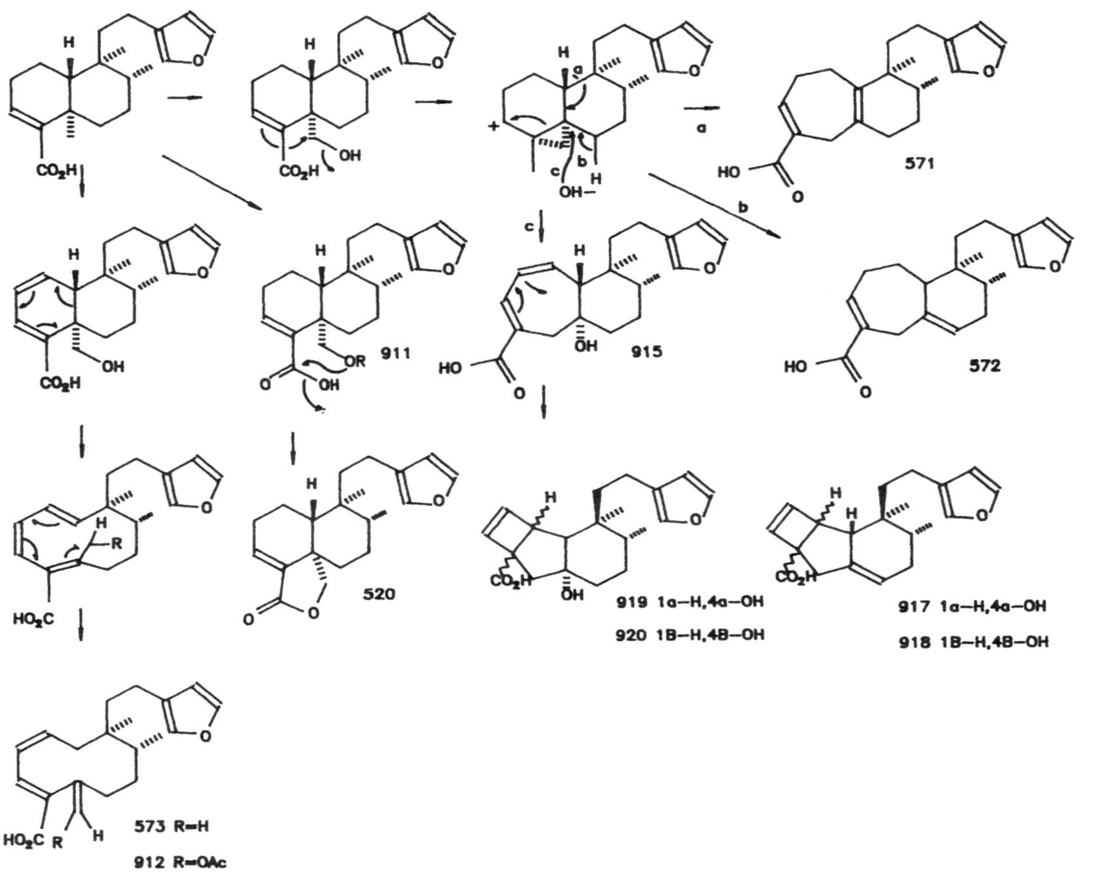

Figure 11. Proposed Biogenetic Routes   423

FIGURE 11. Proposed biogenetic routes for *trans-ent*-clerodanes and other related rearranged skeletons reported from *Conyza* (Astereae, *83*) and *Printzia* (Inuleae, *165*).

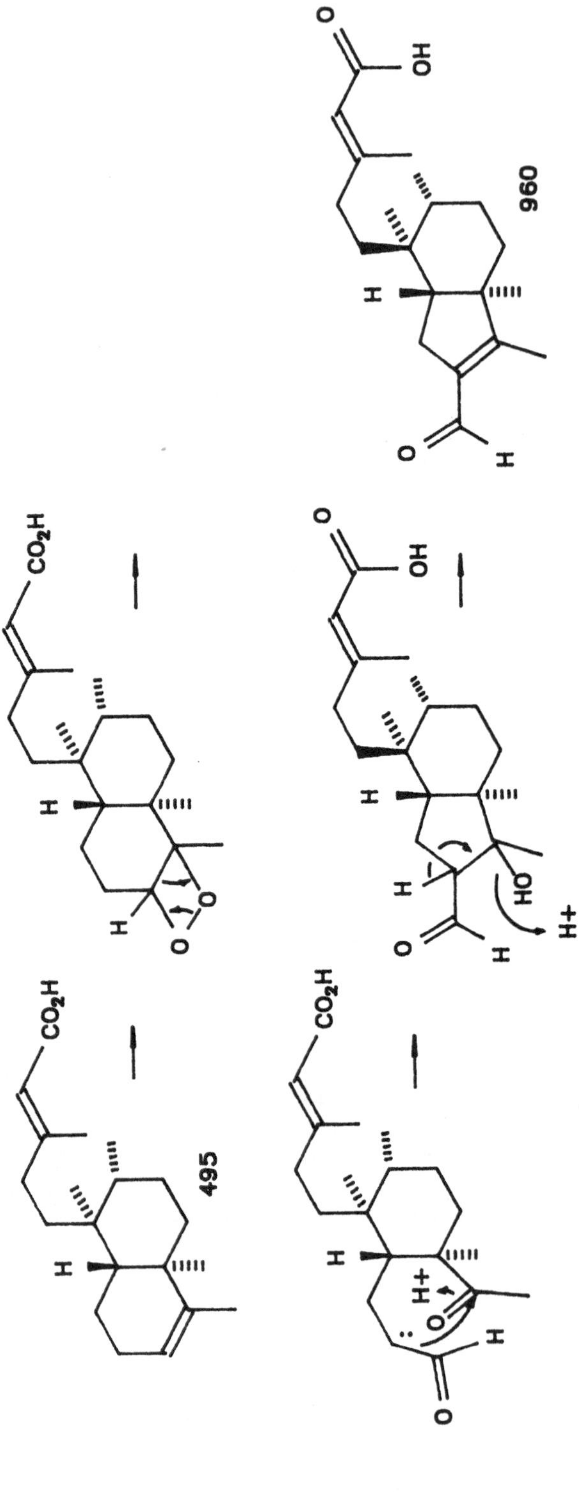

FIGURE 12. A proposed biogenetic route for solidagonane (**960**), a rearranged clerodane reported from *Solidago altissima* (*143*).

Figure 13. Proposed Biogenetic Routes   425

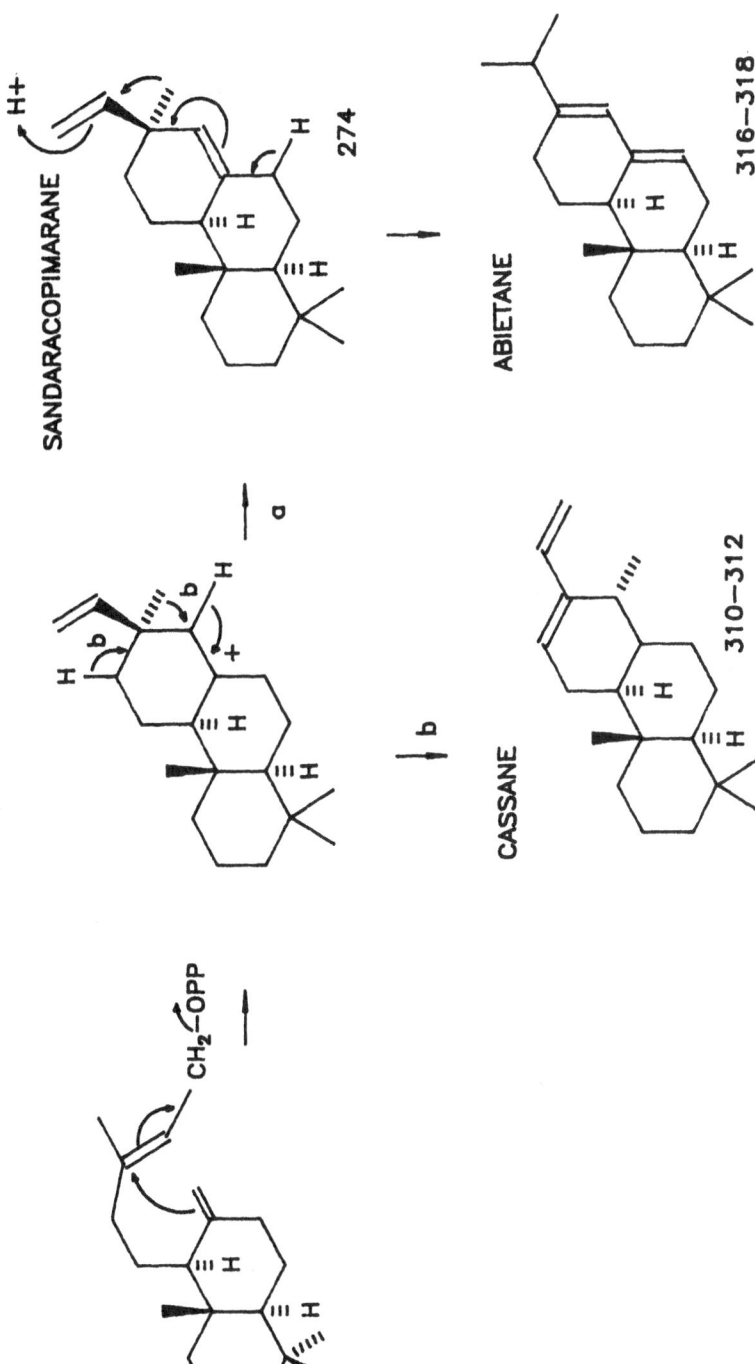

FIGURE 13. Proposed biogenetic routes for sandaracopimaranes (normal-pimaranes) and related skeletons.

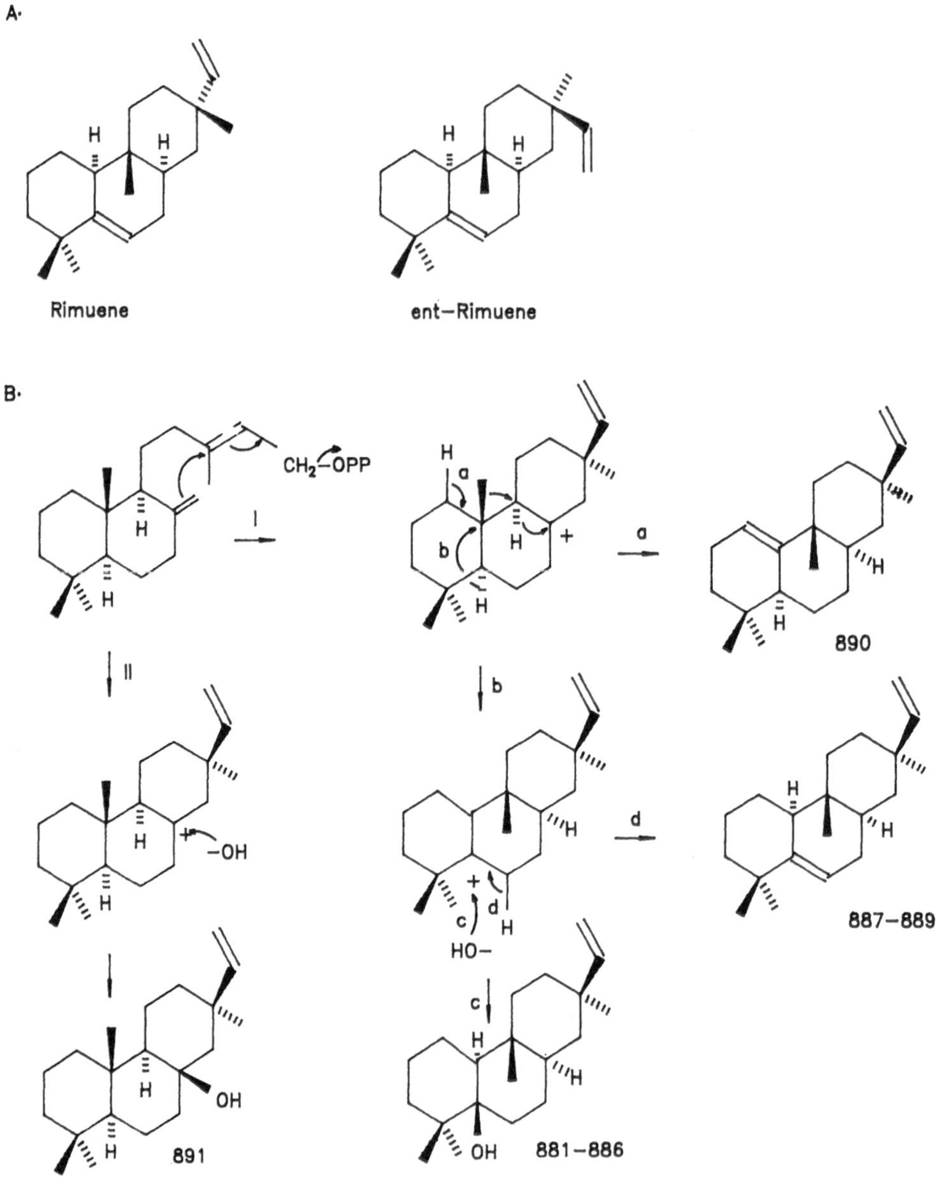

A·

Rimuene   ent—Rimuene

B·

FIGURE 14. Proposed route of rosane biogenesis in *Trichogonia* (Eupatorieae) (*183*): A. an example of C-13 epimeric pair of rosane compounds (note that the prefix "*ent-*" refers not to an *ent*-labdane origin, bot to the C-13 orientivity relationship of this compound pair; B. biogenetic route for sandaracopimarane and rosane constituents of *Trichogonia*.

Figure 15. Proposed Biogenetic Routes 427

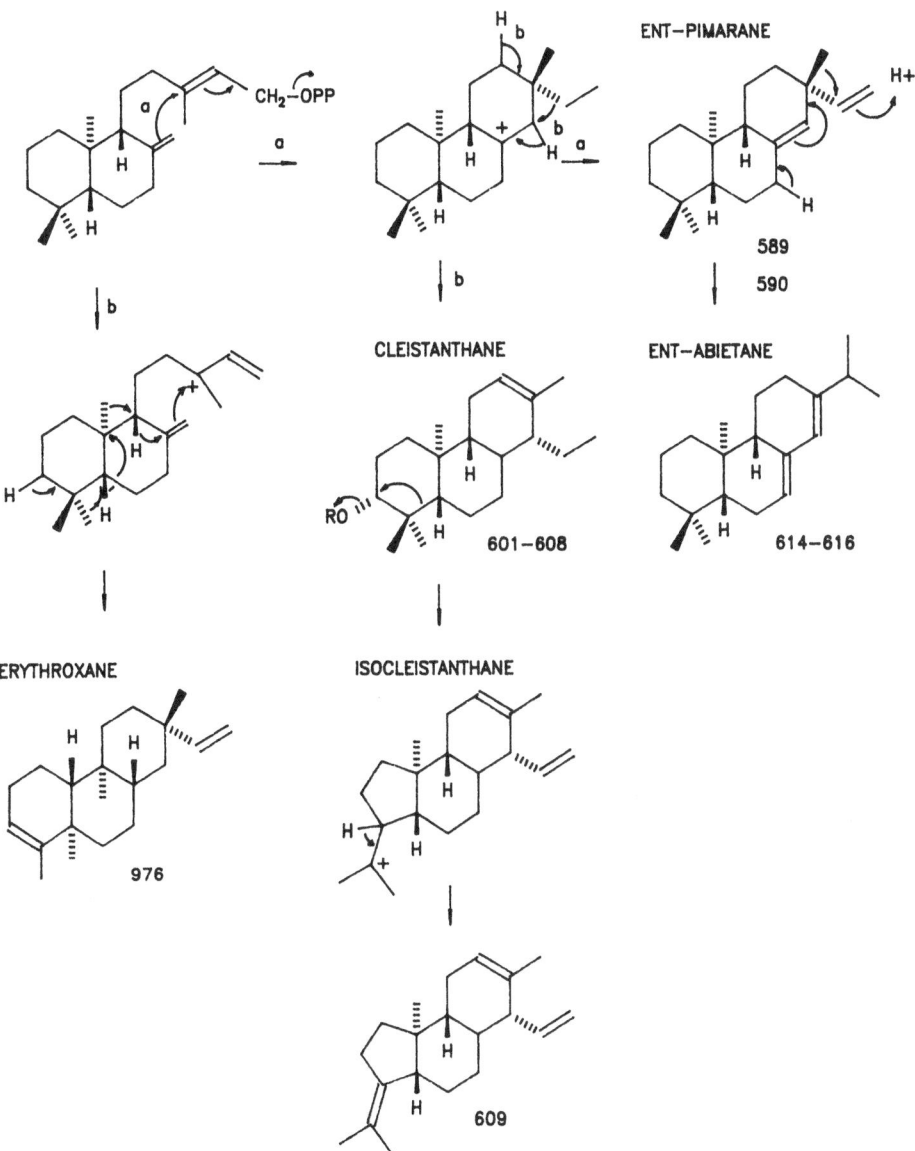

FIGURE 15. Proposed biogenetic routes for tricyclic *ent*-pimarane, *ent*-abietane, erythroxane, cleistanthane and isocleistanthane diterpene skeletons derived from *ent*-labdane precursors.

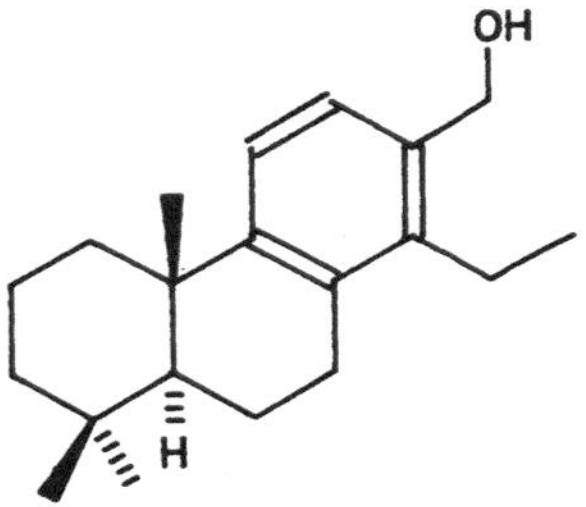

Vellozia (Velloziaceae) 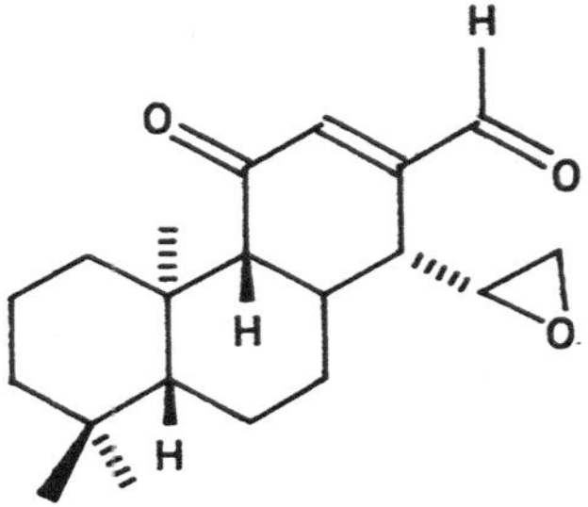   **602 Brickellia (Compositae)**

FIGURE 16. Cleistanthanes from *Brickellia* (Compositae) and *Vellozia* (Velloziaceae).

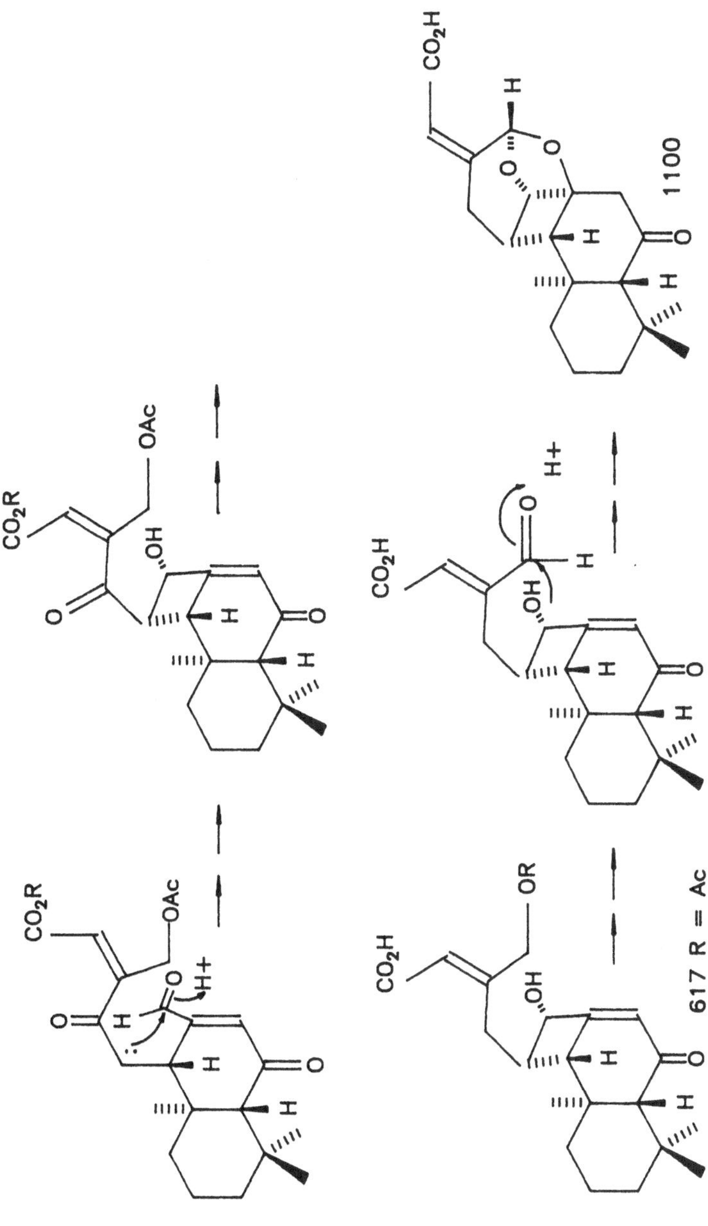

Figure 17. Proposed Biogenesis of Acritoconfertane   429

FIGURE 17. Proposed biogenesis of acritoconfertane (617) and acritoconfertane-acetal (1100) constituents of *Acritopappus confertus (184)*.

3. Biogenesis

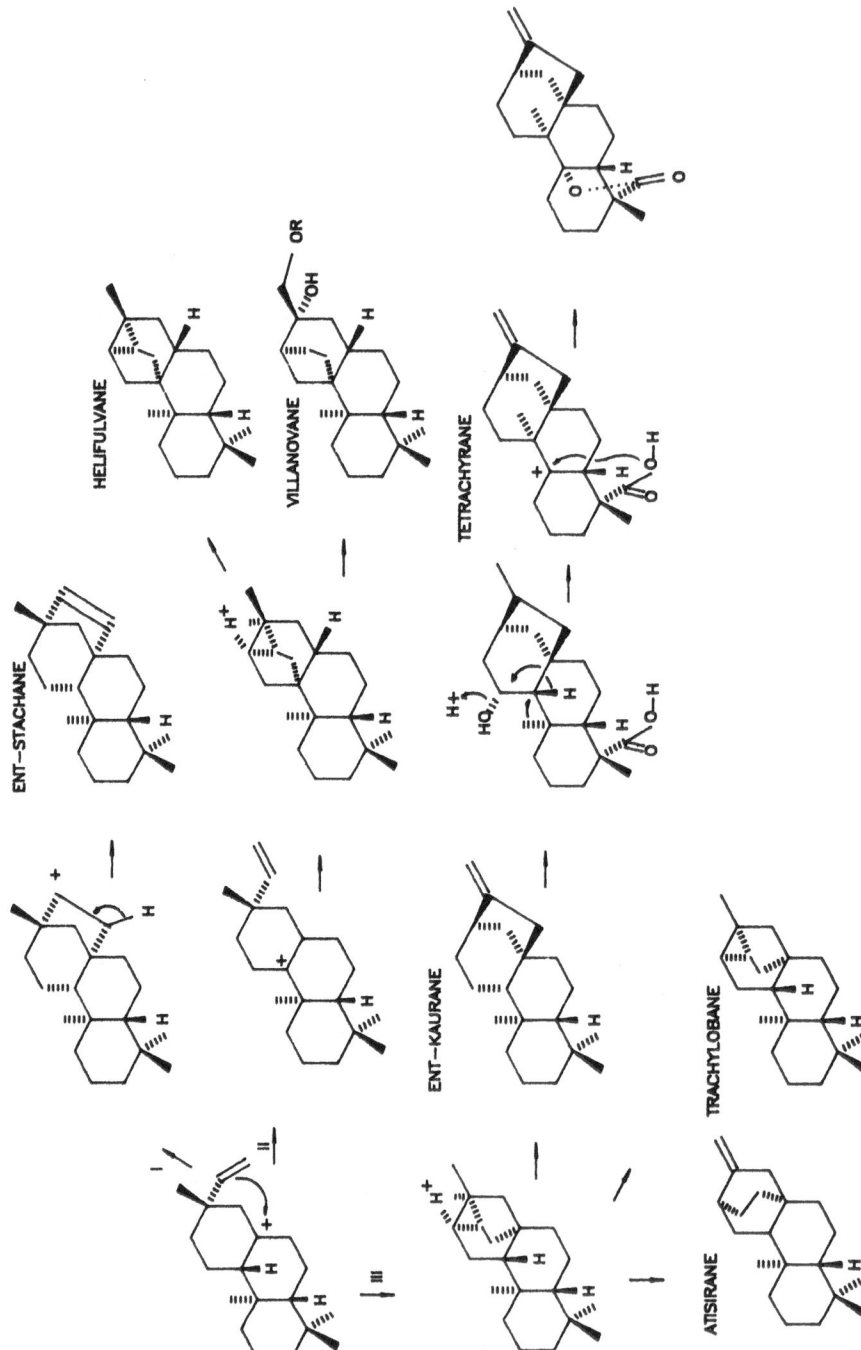

FIGURE 18. Tetracarbocyclic skeletons of *ent*-labdane origin and their proposed biogenetic relationship (in part *208*).

Diterpene Distribution: Compositae

Green plants possess (1) geranylgeraniol, the required precursor for both carotenoid and gibberellin biosynthesis, (2) the enzymatic machinery to cyclize geranylgeraniol (or its isomer) to *ent*-kaurene, and (3) the enzymes for the oxidation of *ent*-kaurene and other diterpene substrates (*420*). Normally, the steady-state concentration of the products of these pathways falls below the level of detectability of all but the most sensitive analytical instruments. By comparison, the compounds that are the focus of this review constitute an "abnormal" biosynthetic excess detectable in the form of resins. Often, general plant metabolism is isolated from these biologically active resins by the comparmentation of resin biosynthesis and storage within glandular trichomes.

The production of copious trichome resin products raises the question of whether or not these resins are adaptive. Various proposals have been advanced regarding the defensive role of such compounds and the link between their structural variability and the coevolutionary process. However, there is little evidence linking these compound's biological function to the fine detail of their structural variability. The extent to which the observed chemical variability is an expression of past selection for specific novel chemical deterrents rather than the product of near-random activation of alternative terpenoid biosynthetic pathways is not known. Thus, in looking at the diterpene variability within the Compositae, care should be exercised in automatically associating a discrete adaptive function with each novel chemistry.

Whatever the cause of the variability, chemical variation patterns correlate with taxonomic boundaries and may prove useful as taxonomic characters. Consequently, chemical differences between taxa may be useful chemical indicators of phylogenetic relationships. As a corollary, our understanding of the factors governing terpenoid biogenesis benefits from superimposing a plant taxonomic framework on the discussion of diterpene distribution and the underlying biosynthetic diversity.

IV.1. Subfamilial Distribution Patterns

The Compositae is so well differentiated from other families that any claims concerning its affinities are difficult to defend. In contrast, its own internal boundaries separating tribes and genera are ill defined, due to complex and overlapping morphological and ecological diversity (233). No taxonomic problem better exemplifies these poorly demarcated internal boundaries than the definition of subfamilies. On this subject, Cronquist noted (233)...

The classical organization of one subfamily for the Lactuceae, and one for everything else, remains the best such expression of the pattern of diversity within the family. Perhaps three subfamilies could be recognized, one for the Lactuceae, one for the radiate [flowering head type with central flowers displaying regular tubular corollas and marginal flowers in which part of the corolla is prolonged into a strap-shaped ligule] tribes, and one for the discoid [flowering head type in which all flowers have regular tubular corollas] tribes plus the Mutisieae...

Although alternative classifications have been proposed (226), the findings of the recent study of subfamilial boundaries by Wagenitz (488) were largely compatible with Cronquist's view. In the Wagenitz phylogenetic proposal, the Compositae was divided into two subfamilies, the Asteroideae and Cichorioideae. The Asteroideae was further subdivided into two groups (Table 4 [p. 438]). A major area of uncertainty for Wagenitz was the placement of the Anthemideae in one of these two Asteroideae groups. He opted for group 1, but admitted that with the exception of a pollen anatomical feature, this tribe could be equally well situated in group 2. The most unusual aspect of his scheme was the assignment of the Vernonieae and Eupatorieae to different groups (488). Traditionally these two tribes had been associated by their lack of ray-flowers and their purple flower color. Wagenitz considered the Cichorioideae to be more closely related to group 1 than to group 2.

Distributions of three major natural products groups, sesquiterpene lactones (458), benzofurans/chromenes(benzopyrans) (429) and diterpenes, were superimposed on the Wagenitz arrangement of the Compositae (Table 4). While sesquiterpene lactones appear to be well-distributed across the tribes, both diterpenes and benzofurans and pyrans are much more heavily concentrated in Wagenitz's group 2. Their presence in the group 1 taxa has only rarely been reported. In support of his separation of the Vernonieae and Eupatorieae, the former tribe, despite intensive chemical investigation, has yielded no reports of either diterpenes or bezofurans/pyrans, while the latter is a major producer of both.

A tribe with uncertain affinities based on nonchemical characters, the Anthemideae, displays a closer chemical link with group 1. Intensive chemical analysis of this tribe failed to produce reports of either diterpenes of benzofurans/pyrans. Finally, the chemistry of the Lactuceae (Cichorioideae) more closely resembles that of the group 1 tribes, a finding which is in agreement with Wagenitz.

IV.2. Tribal Affinities

Patterns of diterpene variation are relevant as taxonomic characters to relationships only within the Wagenitz group 2 tribes. Within this group, diterpene patterns distinguish (1) the Heliantheae because of its *ent*-kaurane-based polycyclic profile, (2) the Calenduleae based on its sandaracopimarane chemistry, and (3) the Eupatorieae and Astereae based on their complex labdane/clerodane bicyclic profile (Table 5 [pp. 439–440]).

IV.2.1. Heliantheae

Natural products variability within the Heliantheae usually parallels the extensive variability of morphological features. As noted earlier (*458*), the sesquiterpene lactones display as much variability within the Heliantheae as exists within the entire family. When viewed against this background, the diterpene chemistry of the Heliantheae seems remarkably uniform. Of the approximately 180 Heliantheae taxa analyzed, only 29 yielded reports of non-polycyclic diterpenes. These 29 reports included 16 of linear diterpenes, 9 or normal-labdanes, 1 of *ent*-labdanes, 2 of *trans*-clerodanes, and 2 of *ent*-pimaranes (Table 5). The picture that emerges for this tribe is one of a fairly uniform kaurane-based polycyclic chemistry and a small group of exceptions, the majority of which contain linear homologs of the geranylgeraniol precursor. As yet no significant structural novelty is reported for any diterpene group other than the polycyclics.

III.2.2. Calenduleae

The Calenduleae is the source for most sandaracopimarane (normal-pimarane) reports (Table 5). Of the 24 investigated taxa, only two produced exclusively non-sandaracopimarane chemistries; *Dimorphoteca aurantiaca* and *D. pseudoaurantiaca* gave *ent*-stachanes. A possibly misleading report (*78*) included a series of *Garuleum sonchifolium* structures drawn as *ent*-pimaranes, but referred to as sandaracopimaranes. As two of these structures were described as being previously reported from *G. pinnatifidum*, it was concluded that the structures were drawn incorrectly. Bohlmann and Grenz (*79*) suggested that this tribe was divisible into two groups, a sandaracopimarane-producing group (*Dimorphotecha, Osteosperum, Garuleum* and *Chrysanthe-moides*) and a smaller sandaracopimarane-deficient group (*Calenula* and *Castalis*). Investigations of *Calendula* and *Castalis* taxa resulted in the detection of no diterpenes (*79*). *Dimorphotheca* yielded sandaracopimaranes, but as mentioned above two species produced *ent*-stachanes.

IV.2.3. Euaptorieae and Astereae

The similar diterpene profiles linking the Astereae and Eupatorieae are notable because of their relevance to the preliminary outcome of a recent phylogenetic analysis of the Compositae. The preliminary working cladogram (phylogenetic tree) resulting from this analysis indicated that these two tribes were a major monophyletic lineage within the family (K. Bremer, pers. commun.). Parallels in the diterpene chemistries of these tribes adds to the set of characters shared by them.

IV.3. Infrageneric Distribution Patterns

As yet, no systematic study focusing on generic-level diterpenoid distribution patterns has been reported for the Compositae. Consequently, little is known about how diterpenes vary within groups of related species. As a partial remedy for this deficiency, diterpene interspecific variation is described for representative genera of the major diterpene-producing tribes: Eupatorieae (*Brickellia*, *Stevia*), Astereae (*Baccharis*), Inuleae (*Helichrysum*), and Heliantheae (*Helianthus*, *Montanoa*).

Diterpenoid distribution was contrasted to that of other terpenoids for the two genera, *Baccharis* and *Helianthus*. These two taxa were selected for a more detailed examination, because their chemistries represented the two major types of diterpene complements: a bicyclic labdane/clerodane type (*Baccharis*) and a polycyclic type (*Helianthus*).

IV.3.1. *Brickellia*

Most of the approximately ninety *Brickellia* (Eupatorieae: Alomiinae) species inhabit the North American deserts, although three species are South American disjuncts. The diterpene chemistries of the ten investigated taxa (Table 6 [p. 441]; Figure 19 [pp. 457–461]) are remarkably uniform. Normal-labdanes often characterized by novel $2\alpha,3\alpha$-hydroxylation/esterification occur in all diterpene-producing species. In *B. eupatoriedes*, these $2\alpha,3\alpha$-substituted normal-labdanes co-occurred with cleistanthanes and other *ent*-pimarane-derived skeletal types. Production of normal-labdanes together with tri- and polycyclic *ent*-labdane-derived skeletons is an often-repeated theme in the diterpene chemistries of Compositae taxa.

IV.3.2. *Stevia*

The 150 to 300 *Stevia* species (Eupatorieae: Piqueriinae) range from southwestern North America to northern Argentina. Seventy of its approximately ninety North American species occur in Mexico (*464*). Chemical investigation of this genus has been hastened by the discovery of high concentrations of sweet-tasting *ent*-kauranoid glycosides in several of the species, one of which, *Stevia rebaudiana*, is currently used as a commercial source of sweetening agents in Japan and Brazil. Despite the common use of these compounds for sweetening pickled vegetables, sea foods, soft drinks, soy sauce, and so on, questions have been raised about possible health risks associated with the consumption of the diterpene aglycone, steviol (*368*).

Like *Brickellia*, *Stevia* species produces mostly mixtures of normal-labdanes and further-cyclized *ent*-labdane-derived compounds, specifically *ent*-kauranes. In several species, only *ent*-labdanes or their clerodane derivative were reported (Table 7 [pp. 442–443]; Figures 20 [pp. 462–463] and 21 [pp. 464–466]). Three of the *ent*-kaurane-producing species, *S. rebaudiana*, *S. ovata* and *S.*

paniculata, produce glycosides in which glycosidic groups of varying complexity are attached at one or more sites, specifically C-13, C-15, or C-19. The distribution of such diterpene glycosides inside and outside this genus is probably underestimated due to the lipophilics-specific extraction methods used in most laboratories.

IV.3.3. *Baccharis*

Centered in South America, the approximately four hundred perennial herbaceous or shrubby *Baccharis* species extend in their range into North America (*292*). Their closest relatives most likely include *Archibaccharis* and *Conyza* (A. Cronquist, pers. commun.). The dominant lipophilic constituents of the leaf extracts from 54 taxa are diterpenes, triterpenes, sesquiterpenes, flavonoids, and other phenolics (Table 8 [pp. 443–446]; Figure 22 [pp. 467–468]). Although *ent*-clerodanes are the most frequently reported diterpene components, labdanes and tetracyclic *ent*-kaurenes and *ent*-stachanes are common (Figure 22).

No clear variation pattern emerges from the examination of the diterpene distribution within this genus. Eighteen of the investigated collections failed to produce any detectable diterpenes, yielding instead various combinations of sesqui- and triterpenoids and phenolics. No clear-cut mutual exclusivity is evident in the distribution of diterpenes and these other terpenoid classes.

The significance of individual negative diterpene reports is difficult to estimate. This difficulty is highlighted by the observation that of the three taxa that have been examined twice, *Baccharis ramoisissima*, *B. tucumanensis*, and *B. alaternoides*, two yielded detectable diterpenes in only one of the two investigations. Given that little is known about (1) factors governing the plant developmental stage at which diterpene production is initiated and (2) factors controlling the rate of production, data for these two taxa suggest that species reports of diterpene absence be treated cautiously.

Looking more closely at the diterpene-producing species, it is apparent that *Baccharis* produces relatively more *ent*-labdanes and related skeletons than were found in *Brickellia* and *Stevia* (Table 9, [pp. 447–449], Figures 23 [p. 469], 24 [pp. 470–473], and 25 [pp. 474–475]). The most common of these skeletons is the $5\alpha,10\beta$-*trans-ent*-clerodane. Among the *ent*-skeletal classes, there is a rough mutual exclusivity of the *trans*-clerodanes and the polycyclic compounds (e.g., *ent*-kaurenes). Normal-labdanes and derivatives occur rarely within this genus.

IV.3.4. *Helichrysum*

The large genus *Helichrysum* (approx. five hundred species; Inuleae: Gnaphaliinae) extends through Eurasia, Africa, Madagascar, Australia and New Zealand. As in *Baccharis*, many investigated *Helichrysum* species failed to yield detectable quantities of diterpenes (*171*). Of the diterpene-producing species, the majority produced *ent*-labdane-derived compounds of the kaurane, trachylobane, helifulvane, and atisirane polycyclic types (Table 10 [pp. 449–

451], Figures 26 [pp. 476–477], 27 [pp. 478–479], and 28 [pp. 480–481]). Other species produced chemistries combining novel unicyclic geranylter-pinenes, related acyclic compounds, normal-labdanes, and *ent*-labdanes. In one of two separate reports regarding the same species, *H. albirosulatum*, (*171*), *ent*-labdanes were identified, although in the other (*180*) normal-labdanes were found. The report of a macrocyclic diterpene (isocembrene) from *H. subfalca-tum* is the only such report from the Compositae.

IV.3.5. *Helianthus*

The distribution of the New World sunflower genus, *Helianthus* (fifty species; Heliantheae: Helianthinae) is centered in North America. It is very closely related to *Viguiera*, and in fact the generic distinctions between the two genera break down in several species (*316*). Many widespread *Helianthus* taxa are the products of hybridization. A summary of the diterpene variability within *Helianthus* (Table 11 [pp. 452–455], Figure 29 [pp. 482–483]) shows that while diterpenes have not been reported from all investigated taxa, many plants produce resins containing both sesquiterpene lactones and diterpenes. Several species (e.g., *H. glaucophyllus* and *H. microcephalus*) produce resins that are rich in sesquiterpene lactones (*249*) but lack diterpenes, while others (e.g. *H. occidentalis* and *H. rigidus*) produce abundant diterpenoids and no detectable sesquiterpene lactones. Chemical profiles of diterpene-producing *Viguiera* and *Helianthus* species display many similarities (Table 11).

IV.3.6. *Montanoa*

Montanoa (25 species; Heliantheae:Montanoinae) ranges from northern Mex-ico to Peru, with the greatest species concentration occurring in southern Mexico and Guatemala. Although few diterpenes are reported from this genus, their pharmacological value justifies including a discussion of *Montanoa* diterpenes. The focus of attention has been the novel oxepane diterpenes (Figure 30 [p. 484]) identified as the abortifacient agent in the medicinal tea brewed from *M. tomentosa* (*419, 382*) (See Chapter 5). Oxepanes have been reported from seven of the nine investigated *Montanoa* taxa (Table 12 [p. 456]; Figure 30). The only other reported diterpenes are *ent*-stachanes, kauranes and related structures. Typically, oxepanes are present in the extract in lower concentrations than the co-occurring sesquiterpene lactones.

IV.3.7. Interspecific Variation: An Overview

In *Baccharis*, *Helichrysum*, and other genera detectable levels of diterpenes may typify only a fraction of the species. The absence of detectable quantities of diterpenes, a trait which can characterize an entire tribe (e.g., Vernonieae) or groups of genera within tribes, seems also to appear within groups of related species some other members of which produce diterpenes. As noted in

Baccharis, some collections of a species may yield diterpenes while others do not.

This "spotty" and sometimes unpredictable distribution can have many possible explanations. Any environmental or developmental factor which disrupts the proper functioning of glandular tissues can affect diterpene production. Another important consideration is the central role the diterpene precursor, geranylgeranyl pyrophosphate, plays in terpenoid biogenesis. Its biosynthesis is dependent on a supply of farnesyl pyrophosphate (the C_{15} precursor of both sesqui- and triterpenes) and isopentenyl pyrophosphate (C_5). The existence of competing demands for the C_{15} precursor is suggested by instances in which sesquiterpenes (e.g., hydrocarbons, alcohols or lactones) and/or triterpenes characterize extracts lacking diterpenes. Additionally, geranylgeraniol is the precursor of phytol, a necessary component of chlorophyll, and tetraterpenoids (C_{40}), such as carotenoids. Any of the competing demands for the C_{20}, C_{15}, or C_5 precursors could effect diterpene production.

In summary, several distributional trends characterize the diterpene-producing species of the Compositae genera:

1. Representatives of both the normal- and *ent*-series occur within genera and often within the same species.
2. When normal- and *ent*-series structures co-occur it is usually as a mixture of normal-labdanes and some tri- or polycyclic *ent*-labdane derivatives. Normal-series structures infrequently exist as tri- or polycyclics; the most common of such structures, the normal-pimaranes (sandaracopimaranes), achieve prominence only within the small tribe Calenduleae.
3. Structural novelty (e.g., oxpanes of *Montanoa*, $2\alpha,3\alpha$-dihydroxylation of *Brickellia*) can be diagnostic for groups of related species.
4. Despite early statements regarding the rarity of linear diterpenes in plants (*279*), linear diterpenes and their hetero- and carbocyclic derivatives are hardly rare within genera of the Compositae.

TABLE 4. Distribution of sesquiterpene lactones (SQTL), diterpenes (DIT), chromenes (CHR), and benzofurans (BENZF) in the Compositae tribes organized according to Wagenitz (*488*). Sesquiterpene lactone data are from *458*. Chromene and benzofuran data are from *429*.

Tribes[1]	Natural Product Classes			Head Type	
	SQTL	DIT	CHR & BENZF	DISC	RAD[5]
Asteroideae:					
Group 1:					
Vernonieae	216[3]	_[4]	_[4]	+	
Liabeae	16	-	-		+
Mutisieae	29	1	2	+	
Cynareae[2]	132	1	-	+	
Arctoteae	13	1	-		+
Anthemideae	412	-	-		+
Group 2:					
Eupatorieae	184	118	65	+	
Astereae	7	96	15		+
Heliantheae	892	192	33		+
Senecioneae	744	10	30		+
Calenduleae	1	23	1		+
Inuleae	114	29	19		+
Cichorioideae:					
Lactuceae	30	-	-	ligulate	

[1] Tribes organized according to Wagenitz (488).

[2] Cynareae includes Wagenitz's tribes Cardueae and Echinopeae.

[3] Number of reports of compounds belonging to this chemical class in taxa of this tribe.

[4] No. of taxa reported to produce compounds of this chemical class.

[5] Flowering heads are either discoid or radiate.

Table 5. Number of Taxa   439

TABLE 5. The Number of taxa (organized by tribes) reported to produce major diterpene skeletal types.

Diterpene Skel. No.	VE	EU	AS	IN	HE	AN	SE	CA	AR	CY	MU	LA	LI	AR
Linear														
1.		12		1	5									
2.		8	1		3									
Cyclic														
I. Normal														
16.		29	9	3	9		1							
17.			17											
18.		3	2		1									
19.		2			5									
20.			2											
21.		1	1											
31.		1	7	1										
32.		2	1		3		2	20						
34.								1						
35.		2			1									
37.			2	1										
II. ent-														
38.		18	10	2	1									
39.		1												
46.		7	16		1							1		
50.			5											
52.		2			5		2							
55.			4	1										
58.		30	17	17	115		5			1	1			
64.			2	3	9			2						
66.					24									

Representatives of the major diterpene skeletal types characteristic of the Heliantheae, Calenduleae, Astereae and Eupatorieae.

TABLE 5. *(contd.).*

HELIANTHEAE

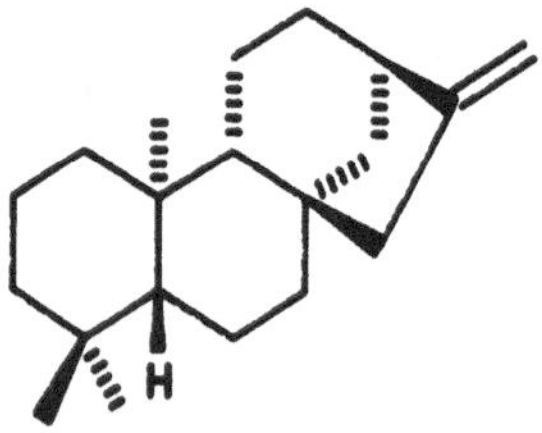

ent—Kaurane

CALENDULEAE

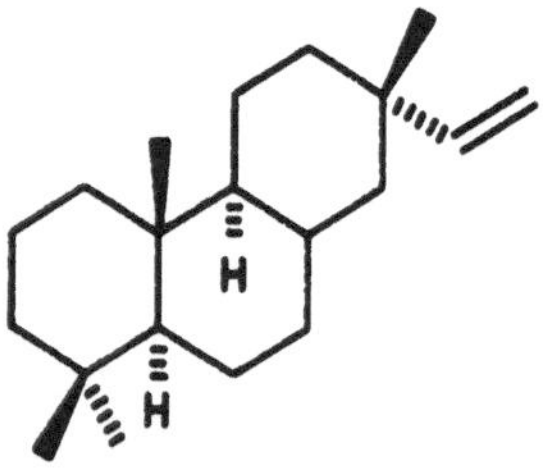

Sandaracopimarane

(normal—pimarane)

ASTEREAE

EUPATORIEAE

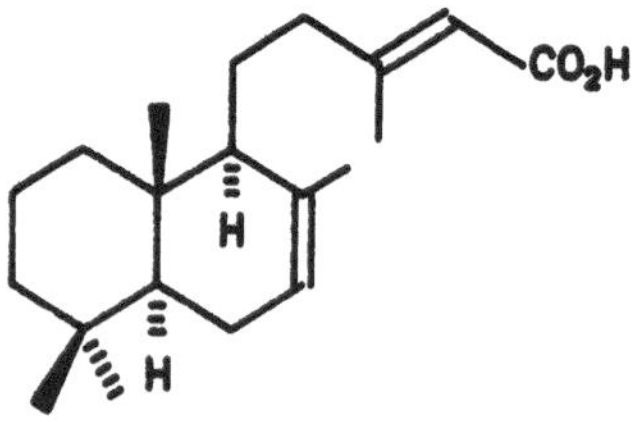

normal—Labdane

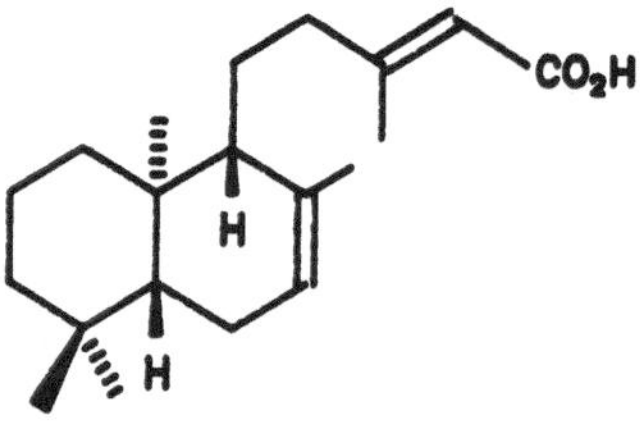

ent—Labdane

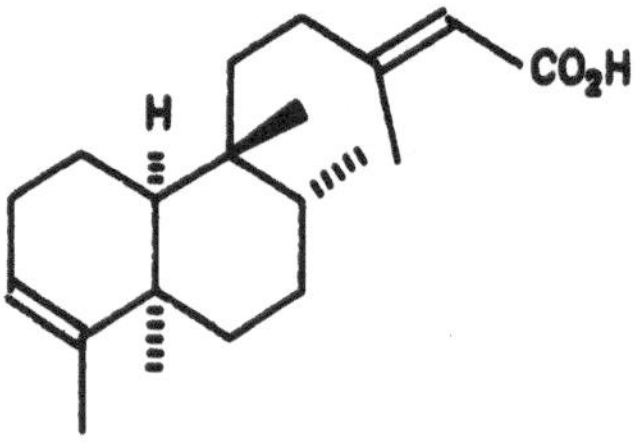

normal—Clerodane

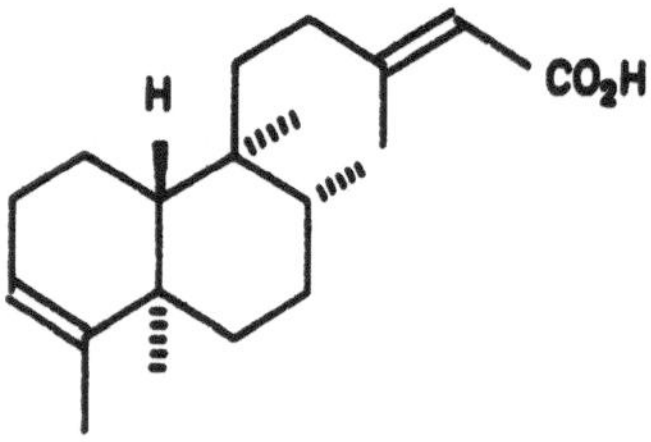

ent—Clerodane

Table 6. Distribution of Diterpenes   441

TABLE 6. Distribution of diterpenes in *Brickellia* (Eupatorieae) species.

| | | | Brickellia Diterpene Cyclic Classes | | | | | |
Species	Linear A./Uni.	Normal Bi. (L)	Bi. (C)	Tri.	Ent- Bi. (L)	Bi. (C)	Tri.	Poly.
B. annulosa						511		
B. argyrolepis		101 102 122 123						
B. corymbosa		106 108						
B. diffusa		97 215 216 217 218						
B. eupatoriedes		78 98 99 103 104 105 126 127 128					601 602 603 604 605 606 607 608 609 610	
B. paniculata		126						
B. squarrosa		106 107 108 109						
B. vernicosa		991 992 993 994 995						
B. veroni- caefolia		124 125 129 220						
B. veroni- caefolia A. Gray var. typica		129						

TABLE 7. Distribution of diterpenes in *Stevia* species.

Species	Linear A./Uni.	Normal Bi. (L)	Normal Bi. (C)	Normal Tri.	Ent- Bi. (L)	Ent- Bi. (C)	Ent- Tri.	Poly.
S. berlandiera		145 1021 1022 1023						
S. bupthalmiflora								649 650 651 652
S. eupatoria								641 998
S. jaliscensis		118						
S. lemmonia	1020							
S. lucida Lag. var. Bipontini					415			
S. monardaefolia		143 153						625 649
S. myriadenia	3 23				421	493 494		625 650 651 774
S. origanoides					428 429 808 809 810 811			
S. ovata								895 896 897 898 899
S. paniculata								895 896 897 898 899
S. polycephala						570		
S. rebaudiana		195			428 429			682 683 684 685

Table 8. Major Natural Products   443

TABLE 7. (*contd.*).

Species	Linear A./Uni.	Normal Bi. (L)	Normal Bi. (C)	Normal Tri.	Ent- Bi. (L)	Ent- Bi. (C)	Ent- Tri.	Poly.
								686
								687
								688
								689
S. sali-cifolia		121 1024 1025 1026			154[abs. ster. unknown] 429			
S. setifera								625

TABLE 8. Major natural products reported from *Baccharis* (Astereae).

Species	Terpenoid C20	C30	C15	Flavonoid	Phenolic	Acetylenic
B. alaternoides H.B.K. (1) Peru	6	11	14	16		
B. alaternoides H.B.K. (2) Peru				16		
B. articulata (Lam.)Pers. Arg.	5,6					
B. caespitosa (Ruiz and Pavón) Pers. Ecua., Peru, Bol.						
B. calvescens DC. Braz.	6	10,13	14			
B. cassinaefolia DC. Braz.	4	11,12,13	14			20
B. chilco H.B.K. (1) Col.	4,6	10				
B. chilco H.B.K. (2) Col., Peru		13	14			
B. concinna Barroso	8	12,13	14	16		

TABLE 8. (*contd.*).

Species	Terpenoid			Flavonoid	Phenolic	Acetylenic
	C20	C30	C15			
B. conferta H.B.K. Mex.	6					
B. eggersii Hieron. Peru	3	11	14			
B. genistelloides (Lam.)Pers. Col., Ecua., Peru	1,6		14			
B. grandicapitu-lata Hieron. Peru	6	11	14			
B. halimifolia		11				
B. helichryi-soides DC. Braz.					19	
B. heterophylla H.B.K.		10				
B. hutchisonii Cuatr. Peru	6	11	14			
B. intermixta Gardn. Braz.	2,8		14			
B. juncea (Lehrn.)Desf. Braz.					18	
B. kingii Cuatr. Peru	6		14	17		
B. latifolia (Ruiz & Pavón) Pers. Bol., Ecua.		11,13	14,15			
B. leptocephala DC. Braz.		11	14	16		
B. microcephala (Less.)DC. Para.	6		14			
B. minutiflora Braz.	8	12	14			

Table 8. Major Natural Products   445

TABLE 8. (*contd.*).

Species	Terpenoid			Flavonoid	Phenolic	Acetylenic
	C20	C30	C15			
B. nitida (Ruiz & Pavón)Pers. Peru	4	10	14			
B. oxydonta DC. Braz.	2	13	14	16		
B. pedunculata (Miller)Cabrera Ven., Peru		11	14			20
B. polyphylla Gardn. Braz.	8		14			
B. pylicoides H.B.K.	8	10	14			
B. quitensis H.B.K.	8	12	14			
B. ramoisissima Gardn. (1) Braz.		12	14	16		
B. ramoisissima Gardn. (2) Braz.	8				18	
B. reticularia DC. Braz.		11,12,13	14	16		
B. salicifolia (Ruiz & Pavón) Pers. Col., Arg., Chile	6	10,11		17		
B. salzmannii DC. Braz.	8	12	14			
B. sarothroides A. Gray	6					
B. scandens R. & P. Peru			14			
B. scoparia (L.)Sw. Jamaica	3,5	10,11	14			
B. serrulata (Lam.)Pers. Braz.		13	14	16		

TABLE 8. (*contd.*).

Species	Terpenoid			Flavonoid	Phenolic	Acetylenic
	C20	C30	C15			
B. sternbergiana Steud. Peru	1,2		14	16	18	
B. subdentata DC. Braz.	4	10	14			
B. tola Chile	2,9	10				
B. tricuneata var. lineata Cuatr.	6	10				
B. tricuneata var. ruiziana Cuatr. Col, Ven.					21	
B.tricuneata (L.f)Pers. var.tricuneata Col., Ven.	4,5,7			17	19	
B. trimera Braz.	6			17		
B. trinervis (Lam.)Pers. Braz., Mex.		12,13	14			20
B. trinervis var. rhexoides (H.B.K)Baker Bol., Ecua.				17		20
B. truncata	8	12	14	16		
B. tucumanensis H. & A. (1) Arg.	6					
B. tucumanensis H. & A. (2) Arg.	2 or 3			17		
B. ulcina H. & A. Bol., Ecua.			14			
B. vaccinoides H.B.K.	6	10				
B. varians Gardn. Braz.				16		

Table 9. Distribution of Diterpenes   447

TABLE 9. Distribution of diterpenes reported from *Baccharis* (Astereae).

Species	Linear A./Uni.	Normal Bi. (L)	Normal Bi. (C)	Normal Tri.	Ent- Bi. (L)	Ent- Bi. (C)	Tri.	Poly.
B. alater-noides						484 569		
B. articulata						510 527		
B. calvescens						518		
B. cassi-naefolia						531 532 533		
B. chilco						529		
B. concinna (concinnia)								646 649 625
B. conferta						521		
B. eggersii		965 966						
B. geni-stelloides						561		
B. grandi-capitulata						963		
B. hutchisonii						507 967		
B. intermixta					363 364			625 646
B. kingii						846		
B. micro-cephala						561		
B. minutiflora								618 619 621 624 625 697 698 713 720

TABLE 9. (*contd.*).

| | Linear | Normal | | | Ent- | | | |
Species	A./Uni.	Bi. (L)	Bi. (C)	Tri.	Bi. (L)	Bi. (C)	Tri.	Poly.
								729
								730
								731
								732
								733
								734
								735
								736
								737
								738
								739
B. nitida						968		
B. oxydonta					442			
					443			
					444			
					445			
					446			
B. polyphylla						569		625
								646
								649
								661
B. quitensis								699
								740
B. ramosissima								624
								625
B. salicifolia						518[1033]		
						904		
						1070		
						1093		
B. salzmannii								625
								650
B. sarothroides						518[1033]		
						1034		
B. scoparia	964					969		
						970		
						1061		
B. stern-bergiana					331			
					332			
					333			
					334			
					357			
					359			

Baccharis Diterpene Cyclic Classes

Table 10. Distribution of Diterpenes   449

TABLE 9. (*contd.*).

Species	Linear A./Uni.	Normal Bi. (L)	Bi. (C)	Tri.	Ent- Bi. (L)	Bi. (C)	Tri.	Poly.
B. subdentata						529		
B. tola					455			765 766 771 772 779 780 782
B. tricuneata (L. f.)Pers. var. tricuneata						509 527 528 529 530		
B. tricuneata var. lineata						508 518 520		
B. trimera						559 562 563		
B. truncata (tricuneata?)						565		624 625 648 650
B. tucumanensis	110					962		
B. vaccinoides						518[1033]		

TABLE 10. Distribution of diterpenes reported from *Helichrysum* (Inuleae).

Species	Linear A./Uni.	Normal Bi. (L)	Bi. (C)	Tri.	Ent- Bi. (L)	Bi. (C)	Tri.	Poly.
H. acutatum	69 70 71 72 1101							

TABLE 10. (*contd.*).

		Helichrysum Diterpene Cyclic Classes						
	Linear	Normal			Ent-			
Species	A./Uni.	Bi. (L)	Bi. (C)	Tri.	Bi. (L)	Bi. (C)	Tri.	Poly.
H. albirosulatum		1105 1106 1107			447 448 449			
H. argentissimum								625
H. aureum (Houtt.) Merile var. monocephalum								625 629 640 690
H. bellum								625
H. calliconum	65 66 67							
H. chiono-sphaerum							614	625 783 796 797
H. confertum		132 137 996 1102 1103 1104						
H. cooperi								625 629 640
H. dendroideum								619 623 626 698 764 765 767 770 771
H. diosmifolium								728
H. fulvum								625 774 791 798 799

Table 10. Distribution of Diterpenes 451

TABLE 10. (*contd.*).

Species	Linear A./Uni.	Normal Bi. (L)	Bi. (C)	Tri.	Ent- Bi. (L)	Bi. (C)	Tri.	Poly.
H. heterolasium								619 624 629 640 648
H. Krebsianum	32 70							
H. miconiifolium								624 625
H. mundii	23							
H. nudifolium	68							
H. odoratissimum	68							
H. pallidum								624 625 640 720 722
H. pinifolium					447			
H. platypterum								690
H. refluxum							976	625 774 791
H. ruderale								640 690
H. subfacultatum	23							
H. subfalcatum	73							
H. suterlandii					430			
H. vernum					447			625

Helichrysum Diterpene Cyclic Classes

TABLE 11. Major di- and sesquiterpenoids of *Helianthus* (Heliantheae).

Species	KA[1]	Terpenoids										
		C20								C15		
		LA	PI	ST	TE	IS	TR	AT	VI	GE	HE	FU
HELIANTHUS												
1.HELIAN-THUS												
H. annuus	15,18[2] 19,21-23 28,29					35	38-40	42,43		47		49
H. ano-malus												
H. argo-phyllus	18,33						39			47		
H. bolan-deri												
H. debilis subsp. cucu-merifolius	23,28,29					35-37						
H. debilis subsp. debilis	23,28,29				34		38					
H. deserti-cola												
H. neglectus												
H. niveus subsp. canescens	23,28						39			47		
H. niveus subsp. niveuś										47	48	
H. para-doxus												
H. agrestis												
H. arizo-nensis	33						39					
H. cili-aris	18,21,33						39					49
H. láci-niatus	18,21, 23,33			16, 17			39	42				
H. cusic-kii	28						39					

Table 11. Major di- and Sesquiterpenoids 453

TABLE 11. (*contd.*).

		Terpenoids										
		C20								C15		
Species	KA[1]	LA	PI	ST	TE	IS	TR	AT	VI	GE	HE	FU
H. gracilentus										47	48	49
H. pumilus										47		
H. californicus							39			47		
H. decapetalus	23,26	1-3					38	45		47		
H. divaricatus										47		
H. eggertii												
H. giganteus	23,24,26 28,29	5					38			47		
H. grosseserratus	24,28					35	39			47		49
H. hirsutus	23,26			11, 12			38			47		49
H. maximiliani	24	3, 5								47		49 (50)
H. mollis										47		
H. nuttallii	23,28					35	39					49
H. resinosus										47		
H. salicifolius	18,33						39			47		
H. schweinitzii												49
H. strumosus				7,8								49
H. tuberosus	23,24 26,29	3,5										48
H. glaucophyllus												(50)
H. laevigatus												
H. microcephalus												(50)

Table 11. (*contd.*).

| | | | | | Terpenoids | | | | | | | |
| | | C20 | | | | | | | | C15 | | |
Species	KA[1]	LA	PI	ST	TE	IS	TR	AT	VI	GE	HE	FU
H. porteri												
H. smithii												
H. atrorubens												
H. occidentalis	21,23,28, 29-31	3,5				35-37	38, 39	44				
H. rigidus	23,26,33						38, 39					
H. silphioides												
H. angustifolius	23,24, 28,33	3,5										49
H. carnosus											48	
H. floridanus												
H. heterophyllus	23,24 25,27										48	49 (50)
H. longifolius												
H. radula	19-21, 23,26,29, 31-33					35, 36	41					
H. simulans	23,26,28, 29,31					35, 36					48	
V. bishopii	18,23,24	4,6			15		38					
V. cordifolia	23											
V. dentata	23,24, 28,29	6					38					
V. excelsa	23-25, 27,31											49
V. grammatoglossa	23				13-15							
V. insignis	23,24				15			43				

Table 11. Major di- and Sesquiterpenoids 455

TABLE 11. (*contd.*).

Species	KA[1]	C20							VI	C15		
		LA	PI	ST	TE	IS	TR	AT		GE	HE	FU
V. lanceolata							38					
V. linearis											48	49
V. maculata	23,28,34											
V. oaxacana	23											
V. pazensis	29						38		46			
V. porteri	23,28,29					35						
V. procumbens	23,24						38					
V. quinqueradiata	29											
V. stenoloba	23,29				34						48	

1 KA = Kaurane
  LA = <u>ent</u>-Labdane
  PI = <u>ent</u>-Pimarane
  ST = <u>ent</u>-Stachane
  TE = Tetrachyrane
  IS = Isokaurane
  TR = Trachylobane
  AT = Atisirane
  VI = Villanovane

  GE = Germacrolide
  HE = Heliangolide
  FU = Furanoheliangolide

2 See Figure **19** for representative structures

TABLE 12. Distribution of diterpenes reported from *Montanoa* (Heliantheae).

	Montanoa Diterpene Cyclic Classes							
	Linear	Normal			Ent-			
Species	A./Uni.	Bi. (L)	Bi. (C)	Tri.	Bi. (L)	Bi. (C)	Tri.	Poly.
M. atri- plicifolia								625
M. frutescens	56[homolog]							
M. leucantha subsp. arborescens	55							
M. leucantha subsp. leucantha	55							
M. mollissima	55							
M. pteropoda								618 619 624 625 649 652 655 690 694 773
M. tomentosa subsp. microcephala	55							
M. tomentosa subsp. tomentosa	55 57 1094 1095 1096 1097 1098 1099							
M. tomentosa subsp. xanthiifolia	56							800

Figure 19. Structures of Normal-Labdanes 457

SCHEME A

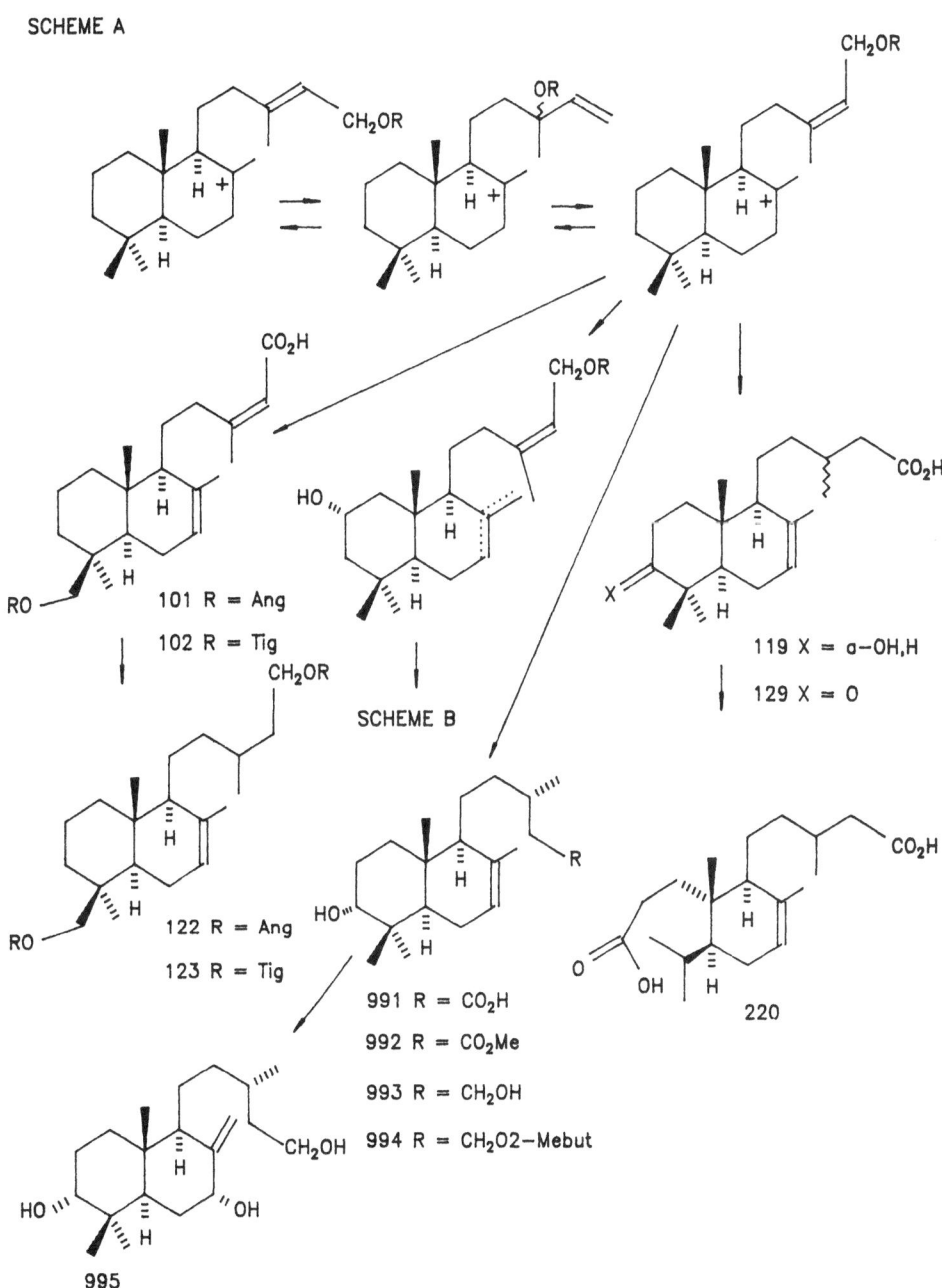

FIGURE 19. Structures of normal-labdanes reported from *Brickellia* (Eupatorieae): compounds with common substitutional patterns (Scheme A) and with unusual 2α,3α-dihydroxy-pattern (Scheme B). Structures of *Brickellia ent*-clerodanes [511], cleistanthanes, etc. [601–609] (Scheme C).

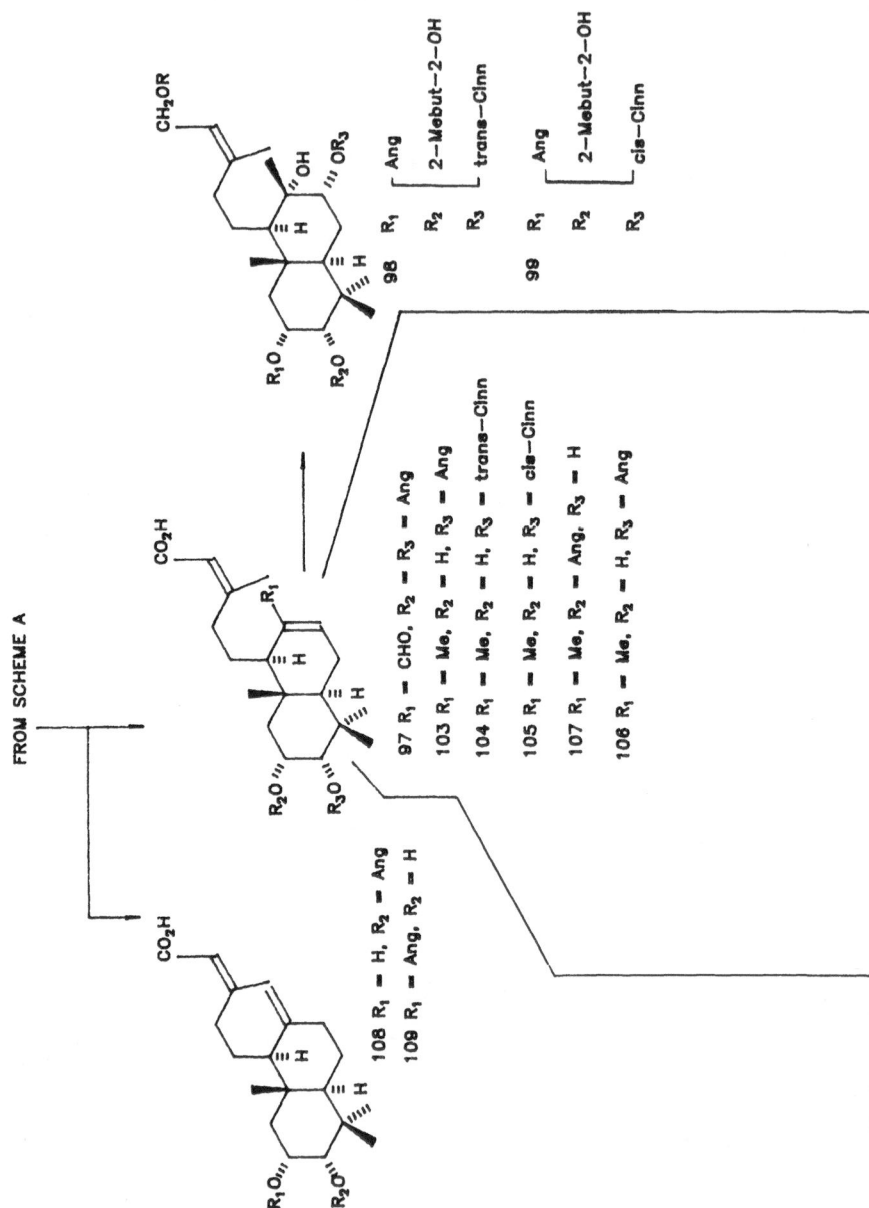

SCHEME B

FROM SCHEME A

108 R₁ = H, R₂ = Ang
109 R₁ = Ang, R₂ = H

97 R₁ = CHO, R₂ = R₃ = Ang
103 R₁ = Me, R₂ = H, R₃ = Ang
104 R₁ = Me, R₂ = H, R₃ = trans-Clnn
105 R₁ = Me, R₂ = H, R₃ = cis-Clnn
107 R₁ = Me, R₂ = Ang, R₃ = H
106 R₁ = Me, R₂ = H, R₃ = Ang

98   R₁   Ang
      R₂   2-Mebut-2-OH
      R₃   trans-Clnn

99   R₁   Ang
      R₂   2-Mebut-2-OH
      R₃   cis-Clnn

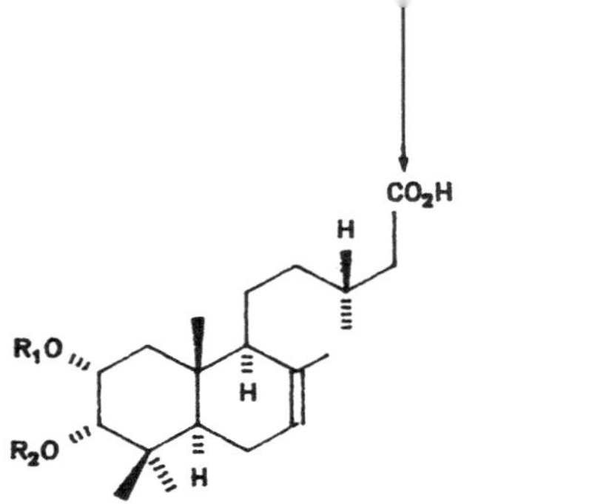

Figure 19. Structures of Normal-Labdanes   459

218 $R_1 = H$, $R_2 = Ang$

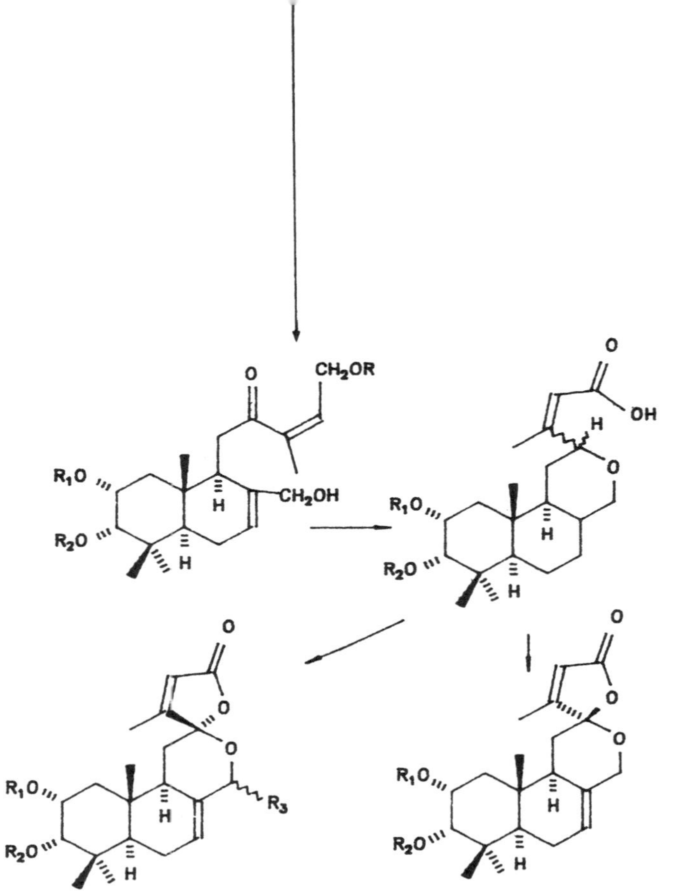

215 $R_1 = R_2 = Ang$, $R_3 = H$
216 $R_1 = R_2 = Ang$, $R_3 = \alpha-OH$
217 $R_1 = R_2 = Ang$, $R_3 = \beta-OH$

124 $R_1 = R_2 = H$
125 $R_1 = H$, $R_2 = 2-Mebut-2-OH$
126 $R_1 = H$, $R_2 = Ang$

C−13 Orientivity Undefined:

127 $R_1 = H$, $R_2 = trans-Cinn$
128 $R_1 = H$, $R_2 = cis-Cinn$

FIGURE 19. (contd.).

SCHEME C

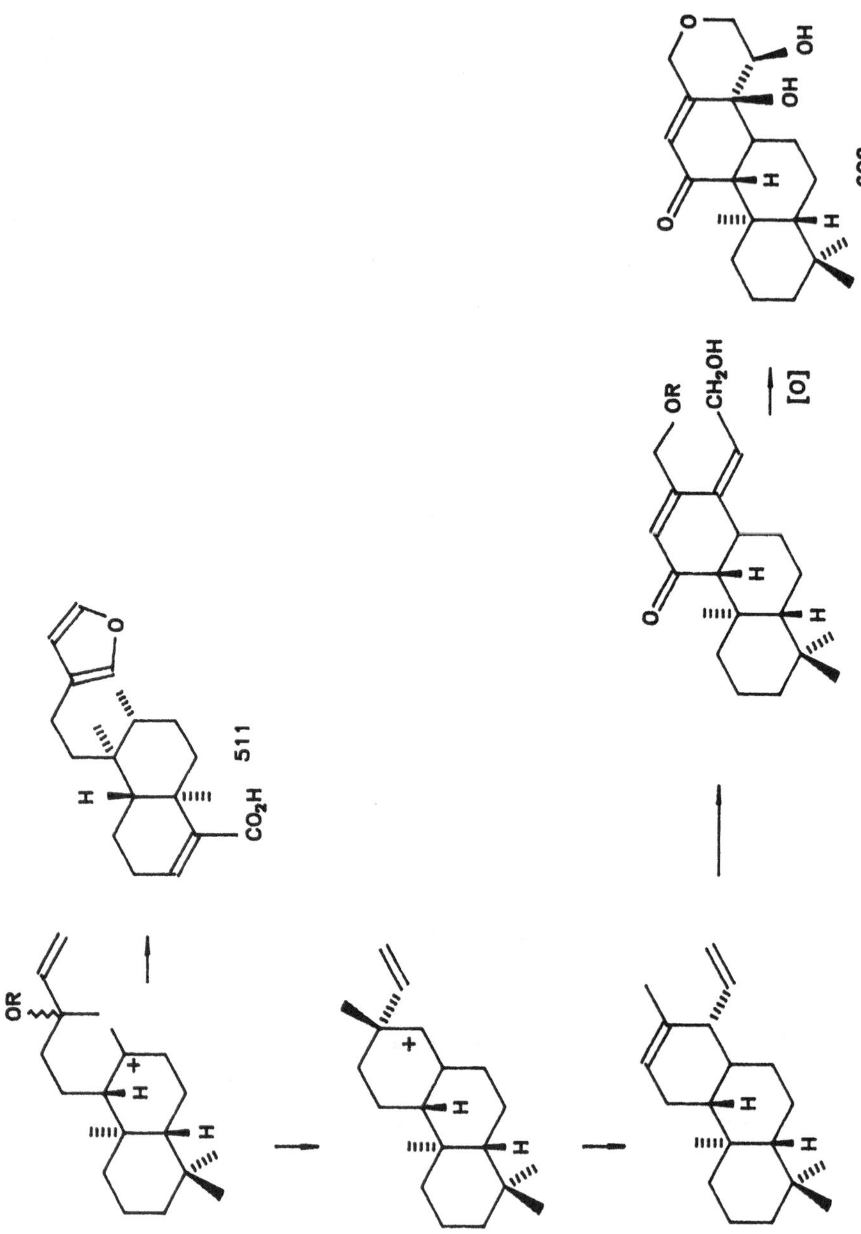

Figure 19. Structures of Normal-Labdanes   461

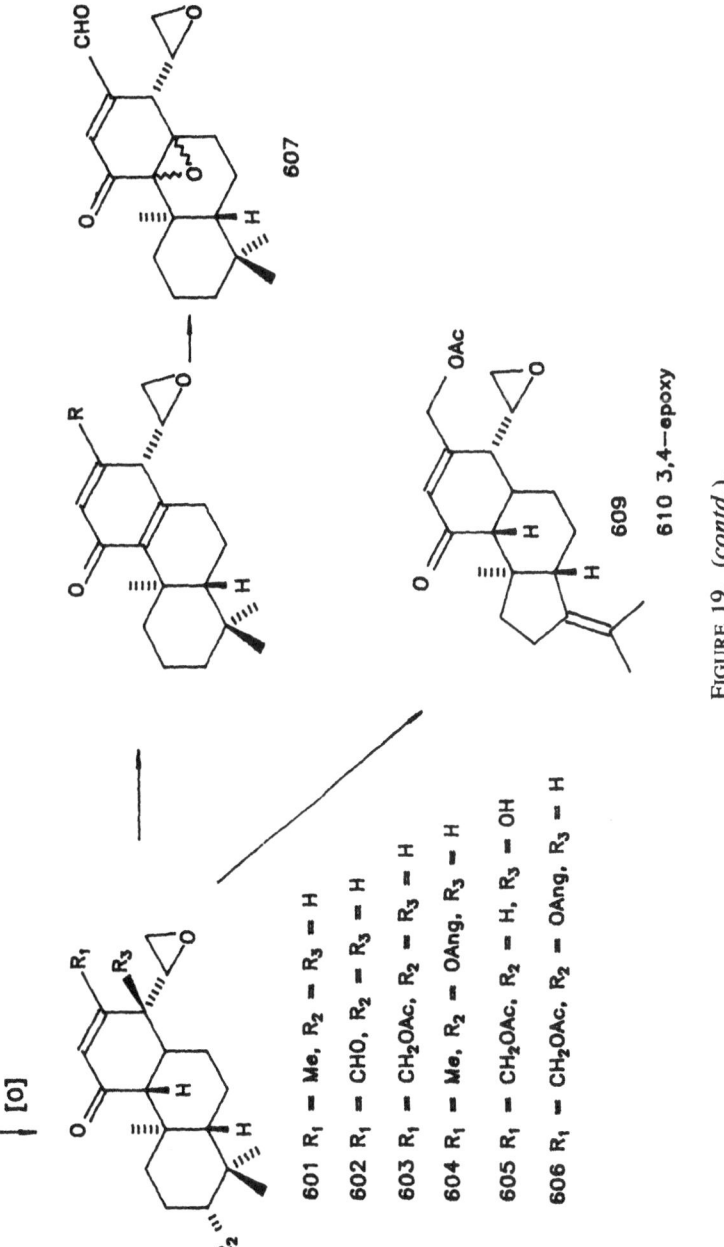

601 R_1 = Me, R_2 = R_3 = H

602 R_1 = CHO, R_2 = R_3 = H

603 R_1 = CH_2OAc, R_2 = R_3 = H

604 R_1 = Me, R_2 = OAng, R_3 = H

605 R_1 = CH_2OAc, R_2 = H, R_3 = OH

606 R_1 = CH_2OAc, R_2 = OAng, R_3 = H

610 3,4–epoxy

FIGURE 19. (contd.).

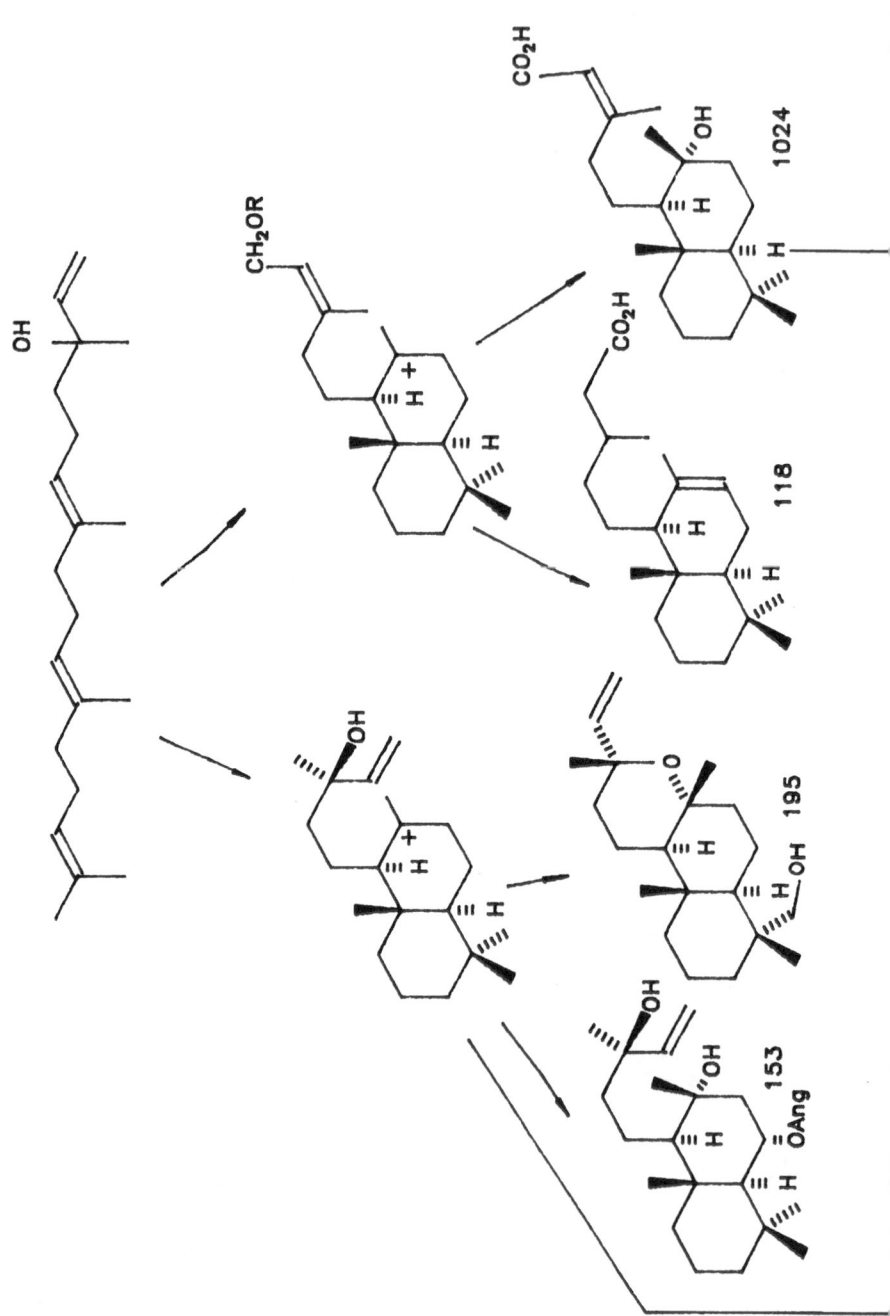

Figure 20. Structures of Normal-Labdane and Related Diterpenes 463

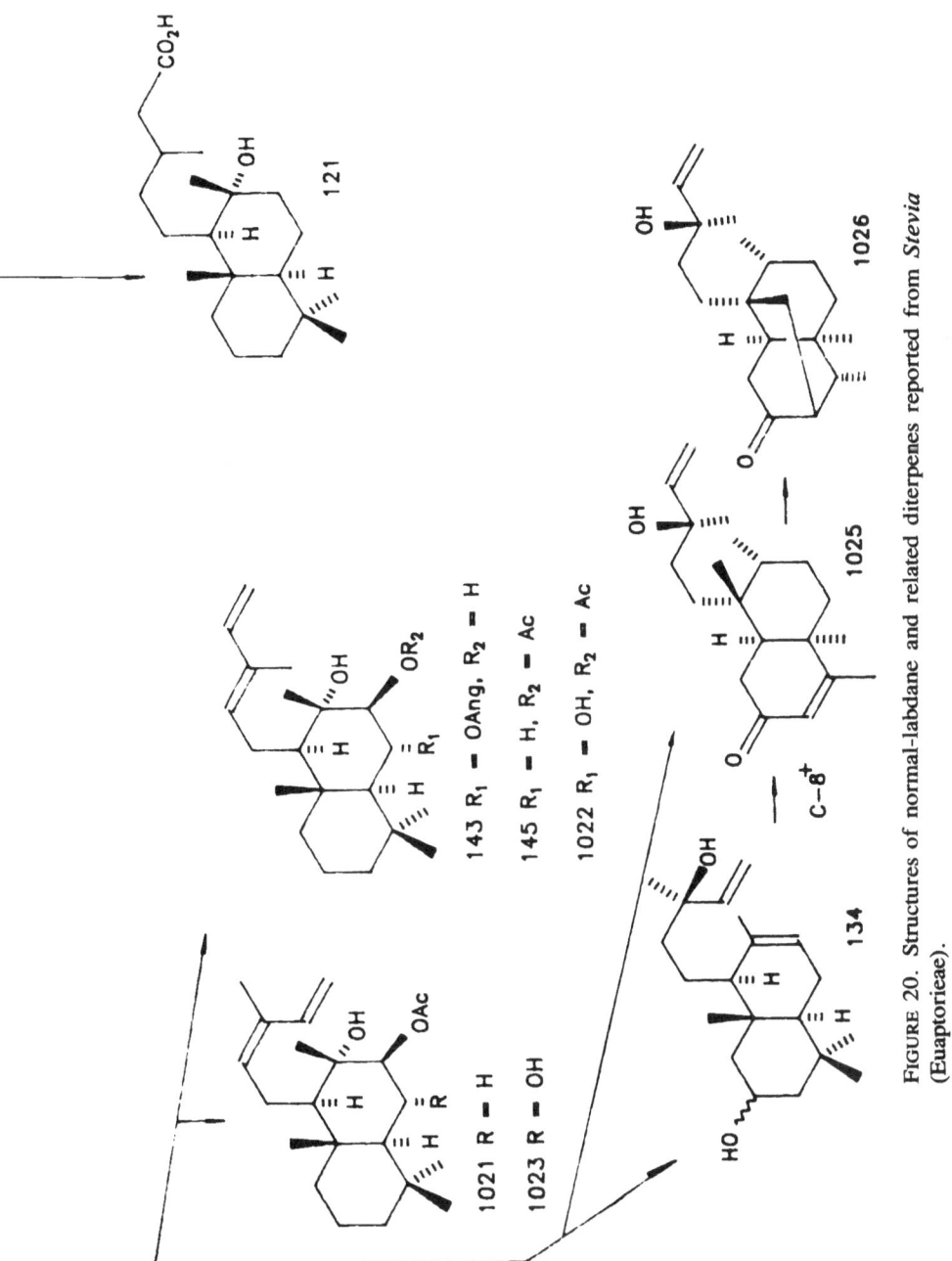

FIGURE 20. Structures of normal-labdane and related diterpenes reported from *Stevia* (Euaptorieae).

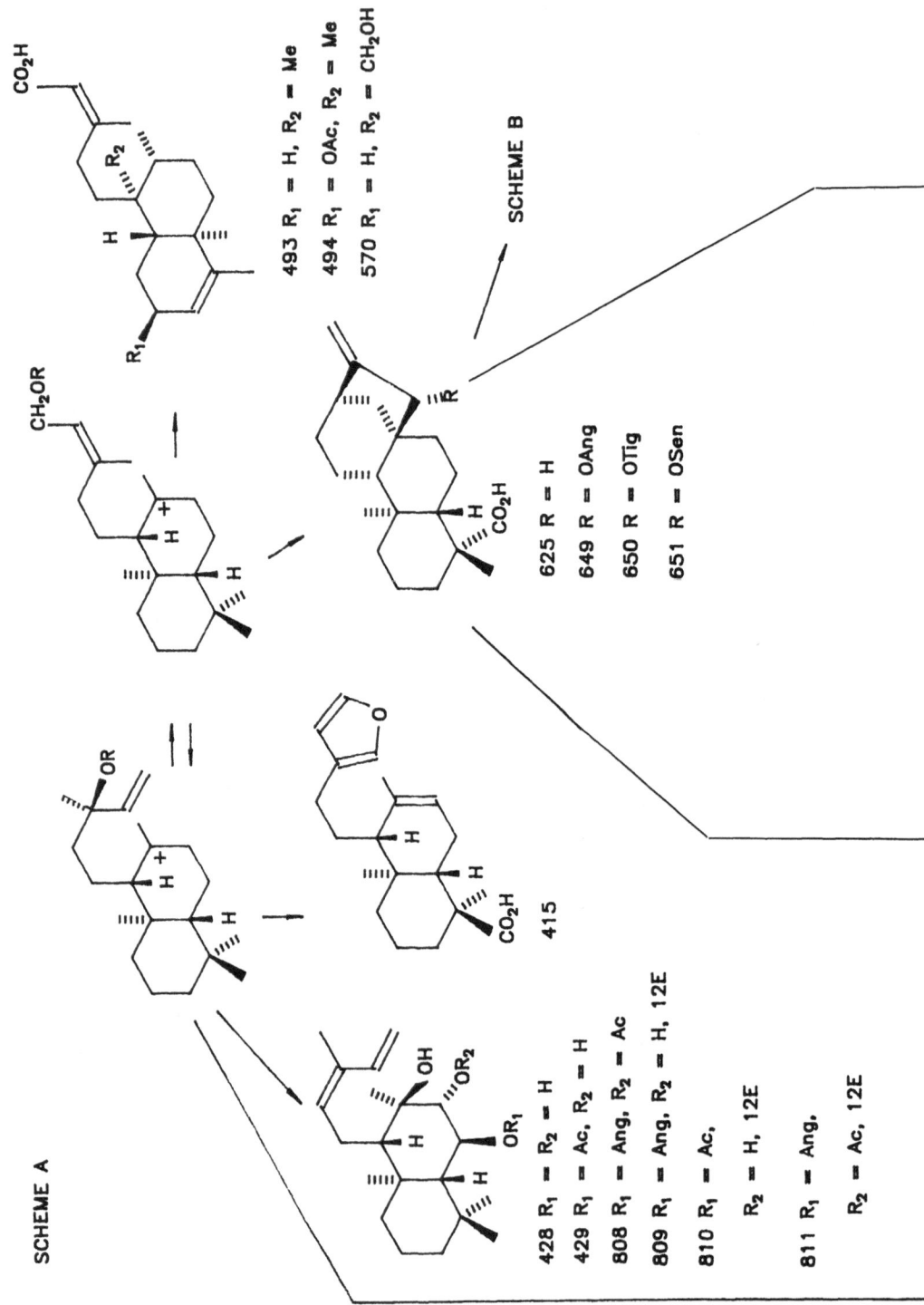

SCHEME A

493 R₁ = H, R₂ = Me
494 R₁ = OAc, R₂ = Me
570 R₁ = H, R₂ = CH₂OH

SCHEME B

625 R = H
649 R = OAng
650 R = OTig
651 R = OSen

415

428 R₁ = R₂ = H
429 R₁ = Ac, R₂ = H
808 R₁ = Ang, R₂ = Ac
809 R₁ = Ang, R₂ = H, 12E
810 R₁ = Ac,
        R₂ = H, 12E
811 R₁ = Ang,
        R₂ = Ac, 12E

Figure 21. Structures of *ent*-Labdane and *ent*-Kauranoid Diterpenes 465

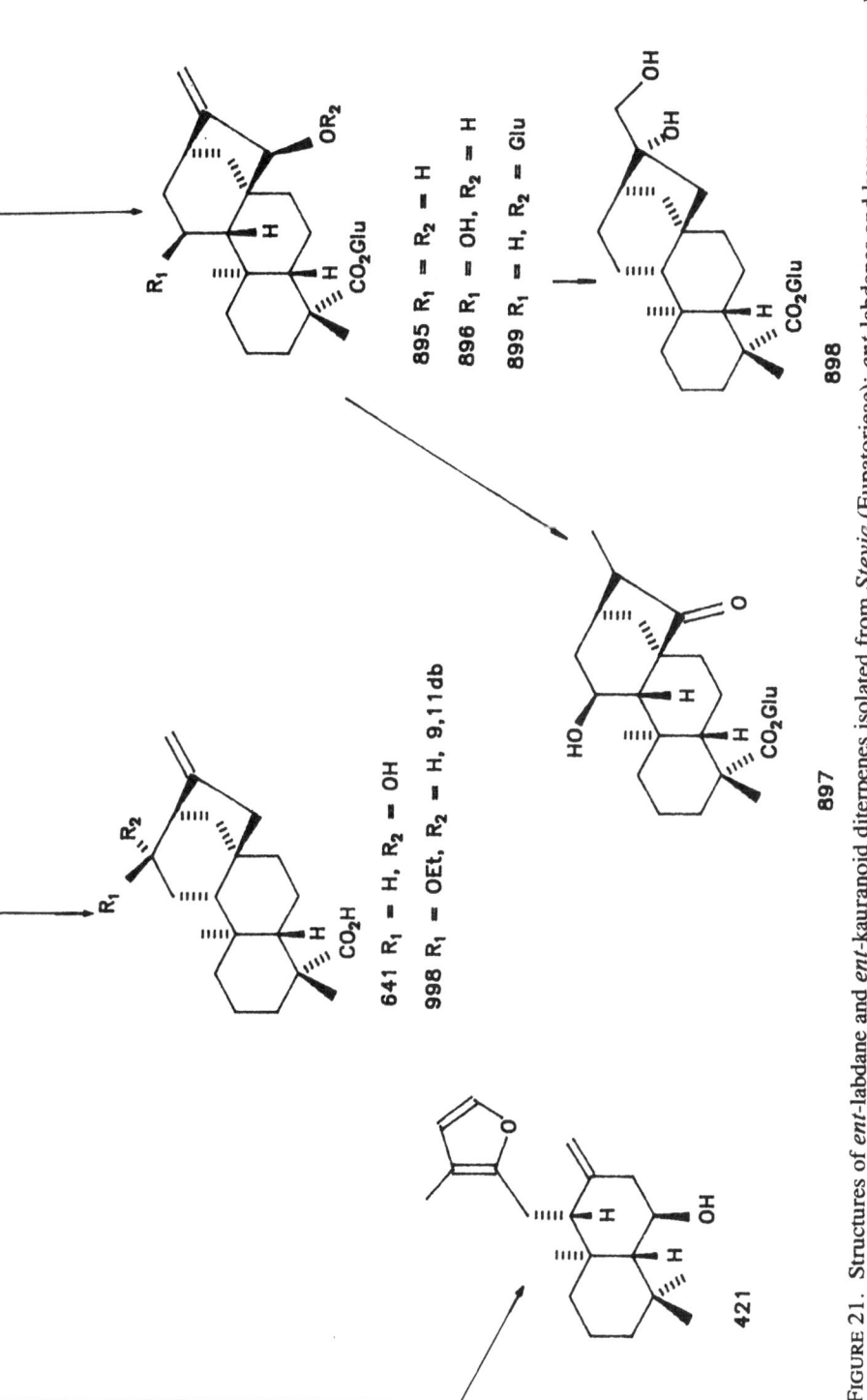

895 R₁ = R₂ = H

896 R₁ = OH, R₂ = H

899 R₁ = H, R₂ = Glu

898

897

641 R₁ = H, R₂ = OH

998 R₁ = OEt, R₂ = H, 9,11db

421

FIGURE 21. Structures of *ent*-labdane and *ent*-kauranoid diterpenes isolated from *Stevia* (Eupatorieae): *ent*-labdanes and kaurene mono- and di-glycosides (Scheme A); C-13-kaurene-glycosides (Scheme B).

SCHEME B

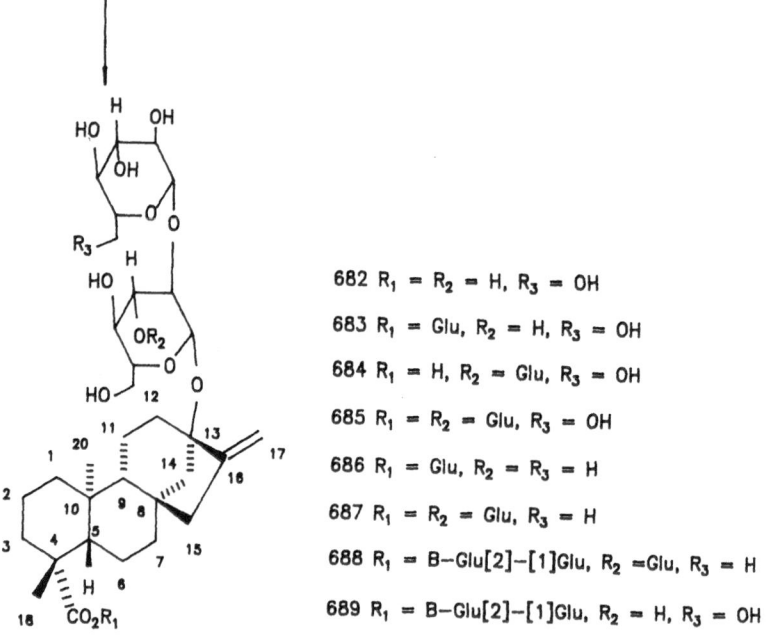

682 $R_1 = R_2 = H$, $R_3 = OH$

683 $R_1 = Glu$, $R_2 = H$, $R_3 = OH$

684 $R_1 = H$, $R_2 = Glu$, $R_3 = OH$

685 $R_1 = R_2 = Glu$, $R_3 = OH$

686 $R_1 = Glu$, $R_2 = R_3 = H$

687 $R_1 = R_2 = Glu$, $R_3 = H$

688 $R_1 = B-Glu[2]-[1]Glu$, $R_2 = Glu$, $R_3 = H$

689 $R_1 = B-Glu[2]-[1]Glu$, $R_2 = H$, $R_3 = OH$

FIGURE 21. (*contd.*).

Figure 22. Representative Structures 467

DITERPENES:

LINEAR

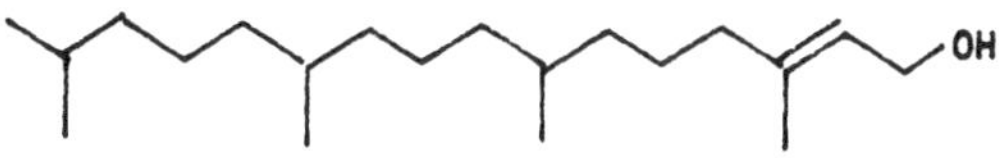

1· Phytol (59)

LABDANE

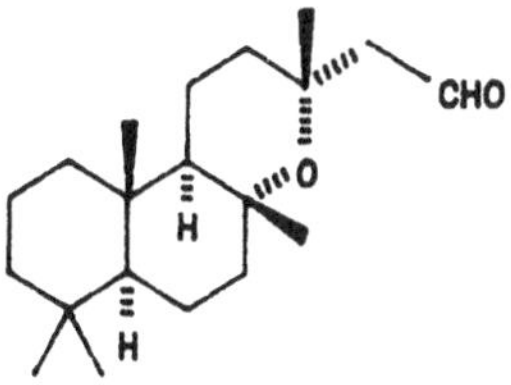

3· normal—Labdane

(202,964,etc·)

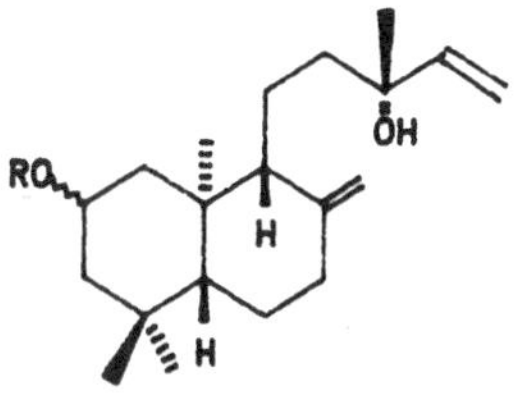

2· ent—Labdane (442—445,etc·)

CLERODANES
(TRANS—ENT—)

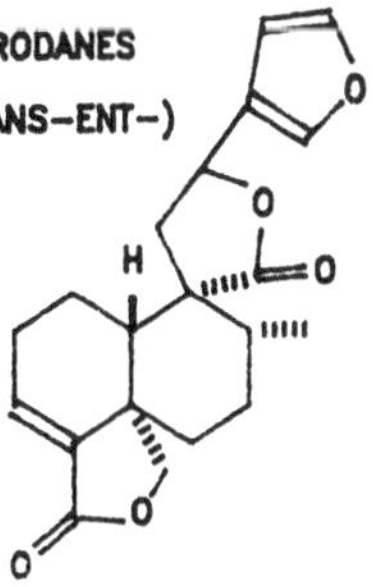

4· (529—533)

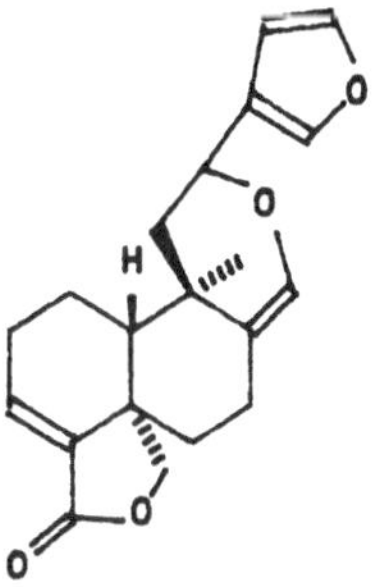

5· (527)

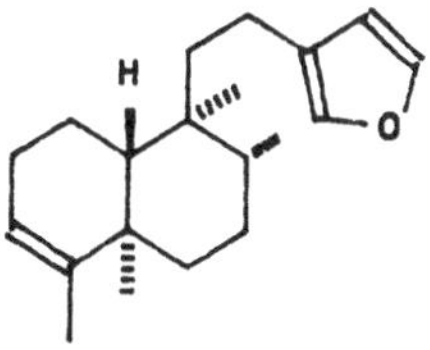

6· (484,508—510,etc·)

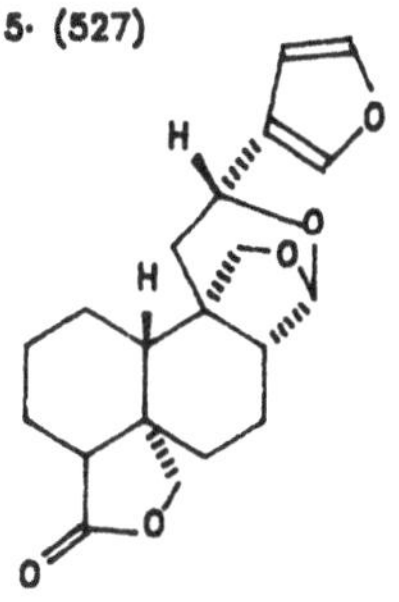

7· (528)

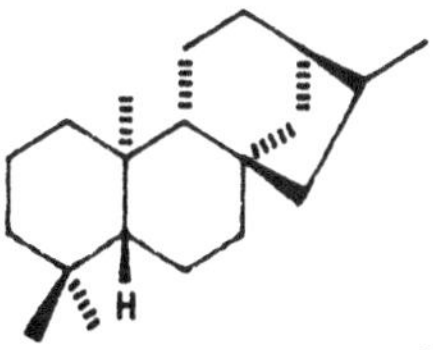

8· ent—Kaurane
(619,624,648—651,etc·)

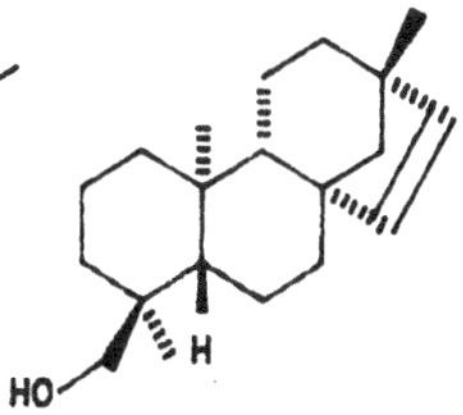

9· ent—Stachane
(765,766,771,772,779)

FIGURE 22. Representative structures of di-, and tri-, and sesquiterpenes, flavonoids, acetophenones, coumarins, and polyacetylenes reported from *Baccharis* (Astereae). The numbers assigned to the structures correspond to the numbers listed in Table 8.

TRITERPENES:

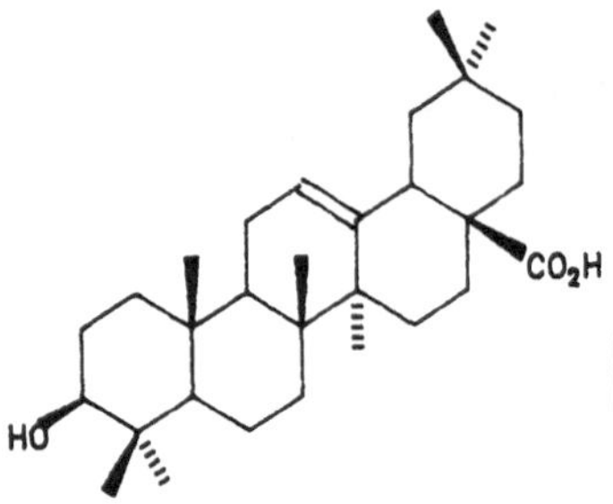

10· Oleanolic Acid—type 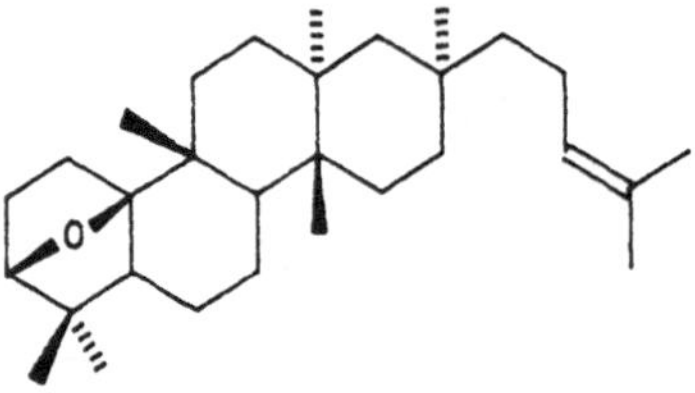   11·

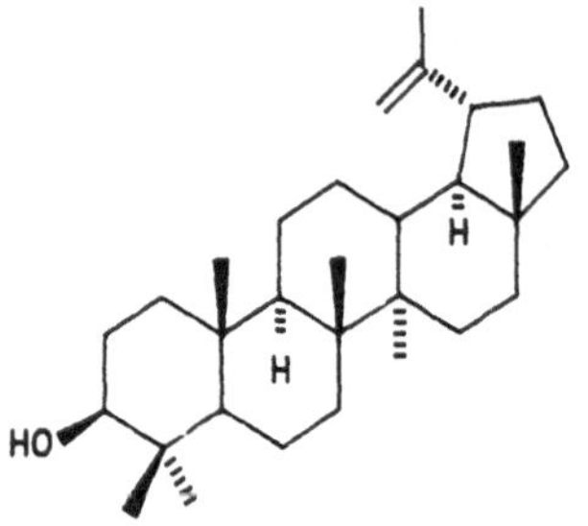

12· Lupeol—type 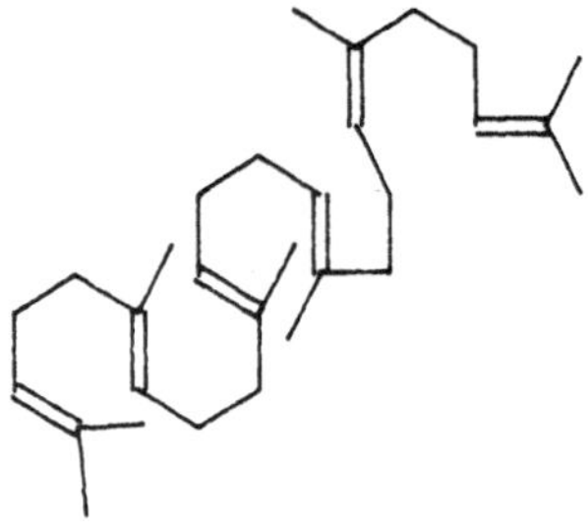   13· Squalene

SESQUITERPENES:

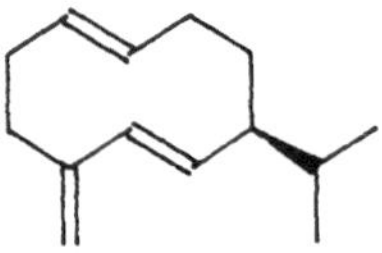

14· Germacrene—D—type 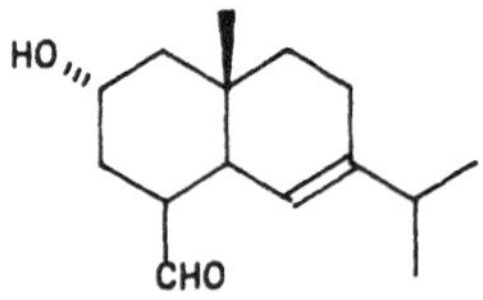   15· Eudesmane—type

FLAVONOIDS:

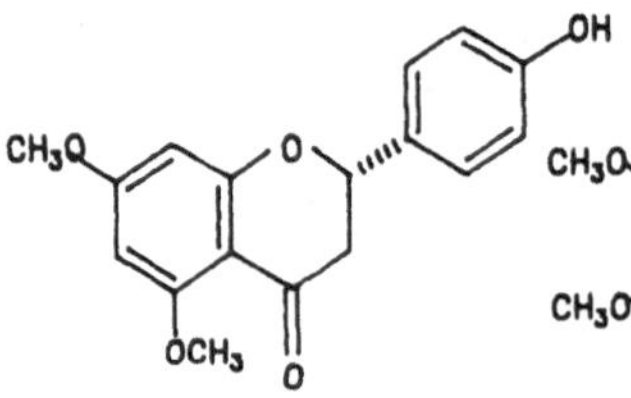

16· Flavonone—type 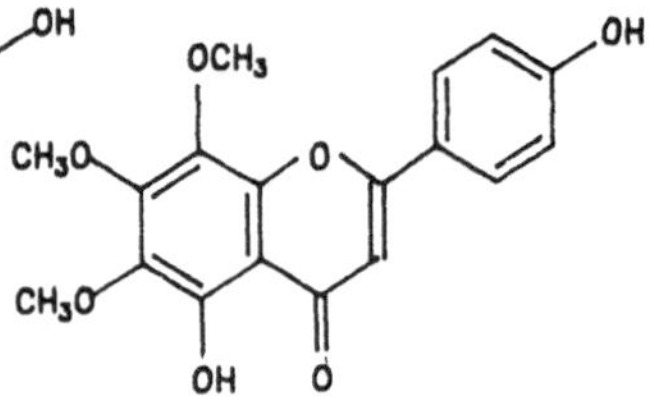   17· Flavone—type

FIGURE 22. (*contd.*).

Figure 23. Structures of Normal-Labdanes 469

ACETOPHENONE:

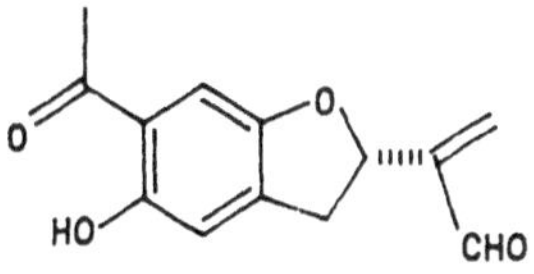

18·

COUMARINS: P–COUMARIC ACIDS

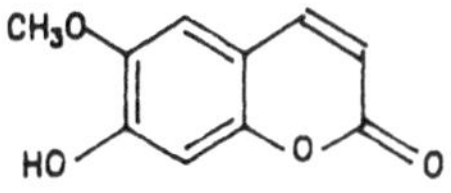

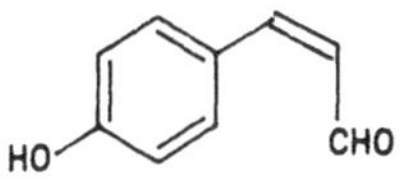

19·   21·

POLYACETYLENES:

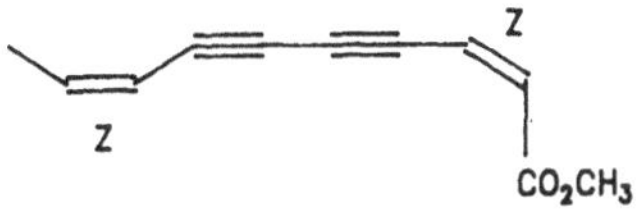

FIGURE 22. (*contd.*).

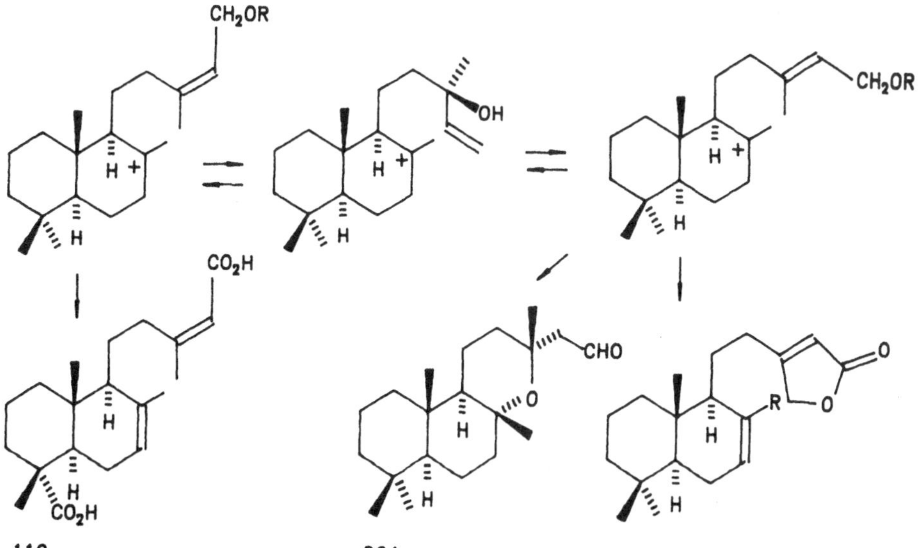

110   964   965 R = CH₂OH
                                       966 R = CHO

FIGURE 23. Structures of normal-labdanes reported from *Baccharis* (Astereae).

SCHEME A

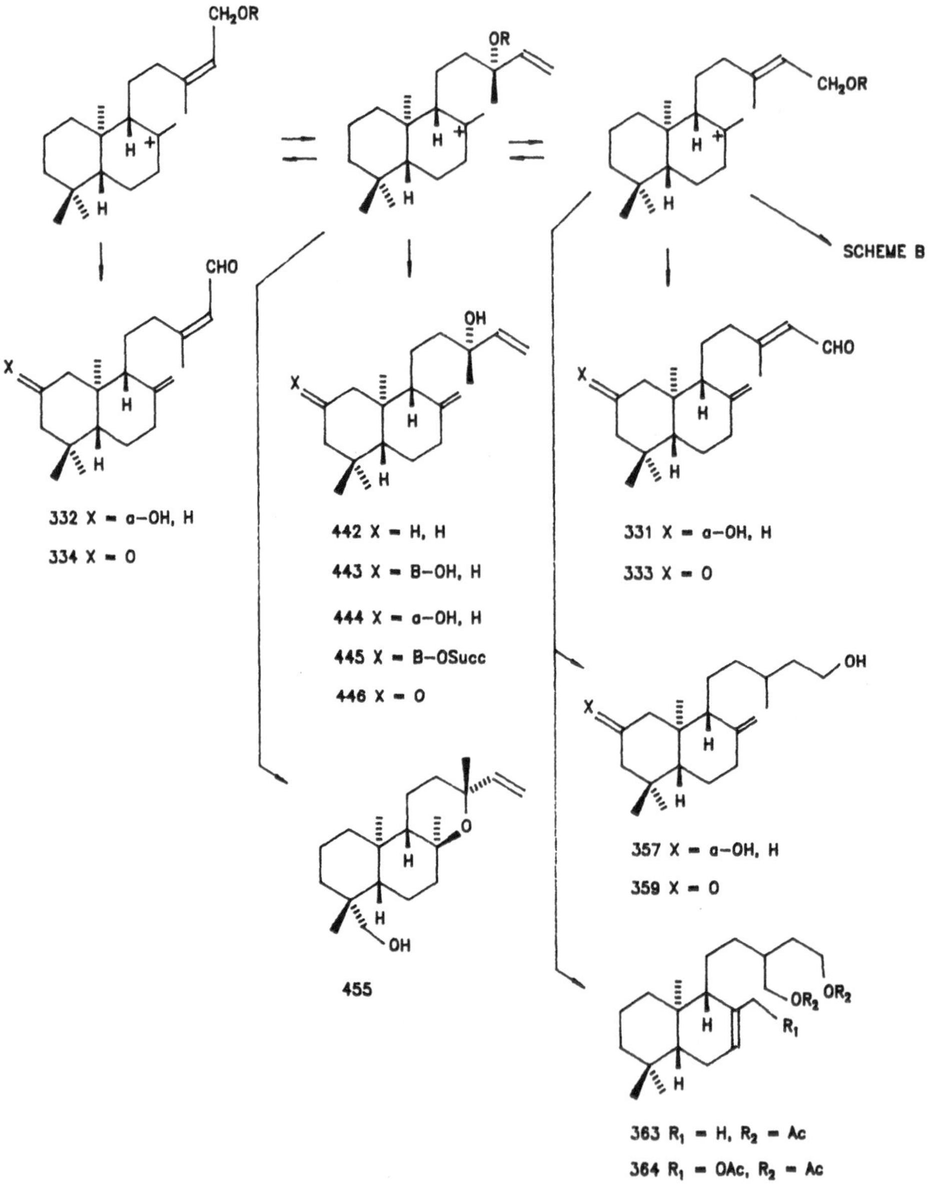
CH₂OR
OR
CH₂OR
SCHEME B
CHO
OH
CHO
X
332 X = α-OH, H
334 X = O
442 X = H, H
443 X = β-OH, H
444 X = α-OH, H
445 X = β-OSucc
446 X = O
331 X = α-OH, H
333 X = O
OH
357 X = α-OH, H
359 X = O
OH
455
OR₂ OR₂
R₁
363 R₁ = H, R₂ = Ac
364 R₁ = OAc, R₂ = Ac

FIGURE 24. Structures of *ent*-labdanes (Scheme A) and related *ent*-clerodanes (Schemes B, C, and D) reported from *Baccharis* (Astereae).

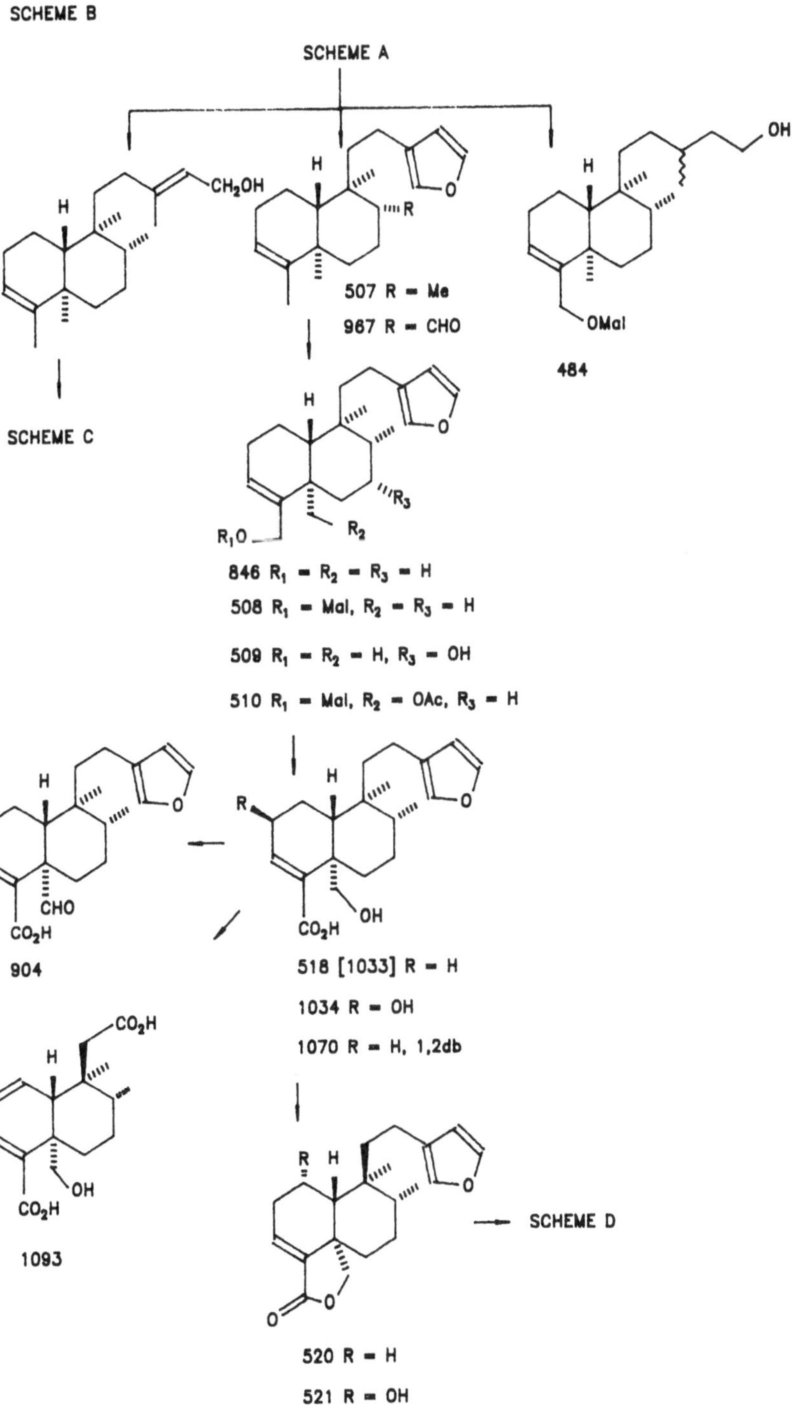

Figure 24. Structures of *ent*-Labdanes and *ent*-Clerodanes 471

SCHEME B

SCHEME A

SCHEME C

507 R = Me
967 R = CHO

484

846 R₁ = R₂ = R₃ = H

846 R$_1$ = R$_2$ = R$_3$ = H

508 R$_1$ = Mal, R$_2$ = R$_3$ = H

509 R$_1$ = R$_2$ = H, R$_3$ = OH

510 R$_1$ = Mal, R$_2$ = OAc, R$_3$ = H

904

518 [1033] R = H

1034 R = OH

1070 R = H, 1,2db

1093

520 R = H

521 R = OH

SCHEME D

FIGURE 24. (*contd.*).

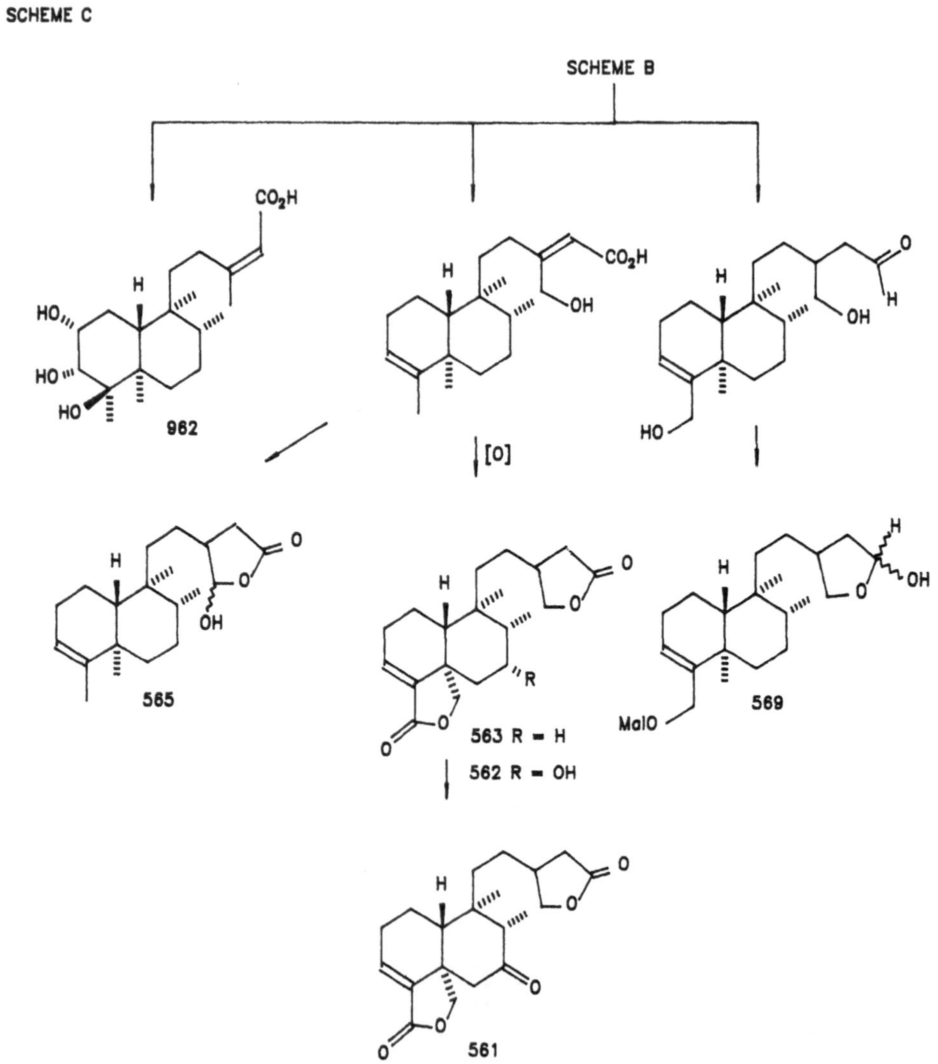

SCHEME C

SCHEME B

FIGURE 24. (contd.).

Figure 24. Structures of *ent*-Labdanes and *ent*-Clerodanes   473

SCHEME D

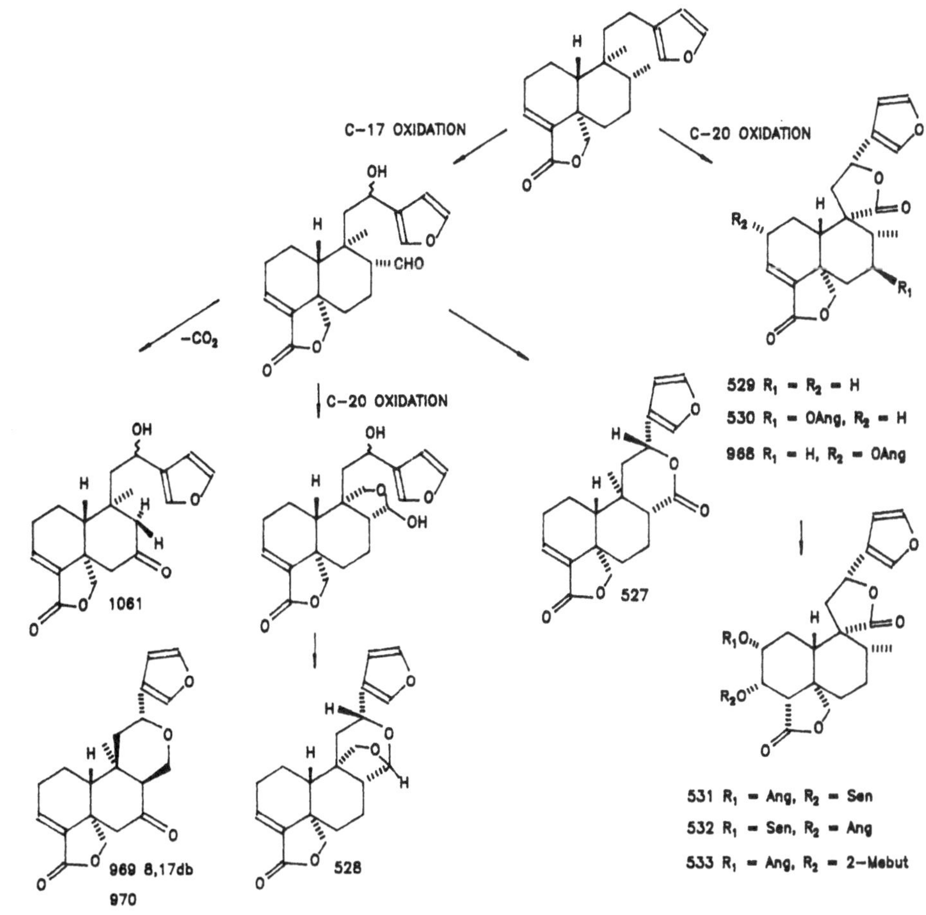

C-17 OXIDATION

C-20 OXIDATION

-CO₂

C-20 OXIDATION

529 R₁ = R₂ = H
530 R₁ = OAng, R₂ = H
968 R₁ = H, R₂ = OAng

1061

527

531 R₁ = Ang, R₂ = Sen
532 R₁ = Sen, R₂ = Ang
533 R₁ = Ang, R₂ = 2-Mebut

969 8,17db
970

528

FIGURE 24. (*contd.*).

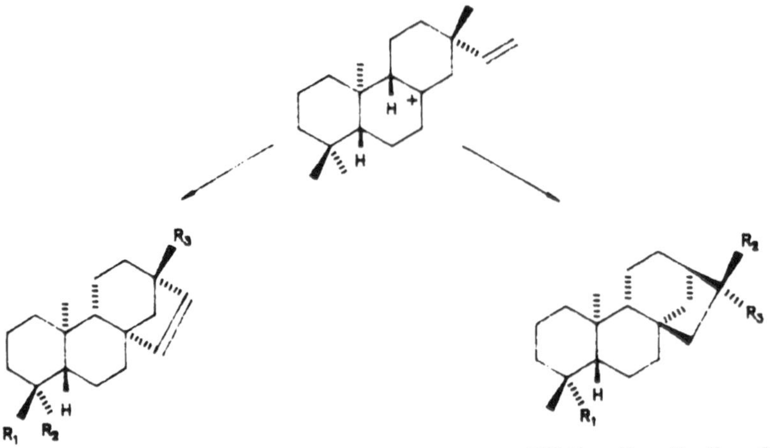

765 R₁ — Me, R₂ — CH₂OH, R₃ — Me
766 R₁ — CH₂OH, R₂ — R₃ — Me
771 R₁ — Me, R₂ — R₃ — CH₂OH
772 R₁ — R₂ — CH₂OH, R₃ — Me
782 R₁ — Me, R₂ — OH, R₃ — Me

720 R₁ — R₂ — Me, R₃ — OH
729 R₁ — Me, R₂ — H, R₃ — CHO
730 R₁ — CH₂OH, R₂ — H, R₃ — CHO
731 R₁ — Me, R₂ — H, R₃ — CO₂H
732 R₁ — CH₂OH, R₂ — H, R₃ — CO₂H
736 R₁ — CH₂OH, R₂ — H, R₃ — COMe
737 R₁ — CHO, R₂ — H, R₃ — CHO
738 R₁ — CO₂H, R₂ — H, R₃ — CH₂OH

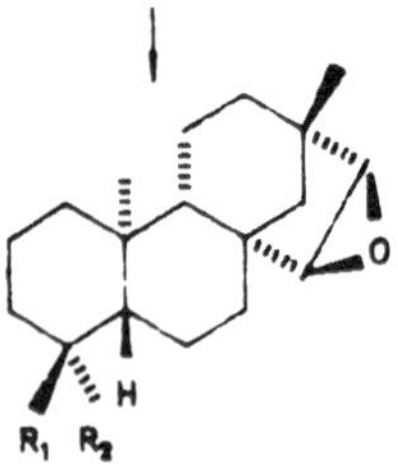

779 R₁ — CH₂OH, R₂ — Me
780 R₁ — Me, R₂ — CH₂OH

FIGURE 25. Structures of *ent*-stachane and *ent*-kauranoid diterpenes isolated from *Baccharis* (Astereae).

Figure 25. Structures of *ent*-Stachane and *ent*-Kauranoid Diterpenes   475

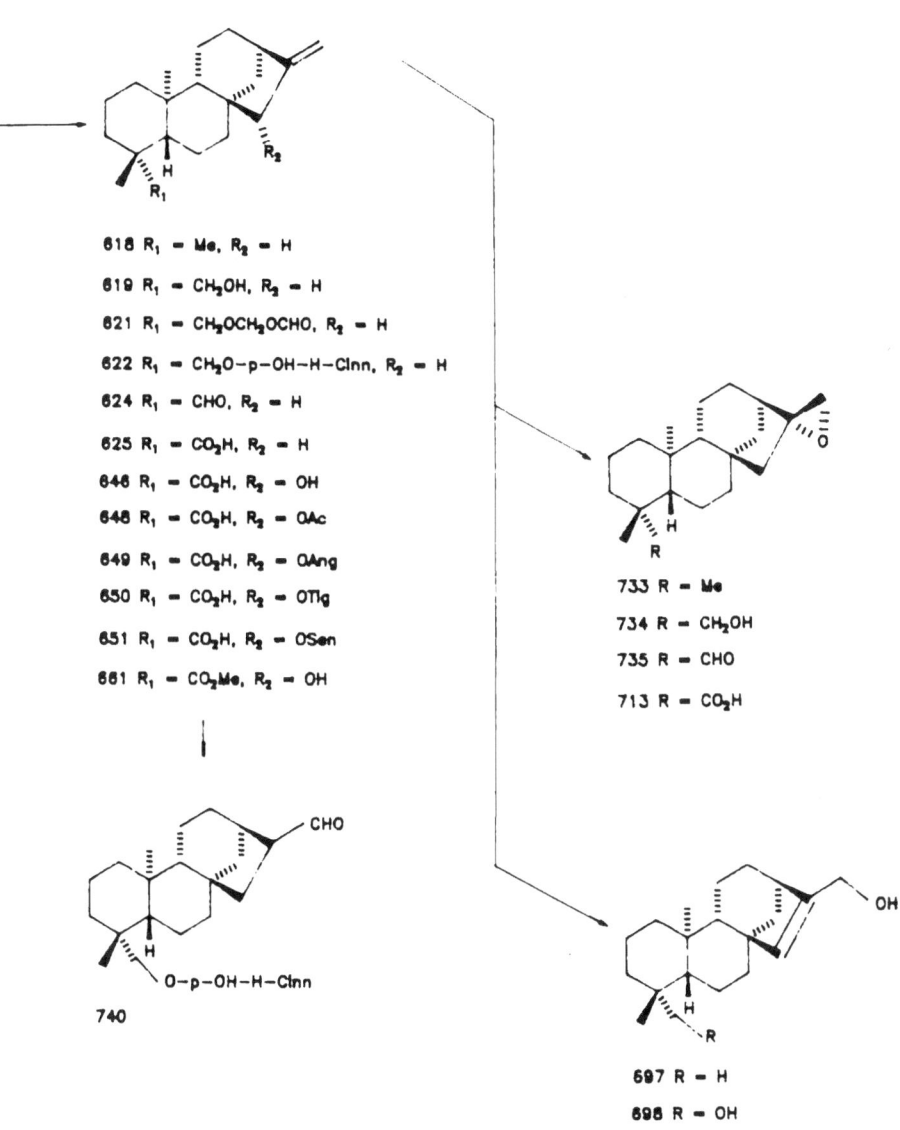

618 R$_1$ — Me, R$_2$ — H

619 R$_1$ — CH$_2$OH, R$_2$ — H

621 R$_1$ — CH$_2$OCH$_2$OCHO, R$_2$ — H

622 R$_1$ — CH$_2$O–p–OH–H–Cinn, R$_2$ — H

624 R$_1$ — CHO, R$_2$ — H

625 R$_1$ — CO$_2$H, R$_2$ — H

646 R$_1$ — CO$_2$H, R$_2$ — OH

648 R$_1$ — CO$_2$H, R$_2$ — OAc

649 R$_1$ — CO$_2$H, R$_2$ — OAng

650 R$_1$ — CO$_2$H, R$_2$ — OTig

651 R$_1$ — CO$_2$H, R$_2$ — OSen

661 R$_1$ — CO$_2$Me, R$_2$ — OH

733 R — Me

734 R — CH$_2$OH

735 R — CHO

713 R — CO$_2$H

CHO

O–p–OH–H–Cinn

740

OH

697 R — H

698 R — OH

699 R — O–p–OH–H–Cinn

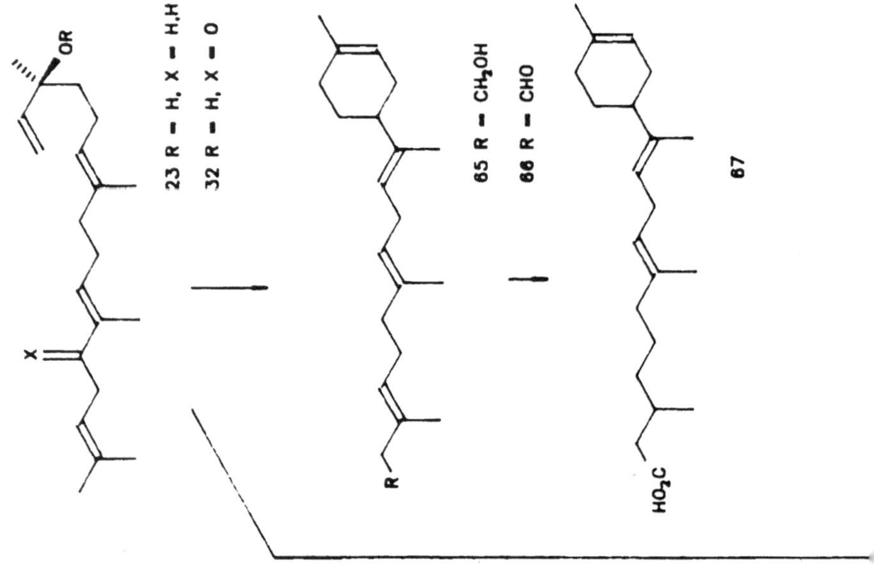
23 R = H, X = H,H
32 R = H, X = O
OR
X
65 R = CH2OH
66 R = CHO
R
67
HO2C

Figure 26. Structures of Diterpenes 477

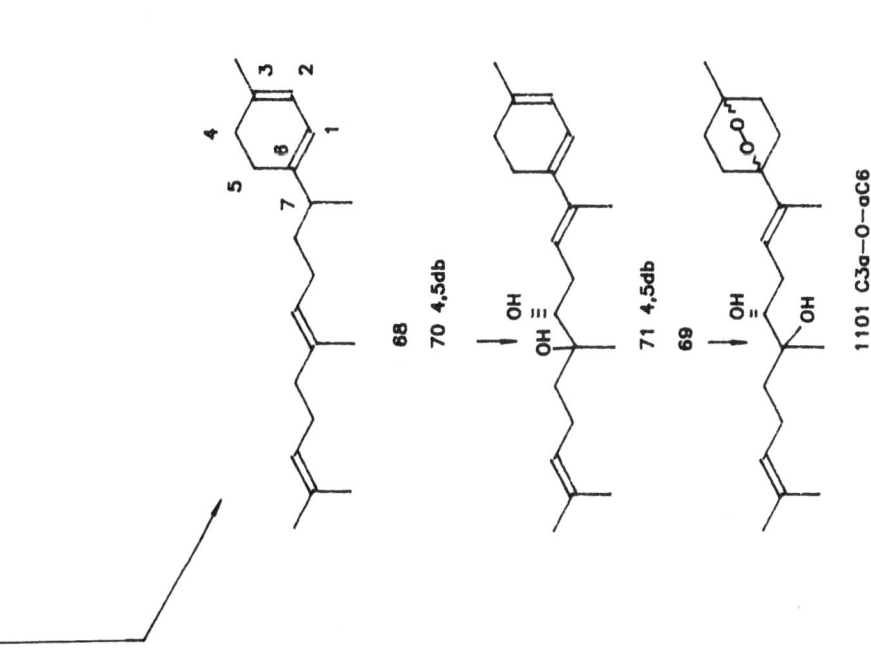

FIGURE 26. Structures of linear (**23**, **32**), unicyclic (**65–72**, **1101**), and macrocyclic (isocembrene, **73**) diterpenes reported from *Helichrysum* (Inuleae).

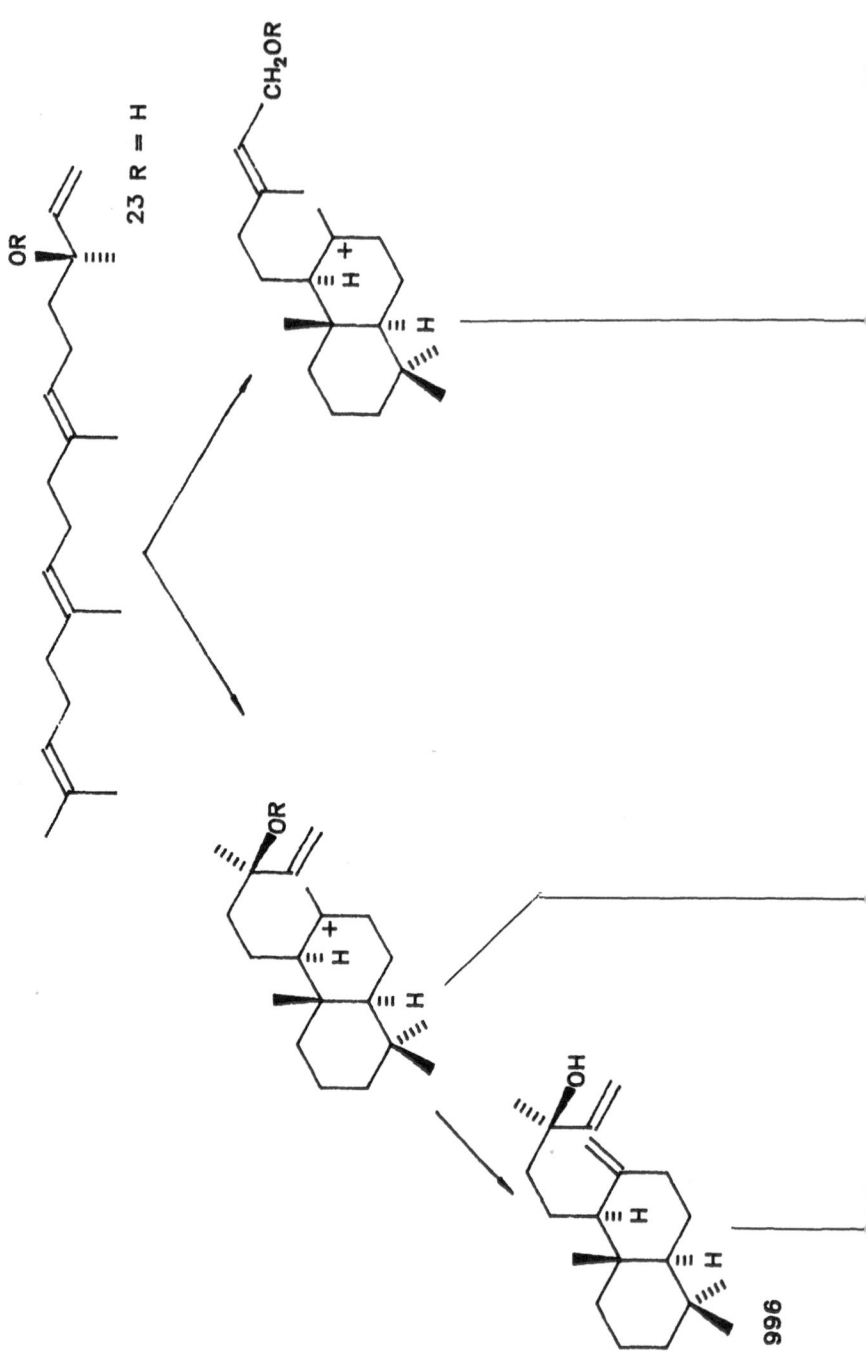

23 R = H

Figure 27. Structures of Normal-Labdane Diterpenes 479

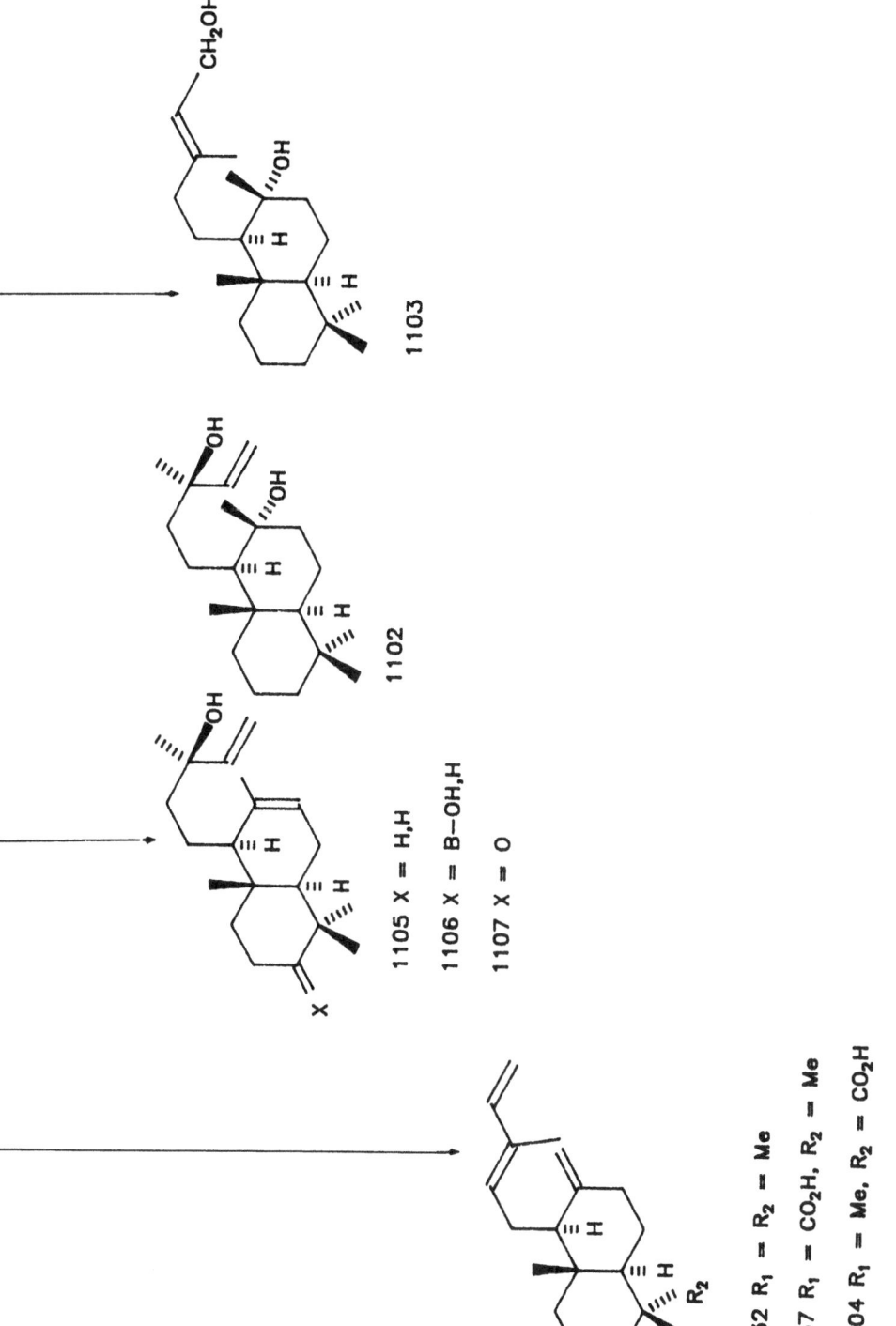

1103

1102

1105 X = H,H
1106 X = B-OH,H
1107 X = O

132 R₁ = R₂ = Me

137 R₁ = CO₂H, R₂ = Me

1104 R₁ = Me, R₂ = CO₂H

FIGURE 27. Structures of normal-labdane diterpenes isolated from *Helichrysum* (Inuleae).

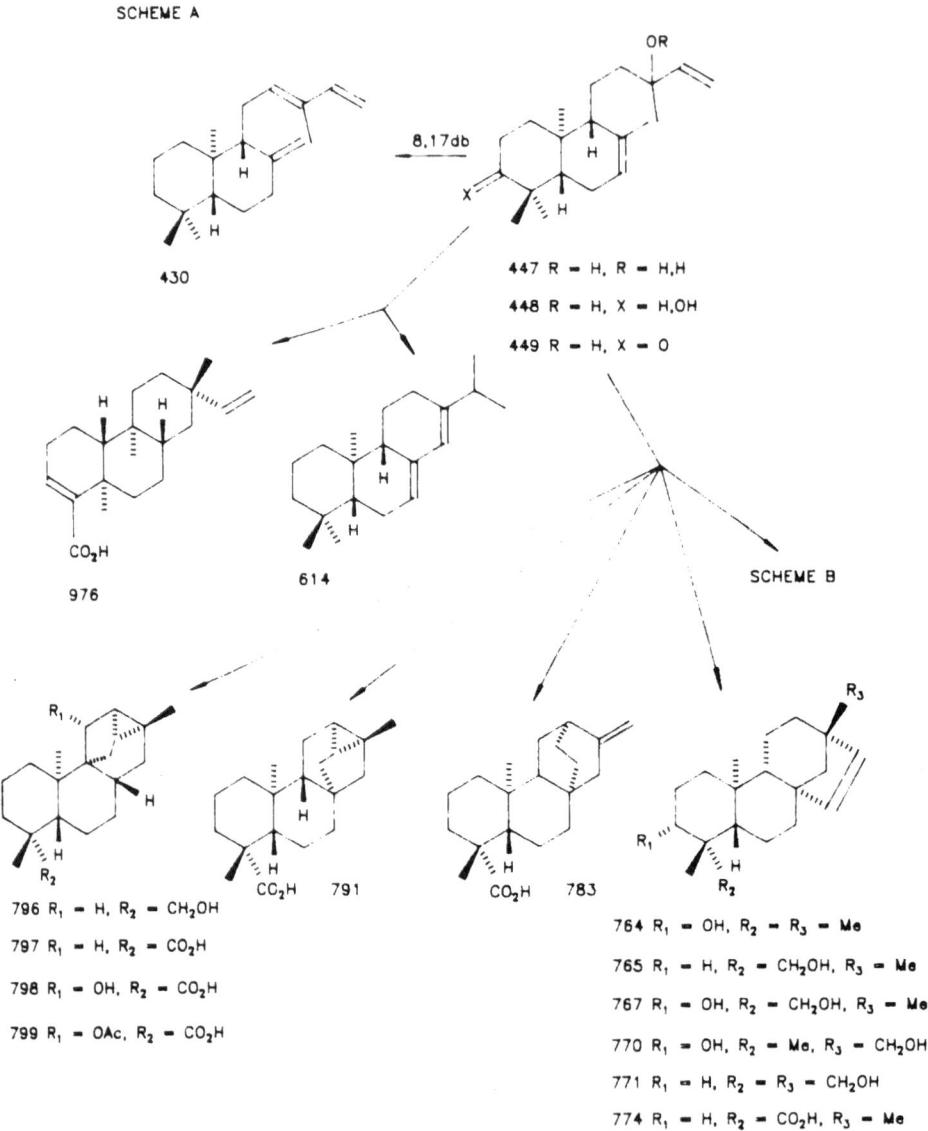

FIGURE 28. Structures of *ent*-labdane (**430, 447–449**), *ent*-pimarane (**976**), *ent*-abietane (**614**), *ent*-helifulvane (**796–799**), *ent*-trachylobane (**791**), *ent*-atisirane (**783**), *ent*-stachane (**764, 765, 767, 770, 771, 774**) (Scheme A), and *ent*-kauranoid (Scheme B) diterpenes reported from *Helichrysum* (Inuleae).

Figure 28. Structures of *ent*-Labdane and Other Diterpenes   481

SCHEME B

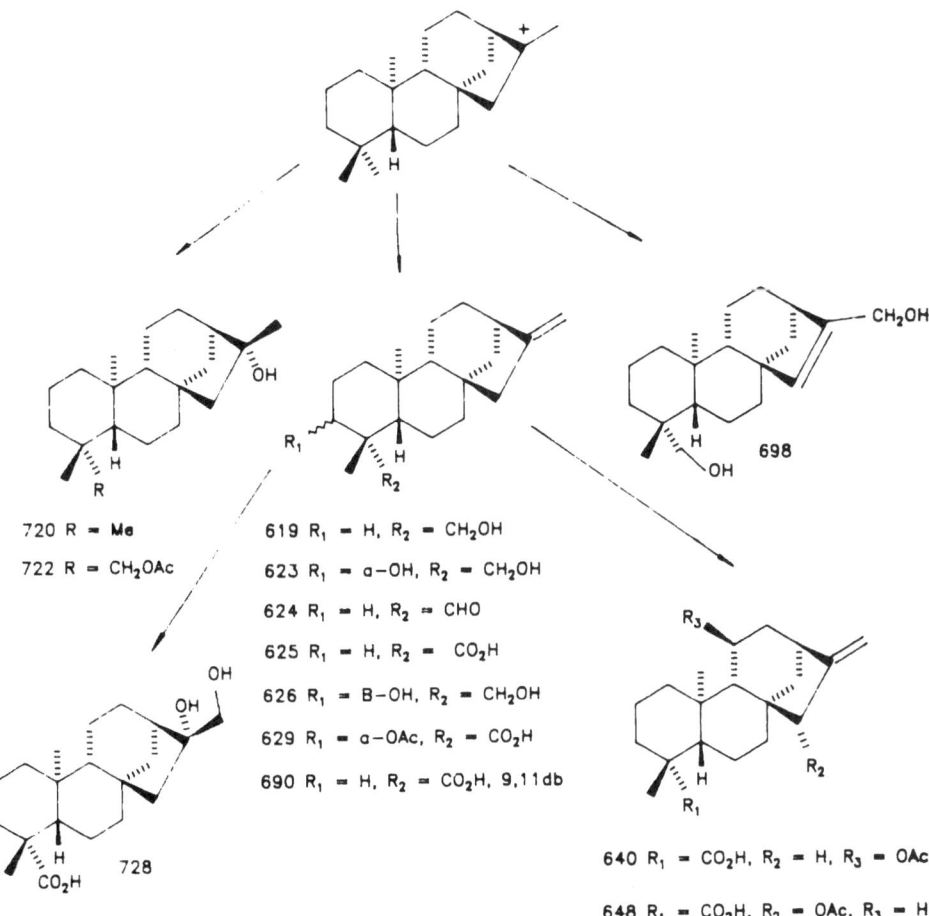

720 R = Me

722 R = CH₂OAc

619 R₁ = H, R₂ = CH₂OH

623 R₁ = α-OH, R₂ = CH₂OH

624 R₁ = H, R₂ = CHO

625 R₁ = H, R₂ = CO₂H

626 R₁ = β-OH, R₂ = CH₂OH

629 R₁ = α-OAc, R₂ = CO₂H

690 R₁ = H, R₂ = CO₂H, 9,11db

698

728

640 R₁ = CO₂H, R₂ = H, R₃ = OAc

648 R₁ = CO₂H, R₂ = OAc, R₃ = H

ENT–LABDANES

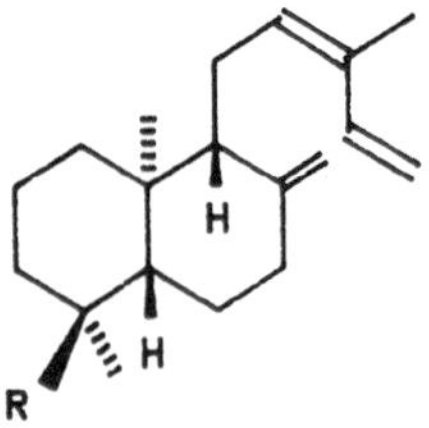

1· R = CH₃

2· R = CHO

3· R = CO₂H

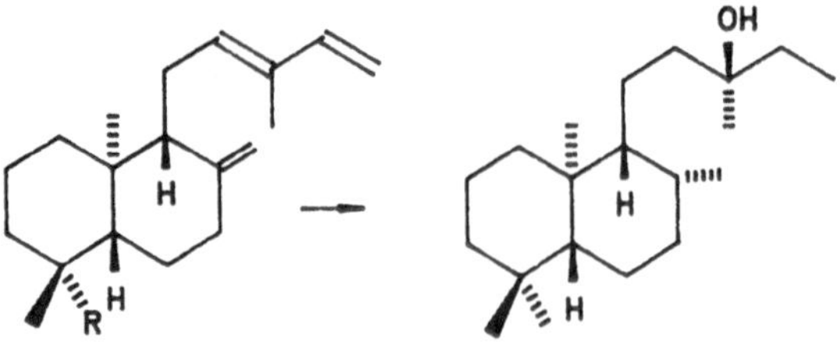

4· R = CH₃

5· R = CO₂H

6·

ENT–PIMARANES

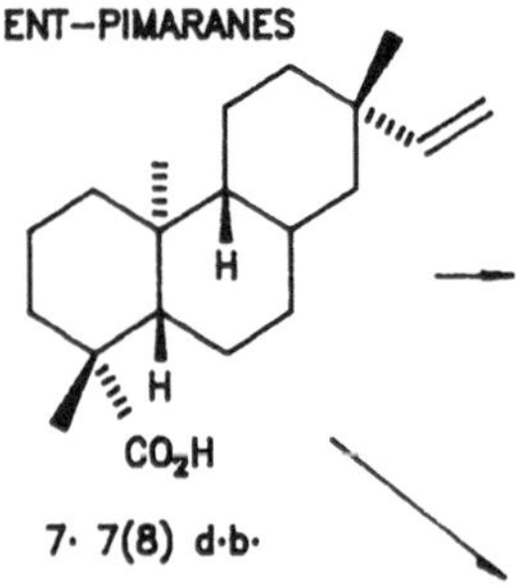

7· 7(8) d·b·

8· 8(9) d·b·

9· 9(11) d·b·

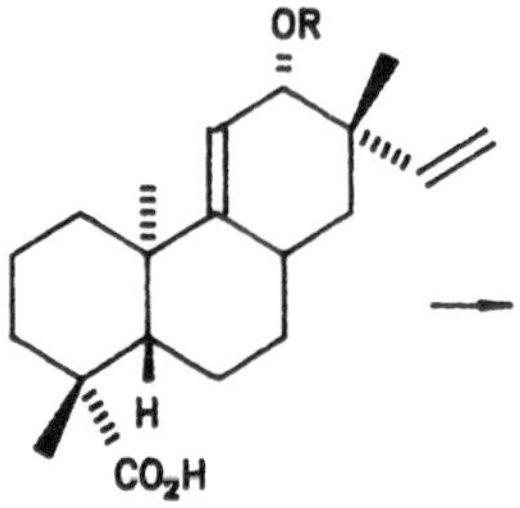

11·

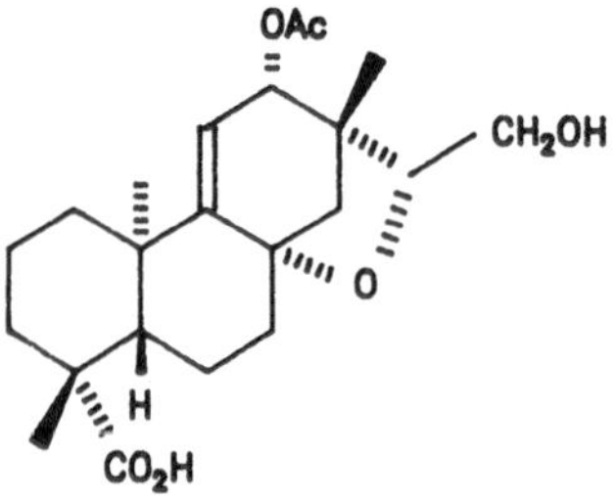

12·

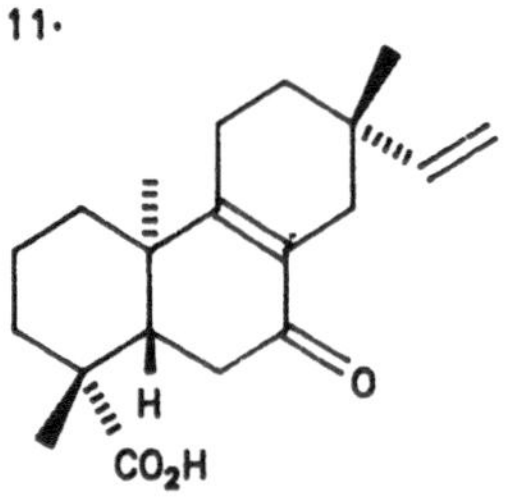

10·

ENT–STACHANES

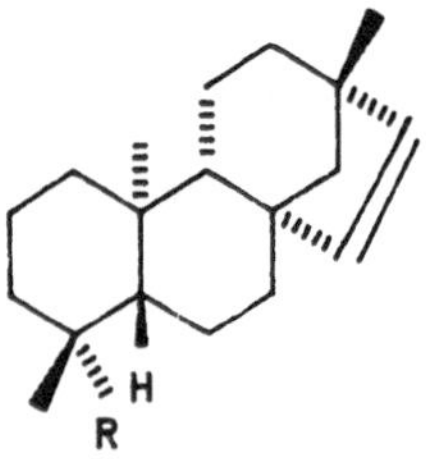

13· R = CH₂OH

14· R = CHO

15· R = CO₂H

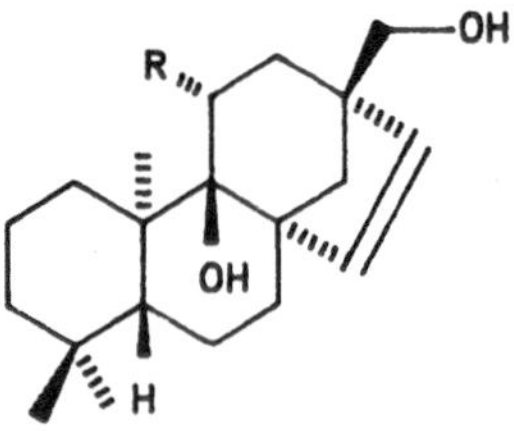

16· R = H

17· R = OH

FIGURE 29. Major diterpenes isolated from *Helianthus* (Heliantheae). The numbers assigned to the compounds correspond to the numbers listed in Table 11.

Figure 29. Major Diterpenes 483

ENT-KAURANES

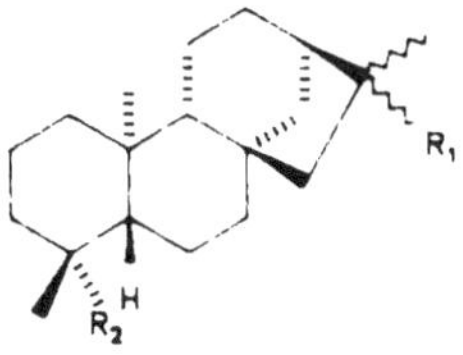

18. $R_1 = a-OH$, $R_2 = CH_3$
19. $R_1 = a-OH$, $R_2 = CH_2OH$
20. $R_1 = a-OH$, $R_2 = CO_2H$
21. $R_1 = B-OH$, $R_2 = CH_3$
22. $R_1 = B-OH$, $R_2 = CO_2H$

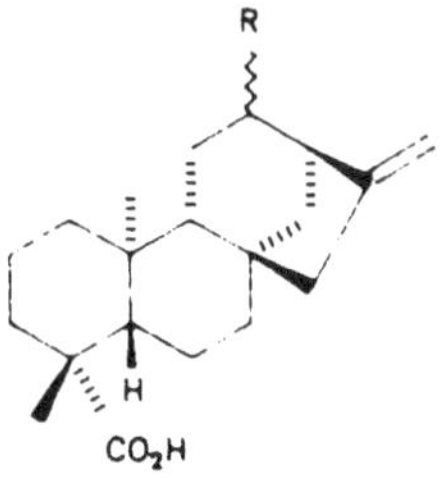

23. R = H
24. 9(11)d·b·, R = H
25. 9(11)d·b·, R = B-OH
26. 9(11)d·b·, R = a-OH
27. 9(11)d·b·, R = =O

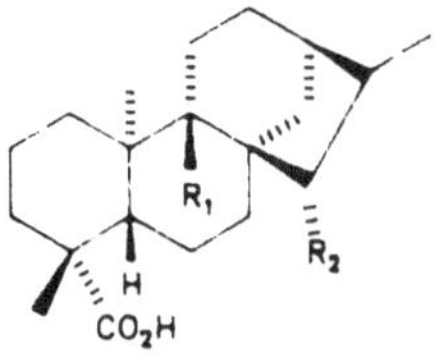

28. $R_1 = H$, $R_2 = OH$
29. $R_1 = H$, $R_2 = OAng,OTig$, etc·
30. $R_1 = R_2 = OH$

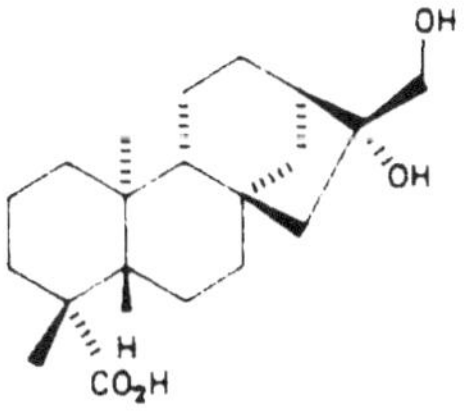

31.

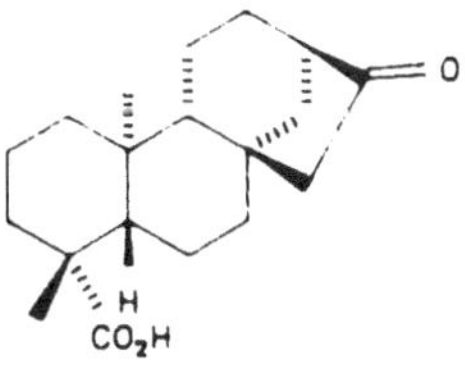

32.

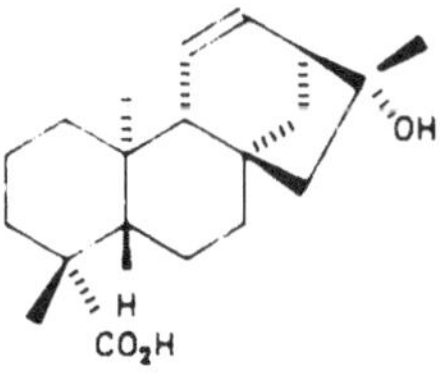

33.

TETRACHYRANE

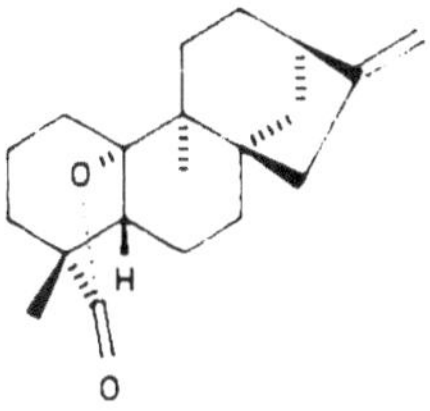

34.

ISOKAURANE

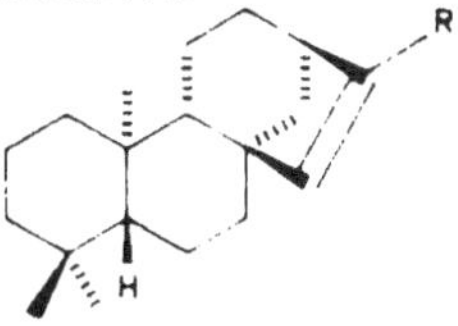

35. R = CH_2OH
36. R = CHO
37. R = CO_2H

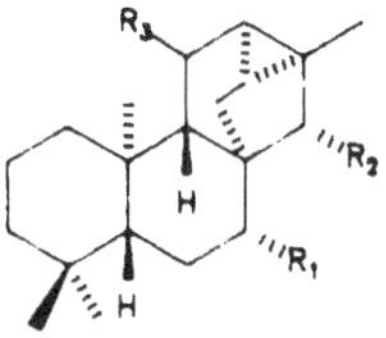

38. $R_1 = R_2 = R_3 = H$
39. $R_1 = OH$, $R_2 = R_3 = H$
40. $R_1 = H$, $R_2 = OH$, $R_3 = H$
41. $R_1 = R_2 = H$, $R_3 = =O$

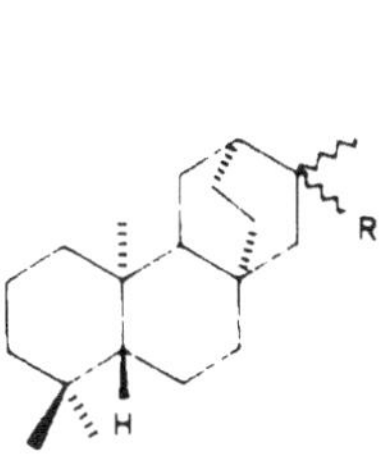

42. R = a-OH
43. R = B-OH

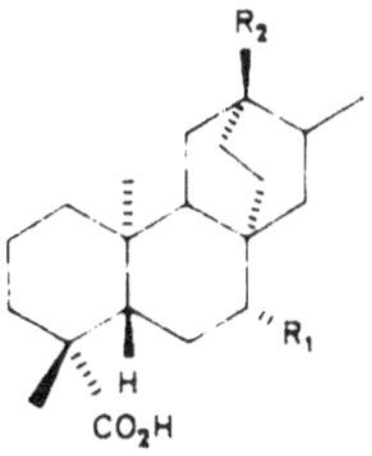

44. $R_1 = OH$, $R_2 = H$
45. $R_1 = H$, $R_2 = OH$

FIGURE 29. (contd.).

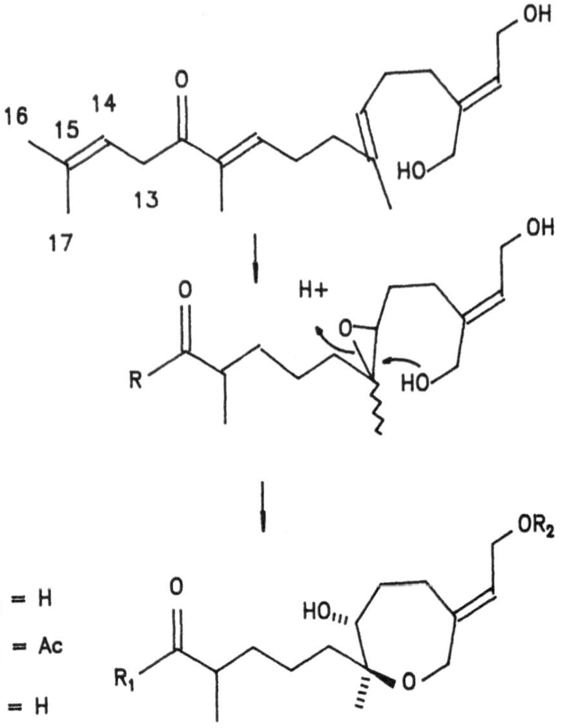

1094 R = D

1095 R = B

1096 R = A

55 R_1 = A, R_2 = H

56 R_1 = B, R_2 = Ac

57 R_1 = E, R_2 = H

1097 R_1 = D, R_2 = H

1098 R_1 = B, R_2 = H

1099 R_1 = C, R_2 = H

A = 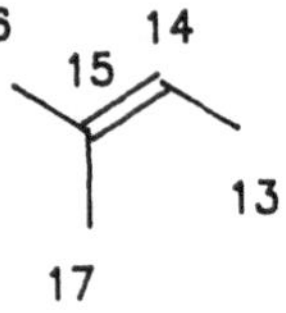

C = 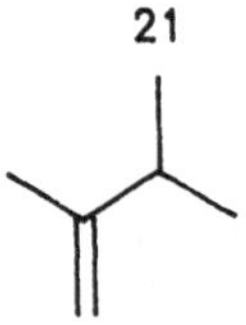

E = 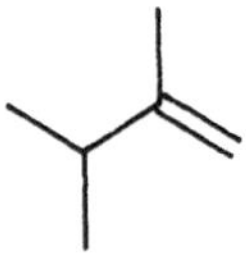

B = 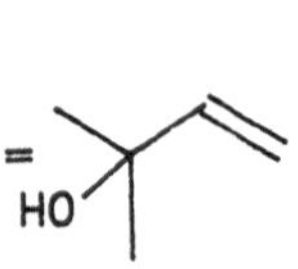

D = 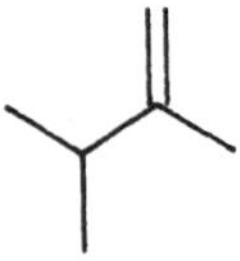

FIGURE 30. Structures of oxepane diterpenes (**55–57**, **1097–1099**) and their linear precursor-like compounds (**1094–1096**) reported from *Montanoa* (Heliantheae).

CHAPTER 5

Biological Activity of Diterpenes

V.1. Introduction

Various biological activities (Table 13 [pp. 489–490]) have been reported for plant diterpenes, principally from members of the Ericaceae, Euphorbiaceae, Lamiaceae, and Compositae. Each of the major diterpenoid classes from this last family will be discussed below in terms of its biological properties.

V.2. Linear Diterpenes

Linear diterpenes derived from geranylnerol, geranylgeraniol, and geranyllinalool occur commonly within the Compositae. The most diverse group of linear diterpenes is derived from geranylnerol and includes lactonic, oxepane, and furanoid members as well as unicyclic and macrocyclic derivatives. In addition to the previously mentioned *Montanoa* oxepane diterpene series, oxepanes and lactones have been detected in *Acanthosperum*, *Ichthyothere*, *Melampodium*, and other genera mostly within the Heliantheae.

For hundreds of years, a tea brewed from the leaves of *Montanoa tomentosa* has been used medicinally to induce labor and as an antifertility agent (*419, 382*). The *M. tomentosa* extract possessed abortifacient, labor induction, and menstruation induction effects (*372, 297*). Zoapatanol (55) appears to be the tea component responsible for most of the activity (*368*). In addition to zoapatanol, several other oxepane diterpenes (**56, 57, 1097, 1098, 1099**) and their uncyclized precursors (**1094, 1095, 1096**) were isolated from this species. Commercial interest in these compounds has led to the synthesis of analogous structures sharing similar pharmacological properties (*353*).

V.3. Bicyclic Diterpenes

Grindelane normal-labdanes **185** and **188** isolated from *Grindelia*, *Chrysothamnus*, *Solidago*, and *Haplopappus* displayed significant antifeedant activity against the Colorado potato beetle (*440*) and the aphid *Schizaphis graminum*

(*441*). The anatomical distribution of grindelanes was examined in *Grindelia camporum* and the development of anatomically defined chemistries was traced over time. This work led to the observations that most diterpene resin accumulated on the plant surface and the quantity of accumulated cuticular resin was positively correlated with the number of multicellular epidermal resin glands (*340*).

Surface resin production in *G. camporum* and other resin-producing xerophytic plants (e.g., *Chrysothamnus*, *Gutierrezia* and *Xanthocephalum*) may serve at least three (not mutually-exclusive) functions: (1) when coating the leaf surface, the resin makes the cuticle less permeable to water and thus reduces water loss, (2) the resin coating increases the reflectance of the leaf surface thereby reducing the amount of solar radiation penetrating the surface and, consequently, lowering the internal temperature, and (3) the resin deters herbivores and/or plant pathogens (*340*).

Although the plant source, *Nicotiana glutinosa*, is a member of the Solanaceae not the Compositae, the major leaf-surface-resin normal-labdane constitutents, sclareol (**1102**) and 13-epi-sclareol, either have been isolated from the Compositae (sclareol) or resemble other normal-labdanes isolated from this family. When tested against a variety of fungi, these *Nicotiana* compounds significantly inhibited the radial growth rate of common fungal colonies. Radial growth was apparently retarded by the effect of diterpenes on the growth-regulatory processes controlling fungal branching (*24*).

Solidago clerodanes are frequently referred to as the "bitter principal" responsible for the characteristic taste of the root tissue. For example, solidagonic acid (**500**) was described as a bitter principle from the roots of *Solidago altissima* (*365*).

V.4. Tricyclic Diterpenes

When each of the diterpene resin acids, pimaric acid, dihydropimaric acid, sandaracopimaric acid, isopimaric acid, neoabietic acid, levopimaric acid, palustric acid, and abietic acid (Figure 31 [p. 491]) was included in the artificial diet of the pink bollworm, *Pectinophora gossypiella*, larval growth was inhibited. In order to ascertain the importance of the compounds' C-19 acid function in growth inhibition, methyl isopimarate and isopimarol, (Figure 31) were tested under the same conditions and found to be substantially less active than the free acid. Isopimarol had virtually no effect on larval growth.

Resemblance of the tricyclic diterpene skeleton to the ABC-ring system of common steroids prompted the suggestion that the diterpenes interfered with the pink bollworm's steroid metabolism. Insect hormones are produced from dietary phytosterols which are absorbed, transported, and biochemically altered. Interference by the diterpene resin acids at any stage of this process could significantly reduce larval viability. To test this hypothesis, feeding studies were conducted using a levopimaric acid test diet with added cholesterol (addition of cholesterol diminishes the insect's requirement for phytosterols). In comparison to the control diet that contained 0.2 percent levopimaric acid and no cholesterol, test diets containing levopimaric acid and steadily incremented cholesterol concentrations from 0.05 to 1.0 percent produced corresponding increments (increases) in larval growth. By 1.0 percent cholesterol, the larval weight had reached 86 percent of larvae grown on an artificial diet containing no levopimaric acid, suggesting that added cholesterol canceled the deleterious effect of the diterpene. It was concluded that reduced larval growth resulted from the antagonism between resin acids and steroids and that the insect's hormonal system was affected (*258*).

V.5. Tetracyclic Diterpenes

Feeding tests using kaur-16-en-19-oic acid (**625**) and trachyloban-19-oic acid (**791**) from the florets of sunflower (*Helianthus annuus*) varieties resistant to the sunflower moth, *Homoeosoma electellum*, produced significant reduction in this insect's larval growth (*491*). Resistance of these varieties to seed predation by the sunflower moth was attributed to the relatively high concentration of these compounds in the sunflower's seed-bearing florets. Further studies (*332*) showed that when other *Helianthus* diterpenes, ciliaric acid (**792**) and angelylgrandifloric acid (**649**), were incorporated (one percent) in the diet of sunflower moth larvae, the larvae experienced a significantly higher level of mortality than those larvae feeding on the diet lacking the diterpenes. The compounds did not affect the length of the development period but did appreciably reduce the size of seven-day-old larvae.

Kaur-16-en-19-oic acid (**625**) obtained from *Mikania monagasensis* inhibited the growth of *Staphylococcus aureus* and *Candida albicans* (*384*). In the latter case, this diterpene was appreciably less active than the co-occurring sesquiterpene lactone, mikanolide.

Diterpenes of *Wedelia* (Heliantheae) were responsible for this plant's toxicity to sheep and cattle. The first of two kaurenoid glycosides, wedeloside (**679**) and atractyloside (**681**), caused the severe loss of sheep in Queensland, Australia, due to grazing on *Wedelia asperrima* (*408*), while the second proved fatal to South American cattle grazing on *W. glauca* (*456*). Investigation of *W. buphthalmiflora*, another South American shrub toxic to cattle, (using extraction methods unlikely to isolate kaurenoid glycosides), revealed a complex chemistry consisting of kaur-9(11),16-dien-19-oic acid (**690**), kaur-16-en-19-oic acid (**625**), and a variety of grandifloric acid (**646**) esters (*455*). One of these esters, 15α-tiglinoyloxy-kaur-16-en-19-oic acid (**650**), isolated from *W. scaberrima* was miracidal (toxic to the miracidia stage of *Schistosoma mansoni*). Kaur-16-en-19-oic acid was molluscicidal against *Biomphalaria glabata* (*476*).

The sulfated glycoside, carboxyatractyloside (**678**), was reported from two sources, *Xanthium strumarium* (*232*) and *Atractylis gummifera* (*239*). This hypoglycemic compound strongly inhibits translocation of adenine nucleotides across the mitochondrial membrane and shows an *in vivo* toxicity ten times greater than that of atractyloside (**681**, which co-occurs in *A. gummifera*) (*239*). Carboxyatractyloside was responsible for cocklebur poisoning of pigs in the southern United States (*238*). Later work on the inhibitors responsible for the dormancy properties of *Xanthium* seeds identified carboxyatractyloside as a potent plant growth inhibitor (*238*).

Sweet-tasting *ent*-kaurene glycosides, stevioside (**683**), rebaudiosides A (**685**), B (**684**), C, D (**688**) and E (**689**), dulcoside A (**686**), and steviolbioside (**682**) (See Figure 21), have been isolated from the New World genus *Stevia* (*359, 464*). Eighteen sweet-tasting *Stevia* taxa have been identified from the total of 121 tested taxa (*464*). Stevioside, which is estimated to be three hundred-fold sweeter than sucrose, is produced commercially as a sweetening agent for desserts, soft drinks, sea foods, pickled vegetables, and so on. Although no evidence indicates that stevioside is harmful to humans, the likely hydrolysis product, steviol (Figure 32 [p. 492]), is metabolized to mutagenic structures by human enzymes and appears to covalently interact with DNA. When steviol was incubated in the presence of rat-liver microsomes (S-9), the principal metabolic product was 15α-hydroxy-steviol. Testing of this latter compound did not reveal any mutagenicity. However, 15-oxo-steviol, a closely-related derivative (Figure 32) was a direct-acting mutagen and a bactericide (*425*).

The role of diterpenes in the resistance of four *Solidago* (Astereae) species to the beetle *Trirhabda* (Coleoptera) is under investigation (*368*). Preliminary results indicate that three of the four kauranoids isolated from the leaves of *S. nemoralis* and *S. altissima*, *ent*-kauran-16α-ol (**720**), 15α-hydroxy-*ent*-kaur-16-en-19-oic acid (**646**), and 17-hydroxy-*ent*-kaur-15-en-19-oic acid reduced *Trirhabda* larval growth by 40 percent and adult growth by 32 to 49 percent. The fourth, *ent*-kaur-16-en-19-oic acid (**625**), had no deterrent effect on the larvae and actually stimulated feeding by adults.

The diterpene alkaloids, lycoctonine (**1009**) and anthranoyllycoctonine (**1010**) appear to be responsible for the blood pressure depressant, smooth muscle relaxant and curareform properties of the source plant, *Inula royleana* (*34*).

Table 13. Plant Diterpene Biological Activities   489

TABLE 13. Plant diterpene biological activities.

Activity Compound	Source	Family	Reference
Antifeedant			
ajugarin I-III	Ajuga remota	Lamiaceae	305,306
grindelanes	Chrysothamnus nauseosus	Compositae	340
	Grindelia (7 spp.)	"	"
	Haplopappus sp.	"	"
ciliaric acid	Helianthus sp.	"	306
kalmitoxins	Kalmia latifolia	Ericaceae	259
grayanotoxins	" "	"	"
kaurenoic acids	Helianthus annuus	Compositae	258,306
trachylobanic acids	" "	"	" "
kaurenoid glycosides	Wedelia spp.	Compositae	408,455,456
	Xanthium strumarium	"	238
	Atractylis gummifera	"	239
teucjaponin A	Teucrium japonicum	Lamiaceae	304
Antifungal			
casbene	Ricinus communis	Euphorbiaceae	261
pseudolaric acid	Pseudolarix kaempferi	Pinaceae	306
sclareol,etc.	Nicotiana glutinosa	Solanaceae	24
Fish poison			
eremone	Emerocarpus setigerus		304
hautriwaic acid	"	"	"
Insecticidal			
ajugarin IV	Ajuga remota	Lamiaceae	305
Molluscicidal, etc.			
kaurenoic acid esters	Wedelia scaberria	Compositae	476
Antibacterial			
longikaurins	Rabdosia longituba	Lamiaceae	306
kaurenoic acid	Mikania monagasensis	Compositae	384

TABLE 13. (*contd.*).

Activity Compound	Source	Family	Reference
Antibiotic			
przewaquinone	Salvia przewalskii	Lamiaceae	304
sonderianol	Croton sonderianus	Euphorbiaceae	305
Antimicrobial			
7,18-dihydroxy-sandaracopimaradiene	Iboza riparia	Lamiaceae	304
Curareform			
lycoctonine,etc.	Inula royleana	Compositae	34
Antiviral			
bacchotricuneatin	Baccharis tricuneata	"	490
Oxytocic			
zoapatlin, etc.	Montanoa tomentosa	"	297,372
Plant growth inhibitor			
podolactone	Coniferae spp.	Coniferae	304
carboxyatrac-tyloside	Xanthium strumarium	Compositae	238
Sweetening agent			
kaurenoid glycosides	Stevia (18 spp.)	"	464

Figure 31. Diterpene Resin Acids   491

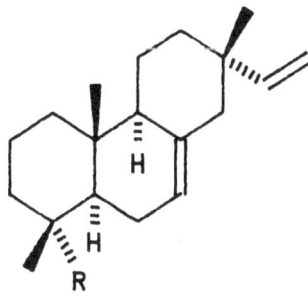

15   16

Pimaric Acid 15,16db

Dihydropimaric Acid

Sandaracopimaric Acid

Isopimarol R = CH₂OH

Isopimaric Acid R = CO₂H

Methyl Isopimarate R = CO₂Me

Neoabietic Acid

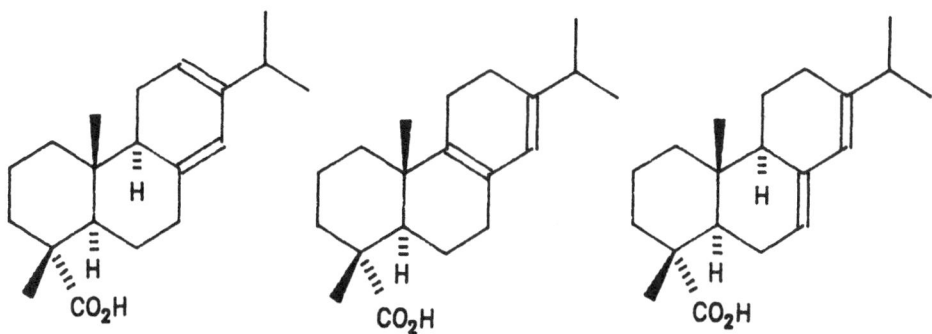

Levopimaric Acid   Palustric Acid   Abietic Acid

FIGURE 31. Diterpene resin acids tested for biological activity against pink bollworm (*Pectinophora gossypiella*) larvae.

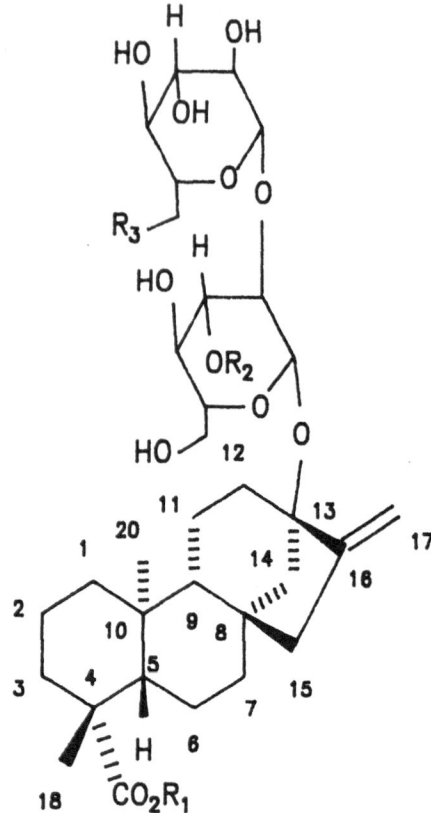

Stevioside R$_1$ = Glu, R$_2$ = H, R$_3$ = OH

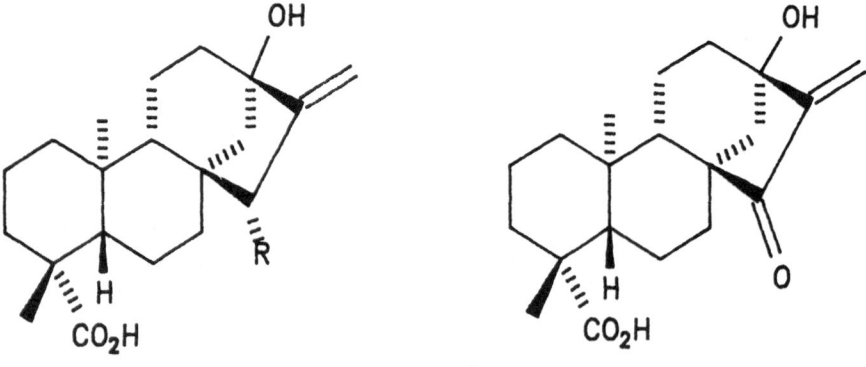

Steviol R = H

15α−Hydroxy−steviol R = OH

15−Oxo−steviol

FIGURE 32. *Stevia* diterpene, stevioside, its aglycone, steviol, and a mutagenic derivative, 15-oxo-steviol.

CHAPTER 6

Diterpene Analysis with Emphasis on Clerodanes

VI.1. Clerodanes

VI.1.1. Major Clerodane Types

VI.1.1.1. trans-ent-*Clerodanes (5α,10β-trans-Clerodanes)*

Early work on Compositae diterpenes focused on *Solidago* and the normal-labdane, solidagenone (Figure 33 [pp. 513–514]; Structure **158**) (*9, 16, 17*). Later, *ent*-clerodanes (kolavanes) were found to co-occur with the labdanes (*364, 365*). The first identified novel clerodane, solidagonic acid (Figure 33; Structure **500**), was shown by chemical interconversion (*364, 365*; Figure 33) to share the orientivity of the major chiral centers, C-5, C-8, C-9, and C10, with the co-occurring clerodanes of known stereochemistry, kolavenic acid (**495**) and kolavenol (Figure 33; Structure **475**).

Structural correlation of solidagonic acid with **495** and **475** was made feasible by the recent assignment by Misra and coworkers (*391*) of absolute stereochemistry to the chiral centers of the *Hardwickia pinnata* clerodanes, (-)-hardwickiic acid (**511**), kolavenic acid, and kolavenol (Figure 34 [p. 515]). Their pivotal work on clerodane stereochemistry proceeded as described below:

1. Chemical degradation of 511 led to a series of products that permitted assignment of the ring fusion configuration. The CD spectrum of derivative **511c** produced a strong Cotton effect which was almost a mirror-image of that produced by the steroid, 3β-hydroxy-5α-androstan-17-one. This suggested that the two rings of **511c** were *trans*-fused, but with an absolute stereochemistry opposite to that of the steroid's CD rings.
2. Orientivity at C-9 was established from the **511d** ^{1}H-NMR data which failed to indicate a shielding effect of the C-9 carbomethoxy group on the C-5 methyl. An α-oriented carbomethoxy group would have a 1,3-diaxial relationship with the C-5 methyl which would then be subject to deshielding. The absence of deshielding indicated a β-oriented sidechain and an α-oriented methyl at C-9.
3. The paramagnetic ^{1}H-NMR chemical shift (δ1.19) of the **511f** C-8 methyl signal (H-17) relative to the shift observed for the parent compound and related structures (c. δ0.83) was attributed to the deshielding influence of the

keto group. Misra argued that since a carbonyl group can deshield only those protons that lie in its plane, the C-8 methyl must be equatorial (α-oriented). The absolute stereochemistry of kolavenic acid and kolavenol follow from their relationship with (-)-hardwickiic acid (*391*).

Unfortunately, the early assumption of (-)-hardwickiic acid-based biogenetic uniformity among diterpenes of related *Solidago* species led to several incorrect stereochemical assignments: Anthonsen and McCrindle (*18*) isolated from *Solidago elongata* a series of clerodanes that co-occurred with kolavenol, kolavelool (**479**), kolavenic acid plus several angelate and acetate ester derivatives (**480, 495, 498**). Based on ¹H-NMR data that paralleled those of the kolavane series, this new series of five clerodanes, elongatolides A–E, were also assigned *trans-ent*-clerodane absolute stereochemistries (Figure 35 [pp. 516–517]). Two other clerodanes from *S. shortii* (**556** and **557**) were assigned the same stereochemistry (*9*).

Diterpenoids from *S. altissima*, first identified as a series of *trans-ent*-clerodanes, solidagolactones I–VI (*413*), were in some instances identical to members of the previously reported elongatolide series (Figure 35). Solidagolac-tone (Figure 35; Structure **543**), the only member of this series lacking a C-6 oxygen function, was shown via reduction/methylation to be identical to a kolavenic acid derivative. Consequently, the absolute stereochemistry of the kolavanes was assigned to **543** as well as the other six *S. altissima* clerodanes, solidagolactones II–VII.

Co-occurrence with (-)-hardwickiic acid of known absolute stereochemistry (*391*) also led to the assignment of its stereochemistry to the clerodane constituents of *S. juncea* (*318*).

VI.1.1.2. cis-normal-*Clerodanes (5α, 10α-cis-Clerodanes)*

The first non-*trans*-fused clerodane (*235*) from the Compositae was identified by X-ray as a C5α,C10α-*cis*-clerodane, gutierolide (**270**). Perhaps prompted by this finding and the growing number of such *cis*-clerodanes reported from other plant families, the absolute stereochemistries of the next reported pair of *Solidago* clerodanes (*15*), solidagoic acids A (**239**), and B (**244**), and a related series of neutral clerodanes (*317*) were studied exhaustively. These structures were assigned a 5α, 10α-*cis*-ring junction indicating a normal-labdane origin. Initially, an attempt was made to correlate structure **239** with the established stereochemistry of (-)-hardwickiic acid (**511**; Figure 36A [p. 518]). Chemical conversion of **239** to **239a** and the corresponding conversion of **511** to **507** permitted the comparison of the properties of the two constitutionally equivalent derivatives, **239a** and **507**. Significant differences in their ¹H-NMR chemical shifts proved the nonequivalence of the two parent structures, **239** and (-)-hardwickiic acid (Figure 36A).

Attempts were then made to correlate **239** with a *cis*-clerodane of known absolute stereochemistry [revised as per *15*], plathyterpenone (**1075**). Compari-son of the CD results for **239b** (a derivative of **239**) and **1075** suggested a *cis*-AB ring-fusion at C-5 and C-10 identical to the revised **1075** (Figure 36B [p. 518]).

The *Solidago shortii* clerodanes reported initially as *trans-ent*-clerodanes (See above) were reinvestigated (*386*) as part of the *S. arguta* study. Preparation of the 3,4-β-epoxide of 5α,10α-*cis*-fused solidagoic acid A produced a structure with ^{1}H-NMR properties strikingly similar to those of a *S. shortii* compound, **556**. It was then suggested that this and other *S. shortii* structures be revised to the corresponding *cis*-normal-clerodanes.

Reinvestigation of the above-mentioned *Solidago elongata* and *S. altissima* C-6 oxygenated clerodanes was also prompted by the identification of tricyclosolidagolactone (Figure 37 [p. 519]; Structure **947**), a *cis*-normal-clerodane (5α,10α-*cis*-fused) from *S. altissima* (*500*). This tricyclic clerodane contained a cis-fused AB ring system. Niwa and Yamamura proposed that the biogenesis of **947** (Figure 37) involved a precursor that was constitutionally identical to the first-proposed structure for solidagolactone V (Figure 35; Structure **262**) but belonging to the *cis*-normal-clerodane class (*402*). They reinvestigated the *S. altissima* clerodanes in order to resolve this inconsistency. Consequently, the structures of solidagolactones II–VII and the corresponding elongatolides were revised to *cis*-normal-clerodanes (Figure 35). Confirmation of the absolute stereochemistry of the solidagolactone series was provided by X-ray analysis of the C-6 p-bromobenzoate derivative of desacylsolidagolactone VIII (*401*).

VI.1.1.3. cis-ent-*Clerodanes (5β,10β-cis-Clerodanes)*

A *Solidago arguta* report identified a mixture of seven clerodanes, one *trans-ent*-clerodane (**507**) and six novel 5β,10β-*cis-ent*-clerodanes, **578–583** (*7*). The absolute stereochemistry assigned these structures was confirmed by X-ray analysis (*263*) and by chemical interconversion (*386*). First, a nonheavy atom X-ray analysis of **583** established its relative stereochemistry (Figure 38 [p. 519]). The lactone **583** was transformed through reduction and acetylation into the co-occurring compounds **581** and **582**, respectively. Then, a study of optical data for *S. arguta* compound **578** (Figure 39 [p. 520]) confirmed the absolute stereochemistry for these structures as well as other compounds of similar properties. Haplopappic acid (**576**), isolated from *Haplopappus angustifolius* and *H. foliosus* (*462*), and the series of cistodioic acid-related structures from *Cistus monspeliensis* (Cistaceae; *33, 469*) were correlated through a series of derivatives with these *Solidago* structures of established stereochemistry (Figure 39).

CD (circular dichroism) and other optical data indicated that haplopappic acid, the *S. arguta* clerodanes and the cistodioic acid-related structure series had a common relative configuration that was largely opposite that of solidagoic acid A (**239**) and B (**244**). The two enones, **576c** and **578b**, derived as indicated in Figure 39 have positive Cotton effects for the n → π^* transition of almost identical molecular amplitudes (α +65 and +64, respectively). The two enones derived from solidagoic acid A had an opposite and weaker Cotton effect (α − 46 and − 33) compared to those observed for **576c**, **578b** and related structures.

Despite opposite orientations at C-5, C-9, and C-10, $5\beta,10\beta$- and $5\alpha,10\alpha$-*cis*-clerodanes are epimeric rather than enantiomeric because both have α-methyls at C-8.

VI.1.1.4. trans-*normal-Clerodanes (5β,10α-trans-Clerodanes)*

Of the four possible AB-ring junction orientations, three are well-documented among the clerodanes of the Compositae. Identification of the fourth, a $5\beta,10\alpha$-*trans*-normal-clerodane, is at best tentative. In the first report of *trans*-normal clerodanes, structures **236** and **237** from *Bedfordia salicina* were illustrated (*122*) without specifying orientation of the methyl groups (Figure 40 [p. 521]). Although constitutionally identical to kolavenol (**475**), **237** differed in its spectral properties. In particular, the optical rotations for **237** and **475** were opposite ([α]$_D$ = +7.9 and [α]$_D$ = −57.1, respectively, *365*). The authors suggested that the data for **236** and **237** were consistent with structures bearing C-5β-, C-8β-, and C-9β-oriented methyls. Nevertheless, the positive optical rotation and methyl chemical shifts (e.g., **237**: H-17,δ0.79; H-18,δ1.69; H-19,δ1.05; H-20,0.82) for **236** and **237** are also consistent with the *cis-ent*-clerodanes (Table 14 [pp. 507–510]).

Koanophyllon admantium compounds **245** and **250** were depicted as *trans*-normal-clerodanes with β-oriented C-8 methyls (Figure 40; *42*). However, the structures were accompanied by the statement that the absolute configuration of these diterpenes was not determined, raising the possibility that **245** and **250** were (-)-hardwickiic acid and its 1,2-dehydro-derivative, respectively. The *Koanophyllon* clerodanes shared with (-)-hardwickiic acid a negative optical rotation. The confusion regarding these structures was heightened by the report of these two structures from *Centipeda* (*128*) and **251**, the methylester of **250**, from *Nidorella* (*72*) in which the structures were displayed with C-8α-methyls (The C-8α-methyl epimers of **245** and **250** are treated here as **249** and **248**, respectively). Although **251** was reported as a methylester (nidoreseda acid methyl ester) from *Nidorella*, when this structure report was cited in the *Centipeda* article the structure was described and drawn as the free acid.

The *Centipeda* compound **249** assignment was based on correlation with the *Solidago* (*286*) *trans-ent*-clerodane, (-)-hardwickiic acid. The discrepancy between the structure illustrated for **249** and that of (-)-hardwickiic acid was not discussed. Possibly, the above-described confusion results from incorrectly illustrated structures for **248**, **249**, and **251**. Because **249** was described as being previously reported by Misra et al. (*391*), it is assumed that this structure is actually (-)-hardwickiic acid and is probably identical to **245**. The other structures, **248** and **250**, are presumably the same compound and identical to 1,2-dehydro-(-)-hardwickiic acid. Very likely, **248** and **249** are the same as **250** and **245**, respectively.

A previously characterized clerodane, **238**, and two novel clerodanes, **253** and **254**, were reported as $5\beta,10\alpha$-*trans*-clerodanes from *Nidorella* (*72*). The citation for the known compound was a *Solidago* article (*317*) in which the structure was drawn differently (as **1064**, a 5α,10α-*cis*-clerodane). Structures **238** and **1064** differ in that their C-5 hydroxymethylenes are β- and α-oriented, respectively (Figure 40). The authors noted that the absolute stereochemistry of these compounds was not determined and that the displayed structures were only tentative. Possibly, the authors intended all three of these structures (**238**, **253**, and **254**) to be shown with the more common configuration of **1064**.

Based on the above-described reports, there is no well-established example of 5β,10α-*trans*-clerodanes in the Compositae.

The previously mentioned mix-up regarding the *Nidorella* compounds has exacerbated an already confused situation regarding the name(s) and structure(s) proposed for the 5,10-seco-clerodane structures. The first report of a compound with these properties from *Conyza stricta* (*460*) incorrectly assigned the structure of Δ¹-polyalthic acid (Figure 40) to the compound, conyzic acid. A second compound, secoreseda acid, from *Nidorella resedifolia* (*72*) (Figure 40) was assigned a relative (but not absolute) stereochemistry, based on analogy with co-occurring compounds. Concurrently, a diterpene isolated from *Conyza stricta*, strictic acid, (*470*) was assigned a structure epimeric at C-9 to that first proposed for secoreseda acid. Comparison of the properties of these three compounds led to the subsequent proposal that conyzic acid, secoreseda acid, and strictic acid were identical (*376*) and that the probable structure was that of secoreseda acid (Figure 41 [p. 522]; Structure 2). The C-9 methyl was assigned a β-orientation because secoreseda acid co-occurred with compounds with β-oriented C-9 methyls. Later, when secoreseda acid was reported from *Conyza scabrida*, the structure was drawn as a *trans-ent*-clerodane derivative (Figure 41; Structure 3; *83*). The assigned stereochemistry was at least in part based on the co-occurrence with biogenetically closely related *trans-ent*-clerodanes (see Figure 11). This revision of the secoreseda acid structure was published without comment by the authors (*83*). However, this revised version has the same absolute stereochemistry as that originally assigned to strictic acid (*470*) and is currently the best guess at the structure of conyzic acid/strictic acid/secoreseda acid.

VI.1.2. ¹H-NMR Spectrometry of Clerodanes

As suggested by the preceding discussion, assignment of orientivity at the four clerodane asymmetric centers, C-5, C-8, C-9, and C-10, is likely the most problematic aspect of structure elucidation. The co-occurrence in the same species of clerodanes representing different ring-junction types erodes the reliability of biogenetic arguments in assigning stereochemistry based on stereochemical properties of compounds previously isolated from the same species or from related species. Given the range of biogenetically feasible structure candidates, assignment of stereochemistry may require choosing between pairs of enantiomeric structures or between constitutionally identical structures that are epimeric at one or more centers. Such a problem greatly complicates the assignment of absolute stereochemistry.

VI.1.2.1. General Properties of Methyl Chemical Shifts

A reference point for the evaluation of the ^{1}H-NMR chemical shifts of the clerodanes is the study initiated by Zürcher on the steroid ^{1}H-NMR properties. Zürcher examined the effects of individual substituents on the chemical shifts of the C-10 and C-13 angular methyls of the four androstane skeletal types (504, 505; Figure 42A [p. 523]). His findings became codified as the "Zürcher Rules." Chemical shifts for the four reference compounds (Figure 42A) demonstrate that angular methyl shift values depend in part on the cis or trans nature of the ring junction.

Other results indicate that the largest deshielding effect on angular methyls results from substituents that have a 1,3-syn-diaxial relationship with the methyl group. Comparable substituents that are equatorial at these positions have little influence on the methyl chemical shifts. In general, substituents on carbons that are vicinal to the methyl-bearing ring carbon display the second greatest deshielding influence.

Similar patterns can be observed in the constitutionally equivalent representatives of the three clerodane types (Figure 43 [p. 524]):

1. The 5α,10β-trans-representative (Figure 43; Structure C-1) with its trans-AB ring junction and the syn-diaxial arrangement of the C-9 and C-5 methyls is comparable to the 5α-14α-androstane structure (Figure 42A; Structure 1): Its relatively upfield C-5 and C-9 methyl chemical shifts are largely a consequence of the methyls's axial relationship to the ring system.
2. The introduction of the A-B and/or C-D cis-ring junction into the last three representative androstane skeletons results in an appreciable downfield shift of the angular methyl situated at the cis-junction. Relative to the 5α-10β-trans-clerodane structure, C-5 methyls of both the 5α,10α-, and 5β,10β-cis-representatives display similar downfield shift effects (Figure 43; 5α,10β:δ1.00; 5α,10α:δ1.17; 5β,10β:δ1.06).
3. In both the 5α,10β-trans- and 5β,10β-cis-clerodanes, the C-9 methyl adopts an axial orientation that is trans to H-10. Both produce chemical shifts (δ0.74 and 0.84, respectively) that are upfield from the shift of the equatorial C-9 methyl of the 5α,10α-cis-structure (δ1.08). Despite the flexibility of the 5α,10α-cis-decalin system, evidence (discussed below) suggests that for this system the conformation with the C-9 equatorial methyl is generally favored.

The usefulness of ^{1}H-NMR chemical shifts in differentiating the two major clerodane isomeric types (5α,10β-trans- and 5α,10α-cis-fused) was first noted in the comparison of the chemical shifts of the solidagoic acid A (239) derivative (239a) and the (-)-hardwickiic acid (511) derivative (507) (Figure 36A). Analysis of additional structures of the two isomeric types demonstrated that the most useful ^{1}H-NMR parameters for assigning stereochemistry were the C-5-, C-8-, and C-9-methyl chemical shifts. Niwa and Yamamura (402) observed that the C-8-methyl doublet (δ0.90) of 239a appeared upfield from the C-9 methyl singlet (δ1.08), while the C-9 methyl singlet (δ0.74) of 507 occurred upfield from the C-8 methyl doublet (δ0.86). These differences were attributed to the presence of an equatorial C-9 methyl in **239a** and an axial C-9 methyl in **507**.

Based on examination of compiled C-4, C-5, C-8, and C-9 methyl chemical shifts (Table 14), with very few exceptions the relative positions of C-8 and C-9 methyl chemical shifts observed by Niwa and Yamamura characterize all listed compounds of the two major clerodane types.

VI.1.2.2. ¹H-NMR Properties of cis-normal-Clerodanes (5α,10α-cis-)

VI.1.2.2.1. C-6 Oxygenation

The C-9 methyl (H-20) chemical shift generalization made by Niwa and Yamamura (*402*) for the 5α,10α-*cis*-clerodanes seems to be consistent for all compounds listed in Table 14. The compounds with the most divergent H-20 chemical shifts, **262** (δ1.21), **265** (δ1.17), and the synthetic product **262a** (δ1.24) (*402*), represent all the listed structures with C-6 keto functions. While no H-20 deshielding is evident for C-6 hydroxyl- or ester-containing structures, such deshielding is pronounced for the C-6 carbonyl-bearing compounds (Figure 44 [p. 525]).

VI.1.2.2.2. Conformational Variability

¹H-NMR data support the view that two basic conformations, I and II, (or some minor variant of I or II in which, for example, the A-ring exists as a flattened half-boat form or some twist analog) occur within the 5α-10α-*cis*-clerodanes (Figures 45A [p. 526] and 46 [p. 527]).

Individual circumstances tend to favor one conformation or the other. For instance, the presumed conformation for hemiacetal or lactone ring-containing compounds with C-18,19-oxido linkages (Figure 45B [p. 526]) is conformation I because the closure of the hemiacetal or lactone ring imposes less constraint on the decalin system in this conformation. Compound **1068** (Figure 45B) is characterized by a H-3 ¹H-NMR signal that is relatively narrow ($w_{1/2} = 3$Hz), in agreement with the proposed conformation (*317*). However, as will be described below not all such C-18,C-19-heterocyclic structures agree with this interpretation.

The assignment of conformation was more problematic for compounds lacking C-18,19-oxido linkages. Based on a variety of data (*15*), Anthonsen and coworkers proposed that conformation II with the A-ring flattened into a half-boat was more reasonable than conformation I. They argued that their proposal was supported by the C-9 methyl (H-20) chemical shift for compounds with the A-ring substituents shown in Structure 4 of Figure 45A. As drawn, the C-9 methyl is directed over the enone function and is shielded by it. In compounds lacking the enone, the methyl produces a signal at approximately δ1.05. However, when the enone is present the methyl signal appears at about δ0.85 (*15*).

These results, in particular those for structure **239b**, suggest that the shielding zone of the ketone (in the absence of α,β-unsaturation) alone is sufficient to cause the observed H-20 chemical shift. Other examples of this shielding effect include structures **239d**, **239b**, and **239c** (Table 14; Figure 60 [pp. 541–544]).

A derivative, **263a**, prepared from a variety of sources (solidagolactones II and III and **263**) by Goswami and coworkers, displayed methyl chemical shifts (Table 14) that were inconsistent with the above-described shielding effect (*291*). In depicting the conformation of the related structures **268** and **269**, Goswami and associates (*291*) opted for the conformation that was most consistent with their results (Figure 45C [p. 527]). As in conformation I, the C-9 methyl was equatorial, and the C-6 ester was axial.

The results reported by Anthonsen and coworkers (*15*) and Goswami and associates (*291*) appear on the surface to be contradictory. However, these and other results (described below) are made compatible by assigning the favored conformation according to "rules" based on several alternative A and B ring substitutional patterns. If R_1 (Figure 46) is a keto function and R_2 is a pair of hydrogens, then conformation II is favored. However, the results suggest that if R_2 is an oxygen substituent the favored conformation is I regardless of the R_1 substituent. Given that the majority of $5\alpha,10\alpha$-*cis*-clerodanes have C-6 oxygen substituents, the most common conformation is expected to be I. If R_2 is a keto function, maximum deshielding of the C-9 methyl is expected in conformation I (or some minor variant of this conformation). This effect is pronounced in the C-9 methyl chemical shifts of **262**, **262a**, and **265** ($\delta 1.21$, 1.24, and 1.17, respectively).

These "rules" are consistent with the data reported by Goswami and coworkers (*291*) and with data reported for the two structures, **1066** and **244a**, (Figure 47 [p. 528]). However, not all results are consistent with the generalization that conformation I should be favored in structures bearing a C-18,C-19-heterocyclic ring. The comparison of the methyl chemical shifts for **1066** and **244a** indicate a sizable difference in the chemical shifts of H-20 ($\delta 1.03$ and 0.84, respectively, Figure 47A). Partly, this difference can be attributed to the occurrence of the tetrahydrofuryl moiety in **244a** and the furyl moiety in **1066**. However, comparison of similar structures differing in this manner (Figure 47B) reveals that the deshielding effect of the furyl moiety on H-20 is approximately $\delta 0.07$–0.08 greater than that of the tetrahydrofuryl moiety. The difference in H-20 chemical shift between **1066** and **244a**, $\delta 0.19$, suggests that some factor other than the furyl moiety is contributing to this difference. The only other structural difference is the C-2 keto function. If the C-2 keto group is responsible for this evident shielding effect of H-20, then conformation II is plausibly assigned to **244a**. This result is consistent with the proposed "rule" that if $R_1 = O$ and $R_2 = H,H$ (Figure 46) then conformation II is favored.

The "rules" for the interpretation of conformational variability are further supported by the data for the two **1025** conformations detected at $-50°C$ (Figure 48 [p. 529]; *170*). While the room temperature CDCl$_3$ spectrum was difficult to interpret, two distinct sets of signals were obtained at $-50°C$. The first (**1025a**; Figure 48) was characterized by a B-ring conformation in which the C-6 acetate adopted an axial position and the C-9 methyl was equatorially oriented. This corresponds most closely to conformation I described above. The C-9 methyl chemical shift ($\delta 0.93$) is in agreement with this equatorial position. In the second conformation (**1025b**; Figure 48), the C-6 acetate adopts an equatorial orientation and the C-9 methyl is axially oriented (conformation II). The upfield position of the C-9 methyl signal ($\delta 0.73$) is consistent with the shielding effect of the C-2 keto on the axial methyl. These assignments are corroborated by the H-6 coupling constants. For **1025a**, the H-6 signal appears as a broad singlet, while the H-6 signal for **1025b** is a doublet ($J = 12 Hz$), values which are consistent with the proposed conformations. In compliance with the above "rules," this structure which possesses a C-6 β-oxygen substituent adopts conformation I as the dominant conformation at room temperature (Figure 48).

Several other lines of evidence were developed for the assignment of conformation I to C-6 oxygen-substituent-bearing structures. In the first, Eu(dpm)$_3$-induced shift results for C-6 hydroxy-representatives of $5\beta,10\beta$- and 5α-10α-clerodanes supported the two conformations indicated in Figure 49 (p. 530). The normalized ratio for the $5\beta,10\beta$-fused structure (Figure 49; *386*) was interpreted as support for a steroidlike conformation. The C-19 value for the $5\alpha,10\alpha$-fused structure (5.3) was indicative of an *anti-trans* relation of C-19 to the axial C-6 hydroxyl and of a nonsteroidal conformation (*378*).

The second line of evidence was based on the chemical transformations of **545** to yield two products, **545a** and **545b** (Figure 50 [p. 531]). The formation of a compound with an ether bond between C-3 and C-6 (**545b**) was considered proof of a $5\alpha,10\alpha$-*cis*-ring junction and a nonsteroidal conformation (*378*).

VI.1.2.2.3. 3,4-Epoxides

The 3,4-epoxide functions of clerodanes can adopt either an α- or β-oriented *cis* configuration. As this represents one more complication in the already difficult task of assigning absolute stereochemistry, it is not surprising that epoxide orientation within the elongatolide (solidagolactone) series of $5\alpha,10\alpha$-*cis*-clerodanes has been misassigned. Elongatolides D (solidagolactone VI, *413*; **553**) and E (solidagolactone VII, *413*; **260**) were the first reported 3,4-epoxide-containing clerodanes of this type (currently accepted structures shown in Figure 51A [p. 532]). Although they were initially described as *trans*-clerodanes with unspecified epoxide orientation (Figure 35; *18*), Niwa and Yamamura (*402*) revised these structures to $5\alpha,10\alpha$-*cis*-clerodanes with α-epoxides (Figure 51). Their assignment was based on a series of transformations involving the furyl derivative of **260** (Fig. 51B [p. 532]). They argued that the formation of **260b** and **260c** suggested that the C-4 tertiary hydroxyl group must be axially oriented which would be compatible with the original presence of an α-epoxide ring in **260**. However, the conformational flexibility of the $5\alpha,10\alpha$-clerodane skeleton provides a variety of reasonable conformations: In the instance of the C-4 α-hydroxyl, conformations with either the axial or equatorial configuration of this functionality can reasonably be proposed. Likewise for the C-4 β-hydroxyl option, there are two possible orientations (axial or equatorial).

The structural questions relating to **553** and **260** were finally resolved by Nishino et al. (*401*). Epoxidation of solidagolactone IV (**545**) yielded (in order of abundance) two products, the β- and α-epoxides (Figure 52B [p. 534]). The observed ratio of products was in agreement with the usual mechanism for the epoxidation of homoallylic alcohols: In the presence of a C-6 β-OH group, the epoxide of the major product should have the same configuration as the hydroxyl group. Data for the major product (β-epoxide) included an IR absorption at 3470 cm^{-1} (unaffected by dilution) indicative of hydrogen bonding and no absorption at 3600 cm^{-1} which, if present, would suggest a free OH. The IR spectrum of the minor product (α-epoxide) included an absorption at 3600 cm^{-1} and only a weak absorption at 3475 cm^{-1} which was affected by dilution. Further evidence for the hydrogen bonding of the epoxide oxygen and the C-6 hydroxyl was the presence in the ^{1}H-NMR spectrum of the β-epoxide-containing structure of a sharp singlet at $\delta5.41$ (OH proton), whereas the OH proton of the α-epoxide occurred as a multiplet at approximately $\delta1.8$. Because this intramolecular hydrogen bonding is only possible if the 3,4-epoxide is β-oriented, the major product was assigned this configuration. The β-epoxide was correlated with solidagolactone VII (**260**) and VIII (**988**) by a set of reactions (Figure 51C [p. 533]) in which **260** and **988** and the p-bromobenzoate derivative (Figure 51C; Structure 5), were transformed to the corresponding furyl derivatives and through hydrolysis to the common product (Figure 51C; Structure 4). More importantly, the other compounds were correlated with the p-bromobenzoate derivative (Figure 51C; Structure 5) for which X-ray analysis provided the absolute configuration. Consequently, the α-epoxide structure proposed for solidagolactone VII (and presumably solidagolactone VI) was revised to a β-epoxide.

An unexpected result of the study by Nashino and his coworkers was the ratio of products in the solidagolactone IV epoxidation reaction (Figure 52B), 1.7(β-epoxide):1.0(α-epoxide). Given that the homoallylic hydroxyl system should favor an attack on the same side of the molecule as the hydroxyl function (β), the surprisingly nonskewed observed ratio suggests that in the absence of this C-6 hydroxyl, attack from the α-face of the A-ring is favored. Support for this proposal came from an apparently unlikely source (*291*) in which just the opposite assumption was made. Solidagolactones VII (**260**) and VIII (**988**) were isolated by Goswami and associates from *Solidago virgaurea* and correctly identified as $5\alpha,10\alpha$-*cis*-clerodanes. The epoxide was assigned an α-orientation in agreement with the previously published findings of Niwa and Yamamura (*402*). Treatment of the solidagolactone II (**258**) and III (**259**) mixture with m-chloroperbenzoic acid yielded an angelate/tiglate pair of epoxide products that were isomeric to **260** and **988** (Figure 52C [p. 535]). The authors assumed that this angelate/tiglate pair possessed a β-epoxide because the H-3's chemical shift ($\delta3.04$) and coupling constant (d, J = 4Hz) were distinct from those ($\delta2.74$, broad singlet) of the naturally occurring pair of structures (**260** and **988**) to which they had previously assigned an α-epoxide. However, the absolute configuration of solidagolactone VII was subsequently established and revised to a β-epoxide (361). Given that Goswami and coworkers originally misassigned the epoxide of **260** and **988**, then the epoxide of the isomeric transformation product (Figure 35C) must have just the opposite configuration (α) to the one originally assigned (β). The similarity of the ^{1}H-NMR data for the H-3 proton of the epoxidation product (d, J = 4Hz) and the α-epoxide synthesized by Nishino et al. (d, J = 5.1Hz) supports this conclusion. The observation that this α-epoxide was the major product (no other products were identified) suggests that in the absence of the C-6 β-hydroxyl, epoxidation occurs most favorably from the α-face. A similar epoxidation reaction (Figure 52D [p. 535]) involving the C-6 acetate of solidagolactone IV (**545**, *413*) yielded one detectable product (probably an α-epoxide) characterized by a doublet H-3 signal with J = 4.5Hz.

In another epoxidation reaction (Figure 52A), a transformation product of solidagoic acid, **239**, (*15*) yielded two epimeric epoxides (*386*). Differentiating the two epimers was attempted by the further transformation of the β-epoxide (Figure 53 [p. 535]) (*386*). ^{1}H-NMR evidence of an axial (β) hydroxyl function at C-3 in the product of this Lewis acid treatment was used as evidence in the assignment of spectroscopic properties to each epoxide. Two aspects of the assignments appear contradictory to previously reported findings. First, the ratio of products, 3(β-epoxide):1(α-epoxide), indicates that in this instance attack from the β-face of the 5α-10α-*cis*-clerodane is favored. Second, the H-3 coupling constants reported for the two epimers are opposite those assigned by Nishino et al. (Figure 52B). Accordingly, it appears that the two epoxide products (Figure 52A) should be revised such that the major product is the α-epoxide. However, caution should be exercised in concluding that the reaction-A products are misassigned. There are several reasons for caution. The absence of a C-6 oxygen function may affect the conformation of the products of reaction A, with the result that coupling constants differ from those of C-6 oxygenated compounds. Further, the effect of the absence of the C-6 function on the likelihood of β-face attack during epoxidation is not well understood.

A recent definitive study of the stereochemistry of C-6 substituted 5α-10α-*cis*-clerodanes (*483*) confirmed the above interpretation of the results for the epoxidation of the C-6 β-acetate and formally revised the epoxide configurations (described above) originally assigned by Niwa and Yamamura (*402*) and Goswami et al. (*291*). The authors also used α- and β-epoxide analysis results to determine the feasibility of applying Tori equation calculations to the prediction of H-3 coupling constants (Figure 54 [p. 536]). Dihedral bond angles for H3-C3-C2-Hβ-2 and H3-C3-C2-Hα-2 for the β-epoxide were known and those for the α-epoxide were estimated based on an anticipated change of 18° in this angle during epoxidation of the olefin. In each instance the calculated coupling constants closely approximated the observed values.

VI.1.2.3. ^{1}H-NMR Properties of trans-ent-Clerodanes

The axial configuration of the C-9 methyl (H-20) results in a relatively high-field ^{1}H-NMR chemical shift for this signal. Departures from the typical value of c. δ0.75 for H-20 can usually be attributed to the deshielding effect of an oxygen function at one of several possible positions:

VI.1.2.3.1. C-7α-OR

The ^{1}H-NMR spectra of C-7α-hydroxyl-bearing compounds **509**, **559**, and **562** and C-7α-acetoxyl-bearing **500** and **509a** produce H-20 singlets at δ1.02, 0.92, 0.86, 0.97, and 1.02, respectively (Figure 55 [p. 537]). Spectra of compounds **509**, **500**, and **509a** display H-19 singlets at δ1.36, 1.17, and 1.34, respectively. The spectra of the C-7 keto derivatives of **500** and **509** are characterized by H-19 and H-20 singlets that are δ0.2–0.3 upfield from the parent compounds (Figure 55). These observed chemical shifts are attributable to the 1,3,5-triaxial relationship of the C-5 methyl, the C-9 methyl, and the C-7α-hydroxyl or ester function. The interaction between these substituents is shown in Figure 56 (p. 538) from data supplied by Ohsuka et al. (*412*).

VI.1.2.3.2. C-12-OR

The C-12 hydroxyl function is the probable cause of the relatively downfield position of the H-20 singlet (δ0.94) in the spectrum of **1061**.

VI.1.2.3.3. C-19-OR and Aldehyde

Introduction of a hydroxyl function at C-19 has minimal effect on the chemical shift of H-20 (Table 14 and Figure 57 [p. 539]). However, attachment of increasingly bulky ester functions (e.g., angelate) at C-19 appears to result in modest deshielding of H-20. Structure **904** which differs from the related compounds **900**, **902**, **903**, and **1033** (hautriwaic acid) (Figure 57) by the further oxidation of C-19 to an aldehyde is characterized by an H-20 singlet at δ0.60. The considerable shielding of this methyl is attributed to the aldehyde (C-19) which is 1,3 diaxially oriented with respect to the C-10 methyl (H-20).

VI.1.2.3.4. C-17 Aldehyde

Oxidation of the C-8 methyl (H-17) in compound **967** appears to have an appreciable deshielding effect on H-20 (δ1.00).

VI.1.2.4. cis-ent-*Clerodanes*

McCrindle and coworkers (*386*) have proposed a conformation for the 5β,10β-*cis*-clerodanes (Figure 58 [p. 539]). The H-20 chemical shifts for the members of this series (Table 14) are consistent with the depicted axial arrangement for this methyl. A comparison of the values reported for the cistodioic acid series of compounds (Figure 59 [p. 540]; *33, 469*) and those of the Compositae (Table 14) suggest that the range of observed C-9 methyl chemical shifts results largely from differences in the sidechain moiety. A saturated sidechain is correlated with the relatively upfield values, while the furyl sidechain is associated with the most downfield.

VI.1.3. ^{13}C-NMR Spectrometry of Clerodanes

Reviews of diterpene ^{13}C-NMR data have focused in one instance on all reported diterpenes (*494*), in the second instance on the *trans*-clerodanes (*375*), and in the third (*378*), on the use of these data to differentiate between 5α,10α-*cis*- and *trans*-clerodanes. As this third review deals specifically with the solidagolactone problem, it will be discussed at length here.

Pendant methyl groups of clerodanes display some of the same properties as those previously elucidated for steroids (Figure 42; *504* and *505*). Principally, the configuration of the ring junction influences the chemical shifts of the associated methyls. Comparison of the C-17, C-19, and C-20 chemical shifts of a series of C6-substituted 5α,10α-*cis*-clerodanes with those of several *trans*-clerodanes reveals differences consistent with those observed for *cis*- and *trans*-decalin (Figure 42B) and steroid systems (Table 15 [pp. 510–512]; Figure 61 [pp. 545–549]). In the *cis*-clerodanes, the C-19 signal appears in an area centered on δ25, while in the *trans*-clerodanes it occurs between δ17 and 19. Similarly, the C-20 signal resonates at a lower field (δ21 to 29) than that of the C-20 of the *trans*-clerodanes (δ17 to 19). The C-17 signals of both *cis*- and *trans*-fused compounds display relatively constant chemical shifts near δ15.

The association between ring junction and methyl chemical shift was attributed to the shielding effect from carbons with a τ-*gauche* relationship to the resonating carbon. In *trans*-clerodanes, the number of carbon atoms having a τ-*gauche* interaction with C-19 is larger than the number in *cis*-clerodanes. In *trans*-clerodanes, C-20 has an axial orientation, while in *cis*-clerodanes, this methyl has an equatorial orientation. The relatively constant chemical shift of C-17 in *trans*- and *cis*-clerodanes is attributable to its equatorial orientation in both ring junction types (*378*). Evidently, C-19 and C-20 ^{13}C-NMR chemical shifts are useful in distinguishing *cis*- and *trans*-clerodanes (Figure 61). Just how useful these data will be depends on how many exceptions (if any) will be found to the data in Table 15.

VI.1.4. Conformational Analysis (X-ray)

Quantitative conformational analysis of *trans*-clerodanes (*256*) and 5α,10α-*cis*-clerodanes (*378*) has been conducted on X-ray diffraction results.

VI.2. Other Bicarbocyclics

^{13}C-NMR studies of labdanes have focused in one instance on the normal-labdanes, manoyl oxide (**272**), manool (**996**), sclareol, and marrubiin (**4**), and in a second on the C-2β- and C-3α-hydroxylated *ent*-labdanes (**424**) reported from a non-Composite, *Phlogacanthus thyrsiflorus*. The results of the first study and other unpublished ^{13}C-NMR findings for normal-labdanes were discussed in detail by Wehrli and Nishida (*494*).

VI.3. Tricarbocyclics

Early ^{1}H-NMR work on the resin acids, sandaracopimaric, pimaric, isopimaric acids and their dehydro- and tetrahydro- derivatives, applied chemical shift calculation methods to the prediction of C-17, C-18, and C-20 methyl chemical shifts (*19, 20*). ^{13}C-NMR results for several other normal-pimaranes were recently reviewed (*494*).

Investigation of ^{1}H-NMR properties of the sandaracopimaranes isolated from *Garuleum* and *Osteosperum* (*156*) led to the proposal of a conformation for these structures in which the A-ring adopted a twist form (Figure 62 [p. 550]). Evidence for this conformation was provided by the comparison of the C-4β-methyl chemical shifts for sandaracopimar-15-en-8β-ol and its C-6β-hydroxy and acetoxy derivatives (Structures 1, 3 and 4, respectively; Figure 62; Table 16 [p. 512]). The C-6β-hydroxyl group had a strong deshielding effect on the C-4β-methyl, an effect which was diminished by introduction of an acetyl group. This effect agrees with the proposed 1,3-*syn*-diaxial relationship between the C-6 hydroxyl and the C-4β-methyl. The deshielding effect of ths C-6 hydroxyl on the C-10 methyl suggests that the same 1,3-*syn*-diaxial relationship exists for these substituents. The introduction of a C-11β-hydroxyl group (Structures 11 and 12; Figure 62) displayed the expected deshielding effect on the C-10 and C-13 methyls (Table 16). However, equatorially oriented hydroxyl groups at C-11 and C-12 (e.g., Structures 5, 7, and 10; Figure 62) had much less effect on neighboring protons.

Detailed ^{13}C-NMR results for ferruginol (**319**) and other aromatic abietane compounds were used to indicate the stereochemistry at C-4 for compounds possessing either an equatorial or axial carbinol group (*494*).

The only review of ^{13}C-NMR results for the cassane series focused on a group of compounds from a non-Compositae source (*Caesalpinia bonducella*) that were substitutionally distinct from those reported from *Osteospermum* (*231*).

VI.4. Tetracarbocyclics and Related Skeletons

Pre-1978 ^{13}C-NMR analyses of *ent*-kauranes (*357, 310, 309, 449*) have been reviewed (*494*). Another report (*502*) included complete data for the *Stevia ent*-kaurene glucosides, stevioside (**683**), rebaudioside A (**685**), and paniculosides I-III (**895–897**). Recently, a detailed study of kaurenoic acid derivatives (*345*) sought to compare the ^{13}C-NMR chemical shifts of *ent*-kaur-16-en-19-oic acid (**625**) with those of hydroxy derivatives at C-6, C-7, C-11, C-12, and C-15 and with kaurenolide. Another study (*265*) reported the results of complexation of *ent*-kaurane polyols with boric acid. Compounds containing 1,2- or/and 1,3-glycol systems displayed boric acid-induced shifts in their ^{13}C-NMR spectra together with broadening of signals of carbons attached to or adjacent to the glycol function.

ent-Beyerane and beyerene ^{13}C-NMR data have been reported for a series of C-1- and C-12-substituted structures (*486, 494*) and for a series of C-2- and C-3-substituted compounds (*308*).

The ^{13}C-NMR parameters for trachylobane and C-19 derivatives were consistent with predicted values derived from 9-methyl-*trans*-decalin and tricyclo[2.2.1.0^{2,6}]octane (*21*).

Table 14. ¹H-NMR Chemical Shifts   507

TABLE 14. ¹H-NMR Chemical Shifts (δ) for the C-17, C-18, C-19, and C-20 Methyls of Clerodanes.

Compound No.	H-17	H-18	H-19	H-20
5α,10α-<u>cis</u>				
1. 1025(RT)[1]	0.93	1.95	1.32	0.93
2. 1025b	1.04		1.18	0.73
3. 1025a	0.84		1.29	0.93
4. 239[2]	0.89	1.52	COOH	0.98
5. 239a[3]	0.92	1.62	1.17	1.08
6. 310	0.94	(4.24,3.96)	(3.83,3.50)	1.09
7. 258	0.86	1.56	1.28	1.08
8. 259	0.85	1.56	1.28	1.08
9. 545	0.88	1.70	1.15	1.03
10. 262	0.83	1.50	1.17	1.21
11. 263	0.89	1.64	1.26	1.08
12. 265	0.84	1.57	1.30	1.17
13. 268(Ang)/ 269(Tig)	0.86	0.98	1.33	1.10
14. 261	0.88	1.29	1.21	1.03
15. 260	0.88	1.31	1.22	1.03
16. 260a	1.03	(4.93-5.01)	1.28	1.18
17. 262a	0.87	1.54	1.24	1.24
18. 260b	0.87	1.08	1.16	1.16
19. 260c	0.86	1.23	1.08*	1.00*
20. 260d	0.85	1.26	1.20*	1.03*
21. 263a	0.97		1.37	1.04
22. 1091a	0.93		COOH	0.91
23. 1091	0.92		COOH	0.89
24. 1085	0.98	(4.47,4.95)	COOH	0.98
25. 1062(CCl$_4$)	0.78	1.42	CHO	0.78
26. 1063(")	0.85	CHO	CHO	0.96
27. 1064(")	0.90	1.67	CH$_2$OH	1.08

TABLE 14. (*contd.*).

Compound No.	H-17	H-18	H-19	H-20
28. 1065(")	0.85	(4.16)	1.18*	1.03*
29. 1066(")	0.90	COOR	(4.44,3.80)	1.03
30. 1067(")	0.85	(4.21)	(5.42)	0.97
31. 1068(")	0.81	(4.89)	-CHOH-	0.88
32. 1026	0.96	0.87	0.96	-CH$_2$-(1.58,1.69)
33. 556	0.88	1.23	1.11	0.94
34. 239b	0.87	0.97	1.22	0.90
35. 239c	0.96	1.94	1.28	0.83
36. 239d	1.02	1.95	1.27	0.90
37. 244a	0.82	COOR	(4.43,3.88)	0.84

5α,10β-trans

Compound No.	H-17	H-18	H-19	H-20
38. 1030	1.03	1.86	1.15	0.83
39. 475	0.79	1.55	0.97	0.71
40. 479	0.75	1.55	0.97	0.70
41. 480	0.82	1.53	1.21	0.97
42. 495	0.85	1.60	1.02	0.76
43. 494(C$_6$D$_6$)	1.00	1.52	0.82	0.67
44. 1032	1.06	1.80	1.04	0.80
45. 1031	0.75	1.55	1.21	0.83
46. 1027	0.81	1.63	1.20	0.82
47. 1028	0.83	1.55	1.24	0.84
48. 1029	0.82	1.56	1.19	0.81
49. 962	0.84	1.29	1.10	0.74
50. 507	0.86	1.64(CCl$_4$)	1.00	0.74
51. 1034a	0.86	COOH	(4.55,4.31)	0.79
52. 509	1.04	(4.12)	1.36	1.02
53. 509a	1.05	4.5	1.34	1.02
54. 509b	0.97	(4.49)	1.06	0.75

Table 14. ¹H-NMR Chemical Shifts   509

TABLE 14. (*contd.*).

Compound No.	H-17	H-18	H-19	H-20
55. 846	0.85	(4.23,3.67)	(3.85,3.98)	0.78
56. 1033a	0.86	COOH	(4.59,4.32)	0.80
57. 1034	0.88	COOH	(4.22)	0.76
58. 1034a	0.86	COOH	(4.55)	0.79
59. 1034b	0.86	COOMe	(4.60)	0.81
60. 967	CHO	1.60	1.07	1.00
61. 551b	0.82	1.22	1.02	0.67
62. 1061		COOR	(4.15,4.09)	0.94
63. 543	0.84	1.57	1.01	0.77
64. 563	0.82	COOR	(4.28,3.89)	0.58
65. 562	1.03	COOR	(5.31,3.9)	0.86
66. 559	1.06	COOR	(5.33,3.92)	0.92
67. 561	0.99	COOR	(3.92,4.01)	0.64
68. 562	1.04	COOR	(3.91,5.30)	0.87
69. 969		COOR	(4.09,4.05)	0.69
70. 970		COOR	(4.00,4.10)	0.87
71. 525		COOR	(4.15,4.18)	1.10
72. 526		COOR	(4.29,4.27)	1.13
73. 526a		COOR	(4.13,4.45)	1.16
74. 900	0.89	COOMe	(4.36,4.60)	0.84
75. 902	0.89	COOMe	(4.62,4.54)	0.89
76. 903	0.89	COOMe	(4.62,4.37)	0.85
77. 904	0.86	COOMe	CHO	0.60
78. 500	0.92	1.59	1.17	0.97
79. 500a			0.96	0.72
80. 500b	1.00	1.59	1.25	0.98
81. 500c			1.27	1.01
82. 500d	0.89		0.95	0.71
83. 960	0.81	2.02	0.92	0.85

TABLE 14. (*contd.*).

Compound No.	H-17	H-18	H-19	H-20
5β,10β-<u>cis</u>				
84. 1057	0.74	(4.23,4.11)	1.12	0.76
85. 578(CCl$_4$)	0.83	1.72	1.06	0.84
86. 580(")	0.82	(4.54)	1.20	0.80
87. 1056a(C$_6$D$_6$)	0.70		1.16	0.76
88. 1055a	0.79	(4.58)	1.12	0.80
89. 579(CCl$_4$)	0.80	(4.24)	1.12	0.78
90. 1058	0.76	1.70	(3.36,3.23)	0.77
91. 582(CCl$_4$)	0.83	(4.82)	1.28	0.80
92. 582a(")	0.87	(4.70)	1.15	0.83
93. 581	0.85	(4.22)	1.35	0.80
94. 581a	0.87	1.85	1.21	0.82
95. 585	0.80	COOR	1.20	0.69
96. 584	0.85	COOR	1.26	0.77
97. 577	0.77	COOR	1.24	0.79
98. Cistodioic Acid	0.71	COOR	1.20	0.75
99. Cistodiol- Acetate	0.73		1.07	0.75

[1] See Figure 48 for conformations of compounds **1025a** and **1035b**.

[2] See Figure 1 for structures of non-derivative compounds.

[3] See Figure 60 for derivative compounds and naturally occurring compounds not included in Figure 1.

TABLE 15. ^{13}C NMR Chemical Shifts of Methyl Carbons, C-17, C-19, and C-20, of 5α,10α- and 5α,10β-clerodanes.

Structure	C-17	C-19	C-20	Reference
I. 5α,10α-<u>cis</u>-Clerodanes				
1. [545][1]	15.40	24.18	28.34	<u>483</u>
2. [262]	15.17	21.55	24.24	<u>483</u>
3. [260]	15.75	26.64	27.46	<u>483</u>

Table 15. ¹³C NMR Chemical Shifts   511

TABLE 15. (*contd.*).

Structure	C-17	C-19	C-20	Reference
4. [988]	15.75	26.53	27.46	483
5.	15.75	26.53	28.05	483
6.	15.81	26.47	28.05	483
7.	15.58	25.24	28.58	483
8.	15.52	25.24	28.34	483
9.	15.28	22.37	28.87	483
10.	15.28	24.54	28.11	483
11.	15.05	21.90	28.69	483
12.	15.46	24.13	28.58	483
13.	14.29	21.73	29.28	483
14.	15.23	28.46	26.66	483
15. [1083]	15.5	31.0	17.6	180
16. [1084]	15.7	64.8	17.2	180
17. [1085]	18.3	77.0	22.7	180
18. [1086]	17.7	101.1	22.3	180
19. [1087]	16.7	97.8	21.8	180
20. [1088]	16.4	101.3	21.4	180
21. [1089]	17.0	102.6	22.2	180
22. [1090]	17.0	104.1	23.6	180
23. [1091]	16.8	176.5	21.0	180
II. 5α,10β-trans-Clerodanes				
24. [961]	15.8	19.5	17.8	483
25.	15.5	*	17.6	483
26.	15.6	*	17.6	483
27.	15.5	*	17.3	483
28.	15.3	*	17.3	483
29.	15.4	*	17.8	483
30. [1033]	15.9	65.5	18.6	464
31. [1033 Ac]	15.6	71.8	18.1	464
32. [1034 Ac]	15.6	70.1	17.8	464
33. [1034Me,Ac]	15.6	70.1	17.8	464

TABLE 15. (contd.).

Structure	C-17	C-19	C-20	Reference
34. [1070 Me]	19.9	60.6	15.5	468
35. [1070Me,Ac]	19.9	62.4	15.6	468
36. [1093Me,Ac]	18.9	62.2	15.9	468
37.	15.3	68.1	17.2	341
38.	12.0	72.6	19.0	341
39.	11.9	72.7	19.1	341
40.	14.4	64.2	16.3	341
41.	15.6	70.2	17.2	341
42.	15.4	69.1	16.4	341
43.	15.3	68.1	17.6	341
44.	*	73.4	18.8	341

[1] See Figure 61 for structures of these compounds.

TABLE 16. ^{1}H-NMR Chemical Shifts for the Sandaracopimaranes (Structures 1–12, Figure 62) Isolated from *Garuleum* and *Osteosperum* (156).

	[4α-Me] H-19	[4β-Me] H-18	[10-Me] H-20	[13-Me] H-17	H-15	H-16(c)	H-16(t)
1.*	0.85	0.85	1.00	1.22	5.73	4.77	4.82
2.*	0.83	0.85	0.95	1.01	5.64	4.76	4.79
3.	0.97	1.23	1.31	1.23	5.74	4.83	4.87
4.*	0.96	0.98	1.16	1.27	5.65	4.74	4.79
5.	0.87	0.87	1.18	1.26	5.73	4.81	4.86
6.	0.85	0.88	1.00	1.31	5.67	4.90	4.92
7(1)	0.88	0.88	1.20	1.28	5.75	5.00	5.02
(2)*	0.87	0.87	1.16	1.22	5.77	4.93	4.94
8.	0.89	0.89	1.20	1.34	5.67	4.93	4.94
9.	0.86	0.88	1.11	1.27	5.83	4.96	4.97
10.	0.86	0.88	1.02	1.26	5.72	5.04	5.05
11.	0.96	1.25	1.65	1.40	5.70	4.82	4.84
12.	0.89	0.91	1.42	1.42	5.78	5.13	5.01

* Solvent:CCl$_4$

Figure 33. Chemical Interconversion of Solidagonic Acid   513

A·

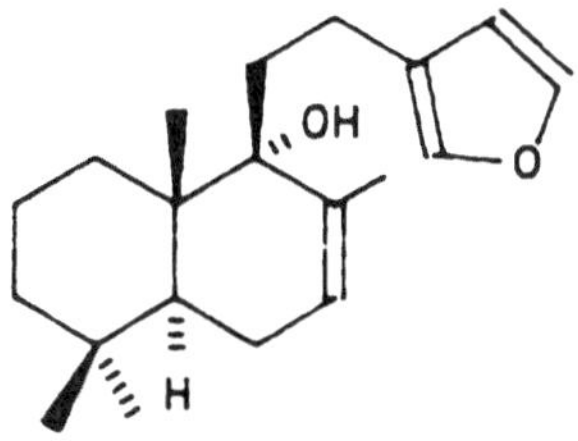

158 Solidagenone

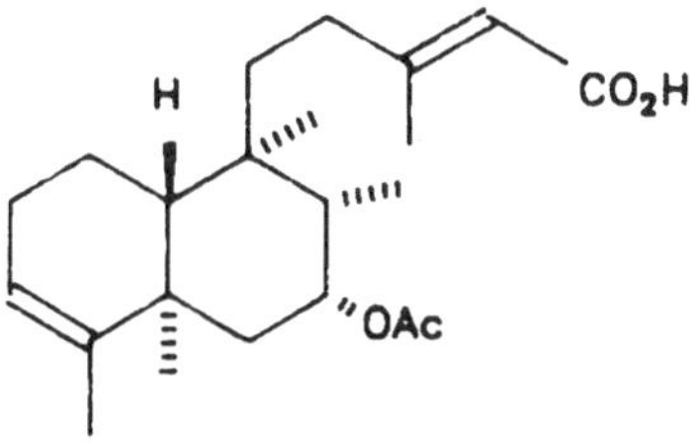

500 Solidagonic Acid

B·

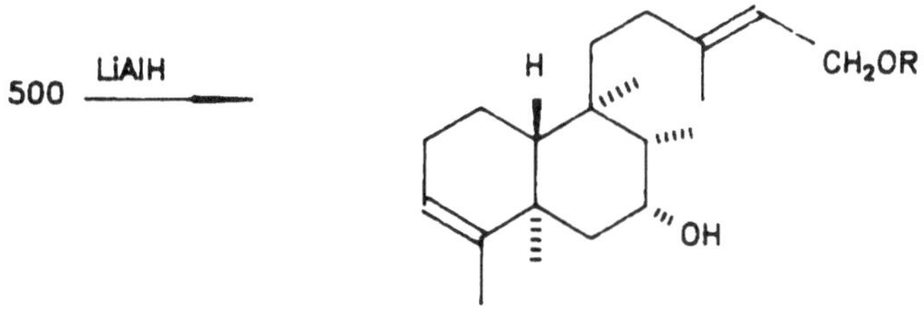

500 $\xrightarrow{\text{LiAlH}}$

500a R = H

500b R = Ac

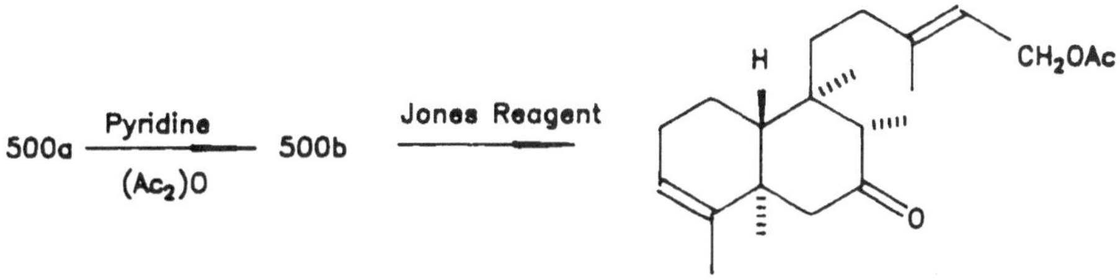

500a $\xrightarrow[\text{(Ac}_2)\text{O}]{\text{Pyridine}}$ 500b $\xrightarrow{\text{Jones Reagent}}$

500c

500c $\xrightarrow[\text{Hydrazine hydrate}]{\text{KOH}}$

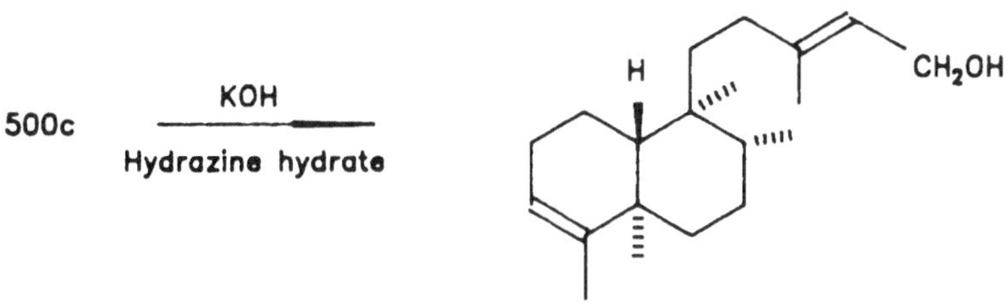

475 Kolavenol

FIGURE 33. Chemical interconversion of solidagonic acid (500) to a compound of established absolute stereochemistry, kolavenol (475) (364, 365).

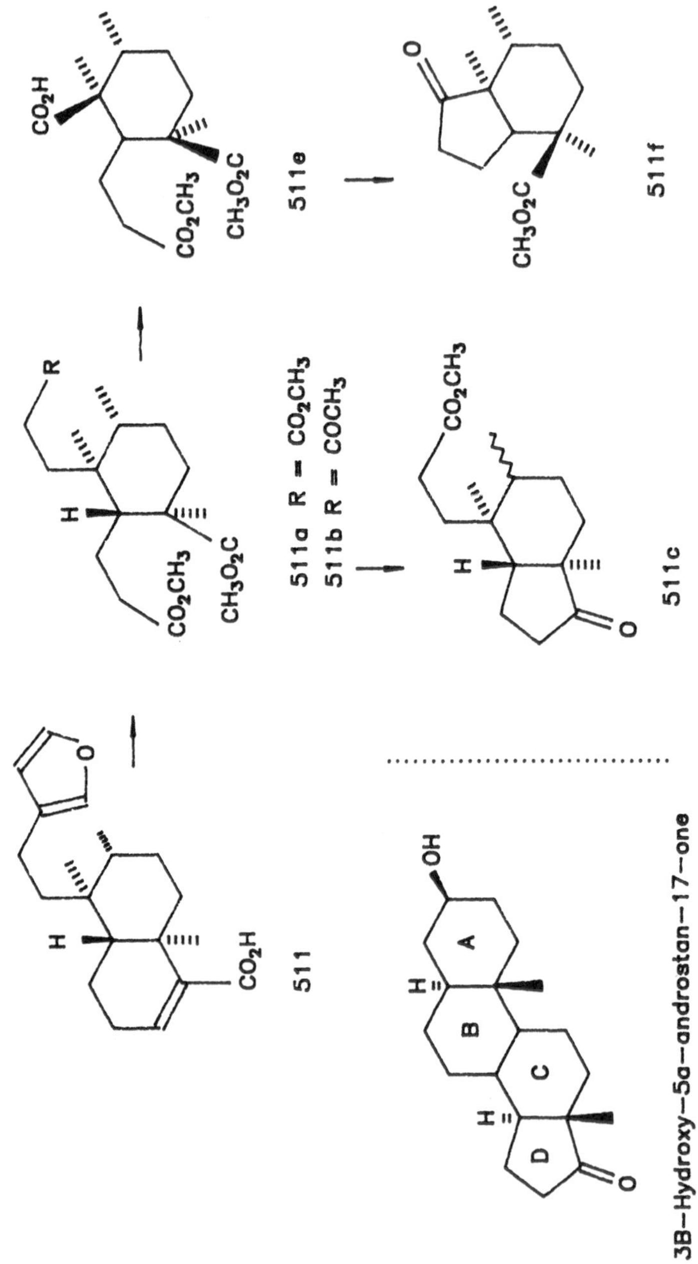

511e

511f

R

511a R = CO₂CH₃
511b R = COCH₃

511c

511

3B-Hydroxy-5α-androstan-17-one

Figure 34. Chemical Degradation of (-)-Hardwickiic Acid   515

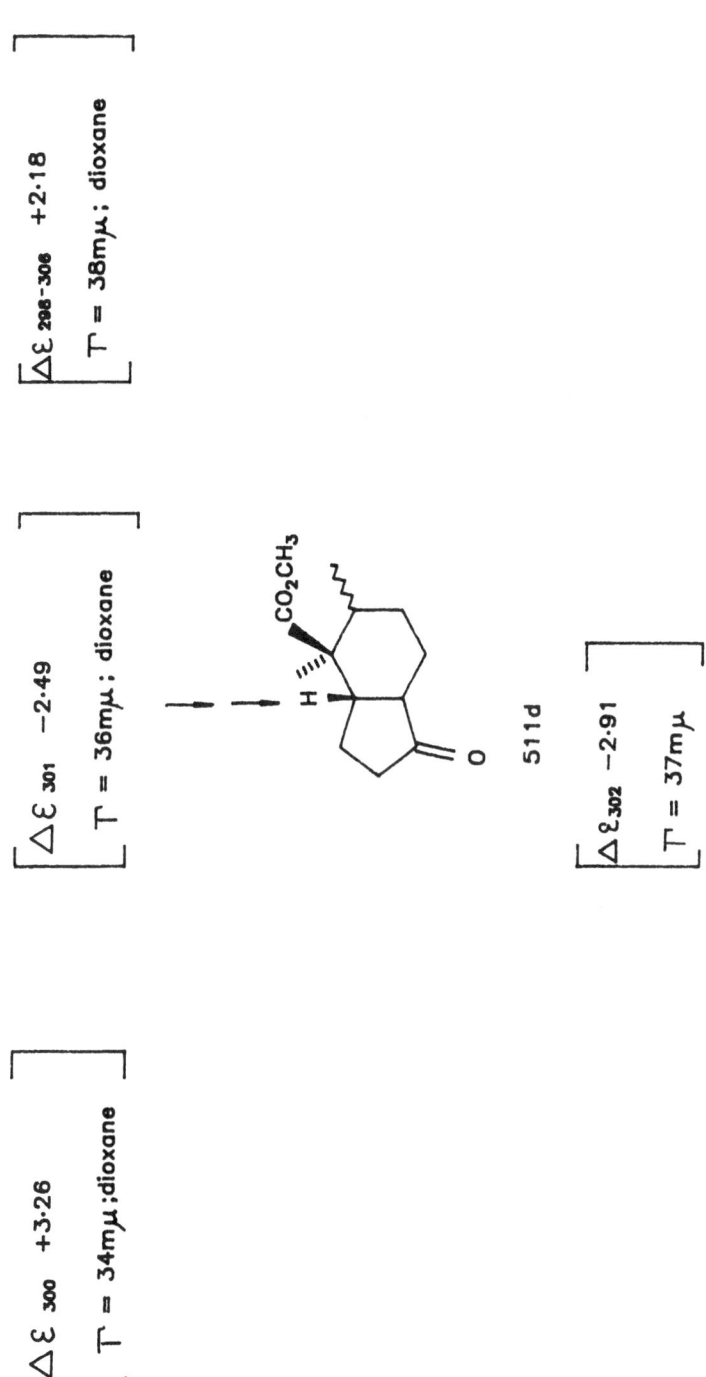

FIGURE 34. Chemical degradation of (-)-hardwickiic acid (**511**) to products permitting the assignment of ring-fusion configuration (*391*).

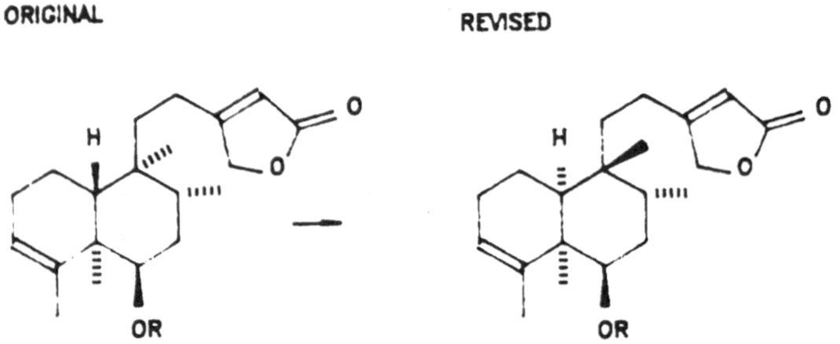

ORIGINAL   REVISED

258 Elongatolide C (R = Ang)
    (Solidagolactone II)

545 Elongatolide A (R = H)
    (Solidagolactone IV)

546 Elongatolide B (R = Ac)

259 Solidagolactone III (R = Tig)

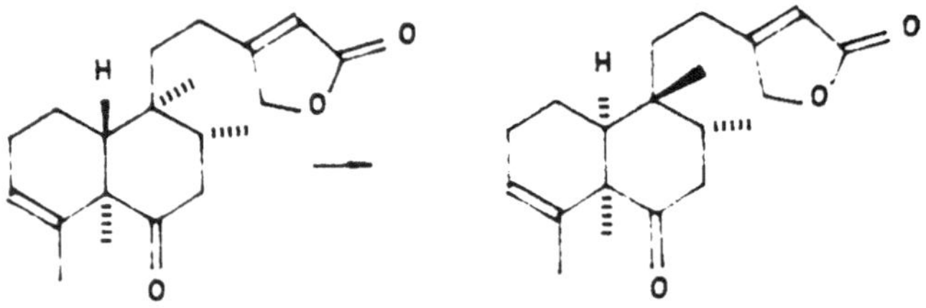

262 Solidagolactone V

FIGURE 35. Original *trans-ent*-clerodane structures assigned as the elongatolide series from *Solidago elongata* (*18*) and the solidagolactone series from *S. altissima* (*413*) and the revised 5α,10α-*cis*-normal-clerodane structures (*401, 402*).

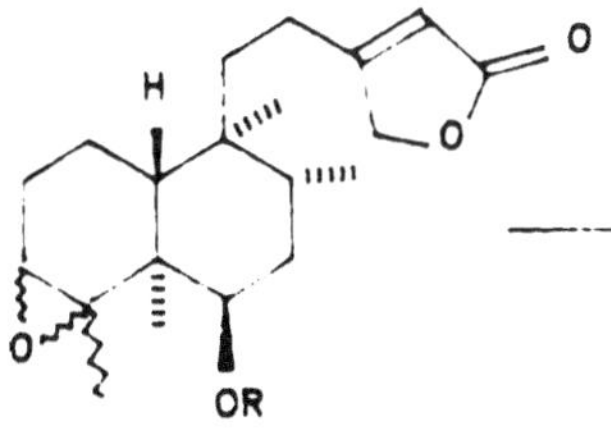

Figure 35. Original *trans-ent*-Clerodane Structures 517

553 Elongatolide D (R = Ac)
(Solidagolactone VI)

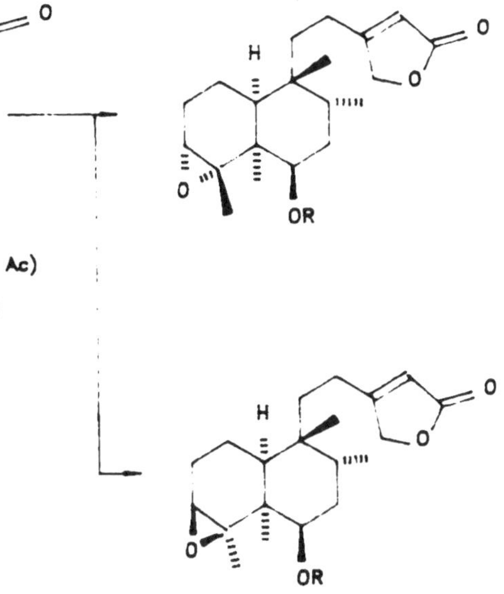

260 Elongatolide E (R = Ang) —— **260 Solidagolactone VII**

988 Solidagolactone VIII (R = Tig)

ORIGINAL

REVISED

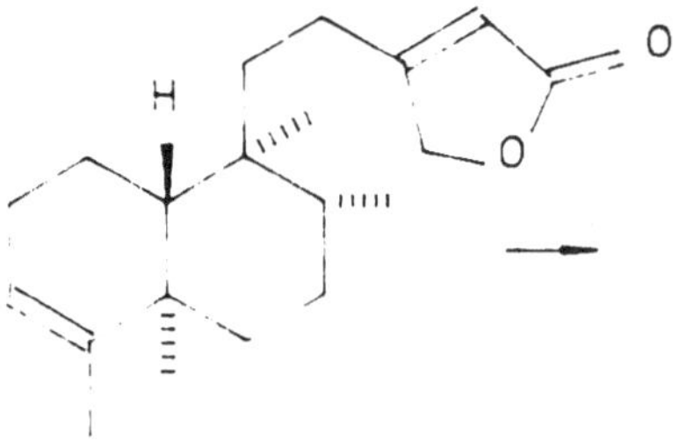

No revision

543 Solidagolactone

FIGURE 35. (*contd.*).

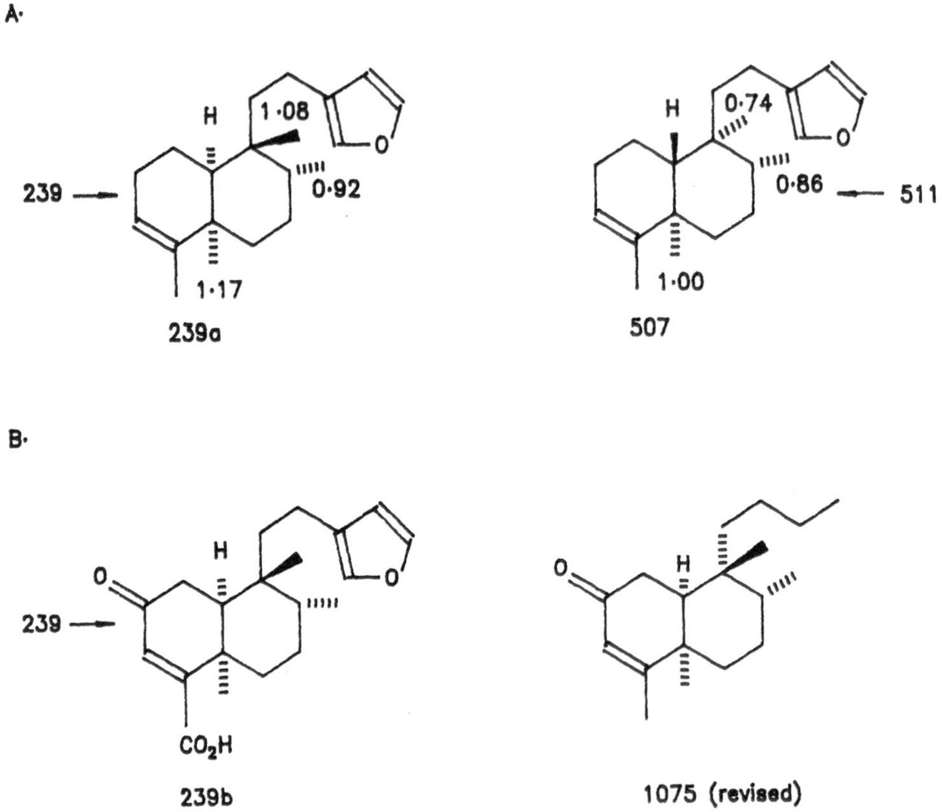

FIGURE 36. Attempts to correlate the structure of solidagoic acid A (239) with those of known cis-clerodanes: A. comparison of the methyl ^{1}H-NMR chemical shifts (δ) of a solidagoic acid A derivative, 239a, to those of a (-)-hardwickiic acid derivative, 507; B. enone derivative (239b) of solidagoic acid A and the revised structure of plathyterpenone used for comparison of CD optical properties.

Figure 38. Relative Stereochemistry of the *cis-ent*-Clerodane   519

262 Solidagolactone V   **947 Tricyclosolidagolactone**

FIGURE 37. Tricyclosolidagolactone (**947**) and possible precursor, solidagolactone V (**262**).

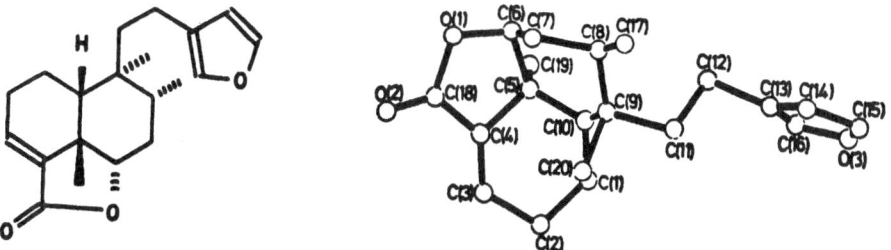

FIGURE 38. Relative stereochemistry of the *cis-ent*-clerodane, **583**, as determined from X-ray analysis.

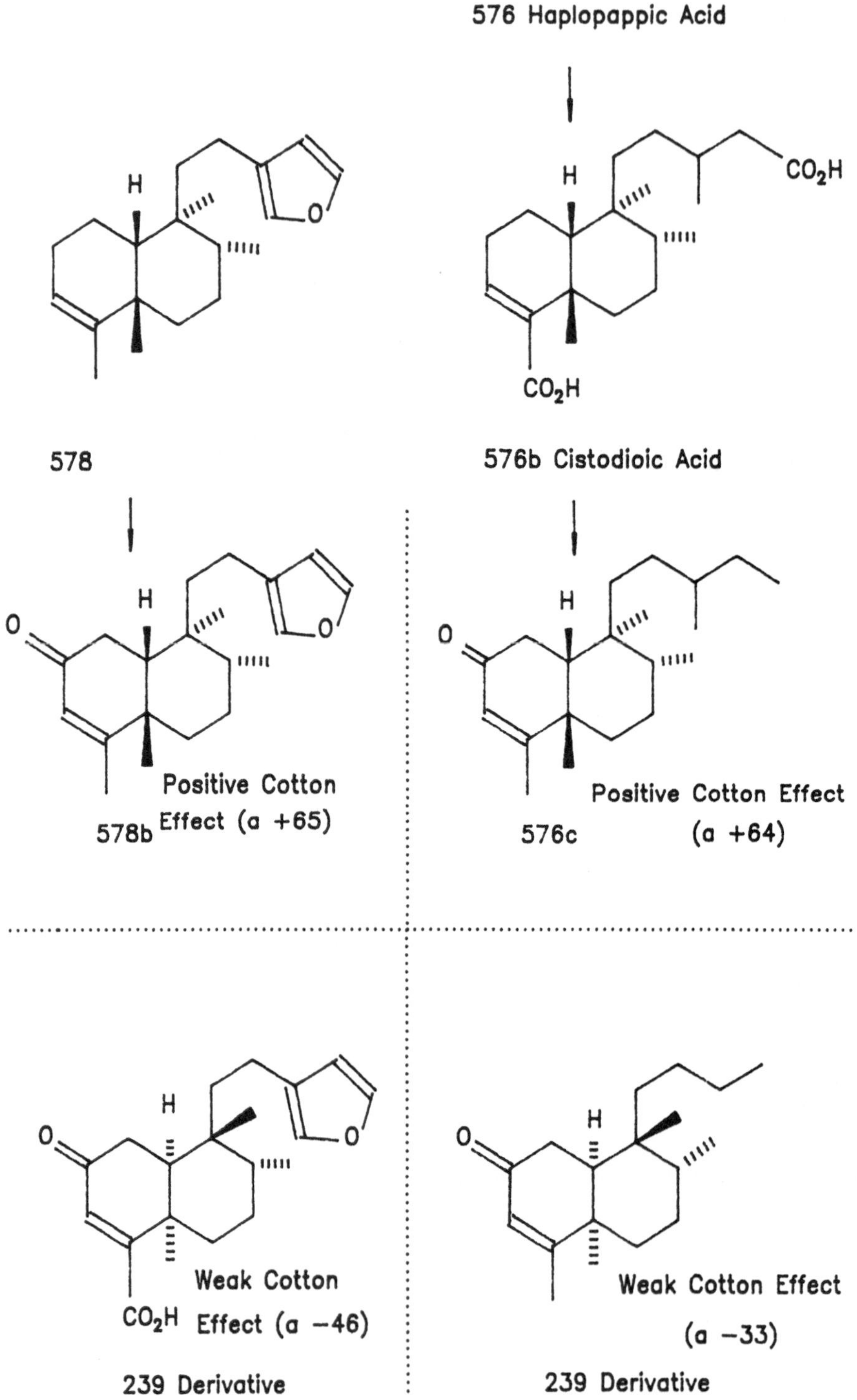

FIGURE 39. Comparison of the optical data obtained from 5β,10β-*cis-ent*-clerodane derivatives of **576** and **578** to that obtained from derivatives of 5α,10α-*cis*-normal clerodane **239** (*237*).

A· Originally Proposed Structures

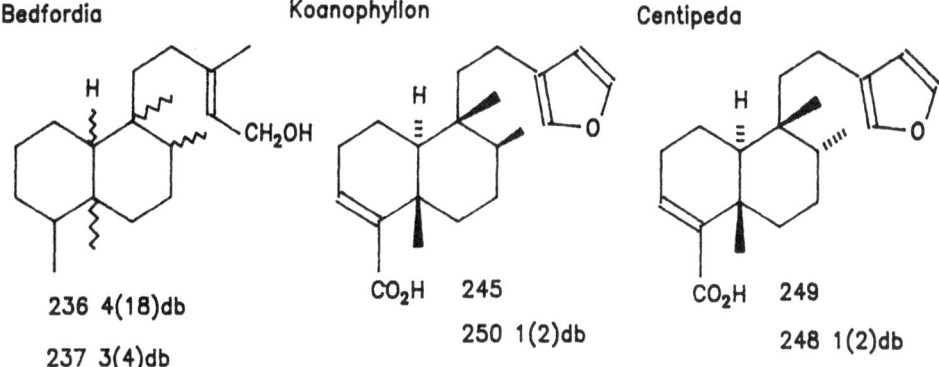

B· Originally Proposed Nidorella Structures

FIGURE 40. Originally proposed *trans*-normal-clerodane structures for compounds isolated from *Bedfordia*, *Koanophyllon*, *Centipeda* and *Nidorella*.

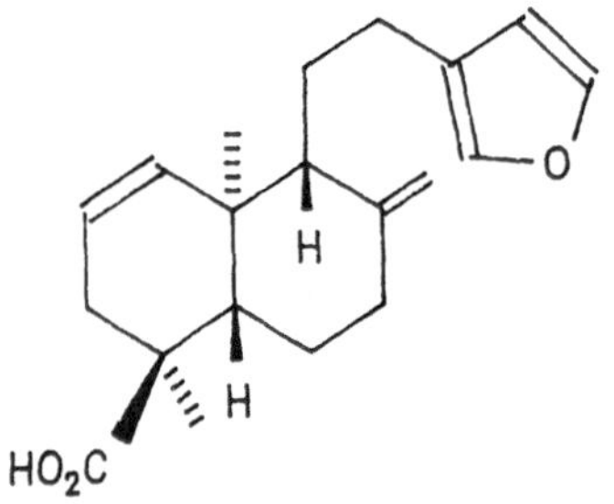

1. 1-Polyalthic acid

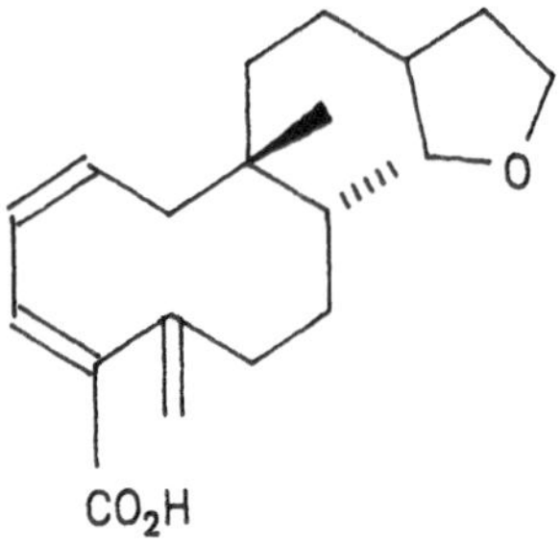

2. Secoreseda acid   (1)

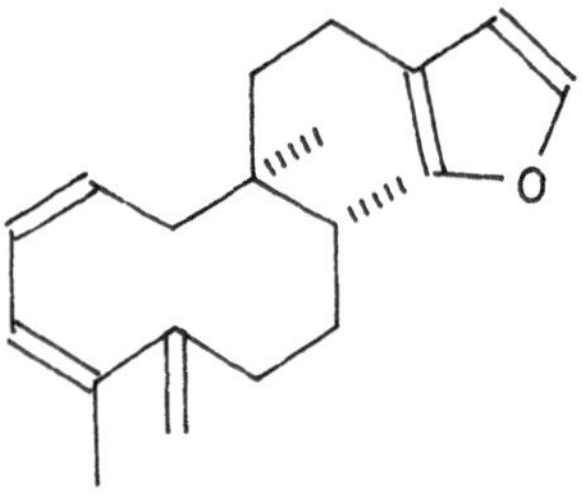

3. Secoreseda acid (2)

FIGURE 41. Proposed structures for the 5,10-seco-clerdanes: (1.) conyzic acid was first assigned the structure of Δ^1-polyalthic acid (411); (2.) first proposed structure of secoreseda acid (61) with C-9 orientivity opposite to that first assigned to strictic acid (420); (3.) second proposed structure of secoreseda acid (72) and by extension to strictic (573) and conyzic acids.

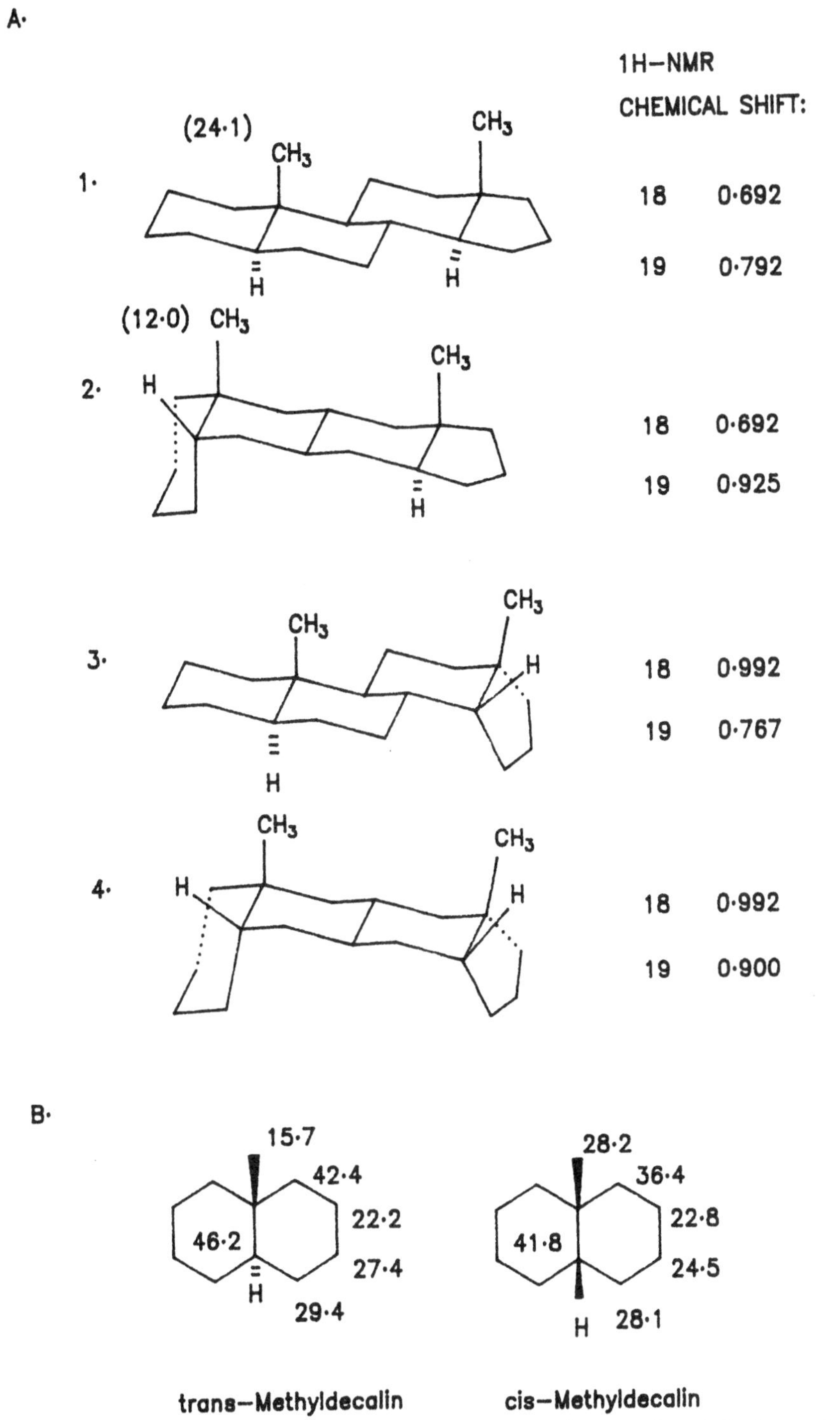

Figure 42. Effect of Ring Junction   523

A.

1H−NMR
CHEMICAL SHIFT:

(24·1)
CH₃
CH₃

1.   18   0·692

                                              19   0·792

(12·0) CH₃
H   CH₃

2.   18   0·692

                                              19   0·925

                        CH₃
              CH₃   H

3.   18   0·992

                                              19   0·767

        H

      CH₃   CH₃
H   H

4.   18   0·992

                                              19   0·900

B.

15·7   28·2
  42·4   36·4
    22·2   22·8
46·2   41·8
    27·4   24·5
 H   H   28·1
  29·4

trans−Methyldecalin   cis−Methyldecalin

FIGURE 42. The effect of ring junction configuration on the ¹H- and ¹³C-NMR chemical shifts of methyls attached at the ring junction in parallel instances: A. androstane steroid skeletal types 1-4; B. *cis-* and *trans*-methyldecalin.

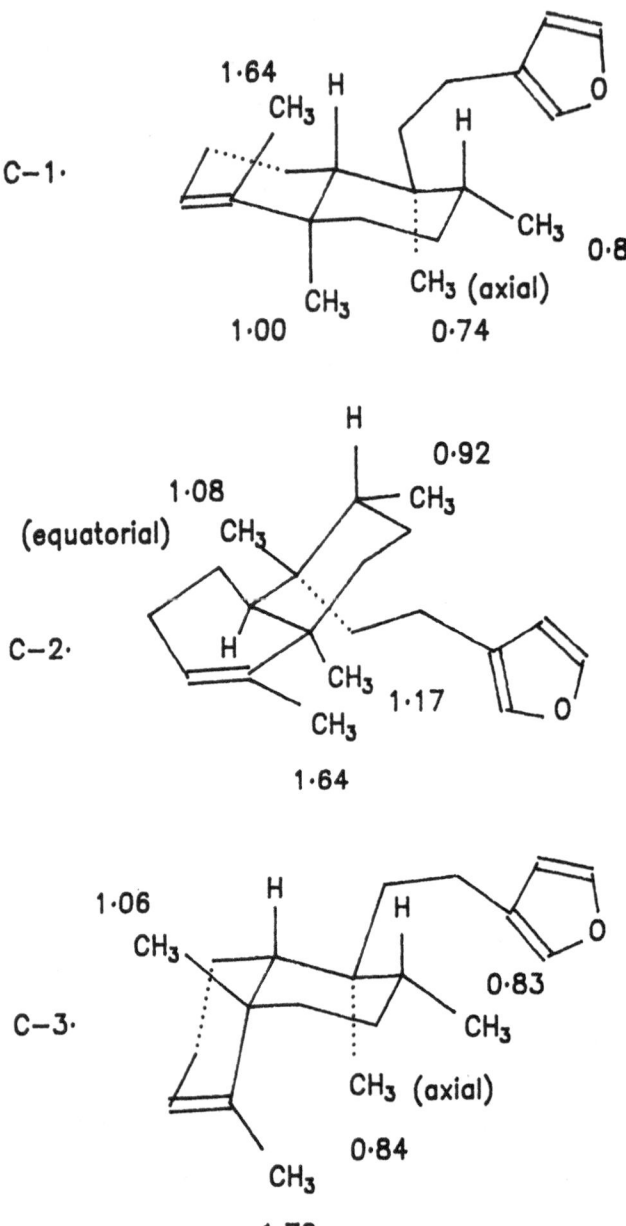

FIGURE 43. ^{1}H-NMR chemical shifts (δ, CDCl$_3$) of the methyls associated with three constitutionally equivalent representatives of the different clerodane ring-junction types, 5α-10β-*trans-ent*-clerodane (C-1), 5α,10α-*cis*-normal-clerodane (C-2) and 5β,10β-*cis-ent*-clerodane (C-3).

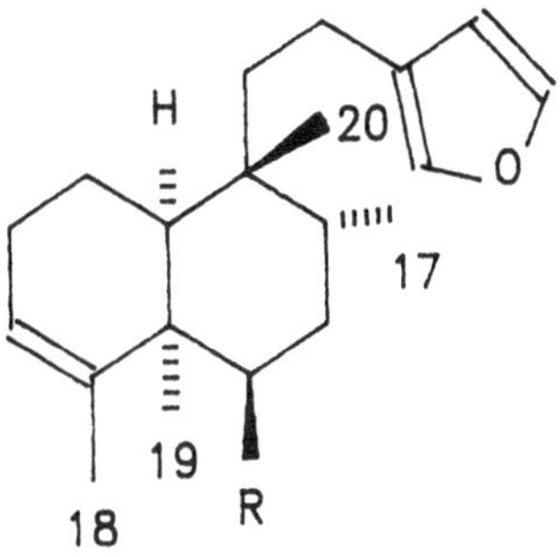

Figure 44. Effect of C-6 Substitution   525

^{1}H-NMR Chemical Shift (δ, CDCl$_3$)

Structure	H-17	H-18	H-19	H-20
1. R = H	0.92	1.64	1.17	1.08
2. R = OH	0.86	1.68	1.21	1.05
3. R = OAng	0.86	1.56	1.28	1.08
4. R = =O	0.87	1.54	1.24	1.24

FIGURE 44. The effect of C-6 substitution on the methyl ^{1}H-NMR chemical shifts of 5α,10α-cis-normal-clerodanes (Structures 1 from 15; Structures 2, 3 and 4 from 402).

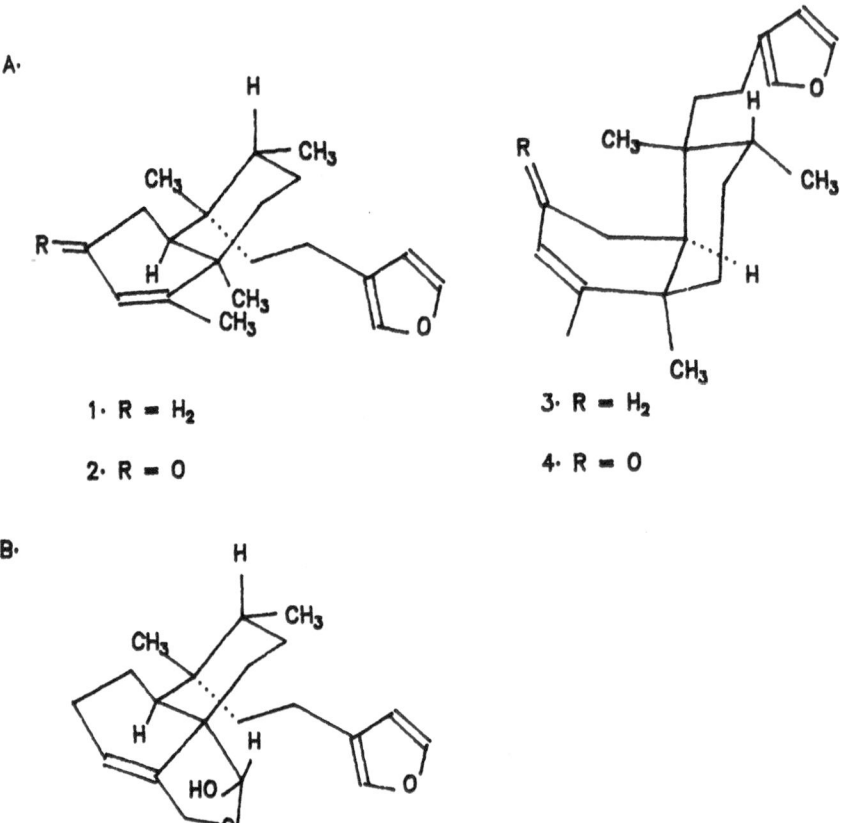

A.

1· R = H₂

2· R = O

3· R = H₂

4· R = O

B.

FIGURE 45. Conformational variation attributed to the 5α-10α-*cis*-normal-clerodanes: A. two favored conformational possibilities (*15*); B. probable conformation of compounds containing a 18,19-oxido linkage; C. the effect of the shielding zone of the C-2 ketone on the ^{1}H-NMR chemical shift of the C-9 methyl (H-20) of *cis*-clerodanes.

Figure 46. Two Conformations   527

C.

H   1·04

O

0·97

1·37

263a

268   R = Ang

269   R = Tig

FIGURE 45. (contd.).

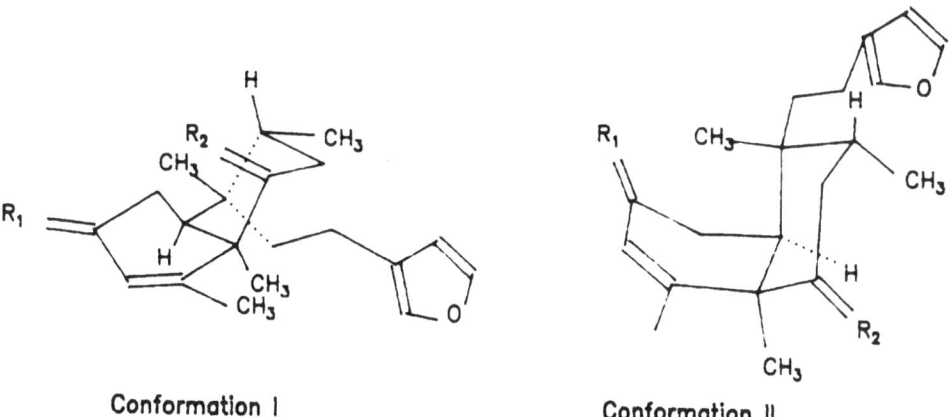

Conformation I   Conformation II

FIGURE 46. Two conformations assignable to $5\alpha,10\alpha$-cis-normal-clerodanes based on the nature of the R_1 and R_2 substituents.

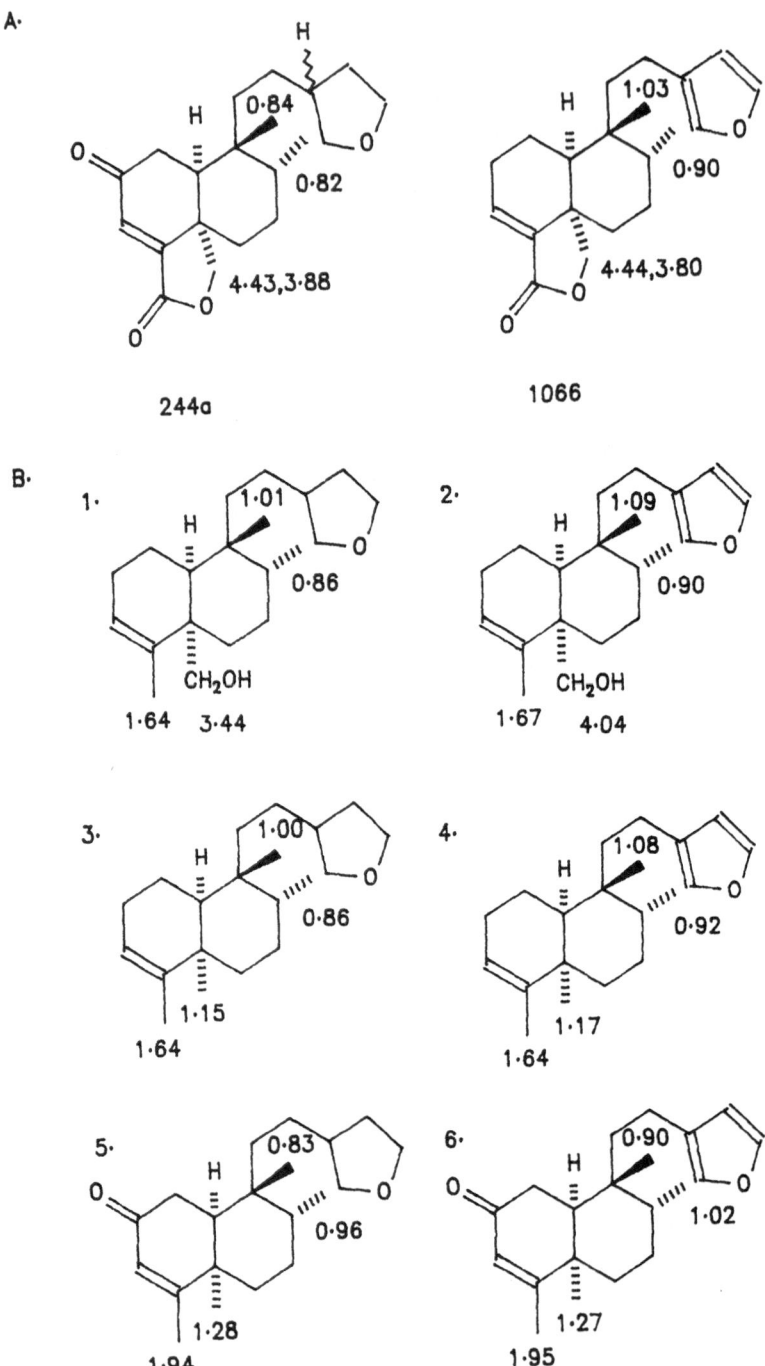

FIGURE 47. The effect of 18,19-oxido linkage on the probability of either conformation I or II: A. comparison of the methyl ^{1}H-NMR chemical shifts (δ, CDCl$_3$) of 244a and 1066; B. comparison of the methyl chemical shifts of tetrahydrofuryl-bearing cis-clerodanes (Structures 1, 3, and 5) and the corresponding furyl compounds (Structures 2, 4, and 6).

Figure 48. Methyl ¹H-NMR Chemical Shifts   529

A· Conformations at −50° C

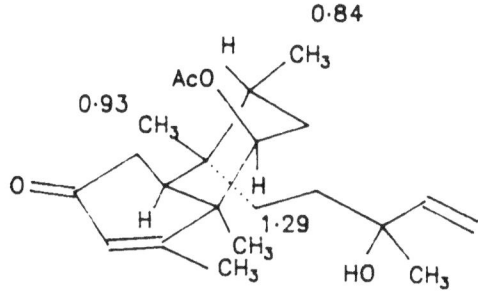

1025a Conformation I (C−6β−acetate: axial)

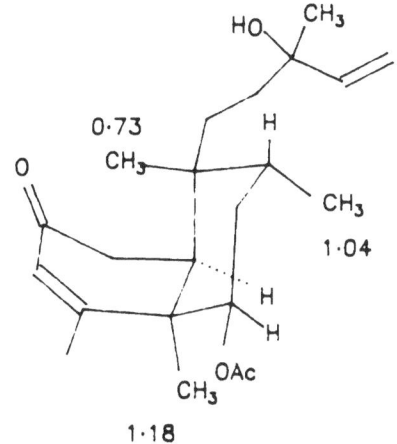

1025b Conformation II (C−6β−acetate: equatorial)

B· Room−Temperature Chemical Shifts:

H−17   0·93

H−19   1·32

H−20   0·93

FIGURE 48. Methyl ¹H-NMR chemical shifts (δ, CDCl₃) for the two conformations (**1025a** and **1025b**) of 1025 observed at −50°C (*170*).

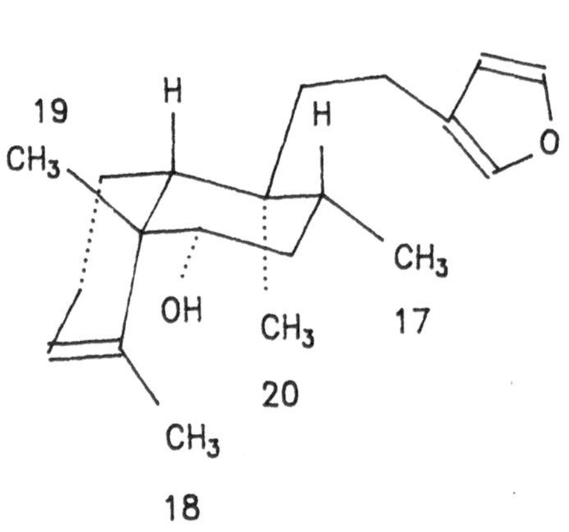

1· 582 Hydrogenolysis product 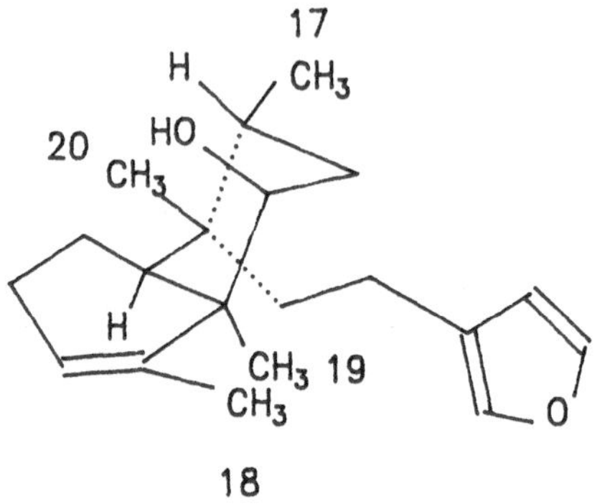   2· 545 Reduction product

Methyl Proton	Eu(pdm)$_3$ Addition-Induced Shift	
	1.	2.
H-18	10.0	3.9
H-19	9.6	5.3
H-17	1.7	2.0
H-20	4.2	3.3

FIGURE 49. Comparison of the Eu(pdm)$_3$ addition-induced ^{1}H-NMR chemical shift results for representative compounds of 5α-10α-*cis*-normal- and 5β,10β-*cis-ent*-clerodanes (*386, 378*).

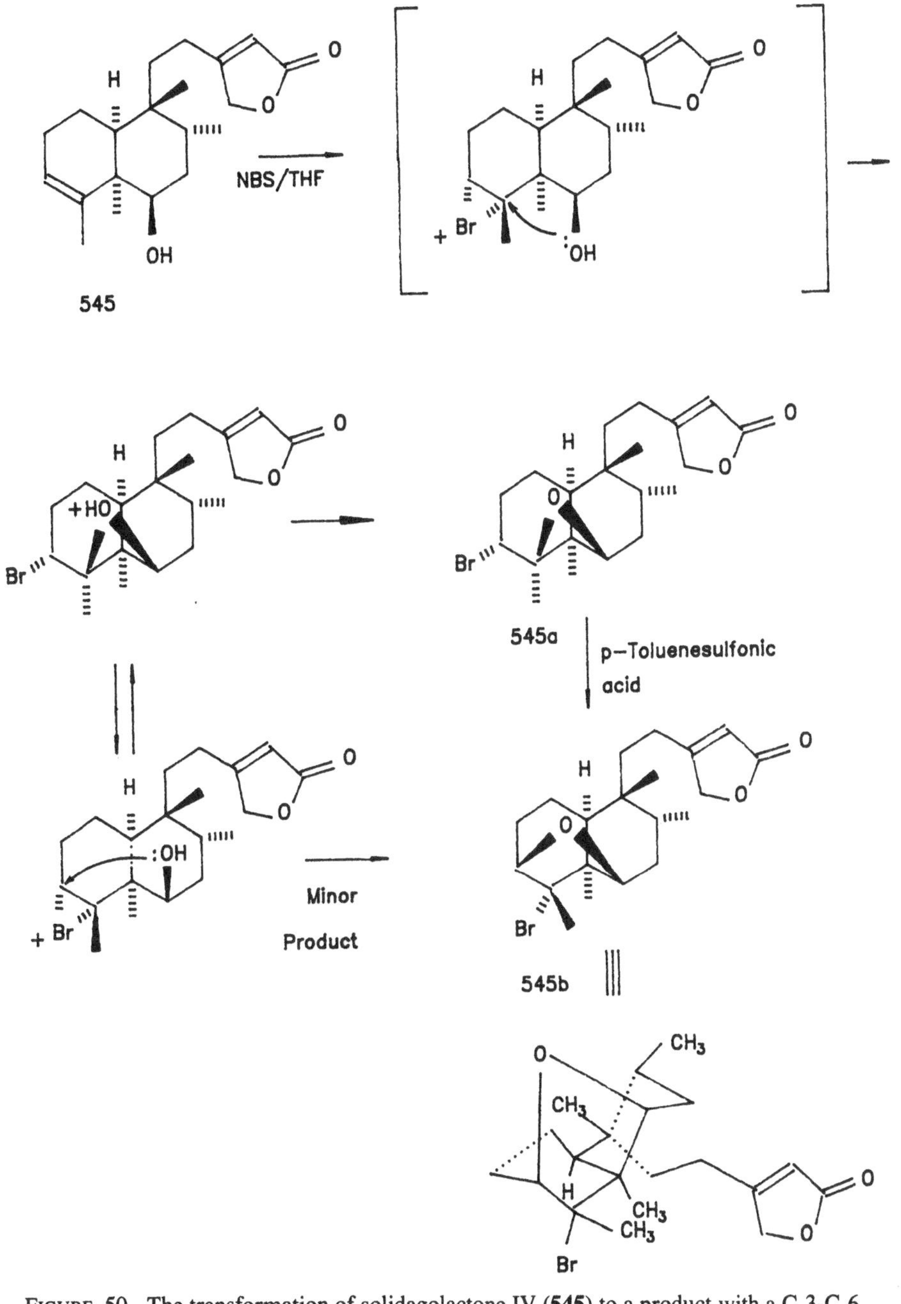

Figure 50. Transformation of Solidagolactone IV   531

FIGURE 50. The transformation of solidagolactone IV (545) to a product with a C-3-C-6 ether function (545b) as partial proof of a *cis*-ring junction (*378*).

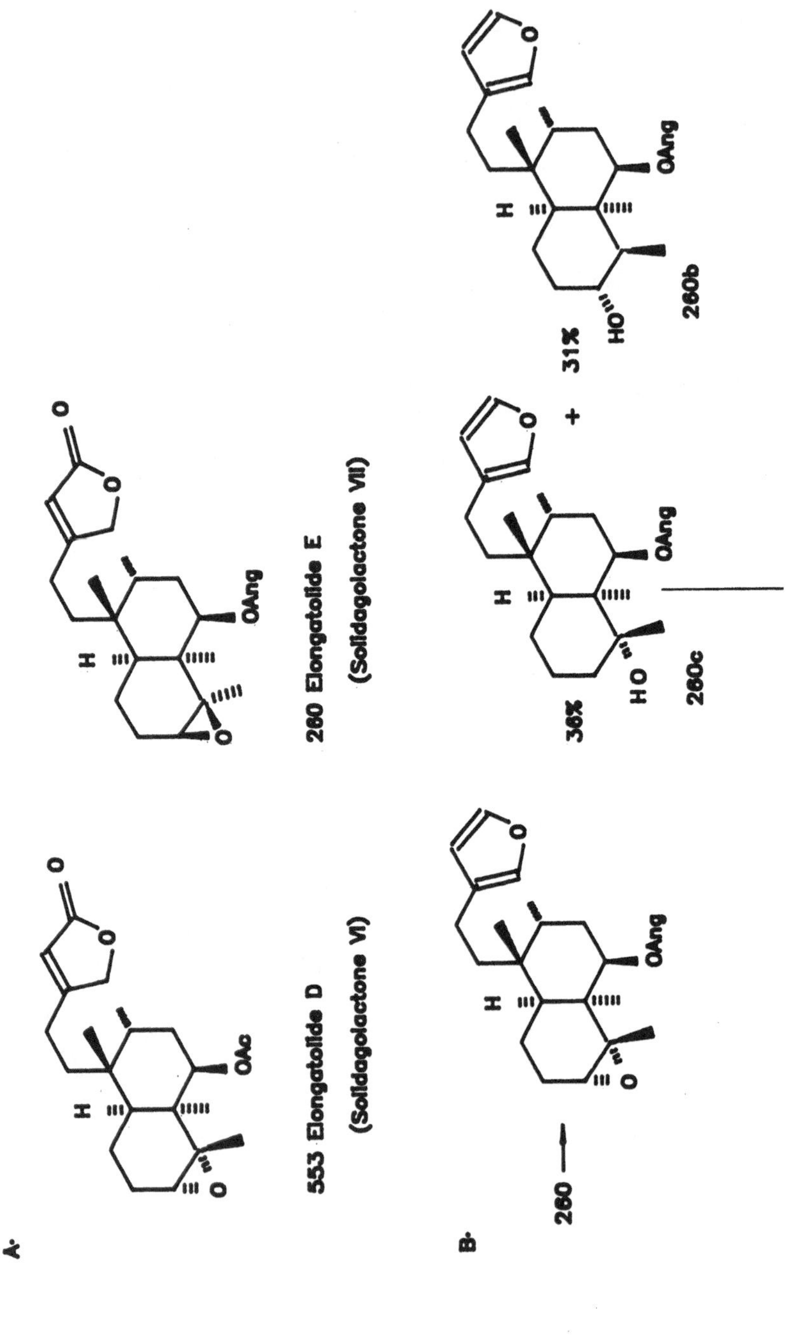
A.
553 Elongatolide D
(Solidagolactone VI)
260 Elongatolide E
(Solidagolactone VII)
B.
260
36%
260c
+
31%
260b
OAc
OAng
HO

Figure 51. Structure Determination of Elongatolides 533

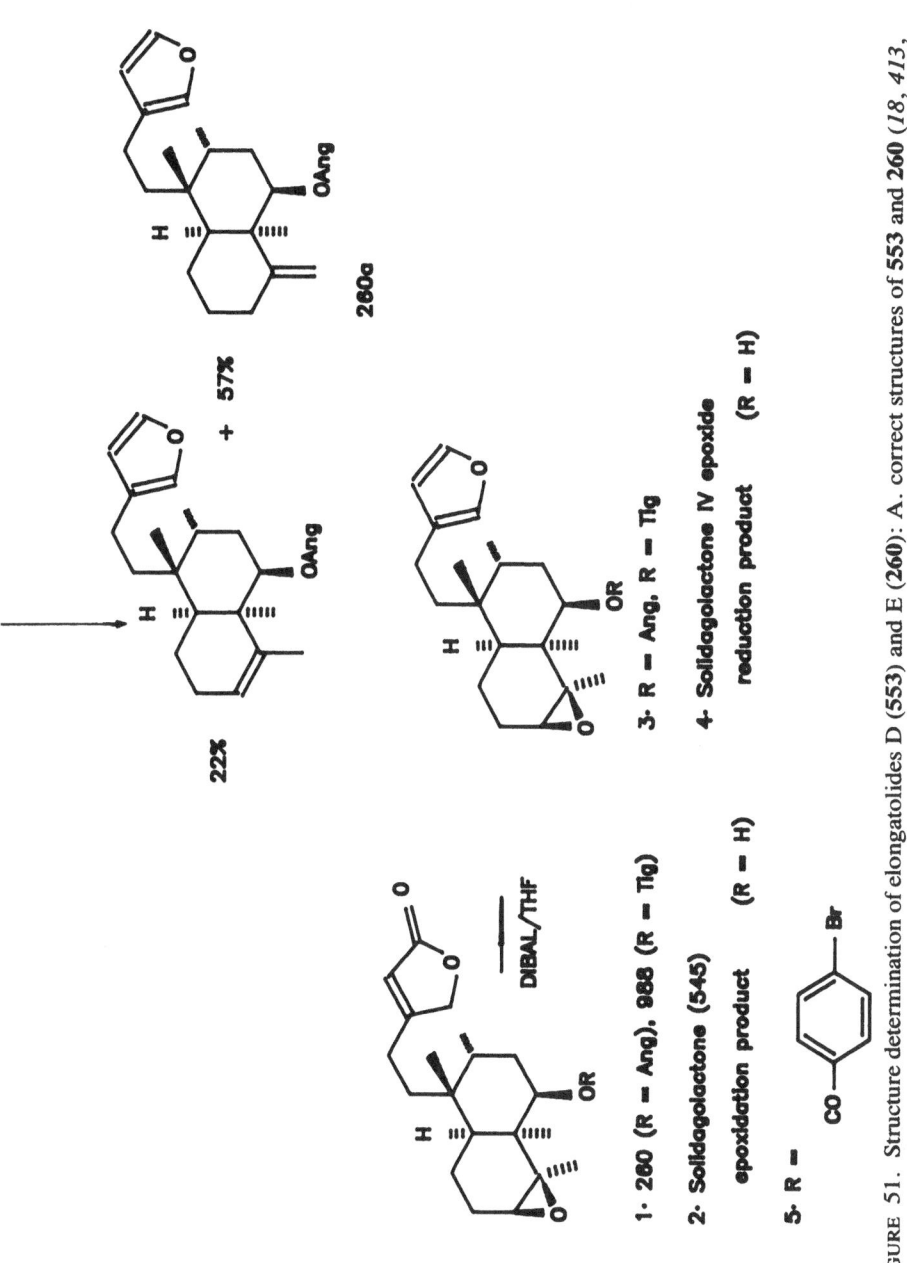

FIGURE 51. Structure determination of elongatolides D (**553**) and E (**260**): A. correct structures of **553** and **260** (*18, 413, 401*); B. incorrect assignment of structure to the products, **260b** and **260c**, of the epoxidation of elongatolide E (**260**) furyl derivative (*402*); C. correlation of the major product of the epoxidation of **545** (Figure 35B.) (Structure 2, R = H) with related butenolide-containing compounds, **260**, **988** and the p-bromobenzoate (Structure 5) (*401*).

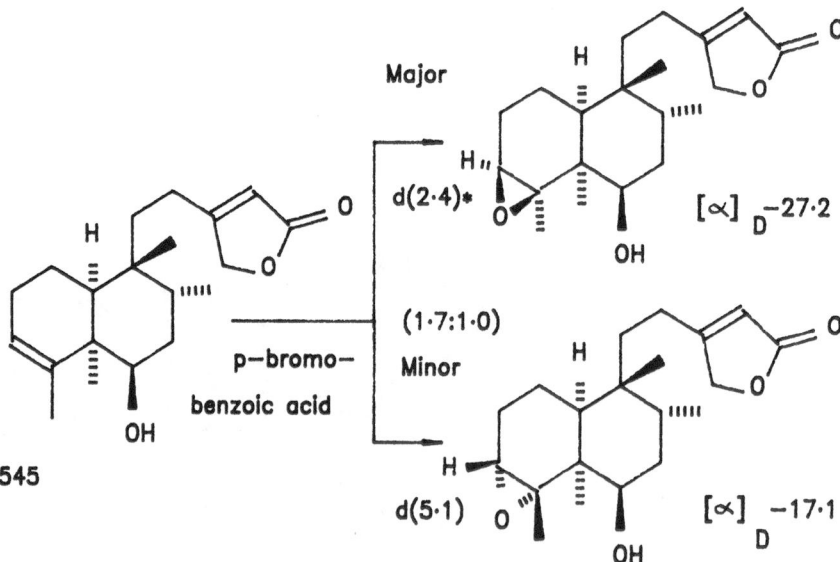

FIGURE 52. The effects of C-6 substitution on the products of 3,4-epoxidation of 5α-10α-*cis*-normal clerodanes: A. epoxidation of **239** (*386*); B. epoxidation of **545** (*401*); C. epoxidation of a **258** and **259** mixture (*291*); D. epoxidation of **546** (*413*).

Figure 53. Transformation of Solidagonic Acid   535

C.

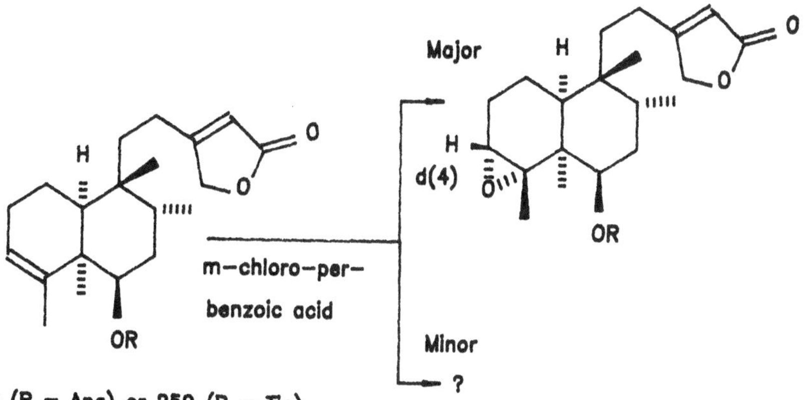

m—chloro—per—
benzoic acid

Major

d(4)

Minor

?

258 (R = Ang) or 259 (R = Tig)

D.

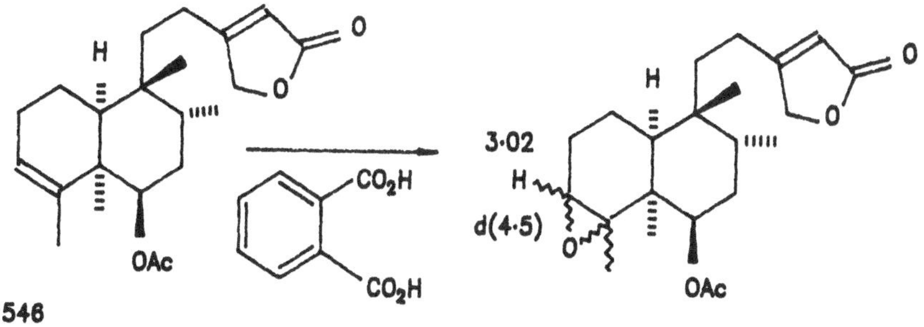

3·02

d(4·5)

546

OAc

FIGURE 52. (contd.).

BFl₃

3·45
HO

m(W₁/₂ = 6)

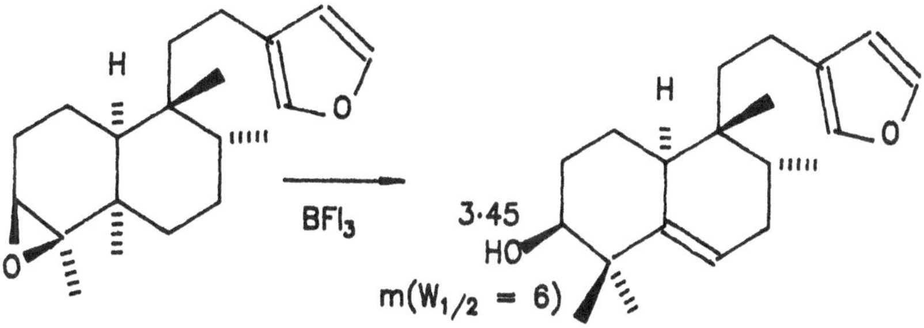

239 Solidagoic Acid A

derivative

C₃—OH axial

FIGURE 53. Transformation of solidagoic acid (239) epoxy-derivative to the 3β-hydroxy product (386).

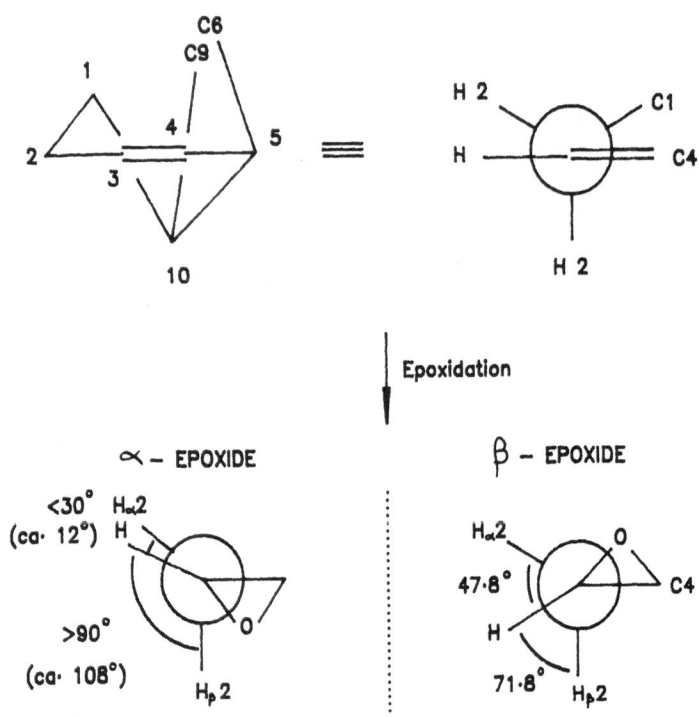

Coupling Constants

<table>
<tr><td>Tori equation
calculation:</td><td>$J_{3,2\alpha}$ = 4.88Hz (12°)</td><td>$J_{3,2\alpha}$ = 0.50Hz (47.8°)</td></tr>
<tr><td></td><td>$J_{3,2\beta}$ = 0.49Hz (108°)</td><td>$J_{3,2\beta}$ = 2.30Hz (71.8°)</td></tr>
<tr><td>Observed:</td><td>$J_{3,2\alpha}$ = 5.1Hz</td><td>$J_{3,2\alpha}$ = 0.0Hz</td></tr>
<tr><td></td><td>$J_{3,2\beta}$ = 0.0Hz</td><td>$J_{3,2\beta}$ = 1.5Hz</td></tr>
</table>

FIGURE 54. The application of Tori equation calculations to the prediction of H-3 coupling constants of 5α-10α-*cis*-normal-clerodane 3,4-α- and β-epoxides (*378*).

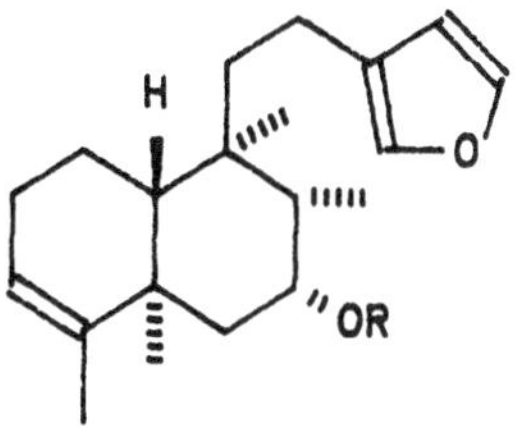

Figure 55. Effect of Different C-7 Substituents   537

509 R = H
509a R = Ac

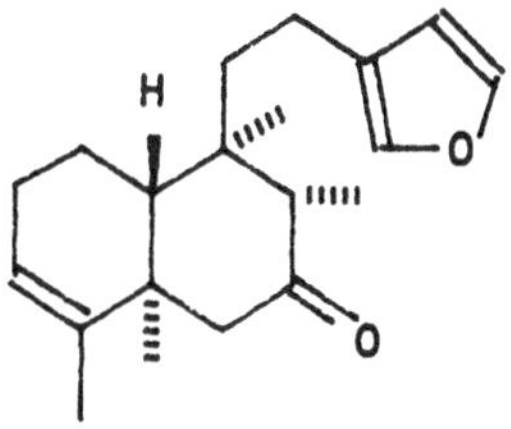

509b

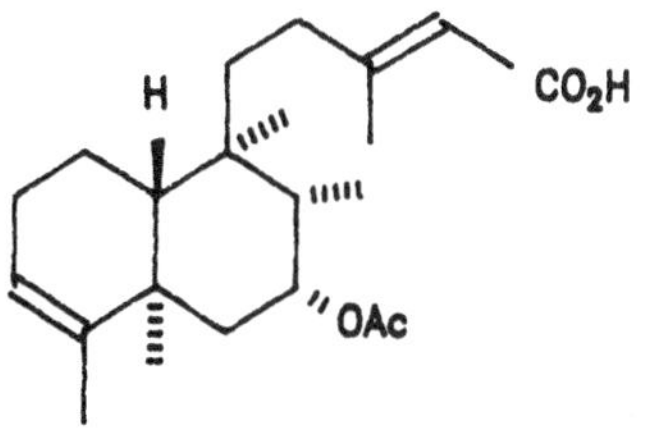

500

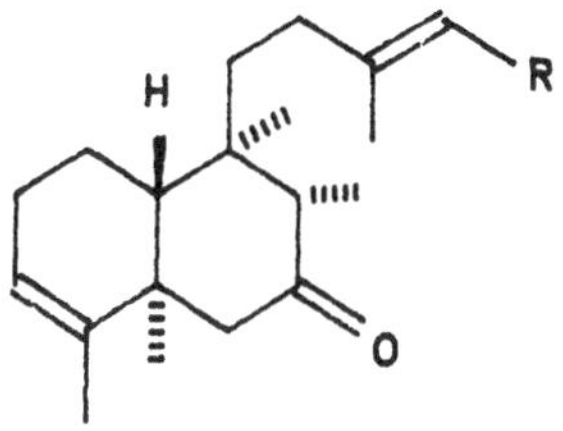

500a R = CH$_2$OAc

500b R = CO$_2$CH$_3$

^{1}H-NMR Chemical Shift (δ, CDCl$_3$)

Structure	H-19	H-20
509	1.36	1.02
509a	1.34	1.02
500	1.17	0.97
509b	1.06	0.75
500a	0.95	0.71
500b	0.96	0.72

FIGURE 55. The effect of different C-7 substituents on the H-19 and H-20 ^{1}H-NMR chemical shifts (δ) of 5α,10β-*trans-ent*-clerodanes.

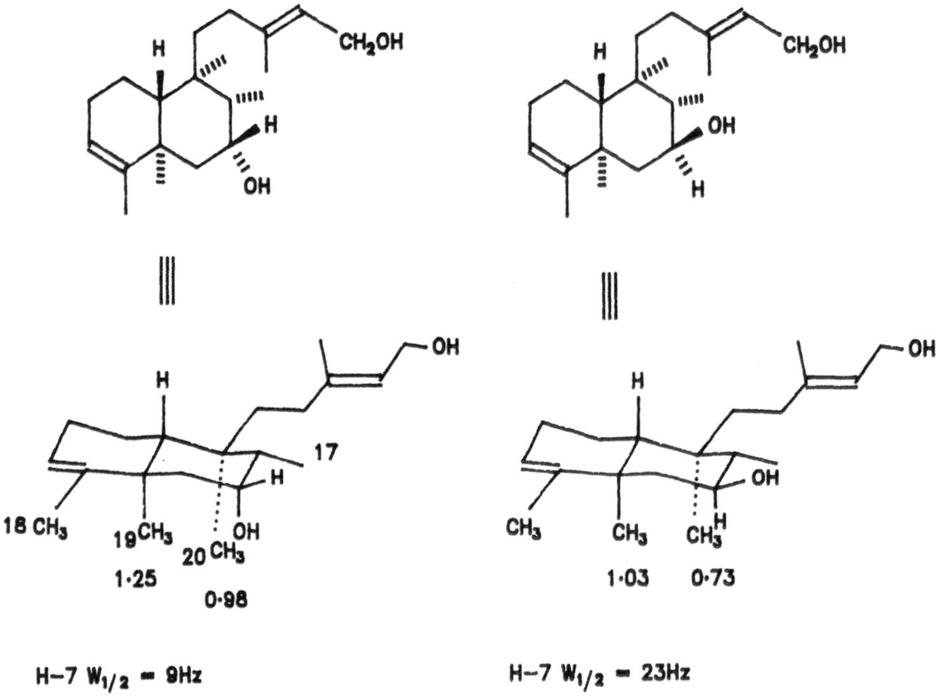

FIGURE 56. The effects of C-7 α- and β-hydroxylation on the ^{1}H-NMR properties of $5\alpha,10\beta$-trans-ent-clerodanes (412).

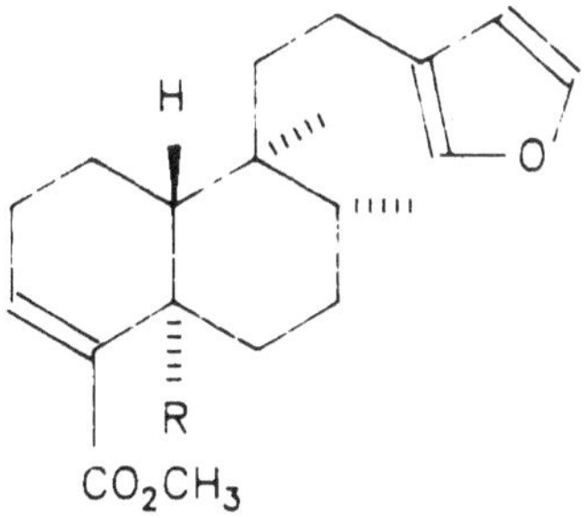

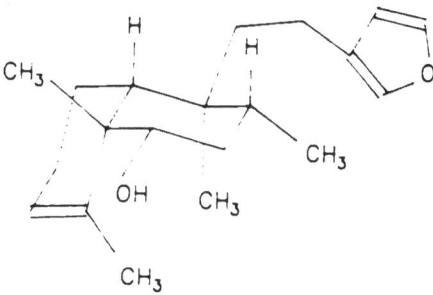

Compound	C19 Substitution	H−20 Chem· Shift
900	CH$_2$OAc	0·84
902	CH$_2$OAng	0·89
903	CH$_2$Oi−Val	0·85
1033	CH$_2$OH	0·79
904	CHO	0·60

FIGURE 57. The effect of different 5α,10β-*trans-ent*-clerodane C-19 substituents on the ¹H-NMR chemical shift (δ) of H-20.

FIGURE 58. Conformation assigned by McCrindle et al. (*386*) to the 5β-10β-*cis-ent*-clerodanes.

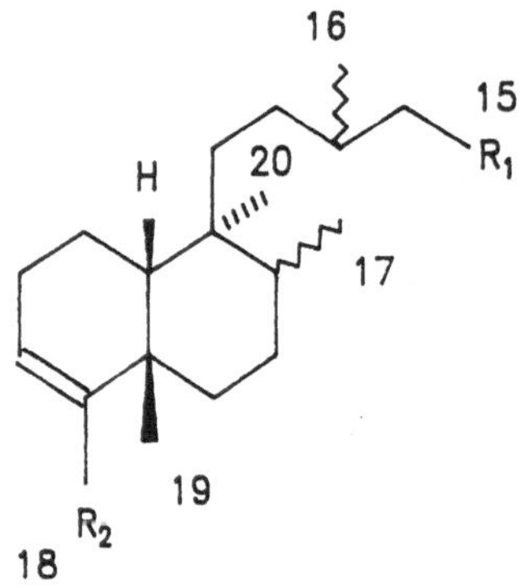

¹H-NMR Chemical Shift(CDCl$_3$)

Structure	H-17	H-18	H-19	H-20
1. R$_1$ = CO$_2$H, R$_2$ = CH$_2$OH	0.74	-	1.12	0.76
2. R^1 = CO$_2$H, R$_2$ = CH$_2$OAc	0.74	-	1.11	0.69
3. R$_1$ = R$_2$ = CH$_2$OH	0.75	-	1.12	0.76
4. R$_1$ = CO$_2$CH$_3$, R$_2$ = CH$_2$OH	0.74	-	1.12	0.76
5. R$_1$ = CH$_2$OH, R$_2$ = CO$_2$CH$_3$	0.75	-	1.22	0.77

FIGURE 59. ^{1}H-NMR chemical shifts (δ, CDCl$_3$) of the methyl protons on cistodioic acid (R$_1$ = R$_2$ = CO$_2$H) derivatives (*33, 469*).

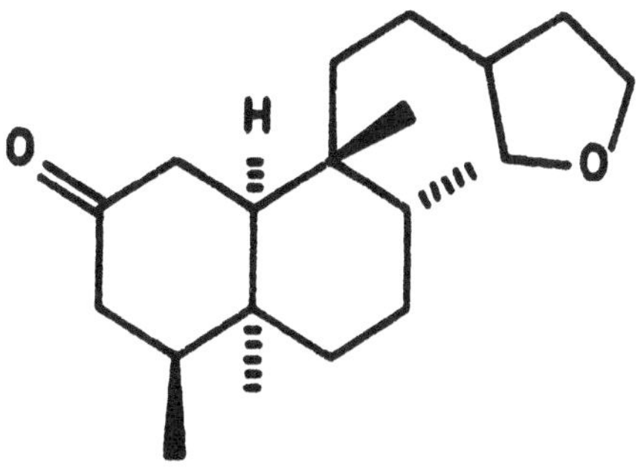

Figure 60. Structures of Clerodane Diterpenes 541

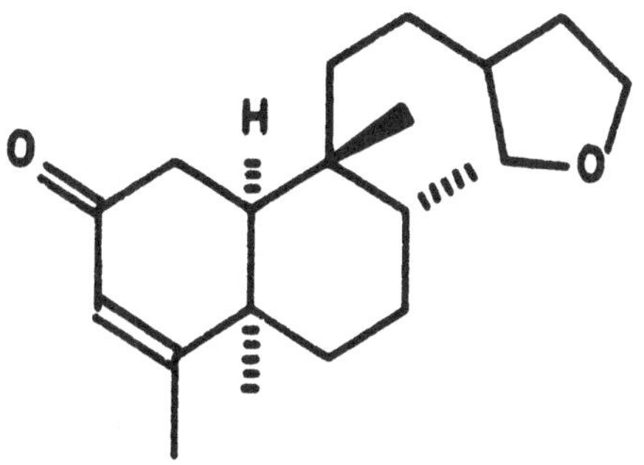

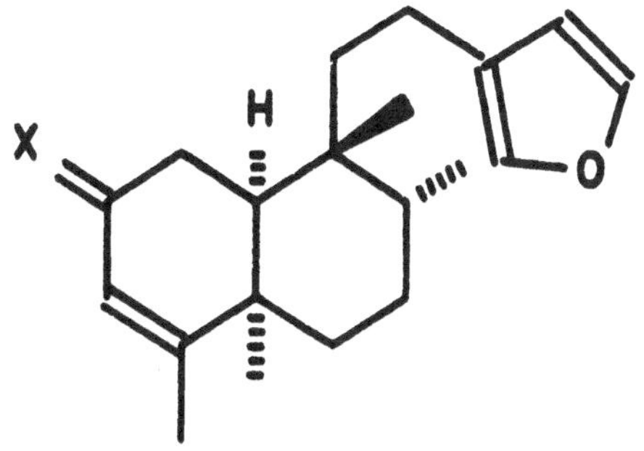

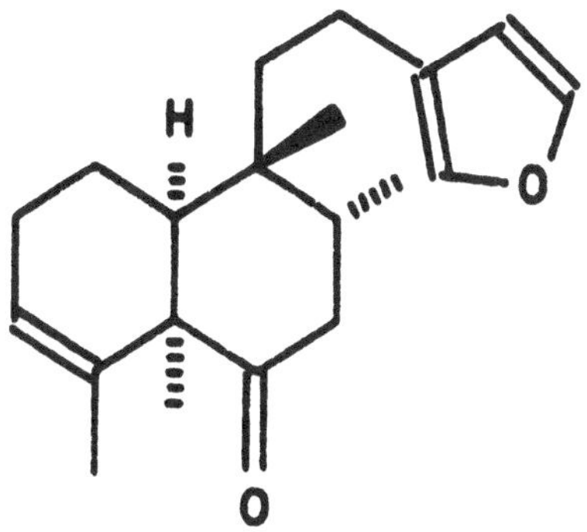

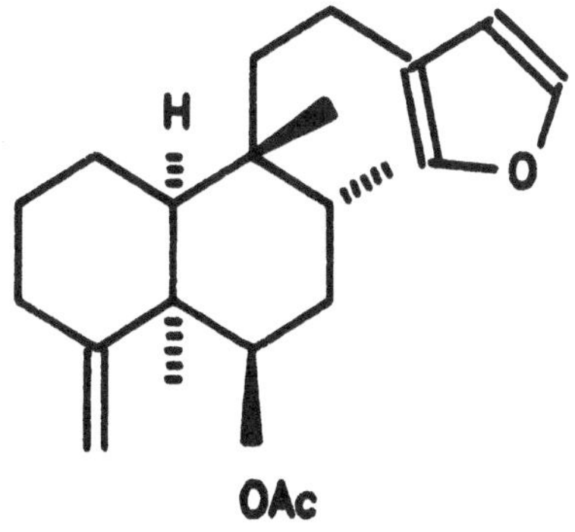

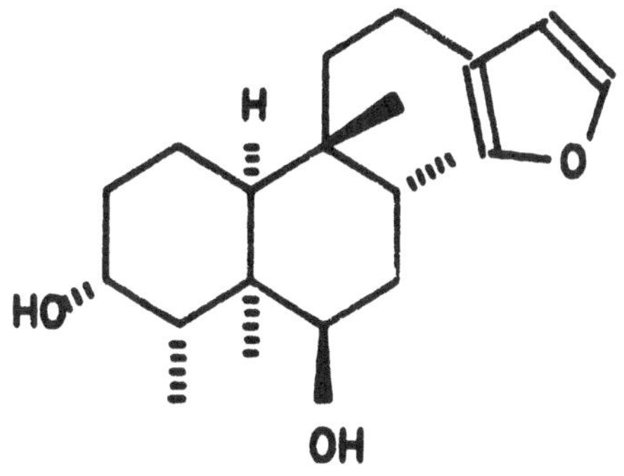

FIGURE 60. Structures of clerodane diterpenes (included in Table 14 compilation of ^{1}H-NMR data) that are derivatives of the natural products listed in Table 2.

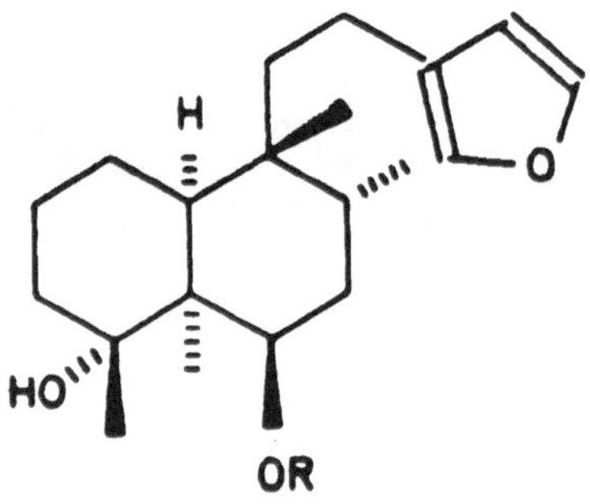

260c R = H
260d R = Ac

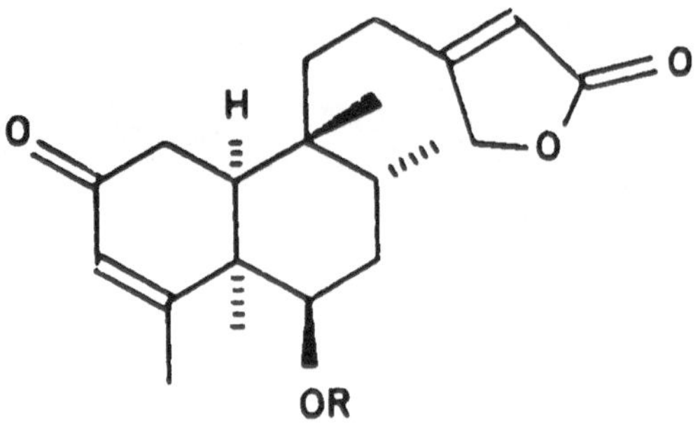

263a R = Ang or Tig

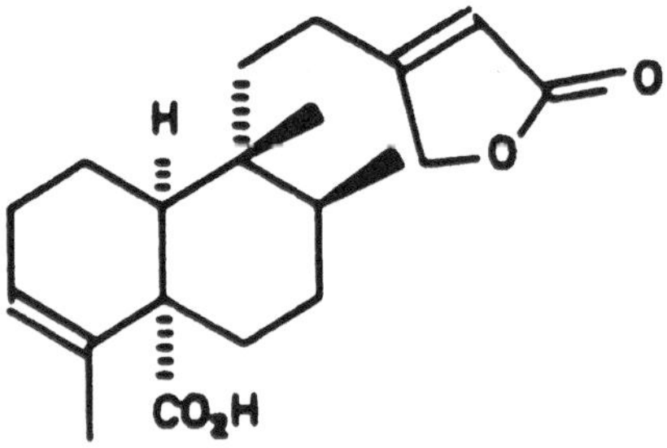

1091a

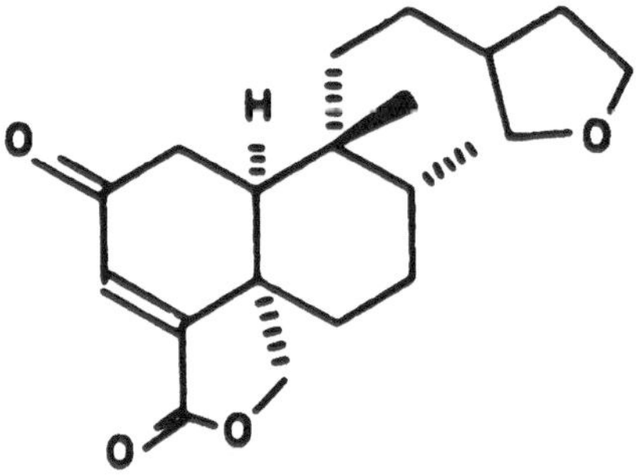

244a

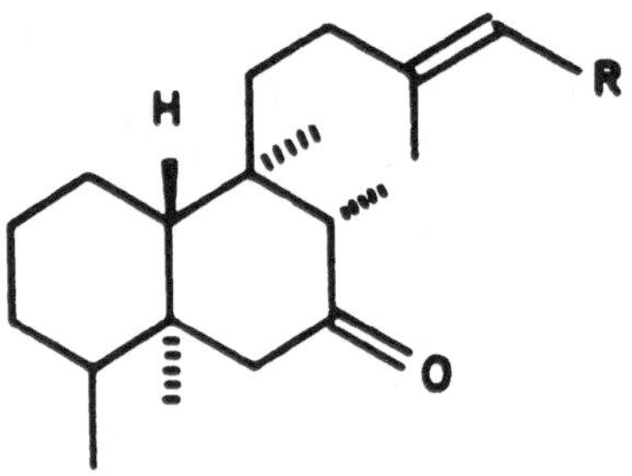

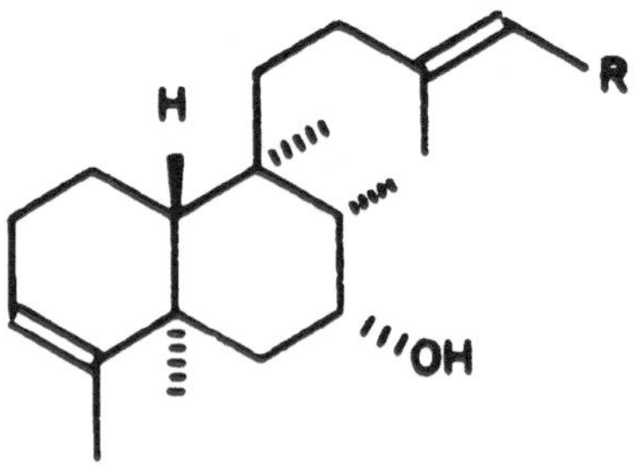

500a R = CO$_2$CH$_3$
500d R = CH$_2$OAc

500b R = CH$_2$OH
500c R = CO$_2$H

FIGURE 60. (contd.).

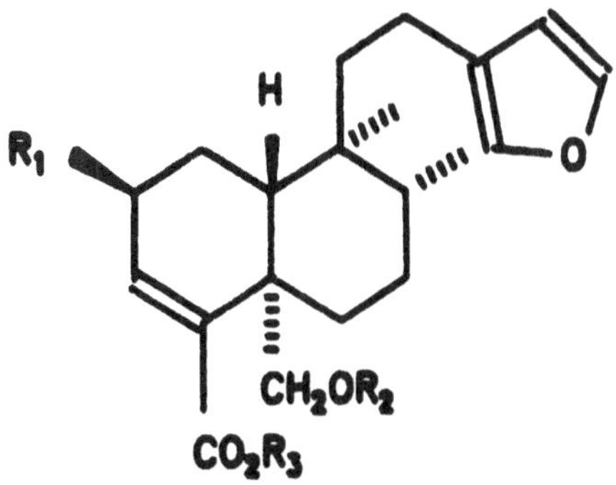

Figure 60. Structures of Clerodane Diterpenes 543

1033a R$_1$ = R$_3$ = H, R$_2$ = Ac

1034a R$_1$ = OAc, R$_2$ = Ac, R$_3$ = H

1034b R$_1$ = OH, R$_2$ = Ac, R$_3$ = CH$_3$

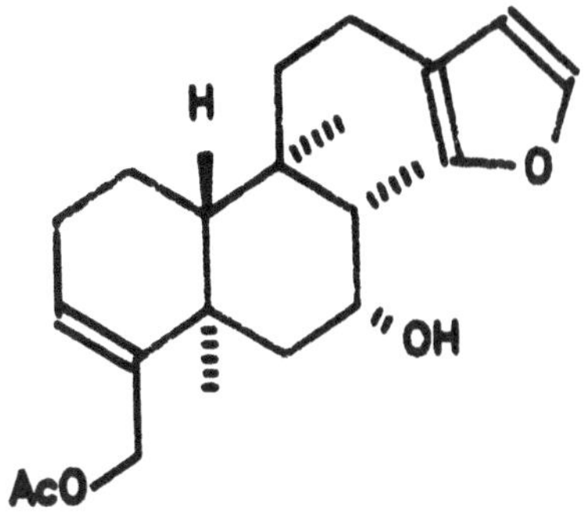

509a

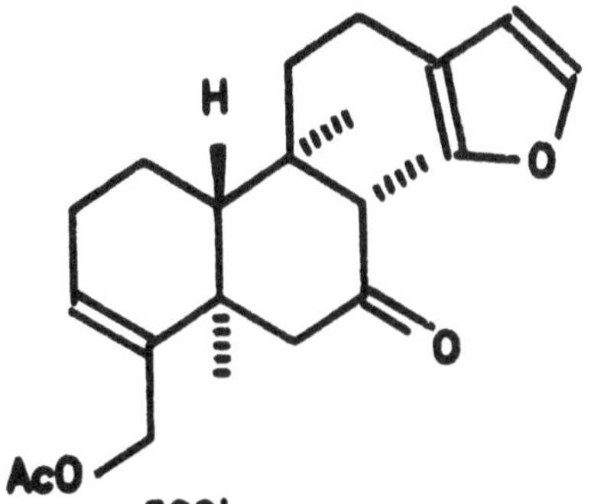

509b

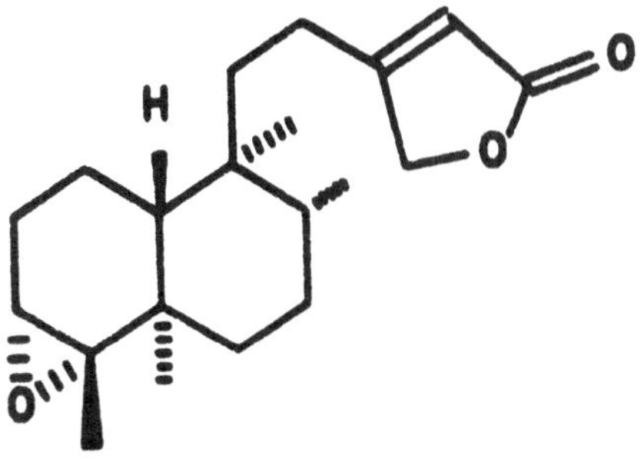

551b

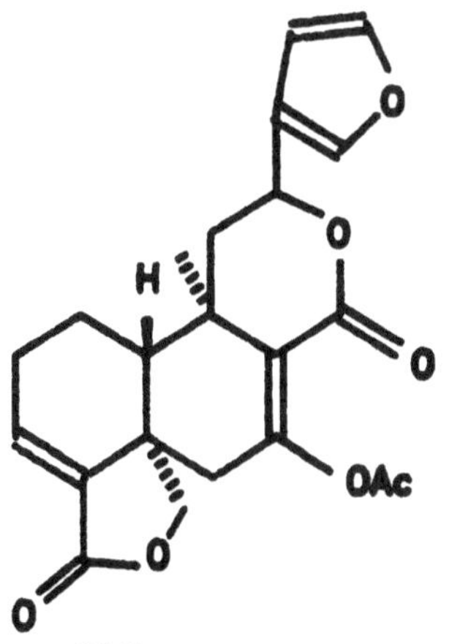

526a

FIGURE 60. (contd.).

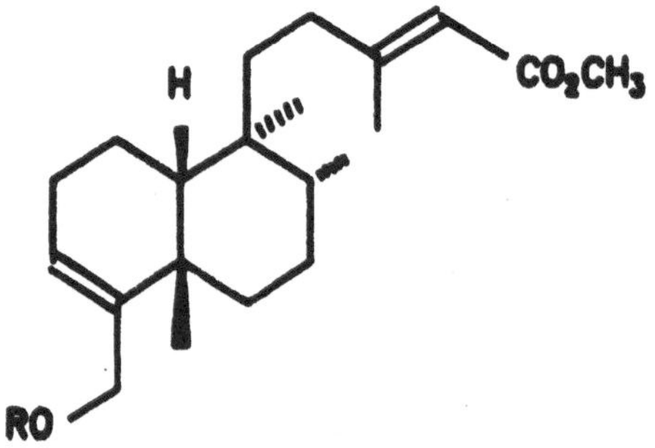

1055a   R = Ac

1056a   R = H

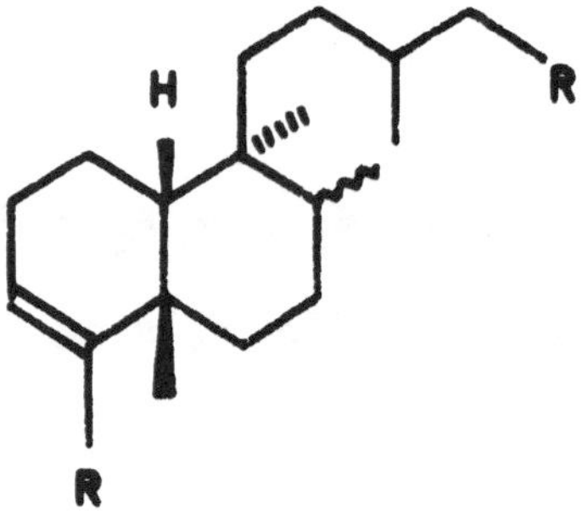

CISTODIOL DIACETATE   R = CH₂OAc

CISTODIOIC ACID   R = CO₂H

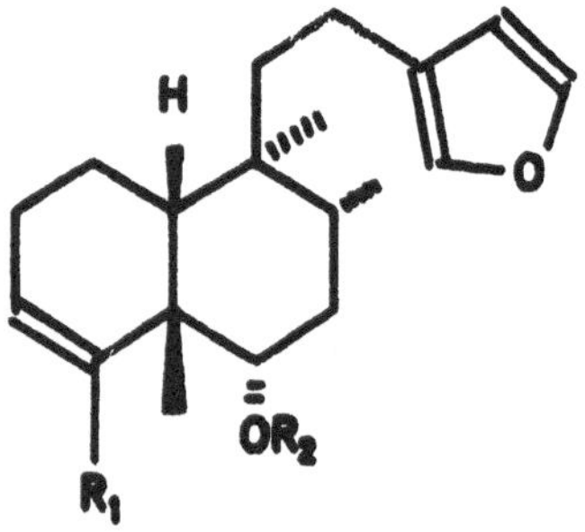

581a   R₁ = CH₃, R₂ = H

582a   R₁ = CH₂OAc, R₂ = Ac

FIGURE 60. (contd.).

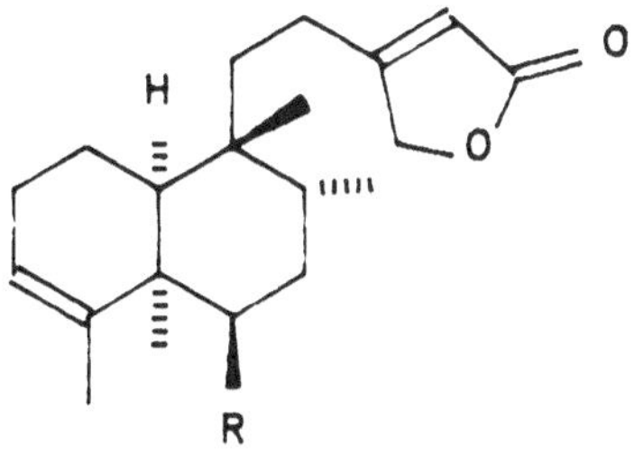

Figure 61. Structures of Clerodane Diterpenes   545

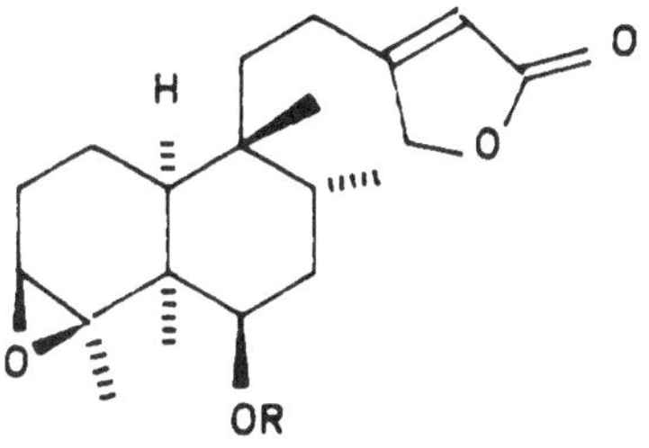

1· (545) R = OH   3· (260) R = Ang

2· (262) R = =O   4· (988) R = Tig

10 R = OAc   8· R = H

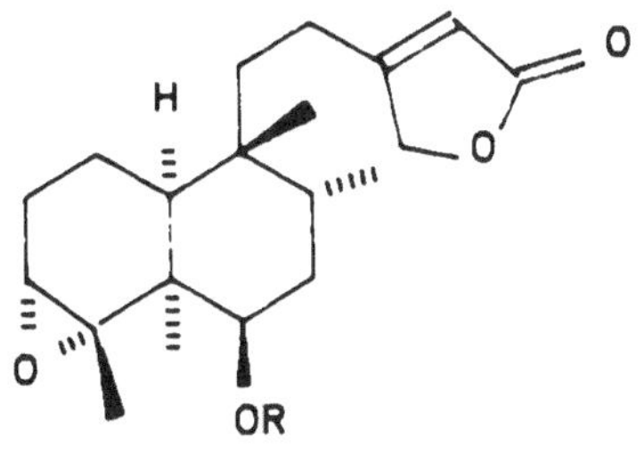

9· R = H   12·

11· R = Ac

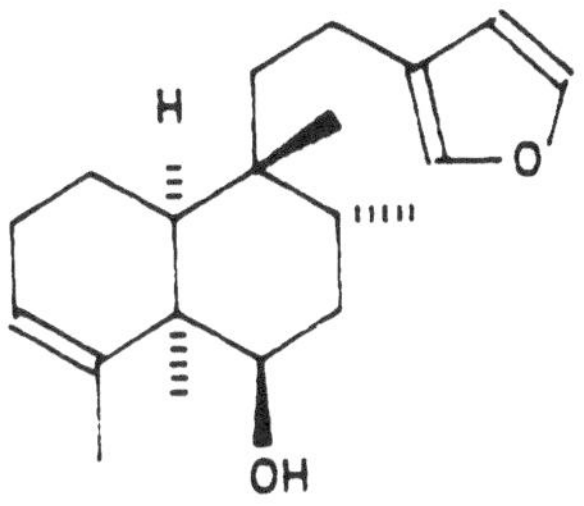

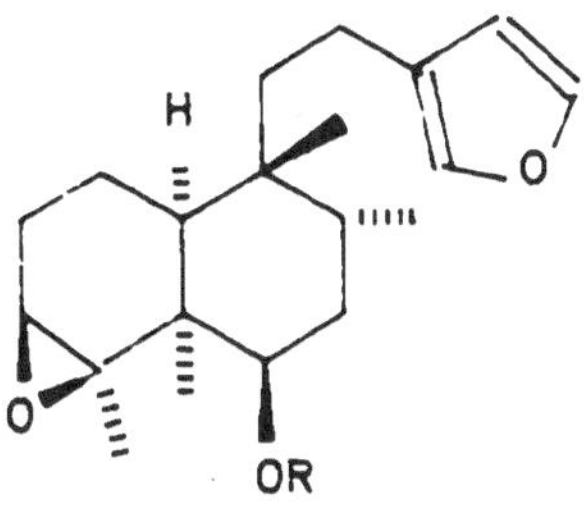

5· R = Ang   13·

6· R = Tig

7· R = H

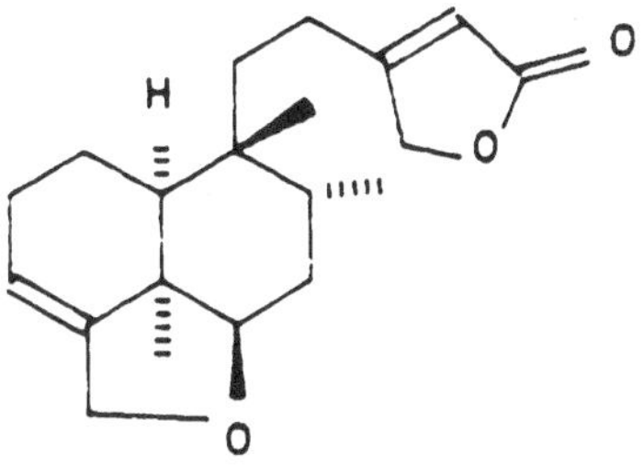

FIGURE 61. Structures of clerodane diterpenes included in the compilation of ^{13}C-NMR data (Table 15).

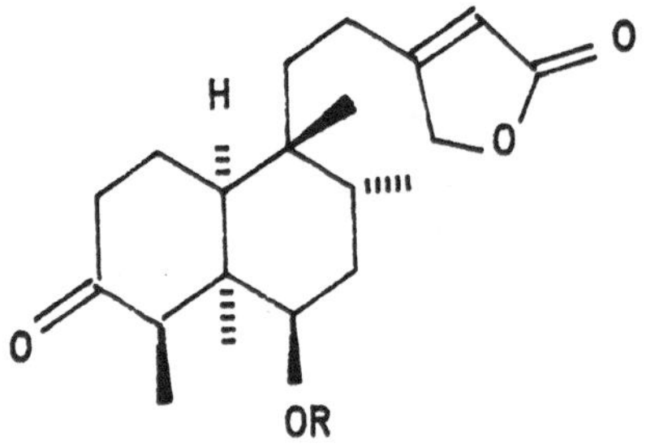

14· R = Tig

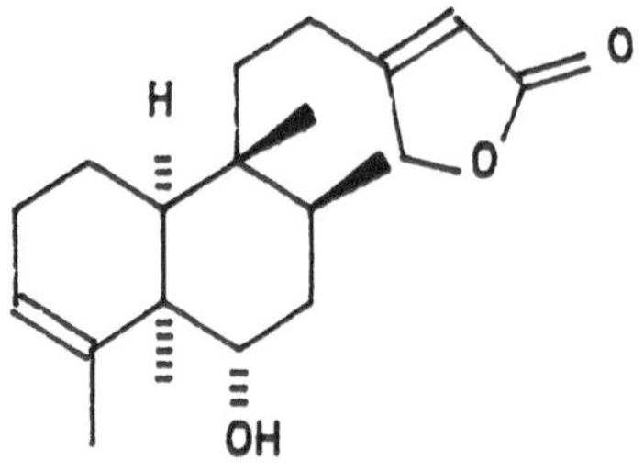

15· (1083)

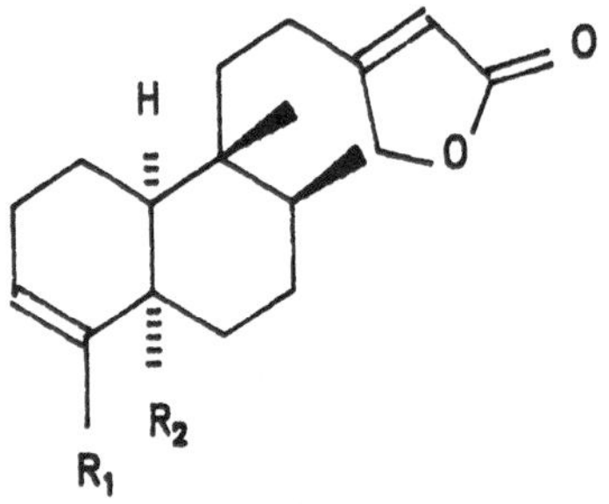

16· (1084) R₁ = R₂ = CH₂OH

16· (1084) R_1 = R_2 = CH_2OH

23· (1091) R_1 = CH_3, R_2 = COO-α-L-Arab

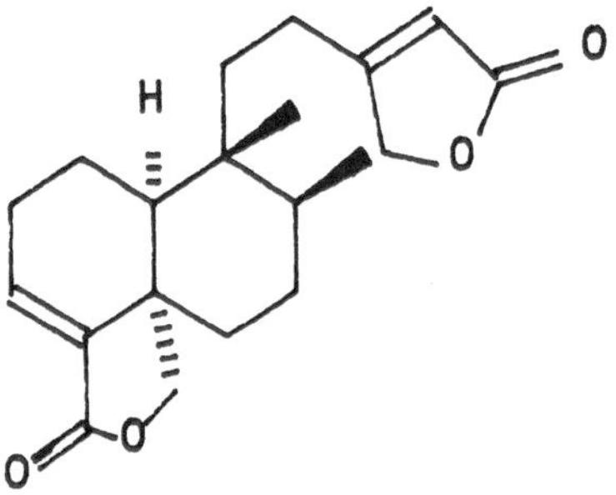

17· (1085)

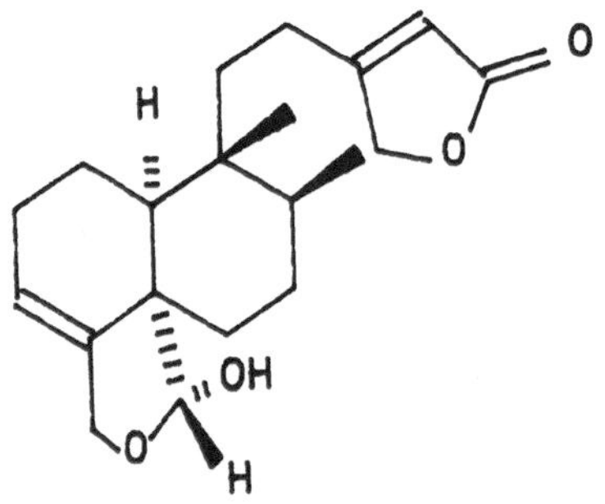

18· (1086)

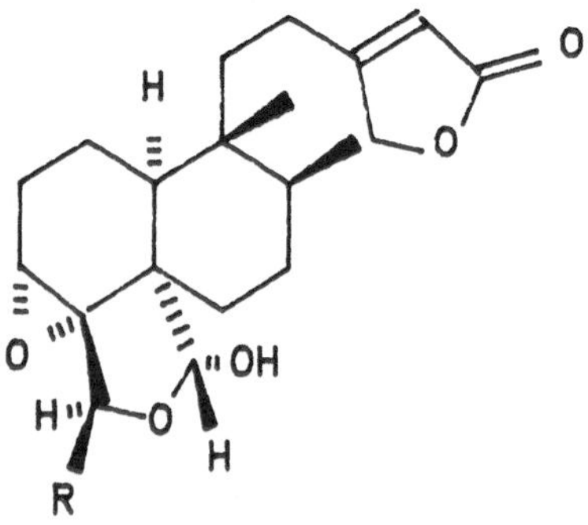

19· (1087) R = OH

21· (1089) R = H

FIGURE 61. (contd.).

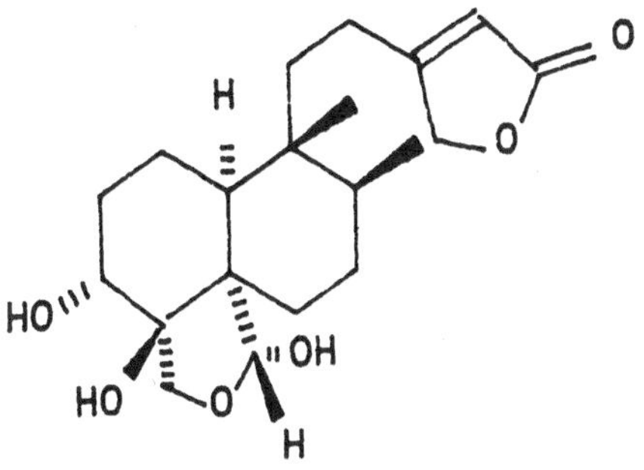

Figure 61. Structures of Clerodane Diterpenes 547

22· (1090)

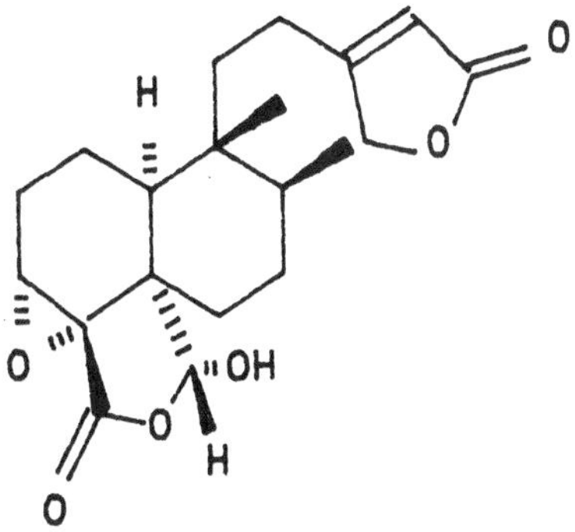

20· (1088)

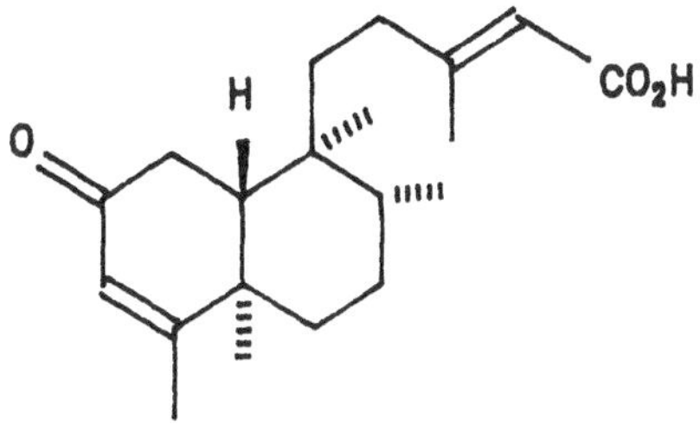

24· (961)

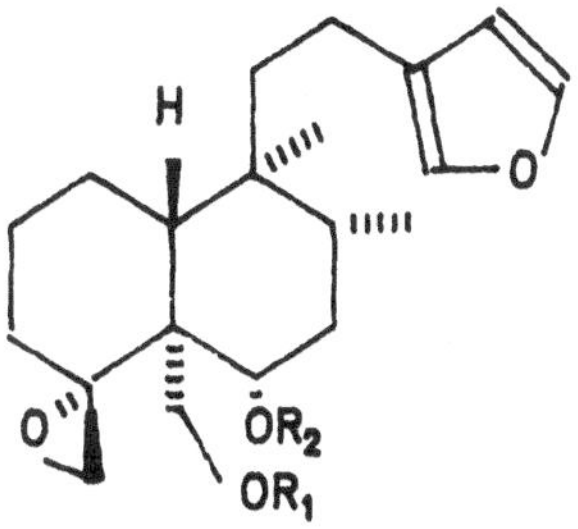

26· Teumassilin, $R_1 = R_2 = H$

27· 6,19—Diacetylteumassilin

$R_1 = R_2 = Ac$

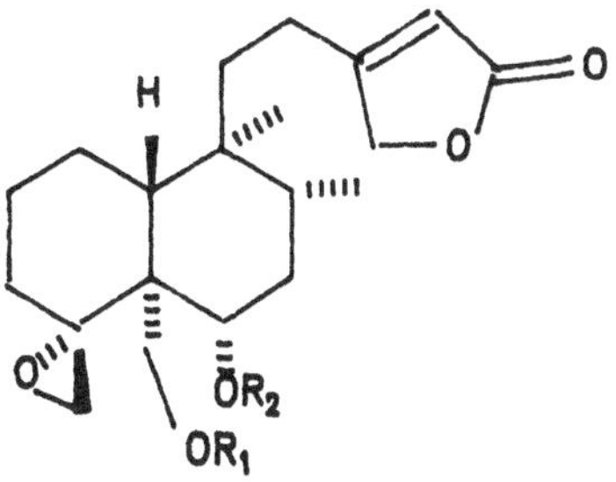

25· Deacetylajugarin—II

$R_1 = R_2 = H$

28· Ajugarin—I, $R_1 = R_2 = Ac$

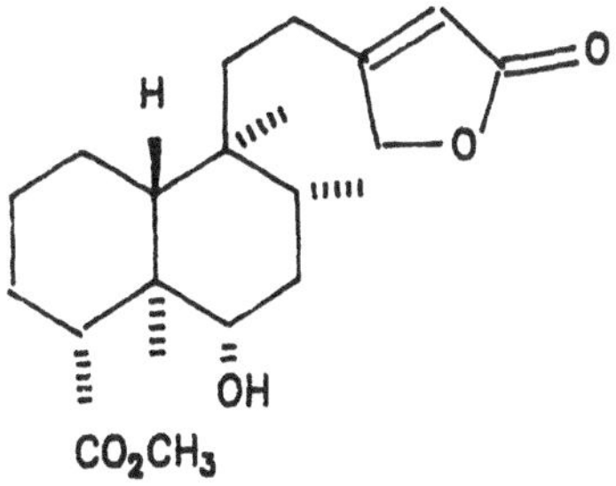

29· Ajugarin—IV

FIGURE 61. (contd.).

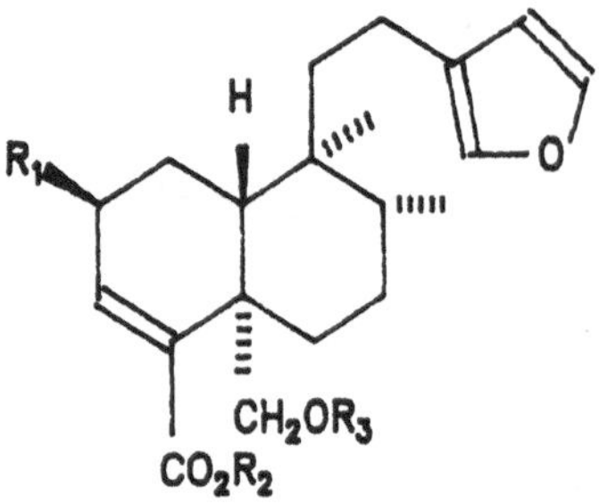

30· (1033) R₁ = R₂ = R₃ = H

31· R₁ = R₂ = H, R₃ = Ac

32· R₁ = OAc, R₂ = H, R₃ = Ac

33· R₁ = OAc, R₂ = CH₃, R₃ = Ac

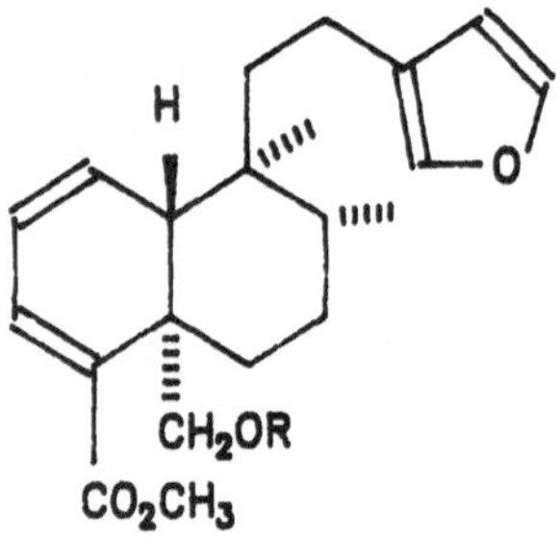

34· R = H

35· R = Ac

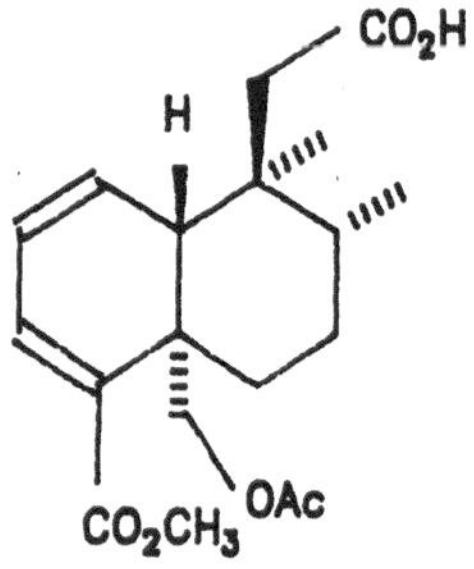

36·

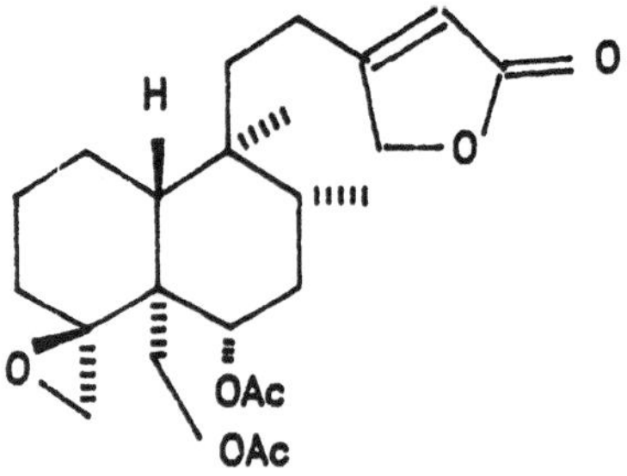

37·

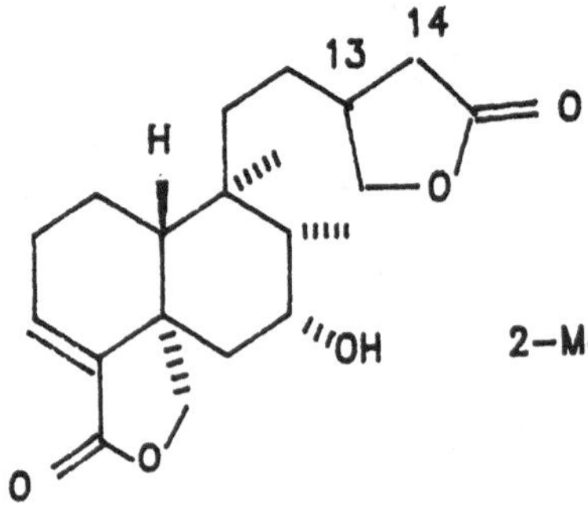

38· 13(14)d·b·

39·

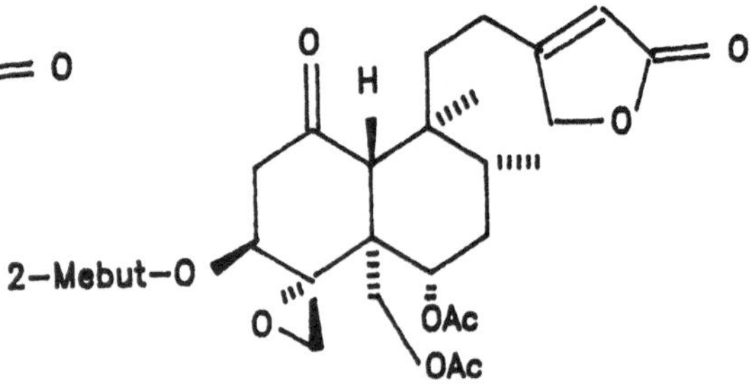

40·

FIGURE 61. (contd.).

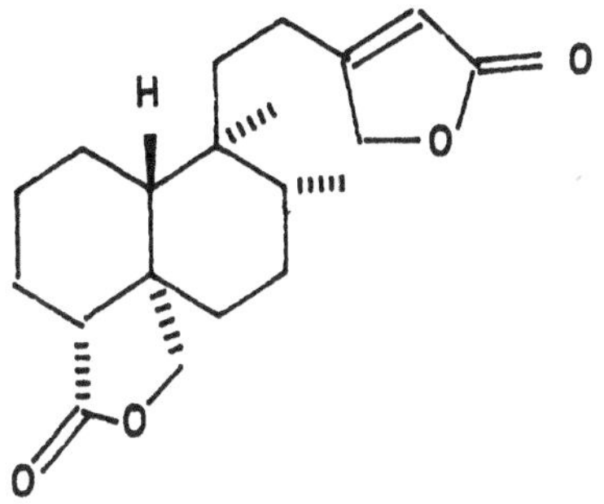

Figure 61. Structures of Clerodane Diterpenes   549

41·

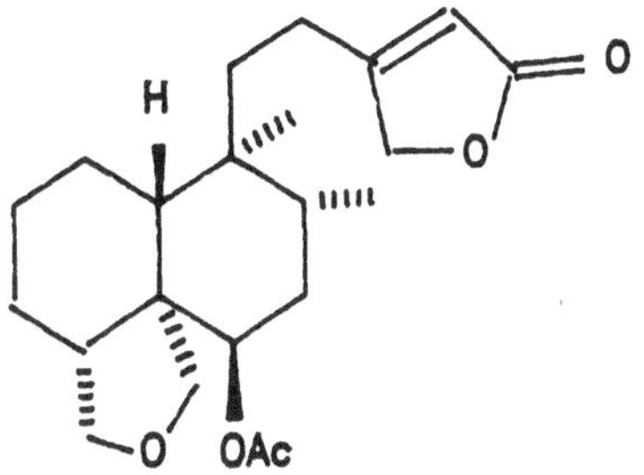

42·

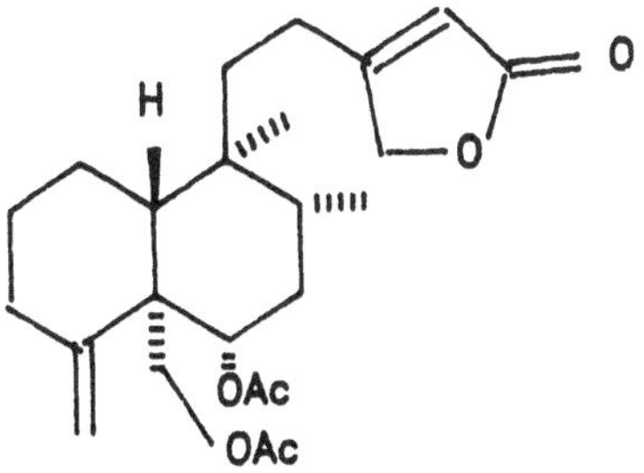

43·

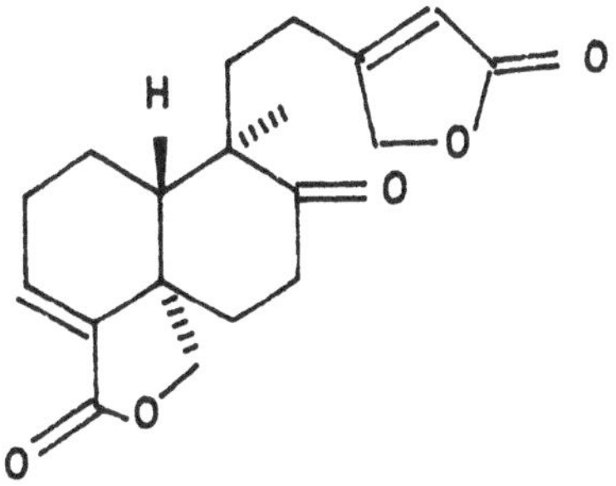

44·

FIGURE 61. (*contd.*).

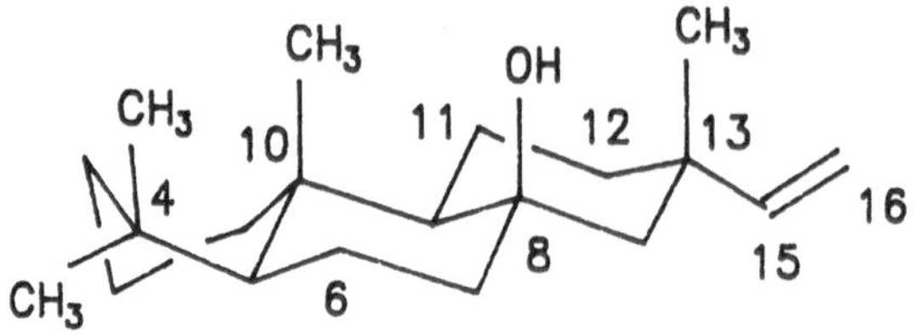

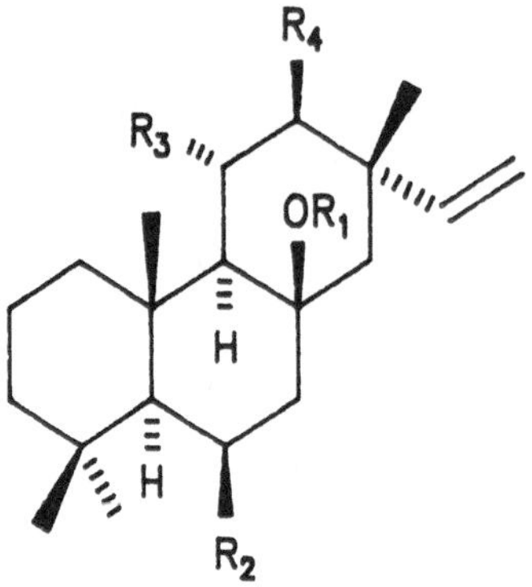

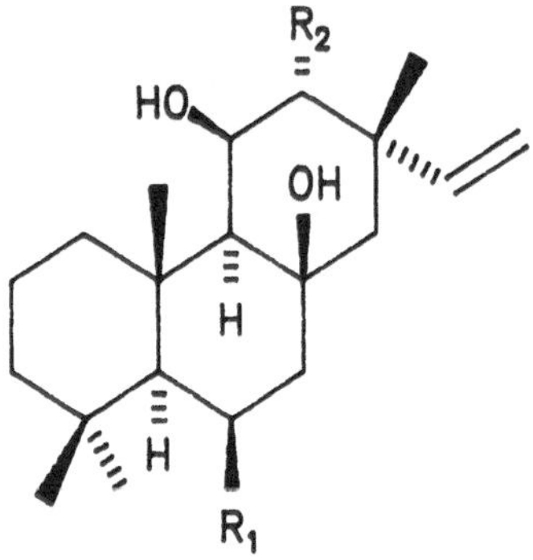

1· [276] R₁ = R₂ = R₃ = R₄ = H

2· R₁ = Ac, R₂ = R₃ = R₄ = H

3· [277] R₁ = R₃ = R₄ = H, R₂ = OH

4· [278] R₁ = R₃ = R₄ = H, R₂ = OAc

5· [279] R₁ = R₂ = R₄ = H, R₃ = OH

6· [284] R₁ = R₂ = R₃ = H, R₄ = OAc

7· R₁ = R₂ = H, R₃ = R₄ = OH

8· [292] R₁ = R₂ = H, R₃ = OH, R₄ = OAc

9· [289] R₁ = R₂ = H, R₃ = OAc, R₄ = OH

10· R₁ = R₂ = R₃ = H, R₄ = OH

11· R₁ = OH, R₂ = H

12· R₁ = H, R₂ = OH

FIGURE 62. Proposed conformation and twelve representative structures of the sandara-copimaranes isolated from *Garuleum* and *Osteospermum* (156). ¹H-NMR data for these structures are included in Table 16.

CHAPTER 7

References

1. Achilladelis, B., and J.R. Hanson: Studies in terpenoid biosynthesis. I. The biosynthesis of metabolites of *Tricothecium roseum*. Phytochem. **7**, 589 (1968).

2. Ahmed, A.A., D.A. Gage, J.S. Calderon, and T.J. Mabry: Labdane diterpenes from *Brickellia vernicosa*. Phytochem. **25**, 1385 (1986).

3. Alieva, S.A., Z.M. Putieva, E.S. Kondratenko, and N.K. Abubakirov: A diterpene glycoside—doronicoside D—from *Doronicum macrophyllum*. Khim. Prir. Soedin. 658 (1977).

4. Almqvist, S.-O., C.R. Enzell, and F.W. Wehrli: Carbon-13 NMR studies of labdane diterpenoids. Acta Chem. Scand **B 29**, 695 (1975).

5. Alvarez, L., R. Mata, G. Delgado, and A. Romo de Vivar: Sesquiterpene lactones from *Viguiera hypargyrea*. Phytochem. **24**, 2973 (1985).

6. Amaro, T.M., and M. Adrien R.: Diterpenoides del *Oxylobus glanduliferus* (Sch.-Bip.) Gray. Rev. Latinoamer. Quim. **13**: 110 (1982).

7. Anderson, A.B., R. McCrindle, and E. Nakamura: Diterpenoids of *Solidago arguta* Ait. The stereochemistry of *cis*-clerodanes. J. Chem. Soc. Chem. Commun. **1974**, 453.

8. Angeles, E., K. Folting, P.A. Grieco, J.C. Huffman, R. Miranda, and M. Salmon: Isolation and structure of stephalic acid, a new clerodane diterpene from *Stevia polycephala*. Phytochem. **21**, 1804 (1982).

9. Anthonsen, T.: A note on the constitution of the diterpene $C_{20}H_{28}O_3$ from *Solidago canadensis* L. Acta Chem. Scand. **20**, 904 (1966).

10. Anthonsen, T., and G. Bergland: The diterpenoids of *Solidago missouriensis* Nutt. Acta Chem. Scand. **24**, 1860 (1970).

11. Anthonsen, T., and G. Bergland: The diterpenoids of some *Solidago* species. Acta Chem. Scand. **25**, 1924 (1971).

12. Anthonsen, T., and G. Bergland: Constitution and stereochemistry of diterpenoids from *Solidago missouriensis* Nutt. Acta Chem. Scand. **27**, 1073 (1973).

13. Anthonsen, T., and S. Chantharasakul: Isolation of *ent*-16-kauren-19-oic acid and *ent*-16-kauren-19-ol from *Abrotanella nivigena* Muell. Acta Chem. Scand. **25**, 1925 (1971).

14. Anthonsen, T., M.S. Henderson, A. Martin, R. McCrindle, and R.D.H. Murray: Furan-containing diterpenoids from *Solidago serotina* Ait. Acta Chem. Scand. **22**, 351 (1968).

15. Anthonsen, T., M.S. Henderson, A. Martin, R.D.H. Murray, R. McCrindle, and D. McMaster: Constituents of *Solidago* species. Part IV. Solidagoic acids

A and B, diterpenoids from *Solidago gigantea* var. *serotina*. Can J. Chem. **51**, 1332 (1973).

16. Anthonsen, T., P.H. McCabe, R. McCrindle, and R.D.H. Murray: The constitution and stereochemistry of solidagenone. Acta Chem. Scand. **21**, 2289 (1967).

17. Anthonsen, T., P.H. McCabe, R. McCrindle, and R.D.H. Murray: Constituents of *Solidago* species I. The constitution and stereochemistry of diterpenoids from *Solidago canadensis* L. Tetrahedron **25**, 2233 (1969).

18. Anthonsen, T., and R. McCrindle: The constitution of diterpenoids from *Solidago elongata* Nutt. Acta Chem. Scand. **23**, 1068 (1969).

19. Apsimon, J.W., W.G. Craig, P.V. DeMarco, D.W. Mathieson, and W.B. Whalley: The n.m.r. spectra of some diterpenes. Chem. Comm. **1966**, 361.

20. Apsimon, J.W., W.G. Craig, P.V. DeMarco, D.W. Mathieson, and W.B. Whalley: The chemical shift—III. The NMR spectra of some diterpenes. Tetrahedron **23**, 2375 (1967).

21. Arnone, A., R. Mondelli, and J. St. Pyrek: 13C NMR Spectroscopy of natural substances. IV. 13C NMR Studies of trachylobane diterpenes: complete carbon assignment. Org. Magn. Reson. **12**, 429 (1979).

22. Arriaga-Giner, F.J., E. Wollenweber, I. Schober, P. Dostal, and S. Braun: 2β-Hautriwaic acid, a clerodane type diterpenoid and other terpenoids from three *Baccharis* species. Phytochem. **25**, 719 (1986).

23. Bahsas, A., and J. Triana: Inulaefol, nuevo furanoditerpeno del *Eupatorium inulaefolium* HBK. VI. Simposium Internacional Quimica de Productos Naturales, Monterrey, N. L., Mexico. April 23–28. **1979**, 15.

24. Bailey, J.A., G.G. Vincent, and R.S. Burden: Diterpenes from *Nicotiana glutinosa* and their effect on fungal growth. J. of General Microbiology. **85**, 57 (1974).

25. Banerjee, S., M. Grenz, J. Jakupovic, and F. Bohlmann: Some alicyclic terpenoids from the tribe Anthemideae. Planta Med. **1985**, 177 (1985).

26. Banerjee, S. and J. Jakupovic: Further diterpenes from *Palafoxia rosea*. Rev. Latinoamer. Quim. **17**, 202 (1986).

27. Banerjee, S., J. Jakupovic, and R.M. King: Weitere derivate des β-jonons aus *Haplopappus fremontii*. Pharmazie **41**, 157 (1986).

28. Barua, R.N., R.P. Sharma, G. Thyagarajan, W. Herz, and S.V. Govindan: New melampolides and darutigenol from *Sigesbeckia orientalis*. Phytochem. **19**, 323 (1980).

29. Baruah. R.N., C. Zdero, F. Bohlmann, R.M. King, and H. Robinson: Some sesqui- and diterpenes from the tribe Eupatorieae. Phytochem. **24**, 2641 (1985).

30. Bastard, J., D. Khac Duc, M. Fetizon, M.J. Francis, P.K. Grant, R.T. Weavers, C. Kaneko, G. Vernon Baddeley, J.-M. Bernaussau, I.R. Burfitt, P.M. Wovkulich, and E. Wenkert: CMR spectroscopy of labdanic diterpenes and related substances. J. Nat. Prod. **47**, 592 (1984).

31. Beale, M.H., J.R. Bearder, J. Macmillan, A. Matsuo, and B.O. Phinney: Diterpene acids from *Helianthus* species and their microbiological conversion by *Gibberella fugikuroi*, Mutant B1–41a. Phytochem. **22**, 875 (1983).

32. Bennett, R.D., E.R. Lieber, and E. Heftmann: Biosynthesis of steviol from (-)-kaurene. Phytochem. **6**, 1107 (1967).

33. Berti, G., O. Livi, and D. Segnini: Cistodiol and Cistodioic acid, diterpenoids with a *cis*-fused clerodane skeleton. Tetrahedron Letters **1970**, 1401.

34. Bhat, S.V., P.S. Kalyanaraman, H. Kohl, and N.J. De Souza: Inuroyleanol and 7-ketoroyleanone, two novel diterpenoids of *Inula royleana* DC. Tetrahedron **31**, 1001 (1975).

35. Bittner, M., and W.H. Watson: Phytochemical study of *Haplopappus ciliatus*. Rev. Latinoamer. Quim. **13**, 24 (1982).

36. Bittner, M.L., V. Zabel, W.B. Smith, and W.H. Watson: A new dihydro-*trans*-clerodane diacid from *Haplopappus ciliatus*. Phytochem. **17**, 1797 (1978).

37. Bjeldanes, L.F., and T.A. Geissman: Constituents of *Helianthus cilaris*. Phytochem. **11**, 327 (1972).

38. Bohlmann, F.: Neues über die chemie der Compositen. Naturwissenschaften **67**, 588 (1980).

39. Bohlmann, F., and W.-R. Abraham: Neue diterpene und weitere inhaltsstoffe aus *Helichrysum calliconum* und *Helichrysum heterolasium*. Phytochem. **18**, 889 (1979).

40. Bohlmann, F., and W.-R. Abraham: Neue diterpene aus *Helichrysum acutatum*. Phytochem. **18**, 1754 (1979).

41. Bohlmann, F. and W.-R. Abraham: Sesquiterpene und acetylenverbindungen aus *Cineraria*-arten. Phytochem. **17**, 1629 (1978).

42. Bohlmann, F., W.-R. Abraham, R.M. King, and H. Robinson: Diterpenes from *Koanophyllon* species. Phytochem. **20**, 1903 (1981).

43. Bohlmann, F., W.-R. Abraham, H. Robinson, and R.M. King: A new labdane derivative and geranylphloroglucinols from *Achyrocline alata*. *Phytochem.* **19**, 2475 (1980).

44. Bohlmann, F., W.-R. Abraham, H. Robinson, and R.M. King: Heliangolides and other constituents from *Bejaranoa semistriata*. Phytochem. **20**, 1639 (1981).

45. Bohlmann, F., W.-R. Abraham, and W.S. Sheldrick: Weitere diterpene mit helifulvan-gerüst und andere inhaltsstoffe aus *Helichrysum chionosphaerum*. Phytochem. **19**, 896 (1980).

46. Bohlmann, F., A. Adler, J. Jakupovic, R.M. King, and H. Robinson: A dimeric germacranolide and other sesquiterpene lactones from *Mikania* species. Phytochem. **21**, 1349 (1982).

47. Bohlmann, F., A. Adler, R.M. King, and H. Robinson. *ent*-Labdanes from *Mikania alvimii*. Phytochem. **21**, 173 (1982).

48. Bohlmann, F., A. Adler, A. Schuster, R.K. Gupta, R.M. King, and H. Robinson: Diterpenes from *Mikania* species. Phytochem. **20**, 1899 (1981).

49. Bohlmann, F., M. Ahmed, N. Borthakur, M. Wallmeyer, J. Jakupovic, R.M. King, and H. Robinson: Diterpenes related to grindelic acid and further constituents from *Grindelia* species. Phytochem. **21**, 167 (1982).

50. Bohlmann, F., M. Ahmed, J. Jakupovic, R.M. King, and H. Robinson: Labdane and dehydronerolidol derivatives from *Brickellia diffusa*. Phytochem. **21**, 691 (1982).

51. Bohlmann, F., M. Ahmed, J. Jakupovic, R.M. King, and H. Robinson: Dimeric sesquiterpene lactones and kolavane derivatives from *Gochnatia paniculata*. Phytochem. **22**, 191 (1983).

52. Bohlmann, F., M. Ahmed, J. Jakupovic, R.M. King, and H. Robinson: Three kolavane derivatives and 2β-angeloyloxy himachanolide from *Acritopappus longifolius*. Rev. Latinoamer. Quim. **15**, 16 (1984).

53. Bohlmann, F., M. Ahmed, R.M. King, and H. Robinson: Labdane and

eudesmane derivatives from *Ageratum fastigiatum*. Phytochem. **20**, 1434 (1981).

54. Bohlmann, F., M. Ahmed, H. Robinson, and R.M. King: A kolavane derivative from *Liatris scariosa*. Phytochem. **20**, 1439 (1981).

55. Bohlmann, F., N. Ates, and J. Jakupovic: Hirsutinolides from South African *Vernonia* species. Phytochem. **22**, 1159 (1983).

56. Bohlmann, F., N. Ates, R.M. King, and H. Robinson: Two sesquiterpenes from *Senecio* species. Phytochem. **22**, 1675 (1983).

57. Bohlmann, F., S. Banerjee, J. Jakupovic, M. Grenz, L.N. Misra, G. Schmeda-Hirschmann, R.M. King, and H. Robinson: Clerodane and labdane diterpenoids from *Baccharis* species. Phytochem. **24**, 511 (1985).

58. Bohlmann, F., S. Banerjee, C. Wolfrum, J. Jakupovic, R.M. King, and H. Robinson: Sesquiterpene lactones, geranylnerol and tremetone derivatives from *Ageratina* species. Phytochem. **24**, 1319 (1985).

59. Bohlmann, F., M. Bapuji, J. Jakupovic, R.M. King, and H. Robinson: Unusual diterpenes from *Brickellia eupatoriedes*. Phytochem. **21**, 181 (1982).

60. Bohlmann, F., M. Bapuji, R.M. King, and H. Robinson: Kolavenol derivatives from *Goyazianthus tetrastichus*. Phytochem. **21**, 939 (1982).

61. Bohlmann, F., R.N. Baruah, R.M. King, and H. Robinson: Alicyclic diterpenes from *Cronquistianthus bishopii*. Planta Med. **1985** 167 (1985).

62. Bohlmann, F., N. Borthakur, R.M. King, and H. Robinson: Dihydrodendroidinic acid from *Pleurocoronis pluriseta*. Phytochem. **20**, 2433 (1981).

63. Bohlmann, F., T. Burkhardt, and C. Zdero: Naturally Occurring Acetylenes. Academic Press. London and New York. 1973.

64. Bohlmann, F., V. Castro, and J. Jakupovic: Germacra-1(10),4-dien-*cis*-6,12-olides and elemanolides from *Montanoa atriplicifolia*. Phytochem. **22**, 1223 (1983).

65. Bohlmann, F., T. Chau-Thi, P. Singh, and J. Jakupovic: Alicyclic diterpenes from *Solidago* species. Planta Med. **1985**, 487 (1985).

66. Bohlmann, F., and H. Czerson: Neue labdan- und pimaren- derivate aus *Palafoxia rosea*. Phytochem. **18**, 115 (1979).

67. Bohlmann, F., A.K. Dhar, J. Jakupovic, R.M. King, and H. Robinson: Two sesquiterpene lactones with an additional propiolactone ring from *Disynaphia halimifolia*. Phytochem. **20**, 1077 (1981).

68. Bohlmann, F., K. Dhar, J. Jakupovic, R.M. King, and H. Robinson: A caryophylene derivative from *Fleischmannia pycnocephaloides*. Phytochem. **20**, 1425 (1981).

69. Bohlmann, F., L.N. Dutta, W. Dorner, R.M. King, and H. Robinson: Zwei neue guajanolide sowie weitere longipinenester aus *Stevia*-arten. Phytochem. **18**, 673 (1979).

70. Bohlmann, F., L. Dutta, H. Robinson, and R.M. King: Neue labdan-derivate aus *Chrysothamnus nauseusus*. Phytochem. **18**, 1889 (1979).

71. Bohlmann, F., D. Ehlers, and C. Zdero: Einige neue furanoeremophilane aus *Senecio* arten. Phytochem. **17**, 467 (1978).

72. Bohlmann, F., and U. Fritz: Neue diterpene und acetylenverbindungen aus *Nidorella*-arten. Phytochem. **17**, 1769 (1978).

73. Bohlmann, F., U. Fritz, H. Robinson, and R.M. King: Sesquiterpene and diterpene derivatives from *Solidago* species. Phytochem. **19**, 2655 (1980).

74. Bohlmann, F., U. Fritz, H. Robinson, and R.M. King: Isosesquicaren aus

Haplopappus tenuisectus. Phytochem. **18**, 1749 (1979).

75. Bohlmann, F., T. Gerke, J. Jakupovic, N. Borthakur, R.M. King, and H. Robinson: Diterpene lactones and other constituents from *Wedelia* and *Aspilia* species. Phytochem. **23**, 1673 (1984).

76. Bohlmann, F., and M. Grenz: Notiz über die isolierung von hautriwa-säure aus *Conyza ivaefolium* Less. Chem. Ber. **105**, 3123 (1972).

77. Bohlmann, F., and M. Grenz: Constituents from representatives of the *Eupatorium* group. Chem. Ber. **110**, 1321 (1977).

78. Bohlmann, F., and M. Grenz: Atisiren-derivate aus *Garuleum sonchifolium* (DC.)T. Norl. Chem. Ber. **111**, 1509 (1978).

79. Bohlmann, F., and M. Grenz: Diterpene vom sandarocopimaren-typ aus *Chrysanthemoides*-arten. Phytochem. **18**, 683 (1979).

80. Bohlmann, F., M. Grenz, A.K. Dhar, and M. Goodman: Labdane derivatives and flavones from *Gutierrezia dracunculoides*. Phytochem. **20**, 105 (1981).

81. Bohlmann, F., M. Grenz, J. Jakupovic, R.M. King, and H. Robinson: New labdane derivatives and other consituents from *Fleischmannia viscidipes*. Rev. Latinoamer. Quim. **15**, 1 (1984).

82. Bohlmann, F., M. Grenz, and H. Schwarz: Neue diterpene aus *Hinterhubera imbricata* Cuatr. Chem. Ber. **106**, 2579 (1973).

83. Bohlmann, F., M. Grenz, P. Wegner, and J. Jakupovic: Clerodan-derivate und neuartige diterpene aus *Conyza scabrida* DC. Liebigs Ann. Chem. **1983**, 2008.

84. Bohlmann, F., R.K. Gupta, R.M. King, and H. Robinson: A clerodane lactone and a tremetone derivative from *Bahianthus viscidus*. Phytochem. **20**, 331 (1981).

85. Bohlmann, F., R.K. Gupta, H. Robinson, and R.M. King: Labdane derivatives and a himachalanolide from *Acritopappus longifolius*. Phytochem. **20**, 275 (1981).

86. Bohlmann, F., L. Hartono, and J. Jakupovic: A diterpene related to erythroxydiol from *Helichrysum refluxum*. Phytochem. **24**, 611 (1985).

87. Bohlmann, F., L. Hartono, C. Zdero, and J. Jakupovic: Constituents of the genus *Oxylobus*. Phytochem. **24**, 1111 (1985).

88. Bohlmann, F., and H. Hoffmann: Further amides from *Echinecea purpurea*. Phytochem. **22**, 1173 (1983).

89. Bohlmann, F., and J. Jakupovic: Über neue chromene und andere inhaltsstoffe von *Lagascea rigida*. Phytochem. **17**, 1677 (1978).

90. Bohlmann, F., and J. Jakupovic: Neue labdan-derivate und andere inhaltsstoffe aus *Relhania acerosa*. Phytochem. **18**, 631 (1979).

91. Bohlmann, F., and J. Jakupovic: Neue labdan-derivate und sesquiterpene aus *Silphium*-arten. Phytochem. **18**, 1987 (1979).

92. Bohlmann, F., J. Jakupovic, M. Ahmed, M. Grenz, H. Suding, H. Robinson, and R.M. King: Germacranolides and diterpenes from *Viguiera* species. Phytochem. **20**, 113 (1981).

93. Bohlmann, F., J. Jakupovic, M. Ahmed, M. Wallmeyer, H. Robinson, and R.M. King: Labdane derivatives from *Hemizonia* species. Phytochem. **20**, 2383 (1981).

94. Bohlmann, F., J. Jakupovic, N. Ates, A. Schuster, J. Pickardt, R.M. King, and H. Robinson: Steiractinolide, eine neue gruppe von sesquiterpenlactonen. Liebigs Ann. Chem. **1983**, 962 (1983).

95. Bohlmann, F., J. Jakupovic, A.K. Dhar, R.M. King, and H. Robinson: Heliangolides and diterpenes from *Hartwrightia floridana*. Phytochem. **20**, 843 (1981).

96. Bohlmann, F., J. Jakupovic, A.K. Dhar, R.M. King, and H. Robinson: Two sesquiterpene and three diterpene lactones from *Acanthospermum australe*. Phytochem. **20**, 1081 (1981).

97. Bohlmann, F., J. Jakupovic, M. Hashemi-Nejad, and S. Huneck: Clerodane diterpenoids from *Aster alpinus*. Phytochem. **24**, 608 (1985).

98. Bohlmann, F., J. Jakupovic, R.M. King, and H. Robinson: Neue *ent*-atisiren- und *ent*-kaurensäure-derivate aus *Helianthus*-arten. Phytochem. **19**, 863 (1980).

99. Bohlmann, F., J. Jakupovic, R.M. King, and H. Robinson: New germacranolides, guaianolides and rearranged guaianolides from *Lasiolaena santosii*. Phytochem. **20**, 1613 (1981).

100. Bohlmann, F., J. Jakupovic, R.M. King, and H. Robinson: New labdane derivatives from *Madia sativa*. Phytochem. **21**, 1103 (1982).

101. Bohlmann, F., J. Jakupovic, H. Robinson, and R.M. King: Neue diterpene aus *Schkuhria*-arten. Phytochem. **19**, 881 (1980).

102. Bohlmann, F., J. Jakupovic, H. Robinson, and R.M. King: Diterpenes and other constituents of *Morithamnus crassus*. Phytochem. **19**, 2769 (1980).

103. Bohlmann, F., J. Jakupovic, and A. Schuster: Germacranolides from *Perymenium klattianum* and *Perymeniopsis ovalifolia*. Phytochem. **24**, 495 (1985).

104. Bohlmann, F., J. Jakupovic, and A. Schuster: 8-Hydroxypegolettiolide, a sesquiterpene lactone with a new carbon skeleton and further constituents from *Pegolettia senegalensis*. Phytochem. **22**, 1637 (1983).

105. Bohlmann, F., J. Jakupovic, A. Schuster, R.M. King and H. Robinson: Germacranolide, hydroxyverbenon und *ent*-kaur-15(16)-en-17,19-disäure aus *Helianthus occidentalis* var. *dowellianus*. Planta Med. **1984**, 202 (1984).

106. Bohlmann, F., J. Jakupovic, A. Schuster, R.M. King, and H. Robinson: Guaianolides and homoditerpenes from *Lasiolena morii*. Phytochem. **21**, 161 (1982).

107. Bohlmann, F., J. Jakupovic, A. Schuster, R.M. King, and H. Robinson: New melampolides, kaurene derivatives and other constituents from *Ichthyothere* species. Phytochem. **21**, 2317 (1982).

108. Bohlmann, F., J. Jakupovic, A. Schuster, R.M. King, and H. Robinson: Homogeranylnerol derivatives and a melampolide from *Smallanthus glabratus*. Phytochem. **24**, 1309 (1985).

109. Bohlmann, F., J. Jakupovic, C. Zdero, R.M. King, and H. Robinson: Neue melampolide und *cis*,*cis*-germacranolide aus vertretern der subtribus Melampodiinae. Phytochem. **18**, 625 (1979).

110. Bohlmann, F., W. Knauf, M. Grenz, and M.A. Lane: Ein neues diterpen aus *Xanthocephalum linearifolium*. Phytochem. **18**, 2040 (1979).

111. Bohlmann, F., W. Knauf, R.M. King, and H. Robinson: Ein neues diterpen und weitere inhaltsstoffe aus *Baccharis*-arten. Phytochem. **18**, 1011 (1979).

112. Bohlamnn, F., and K.-H. Knoll: Neue pimardien-derivate aus *Othonna*-arten. Phytochem. **15**, 1072 (1976).

113. Bohlmann, F., K.-H. Knoll, R.M. King, and H. Robinson: Neue α-santalen- und labdan-derivate aus *Ayapana amygdalina*. Phytochem. **18**, 1997 (1979).

114. Bohlmann, F., K.-H. Knoll, H. Robinson, and R.M. King: Neue kauren-

derivate und melampolide aus *Smallanthus uvedalia*. Phytochem. **19**, 107 (1980).

115. Bohlmann, F., K.-H. Knoll, H. Robinson, and R.M. King: Neue eudesmanolide aus *Steiractinia mollis*. Phytochem. **19**, 971 (1980).

116. Bohlmann, F., W. Kramp, M. Grenz, H. Robinson, and R.M. King: Diterpenes from *Baccharis* species. Phytochem. **20**, 1907 (1981).

117. Bohlmann, F., W. Kramp, J. Jakupovic, H. Robinson, and R.M. King: Diterpenes from *Baccharis* species. Phytochem. **21**, 399 (1982).

118. Bohlmann, F., and N. Le Van: Über neue diterpene aus *Dimorphotheca pluvialis* Moench. Chem. Ber. **109**, 1446 (1976).

119. Bohlmann, F., and N. Le Van: Sesquiterpenlactone und polyine aus der gattung *Arctotis*. Phytochem. **16**, 487 (1977).

120. Bohlmann, F., and N. Le Van: Neuen kaurensäure-derivate aus Wedelia-arten. Phytochem. **16**, 579 (1977).

121. Bohlmann, F., and N. Le Van: Neue guajanolide aus *Podachaenium eminens*. Phytochem. **16**, 1304 (1977).

122. Bohlmann, F., and N. Le Van: Neue sesqui- und diterpene aus *Bedfordia salicina*. Phytochem. **17**, 1173 (1978).

123. Bohlmann, F. and N. Le Van: Neue kaurensäure-derivate und germacranolide aus *Montanoa pteropoda*. Phytochem. **17**, 1957 (1978).

124. Bohlmann, F., and M. Lonitz: Neue sadaracopimardien-derivate, sesquiterpene und sesquiterpenlactone aus *Zexmenia*-arten. Chem. Ber. **111**, 843 (1978).

125. Bohlmann, F., and M. Lonitz: Über die inhaltsstoffe von *Zexmenia gnaphaloides* und die synthese von valerenan-derivaten. Chem. Ber. **113**, 2410 (1980).

126. Bohlmann, F., G.-W. Ludwig, J. Jakupovic, R.M. King, and H. Robinson: A daucanolide and further farnesene derivatives from *Ageratum fastigiatum*. Phytochem. **22**, 983 (1983).

127. Bohlmann, F., and P.K. Mahanta: Kaurenic acid derivatives from *Adenostemma caffrum*. Phytochem. **17**, 814, (1978).

128. Bohlmann, F., and P.K. Mahanta: Ein neue diterpensäure aus *Centipeda orbicularis*. Phytochem. **18**, 1067 (1979).

129. Bohlmann, F., L.N. Misra, and J. Jakupovic: Pseudoguaianolides and carabrone derivatives from *Loxothysanus sinuatus*. Phytochem. **24**, 1021 (1985).

130. Bohlmann, F., L.N. Misra, J. Jakupovic, R.M. King, and H. Robinson: Guaianolides, heliangolides, diterpenes and cycloartenol derivatives from *Balsamorhiza sagittata*. *Phytochem*. **24**, 2029 (1985).

131. Bohlmann, F., A.A. Natu, and P.K. Mahanta: Neue diterpene und germacranolide aus *Mikania*-arten. Phytochem. **17**, 483 (1978).

132. Bohlmann, F., and N. Rao: Neue hydroxyacetophenon-derivate aus *Espeletia schultzii*. Chem. Ber. **106**, 3035 (1973).

133. Bohlmann, F., H. Robinson, and R.M. King: Chemotaxonomy of the genus *Plagiocheilus*. Phytochem. **19**, 2235 (1980).

134. Bohlmann, F., E. Rosenberg, R.M. King, and H. Robinson: Neue labdanderivate aus *Aristeguietia buddleaefolia*. Phytochem. **19**, 977 (1980).

135. Bohlmann, F., C. Scheidges, R.M. King, and H. Robinson: Five labdane derivatives from *Koanophyllon conglobatum*. Phytochem. **23**, 1190 (1984).

136. Bohlmann, F., C. Scheidges, C. Zdero, R.M. King, and H. Robinson: *Ent*-labdanes from *Baccharis sternbergiana*. Phytochem. **23**, 1109 (1984).

137. Bohlmann, F., G. Schmeda-Hirschmann, and J. Jakupovic: Heliangolides and germacranolides from *Disynaphia multicrenulata*. Phytochem. **23**, 1435 (1984).

138. Bohlmann, F., G. Schmeda-Hirschmann, and J. Jakupovic: Neue melampolide aus *Acanthosperum australe*. Planta Med. **1984**, 37 (1984).

139. Bohlmann, F., G. Schmeda-Hirschmann, and J. Jakupovic: Nor-*ent*-labdan derivate aus *Austroeupatorium inulaefolium*. Planta Med. **1984**, 199 (1984).

140. Bohlmann, F., P. Singh, and J. Jakupovic: Further ineupatorolide-like germacranolides from *Inula cuspidata*. Phytochem. **21**, 157 (1982).

141. Bohlmann, F., P. Singh, J. Jakupovic, R.M. King, and H. Robinson: Eudesmanolides from *Dimerostemma brasilianum*. Phytochem. **21**, 1343 (1982).

142. Bohlmann, F., P. Singh, J. Jakupovic, H. Robinson, and R.M. King: An epoxygermacranolide and further consitituents from *Mikania* species. Phytochem. **21**, 705 (1982).

143. Bohlmann, F., P. Singh, R.K. Singh, K.C. Joshi, and J. Jakupovic: A diterpene with a new carbon skeleton from *Solidago altissima*. Phytochem. **24**, 1114 (1985).

144. Bohlmann, F., H. Suding, J. Cuatrecasas, R.M. King, and H. Robinson: Neue diterpene aus der subtribus Espeletiinae. Phytochem. **19**, 267 (1980).

145. Bohlmann, F., H. Suding, J. Cuatrecasas, H. Robinson, and R.M. King: Tricyclic sesquiterpenes and further diterpenes from *Espeletiopsis* species. Phytochem. **19**, 2399 (1980).

146. Bohlmann, F., A. Suwita, R.M. King, and H. Robinson: Neue *ent*-labdan-derivate aus *Austroeupatorium chaparense*. Phytochem. **19**, 111 (1980).

147. Bohlmann, F., A. Suwita, and T.J. Mabry: New labdane derivatives and further constituents of *Brickellia* species. Phytochem. **17**, 763 (1978).

148. Bohlmann, F., A. Suwita, H. Robinson, and R.M. King: Six guaianolides from *Stylotrichium rotundifolium*. Phytochem. **20**, 1887 (1981).

149. Bohlmann, F., K. Umemoto, and J. Jakupovic: Pseudoguaianolides related to confertin from *Stevia isomeca*. Phytochem. **24**, 1017 (1985).

150. Bohlmann, F., M. Wallmeyer, and J. Jakupovic: A new seco-labdane derivative from *Athrixia elata*. Phytochrem. **21**, 1806 (1982).

151. Bohlmann, F., M. Wallmeyer, J. Jakupovic, and J. Ziesche: Diterpenes and sesquiterpenes from *Osteospermum* species, Phytochem. **22**, 1645 (1983).

152. Bohlmann, F., M. Wallmeyer, R.M. King, and H. Robinson: 2-Oxo-labda-8(17),13-dien-15-ol from *Ophryosporus chilca*. Phytochem. **23**, 1513 (1984).

153. Bohlmann, F., and P. Wegner: *Ent*-beyer-15-ene derivatives from *Nidorella anomala*. Phytochem. **21**, 1175 (1982).

154. Bohlmann, F., and P. Wegner: Three diterpenes from *Conyza podocephala*. Phytochem. **21**, 1693 (1982).

155. Bohlmann, F., P. Wegner, and J. Jakupovic: Unusual diterpenes and sesquitepene xylosides from *Nidorella hottentotica*. Phytochem. **21**, 1109 (1982).

156. Bohlmann, F., G. Weickgenannt, and C. Zdero: Neue diterpene aus der tribus Calendulaceae (sic.). Chem. Ber. **106**, 826 (1973).

157. Bohlmann, F., and C. Zdero: Notiz über ein weiteres diterpen aus *Osteospermum subulatum* DC. Chem. Ber. **108**, 362 (1975).

158. Bohlmann, F., and C. Zdero: Über inhaltsstoffe der gattung *Brickellia*. Chem. Ber. **109**, 1436 (1976).

159. Bohlmann, F., and C. Zdero: Über ein neues diterpen aus Melampodium perfoliatum (Carv.)A. Gray. Chem. Ber. **109**, 1670 (1976).

160. Bohlmann, F., and C. Zdero: Neue terpen-inhaltsstoffe aus *Verbesina*-arten. Phytochem. **15**, 1310 (1976).

161. Bohlmann, F., and C. Zdero: Inhaltsstoffe der gattung *Polymnia*. Phytochem. **16**, 492 (1977).

162. Bohlmann, F., and C. Zdero: Ein neue diterpensäure aus *Perymenium ecuadoricum*. Phytochem. **16**, 786 (1977).

163. Bohlmann, F., and C. Zdero: Ein neues clerodan-derivat sowie weitere inhaltsstoffe aus der gattung *Macowania*. Phytochem. **16**, 1583 (1977).

164. Bohlmann, F., and C. Zdero: Neue norkauren- und thymol derivate aus *Athrixia*-arten. Phytochem. **16**, 1773 (1977).

165. Bohlmann, F., and C. Zdero: Diterpenes with a new carbon skeleton from *Printzia laxa*. Phytochem. **17**, 487 (1978).

166. Bohlmann, F., and C. Zdero: Ein neues kaurensäure und ein euparin-derivat aus *Oyedaea*-arten. Phytochem. **18**, 492 (1979).

167. Bohlmann, F., and C. Zdero: Neue phloroglucin—derivate aus *Helichrysum natalitium* und *Helichrysum bellum*. Phytochem. **18**. 641 (1979).

168. Bohlmann, F., and C. Zdero: Korrektur der konstitutionen von diterpenen aus *Palafoxia rosea*. Phytochem. **18**, 2038 (1979).

169. Bohlmann, F., and C. Zdero: Sandaracopimarene derivatives from *Senecio subrubriflorus*. Phytochem. **21**, 1697 (1982).

170. Bohlmann, F. and C. Zdero: Stevisalicinon, ein neuer Diterpentyp, sowie weitere inhaltsstoffe aus *Stevia*-arten. Liebigs Ann. Chem. 1764 (**1985**).

171. Bohlmann, F., C. Zdero, W.-R. Abraham, A. Suwita, and M. Grenz: Neue diterpene und neue dihydrochalkon-derivate sowie weitere inhaltsstoffe aus *Helichrysum*-arten. Phytochem. **19**, 873 (1980).

172. Bohlmann, F., C. Zdero, and M. Ahmed: New sesquiterpene lactones, geranyllinalol derivatives and other constituents from *Geigeria* species. Phytochem. **21**, 1679 (1982).

173. Bohlmann, F., C. Zdero, D. Berger, A. Suwita, P. Mahanta, and C. Jeffrey: Neue furanoeremophilane und weitere inhaltsstoffe aus südafrikanischen *Senecio-arten. Phytochem.* **18**, 79 (1979).

174. Bohlmann, F., C. Zdero, J. Cuatrecasas, R.M. King, and H. Robinson: Neue sesquiterpene und norditerpene aus vertretern der gattung *Libanothamnus*. Phytochem. **19**, 1145 (1980).

175. Bohlmann, F., C. Zdero, L. Fiedler, H. Robinson, and R.M. King: A labdane derivative from *Chromolaena collina* and a *p*-hydroxyacetophenone derivative from *Stomatanthes corumbensis*. Phytochem. **20**, 1141 (1981).

176. Bohlmann, F., C. Zdero, and M. Grenz: New sesquiterpenes of the genus *Othonna*. Chem. Ber. **107**, 3928 (1974).

177. Bohlmann, F., C. Zdero, and M. Grenz: Weitere inhaltsstoffe aus vertretern der *Eupatorium*-gruppe. Chem. Ber. **110**, 1034 (1977).

178. Bohlmann, F., C. Zdero, M. Grenz, A.K. Dhar, H. Robinson, and R.M. King:

Five diterpenes and other constituents from nine *Baccharis* species. Phytochem. **20**, 281 (1981).

179. Bohlmann, F., C. Zdero, R.K. Gupta, R.M. King, and H. Robinson: Diterpenes and teranorditerpenes from *Acritopappus* species. Phytochem. **19**, 2695 (1980).

180. Bohlmann, F., C. Zdero, E. Hoffmann, P.K. Mahanta, and W. Dorner: Neue diterpene und sesquiterpene aus südafrikanischen *Helichrysum*-arten. Phytochem. **17**, 1917 (1978).

181. Bohlmann, F., C. Zdero, and S. Huneck: Diterpenes from *Heteropappus altaicus*. Phytochem. **24**, 1027 (1985).

182. Bohlmann, F., C. Zdero, J. Jakupovic, N. Ates, R.M. King, and H. Robinson: Steiractinolide aus *Aspilia* und *Wedelia*-arten. Liebigs Ann. Chem. **1983**, 1257 (1983).

183. Bohlmann, F., C. Zdero, J. Jakupovic, T. Gerke, M. Wallmeyer, R.M. King, and H. Robinson: Neue sesquiterpenlactone und rosan-derivate aus *Trichogonia*-arten. Liebigs Ann. Chem. **1984**, 162.

184. Bohlmann, F., C. Zdero, J. Jakupovic, R.M. King, and H. Robinson: Diterpenes from *Acritopappus confertus*. Phytochem. **22**, 2243 (1983).

185. Bohlmann, F., C. Zdero, R.M. King, and H. Robinson: Neue labdan-derivate aus *Carterothamnus anomalochaeta*. Phytochem. **18**, 621 (1979).

186. Bohlmann, F., C. Zdero, R.M. King, and H. Robinson: Ein neues labdan-derivat und neue thymol-derivate aus *Bishovia boliviensis*. Phytochem. **18**, 1234 (1979).

187. Bohlmann, F., C. Zdero, R.M. King, and H. Robinson: Neue labdan-derivate aus *Gutierrezia*-arten. Phytochem. **18**, 1533 (1979).

188. Bohlmann, F., C. Zdero, R.M. King, and H. Robinson: Ein neues germacran-8,12-olid und neue diterpene aus *Polymnia canadensis*. Phytochem. **19**, 115 (1980).

189. Bohlmann, F., C. Zdero, R.M. King, and H. Robinson: Germacranolides, a guaianolide with a β-lactone ring and further constituents from *Grazielia* species. Phytochem. **20**, 1069 (1981).

190. Bohlmann, F., C. Zdero, R.M. King, and H. Robinson: Thirteen kolavane derivatives from *Symphyopappus* species. Phytochem. **20**, 1657 (1981).

191. Bohlmann, F., C. Zdero, R.M. King, and H. Robinson: Sesquiterpenes, guaianolides and diterpenes from *Stevia myriadenia*. Phytochem. **21**, 2021 (1982).

192. Bohlmann, F., C. Zdero, R.M. King, and H. Robinson: New germacranolides and other constituents from *Trichogoniopsis morii*. Phytochem. **21**, 2035 (1982).

193. Bohlmann, F., C. Zdero, R.M. King, and H. Robinson: Eudesmanolides and kaurane derivatives from *Wedelia hookeriana*. Phytochem. **21**, 2329 (1982).

194. Bohlmann, F., C. Zdero, R.M. King, and H. Robinson: New labdan-derivate aus *Aristeguietia pseudoarborea*. Liebigs Ann. Chem. **1983**, 2127.

195. Bohlmann, F., C. Zdero, R.M. King, and H. Robinson: Epoxycannabinolid und diterpene mit neuem kohlenstoffgerüst aus *Villanova titicaensis*. Liebigs Ann. Chem. **1984**, 250.

196. Bohlmann, F., C. Zdero, R.M. King, and H. Robinson: Kingidiol, a kolavane derivative from *Baccharis kingii*. Phytochem. **23**, 1511 (1984).

197. Bohlmann, F., C. Zdero, R.M. King, and H. Robinson: Gutierrezial and

further diterpenes from *Gutierrezia sarothrae*. Phytochem. **23**, 2007 (1984).

198. Bohlmann, F., C. Zdero, R.M. King, and H. Robinson: A further geranylnerol derivative from *Helianthopsis* species. Phytochem. **24**, 1108 (1985).

199. Bohlmann, F., C. Zdero, and P. Mahanta: Neue diterpene aus *Dimorphotheca*- und *Viguiera*-arten. Phytochem. **16**, 1073 (1977).

200. Bohlmann, F., C. Zdero, H. Robinson, and R.M. King: Ein neues germacrenderivat sowie ein diterpenmalonat aus *Baccharis* arten. Phytochem. **18**, 1993 (1979).

201. Bohlmann, F., C. Zdero, H. Robinson, and R.M. King: 15α-Methylacryloyloxy-*ent*-kaurenic acid from *Ichthyothere* species. Phytochem. **20**, 522 (1981).

202. Bohlmann, F., C. Zdero, H. Robinson, and R.M. King: A diterpene, a sesquiterpene quinone and flavanones from *Wyethia helenioides*. Phytochem. **20**, 2245 (1981).

203. Bohlmann, F., C. Zdero, H. Robinson, and R.M. King: Labdane derivatives from *Planaltoa lychnophoroides*. Phytochem. **21**, 465 (1982).

204. Bohlmann, F., C. Zdero, H. Robinson, and R.M. King: germacranolides from *Lychnophora* species. Phytochem. **21**, 1087 (1982).

205. Bohlmann, F., C. Zdero, G. Schmeda-Hirschmann, J. Jakupovic, V. Castro, R.M. King, and H. Robinson: Heliangolide, trachyloban und villanovan-derivate aus *Viguiera*-arten. Liebigs. Ann. Chem. **1984**, 495 (1984).

206. Bohlmann, F., C. Zdero, and S. Schoneweiss: Über die Inhaltsstoffe aus *Stevia*-arten. Chem. Ber **109**, 3366 (1976).

207. Bohlmann, F., C. Zdero, and B. L. Turner: Guaianolides and heliangolides from *Hymenopappus newberryi*. Phytochem. **23**, 1055 (1984).

208. Bohlmann, F., C. Zdero, R. Zeisberg, and W.S. Sheldrick: Helifulvanolsaure -ein neues diterpen mit anomalem kohlenstoffgerust aus *Helichrysum fulvum*. Phytochem. **18**, 1359 (1979).

209. Bohlmann, F., and J. Ziesche: Neue germacren-derivate aus *Senecio*-arten. Phytochem. **18**, 1489 (1979).

210. Bohlmann, F., and J. Ziesche: Neue diterpene aus *Gnaphalium*-arten. Phytochem. **19**, 71 (1980).

211. Bohlmann, F., J. Ziesche, R.M. King, and H. Robinson: Neuemelampolide aus *Smallanthus fruticosus*. Phytochem. **19**, 973 (1980).

212. Bohlmann, F., J. Ziesche, R.M. King, and H. Robinson: Seven furanoeremophilanes from three *Senecio* species. Phytochem. **19**, 2675 (1980).

213. Bohlmann, F., J. Ziesche, R.M. King, and H. Robinson: Eudesmanolides and diterpenes from *Wedelia trilobata* and an *ent*-kaurenic acid derivative from *Aspilia parvifolia*. Phytochem. **20**, 751 (1981).

214. Bohlmann, F., J. Ziesche, R.M. King, and H. Robinson: Eudesmanolides and other constituents from *Dimerostemma asperatum*. Phytochem. **20**, 1335 (1981).

215. Bohlmann, F., J. Ziesche, R.M. King, and H. Robinson: Eudesmanolides, guaianolides, germacranolides and elemanolides from *Zinnia* species. Phytochem. **20**, 1623 (1981).

216. Bohlmann, F., J. Ziesche, R.M. King, and H. Robinson: A pseudoguaianolide and a hydroxygeranylnerol from *Kingianthus paradoxus*. Phytochem. **20**, 1146 (1981).

217. Bohlmann, F., P. Zitzkowski, A. Suwita, and L. Fiedler: *cis*-Kolaveninsäure und weitere inhaltsstoffe aus vertretern der tribus Eupatorieae. Phytochem. **17**, 2101 (1978).

218. Brieskorn, C.H., and E. Pohlmann: Diterpene vom kaurantyp aus der composite *Espeletia schultzii* (Wedd.). Tetrahedron Lett., 566 (1968).

219. Brieskorn, C.H., and E. Pohlmann: Kaurandien-(9(11).16)-saure-(19) und 15α-acetoxy-kauren-(16)-saure-(19). Chem. Ber. **102**, 2621 (1969).

220. Brunn, T., L.M. Jackman, and E. Stenhagen: Grindelic and oxygrindelic acids. Acta. Chem. Scand. **16**, 1675 (1962).

221. Calderon, J.S., and J.R. Puig: Diterpenos tipo labdano de *Stevia origanoides*. X. Simposium Internacional Quimica de Productos Naturales, Monterrey, N. L., Mexico. April 28–30, 1983. p. 6.

222. Calderon, J.S., Quijano, L., Cristia, M., Gomez, F., and T. Rios: Labdane diterpenes from *Brickellia veronicaefolia*. Phytochem. **22**, 1783 (1983).

223. Calderon, J.S., L. Quijano, M. Garduno, F. Gomez, and T. Rios: 2α-iso-valeroyloxy-eperuic acid, a diterpene from *Eupatorium petiolare*. Phytochem. **22**, 2617 (1983).

224. Canonica, L., B. Rindone, and C. Scolastico: Structure and stereochemistry of psiadiol, a new diterpenoid. Tetrahedron Lett. 2639 (1967).

225. Canonica, L., B. Rindone, and C. Scolastico: A new diterpene with pimarane skeleton. Tetrahedron Lett. 4801 (1969).

226. Carlquist, S.: Structure and ontogeny of glandular trichomes of *Madinae* (Compositae). Amer. J. Bot. **45**, 675 (1958).

227. Castro, V. and J. Jakupovic: Two further 6,12-*cis*-germacranolides from *Montanoa tomentosa* subsp. *xanthiifolia*. Phytochem. **24**, 2449 (1985).

228. Castro, V. and J. Jakupovic: 4β,19-Epoxy-norkaurene and other diterpenes from *Mikania banisteriae*. Phytochem. **24**, 2450 (1985).

229. Ceccarelli, N., R. Lorenzi, and A. Alpi: Kaurene and kaurenol biosynthesis in cell-free systems of *Phaseolus coccineus* suspensor. Phytochem. **18**, 1657 (1979).

230. Cheng, P.C., C. Hufford, and N.J. Doorenbos: Isolation of 11-hydroxylated kauranic acids from *Adenostemma lavenia*. J. Nat. Prod. **42**, 183 (1979).

231. Connolly, J.D., F. Orsini, F. Pelizzoni, and G. Ricca: 13C NMR spectra of cassane diterpenoids. The stereochemistry of the caesalpins. Org. Magn. Reson. **17**, 163 (1981).

232. Craig, J.C., M.L. Mole, S. Billets, and F. El-Feraly: Isolation and identification of the hypoglycemic agent, carboxyatractylate from *Xanthium strumarium*. Phytochem. **15**, 1178 (1976).

233. Cronquist, A.: The Compositae revisited. Brittonia **29**, 137 (1977).

234. Croteau, R., and M.A. Johnson: Biosynthesis of terpenoids in glandular trichomes. In: E. Rodriguez (ed.). Biology and Chemistry of Plant Trichomes. Plenum Press (1984).

235. Cruse, W.B.T., M.N.G. James, and Ali A. Al-Shamma, J.K. Beal, and R.W. Doskotch: The molecular structure of gutierolide, a novel chloro-diterpenoid lactone. J. Chem. Soc. Chem. Commun. 1278 (1971).

236. Cuatrecasas, T.: Prima Flora Colombiana. 3. Compositae—Astereae. Webbia **24**, 1 (1969).

237. Cuevas, L.A., F. Garcia Jimenez, and A. Romo de Vivar: Estructura de la estenolobina. Rev. Latinoamer. Quim. **3**, 22 (1972).

238. Cutler, H.G.: Carboxyatractyloside: A compound from *Xanthium strumarium* and *Atractylis gummifera* with plant growth inhibiting properties. The probable "Inhibitor A." J. Nat. Prod. **46**, 609 (1983).

239. Danieli, B., E. Bombardelli, A. Bonati, and B. Gabetta: Structure of the diterpenoid carboxyatractyloside. Phytochem. **11**, 3501 (1972).
240. Delgado, G., L. Alvarez, and A. Romo de Vivar: Terpenoids and a flavan-3-ol from *Viguiera quinqueradiata*. Phytochem. **23**, 675 (1984).
241. Delgado, G., L. Alvarez, and A. Romo de Vivar: 15-Oxo-zoapatlin, a diterpene lactone from *Viguiera maculata*. Phytochem. **23**, 2674 (1984).
242. Delgado, G., H. Candenas, G. Pelaez, and A. Romo de Vivar: Terpenoids from *Viguiera excelsa* and *Viguiera oaxacana*. J. Nat. Prod. **47**, 1042 (1984).
243. Delgado, G., and A. Romo de Vivar: *ent*-Kaurenoid methyesters from *Viguiera stenoloba*, structural revision of stenolobin and its biomimetic conversion to zoapatlin. Chem. Lett. 1237 (1984).
244. Delgado, G., A. Romo de Vivar, J. Cardenas, R. Pereda-Miranda, and E. Huerta: *ent*-Beyerene and *ent*-atisene-diterpenes from *Viguiera insignis*. Phytochem. **23**, 2285 (1984).
245. Delgado, G., A. Romo de Vivar, A. Ortega, J. Cardenas, and E. O. Schlemper: Diterpenoids from *Viguiera insignis*. Phytochem. **22**, 1227 (1983).
246. Dell, B., and A.J. McComb: Resin production and glandular hairs in *Beyeria viscosa* (Labill.) Miq. (Euphorbiaceae). Aust. J. Bot. **25**, 195 (1974).
247. Dell, B., and A.J. McComb: Glandular hairs, resin production, and habitat of *Newcastelia viscida* E. Pritzel (Dicrastylidaceae). Aust. J. Bot. **23**, 373 (1975).
248. Dominguez, X.A.: Eupatorieae—chemical review, In V.H. Heywood, J.B. Harborne and B.L. Turner (eds.). The biology and chemistry of the Compositae. vol. 1, 487–502, Academic Press, London (1977).
249. Dominguez, X.A., D. Butruille, I. Sandler, and G. Vazquez: Identification de la solidagenona en *Solidago sempervirens*. Rev. Latinoamer. Quim. **6**, 159 (1975).
250. Dominquez, X.A., C. Cisneros, E. Guajardo, T., R. Villarreal, and A. Zamudio: Diterpenoids from *Palafoxia rosea*. Phytochem. **14**, 1665 (1975).
251. Dominquez, X.A., C. Cisneros, R. Guajardo, R. Villarreal, V. Zebel, and W.H. Watson: Jesromotetrol, a new diterpene from *Palafoxia rosea* (Compositae). Rev. Latinoamer. Quim. **9**, 99 (1978).
252. Dominquez, X.A., and S. Jimenez Jimennez: Aislamiento y estructuras del julslimtetrol, *nor*-julslimdiolona y la isocumambrina, metabolitos secundarius del *Croptilon divaricatum* (Compuesta). Rev. Latinoamer. Quim. **3**, 177 (1973).
253. Dominguez, X.A., J. Verde S., N.E. Guerra R., E. Ellenmaurer, and J. Jakupovic: A *nor*-diterpene and other constituents from *Isocoma coronopifolia*. Phytochem. **25**, 2893 (1986).
254. Edwards, O.E., G. Feniak, and M. Los: Diterpenoid quinones of *Inula royleana* D.C. Can. J. Chem. **40**, 1540 (1962).
255. Edwards, O.E., and M.N. Rodger: The alkaloids of *Inula royleana*. Can. J. Chem. **37**, 1187 (1959).
256. Eguren, L., A. Perales, J. Fayos, B. Rodriquez, G. Savona, and F. Piozzi: New neoclerodane diterpenoid containing an oxetane ring isolated from *Teucrium chamaedrys*. X-ray structure determination. J. Org. Chem. **47**, 4157 (1982).
257. El-Emary, N.A., G. Kusano, and T. Takemoto: Kaurenoids from *Cacalia bulbifera*. Phytochem. **14**, 1660 (1975).
258. Elliger, C.A., D.F. Zinkel, B.G. Chan, and A.C. Waiss, Jr.: Diterpene acids as larval growth inhibitors. Experientia **32**, 1364 (1976).

259. El-Naggar, S.F., R.W. Doskotch, T.M. O'Dell, and L. Girard: Antifeedant diterpenes for the gypsy moth larvae from *Kalmia latifolia*: Isolation and characterization of ten grayanoids. J. Nat. Prod. **43**, 617 (1980).

260. Evans, F.J., and C.J. Soper: The tigliane, daphnane and ingenane diterpenes, their chemistry, distribution and biological activities. A review. Lloydia **41**, 193 (1978).

261. Evans, F.J., and S.E. Taylor: Pro-inflammatory, tumour-promoting and anti-tumour diterpenes of the plant families Euphorbiaceae and Thymelaeaceae. Fortschritte Chem. Org. Naturst. **44**, 1 (1984).

262. Fahn, A.: Secretory Tissues in Plants. Chap. 9 (158–222), Tissues secreting lipophilic substances. Academic Press, London, New York, and San Francisco.

263. Ferguson, G., W.C. Marsh, R. McCrindle, and E. Nakamura: Stereochemistry of clerodanes. X-ray structure of a key diterpenoid from *Solidago arguta*. J. Chem. Soc. Chem. Commun. 299 (1975).

264. Ferguson, G., R. McCrindle, S.T. Murphy, and M. Parvez: Further diterpenoid constituents of *Helianthus annuus* L. Crystal and molecular structure of methyl *ent*-15β-hydroxy-trachyloban-19-oate. J. Chem. Res. **5**, 200 (1982).

265. Fernandez-Gadea, F., M.L. Jimeno, and B. Rodriguez: Complexation of hydroxylated *ent*-kaurane diterpenoids with boric acid as an aid to the assignment of the 13C NMR spectra. Org. Magn. Reson. **22**, 515 (1984).

266. Fischer, N.H., E.J. Olivier, and H.D. Fischer: The biogenesis and chemistry of sesquiterpene lactones, In: W. Hertz, H. Grisebach and G. W. Kirby (eds.). Progress in the chemistry of organic natural products, 48–390. Spring-Verlag, New York (1979).

267. Fujita, E., K. Fuji, Y. Nagao, M. Node, and M. Ochiai: Bull. Inst. Chem. Res., Kyoto Univ., **61**, 142 (1983).

268. Gage, D.A.: Chemical data and their bearing upon generic delineations in the Eupatorieae. Taxon **34**, 61 (1985).

269. Gage, D.A.: A Chemosystematic Study of the genus *Piptothrix* (Compositae, Eupatorieae). Doctoral Dissertation. The University of Texas at Austin (1986).

270. Ganguly, S.N., T. Ganguly, and S.M. Sircar: Gibberellins of *Enhydra fluctuans*. Phytochem. **11**, 3433 (1972).

271. Gao, F., M. Leidig, and T.J. Mabry: *ent*-Labdane derivatives from *Gutierrezia grandis*. Phytochem. **24**, 1541 (1985).

272. Gao, F., M. Leidig, and T.J. Mabry: Diterpene glycosides from *Gutierrezia sphaerocephala*. Phytochem. **25**, 1371 (1986).

273. Gao, F., and T.J. Mabry: An acyclic diterpene from *Viguiera deltoidea*. Phytochem. **24**, 3061 (1985).

274. Gao, F., and T.J. Mabry: Ten *cis*-clerodane-type diterpene lactones from *Gutierrezia texana*. Phytochem. **26**, 209 (1987).

275. Gao, F., M. Miski, D.A. Gage, and T.J. Mabry: Terpenoid constituents of *Viguiera dentata*. J. Nat. Prod. **48**, 316 (1985).

276. Gao, F., M. Miski, D.A. Gage, J.A. Norris, and T.J. Mabry: Terpenoids from *Viguiera potosina*. J. Nat. Prod. **48**, 489 (1985).

277. Gao, F., H. Wang, and T.J. Mabry: Sesquiterpene lactones and flavonoids from *Helianthus* species. J. Nat. Prod. **50**, 23 (1987).

278. Garcia, E.E., E. Guerreiro, and P. Joseph-Nathan: *ent*-Pimaradiene diterpenes from *Gochnatia glutinosa*. Phytochem. **24**, 3059 (1985).

279. Geissman, T.A., and D.H.G. Crout: Organic Chemistry of Secondary Plant Metabolism, Chap. 10 (Diterpenoid Compounds) pp. 291–311. Freeman, Cooper and Company, San Francisco.

280. Gershenzon, J., and T.J. Mabry: Sesquiterpene lactones from a Texas population of *Helianthus maximiliani*. Phytochem. **23**, 1959 (1984).

281. Gershenzon, J., and T.J. Mabry: Furanoheliangolides from *Helianthus schweinitzii*. Phytochem. **23**, 2557 (1984).

282. Gershenzon, J., T.J. Mabry, J.D. Korp, and I. Bernal: Germacranolides from *Helianthus californicus*. Phytochem. **23**, 2561.

283. Gianello, J.C., and O.S. Giordano: Barticulidiol, un nuevo furano diterpeno aislado de *Baccharis articulata* Lam (Persoon). Rev. Lationamer. Quim. **13**, 76 (1982).

284. Givovich, A., A. San-Martin, and M. Castillo: Neo-clerodane diterpenoids from *Baccharis incarum*. Phytochem. **25**, 2829 (1986).

285. Glasby, J.S.: Encyclopaedia of the Terpenoids, Vols. 1 and 2. Wiley-Interscience Publication, Chichester and New York (1982).

286. Gomez, G.F., J.S. Calderon, L., Quijano, and T. Rios: Acido brickelico C, un neuvo diterpeno aislado de *Brickellia* sp. VIII. Simposium Internacional Quimica de Productos Naturales, Monterrey, N. L., Mexico. March 26–28, 1981.

287. Gomez, F., L. Quijano, J.S. Calderon, and T. Rios: 3α-Angeloyloxy-2α-hydroxy-cativic acid, a new diterpene from *Brickellia paniculata*. Phytochem. **22**, 1292 (1983).

288. Gonzalez, A.G., J.M. Artega, B.M. Fraga, M.G. Hernandez, and J. Fayos: The structure of jhanilactone. Experientia **34**, 554 (1978).

289. Gonzalez, A.G., J.M. Artega, J.L. Breton, and B.M. Fraga: Five new labdane diterpene oxides from *Eupatorium jhanii*. Phytochem. **16**, 107 (1977).

290. Gonzalez, A.G., J.J. Mendoza, J.G. Luis, A.G. Ravelo, X.A. Dominguez, and G. Cano: Diterpenes from *Palafoxia texana*. Phytochem. **24**, 3056 (1985).

291. Goswami, A., R.N. Barua, R.P. Sharma, J.N. Baruah, P. Kulanthaivel, and W. Herz: Clerodanes from *Solidago virgaurea*. Phytochem. **23**, 837 (1984).

292. Grau, J.: Astereae—systematic review. In: The Biology and Chemistry of the Compositae (Heywood, V.H., J.B. Harborne, and B.L. Turner, eds.). New York and London: Academic Press. 1977.

293. Guerreiro, E., J. Kavka, and O.S. Giordano: Acido 1-hidroxigrindelico en *Grindelia pulchella*. Rev. Latinoamer. Quim. **13**, 72 (1982).

294. Guerreiro, E., J. Kavka, J.R. Saad, M.A. Oriental, and O.S. Giordano: Acidos diterpenicos en *Grindelia pulchella* and *G. chiloensis* Cabr. Rev. Latinoamer. Quim. **12**, 77 (1981).

295. Guerrero, C., and A. Romo de Vivar: Estructura y estereoquimica de la bacchofertina; diterpeno aislado de *Baccharis conferrta* H.B.K. Rev. Latinoamer. Quim. **4**, 178 (1973).

296. Gutierrez, A.B., J.C. Oberti, P. Kulanthaivel, and W. Herz: Sesquiterpene lactones and diterpenes from *Mikania periplocifolia*. Phytochem. **24**, 2967 (1985).

297. Hahn, D.W., E.W. Ericson, M.T. Lai, and A. Probst: Antifertility activity of *Montanoa tomentosa* (zoapatle). Contraception **23**, 133 (1981).

298. Hänsel, R., E.-M. Cybulski, B. Çubukçu, A.H. Meriçli, F. Bohlmann, and C. Zdero: Neue pyron-derivate aus *Helichrysum*-arten. Phytochem. **19**, 639 (1980).

299. Hanson, J.R.: The biosynthesis of the diterpenes. In: Progress in the Chemistry of Organic Natural Products (Herz, W., H. Grisebach, and G. W. Kirby, eds.). Wien and New York: Springer-Verlag. 1971.

300. Hanson, J.R.: The bicyclic diterpenes. In: Progress in Phytochemistry (Reinhold, L., and Y. Liwschitz, eds.). London: Interscience Publishers of John Wiley and Sons. 1972.

301. Hanson, J.R.: The di- and sesquiterpenes. In: Chemistry of Terpenes and Terpenoids (Newman, A.A., ed.). London and New York: Academic Press. 1972.

302. Hanson, J.R. (ed.): Terpenoids and Steroids (Specialist Periodical Reports), Vol. 12. London: The Royal Society of Chemistry. 1983.

303. Hanson, J.R.: The biosynthesis of C_5-C_{20} terpenoid compounds. Natural Products Reports **1**, 443 (1984).

304. Hanson, J.R.: Diterpenoids. Natural Products Reports **1**, 171 (1984).

305. Hanson, J.R.: Diterpenoids. Natural Products Reports **1**, 339 (1984).

306. Hanson, J.R.: Diterpenoids. Natural Products Reports **1**, 533 (1984).

307. Hanson, J.R.: Diterpenoids. Natural Products Reports **3**, 307 (1986).

308. Hanson, J.R.: The chemistry of the tetracyclic diterpenoids—X. Some beyerene 2 and 3-alcohols. Tetrahedron **26**, 2711 (1970).

309. Hanson, J.R., G. Savona, and M. Siverns: 13C Nuclear magnetic resonance spectra and microbiological hydroxylation of 7α- and 7β-hydroxykaurenolide. J.C.S. Perkin I **1974**, 2002.

310. Hanson, J.R., M. Siverns, F. Piozzi, and G. Savona: The 13C Nuclear magnetic resonance spectra of kauranoid diterpenes. J.C.S. Perkin I **1976**, 114.

311. Hanson, J.R., and A.F. White: Studies in terpenoid biosynthesis II. The biosynthesis of steviol. Phytochem. **7**, 595 (1968).

312. Harborne, J.B.: Inuleae—chemical review. In: The Biology and Chemistry of the Compositae (Heywood, V.H., J.B. Harborne and B.L. Turner, eds.). New York and London: Academic Press. 1977.

313. Hedden, P., and J.E. Grabbe: Kaurenolide biosynthesis in a cell-free system from *Cucurbita maxima* seeds. Phytochem. **20**, 1011 (1981).

314. Hegnauer, R.: Compositae. In: Chemotaxonomie der Pflanzen, Vol. 3. Basel: Birkhäuser Verlag. 1964.

315. Hegnauer, R.: The chemistry of the Compositae. In: The Biology and Chemistry of the Compositae (Heywood, V.H., J.B. Harborne, and B.L. Turner, eds.). New York and London: Academic Press. 1977.

316. Heiser, C.B., D.M. Smith, S.B. Clevenger, and W.C. Martin: The North American Sunflowers (*Helianthus*). Bull. Torr. Bot. Club **22**, 1 (1969).

317. Henderson, M.S., R. McCrindle, and D. McMaster: Constituents of *Solidago* species. Part V. Non-acidic diterpenoids from *Solidago gigantea* var. *serotina*. Can. J. Chem. **51**, 1346 (1973).

318. Henderson, M.S., R.D.H. Murray, R. McCrindle, and D. McMaster: Constituents of *Solidago* species. Part III. The constitution of diterpenoids from *Solidago juncea* Ait. Can. J. Chem. **51**, 1322 (1973).

319. Herz, W.: Astereae—chemical review. In: The Biology and Chemistry of the Compositae (Heywood, V.H., J.B. Harborne, and B.L. Turner, eds.). New York and London: Academic Press. 1977.

320. Herz, W., S.V. Bhat, and R. Murari: The diterpene darutigenol from *Palafoxia arida*. Phytochem. **17**, 1060 (1978).

321. Herz, W., and M. Bruno: Heliangolides, kauranes and other constituents of *Helianthus heterophyllus*. Phytochem. **25**, 1913 (1986).

322. Herz, W., D. Gage, and N. Kumar: Damsinic acid and ambrosanolides from vegetative *Ambrosia hispida*. Phytochem. **20**, 1601 (1981).

323. Herz, W., S.V. Govindan, and J.F. Blount: Tetracyclic analogues of the rosane lactones from *Eupatorium album*. J. Org. Chem. **44**, 2999 (1979).

324. Herz, W., S.V. Govindan, and K. Watanabe: Diterpenes of *Helianthus rigidus* and *H. salicifolius*. Phytochem. **21**, 946 (1982).

325. Herz, W., and P. Kulanthaivel: Eudesmanolides and *ent*-pimaranes from *Liatris laevigata*. Phytochem. **22**, 715 (1983).

326. Herz, W., and P. Kulanthaivel: *Ent*-kaurenes and trachylobanes from *Helianthus radula*. Phytochem. **22**, 2543 (1983).

327. Herz, W., and P. Kulanthaivel: *Ent*-pimaranes, *ent*-kauranes, heliangolides and other constituents of three *Helianthus* species. Phytochem. **23**, 1453 (1984).

328. Herz, W., and P. Kulanthaivel: *Ent*-kauranes and 10α-methyl-eudesman-8αH,12-olides from *Wedelia calycina* and *Wedelia hispida*. Phytochem. **23**, 2271 (1984).

329. Herz, W., and P. Kulanthaivel: Trihydroxy-C_{18}-acids and a labdane from *Rudbeckia fulgida*. Phytochem. **24**, 89 (1985).

330. Herz, W., and P. Kulanthaivel: Diterpenes and sesquiterpene lactones from *Mikania congesta*. Phytochem. **24**, 1761 (1985).

331. Herz, W., and P. Kulanthaivel: Diterpenes from *Viguiera porteri*. J. Nat. Prod. **48**, 676 (1985).

332. Herz, W., P. Kulanthaivel, and K. Watanabe: *Ent*-kauranes and other constituents of three *Helianthus* species. Phytochem. **22**, 2021 (1983).

333. Herz, W., and N. Kumar: Sesquiterpene lactones from *Helianthus grosseserratus*. Phytochem. **20**, 99 (1981).

334. Herz, W., A.-M. Pilotti, A.-C. Söderholm, I.K. Shuhama, and W. Vichnewski: New *ent*-clerodane-type diterpenoids from *Baccharis trimera*. J. Org. Chem. **42**, 3913 (1977).

335. Herz, W., and R.P. Sharma: Ligantrol and ligantrol monoacetate, two new linear polyoxgenated diterpenes from *Liatris elegans*. J. Org. Chem. **40**, 192 (1975).

336. Herz, W., and R.P. Sharma: New hydroxylated *ent*-kauranoic acids from *Eupatorium album*. J. Org. Chem. **41**, 1021 (1976).

337. Herz, W., K. Watanabe, P. Kulanthaivel, and J.F. Blount: Cycloartanes from *Lindheimera texana*. Phytochem. **24**, 2645 (1985).

338. Hochmannova, J., L. Novotny, and V. Herout: Hydrocarbons from *Petasites albus* (L.)Gaertn. rhizomes. Coll. Czech. Chem. Commun. **27**, 2711 (1962).

339. Hoeneisen, M. and M. Silva: Quimica y actividad biologica de algunas Compositae chilenas. Rev. Latinoamer. Quim. **17**, 24 (1986).

340. Hoffmann, J.J., B.E. Kingsolver, S.P. McLaughlin, and B.N. Timmerman:

Production of resins by arid-adapted Astereae. In: Recent Advances in Phytochemistry (Timmermann, B.N., C. Steelink, and F.A. Loewus, eds.), Vol. 18, p. 251. New York and London: Plenum Press. 1984.

341. Hoffmann, J.J., S.P. McLaughlin, S.D. Jolad, K.H. Schram, M.S. Tempesta, and R.B. Bates: Constituents of *Chrysothamnus paniculatus* (Compositae): Chrysothame, a new diterpene, and 6-oxogrindelic acid. J. Org. Chem. **47**, 1725 (1982).

342. Hosozawa, S., N. Kato, and K. Munakata: Antifeeding active substances for insects in *Caryopteris divaricata* Maxim. Agr. Biol. Chem. **38**, 823 (1974).

343. Hubert, T.D., D.F. Wiemer: Ant-repellent terpenoids from *Melampodium divaricatum*. Phytochem. **24**, 1197 (1985).

344. Huneck, S., Y. Asakawa, Z. Taira, A.F. Cameron, J.D. Connolly, and D.S. Rycroft: Gymnocolin, a new *cis*-clerodane diterpenoid from the liverwort, *Gymnocolea inflata*. Crystal structure analysis. Tetrahedron Lett. **24**, 115 (1983).

345. Hutchison, M., P. Lewer, and J. Macmillan: Carbon-13 nuclear magnetic resonance spectra of eighteen derivatives of *ent*-kaur-16-en-19-oic acid. J. Chem. Soc. Perkin Trans. I **1984**, 2363.

346. Jakupovic, J., S. Banerjee, F. Bohlmann, R.M. King, and H. Robinson: New diterpenes from *Chiliotrichium rosmarinifolium* and Nardophyllum lanatum. Tetrahedron **42**, 1305 (1986).

347. Jakupovic, J., R.N. Baruah, F. Bohlmann, R.M. King, and H. Robinson: New alicyclic diterpenes and *ent*-labdanes from *Guterrezia solbrigii*. Tetrahedron **41**, 4537 (1985).

348. Jakupovic, J., R.N. Baruah, C. Zdero, F. Eid, V.P. Pathak, T.V. Chau-Thi, F. Bohlmann, R.M. King, and H. Robinson: Further diterpenes from plants of the Compositae, subtribe Solidagininae. Phytochem. **25**, 1873 (1986).

349. Jakupovic, J., E. Ellmauerer, F. Bohlmann, R.M. King, and H. Robinson: *ent*-Labdanes from *Austrobrickellia patens*. Phytochem. **25**, 1927 (1986).

350. Jakupovic, J., E. Ellmauerer, F. Bohlmann, A. Whittemore, and D.A. Gage: Diterpenes from *Eupatorium turbinatum*. Phytochem. **25**, 2677 (1986).

351. Jakupovic, J., J. Kuhnke, A. Schuster, M.A. Metwally and F. Bohlmann: Phloroglucinol derivatives and other constituents from South American *Helichrysum* species. Phytochem. **25**, 1133 (1986).

352. Jeffries, P.R., J.R. Knox, K.R. Price, B. Scaf: Constituents of the tumor-inhibitory extract of *Olearia muelleri*. Aust. J. Chem. **27**, 221 (1974).

353. Jiang, J.B., M.J. Urbanski, and Z.G. Hajos: Total synthesis of dioxane analogues related to zoapatanol. J. Org. Chem. **48**, 2001 (1983).

354. Jiménez, E.M., and M. Gonzalez de la Parra: nuevo alcohol diterpenico aislado de *Piqueria trinervia* Cav. Rev. Latinoamer. Quim. **14**, 20 (1983).

355. Kaneda, N., H. Kohda, K. Yamasaki, O. Tanaka, and K. Nishi: Paniculosides I-V. Diterpene glucosides from *Stevia ovata* Lag. Chem. Pharm. Bull. **26**, 2266 (1978).

356. Kanojia, R.M., M.P. Wachter, S.D. Levine, R.E. Adams, R. Chen, E. Chin, M.L. Cotter, A.F. Hirsch, R. Huettemann, V.V. Kane, P. Ostrowski, and C.J. Shaw: Isolation and structural elucidation of zoapatanol and montanol, novel oxepane diterpenoids from the Mexican plant zoapatle (*Montanoa tomentosa*). J. Org. Chem. **47**, 1310 (1982).

357. Khac Manh Duc, D., M. Fetizon, S. Lazare, P.K. Grant, M.J. Nicholls,

H.T.L. Liau, M.J. Francis, J. Poisson, J.-M. Bernassau, N.F. Roque, P.M. Wovkulich, and E. Wenkert: 13C NMR spectroscopy of tetracarbocyclic diterpenes and related substances. Tetrahedron **37**, 2371 (1981).

358. Kim, J.H., K.D. Han, K. Yamasaki, and O. Tanaka: Darutoside, a diterpenoid from *Siegesbeckia pubescens* and its structure revision. Phytochem. **18**, 894 (1979).

359. Kinghorn, A.D., D.D. Soejarto, N.P.D. Nanayakhara, C.M. Compadre, H.C. Makapugay, J.M. Hovanec-Brown, P.J. Medon, and S.K. Kamath: A phytochemical screening procedure for sweet *ent*-kaurene glycosides in the genus *Stevia*. J. Nat. Prod. **47**, 439 (1984).

360. Knudsen, F., W. Vilegas, F. Oliveira, and N.F. Roque: Pimaradiene diterpenes from *Mikania triangularis*. Phytochem. **25**, 1240 (1986).

361. Kobayashi, M., S. Horikawa, I.H. Degrandi, J. Ueno, and H. Mitsuhashi: Dulcosides A and B, new diterpene glycosides from *Stevia rebaudiana*. Phytochem. **16**, 1405 (1977).

362. Kohda, H., R. Kasai, K. Yamasaki, K. Murakami, and O. Tanaka: New sweet diterpene glycosides from *Stevia rebaudiana*. Phytochem. **15**, 981 (1976).

363. Kohda, H., O. Tanaka, and K. Nishi: Diterpene—glycosides of *Stevia paniculata* Lag.: Structures of aglycones. Chem. Pharm. Bull. **24**, 1040 (1976).

364. Kusumoto, S., T. Okazaki, A. Ohsuka, and M. Kotake: The structure of solidagonic acid, a bitter principle of goldenrod. Tetrahedron Lett. **40**, 4325 (1968).

365. Kusumoto, S., T. Okazaki, A. Ohsuka, and M. Kotake: The structure and stereochemistry of solidagonic acid. Bulletin of the Chemical Society of Japan. **42**, 812 (1969).

366. Lassak, E.V., and J.T. Pinhey: The constituents of some *Helichrysum* species (family Compositae). Aust. J. Chem. **21**, 1927 (1968).

367. Lee, E.F., J. Gershenzon, and T.J. Mabry: Terpenoids of *Helianthus nuttallii*. J. Nat. Prod. **47**, 1021 (1984).

368. Le Quesne, P.W., G.A. Cooper-Driver, M. Villani, M.N. Do, P.A. Morrow, and D.A. Tonkin: Biologically active diterpenes from *Solidago* species. Page 271 in: A. Rahman and P. W. Le Quesne (eds.). New Trends in Natural Products Chemistry. Studies in Organic Chemistry, Vol. 26. Elsevier Scientific Publishers. Amsterdam (1986).

369. Lequesne, P.W., V. Honkan, K.D. Onan, P.A. Morrow, and D. Tonkyn: Oxidized kaurene derivatives from leaves of *Solidago missouriensis* and *S. rigida*. Phytochem. **24**, 1785 (1985).

370. Le Van, N., and T.V. Cuong Pham: Viscidic acid A and B, two *ent*-labdane derivatives from *Chrysothamnus viscidiflorus*. Phytochem. **19**, 1971 (1980).

371. Levan, N., and N.J. Fischer: Three new melampolide sesquiterpenes, polymatin A, B, and C, from *Polymnia maculata*. Phytochem. **18**, 851 (1979).

372. Levine, S.D., D.W. Hahn, M.L. Cotter, F.C. Greenslade, R.M. Kanojia, S.A. Pasquale, M. Wachter, and J.L. McGuire: The Mexican plant zoapatle *(Montanoa tomentosa)* in reproductive medicine. Past, present and future. Journal of Reproductive Medicine **26**, 524 (1981).

373. Lloyd, H.A., S.L. Evans, and H.M. Fales: Terpene alcohols of *Helichrysum*

dendroideum. (1978).

374. Lloyd, H.A., and H.M. Fales: Terpene alcohols of *Helichrysum dendroideum.* Tetrahedron Lett. **48**, 4891 (1967).

375. Luteijn, J.M., A. Van Veldhuizen, and A. De Groot: The assignment of ^{13}C NMR shift data in clerodanes and related structures. Organic Magnetic Resonance **19**, 95 (1982).

376. Mahanto, S.B., A.K. Sen, P.C. Mazumbar, and K. Yamasaki: Diterpenes of *Conyza stricta*, identification of conyzic acid, seconidoresedasäure and strictic acid. Phytochem. **20**, 850 (1981).

377. Manabe, S., N. Enoki, and C. Nishino: Application of the CD homoallylic benzoate method as a chiroptical tool for determination of absolute configuration. Tetrahedron Lett. **26**, 2213 (1985).

378. Manabe, S. and C. Nishino: Stereochemistry of *cis*-clerodane diterpenes. Tetrahedron **42**, 3461 (1986).

379. Manchand, P.S., J.F. Blount, T. McCabe, and J. Clardy: Structure of evillosin, a novel labdane diterpenoid lactone from *Eupatorium villosum* Sw. J. Org. Chem. **44**, 1322 (1979).

380. Mangoni, L., and M. Belardini: Sui componenti della *Grindelia robusta.* Nota II. Acido 6-oxo-grindelico. Gass. Chim. Ital. **92**, 983 (1962).

381. Mangoni, L., and M. Belardini: Sui componenti della *Grindelia robusta.* Nota III. Acido 7α,8α-oxido-diidrogrindelico. Gass. Chim. Ital. **92**, 995 (1962).

382. Marcelle, G.B., N. Bunyapraphatsara, G.A. Cordell, H.H.S. Fong, K.C. Nicolaou, and R.E. Zipkin: Studies on Zoapatle I. The extraction of zoapatle (*Montanoa tomentosa*) and the identification of 21-normontanol as the initial decomposition product of zoapatanol. J. Nat. Prod. **48**, 739 (1985).

383. Mathur, S.B., and C.M. Fermin: Terpenes of *Mikania mongenansis.* Phytochem. **12**, 226 (1973).

384. Mathur, S.B., P. Garcia Tello, C.M. Fermin, and V. Mora-Arellano: Terpenoids of *Mikania monagasensis* and their biological activities. Rev. Latinoamer. Quim. **6**, 201 (1975).

385. McCrindle, R., and E. Nakamura: Constituents of *Solidago* species. Part VI. The constitution of diterpenoids from a chemically distinct variety of *Solidago serotina.* Can. J. Chem **52**, 2029 (1974).

386. McCrindle, R., E. Nakamura, and A.B. Anderson: Constituents of *Solidago* species. Part VII. Constitution and stereochemistry of the *cis*-clerodanes from *Solidago arguta* Ait. and of related diterpenoids. J. Chem. Soc. Perkin Trans. I. **15**, 1590 (1976).

387. Melek, F.R., D.A. Gage, J. Gershenzon, and T.J. Mabry: Sesquiterpene lactone and diterpene constituents of *Helianthus annuus.* Phytochem. **24**, 1537 (1985).

388. Merxmüller, H., P. Leins, and H. Roessler: Inuleae—systematic review. In: The Biology and Chemistry of the Compositae (Heywood, V.H., J.B. Harborne, and B.L. Turner, eds.), Vol. 1. London: Academic Press.

389. Michie, M.J., and W.W. Reid: Biosynthesis of complex terpenes in the leaf cuticle and trichomes of *Nicotiana tabacum.* Nature **218**, 578 (1968).

390. Misra, L.N., J. Jakupovic, F. Bohlmann, and G. Schmeda-Hirschmann: Isodaucane derivatives, norsesquiterpenes and clerodanes from *Chromolaena laevigata.* Tetrahedron **41**, 5353 (1985).

391. Misra, R., R.C. Pandey, and Sukh Dev: The absolute stereochemistry of hardwickiic acid and its congeners. Tetrahedron Letters **1968**, 2681.

392. Mitscher, L.A., G.S.R. Rao, T. Veysoglu, S. Drake, and T. Haas: Isolation and identification of trachyloban-19-oic and (-)-kaur-16-en-19-oic acids as antimicrobial agents from the prairie sunflower, *Helianthus annuus*. J. Nat. Prod. **46**, 745 (1983).

393. Miyakado, M., N. Ohno, H. Yoshioka, T.J. Mabry, and T. Whiffin: Gymnospermin: A new labdantriol from *Gymnosperma glutinosa*. Phytochem. **13**, 189 (1974).

394. Morales Méndez, A., A. Usubillaga, A.K. Banerjee, and T. Nakano: Studies on the constituents of *Espeletia weddellii*. Planta Med. **24**, 243 (1973).

395. Mossa, J.R., J.M. Cassady, M.D. Antoun, S.R. Byrn, A.T. McKenzie, J.F. Kozlowski, and P. Main: Saudin, a hypoglycemic diterpenoid with a novel 6,7-secolabdane carbon skeleton from *Cluytia richardiana*. J. Org. Chem. **50**, 916 (1985).

396. Muradian, J., M. Motidome, and P.C. Ferreira: Flavonols and (-)-kaur-16-en-19-oic acid from *Mikania hirsutissima* DC. var. *hirsutissima*. Rev. Latinoamer. Quim. **8**, 88 (1977).

397. Murakami, T., T. Isa, and T. Satake: Eine neuuntersuchung der inhaltsstoffe von *Siegesbeckia pubescens* Makino. Tetrahedron Lett. **50**, 4991 (1973).

398. Nair, A.G.R., S. Subramanan, F. Bohlmann, S. Schöneweiss, and T.J. Mabry: A new diterpene galactoside from *Acanthospermum hispidum*. Phytochem. **15**, 1776 (1976).

399. Nakano, T., and A. Martin: Synthesis of labda-8(17),13(14)-diene-15,16-olide and 15,16-epoxy-labda-8(17),13(16)-14-triene and their rearrangement to clerodane derivatives. Tetrahedron **38**, 1217 (1982).

400. Nicholas, H.J.: Biosynthesis and metabolism of [^{14}C] sclareol. Biochem. Biophys. Acta **84**, 80 (1964).

401. Nishino, C., S. Manabe, M. Kazui, and T. Matsuzaki: Piscicidal *cis*-clerodane diterpenes from *Solidago altissima* L.: Absolute configurations of $5\alpha,10\alpha$-*cis*-clerodanes. Tetrahedron Lett. **25**, 2809 (1984).

402. Niwa, M., and S. Yamamura: Stereostructures of several solidagolactones (elongatolides). Tetrahedron Lett. **22**, 2789 (1981).

403. Nurmukhamedova, M.R., N.D. Abdullaev, and G.P. Sidyakin: Diterpenoids of *Pulicaria salviifolia*. II. Structure of salvicin. Khim. Prir. Soedin. 299 (1986).

404. Nurmukhamedova, M.R., S.Z. Kasymov, N.D. Abdullaev, G.P. Sidyakin, and M.R. Yagudaev: Diterpenoids of *Pulicaria salviifolia* I. Structures of salvin and salvinin. Khim. Prir. Soedin. 201 (1985).

405. Oberti, J.C., A.B. Pomilio, and E.G. Gros: Diterpenes and sterols from *Wedelia glauca*. Phytochem. **19**, 2051 (1980).

406. Oberti, J.C., V.E. Sosa, P. Kulanthaivel, and W. Herz: *Ent*-norlabdane triols from *Austroeupatorium inulaefolium*. Phytochem. **23**, 2003 (1984).

407. Oehlschlager, A.C., and G. Ourisson: A comparison of *in vivo* and *in vitro* skeletal transformations of diterpenes. In: Terpenoids in Plants (Pridham, J.B., ed.). London and New York: Academic Press. 1967.

408. Oelrichs, P.P., P.J. Vallely, J.K. MacLeod, and I.A.S. Lewis: Isolation of a new potential antitumor compound from *Wedelia asperrima*. J. Nat. Prod. **43**,

414 (1980).

409. Ohno, N., J. Geshenzon, P. Newman, and T.J. Mabry: Diterpene carboxylic acids and a heliangolide from *Helianthus angustifolius*. Phytochem. **20**, 2393 (1981).

410. Ohno, N., and T.J. Mabry: Sesquiterpene lactones and diterpene carboxylic acids in *Helianthus niveus* subspecies *canescens*. Phytochem. **19**, 609 (1980).

411. Ohno, N., T.J. Mabry, V. Zabel, and W.H. Watson: Tetrachyrin, a new rearranged kaurenoid lactone, and diterpene acid from *Tetrachryon orizabaensis* and *Helianthus debilis*. Phytochem. **18**, 1687 (1979).

412. Ohsuka, A., S. Kusumoto, and M. Kotake: The steric structures of diterpene carboxylic acids from *Solidago altissima* L. Nippon Kagaku Kaishi **1973**, 590.

413. Okazaki, T., A. Ohsuka, and M. Kotake: Structure of diterpene lactones from *Solidago altissima* L. Nippon Kagaku Kaishi **1973**, 584.

414. Oriental, M.A., E. Guerreiro, and O.S. Giordano: Diterpene acids from *Grindelia aegialitis*. Rev. Latinoamer. Quim. **15**, 73 (1984).

415. Ortega, A., R. Martinez, and C.L. Garcia: Los diterpenos de *Stevia salicifolia* Cav. Estructura del stevinsol and salicifoliol. Rev. Latinoamer. Quim. **11**, 45 (1980).

416. Ortega, A., F.J. Morales, and M. Salmon: Kaurenic acid derivatives from *Stevia eupatoria*. Phytochem. **24**, 1850 (1985).

417. Oshima, Y., G.A. Cordell, and H.H.S. Fong: Oxepane diterpenes from *Montanoa tomentosa*. Phytochem. **25**, 2567 (1986).

418. Oshima, Y., J. Saito, and H. Hikino: Sterebins A, B, C, and D, bisnorditerpenoids of *Stevia rebaudiana* leaves. Tetrahedron **42**, 6443 (1986).

419. Oshima, Y., S.-M. Wong, C. Konno, G.A. Cordell, D.P. Waller, D.D. Soejarto, and H.H.S. Fong: Studies on Zoapatle, II. Leucantholide, a novel sesquiterpene lactone from *Montanoa leucantha* ssp. *leucantha*. J. Nat. Prod. **49**, 313 (1986).

420. Ourisson, G.: Some aspects of the distribution of diterpenes in plants. In: Chemistry in Botanical Classification (Nobel Foundation, Stockholm), Nobel Symposium **25**. New York and London: Academic Press. 1973).

421. Pakrashi, S.C., P.P. Ghosh Dastidar, and E. Ali: Newer (-)-kaurene derivatives from *Enhydra fluctuans* Lour. Indian J. Chem. **9**, 84 (1971).

422. Pakrashi, S.C., P.P. Ghosh Dastidar, and S.K. Gupta: Diterpenoids from *Enhydra fluctuans*. Phytochem. **9**, 459 (1970).

423. Pandey, U.C., A.K. Singhal, N.C. Barua, R.P. Sharma, J.N. Baruah, K. Watanabe, P. Kulanthaivel, and W. Herz: Stereochemistry of strictic acid and related furanoditerpenes from *Conyza japonica* and *Grangea maderaspatana*. Phytochem. **23**, 391 (1984).

424. Patra, A., A.K. Mitra, S. Biswas, C.D. Gupta, T.K. Chatterjee, K. Basu, and A.K. Barua: Carbon-13 NMR spectra of some labdane diterpenoids. Org. Magn. Reson. **17**, 301 (1981).

425. Pezzuto, J.M.: Chemistry, metabolism and biological activity of steviol (*ent*-13-hydroxykaur-16-en-19-oic acid), the aglycone of stevioside. Page 371 in A. Rahman and P.W. Le Quesne (eds.). New Trends in Natural Products Chemistry. Studies in Organic Chemistry, Vol. 26. Elsevier Scientific Publishers. Amsterdam (1986).

426. Pinhey, J.T., R.F. Simpson, and J.L. Batey: The constituents of *Olearia*

heterocarpa. The structure of olearin, a diterpene dilactone of the cascarillin group. Aust. J. Chem. **25**, 2621 (1972).

427. Pinto, A.C., M.L. Patitucci, R.S. Dasilva, P.P.S. Queiroz, and A. Kelecom: Pimarane and cleistanthane diterpenes from Velloziaceae: absolute configuration and biomimetic conversion. Tetrahedron **39**, 3351 (1983).

428. Piozzi, F., S. Passannanti, and M.P. Paternostro: Kaurenoid diterpenes in *Espeletia grandiflora*. Phytochem. **10**, 1164 (1971).

429. Proksch, P., and E. Rodriguez: Chromenes and benzofurans of the Asteraceae, their chemistry and biological significance. Phytochem. **22**, 2335 (1983).

430. Purushothaman, K.K., A. Sarada, A. Saraswathy, and J.D. Connolly: Sempervirenic acid, a diterpene acid from *Solidago sempervirens*. Phytochem. **22**, 1042 (1983).

431. Quijano, L., J.S. Calderon, F. Gomez G., V. Rosario M., and T. Rios: Oxepane diterpenoids and sesquiterpene lactones from "zoapatle" (*Montanoa tomentosa*), a Mexican plant with oxytocic activity. Phytochem. **24**, 2337 (1985).

432. Quijano, L., J.S. Calderon, F. Gomez G., V. Rosario, and T. Rios: Acyclic precursors of the uterotonic oxepane diterpenoids of "zoapatle" (*Montanoa tomentosa*). Phytochem. **24**, 2741 (1985).

433. Quijano, L., J.S. Calderon, F. Gomez, J.L. Vega, and T. Rios: Diterpenes from *Stevia monardaefolia*. Phytochem. **21**, 1369 (1982).

434. Quijano, L., and N.H. Fischer: Sesquiterpene and diterpene lactones from *Melampodium longipilum*. Phytochem. **23**, 829 (1984).

435. Quijano, L., and N.H. Fischer: Two oxepane diterpene lactones from *Melampodium diffusum*. Phytochem. **23**, 833 (1984).

436. Quijano, L., F.R. Fronczek, and N.H. Fischer: Sesquiterpene and diterpene lactones from *Melampodium leucanthum* and the molecular structure of 4(5)-dihydromelampodin-B. Phytochem. **24**, 1747 (1985).

437. Railton, I.D., B. Fellows, and C.A. West: *Ent*-kaurene synthesis in chloroplasts from higher plants. Phytochem. **23**, 1261 (1984).

438. Robinson, H., and R.M. King: Eupatorieae—systematic review. In: The Biology and Chemistry of the Compositae (Heywood, V.H., J.B. Harborne, and B.L. Turner, eds.), Vol. 1. London: Academic Press. 1977.

439. Rogers, L.J., S.P.J. Shah, and T.W. Goodwin: Compartmentation of biosynthesis of terpenoids in green plants. Photosynthetica **2**, 184 (1968).

440. Rose, A.F.: Grindelane diterpenoids from *Chrysothamnus nauseosus*. Phytochem. **19**, 2689 (1980).

441. Rose, A.F., K.C. Jones, W.F. Haddon, and D.L. Dreyer: Grindelane diterpenoid acids from *Grindelia humilis*: Feeding deterrency of diterpene acids towards aphids. Phytochem. **20**, 2249 (1981).

442. Rossomando, P.C., O.S. Giordano, J. Espiñeira, and P. Joseph-Nathan: A diterpene acid from *Baccharis tucumanensis*. Phytochem. **24**, 787 (1985).

443. Rustaiyan, A., E. Simozar, A. Ahmadi, M. Grenz, and F. Bohlmann: A hardwickiic acid derivative from *Pulicaria gnaphalodes*. Phytochem. **20**, 2772 (1981).

444. Ruzicka, L.: The isoprene rule and the biogenesis of terpenic compounds. Experientia **9**, 357 (1953).

445. Ruzicka, L., A. Eschenmoser, and H. Heusser: Biogenesis of steroids and terpenic compounds. Experientia **9**, 362 (1953).

446. St. Pyrek: Neutral diterpenoids of *Helianthus annuus*. J. Nat. Prod. **47**, 822 (1984).

447. Sakamoto, I., K. Yamasaki, and O. Tanaka: Application of ^{13}C NMR spectroscopy to chemistry of plant glycosides: rebaudiosides-D and -E, new sweet diterpene-glucosides of *Stevia rebaudiana* Bertoni. Chem. Pharm. Bull. **25**, 3437 (1977).

448. Salmon, M., A. Ortega, G. Garcia de la Mora, and E. Angeles. A diterpenic acid from *Stevia lucida*. Phytochem. **22**, 1512 (1983).

449. San-Martin, A., A. Givovich, and M. Castillo: Neo-clerodane diterpenoids from *Baccharis incarum*. Phytochem. **25**, 264 (1986).

450. San-Martin, A., J. Rovirosa, R. Becker, and M. Castillo: Diterpenoids from *Baccharis tola*. Phytochem. **19**, 1985 (1980).

451. San-Martin, A., J. Rovirosa, and M. Castillo: Diterpenoids from *Baccharis tola*. Phytochem. **22**, 1461 (1983).

452. San-Martin, A., J. Rovirosa, C. Labbe, A. Givovich, M. Mahu and M. Castillo: Neo-clerodane diterpenoids from *Baccharis rhomboidalis*. Phytochem. **25**, 1393 (1986).

453. Schmeda-Hirschmann, G., J. Jakupovic, V.P. Pathak, and F. Bohlmann: Balansolide and other sesquiterpene lactones from *Bejaranoa balansae*. Phytochem. **25**, 2167 (1986).

454. Schmeda-Hirschmann, G., C. Zdero, R.N. Baruah, and F. Bohlmann: Further sesquiterpene lactones from *Calea* and *Viguiera* species. Phytochem. **24**, 2019 (1985).

455. Schteingart, C.D., and A.B. Pomilio: Terpenoids from *Wedelia bupthalmiflora*. Phytochem. **20**, 2589 (1981).

456. Schteingart, C.D. and A.B. Pomilio: Atractyloside, toxic compound from *Wedelia glauca*. J. Nat. Prod. **47**, 1046 (1984).

457. Scott, A.I., F. McCapra, F. Comer, S.A. Sutherland, D.W. Young, G.A. Sim, and G. Ferguson: Stereochemistry of the diterpenoids—IV. Structure and stereochemistry of some polycyclic diterpenoids. Tetrahedron **20**, 1339 (1964).

458. Seaman, F.: Sesquiterpene lactones as taxonomic characters in the Asteraceae. Bot. Rev. **48**, 121 (1982).

459. Seaman, F., A.J. Malcolm, and N.H. Fischer: Tomexanthin, an oxepane diterpene from *Montanoa tomentosa*. Phytochem. **23**, 464 (1984).

460. Sen, A.K., S.B. Mahato, and N.L. Dutta: Conyzic acid, a new diterpene resin from *Conyza stricta*. Indian J. Chem. **13**, 504 (1975).

461. Sholichin, M., K. Yamasaki, R. Miyama, R. Yahara, and O. Tanaka: Labdane-type diterpenes from *Stevia rebaudiana*. Phytochem. **19**, 326 (1980).

462. Silva, M., and P.G. Sammes: A new diterpenic acid and other constituents of *Haplopappus foliosus* and *H. angustifolius*. Phytochem. **12**, 1755 (1973).

463. Singh, P., M.C. Sharma, K.C. Joshi, and F. Bohlmann: Diterpenes derived from clerodanes from *Pulicaria angustifolia*. Phytochem. **24**, 190 (1985).

464. Soejarto, D.D., A.D. Kinghorn, and N.R. Farnsworth: Potential sweetening agents of plant origin. III. Organoleptic evaluation of *Stevia* leaf herbarium samples for sweetness. J. Nat. Prod. **45**, 590 (1982).

465. Sorensen, N.A.: Polyacetylenes and conservation of chemical characters in the Compositae. In: The Biology and Chemistry of the Compositae (Heywood,

V.H., J.B. Harborne, and B.L. Turner, eds.), Vol. 1. London: Academic Press. 1977.

466. Stapel, G., H.G. Menssen, and G. Snatzke: Isolierung und strukturaufklärung von zwei diterpenen aus *Baccharis articulata*. Planta Med. **38**, 366. (1980).

467. Stuessy, T.F.: Heliantheae—systematic review. In: The Biology and Chemistry of the Compositae (Heywood, V.H., J.B. Harborne, and B.L. Turner, eds.), Vol. 1. London: Academic Press. 1977.

468. Swain, T., and C.A. Williams: Heliantheae—chemical review. In: The Biology and Chemistry of the Compositae (Heywood, V.H., J.B. Harborne, and B.L. Turner, eds.), Vol. 1. London: Academic Press. 1977.

469. Tabacik, C., and M. Bard: Etude chimio-taxonomique dans le genre *Cistus*. Phytochem. **10**, 3093 (1971).

470. Tandon, S., and R.P. Rastogi: Strictic acid, a novel diterpene from *Conyza stricta*. Phytochem. **18**, 494 (1979).

471. Timmermann, B.N., J.J. Hoffmann, S.D. Jolad, R.B. Bates, and T.J. Siahaan: Diterpenoids and flavonoids from *Grindelia discoidea*. Phytochem. **25**, 723 (1986).

472. Timmermann, B.N., J.J. Hoffmann, S.D. Jolad, R.B. Bates, and T.J. Siahaan: Labdane diterpenoids from *Grindelia discoidea*. Phytochem. **25**, 1389 (1986).

473. Timmermann, B.N., J.J. Hoffmann, S.D. Jolad, and K.H. Schram: Grindelane diterperoids from *Grindelia squarrosa* and *G. camporum*. Phytochem. **24**, 1031 (1985).

474. Timmermann, B.N., J.J. Hoffmann, S.D. Jolad, K.H. Schram, R.E. Klenck, and R.B. Bates: Constituents of *Chrysothamnus paniculatus* (Compositae). 2. Chrysolic acid, a new labdane-derived diterpene with an aromatic B ring. J. Org. Chem. **47**, 4114 (1982).

475. Timmermann, B.N., D.J. Luzbetak, J.J. Hoffmann, S.D. Jolad, K.H. Schram, R.B. Bates, and R.E. Klenck: Grindelane diterpenoids from *Grindelia camporum* and *Chrysothamnus paniculatus*. Phytochem. **22**, 523 (1983).

476. Tomassini, T.C.B., and M.E.O. Matos: On the natural occurrence of 15α-tiglinoyloxy-kaur-16-en-19-oic acid. Phytochem. **18**, 663 (1979).

477. Tonn, C.E., P.C. Rossomando, and O.S. Giordano: Batudioic acid and flavonoids from *Baccharis tucumanensis*. Phytochem. **21**, 2599 (1982).

478. Tsankova, E., and F. Bohlmann: A monoterpene from *Aster bakeranus*. Phytochem. **22**, 1285 (1983).

479. Usubillaga, A., J. de Hernandez, N. Perez, and M. Kiriakidis: Kauranoid diterpenes in *Espeletia* species. Phytochem. **12**, 2999 (1973).

480. Usubillaga, A., and A. Morales: Kaurenic acids in *Espeletia* species. Phytochem. **11**, 1856 (1972).

481. Usubillaga, A., and A. Nakano: Kauranoid diterpenes in *Ruilopezia margarita*. Planta Med. **35**, 331 (1978).

482. Vichnewski, W., H. de Freitas Leitao Filho, R. Murari, and W. Herz: Cinnamoylgrandifloric acid from *Mikania oblongifolia*. Phytochem. **16**, 2028 (1977).

483. Vichnewski, W., R. Murari, and W. Herz: New clerodanes from *Symphiopappus itatiayensis*. Phytochem. **18**, 129 (1979).

484. Vis, E., and H. G. Fletcher: Stevioside. IV. Evidence that stevioside is a sophoroside. J. Am. Chem. Soc. **78**, 4709 (1956).

485. Vlad, P.F., N.D. Ungur, and M.N. Koltsa: Cyclization and rearrangements of diterpenoids. I. Synthesis of tetracyclic diterpenoids with a new carbon skeleton from labdanes. Tetrahedron **39**, 3947 (1983).

486. Von Carstenn-Lichterfelde, C., C. Pascual, J. Pons, R.M. Rabanal, B. Rodriguez, and S. Valverde: Carbon-13 NMR spectra of ent-beyerane and ent-beyerene derivatives. Tetrahedron Lett. **1975**, 3569.

487. Waddell, T.G., C.B. Osborne, R. Collins, M.J. Levine, M.C. Cross, J.V. Silverton, H.M. Fales, and E.A. Sokoloski: Erigerol, a new labdane diterpene from *Erigeron philadelphicus*. J. Org. Chem. **48**, 4450 (1983).

488. Wagenitz, G.: Systematics and phylogeny of the Compositae (Asteraceae). Plant Syst. Evol. **125**, 29 (1976).

489. Wagner, H., R. Seitz, V.M. Chari, and H. Lotter: Bacchotrcuneatin A and B, zwei neue diterpenlactone aus *Baccharis tricuneata* var. *tric.* Tetrahedron Lett. **1977**, 3039.

490. Wagner, H., R. Seitz, and H. Lotter: New furanoid ent-clerodanes from *Baccharis tricuneata*. J. Org. Chem. **43**, 3339 (1978).

491. Waiss, A.C., B.G. Chan, C.A. Elliger, V.H. Garrett, E.C. Carlson, and B. Beard: Larvicidal factors contributing to host-plant resistance against sunflower moth. Naturwissenschaften **64**, 341 (1977).

492. Warning, U., F. Bohlmann, H. Sanchez V., E. Del Rio S., and X.A. Dominguez: New constituents of *Baccharis salicifolia*. Rev. Latinoamer. Quim. **17**, 199 (1986).

493. Watanabe, K., N. Ohno, H. Yoshioka, J. Gershenzon, and T.J. Mabry: Sesquiterpene lactones and diterpenoids from *Helianthus argophyllus*. Phytochem. **21**, 709 (1982).

494. Wehrli, F.W., and T. Nishida: The use of carbon-13 nuclear magnetic resonance spectroscopy in natural products chemistry. Fortschritte d. Chem. org. Naturst. **36**, 1 (1979).

495. Wenkert, E.: Structural and biogenetic relationships in the diterpene series. Chem. and Ind. **1955**, 282.

496. Wenkert, E., and Z. Kumazawa: Manool-derived diterpenic hydrocarbons and related products. Chemical Communications **1968**, 140.

497. West, C.A.: Biosynthesis of diterpenes. In: Biosynthesis of Isoprenoid Compounds (Porter, J.W., and S.L. Spurgeon, eds.), Vol. 1, p. 376. New York: John Wiley and Sons. 1981.

498. West, C.A., M.W. Dudley, and M.T. Dueber: Regulation of terpenoid biosynthesis in higher plants. In: Topics in the Biochemistry of Natural Products (Swain, T., and G.R. Waller, eds.). New York and London: Plenum Press.

499. Wollenweber, E., and V.H. Dietz: Occurrence and distribution of free flavonoid aglycones in plants. Phytochem. **20**, 869 (1981).

500. Yamamura, Y., I. Masatoki, M. Niwa, Hasegawa, S. Ohba, and Y. Saito: The isolation and structure of tricyclosolidagolactone, a new diterpene from *Solidago altissima*. Tetrahedron Lett. **22**, 739 (1981).

501. Yamasaki, K., H. Kohda, T. Kobayashi, N. Kaneda, R. Kasai, O. Tanaka, and K. Nishi: Application of 13C nuclear magnetic resonance spectroscopy to chemistry of glycosides: structures of paniculosides-I, -II, -III, -IV, and -V, diterpene glucosides of *Stevia paniculata* Lag. Chem. Pharm. Bull. **25**, 2895 (1977).

502. Yamasaki, K., H. Kohda, T. Kobayashi, R. Kasai, and O. Tanaka: Structures of *Stevia* diterpene-glucosides: Application of 13C NMR. Tetrahedron Lett. **1976**, 1005.

503. Zdero, C., F. Bohlmann, R.M. King, and H. Robinson: Diterpene glycosides and other constituents from Argentinian *Baccharis* species. Phytochem. **25**, 2841 (1986).

504. Zurcher, R.F.: Protonenresonanzspektroskopie und steroidstruktur. I. Das C-19-methylsignal in funktion der substituenten. Helv. Chim. Acta **44**, 1380 (1961).

505. Zurcher, R.F.: Protonenresonanzspektroskopie und steroidstruktur. II. Die lage der C-18- und C-19-methylsignale in abhängigkeit von den substituenten am steroidgerüst. Helv. Chim. Acta **46**, 2054 (1963).

Appendix

APPENDIX. Alphabetical Listing of Molecular Substituents including Ester Sidechains.

Name	Code	Structure
(1,2-dihydroxyethyl)acrylate	(1,2-OH-Et)Acr	
(2α-acetoxyethyl)acrylate	(2-AcEt)Acr	
(2α-hydroxyethyl)acrylate	(2-OH-Et)Acr	
2,3-diacetyloxy-2-methylbutyrate	2-Mebut-2,3-Ac	
2,3-dihydroxy-2-methylbutyrate	2-Mebut-2,3-OH	
2,3-dihydroxyisobutyrate	i-But-2,3-OH	

Appendix. (*contd.*).

Name	Code	Structure
2,3-epoxy-2-methylbutyrate	Epoxyang	
2,3-epoxy-n-butyrate	Epoxy-n-But	
2-hydroxy-2-methylbutyrate	2-Mebut-2-OH	
2-hydroxy-3-acetyloxy-2-methylbutyrate	2-Mebut-2-OH-3-Ac	
2-hydroxy-3-chloro-isobutyrate	i-But-2-OH-3-Cl	
2-hydroxy-3-oxo-2-methylbutyrate	2-Mebut-2-OH-3=O	
2-hydroxy-isobutyrate	i-But-2-OH	
2-methyl-butanoate	2-Mebut	

APPENDIX. (*contd.*).

Name	Code	Structure
3-acetoxy-isovalerate	i-Val-3-Ac	
3-furoate	Fur	
3-hydroxy-2-ethoxy-isobutyrate	i-But-3-OH-2-OEt	
3-hydroxy-2-methylbutyrate	2-Mebut-3-OH	
3-hydroxy-isovalerate	i-Val-3-OH	
3-methyl-2,3-dehydro-valerate	deh-Val-3-Me	
3-methyl-coumarate	3-Me-Coum	
4,5-dihydroxytiglate	Tig-4,5-OH	

APPENDIX. *(contd.).*

Name	Code	Structure
4-[3-hydroxy-stearoyloxy]-tiglate	Tig-4-Stear	
4-acetyloxy-angelate	Ang-4-Ac	
4-hydroxy-5-[5-hydroxy-tiglinoyloxy]-tiglate	Tig-4-OH-5-Tig-5-OH	
4-hydroxy-5-acetyloxytiglate	Tig-4-OH-5-Ac	
4-hydroxy-5-tiglinoyloxy-tiglate	Tig-4-OH-5-Tig	
4-hydroxy-angelate	Ang-4-OH	
4-hydroxy-isobutyrate	i-But-4-OH	

APPENDIX. (*contd.*).

Name	Code	Structure
4-hydroxy-methacrylate	Mac-4-OH	
4-hydroxy-sarracinate	Sar-4-OH	
4-hydroxy-tiglate	Tig-4-OH	
5-[5-hydroxy-tiglinoyloxy]-tiglate	Tig-5-OTig-5-OH	
5-acetyloxy-tiglate	Tig-5-Ac	
5-hydroxy-2,3-dihydro-3-sulfydryl-sarracinate	dih-Sar-5-OH-3-SH	
5-hydroxy-tiglate	Tig-5-OH	
acetate	Ac	

APPENDIX. (*contd.*).

Name	Code	Structure
acetyloxy	OAc	
acetylsarracinate	Sarac	
angelate	Ang	
anthranoate	Anth	
arabinoside	Arab	
arachinate	Arac	$[CH_3(CH_2)_{18}CO-]$
behenate (CO(CH2)20Me	Behen	$[CH_3(CH_2)_{20}CO-]$
benzoate	Benz	

APPENDIX. (*contd.*).

Name	Code	Structure
benzoyloxy	O-Ben	
bernsteinoate (= succinate)	O-bern	
cis-3-S-methylacrylate	cis-Acr-S-Me	
cis-cinnamate	cis-Cinn	
cis-coumarate	cis-Coum	
coumarate	Coum	
epoxymethacrylate	Epoxymac	
epoxysarracinate	Epoxysar	

APPENDIX. (*contd.*).

Name	Code	Structure
formyloxy	OCHO	
β-Galactoside	Gal	
β-D-[6-acetoxyglucopyranoside]	Glu-6-Ac	
β-D-glucopyranoside	Glu	
glycyl	Gly	
hydrogen	H	
hydroxy	OH	
hydrocinnamate	H-Cinn	

APPENDIX. (*contd.*).

Name	Code	Structure
isobutyrate	i-But	
isovalerate	i-Val	
laurate	Dodec	[CH$_3$(CH$_2$)$_{10}$CO-]
linoleate	Linol	
linolenate	Linolen	
malonate	Mal	
methacrylate	Mac	
methoxy	OMe	
methyl	Me	

APPENDIX. (*contd.*).

Name	Code	Structure
methylene	CH2	
methylvalerate	MeVal	
myristate	Tetradec	$[CH_3(CH_2)_{12}CO-]$
p-hydroxy-hydrocinnamate	p-OH-H-Cinn	
p-hydroxy-phenylacetate	p-OH-Phe-Ac	
palmitate	Hexadec Palm	$[CH_3(CH_2)_{14}CO-]$
peroxy	O-OH	
peroxy-bridge	-O-O-	
phenylacetate	Phe-Ac	

APPENDIX. (*contd.*).

Name	Code	Structure
piperidine	Pip	
propionate	Pro	
sarracinate	Sar	
senecioate	Sen	
stearate	Octadec Stear	$[CH_3(CH_2)_{16}CO-]$
stearoyloxy-isovalerate	i-Val-5-O-Stear	
succinate	Succ	
tiglate	Tig	

APPENDIX. (*contd.*).

Name	Code	Structure
<u>trans</u>-3-S-methylacrylate	Trans-Acr-S-Me	
<u>trans</u>-cinnamate	trans-Cinn	

Compound Index

Name	Number	Page Numbers
Abienol	139	15, 73, 246
Abienol, 3β-acetoxy-iso-	1151	17, 75, 249
Abienol, 3β-hydroxy-*ent*-	436	33, 100, 305
Abienol, 6α-hydroxy-7β-acetoxy-12E-	1022	16, 74, 246
Abienol, 6β,7β-dihydroxy-12E-	146	16, 74, 246
Abienol, 6β,7α-dihydroxy-*ent*-	809	33, 100, 305
Abienol, 6β-acetoxy-7α-hydroxy-*ent*-	810	33, 100, 305
Abienol, 6β-angeloyloxy-7α-acetoxy-*ent*-	811	33, 100, 305
Abienol, 6β-hydroxy-12E-	144	16, 74, 246
Abienol, 6β,18-dihydroxy-12E-	147	16, 74, 246
Abienol, 7β-acetoxy-12E-	145	16, 74, 246
Abienol, 7β-acetoxy-6β-hydroxy-12E-	148	16, 74, 246
Abienol, 7β-acetoxy-6β-hydroxy-12α-peroxiiso-	157	16, 74, 249
Abienol, 7β-acetoxy-6β-hydroxy-12β-peroxiiso-	156	16, 74, 249
Abienol, 7β-hydroxy-	140	15, 73, 246
Abienol, 18-hydroxy-	1015	34, 101, 305
Abienol, iso-	155	16, 74, 249
Abieta-7,13(14)-dien-18-oic acid	616	49, 129, 356
Abieta-7,13(14)-diene, 5-acetoxy-	615	49, 129, 356
Abieta-7,13-diene	614	49, 129, 356
Abieta-7,13-diene, 19-succinyloxy-	1054	49, 129, 356
Abietadien-3β-ol, 7,13-	317	27, 94, 280
Abietadien-2α-ol acetate	316	27, 94, 280, 281
Abietadien-3-one, 7,13-	318	27, 94, 281
Acanthoaustralide	40	10, 64, 228
Acanthoaustralide, 1,6-diacetyl-	1124	10, 65, 228
Acanthoaustralide, 17-acetoxy-	43	10, 65, 228
Acanthoaustralide, 17-hydroxy-	42	10, 65, 228
Acanthoaustralide-1-O-acetate	41	10, 65, 228
Acanthospermol-β-galactosidopyranoside	77	12, 71, 238
Acritipappus lactone B	420	37, 106, 313
Acritoconfert-7-en-15-oic acid, 16-acetoxy-17-hydroxy-	617	49, 130, 356

Name	Number	Page Numbers
Acritoconfertic acid [13,14,15,16-Tetra-nor-17-hydroxy-*ent*-labda-7-en-12-oic acid]	472	38, 109, 317
Acritoconfertic acid, 8,16,16,17-bisoxido-	1100	50, 130, 357
Acritopappus acid	575	46, 123, 346
Acritopappus lactone A	419	36, 106, 313
Acritopappusol	574	46, 123, 346
Agathenic acid, 8,17,13,14H-7,8-dehydro-	131	15, 73, 244
Agathenic acid, 8,17H-7,8-dehydro-	130	15, 73, 242
Agathenic acid-16-lactone, 8,17H-7,8-dehydro-	164	19, 81, 257
Ambreinolide, nor-	1156	21, 85, 262
Aristin-19-O-arachinate, 1β-acetoxy-	149	16, 74, 246
Aristin-19-O-*cis*-coumarate, 1β-acetoxy-	151	16, 74, 246
Aristin-19-O-coumarate, 1β-acetoxy-	150	16, 74, 246
Articulin	1237	45, 118, 332
Articulin acetate	1238	45, 118, 332
Aspiliaparviol	1120	
Athrixianone, 16,17-dihydro-16,17-epoxy-	757	56, 140, 373
Athrixianone, 4-carbomethoxy-	756	56, 140, 373
Athrixianone, 4-formyl-	755	56, 140, 373
Athrixianone, 4-hydroxymethyl-	754	56, 139, 373
Athrixic acid, seco-	473	38, 109, 318
Atis-13-en-3β,16α-diol, *ent*-	816	58, 144, 379
Atis-16-en-19-oic acid	783	58, 144, 378
Atis-16-en-19-oic acid, 7α-hydroxy-*ent*-[occidentalic acid]	784	58, 144, 378
Atisan-16α-ol, *ent*-	853	58, 144, 378
Atisan-16β-ol, *ent*-	854	58, 144, 379
Atisirene acid, 11α-acetoxy-[11α-acetoxy-atisirene-20-acid]	785	58, 144, 378
Atisirene acid, 13α-angeloyloxy-	786	58, 144, 378
Atisirene acid, 13α-isobutyryloxy-	788	58, 144, 378
Atisirene acid, 13α-isovaleryloxy-	787	58, 144, 378
Atisirene acid, 14β-acetoxy-9,11-didehydro- [14β-acetoxy-9,11-didehydro-atisirene-20-acid]	790	58, 144, 379
Atisirene acid, 7α-acetoxy-9,11-didehydro- [7α-acetoxy-9,11-didehydroatisirene-20-acid]	789	58, 144, 379
Atractyligenin	758	56, 140, 373
Atractyloside	681	53, 137, 370
Austrochaparol acetate	402	31, 98, 299
Austrochaparol acetate, 18-hydroxy-	408	32, 99, 302
Austrochaparol acetate, 18-oxo-	411	32, 99, 302
Austrochaparol acetate, 19-acetoxy-	409	32, 99, 302
Austrochaparol, 19-oxo-	410	32, 99, 302
Austroeupatorione	1207	38, 110, 318

Name	Number	Page Numbers
Austroeupatorione, 12-desoxo-	1205	38, 110, 318
Austroeupatorione, 12-desoxo-2-desacetyl-	1206	38, 110, 318
Austrofolin	407	32, 99, 301
Austrofolin, 3α,12-dihydroxy-12-desoxo-	413	32, 99, 301
Austrofolin, 3α-hydroxy-	412	32, 99, 301
Austrofolin-12-ol, 12-desoxo-	406	32, 99, 301
Austrofolin-15α-ol, 15,16-dihydro-	417	32, 99, 303
Austrofolin-15β-ol, 15,16-dihydro-	418	32, 99, 303
Austroinulin	428	33, 100, 304
Austroinulin, 6-O-acetyl-[Stevinsol]	429	33, 100, 304
Austroinulin, 6-angelyl-7-acetyl-	808	33, 100, 304
Austroinulin-7-O-acetate, (12Z)-	1023	17, 74, 246
Austroinulin-7-O-acetate, (12Z)-6-desoxy-	1021	16, 74, 246
Ayapanone acid, 3-oxo-	468	37, 108, 316
Ayapanone acid, 3α-hydroxy-	467	37, 108, 316
Bacchalineol	1232	45, 118, 328
Bacchalineol malonate	508	41, 115, 328
Bacchasalicyclic acid	1224	45, 118, 323
Bacchasalicyclic acid-15-O-acetate, 16-hydroxy-	1225	45, 118, 323
Bacchasalicyclic acid-15-O-[1′-β-xylopyranoside]	1226	45, 118, 323
Baccharis incarum compound 1	1227	45, 118, 324
Baccharis incarum neoclerodane 2	1236	45, 118, 332
Baccharis incarum compound 3	1240	45, 118, 334
Baccharis rhomboidalis neoclerodane 1	1241	45, 118, 335
Baccharis rhomboidalis neoclerodane 3	1242	45, 118, 335
Baccharis rhomboidalis neoclerodane 4	1243	45, 118, 335
Baccharis rhomboidalis neoclerodane 5	1229	45, 118, 325
Baccharis rhomboidalis neoclerodane 6	1230	45, 118, 325
Bacchascoparone	1061	46, 123, 346
Bacchofertin	521	42, 116, 330
Bacchomagellin A	1244	45, 120, 342
Bacchomagellin B	1245	45, 121, 342
Bacchotricuneatin A	527	42, 119, 339
Bacchotricuneatin A [Corrected to compound 527 in later publication]	252	42, 119, 339
Bacchotricuneatin A, 1α-hydroxy-	1069	44, 119, 339
Bacchotricuneatin A, 1α-acetyl-	1257	45, 120, 340
Bacchotricuneatin A, 6β-hydroxy-7,8-dehydro-	972	42, 119, 341
Bacchotricuneatin A, 7α-hydroxy-	1258	45, 120, 340
Bacchotricuneatin B	529	42, 120, 342
Bacchotricuneatin B, 2α-angeloyloxy-	968	42, 120, 342
Bacchotricuneatin B, 7β-angeloyloxy-	530	42, 120, 342
Bacchotricuneatin C	528	42, 119, 340
Bacchotricuneatin D	509	41, 115, 328
Bacchotricuneatin, 2α-angeloyloxy-3α-[2-methylbutyryloxy]-3,4βH-	533	42, 120, 342
Bacchotricuneatin, 2α-angeloyloxy-3α-senecioyloxy-3,4βH-	531	42, 120, 342

Name	Number	Page Numbers
Bacchotricuneatin, 2α-senecioyloxy-3α-angeloyloxy-3,4βH-	532	42, 120, 342
Barticulidiol	1233	45, 118, 328
Barticulidiol, 18-acetoxy-19-malonyloxy-	510	41, 115, 328
Batudioic acid	110	14, 72, 242
Bedfordia diterpene alcohol	236	39, 113, 324
Bedfordia diterpene alcohol, iso-	237	47, 124
Beyer-15-en-12α, 19-diol, *ent-*	817	57, 142, 376
Beyer-15-en-17,19-diol, *ent-*[17-hydroxymonogynol]	771	57, 142, 376
Beyer-15-en-18,19-diol, *ent-*	772	57, 142, 376
Beyer-15-en-18-ol, *ent-*	766	57, 142, 375
Beyer-15-en-19-oic acid, 3α-tigloyloxy-*ent-*	776	57, 142, 375
Beyer-15-en-19-oic acid, *ent-*[stach-15-en-19-oic acid; beyerenic acid]	774	57, 142, 375
Beyer-15-en-19-oic acid, 3α-hydroxy-*ent-*	775	57, 142, 375
Beyer-15-en-19-ol, *ent-*[Erythroxylol-A]	765	57, 142, 375
Beyer-15-en-3α, 12β-diol, *ent-*	768	57, 142, 375
Beyer-15-en-3α, 17-diol, *ent-*	770	57, 142, 376
Beyer-15-en-3α, 19-diol, *ent-*	767	57, 142, 375
Beyer-15-en-3α-ol, *ent-*	764	57, 142, 375
Beyer-15-ene, 19-oxo-*ent-*[20-oxostachene]	773	57, 142, 375
Beyer-15-ene, 4α-hydroxy-19-nor-*ent-*	782	57, 143, 378
Beyer-9(11),15-dien-19-oic acid, 3α-tigloyloxy-*ent-*	778	57, 142, 376
Beyer-9(11),15-dien-19-oic acid, *ent-*	777	57, 142, 376
Beyeran-18-ol, 15β,16β-epoxy-*ent-*	779	57, 142, 377
Beyeran-19-ol, 15,16-epoxy-*ent-* [erythroxylol A-14,15-oxide]	780	57, 142, 377
Biformen, 3-oxo-12,13E-	136	15, 73, 245
Biformen, 3α-hydroxy-12,13E-	133	15, 73, 245
Biformen, 3β-hydroxy-12,13E-	134	15, 73, 245
Biformen, 3β-acetoxy-12,13E-	135	15, 73, 245
Biformene, *ent-*	422	33, 100, 303
Biformene, 3α,14,15-trihydroxy-12E, 14,15-dihydro-*ent-*	1186	36, 102, 305
Biformene, 12,13E-*ent-*	430	33, 100, 305
Biformene, 18-oxo-*ent-*	423	33, 100, 303
Biformen, 12,13E-	132	15, 73, 244
Bincatriol	1222	44, 118, 322
Brickellidiffusic acid angelate	97	13, 72, 242
Brickellidiffusic acid spiro ketal lactone, 2α-angeloyloxy-17,O-dihydro-	215	20, 81, 258
Brickellidiffusic acid spiro ketal lactone, 2α-angeloyloxy-17α-hydroxy-17,O-dihydro-	216	20, 82, 258
Brickellidiffusic acid spiro ketal lactone, 2α-hydroxy-17,O-dihydro-	218	20, 82, 258

Name	Number	Page Numbers
Brickellidiffusic acid spiro ketal lactone, 2α-angeloyloxy-17β-hydroxy-17,O-dihydro-	217	20, 82, 258
Carboxyatractyloside	678	53, 137, 369
Carterochaeta acid	211	19, 80, 256
Carterochaetal, 16-acetoxy-14-oxo-13,14H-12,13-dehydro-	213	19, 80, 256
Carterochaetal, 16-oxo-	207	19, 80, 256
Carterochaetal, 16-oxo-13,14H-12,13-dehydro-	212	19, 80, 256
Carterochaetol	210	19, 80, 256
Carterochaetol acetate, 16-acetoxy-	205	19, 80, 256
Carterochaetol, 16-hydroxy-	203	19, 80, 256
Carterochaetol, 16-hydroxy-13,14H-13α,14α-epoxy-	208	19, 81, 257
Carterochaetol, 16-hydroxy-13,14H-13β,14β-epoxy-	209	19, 81, 257
Carterochaetol, 16-oxo-	206	19, 80, 256
Carterothaminotriol	82	13, 71, 240
Carterothaminotriol-14,15-O-diacetate	84	13, 71, 240
Carterothaminotriol-15-O-acetate	83	13, 71, 240
Carterothamnotriol, 14β-	94	13, 72, 247
Carterothamnotriol-14,15-O-diacetate, 14β-	96*	75, 247
Carterothamnotriol-15-O-acetate, 14β-	95*	6, 75, 247
Carterochaetol, 16-acetoxy-	204	19, 80, 256
Cativic acid	118	14, 72, 243
Cativic acid methyl ester, 3α-hydroxy-	992	14, 72, 245
Cativic acid, 18-angeloyloxy-	122	15, 73, 244
Cativic acid, 18-angeloyloxy-13,14-dehydro-	101	13, 72, 241
Cativic acid, 18-tigloyloxy-	123	15, 73, 244
Cativic acid, 18-tigloyloxy-13,14-dehydro-	102	13, 72, 241
Cativic acid, 2-angeloyloxy-13,14Z-dehydro-	100	13, 72, 241
Cativic acid, 2α,3α-[angeloyloxy- & 2-hydroxy-2-methylbutyryloxy]-7α-cis-cinnamoyloxy-	99	13, 72, 243
Cativic acid, 2α,3α-[angeloyloxy-& 2-hydroxy-2-methylbutyryloxy]-7α-trans-cinnamoyloxy-	98	13, 72, 243
Cativic acid, 2α,3α-dihydroxy-	124	15, 71, 238
Cativic acid, 2α-angeloyloxy-3α-hydroxy-(13Z)-13,14-didehydro-	107	14, 72, 242
Cativic acid, 2α-angeloyloxy-3α-hydroxy-(13Z)-13,14-didehydro-8(17)-	109	14, 72, 246
Cativic acid, 2α-hydroxy-3α-[2-hydroxy-2-methylbutyryloxy]-	125	15, 71, 238
Cativic acid, 2β,3β-dihydroxy-ent-	1196	36, 102
Cativic acid, 3-oxo-	129	15, 71, 238
Cativic acid, 3α-angeloyloxy-2α-hydroxy- [Brickellia acid C]	126	15, 73, 238

* Acetylation products of compound 94. These compound numbers were applied to other structures, but these earlier uses were not removed.

Name	Number	Page Numbers
Cativic acid, 3α-angeloyloxy-2α-hydroxy-(13Z)-13,14-didehydro-	106	14, 72, 241
Cativic acid, 3α-angeloyloxy-2α-hydroxy-(13Z)-13,14-didehydro-8(17)-	108	14, 72, 240
Cativic acid, 3α-angeloyloxy-2α-hydroxy-13,14Z-dehydro-	103	14, 72, 241
Cativic acid, 3α-cis-cinnamoyloxy-2α-hydroxy-	128	15, 73, 244
Cativic acid, 3α-cis-cinnamoyloxy-2α-hydroxy-13,14Z-dehydro-	105	14, 72, 241
Cativic acid, 3α-hydroxy-	991	14, 72, 245
Cativic acid, 3α-trans-cinnamoyloxy-2α-hydroxy-	127	15, 73, 244
Cativic acid, 3α-trans-cinnamoyloxy-2α-hydroxy-13,14Z-dehydro-	104	14, 72, 241
Cativin-3,15-dioic acid, 3,4-seco-	220	20, 83, 259
Centipedic acid	60	11, 67, 232
Chiliolide, 3α,5α-dihydroxy-	1212	39, 111, 320
Chiliolide, 3α-hydroxy-5,6-dehydro-	1213	39, 111, 320
Chiliolide, 3α-hydroxy-5β,10β-epoxy-	1214	39, 111, 320
Chiliolide aldehyde, seco-	1215	39, 111, 320
Chiliolide acid, seco-	1216	39, 111, 321
Chiliolide, 19-hydroxy-seco-	1217	39, 112, 321
Chiliolide acid methyl ester, 19-hydroxy-seco-	1218	39, 112, 321
Chiliolide lactone, seco-	1219	39, 112, 321
Chiliotrin, seco-	1220	39, 112, 321
Chiliolide lactone, iso-	1221	39, 113, 321
Chiliomarin	1246	47, 124, 349
Chrysolic acid	271	24, 89, 272
Chrysothane [Strictanonic acid]	223	20, 83, 259
Cleistanth-12-en-11-one, 14β,15β-dihydroxy-16,17-oxido-	608	49, 128, 355
Cleistanth-12-en-11-one, 15,16-epoxy-	601	49, 128, 354
Cleistanth-12-en-11-one, 17-acetoxy-14β-hydroxy-15,16-epoxy-	605	49, 128, 354
Cleistanth-12-en-11-one, 17-acetoxy-15,16-epoxy-	603	49, 128, 354
Cleistanth-12-en-11-one, 17-acetoxy-15,16-epoxy-iso-	609	49, 129, 355
Cleistanth-12-en-11-one, 17-acetoxy-3,4,15,16-diepoxy-iso-	610	49, 129, 355
Cleistanth-12-en-11-one, 17-acetoxy-3α-angeloyloxy-15,16-epoxy-	606	49, 128, 354
Cleistanth-12-en-11-one, 3α-angeloyloxy-15,16-epoxy-	604	49, 128, 354
Cleistanth-12-en-17-al, 11-oxo-8,9,15,16-diepoxy-	607	49, 128, 354
Cleistanth-12-en-17-al, 15,16-epoxy-11-oxo-	602	49, 128, 354
Cleroda-1,3,13(16),14-tetraen-18-oic acid, 15,16-epoxy-ent-	512	41, 115, 322
Cleroda-13(14)-ene-15,16-olide, 3α,4:18,	1087	24, 89, 270

Name	Number	Page Numbers
19-diepoxy-18β,19α-dihydroxy-*cis*-		
Cleroda-13(14)-ene-15,16-olide,	1089	24, 89, 271
3α,4:18,19-diepoxy-19α-hydroxy-*cis*-		
Cleroda-13(14)-ene-15,16-olide,	1090	24, 89, 271
3α,4β,19α-trihydroxy-18,19-epoxy-*cis*-		
Cleroda-13(14)-ene-15,16:18,19-diolide,	1088	24, 89, 270
3α,4-epoxy-19α-hydroxy-*cis*-		
Cleroda-3,13(14)-diene-15,16-olide,	1084	24, 89, 269
18,19-dihydroxy-*cis*-		
Cleroda-3,13(14)-diene-15,16-olide,	1086	24, 89, 270
18,19-epoxy-19α-hydroxy-*cis*-		
Cleroda-3,13(14)-diene-15,16-olide,	1083	24, 89, 269
6α,18-dihydroxy-*cis*-		
Cleroda-3,13(14)-diene-15,16-olide-19-oic ester,	1091	24, 89, 269
19-O-α-L-arabinopyranosyl-*cis*-		
Cleroda-3,13(14)-diene-15,16:18,19-diolide, *cis*-	1085	24, 89, 270
Cleroda-3,13(14)-diene-15,16:18,6α-diolide,	1092	24, 89, 271
2β,6α-dihydroxy-*cis*-		
Cleroda-3,13(16),14-trien-15,16-oxide, 17-oxo-	967	41, 115, 328
Cleroda-3,13(16),14-trien-18,6α-olide,	583	47, 124, 348
15,16-epoxy-*cis-ent*-		
Cleroda-3,13(16),14-triene,15,16-epoxy-*cis*-	578	47, 124, 347
Cleroda-3,13(16),14-triene,	580	47, 124, 347
18-acetoxy-15,16-epoxy-*cis*-		
Cleroda-3,13(16),14-triene,	579	47, 124, 347
18-hydroxy-15,16-epoxy-*cis*-		
Cleroda-3,13(16),14-triene,6α,	581	47, 124, 347
18-dihydroxy-15,16-epoxy-*cis-ent*-		
Cleroda-3,13(16),14-triene,6α-hydroxy-	582	47, 124, 347
18-acetoxy-15,16-epoxy-*cis*-		
Cleroda-3,13E-dien-15-oic acid, 18-acetoxy-*cis*-	1055	47, 124, 348
Cleroda-3,13E-dien-15-oic acid, 18-hydroxy-*cis*-	1056	47, 124, 348
Cleroda-3-en-15,16,18,6α-diolide,	585	47, 124, 348
2β-hydroxy-*cis*-		
Cleroda-3-en-15,16,18,6α-diolide, *cis*-	584	47, 124, 348
Cleroda-3-en-15-oic acid, 18-hydroxy-*cis*-	1057	47, 124, 349
Cleroda-3-en-15-oic acid, 19-hydroxy-	1058	47, 124, 349
Clerodan-15-oic acid, 3,4-dehydro-*cis*-	257	22, 87, 266
Communic acid	137	15, 73, 245
Communic acid, 7α-acetoxy-*trans*-	138	15, 73, 245
Conycephaloide	525	42, 119, 339
Conycephaloide, 12-epi-	969	42, 119, 340
Conycephaloide, 12-epi-8β,17-dihydro-	970	42, 119, 340
Conycephaloide, 7-hydroxy-17-oxo-	526	42, 119, 339
7,8-dehydro-8,17-dihydro-		
Conypododiol	61	11, 67, 232
Conyscabraic acid, 5,6-dehydro-	918	46, 122, 344
Conyscabraic acid, 5,6-dehydroiso-	917	46, 122, 344
Conyscabraic acid, 5α-hydroxy-	920	46, 122, 345

Name	Number	Page Numbers
Conyscabraic acid, 5α-hydroxyiso-	919	46, 122, 344
Conyzic acid	405	32, 99, 301
Cordobic acid	1140	17, 71, 239
Cordobic acid-18-acetate	1141	17, 71, 239
Cordobic acid, 7-epi-	1142	17, 71, 239
Cupressen-18-oic acid	325	28, 95, 284
Dimerobrasiolide	64	11, 68, 233
Daniellic acid	839	31, 99, 300
Daniellic acid, 3α-hydroxy-	840	31, 99, 300
Daniellol	403	31, 98, 299
Daniellol, 18-hydroxy-	841	31, 99, 300
Daniellol, 3α,18-dihydroxy-19-deoxy-	842	32, 99, 300
Darutigenol	591	48, 126, 353
Darutigenol-3α-glucoside	592	48, 126, 353
Dimeroaperatate, methyl-	63	11, 68, 232
Dimeroaperatic acid	62	11, 68, 232
Dimerobrasiolide	64	11, 68, 233
Discoidic acid	1145	17, 75, 244
Doronicoside D	945	53, 137, 369
Dulcoside A	686	53, 139, 371
Dulcoside B	687	53, 139, 371
Elongatolide A [Solidagolactone IV; 6β-hydroxy-solidagolactone]	545	23, 88, 265
Elongatolide B	546	23, 88, 265
Elongatolide D [Solidagolactone VI (structure no. 555)]	553	23, 88, 266
Elongatolide E [Solidagolactone VII]	260	22, 87, 266
Eperuic acid, 2α-iso-valeroyloxy-	358	29, 97, 291
Erigerol [1,6β-dihydroxy-7,8β-epoxydihydrogrendelyl ester of 3,3-dimethylacrylic acid]	461	34, 101, 308
Erythroxa-3,15-dien-18-oic acid	976	50, 131, 357
Erythroxylol A-malonate	769	57, 142, 375
Eupatalbin	805	59, 146, 382
Eupatoralbin	806	59, 146, 382
Evillosin	938	20, 82, 259
Ferruginol	319	27, 94, 281
Geranylascaridol, α-10,11H-10,11-dihydroxy-9-	1101	12, 69, 236
Furosolidagonol	1126	11, 67, 232
Furosolidagonone	1125	11, 67, 232
Furosolidagonone, 13-hydroxy-	1127	11, 67, 232
Furosolidagonone, 1-oxo-2,3-dehydro-1,2,3,20-tetrahydro-	1129	11, 68, 233
Furosolidagonone, 15-peroxy-13,14E-dehydro-14,15-dihydro-	1128	11, 68, 232
Furosolidagonone, 20-oxo-1,2-epoxy-1,2,3,20-tetrahydro-	1137	12, 69, 234
Geranylascaridol, β-10,11H-10,11-dihydroxy-9-	72	12, 69, 235

Name	Number	Page Numbers
Geranylascaridol, α-10,11H-10,11-dihydroxy-9-	1101	12, 69, 236
Geranylcurcumen, 9-	70	12, 69, 235
Geranylcurcumen, 10,11H-10,11-dihydroxy-9-	71	12, 69, 235
Geranylgearanal, 12-oxo-	1119	8, 62, 224
Geranylgeranial	17	8, 62, 223
Geranylgeranial, 4-hydroxy-	18	8, 62, 223
Geranylgeraniol acetate, 12,19-dihydroxy-	12	8, 62, 222
Geranylgeraniol, 12,18-dihydroxy-6,7Z-	22	8, 62, 224
Geranylgeraniol, 12,19-dihydroxy-	11	8, 62, 222
Geranylgeraniol, 14,15-dihydro-4,14,15-trihydroxy-	16	8, 62, 223
Geranylgeraniol, 16-hydroxy-	13	8, 62, 223
Geranylgeraniol, 17-hydroxy-18-acetoxy-	1118	8, 62, 223
Geranylgeraniol, 19-hydroxy-12-oxo-10,11-dihydro-	21	8, 62, 224
Geranylgeraniol, 13-oxo-2,3-dihydro-	20	8, 62, 223
Geranylinalool, 14,15-dihydro-13,14,15,17-teradehydro-	36	9, 63, 226
Geranylgeraniol-20-acid lactone, 18-acetoxy-6,7,10,11-tetrahydro-	1133	11, 68, 233
Geranylgeraniol-20-acid lactone, 17-hydroxy-18,19-diacetoxy-6,7,10,11-tetrahydro-	1134	11, 68, 233
Geranyllinalool	23	8, 62, 224
Geranyllinalool, 12-oxo-	32	9, 63, 225
Geranyllinalool, 12-oxo-15-hydroxy-13,14E-dehydro-10,11,14,15-tetrahydro-	1123	9, 63, 226
Geranyllinalool, 13-hydroxy	24	8, 62, 225
Geranyllinalool, 13-acetoxy-	25	8, 62, 225
Geranyllinalool, 13-acetoxy-5-hydroxy-	31	9, 63, 225
Geranyllinalool, 13-oxo-	33	9, 63, 225
Geranyllinalool, 13-oxo-14,15-epoxy-	34	9, 63, 225
Geranyllinalool, 14,15-dihydro-14,15-dihydroxy-	37	9, 63, 226
Geranyllinalool, 14,15-dihydroxy-9-acetoxy-14,15-dihydro-	39	9, 63, 226
Geranyllinalool, 15-hydroxy-13,14-dehydro-14,15-dihydro-	35	9, 63, 226
Geranyllinalool, 5,14,15-trihydroxy-14,15-dihydro-	38	9, 63, 226
Geranyllinalool, 5,9-diacetoxy-	30	9, 63, 225
Geranyllinalool, 5-hydroxy-	28	8, 63, 225
Geranyllinalool, 9-acetoxy-	27	8, 63, 225
Geranyllinalool, 9-acetoxy-5-hydroxy-	29	9, 63, 225
Geranyllinalool, 9-hydroxy-	26	8, 63, 225
Geranyllinalool, 12-oxo-10,11-dihydro-	1122	9, 63, 225
Geranyllinalool, 13-acetoxy-	1121	9, 63, 225
Geranylneral	5	7, 60, 219
Geranylneral, 12-oxo-	1111	7, 60, 219

Name	Number	Page Numbers
Geranylnerol	1	7, 60, 219
Geranylnerol acetate, 19-acetoxy-15-hydroxy-14-oxo-	1005	7, 61, 221
Geranylnerol, 6,7-epoxy-19-hydroxy-12-oxo-6,7-dihydro-	979	7, 61, 221
Geranylnerol, 8,12,19-trihydroxy-	948	7, 60, 220
Geranylnerol, 9,17-dihydroxy-	1116	8, 61, 220
Geranylnerol, 9,7,19-trihydroxy-	1117	8, 61, 220
Geranylnerol, 12,19,20-trihydroxy-14-methylene-	9	9, 64, 227
Geranylnerol, 12,20-dihydroxy-16-oxo-	8	7, 60, 220
Geranylnerol, 12,20-dihydroxy-19-acetoxy-	15	7, 60, 220
Geranylnerol, 12,20-dihydroxy-19-acetoxy-14-methylene-	10	9, 64, 227
Geranylnerol, 15,19-dihydroxy-14-oxo-	1004	5, 7, 61, 221
Geranylnerol, 16,18,19-trihydroxy-	957	7, 60, 220
Geranylnerol, 17,18-dihydroxy-	1109	7, 60, 219
Geranylnerol, 17,18,20-trihydroxy-	1110	7, 60, 219
Geranylnerol, 18-hydroxy-	2	7, 60, 219
Geranylnerol, 18,19-diacetoxy-17,20-dihydroxy-	1112	8, 60, 220
Geranylnerol, 18,19,20-triacetoxy-17-hydroxy-	1113	8, 60, 220
Geranylnerol, 18-acetoxy-17,20-dihydroxy-	1114	8, 60, 220
Geranylnerol, 19-acetoxy-17,20-dihydroxy-	1115	8, 61, 220
Geranylnerol, 19-acetoxy-20-hydroxy-	984	7, 60, 219
Geranylnerol, 19-hydroxy-12-oxo-	978	7, 60, 221
Geranylnerol, 20-acetoxy-10,11-epoxy-10,11-dihydro-9-hydroxy-	1020	7, 61, 222
Geranylnerol, 17,20-dihydroxy-	4	7, 60, 219
Geranylnerol, 19-hydroxy-13-oxo-	7	7, 60, 220
Geranylnerol, 20-hydroxy-	3	7, 60, 219
Geranylnerol-19-oic acid	1002	7, 60, 219
Geranylnerol-19-oic acid, 1-acetyl-	1003	7, 60, 219
Geranylnerol-6,7-epoxide, 8,12,19-trihydroxy-	949	7, 61, 221
Gerranyllinalool, 13-hydroxy-	24	8, 62, 225
Geranyl-α-terpinene, 9-	68	12, 69, 235
Geranyl-α-terpinene, 10,11H-10,11-dihydro-9-	69	12, 69, 235
Geranyl-α-terpineol, 9-	1138	12, 69, 234
Gnaphalene, 8α,13α-dihydroxy-	586	47, 125, 350
Gnaphala-13(16),14-diene, 8α-hydroxy-	1152	17, 75, 250
Gochnatoic acid-17-O-phenylacetate	522	42, 116, 326
Gochnatol-17-O-acetate	487	40, 114, 326
Grandifloric acid [15α-hydroxy-*ent*-kaur-16-en-19-oic acid]	646	51, 132, 359
Grinde-6,8(17)-dienic acid	166	17, 76, 251
Grindelate, methyl-	1153	18, 77, 252
Grindelate, methyl-4β-hydroxy-6-oxo-19-nor-	1154	20, 83, 260
Grindelic acid	167	17, 76, 251
Grindelic acid, 17-[2-methylbutyryloxy]-	178	18, 76, 252
Grindelic acid, 17-acetoxy-	174	17, 76, 252

Name	Number	Page Numbers
Grindelic acid, 17-grindeloxy-	956	18, 78, 254
Grindelic acid, 17-hydroxy-[Oxygrindelic acid]	173	17, 76, 251
Grindelic acid, 17-isobutyryloxy-	176	18, 76, 252
Grindelic acid, 17-isovaleryloxy-	177	18, 76, 252
Grindelic acid, 17-methoxy-	179	18, 76, 252
Grindelic acid, 17-propionyloxy-	175	17, 76, 252
Grindelic acid, 18-hydroxy-	188	18, 77, 252
Grindelic acid, 19-[2-methylbutyryloxy]-	184	18, 77, 252
Grindelic acid, 19-acetoxy-	181	18, 76, 252
Grindelic acid, 19-acetoxy-6-oxo-	190	18, 77, 252
Grindelic acid, 19-hydroxy-	180	18, 76, 252
Grindelic acid, 19-isobutyryloxy-	183	18, 76, 254
Grindelic acid, 19-isovaleryloxy-	182	18, 76, 252
Grindelic acid, 19-oic-	187	18, 77, 252
Grindelic acid, 19-oxo-	186	18, 77, 252
Grindelic acid, 19-succinyloxy-	185	18, 77, 252
Grindelic acid, 1α-hydroxy-	168	17, 76, 251
Grindelic acid, 3α-hydroxy-	1035	18, 77, 252
Grindelic acid, 3β-hydroxy-	169	17, 76, 251
Grindelic acid, 4α-formyloxy-19-nor-	222	20, 83, 260
Grindelic acid, 4α-hydroxy-19-nor-	221	20, 83, 260
Grindelic acid, 6-oxo-	189	18, 77, 252
Grindelic acid, 6α-formyloxy-	171	17, 76, 251
Grindelic acid, 6α-hydroxy-	170	17, 76, 251
Grindelic acid, 6β-hydroxy-	172	17, 76, 251
Grindelic acid, 7α,8α-epoxy-7,8-dihydro-	192	18, 77, 253
Grindelic acid, 7α-hydroxy-7,8-dihydro-8(17)-dehydro-	165	17, 76, 251
Grindelic acid, 7β,8β-epoxy-7,8-dihydro-	191	18, 77, 253
Grindelic acid, 7β-hydroxy-8(17)-dihydro-7,8-dehydro-	1036	18, 77, 251
Grindelic acid, iso-	955	18, 76, 254
Grindelic acid-15-O-arabinoside	1060,1187	18, 78, 254
Grindelistrictic acid	224	21, 84, 260
Guamaic acid	1164	35, 101, 285
Guteriolide	270	23, 88, 268
Gutierrezial	845	47, 125
Gutierrezianol acid-[2′-methylbutyrate]	160	16, 74, 250
Gutierrezianol acid-isobutyrate	161	16, 74, 250
Gutierrezianol acid-isovalerate	162	16, 74, 250
Gutierrezia sphaerocephala labdane-α-L-arabino-pyranoside 1	95	17, 75, 248
Gutierrezia sphaerocephala labdane-α-L-arabino-pyranoside 2	96	17, 75, 248
Gutierrezia sphaerocephala labdane-β-D-xylo-pyranoside 3	1148	17, 75, 249
Gutierrezia sphaerocephala labdane-β-D-xylo-pyranoside 5	1149	17, 75, 249

Name	Number	Page Numbers
Gutierrezia sphaerocephala labdane-β-D-xylo-pyranoside 6	1150	17, 75, 249
Gutierrezia spathulata *ent*-labdane 4	1194	36, 102, 311
Gutierrezia spathulata *ent*-labdane 5	1195	36, 102, 311
Gutiesolbriolide, 17-hydroxy-	1130	11, 68, 233
Gutiesolbriolide, 10E, 17-hydroxy-	1131	11, 68, 233
Gutiesolbriolide, 17-acetoxy-	1132	11, 68, 233
Gutiesolbriolide, 17-hydroxyiso-	1135	11, 68, 234
Gutiesolbriolide, 17-acetoxyiso-	1136	11, 68, 234
Gymnospermin	80	13, 70, 237
Haplociliatic acid	505	41, 115, 327
Haplopappate, monomethyl-	577	47, 124, 348
Haplopappic acid	576	47, 124, 348
Haplopappus pauciden-tatus compound 27	1208	38, 110, 319
Haplopappus pectinatus *ent*-labdane 13	1175	35, 101, 293
Haplopappus pectinatus *ent*-labdane 14	1176	35, 102, 293
Haplopappus pectinatus *ent*-labdane 15	1177	35, 102, 293
Haplopappus pectinatus *ent*-labdane 17	1178	35, 102, 293
Haplopappus pectinatus *ent*-labdane 16	360	30, 97, 292
Hardwickiic acid lactone, 7-oxo-6α-hydroxy	523	42, 116, 330
Hardwickiic acid, (−)-	511	41, 115, 328
Hardwickiic acid, 12α-[2-methylbutyryloxy]-(−)-	953	41, 115, 329
Hardwickiic acid, 15,16H-15-oxo-	558	43, 116, 332
Hartwrightia acid	462	36, 105, 312
Hautriwaic acid	1033	44, 117, 336
Hautriwaic acid lactone	520	42, 116, 330
Hautriwaic acid, 1,2-dehydro-	1070	44, 117, 322
Hautriwaic acid, 10-O-methyl	901	41, 115, 328
Hautriwaic acid, 13,14,15,16-tetranor-1,2-dehydro	1093	46, 123, 347
Hautriwaic acid, 14α,15α-epoxy-16-oxo-13,14,15,16-tetrahydro-	907	44, 117, 334
Hautriwaic acid, 14β,15β-epoxy-16-oxo-13,14,15,16-tetrahydro-	908	44, 117, 334
Hautriwaic acid, 15α-hydroxy-16-oxo-15,16-dihydro-	905	43, 117, 334
Hautriwaic acid, 15β-hydroxy-16-oxo-15,16-dihydro-	906	44, 117, 334
Hautriwaic acid, 16α-acetoxy-19-O-acetyl-15-oxo-15,16-dihydro-	909	43, 116, 332
Hautriwaic acid, 19-O-acetyl-1,2-dehydro-	911	41, 115, 322
Hautriwaic acid, 19-O-angelyl-	902	41, 115, 328
Hautriwaic acid, 19-O-isovaleryl-	903	41, 115, 328
Hautriwaic acid, 19-acetyl-	900	41, 115, 328
Hautriwaic acid, 19-oxo-	904	41, 115, 328
Hautriwaic acid, 2β-hydroxy-	1034	44, 117, 329
Hautriwaic acid-19-lactone, 12α-hydroxy-	950	42, 116, 330
Hautriwaic acid-19-lactone, 7α,12α-dihydroxy-	951	42, 116, 330

Name	Number	Page Numbers
Hautriwaiic acid, 16β-acetoxy-19-O-acetyl-15-oxo-15,16-dihydro-	910	43, 116, 332
Hebeclinolide	999	59, 149, 384
Helicallen-16-al	66	12, 69, 234
Helicallen-16-oic acid, 14,15-dihydro-	67	12, 69, 234
Helicallen-16-ol	65	12, 69, 234
Helifulvan-19-oic acid	797	59, 146, 382
Helifulvan-19-oic-acid, 11α-acetoxy-	799	59, 146, 382
Helifulvan-19-oic-acid, 11α-hydroxy-	798	59, 146, 382
Helifulvan-19-ol	796	58, 146, 382
Hexa-deca-2E,6E,10E,14-tetraene, 1-hydroxy-11-carboxy-3,7,15-trimethyl-	954	8, 62, 223
Ichthyouleolide	49	10, 66, 229
Ichthyouleolide, 14-hydroxy-15,17-dehydro-14,15-dihydro-	54	10, 66, 230
Ichthyouleolide, 14-peroxy-15,17-dehydro-14,15-dihydro-	53	10, 66, 230
Ichthyouleolide, 15-hydroxy-13,14t-dehydro-14-15-dihydro-	52	10, 66, 229
Ichthyouleolide, 15-peroxy-13,14t-dehydro-14-15-dihydro-	51	10, 66, 229
Ichthyouleolide, 19-acetoxy-	50	10, 66, 229
Imbricatol-angelate, 3-hydroxy-	516	41, 115, 329
Imbricatol-isovalerate, 3-hydroxy-	514	41, 115, 329
Imbricatol-α-methylbutyrate, 3-hydroxy-	515	41, 115, 329
Inulaefol	804	59, 148, 384
Inuroyleanol	320	27, 94, 281
Isoacanthoaustralide, 17-acetoxy-	47	10, 65, 229
Isoacanthoaustralide, 17-hydroxy-	46	10, 65, 229
Isoacanthoaustralide, 1-O-acetate	48	10, 65, 229
Isocembrene, 15-hydroxy-13,18H-	73	12, 69, 236
Jesromotetrol, 3β-acetoxy-	1162	27, 93, 278
Jesromotetrol, 3β,19-diacetoxy-	1161	27, 93, 278
Jhanic acid	940	59, 149, 384
Jhanidiol	197	19, 79, 255
Jhanidiol-19-monoacetate	198	19, 79, 255
Jhanidiol-diacetate	199	19, 79, 255
Jhanilactone	941	21, 85, 262
Jhanol	195	19, 79, 255
Jhanol acetate	196	19, 79, 255
Julslimdiolone, 19-nor-	309	27, 93, 280
Julslimtetrol	308	26, 92, 277
Junceanol W	1072	28, 94, 283
Junceanol X	1073	28, 94, 283
Junceanol Y	1074	28, 94, 283
Junceic acid epoxide	1071	44, 117, 336
Kaurenic acid, 18-angeloyloxy-ent-	939	50, 132, 358
Kaur-11-en-19-oic acid, 16α-hydroxy-ent-	724	55, 135,362

Name	Number	Page Numbers
Kaur-15(16)-en-19-oic acid, 3α-isobutyryloxy-9β-hydroxy-*ent*- [3α-Isobutyryloxy-9β-hydroxy- 9,11-dihydro-polymnia acid]	708	54, 134, 364
Kaur-15(16)-en-19-oic acid, 3α-isovaleryloxy-9β-hydroxy-*ent*- [3α-Isovaleryloxy-9β-hydroxy- 9,11-dihydro-polymnia acid]	707	54, 134, 364
Kaur-15-en-17, 19-dioic acid, *ent*-	703	54, 134, 364
Kaur-15-en-19-al, (−)-	946	54, 134, 364
Kaurenal, 9,11-dehydro-*ent*	942	56, 136, 362
Kaur-15-en-19-oic acid, 17-hydroxy-*ent*-[17-hydroxy-ent-isokaur- 15(16)-en-19-oic acid]	701	54, 134, 364
Kaur-15-en-19-oic acid, 17-oxo-*ent*-	702	54, 134, 364
Kaur-15-en-19-oic acid, *ent*-	700	54, 134, 364
Kaur-15-ene, 17, 19-dihydroxy-*ent*-	698	53, 134, 364
Kaur-15-ene, 17-hydroxy-19- [p-hydroxy-hydrocinnamoyloxy]- [17-hydroxy-19-[p-hydroxy-hydro- cinnamoyloxy]-isokaurene]	699	53, 134, 364
Kaur-15-ene, 17-hydroxy-*ent*-[17-hydroxy-isokaurene]	697	53, 134, 364
Kaur-16-ene, 18-acetoxy-*ent*-	626	50, 132, 358
Kaur-16-en-19-oate, methyl-15α-hydroxy-*ent*- [grandifloric acid methylester]	661	52, 133, 360
Kaur-16-en-19-oic acid glycol ester, *ent*-	627	50, 132, 358
Kaur-16-en-19-oic acid thujanol ester, *ent*-	850	56, 136, 369
Kaur-16-en-19-oic acid, 11β,15β-dihydroxy-*ent*-	671	52, 133, 361
Kaur-16-en-19-oic acid, 11β,12β,15β-trihydroxy-*ent*-	1260	52, 136, 361
Kaur-16-en-19-oic acid, 11β-acetoxy-*ent*-	640	51, 132, 358
Kaur-16-en-19-oic acid, 11β-hydroxy-15-oxo-*ent*-	674	52, 134, 361
Kaur-16-en-19-oic acid, 11β-hydroxy-15β-acetoxy-*ent*-	672	52, 133, 361
Kaur-16-en-19-oic acid, 12α-acetoxy-*ent*-	642	51, 132, 359
Kaur-16-en-19-oic acid, 12α-hydroxy-*ent*-	641	51, 132, 359
Kaur-16-en-19-oic acid, 15α-[2-methylacryloyloxy]-*ent*-	653	51, 133, 359
Kaur-16-en-19-oic acid, 15α-acetoxy-*ent*-	648	51, 132, 359
Kaur-16-en-19-oic acid, 15α-angeloyloxy-*ent*-	649	51, 133, 359
Kaur-16-en-19-oic acid, 15α-benzoyloxy-*ent*-	657	51, 133, 359
Kaur-16-en-19-oic acid, 15α-cinnamoyloxy-*ent*- [cinnamoyloxy-grandifloric acid]	655	51, 133, 359
Kaur-16-en-19-oic acid, 15α-isobutyroyloxy-*ent*-	654	51, 133, 359
Kaur-16-en-19-oic acid, 15α-isovaleryloxy-*ent*-	652	51, 133, 359

Name	Number	Page Numbers
Kaur-16-en-19-oic acid, 15α-senecioyloxy-*ent*-	651	51, 133, 359
Kaur-16-en-19-oic acid, 15α-tigloyloxy-*ent*-	650	51, 133, 359
Kaur-16-en-19-oic acid, 15α[2′-methylbutyryloxy]-*ent*-	892	52, 133, 359
Kaur-16-en-19-oic acid, 15β-angeloyloxy-*ent*-	658	52, 133, 359
Kaur-16-en-19-oic acid, 15β-hydroxy-11-oxo-*ent*-	673	52, 133, 361
Kaur-16-en-19-oic acid, 15β-hydroxy-*ent*-	647	51, 132, 359
Kaur-16-en-19-oic acid, 15β-senecioyloxy-*ent*-	660	51, 133, 359
Kaur-16-en-19-oic acid, 15β-tigloyloxy-*ent*-	659	51, 133, 359
Kaur-16-en-19-oic acid, 18-angeloyloxy-*ent*-	643	51, 132, 358
Kaur-16-en-19-oic acid, 18-isovaleryloxy-*ent*-	645	51, 132, 358
Kaur-16-en-19-oic acid, 18-senecioyloxy-*ent*-	644	51, 132, 358
Kaur-16-en-19-oic acid, 3α-acetoxy-*ent*-	629	50, 132, 358
Kaur-16-en-19-oic acid, 3α-angeloyloxy-9β-hydroxy-*ent*-	662	52, 133, 359
Kaur-16-en-19-oic acid, 3α-angeloyloxy-*ent*-	630	50, 132, 358
Kaur-16-en-19-oic acid, 3α-cinnamoyloxy-9β-hydroxy-*ent*-	663	52, 133, 360
Kaur-16-en-19-oic acid, 3α-cinnamoyloxy-*ent*-	632	50, 132, 358
Kaur-16-en-19-oic acid, 3α-hydroxy-*ent*-	628	50, 132, 358
Kaur-16-en-19-oic acid, 3α-isovaleryloxy-*ent*-	634	51, 132, 358
Kaur-16-en-19-oic acid, 3α-senecioyloxy-*ent*-	633	51, 132, 358
Kaur-16-en-19-oic acid, 3α-tigloyloxy-*ent*-	631	50, 132, 358
Kaur-16-en-19-oic acid, 3β,9β-dihydroxy-15α-angeloyloxy-*ent*-	675	52, 134, 360
Kaur-16-en-19-oic acid, 3β,9β-dihydroxy-15α-senecioyloxy-*ent*-	676	52, 134, 360
Kaur-16-en-19-oic acid, 3β,9β-dihydroxy-15α-tigloyloxy-*ent*-	677	52, 134, 360
Kaur-16-en-19-oic acid, 9β,11β-dihydroxy-15α-angeloyloxy-*ent*-	869	52, 134, 361
Kaur-16-en-19-oic acid, 9β,15α-dihydroxy-*ent*- [grandifloric acid, 9β-hydroxy]	664	52, 133, 360
Kaur-16-en-19-oic acid, 9β-hydroxy-15α-[3′-hydroxy-2′-methylbutyryloxy]-*ent*-	819	52, 133, 360
Kaur-16-en-19-oic acid, 9β-hydroxy-15α-acetoxy-*ent*-	667	52, 133, 360
Kaur-16-en-19-oic acid, 9β-hydroxy-15α-angeloyloxy-*ent*-	665	52, 133, 360
Kaur-16-en-19-oic acid, 9β-hydroxy-15α-cinnamoyloxy-*ent*-	670	52, 133, 360
Kaur-16-en-19-oic acid, 9β-hydroxy-15α-isovaleryloxy-*ent*-	669	52, 133, 360
Kaur-16-en-19-oic acid, 9β-hydroxy-15α-senecioyloxy-*ent*-	668	52, 133, 360
Kaur-16-en-19-oic acid, 9β-hydroxy-15α-tigloyloxy-*ent*-	666	52, 133, 360

Name	Number	Page Numbers
Kaur-16-en-19-oic acid, 9β,13β-dihydroxy-15α-angeloyloxy-	636	51, 132, 360
Kaur-16-en-19-oic acid, 3β-angeloyloxy-*ent*-	635	51, 132, 358
Kaur-16-en-19-oic acid, 7β-hydroxy-(−)-	1000	51, 132, 358
Kaur-16-en-19-oic acid, 17-carbomethoxy-*ent*-	812	52, 133, 368
Kaur-16-en-19-oic acid, 9(11)-dehydro-15α-cinnamoyloxy-*ent*-	943	52, 136, 363
Kaur-16-en-3α,19-diol, (−)-	623	50, 132, 358
Kaur-16-ene, 19-hydroxy-*ent*-	619	50, 131, 357
Kaur-16-ene, 18-nor-*ent*-	752	56, 139, 372
Kaur-16-ene, 19-acetoxy-*ent*-	620	50, 131, 357
Kaur-16-ene, 19-nor-*ent*-	750	56, 140, 373
Kaur-16-ene, 19-oxo-*ent*-	624	50, 132, 358
Kaur-16-ene, 4α-hydroxy-19-nor-*ent*- [Ruilopeziol]	751	56, 140, 374
Kaur-16-ene, 4β-hydroxy-18-nor-*ent*-[4-epi-ruilopeziol]	753	56, 139, 372
Kaur-16-ene, *ent*-	618	50, 131, 357
Kaur-9(11),15(16)-dien-19-oic acid, 3α-isobutyryloxy-*ent*- [3α-isobutyryloxy-polymnia acid]	705	54, 134, 364
Kaur-9(11),15(16)-dien-19-oic acid, 3α-isovaleryloxy-*ent*- [3α-isovaleryloxy-polymnia acid]	704	54, 134, 364
Kaur-9(11),15(16)-dien-19-oic acid, 3α-tiglyloxy-*ent*- [3α-tiglyloxy-polymnia acid]	706	54, 134, 364
Kaur-9(11),16-dien-19-oic acid, 12-oxo-*ent*-	695	53, 134, 363
Kaur-9(11),16-dien-19-oic acid, 12β-ethoxy-*ent*-	998	53, 134, 363
Kaur-9(11),16-dien-19-oic acid, 12β-hydroxy-*ent*-	694	53, 134, 363
Kaur-9(11),16-dien-19-oic acid, 15α-hydroxy-*ent*-	696	53, 134, 363
Kaur-9(11),16-dien-19-oic acid, 2β-hydroxy-*ent*-	691	53, 134, 363
Kaur-9(11),16-dien-19-oic acid, 3β-hydroxy-*ent*-	692	53, 134, 362
Kaur-9(11),16-dien-19-oic acid, 7β-hydroxy-*ent*-	693	53, 134, 363
Kaur-9(11),16-dien-19-oic acid, *ent*-	690	53, 134, 363
Kaur-9(11)-en-19-oic acid, 16α,17-dihydroxy-*ent*-	739	55, 136, 368
Kauran-16β-ol, (−)-	1001	54, 135, 366
Kauran-19-al, 16α-hydroxy-*ent*-	1255	56, 136, 362
Kauran-17,19-dial, 16β-H-*ent*-	737	55, 135, 367
Kauran-17,19-dioic acid, 16α(−)-	990	55, 136, 367
Kauran-17,19-dioic acid, 16α-H-*ent*-	925	55, 135, 367
Kauran-17-al, 16β-H-*ent*-	729	55, 135, 367
Kauran-17-al, 18-[p-hydroxyhydro-cinnamoyloxy]-*ent*-	740	55, 136, 367
Kauran-17-al, 19-hydroxy-16β-H-*ent*-	730	55, 135, 367
Kauran-17-oic acid, 16β-H-*ent*-	731	55, 135, 367
Kauran-17-oic acid, 19-hydroxy-16β-H-*ent*-	732	55, 135, 367
Kauran-19-al, 16α,17-epoxy-*ent*-	735	54, 135, 365

Name	Number	Page Numbers
Kauran-19-al, 16β,17-epoxy-*ent*-	712	54, 135, 365
Kauran-19-al-17-oic acid, 16α(−)-	989	55, 136, 367
Kauran-19-oic acid, (16R)-11α-hydroxy-15-oxo-*ent*-	748	56, 136, 368
Kauran-19-oic acid, 11β-hydroxy-15-oxo-16α-H-*ent*-	747	55, 136, 368
Kauran-19-oic acid, 15α,16α-epoxy-17-hydroxy-*ent*-	746	55, 136, 368
Kauran-19-oic acid, 15α-acetoxy-16,17-epoxy-*ent*-	716	54, 135, 366
Kauran-19-oic acid, 15α-angeloyloxy-16,17-epoxy-*ent*- [Perymenium acid]	715	54, 135, 365
Kauran-19-oic acid, 15β-angeloyloxy-16,17-epoxy-*ent*-	717	54, 135, 366
Kauran-19-oic acid, 15β-tigloyloxy-16,17-epoxy-*ent*-	718	54, 135, 366
Kauran-19-oic acid, 16α,17-epoxy-*ent*-	713	54, 135, 365
Kauran-19-oic acid, 16α-hydroxy-*ent*-	723	55, 135, 362
Kauran-19-oic acid, 16β,17-dihydroxy-*ent*-	822	55, 135, 366
Kauran-19-oic acid, 16β,17-epoxy-*ent*-	714	54, 135, 365
Kauran-19-oic acid, 17-[n-dodecanoyloxy]-16β-H-*ent*-	742	55, 135, 367
Kauran-19-oic acid, 17-[n-hexadecanoyloxy]-16β-H-*ent*-	744	55, 136, 367
Kauran-19-oic acid, 17-[n-octadecanoyloxy]-16β-H-*ent*-	745	55, 136, 368
Kauran-19-oic acid, 17-[n-tetra-dodecanoyloxy]-16β-H-*ent*-	743	55, 136, 367
Kauran-19-oic acid, 17-hydroxy-16α-H-*ent*-	924	55, 135, 367
Kauran-19-oic acid, 17-hydroxy-16β-H-*ent*-	738	55, 135, 367
Kauran-19-oic acid, 17-isobutyrloyloxy-16β-H-*ent*-	741	55, 136, 367
Kauran-19-oic acid, 17-nor-16-oxo-*ent*-	749	56, 139, 372
Kauran-19-oic acid, 16α,17-dihydroxy-*ent*-	728	55, 135, 362
Kauran-19-ol, 16α,17-epoxy-*ent*-	734	54, 134, 365
Kaurane, 16α,17,19-trihydroxy-*ent*-	726	55, 135, 362
Kaurane, 16α,17-dihydroxy-3-oxo-*ent*-[Abbeokutone]	727	55, 135, 366
Kaurane, 16α,17-dihydroxy-*ent*- [17-hydroxy-*ent*-kauranol]	725	55, 135, 362
Kaurane, 16α,17-epoxy-*ent*-	733	54, 134, 365
Kaurane, 16α-acetyl-19-hydroxy-16-desmethyl-*ent*-	736	55, 135, 367
Kaurane, 16α-hydroxy-19-acetoxy-*ent*-	722	55, 135, 362
Kaurane, 16α-hydroxy-*ent*-[(−)-kauranol]	720	54, 135, 362
Kaurane, 16β,19-dihydroxy-*ent*- [*ent*-kauran-16β,19-diol]	852	55, 135, 366

Name	Number	Page Numbers
Kaurane, 16β-hydroxy-*ent*-[*ent*-kauran-16β-ol]	851	55, 135, 366
Kaurane, 19-angeloyloxy-16α,17-epoxy-*ent*-	710	54, 134, 365
Kaurane, 3α-angeloyloxy-16α,17-epoxy-*ent*-	709	54, 134, 365
Kaurane, 3α-angeloyloxy-19-oxo-16α,17-epoxy-*ent*-	711	54, 134, 365
Kaurane, 3β-acetoxy-16α-hydroxy-*ent*-	721	54, 135, 362
Kauranol, *ent*-[(−)-kauranol]	997	50, 131, 362
Kauren-16-en-19-oic acid, 15α-[2,3-epoxy-2-methylbutyloxy]-*ent*-	656	51, 133, 359
Kauren-16-en-19-oic acid, 15α-[2′,3′-dihydroxy-2-methylbutyryloxy]-*ent*-	818	52, 133, 359
Kauren-18-oic acid, *ent*-	926	50, 132, 358
Kauren-19-oic acid, 12β-hydroxy-16α,17-epoxy-16,17-dihydro-9(11)-dehydro-*ent*-	719	54, 135, 366
Kauren-19-oic acid, *ent*-	625	50, 132, 358
Kaurene, 4β,19-epoxy-18-nor-*ent*-	813	56, 140, 372
Kaurene, 19-[p-hydroxy-hydrocinnamoyloxy]-*ent*-	622	50, 131, 358
Kaurene, 19-formyloxy-methylenoxy-*ent*-	621	50, 131, 357
Kingidiol [Barticulidiol]	846	41, 115, 328
Kingidiol, 19-O-malonyl-	1231	45, 118, 328
Kirenol	589	48, 126, 351
Koanoadmantic acid	19	8, 62, 223
Koanolabda-12E,14-diene, seco-	219	20, 82, 259
Koanophyllic acid A	233	21, 86, 263
Koanophyllic acid B	232	21, 86, 263
Koanophyllic acid C	234	21, 86, 263
Koanophyllic acid D	235	22, 86, 264
Kolav-13(16),14-diene, 3α,4α,15,16-diepoxy-	1076	44, 117, 336
Kolav-13-en-15-oic acid lactone, 16,3α,4β-trihydroxy	1082	44, 118, 338
Kolav-13-en-15-oic acid lactone, 16-hydroxy-3-oxo-	1081	44, 117, 338
Kolav-13-en-15-oic acid lactone, 16-hydroxy-3α,4α-epoxy-	1080	44, 117, 338
Kolav-13E-en-15-al, 3-oxo-	1079	44, 117, 337
Kolav-13E-en-15-al, 3α,4α-epoxy-	1077	44, 117, 337
Kolav-13Z-en-15-al, 3-oxo-	1078	44, 117, 337
Kolav-13Z-en-15-al, 3α,4α-epoxy-	1075	44, 117, 337
Kolav-3,13(16),14-trien-18-oic acid, 2β-hydroxy-15,16-epoxy-*ent*-	513	41, 115, 329
Kolav-3,13(16),14-trien-20-oic acid, 15,16-epoxy-*ent*-[Junceic acid]	524	41, 116, 329
Kolav-3,13(16),14-trien-18-oic acid, 15,16-epoxy-19-hydroxy-*ent*- [Hautriwaic acid]	518	41, 115, 328
Kolav-3,13(16),14-trien-18-oic acid, 15,16-epoxy-2α,19-dihydroxy-*ent*-	519	41, 115, 329

Name	Number	Page Numbers
Kolav-3,13(16),14-triene, 15,16-epoxy-	507	41, 115, 328
Kolav-3-en-15-oic acid lactone, 16,16-dihydroxy-	565	43, 117, 333
Kolav-3-en-15-oic acid, 13,14-dihydro-	503	41, 115, 325
Kolav-3-en-15-oic acid, 16-oxo-	542	43, 116, 331
Kolav-3-en-15-oic acid, 18-acetoxy-	504	41, 115, 325
Kolav-3-en-15-ol, 6α,18-dihydroxy-17-acetoxy-	488	40, 114, 325
Kolav-3-en-15-ol, 6α,18-dihydroxy-17-phenylacetoxy-	489	40, 114, 325
Kolav-3-en-15-ol, 6α-hydroxy-17-acetoxy-18-oxo-	491	40, 114, 325
Kolav-3-en-15-ol, 6α-hydroxy-17-phenylacetoxy-18-oxo-	490	40, 114, 325
Kolav-3-ene, 18-malonyloxy-15-hydroxy-15,16-epoxy-	569	44, 117, 335
Kolava-3,13E-dien-15-oic acid, 16-hydroxy-	865	42, 116, 331
Kolava-3,13Z-dien-15-oic acid lactone, 16,16-dihydroxy-	864	43, 116, 332
Kolavan-2-on-15-oic acid, 3β,4β-epoxy-	506	41, 115, 328
Kolavelool [Kolavenool]	479	40, 114, 324
Kolavelool, 6β-angeloyloxy- [6β-Angeloyloxy-kolavenool]	480	40, 114, 324
Kolaven-15,18-dioic acid dilactone, 16,19-dihydroxy-13,14-dihydro-	563	43, 117, 333
Kolaven-15,18-dioic acid dilactone, 7-oxo-16,19-dihydroxy-13,14-dihydro-	561	43, 116, 333
Kolaven-15,18-dioic acid dilactone, 7α,16,19-trihydroxy-	559	43, 116, 332
Kolaven-15,18-dioic acid dilactone, 7α,16,19-trihydroxy-13,14-dihydro-	562	43, 117, 333
Kolaven-15-al, 17-hydroxy-13,14-dihydro-	492	40, 114, 325
Kolaven-15-al, 17-hydroxy-13E-	535	42, 116, 323
Kolaven-15-al, 17-hydroxy-13Z-	534	42, 116, 326
Kolaven-15-oic acid lactone, 16-hydroxy-13,14-dihydro-	564	43, 117, 333
Kolaven-15-oic acid lactone, 16-hydroxy-16-methoxy-18-acetoxy-	963	43, 117, 333
Kolaven-16-oic acid lactone, 15-hydroxy-2-oxo-	568	44, 117, 335
Kolaven-16-oic acid lactone, 2β,15-dihydroxy-	567	44, 117, 334
Kolavenic acid lactone, 16,18-dihydroxy-	550	43, 116, 332
Kolavenic acid lactone, 16-hydroxy-3,4-epoxy-[Structure was revised to 5α,10α-cis-]	556	23, 88, 266
Kolavenic acid lactone, 2α,16-dihydroxy-	544	43, 116, 332
Kolavenic acid lactone, 7β-angeloyloxy-16-hydroxy-3,4-epoxy- [Revised to 5α,10α-cis-]	557	23, 88, 266
Kolavenic acid, 16-18-dihydroxy-	538	42, 116, 331
Kolavenic acid, 16-acetoxy-18-hydroxy-	539	42, 116, 331
Kolavenic acid, 16-acetoxy-18-oxo-	541	43, 116, 331

Name	Number	Page Numbers
Kolavenic acid, 16-hydroxy-18-oxo-	540	43, 116, 331
Kolavenic acid, 2-oxo-	961	41, 115, 327
Kolavenic acid, 6-angeloyloxy-	498	40, 114, 323
[6α-Angeloyloxy-13,14E-kolavenoic acid]		
Kolavenic acid, 6-tigloyloxy-	499	40, 114, 323
[6α-Tigloyloxy-13,14E-kolavenoic acid]		
Kolavenic acid, 6α,7β-diacetoxy-13Z-	1029	44, 117, 336
Kolavenic acid, 6α,7β-dihydroxy-2-oxo-13E-	1030	44, 117, 336
Kolavenic acid, 6α,7β-dihydroxy-nor-	1032	46, 123, 346
Kolavenic acid, 6α-angeloyloxy-7β-acetoxy-13Z-	1028	44, 117, 336
Kolavenic acid, 6α-isobutyryloxy-7β-acetoxy-13-Z-	1027	44, 117, 335
Kolavenic acid, 6α-isobutyryloxy-7β-angeloyloxy-13,14-dihydro-	1031	44, 117, 336
Kolavenic-15-acid lactone, 2α,16-dihydroxy-3α,4α-epoxy-	551	23, 43, 265
Kolavenic-15-acid lactone, 6β-angeloyloxy-16-hydroxy-3α,4α-epoxy-	552	23, 43, 265
Kolavenoic acid [13,14E-Kolavenoic acid]	495	40, 114, 323
Kolavenoic acid, 13,14Z-	493	40, 114, 326
Kolavenoic acid, 2β-acetoxy-13,14Z-	494	40, 114, 326
Kolavenoic acid, 6-acetoxy-	496	40, 114, 323
[6β-Acetoxy-13,14E-kolavenoic acid]		
Kolavenoic acid, 6-angeloyloxy-	497	40, 114, 323
[6β-Angeloyloxy-13,14E-kolavenoic acid]		
Kolavenoic acid, 6α,18-dihydroxy-17-acetoxy-	501	40, 115, 327
Kolavenoic acid, 6α,18-dihydroxy-17-phenylacetoxy-	502	40, 115, 327
Kolavenol	475	39, 114, 322
Kolavenol acetate, 16-hydroxy-2α-acetoxy-3β,4β-epoxy-3,4-dihydro-	1228	45, 118, 324
Kolavenol acetate, 2α,16-dihydroxy-3β,4β-epoxy-3,4-dihydro-	478	39, 114, 324
Kolavenol acetate, 2α-hydroxy-3β,4β-epoxy-3,4-dihydro-	477	39, 114, 324
Kolavenol arachidate, 17-hydroxy-13,14-dihydro-	482	40, 114, 325
Kolavenol arachidate,17-hydroxy-13E-	1010	40, 114, 322
Kolavenol behenate, 17-hydroxy-13,14-dihydro-	483	40, 114, 325
Kolavenol behenate, 17-hydroxy-13E-	1011	40, 114, 322
Kolavenol, 17-hydroxy-13,14-dihydro-	481	40, 114, 325
Kolavenol, 17-hydroxy-13E-	476	39, 114, 322
Kolavenol, 17-oxo-13,14-dihydro-	486	40, 114, 325
Kolavenol, 18-succinyloxy-13,14-dihydro-	484	40, 114, 325
Kolavenol, 2α-hydroxy-3α,4α-epoxy-13,14-dihydro-	485	40, 114, 326
Kolavenool, 6β-acetoxy-2-oxo-	1025	23, 88, 268
Labda-8(17),12E,14-triene, 18,19-dihydroxy-ent-	434	33, 100, 305

Name	Number	Page Numbers
Labda-13(14)E-en-15-al,8-hydroxy-*ent*-	349	29, 96, 289
Labda-13(14)E-en-15-ol, 8-hydroxy-*ent*-	348	29, 96, 289
Labda-13(14)Z-en-15-al, 8-hydroxy-*ent*-	350	29, 96, 290
Labda-13(16),14-diene, 3α-angeloyloxy-8-hydroxy-15,16-epoxy-*ent*-	416	32, 99, 302
Labda-13(16),14-diene, 8α-hydroxy-*ent*-	441	33, 100
Labda-13-Z-en-15-oic acid, 6,16-dioxo-8,12-oxido-*ent*-	460	34, 100, 308
Labda-13-en-15-oic acid, 3,17-dioxo-8β-H-*ent*-	341	29, 96, 288
Labda-13E-en-15,17-dioic acid, 3β-H-*ent*-	345	29, 96, 288
Labda-13Z-en-15,17-dioic acid, 8β-H-*ent*-	344	29, 96, 288
Labda-13Z-en-15-oic acid, 3,17-dioxo-8α-H-*ent*-	342	29, 96, 288
Labda-13Z-en-15-oic acid, 3-oxo-8,17-epoxy-*ent*-	346	29, 96, 289
Labda-13Z-en-15-oic acid, 8-oxo-8-desmethyl-*ent*-	1201	37, 108, 316
Labda-13Z-en-15-oic acid, 3,8-dioxo-*ent*-	1202	37, 108, 316
Labda-13Z-en-15-oic acid, 3α, 8α-dihydroxy-17-oxo-8β-H-*ent*-	343	29, 96, 288
Labda-13Z-en-15-oic acid, 3α-hydroxy-17-oxo-8α-H-*ent*-	340	29, 96, 288
Labda-13Z-en-15-oic acid, 3α-hydroxy-17-oxo-8β-H-*ent*-	339	29, 96, 288
Labda-7,13(16),14-trien-15,16-epoxy-18-oic acid, 5β,9βH,10α-*ent*-	415	32, 99, 302
Labda-7,13(16),14-trien-2β,3α-diol, 15,16-epoxy-*ent*-	1188	36, 102, 310
Labda-6,13-diene, 15,16,17-triacetoxy-*ent*-	364	30, 97, 294
Labda-6,13-diene, 15,16-diacetoxy-*ent*-	363	30, 97, 293
Labda-6,13E-dien-15,17-dioic acid, 16-acetoxy-*ent*-	381	30, 98, 296
Labda-6,13E-dien-15-oic acid, 16-acetoxy-17-oxo-*ent*-	380	30, 98, 296
Labda-6,13E-dien-15-oic acid, 16-acetoxy-8-oxo-8-desmethyl-*ent*-	466	37, 108, 316
Labda-6,13E-dien-15-oic acid, 16-acetoxy-8β-formyl-8-desmethyl-*ent*-	465	37, 108, 315
Labda-6,13E-dien-15-oic acid, 8α,16-dihydroxy-*ent*-	379	30, 98, 296
Labda-6,8(17),13-trien-15-oic acid lactone, 16-hydroxy-*ent*-	384	31, 98, 297
Labda-6,8(17),13-trien-15-oic acid, 16-hydroxy-*ent*-	365	30, 97, 294
Labda-6,8(17),13-dien-15-oic acid, *ent*-	356	29, 97, 291
Labda-7(8),13E-diene, 15,16,17-trihydroxy-*ent*-	362	30, 97, 293
Labda-8(17),13-diene, 7β,15,16-trihydroxy-*ent*-	382	30, 98, 296
Labda-7,13-dien-2β,15-diol, 3α-angeloyloxy-*ent*-	1192	36, 102, 311
Labda-7,13E-diene, 2β,3β,15-trihydroxy-*ent*-	1193	36, 102, 311

Name	Number	Page Numbers
Labda-7,13-dien-15-oic acid lactone, 12,16-dihydroxy-*ent*-	387	31, 98, 298
Labda-7,13-dien-15-oic acid lactone, 16,12,16-trihydroxy-*ent*-	390	31, 98, 298
Labda-7,13-dien-15-oic acid lactone, 16,12,16α-trihydroxy-*ent*-	391	31, 98, 298
Labda-7,13-dien-15-oic acid lactone, 16,12,16β-trihydroxy-*ent*-	392	31, 98, 298
Labda-7,13-dien-15-oic acid lactone, 16,16-dihydroxy-6-oxo-*ent*-	394	31, 98, 298
Labda-7,13-dien-15-oic acid lactone, 16,16α-dihydroxy-*ent*-	388	31, 98, 298
Labda-7,13-dien-15-oic acid lactone, 16,16α-dihydroxy-17-oxo-*ent*- [Acritolongifolide A]	395	31, 98, 298
Labda-7,13-dien-15-oic acid lactone, 16,16β-dihydroxy-*ent*-	389	31, 98, 298
Labda-7,13-dien-15-oic acid lactone, 16,16β-dihydroxy-17-oxo-*ent*- [Acritolongifolide B]	396	31, 98, 298
Labda-7,13-dien-15-oic acid lactone, 16-hydroxy-*ent*-	386	30, 98, 298
Labda-7,13-dien-15-oic acid lactone, 16-hydroxy-6-oxo-*ent*-	393	31, 98, 298
Labda-7,13-dien-15-oic acid, 16-acetoxy-17-oxo-*ent*-	371	30, 97, 295
Labda-7,13-dien-15-oic acid, 16-hydroxy-17-methoxy-*ent*-	1180	36, 102, 294
Labda-7,13-dien-15-oic acid, 2β-angeloyloxy-*ent*-[Dendroidinic acid]	355	29, 97, 290
Labda-7,13-dien-16-oic acid lactone, 15,12,15α-trihydroxy-*ent*-	400	31, 98, 299
Labda-7,13-dien-16-oic acid lactone, 15,12,15β-trihydroxy-*ent*-	399	31, 98, 299
Labda-7,13-dien-16-oic acid lactone, 15-hydroxy-*ent*-	398	31, 98, 299
Labda-7,13-diene, 3α,15-diacetoxy-11-hydroxy-*ent*-	353	29, 97, 290
Labda-7,13E-dien-15-oic acid, *ent*-	354	29, 97, 291
Labda-7,13E-dien-15-oic acid, 16,17-dihydroxy-*ent*-	367	30, 97, 294
Labda-7,13E-dien-15-oic acid, 16-acetoxy-6-oxo-*ent*-	377	30, 98, 295
Labda-7,13E-dien-15-oic acid, 16-acetoxy-6α-hydroxy-17-oxo-*ent*-	372	30, 97, 295
Labda-7,13E-dien-15-oic acid, 16-acetoxy-6β-hydroxy-17-oxo-*ent*-	373	30, 97, 295
Labda-7,13E-dien-15-oic acid, 16-hydroxy-*ent*-	366	30, 97, 294
Labda-7,13E-dien-15-oic acid,	369	30, 97, 294

Name	Number	Page Numbers
16-hydroxy-17-oxo-*ent*-		
Labda-7,13E-dien-15-oic acid,	376	30, 98, 295
16-hydroxy-6-oxo-*ent*-		
Labda-7,13E-dien-15-oic acid,	378	30, 98, 295
16-methoxy-6-oxo-*ent*-		
Labda-7,13E-dien-15-oic acid,	370	30, 97, 294
17-hydroxy-16-acetoxy-*ent*-		
Labda-7,13Z-dien-15-oic acid,	368	30, 97, 294
12-hydroxy-16-oxo-*ent*-		
Labda-7,13Z-dien-15-oic acid,	374	30, 97, 295
16-acetoxy-6-oxo-*ent*-		
Labda-7,13Z-dien-15-oic acid, 6,16-dioxo-*ent*-	375	30, 97, 295
Labda-7,14-diene, 13-hydroxy-*ent*-	447	34, 100, 307
Labda-7,14-diene, 13-hydroxy-3-oxo-*ent*-	449	34, 100, 308
Labda-7,14-diene, 3,13-dihydroxy-*ent*-	448	34, 100, 307
Labda-7,13(Z)-dien-6-one, 15-hydroxy-*ent*-	352	29, 96, 290
Labda-7,13(Z)-diene-6β,15-dihydroxy-*ent*-	351	29, 96, 290
Labda-7-en-15-oic acid,	361	30, 97, 292
2β-angeloylocy-*ent*-		
[13,14-Dihydro-dendroidinic acid]		
Labda-7-en-aldehyde,	471	38, 109, 317
nor-*ent*-[13,14,15,16-Tetra-nor-		
2-oxo-ent-labda-7-ene]		
Labda-8(17),12E,14-triene, 18-hydroxy-*ent*-	433	33, 100, 305
Labda-8(17),12E,14-triene, 19-hydroxy-*ent*-	431	33, 100, 305
Labda-8(17),12E,14-triene, 3α,19-dihydroxy-*ent*-	432	33, 100, 305
Labda-8(17),12E,14-triene, 18,19-diacetoxy-*ent*-	1185	36, 102, 305
Labda-8(17),12E-diene, 14-oxo-15-nor-*ent*-	469	38, 108, 316
Labda-8(17),13Z-diene,	1191	36, 102, 311
2β,15,16,18-tetrahydroxy-		
3α-angeloyloxy-*ent*-		
Labda-8(17),13(14)-dien-16-oic acid lactone,	397	31, 98, 299
7β,15-dihydroxy-*ent*-		
Labda-8(17), 13,14E-dien-15-ol, 7β-acetoxy-*ent*-	328	28, 96, 285
Labda-8(17), 13E-dien-15-oic acid,	1189	36, 102, 310
3α,18-dihydroxy-*ent*-		
Labda-8(17), 13E-dien-15-oic acid,	1190	36, 102, 311
3α-angeloyloxy-18-hydroxy-*ent*-		
Labda-8(17),	327	28, 96, 284
13-dien-15-O-bernsteinoate, *ent*-		
Labda-8(17),13-dien-15-O-bernsteinoate,	347	29, 96, 289
8,17-dihydro-8β,17-epoxy-*ent*-		
Labda-8(17),13-dien-15, 16-olide,	1181	36, 102, 297
2β,18-dihydroxy-*ent*-		
Labda-8(17),13-dien-15, 16-olide,	1182	36, 102, 297
3β-angeloyloxy-18-hydroxy-*ent*-		
Labda-8(17),13-dien-15, 16-olide,	1183	36, 102, 297
3β-angeloyloxy-18β-glucopyranosyloxy-*ent*-		
Labda-8(17),13-dien-15, 16-olide,	1184	36, 102, 297

Name	Number	Page Numbers
3α-angeloyloxy-2β,18-dihydroxy-*ent*-		
Labda-8(17),13-dien-15-oic acid lactone,	385	31, 98, 297
16,19-dihydroxy-*ent*-		
Labda-8(17),13-dien-15-ol, *ent*-	326	28, 96, 284
Labda-8(17),13-dien-15-ol, 2-oxo-*ent*-	849	28, 96, 286
Labda-8(17),13-dien-15-ol, 2α-hydroxy-*ent*-	848	28, 96, 285
Labda-8(17),13E-dien-15-acetoxy-18-oic	330	28, 96, 285
acid,*ent*-[Viscidic acid B]		
Labda-8(17),13E-dien-15-al, 2-oxo-*ent*-	333	28, 96, 286
Labda-8(17),13E-dien-15-al, 2α-hydroxy-*ent*-	331	28, 96, 285
Labda-8(17),13E-dien-15-oic acid,	335	28, 96, 285
ent-[E-Copalic acid]		
Labda-8(17),13E-dien-15-ol-18-oic acid,	329	28, 96, 285
ent-[Viscidic acid A]		
Labda-8(17),13E-diene,	844	28, 96, 285
3α-angeloyloxy-15,18-dihydroxy-*ent*-		
Labda-8(17),13Z-dien-15-al, 2-oxo-*ent*-	334	28, 96, 287
Labda-8(17),13Z-dien-15-al, 2α-hydroxy-*ent*-	332	28, 96, 287
Labda-8(17),13Z-dien-15-oic acid,	336	28, 96, 287
ent-[Z-Copalic acid]		
Labda-8(17),13Z-dien-15-oic acid,	338	28, 96, 287
3α,7α-dihydroxy-*ent*-		
Labda-8(17),13Z-dien-15-oic acid,	337	28, 96, 287
3α-hydroxy-*ent*-		
Labda-8(17),14-dien-18-oic acid,	439	33, 104, 306
12α,13αE-epoxy-*ent*-		
Labda-8(17),14-dien-18-oic acid,	437	33, 104, 306
12α,13αZ-epoxy-*ent*-		
Labda-8(17),14-dien-18-oic acid,	440	33, 105, 306
12β,13βE-epoxy-*ent*-		
[12,13-Dehydro-12,13α-epoxy-ozic acid]		
Labda-8(17),14Z-dien-18-oic acid,	438	33, 104, 306
12β,13βZ-epoxy-*ent*-		
Labda-8(17)-en-15-oic acid, 16-oxo-*ent*-	383	30, 98, 296
Labda-8(17)-en-15-oic acid, 2-oxo-*ent*-	861	30, 97, 292
Labda-8(17)-en-15-oic acid,	860	29, 97, 292
2α-[2-methylbutyryloxy]-*ent*-		
Labda-8(17)-en-15-oic acid,	858	29, 97, 292
2α-angeloyloxy-*ent*-		
Labda-8(17)-en-15-oic acid, 2α-hydroxy-*ent*-	857	29, 97, 291
Labda-8(17)-en-15-oic acid, 2α-tigloyloxy-*ent*-	859	29, 97, 292
Labda-8(17)-en-15-ol, 2-oxo-*ent*-	359	30, 97, 292
Labda-8(17)-ene, 13-oxo-14,15-bis-nor-*ent*-	470	38, 108, 317
Labda-8(17)-ene,	975	32, 99 300
18-angeloyloxy-3α-hydroxy-13-furyl-*ent*-		
Labda-8(17)-ene, 2α,15-dihydroxy-*ent*-	357	29, 97, 291
Labda-8(17)-ene,	974	32, 99 300
3α-angeloyloxy-18-hydroxy-13-furyl-*ent*-		

Name	Number	Page Numbers
Labdan-15-oic acid,	863	30, 97, 292
2α-[2-methylbutyryloxy]-8β,17-epoxy-*ent*-		
Labdan-15-oic acid,	862	30, 97, 292
2α-angeloyloxy-8β,17-epoxy-*ent*-		
Labdane triol 1a,18-nor-*ent*-	828	38, 110, 318
Labdane triol 2a,18-nor-*ent*-	829	38, 110, 318
Labdaturbinic acid, friedo-	1211	38, 110, 319
Labd-1(10),13E-dien-15-ol, 2-oxo-*ent*-friedo-	1209	38, 110, 319
Labd-1(10),13E-diene, 2α,15-dihydroxy-	1210	38, 110, 319
ent-friedo-		
Labd-7-ene, 2β,3β,15-trihydroxy-	1197	36, 102, 311
Labd-7-en-2-O-[fuco-pyranoside-4'-O-acetate],	1203	38, 109, 317
14,15-nor-*ent*-		
Labd-7-en-15-oic acid, 4β-hydroxymethyl-	1145	17, 75, 244
Labd-(8,9)-ene, 13R,14R,15-trihydroxy-	87	13, 71, 237
Labd-13-ene, 8α,15-dihydroxy-	1103	12, 71, 239
Labd-14-en-18-oic acid, 7-oxo-9β,13β-epoxy-*ent*-	1019	34, 101, 309
Labd-14-ene,	1017	34, 101, 309
18-hydroxy-7-oxo-9β,13β-epoxy-*ent*-		
Labd-14-ene, 7,18-dioxo-9β,13β-epoxy-*ent*-	1018	34, 101, 309
Labd-7-en-15-oic acid, 2-hydroxy-17-methoxy-	1179	35, 102, 293
Labd-7-en-15-oic acid, 3β-hydroxy-	1146	75, 244
	[= 120]	
Labd-7-en-15-oic acid,	870	15, 73, 244
3β-hydroxy-2α-senecioyloxy-13-		
Labd-7-en-2α,15-diol,13-	871	14, 73, 244
Labd-7-ene, 13R,14R,15-trihydroxy-	86	13, 71, 237
Labd-7-ene,	994	13, 71, 245
15-[2-methylbutyryloyloxy]-3α-hydroxy-		
Labd-7-ene, 2α-hydroxy-15-	872	14, 73, 244
[3,4-dihydroxy-cinnamoyloxy]-13-		
Labd-7-ene, 2α-hydroxy-15-	873	14, 73, 244
[4-hydroxy-3-methoxy-cinnamoyloxy]-13-		
Labd-7-ene, 2α-hydroxy-15-	874	15, 73, 244
[4-hydroxy-cinnamoyloxy]-13-		
Labd-7-ene, 3α,15-dihydroxy-	993	13, 71, 245
Labd-7-13E-dien-2-O-β-fucopyranoside,	1165	35, 102, 285
3α-angeloyloxy-2β,15-dihydroxy-*ent*-		
Labd-7,13E-dien-2-O-β-	1166	35, 102, 285
[fucopyranoside-4'-O-acetate],		
3α-angeloyloxy-2β,15-dihydroxy-*ent*-		
Labd-7,13E-dien-2-O-β-	1167	35, 102, 285
[fucopyranoside-3'-O-acetate],		
3α-angeloyloxy-2β,15-dihydroxy-*ent*-		
Labd-7,13E-dien-2-O-β-	1168	35, 103, 285
[fucopyranoside-4'-O-acetate],		
3α-angeloyloxy-2β,15,15-trihydroxy-*ent*-		
Labd-7,13(16)-dien-2-O-β-	1169	35, 103, 285

Name	Number	Page Numbers
[fucopyranoside-4′-O-acetate], 3α-angeloyloxy-2β,14,15-trihydroxy-*ent*-		
Labd-7,13E-dien-2-O- [rhamnopyranoside-4′-O-acetate], 3α-angeloyloxy-2β,15-dihydroxy-*ent*-	1170	35, 103, 286
Labd-7,13E-dien-2-O-β-xylopyranoside, 3α-angeloyloxy-2β,15-dihydroxy-*ent*-	1171	35, 103, 286
Labd-13E-ene, 2β,8β,15-trihydroxy-*ent*-	1174	35, 101, 289
Labd-13E-en-8-O-β-xylopyranoside, 8β,15-dihydroxy-*ent*-	1172	35, 103, 286
Labd-8(17)-en-15-al-19-oic acid	977	15, 73, 242
Labd-8(17)-en-15-oic acid, 2β-acetoxy-	959	15, 73, 241
Labd-8(17)-ene, 13,14,15-trihydroxy-	89	13, 71, 240
Labd-8(17)-ene, 14,15-dihydroxy-	88	13, 71, 240
Labd-8(17)-ene, 14-acetoxy-15-hydroxy-	92	13, 72, 240
Labd-8(17)-ene, 15-acetoxy-13,14-dihydroxy-	91	13, 72, 240
Labd-8(17)-ene, 15-acetoxy-13-hydroxy-14-propionyloxy-	93	13, 72, 240
Labd-8(17)-ene, 15-acetoxy-14-hydroxy-	90	13, 71, 240
Labd-8(17)-ene, 15-hydroxy-13,14-epoxy-	85	13, 71, 240
Labd-8(17)-ene, 3α,7α,15-trihydroxy-	995	13, 71, 241
Labd-8(17)-en-19-oic acid, 15-hydroxy-*ent*-	1173	35, 101, 287
Labd-8,14-diene, 13,18-dihydroxy-7-oxo-*ent*-	1016	34, 101, 309
Labd-8(17),12E,14-trien-19-oic acid	1104,1147	17, 75, 245
Labda-12(E), 14-diene, 9α,19-dihydroxy-	152	16, 74, 247
Labda-13E-en-15-oic acid, 5β,8β-peroxy-*ent*-	1044	35, 101, 312
Labda-5,13E-dien-15-oic acid, 18-hydroxy-10-desmenthyl-9α-methyl-8β(H)-*ent*-	1059	38, 110, 319
Labda-7(8),13E-dien-15,18-dioic acid, 6β-isovaleryloxy-*ent*-	1043	34, 101, 312
Labda-7(8),13E-dien-15-oic acid, 6β-hydroxy-18-oxo-*ent*-	1042	34, 101, 312
Labda-7(8),13E-diene, 15,18-diacetoxy-2β-hydroxy-*ent*-[or enantiomer]	1012	34, 101, 309
Labda-7(8),13E-diene, 18-acetoxy-2β,15-dihydroxy-*ent*-[or enantiomer]	1013	34, 101, 309
Labda-7,13-dien-15-acid-16-lactone, 17-hydroxy-	965	19, 81, 257
Labda-7,13-dien-15-acid-16-lactone, 17-oxo-	966	20, 81, 257
Labda-7,13E-dien-15-oic acid, 2α-tigloyloxy-	112	14, 72, 242
Labda-7,13E-dien-15-oic acid, 3-oxo-	117	14, 72, 243
Labda-7,13E-dien-15-oic acid, 6β,18-dihydroxy-*ent*-	1045	35, 101, 312
Labda-7,13Z-15-oic acid, 3α-hydroxy-	116	13, 72, 241
Labda-7-en-15,17-dioic acid, *ent*-	1052	35, 101, 293
Labda-7-en-15-oic acid, 17-acetoxy-*ent*-	1047	35, 101, 293
Labda-7-en-15-oic acid, 17-hydroxy-	1046	35, 101, 293
Labda-7-en-15-oic acid, 17-oxo-*ent*-	1048	35, 101, 293

Name	Number	Page Numbers
Labda-7-en-15-oic acid, 19-hydroxy-17-acetoxy-*ent*-	1053	35, 101, 293
Labda-7-en-15-oic acid, 2α,17-dihydroxy-*ent*-	1049	35, 101, 293
Labda-7-en-15-oic acid, 2α-hydroxy-17-acetoxy-*ent*-	1050	35, 101, 293
Labda-7-en-15-oic acid, 2α-hydroxy-17-oxo-*ent*-	1051	35, 101, 293
Labda-7-en-15-oic acid, 3α-hydroxy-	119	14, 72, 243
Labda-7-en-15-oic acid, 3β-hydroxy-	120	14, 73, 243
Labda-8(17),13(16),14-triene, 18-hydroxy-3α-angeloyloxy-15,16-epoxy,*ent*-	1037	34, 101, 310
Labda-8(17),13(16),14-triene, 2β,18-dihydroxy-3α-angeloyloxy-15,16-epoxy,*ent*-	1038	34, 101, 310
Labda-8(17),13-diene	78	12, 71, 239
Labda-8(17),13E-dien-15-oic acid	111	14, 72, 238
Labda-8(17),13E-dien-15-oic acid, 18-hydroxy-*ent*-	1041	34, 101, 285
Labda-8(17),13E-dien-15-oic acid, *ent*-[E-Copalic acid]	1040	34, 101
Labda-8(17),13E-diene, 3α-acetoxy-15-hydroxy-18-tigloyloxy-*ent*- [or enantiomer]	1014	34, 101, 285
Labda-8(17),13E-dien-15,18-dioic acid, ent-[Guamaic acid]	1164	35, 101, 285
Labda-8(17),13Z-diene, 2β,15,16,18-tetrahydroxy-3α-angeloyloxy-*ent*-	1039	34, 101
Labda-8(17)-13-dien-15-ol	76	12, 71, 238
Labda-[7.13E-dien]-15-oic acid, 3α-hydroxy-	114	14, 72, 242
Labda-[7.13E-dien]-15-oic acid, 3β-hydroxy-	115	14, 72, 242
Labdan-15-oic acid, 8α-hydroxy-	958	14, 70, 237
Labdan-8α,15-diol	79	13, 70, 237
Labdan-8α,15-diol, 13-epi-	81	13, 70, 237
Labdan-8α,15-diol, 13βH-	1139	17, 70, 238
Labdanol acid	121	14, 73, 244
Labden-13-one, 14,15-bisnor-8-hydroxy-11E-	1155	21, 85, 262
Labdenol acid, (13Z)-	1024	17, 74, 243
Lambertianic acid, 17-hydroxyiso-	159	16, 74, 250
Lambertianic acid, 7α-hydroxy-	163	16, 74, 250
Ligantrol	74	12, 70, 236
Ligantrol-monoacetate	75	12, 70, 236
Lycaconitine, methyl-	801	59, 148, 383
Lycoctonine [royline; roylene]	802	59, 148, 384
Lycoctonine, anthranoyl-	803	59, 148, 384
Manool, (+)-	996	16, 74, 248
Manool, *ent*-	442	33, 100, 307
Manool, iso-	1105	16, 74, 248
Manool, 2-oxo-*ent*-	446	34, 100, 307
Manool, 2α-hydroxy-*ent*-	444	34, 100, 307

Name	Number	Page Numbers
Manool, 2β-hydroxy-*ent*-	443	34, 100, 307
Manool, 2β-succinyloxy-*ent*-	445	34, 100, 307
Manool, 3β-hydroxy-iso-	1106	16, 74, 248
Manool, 3-oxo-iso-	1107	16, 74, 249
Manoyl oxide	272	18, 79, 255
Manoyl oxide, 13-epi-	273	19, 79, 255
Manoyl oxide, *ent*-	458	37, 107, 315
Manoyl oxide, 13-epi-*ent*- [*ent*-8,13β-Epoxy-14-labdene]	451	37, 106, 314
Manoyl oxide, 3α-acetoxy-*ent*-	1198	37, 107, 314
Manoyl oxide, 19-cinnamoyloxy-3α-hydroxy-*ent*-	1199	37, 107, 314
Manoyl oxide, 3α,19-dihydroxy-*ent*-	1200	107, 314
Manoyl oxide, 19-hydroxy-13-epi-*ent*-	455	37, 107, 314
Manoyl oxide, 2-oxo-13-epi-*ent*-	454	37, 107, 314
Manoyl oxide, 2-oxo-19-hydroxy-13-epi-*ent*-	456	37, 107, 314
Manoyl oxide, 3-oxo-13-epi-*ent*- [*ent*-8,13β-epoxy-14-labden-3-one]	453	37, 107, 314
Manoyl oxide, 3α-hydroxy-*ent*-	452	37, 107, 314
Manoyl oxide, 3β-hydroxy-*ent*-	459	37, 107, 315
Manoyl oxide, 15-oxo-14,15-dihydro-	964	19, 79, 255
Manoyl oxide, 18-benzoyloxy-3β-hydroxy- [18-benzoyl-oxylarcardenasol]	201	19, 79, 255
Manoyl oxide, 18-hydroxy-13-epi-	202	19, 79, 255
Manoyl oxide, 3β-hydroxy-[18-deoxy-lazarcardenasol]	200	19, 79, 255
Melcantholide, 1,6,7-trihydroxy-17-acetoxy-	1006	11, 67, 231
Melfusanolide, 1,10,17-trihydroxy-	45	10, 65, 228
Melfusanolide, 1,10-dihydroxy-17-acetoxy-	44	10, 65, 228
Mikanifuran	58	11, 67, 231
Missourienol A [initially isolated as dehydration product: 3-oxo-abieta-7,13-diene]	611	49, 129, 355
Missourienol B [initially isolated as dehydration product: 3β-hydroxy-abieta-7,13-diene]	612	49, 129, 356
Missourienol C [initially isolated as dehydration product: 2β-acetoxy-abieta-7,13-diene]	613	49, 129, 356
Montanol	57	10, 66, 230
Neophytadiene	59	9, 64, 226
Nidoanomalin	781	57, 143, 377
Nidorella lactone	214	20, 81, 258
Nidorella lactone, seco-	225	21, 84, 261
Nidorella lactone-6-O-[2-methylbutyrate], seco-	228	21, 84, 261
Nidorella lactone-6-O-angelate, 5-deoxy-5-hydroxy-5H-seco-	230	21, 84, 261
Nidorella lactone-6-O-angelate, seco-	226	21, 84, 261
Nidorella lactone-6-O-isobutyrate, seco-	227	21, 84, 261
Nidorella lactone-6-O-isovalerate, seco-	229	21, 84, 261
Nidorellalactone	254	22, 87, 266
Nidorellalactone, iso-	253	22, 87, 265

Name	Number	Page Numbers
Nidorellol	141	15, 73, 246
Nidorellol, 6α-angeloyloxy-	143	15, 73, 246
Nidorellol, 6α-hydroxy-	142	15, 73, 246
Nidoreseda acid methyl ester, seco-	256	22, 87, 264
Nidoreseda acid, seco-[Conyzic acid, strictic acid]	255	22, 87, 264
Nidoresedaic acid, 19-acetoxyseco-	912	46, 122, 345
Nivenolide	843	31, 98, 299
Olearin	560	43, 116, 332
Osteomuricone, 1β,2α-dihydroxy-	311	26, 92, 278
Osteomuricone, 1β,2α-dihydroxy-15,16-dihydro-	312	26, 92, 278
Osteomuricone, 2α-hydroxy-	310	26, 92, 278
Ozic acid, 12,13E- [ent-Labda-8(17),12E,14-trien-18-oic acid]	425	33, 100, 304
Ozic acid, 2-oxo-12,13Z-	427	33, 100, 304
Ozic acid, 2α-hydroxy-12,13Z-	426	33, 100, 303
Ozic acid, cis-	424	33, 100, 303

Ozic acid, 1β-acetoxy-trans-
  [1β-acetoxy-ent-
  Labda-8(17),12E,14-trien-19-oic acid]
	435	33, 100, 305

Palarosan, 3α-hydroxy-13-epi-	304	26, 91, 276
Palarosan, 9α-hydroxy-13-epi-	305	26, 91, 276
Palarosane, 3-oxo-	315	27, 93, 279
Palarosane, 3β,18-dihydroxy-[Jesromotetrol]	314	26, 93, 278
Palarosane, 3α-hydroxy-	306	26, 93, 276
Palarosane, 3β-hydroxy-	313	26, 93, 278
Paniculoside I	895	53, 138, 370
Paniculoside II	896	53, 138, 370
Paniculoside III	897	53, 138, 370
Paniculoside IV	898	53, 138, 370
Paniculoside V	899	53, 138, 371
Pimar-15-en-8β-ol, 13-epi-	891	26, 92, 277
Pimar-9(11)-en-19-oic acid, ent-8(R),15(S)-epoxy-12α-acetoxy	824	49, 128, 354
Pimar-9,11(15)-dien-19-oic acid, (−)-	600	48, 127, 352
Pimar-9,11(15)-dien-19-ol, (−)-	599	48, 127, 352
Pimara-12α,16-diol, 8,15R-epoxy-3-oxo-ent-	598	49, 127, 353
Pimara-3β,12α,16-triol C-15 epimer 10a, 8-15-epoxy-ent-	597	48, 128, 353
Pimara-3β,12α,16-triol C-15-epimer 9a, 8,15-epoxy-ent-	596	48, 127, 353
Pimara-7,15-dien-19-oic acid, ent-	820	48, 126, 351
Pimara-8(14), 15-dien-19-oic acid, ent-	590	48, 126, 351
Pimara-8(14)-en-6β-O-glucoside, 15β,16-dihydroxy-	593	48, 126, 353
Pimara-8(14)-ene, 2β,15β,16,19-trihydroxy-	595	48, 126, 351
Pimara-8(14)-ene, 6β,15β,16-trihydroxy-	594	48, 126, 353
Pimara-8(14)-ene, 6β,15,16,18-tetrahydroxy-ent-	944	48, 126, 353
Pimaradiene-3β,18-diol, ent-8(14), 15-	1248	48, 126, 351

Name	Number	Page Numbers
Pimaradiene-3β,19-diol, *ent*-8(14),15-	1247	48, 126, 351
Pimar-8(9),15-diene, 20-hydroxy-7-oxo-	1249	48, 126, 352
Pimar-8(9),15-diene, 18-hydroxy-*ent*-	1251	48, 127, 352
Pimara-8,15-dien-19-oic acid, 7-oxo-*ent*-	821	48, 126, 352
Pimara-9(11)-en-19-oic acid, 12β-acetoxy-*ent*-	823	48, 126, 352
Pimara-9(11),15-dien-19-ol, *ent*-	1250	48, 126, 352
Pimar-9(11),15-dien malonate, 18-hydroxy-*ent*-	1252	48, 127, 352
Pimar-9(11),15-dien malonate, 19-hydroxy-*ent*-	1253	48, 127, 352
Pimar-9(11),15-diene, 3α,19-dihydroxy-*ent*-	1254	48, 127, 352
Pimar-8(9),15-diene, 20-hydroxy-7-oxo-	1249	48, 126, 352
Polyalthic acid	830	32, 99, 300
Polyalthic acid, 17-oxo-8β,17-dihydro-	838	32, 99, 301
Polyalthic acid, 19-hydroxy-	835	32, 99, 300
Polyalthic acid, 3α-[2-methylbutyryloxy]-	833	32, 99, 300
Polyalthic acid, 3α-angeloyloxy-	832	32, 99, 300
Polyalthic acid, 3α-hydroxy-	831	32, 99, 300
Polyalthic acid, 3α-isobutyryloxy-	834	32, 99, 300
Polyalthic acid, 8α,17-epoxy-8,17-dihydro-	836	32, 99, 300
Polyalthic acid, 8β,17-epoxy-8,17-dihydro-	837	32, 99, 300
Polyalthine, 17-hydroxy-iso-	414	32, 99, 302
Polyalthine, 7β-hydroxy-[Austrochaparol]	401	31, 98, 299
Printziaic acid lactone, 5α-hydroxy-5,10-dihydro-	916	46, 121, 344
Printziaic acid, 5α-hydroxy-1,2-dehydro-5,10-dihydro-	915	46, 121, 343
Printziaic acid, 5α-hydroxy-5,10-dihydro-	913	46, 121, 343
Printziaic acid, 5α-methoxy-5,10-dihydro-	914	46, 121, 343
Printzianic acid	571	45, 121, 343
Printzianic acid, iso-	572	46, 121, 343
Psiadiol	404	31, 98, 300
Pulicaria angustifolia compound 1a	985	46, 122, 345
Pulicaria angustifolia compound 2a	986	46, 122, 345
Pulicaria angustifolia compound 3a	987	46, 122, 346
Pumiloxide, 6β-hydroxy-	421	32, 99, 303
Rebaudioside A	685	53, 139, 371
Rebaudioside B	684	53, 139, 371
Rebaudioside D	688	53, 139, 371
Rebaudioside E	689	53, 139, 371
Relhania acid	231	21, 86, 263
Rimuen-18-yl acetate, 5β-hydroxy-*ent*-	883	27, 93, 279
Rimuen-18-yl tiglate, 5β-hydroxy-*ent*-	885	27, 93, 279
Rimuen-3α,5β,18-triol, *ent*-	886	27, 93, 279
Rimuen-3α,5β-diol, *ent*-	884	27, 93, 279
Rimuen-5β,18-diol, *ent*-	882	27, 93, 279
Rimuen-5β-ol, *ent*-	881	27, 93, 279
Rosa-1(10),15-dien-18-oic acid	890	27, 93, 280
Rosa-5,15-dien-18-yl-acetate, 3α-hydroxy-	889	27, 93, 279
Rosa-5,15-dien-3α,18-diol	887	27, 93, 279
Rosa-5,15-dien-3α-yl acetate, 18-hydroxy-	888	27, 93, 279

Name	Number	Page Numbers
Royleanone	321	27, 94, 282
Royleanone, 7-acetoxy-	323	27, 94, 282
Royleanone, 7-keto-	324	27, 94, 283
Royleanone, dehydro-	322	27, 94, 282
Rugosolide	566	43, 117, 334
Ruilopezia acid	1256	56, 136, 363
Salicifoliol	154	16, 74, 248
Salviarin	973	42, 119, 341
Salviarin, 6α-angeloyloxy-1-oxo-2,3-dihydro-	971	42, 119, 341
Salvic acid	1143	17, 75, 239
Salvic acid, 7α-acetyl-	1144	17, 75, 239
Salvicin	1223	44, 118, 323
Salvin	1234	45, 118, 330
Salvinin	1235	45, 118, 330
Sandaracopima-8(14)-en-6β-D-glucoside, 15,16-dihydroxy-	922	26, 91, 277
Sandaracopimar-15-en-11-one, 12β-acetoxy-8β-hydroxy-	297	26, 91, 274
Sandaracopimar-15-en-11-one, 8β,12α-dihydroxy-	296	25, 91, 274
Sandaracopimar-15-en-11-one, 8β-hydroxy-	283	25, 91, 274
Sandaracopimar-15-en-3β,18-diol	285	25, 91, 274
Sandaracopimar-15-en-6β,8β,11α-triol	298	26, 91, 274
Sandaracopimar-15-en-6β,8β-diol	277	25, 90, 274
Sandaracopimar-15-en-8β,11α-diol	279	25, 90, 274
Sandaracopimar-15-en-8β,11α-diol, 12β-(p-hydroxycinnamoyloxy)-	293	25, 91, 274
Sandaracopimar-15-en-8β,11α-diol, 12β-acetoxy-	292	25, 91, 274
Sandaracopimar-15-en-8β,11α-diol, 12β-acetoxy-[Structure incorrectly drawn = Cmpd No. 292]	588	48, 91, 274
Sandaracopimar-15-en-8β,11α-diol, 6β-acetoxy-	299	26, 91, 274
Sandaracopimar-15-en-8β,12α-diol, 11β-acetoxy-	295	25, 91, 275
Sandaracopimar-15-en-8β,12β-diol, 11α-acetoxy-	289	25, 91, 274
Sandaracopimar-15-en-8β-ol [Published structure incorrectly drawn = Cmpd No. 276]	587	48, 90, 273
Sandaracopimar-15-en-8β-ol, 11α,12β-diacetoxy-	294	25, 91, 274
Sandaracopimar-15-en-8β-ol, 6β-acetoxy-	278	25, 90, 274
Sandaracopimar-15-ene, 20-hydroxy-8β,20-oxido-	287	25, 91, 275
Sandaracopimar-15-ene, 8β,11α,12β-trihydroxy-	288	25, 91, 274
Sandaracopimar-15-ene, 8β,12β-dihydroxy-11α-senecioyloxy-	290	25, 91, 274
Sandaracopimar-15-ene, 8β,12β-dihydroxy-11α-tiglinoyloxy-	291	25, 91, 274
Sandaracopimar-15-ene, 8β,20-dihydroxy-	286	25, 91, 275
Sandaracopimar-15-ene, 8β-hydroxy-	276	25, 90, 273
Sandaracopimar-15-ene, 8β-hydroxy-11α-acetoxy-	282	25, 91, 274
Sandaracopimar-15-ene, 8β-hydroxy-11α-senecioyloxy-	280	25, 90, 274

Name	Number	Page Numbers
Sandaracopimar-15-ene, 8β-hydroxy-11α-tiglinoyloxy-	281	25, 91, 274
Sandaracopimar-15-ene, 8β-hydroxy-12β-acetoxy-	284	25, 91, 274
Sandaracopimar-7,15-dien-18-oic acid	301	26, 91, 273
Sandaracopimar-7,15-dien-18-oic acid, 3β-acetoxy-	303	26, 91, 273
Sandaracopimar-7-ene, 3α,15,16-trihydroxy-13-epi-	306	25, 92, 276
Sandaracopimar-7-ene, 9α,15,16-trihydroxy-	307	26, 92, 276
Sandaracopimar-8(14),15-dien-18-oic acid	300	26, 91, 275
Sandaracopimar-8(14),15-dien-18-oic acid, 3β-acetoxy-	302	26, 91, 275
Sandaracopimar-8(14),15-diene, 1β,11α-diacetoxy-	932	24, 90, 272
Sandaracopimar-8(14), 15-diene, 1β,11α-diacetoxy-6,7-epoxy-	936	25, 90, 273
Sandaracopimar-8(14),15-diene, 1β,11α-diacetoxy-7-oxo-	937	25, 90, 273
Sandaracopimar-8(14),15-diene, 1β,11α-dihydroxy-	931	24, 90, 272
Sandaracopimar-8(14), 15-diene, 1β,11α-dihydroxy-7α-acetoxy-	933	25, 90, 272
Sandaracopimar-8(14), 15-diene, 1β,7α,11α-triacetoxy-	935	25, 90, 272
Sandaracopimar-8(14),15-diene, 1β,7α-dihydroxy-	928	24, 90, 272
Sandaracopimar-8(14),15-diene, 1β-hydroxy-7α-acetoxy-	930	24, 90, 272
Sandaracopimar-8(14),15-diene, 7α-hydroxy-	927	24, 90, 272
Sandaracopimar-8(14),15-diene, 7α-hydroxy-1β,11α-diacetoxy-	934	25, 90, 272
Sandaracopimar-8(14),15-diene, 7α-hydroxy-1β-acetoxy-	929	24, 90, 272
Sandaracopimar-8(14)-en-19-oic acid, 6β,15,16-trihydroxy-	921	26, 91, 276
Sandaracopimar-8(14)-ene, 6α,15,16,18-tetrahydroxy-	923	26, 91, 277
Sandaracopimaradiene	274	24, 90, 272
Sandaracopimaradiene, iso-	275	25, 90, 273
Schkuhriadiol, cyclo-	463	36, 105, 313
Schkuhriadiol-16-O-acetate, cyclo-	464	36, 106, 313
Schkuhrianol, acetate [16-acetoxy-14,15-epoxy-4,15-dihydro-13-epi-*ent*-Manoyl oxide]	457	37, 107, 315
Sclareol	1102	16, 74, 247
Sclareol, 6α-angeloyloxy-	153	16, 74, 247
Sclareol, 8-epi-*ent*	450	34, 100, 308
Sempervirenic acid [3β-acetoxy-labda-7,13-dien-15-oic acid]	113	14, 72, 242

Name	Number	Page Numbers
Smallantha-2Z,10E,13E-triene, 1,19-dihydroxy-6,7-epoxy-12-oxo-	981	9, 64, 227
Smallantha-2Z,10E,14(21)-triene, 1,19-dihydroxy-6,7-epoxy-12-oxo-	983	9, 64, 227
Smallantha-2Z,6Z,10E,13E-tetraene, 1,19-dihydroxy-	980	9, 64, 227
Smallantha-2Z,6Z,10E,14(21)-tetraene, 1,19-dihydroxy-12-oxo-	982	9, 64, 227
Solidagenone	158	16, 74, 250
Solidagenone, 13-epi-dehydroxy-9α,13α-epoxy-13,16-dihydro-	194	18, 77, 253
Solidagenone, 9-dehydroxy-9α,13α-epoxy-13,16-dihydro-	193	18, 77, 253
Solidago alcohol	241	23, 88, 265
Solidago alcohol, 8-epi-	238	22, 87, 264
Solidago aldehyde	240	23, 88, 265
Solidago dialdehyde	243	23, 88, 265
Solidago epoxylactol	247	23, 88, 268
Solidago gigantea compound I	1062	23, 88, 265
Solidago gigantea compound II	1066	23, 88, 269
Solidago gigantea compound III	1063	23, 88, 265
Solidago gigantea compound IV	1064	23, 88, 241
Solidago gigantea compound V	1065	23, 88, 265
Solidago gigantea compound VI	1067	23, 88, 269
Solidago gigantea compound VII	1068	24, 89, 269
Solidago glycol	242	23, 88, 265
Solidago lactol	246	23, 88, 267
Solidago-18-oate, methyl 8-epi-1,2-dehydro- [Nidoreseda acid methyl ester]	251	39, 114, 322
Solidago-18-oic acid [Incorrectly drawn with C-8β-methyl; probably the same as 249]	245	39, 114
Solidago-18-oic acid, 1,2-dehydro- [Incorrectly drawn with C-8β-methyl; probably = 248]	250	39, 114, 322
Solidago-18-oic acid, 8-epi	249	39, 113
Solidago-18-oic acid, 8-epi-1,2-dehydro-	248	39, 113, 322
Solidagoic acid A	239	23, 88, 265
Solidagoic acid B	244	23, 88, 265
Solidagolactone [Soldagolactone I; 16-Hydroxy-kolavenic acid lactone]	543	43, 116, 332
Solidagolactone II [Elongatolide C]	258	22, 87, 265
Solidagolactone II, 2α-hydroxy- Solidago virgaurea compound 4a]	263	22, 87, 265
Solidagolactone II, 3-oxo-3,4-dihydro-	268	23, 88, 267
Solidagolactone II, 3β,4α-dihydroxy-3,4-dihydro-	266	22, 88, 267
Solidagolactone III	259	22, 87, 265
Solidagolactone III, 2α-hydroxy-	264	22, 87, 265

Name	Number	Page Numbers
[Solidago virgaurea compound 4c]		
Solidagolactone III, 3-oxo-3,4-dihydro-	269	23, 88, 267
Solidagolactone III, 3α,4α-epoxy-3,4-dihydro-	261	22, 87, 266
Solidagolactone III, 3β,4α-dihydroxy-3,4-dihydro-	267	22, 88, 267
Solidagolactone V	262	22, 87, 267
Solidagolactone V, 2β-hydroxy-	265	22, 88, 267
Solidagolactone VIII	988	23, 88, 266
Solidagolactone, tricyclo-	947	24, 89, 271
Solidagonal acid	960	47, 125, 349
Solidagonic acid	962	42, 116, 327
Solidagonic acid	500	40, 115, 323
[7α-Acetoxy-13,14E-kolavenoic acid]		
Stenolobin	639	51, 132, 358
Stenolobin, 15α-angeloyloxy-	1007	51, 133, 361
[methyl-15α-angeloyloxy- ent-kaur-16-en-19-oate]		
Stephalic acid	570	44, 119, 338
Sterebin A	1157	21, 85, 262
Sterebin B	1158	21, 85, 262
Sterebin C	1159	21, 85, 262
Sterebin D	1160	21, 85, 262
Steviolbioside	682	53, 138, 371
Stevioside	683	53, 138, 371
Stevisalicinone	1026	28, 95, 284
Strictic acid	573	46, 122, 345
Strictic acid, 12α-[2-methylbutyryloxy]-	952	46, 122, 345
Symphyoreticulic acid, 13E-	537	42, 116, 325
Symphyoreticulic acid, 13Z-	536	42, 116, 326
Terminaloic acid	759	56, 140, 374
Tetrachyrin [Zoapatlin]	800	59, 146, 382
Tomentanol	1099	11, 66, 231
Tomentol	1097	10, 66, 231
Tomentol, pre-	1094	9, 64, 228
Tomexanthin	56	10, 66, 230
Tomexanthol	1098	11, 66, 230
Tomexanthol, pre-	1096	7, 61, 222
Trachyloban-19-al, ent-	855	58, 144, 380
Trachyloban-19-oic acid	791	58, 144, 380
Trachyloban-19-oic acid, 7β-hydroxy-	1259	58, 145, 380
Trachyloban-19-oic acid, 11-oxo-	793	58, 144, 380
Trachyloban-19-oic acid, 11α-[16′α-hydroxy- ent-kaur-11′-en-19′-oyloxy]-	795	58, 145, 381
Trachyloban-19-oic acid, 15α-acetoxy-	825	58, 144, 380
Trachyloban-19-oic acid, 15α-angeloyloxy-	826	58, 144, 380
Trachyloban-19-oic acid, 15α-isobutyryloxy-	893	58, 144, 380
Trachyloban-19-oic acid, 7α-hydroxy-[Ciliarix acid]	792	58, 144, 380

Name	Number	Page Numbers
Trachyloban-19-oic acid, 15α-isovaleryloxy-	827	58, 144, 380
Trachylobanoic acid, 9,11-dehydro-	794	58, 145, 380
Trachyloban-19-oic acid thujanol ester, *ent*-	856	58, 145, 381
Trachyloban-19-oic acid methyl ester	1204	58, 145, 380
Tricyclosolidagolactone	947	24, 89, 271
Trinervinol	807	26, 91, 276
[Isopimara-8(14)-en-3β,15,16,17-tetrol]		
Tucumanoic acid	962	40, 115, 327
Turbinatone	1163	28, 94, 283
Viguieric acid	14	7, 61, 222
Villanovane,	877	59, 147, 383
13α,17-dihydroxy-19-(3-methylvaleryloxy)-		
Villanovane,	875	59, 147, 383
13α,17-dihydroxy-19-isovaleryloxy-		
Villanovane, 17-acetoxy-	878	59, 147, 383
13α-hydroxy-19-(3-methylvaleryloxy)-		
Villanovane,		
17-acetoxy-13α-hydroxy-19-isovaleryloxy-	876	59, 147, 383
Villanovane, 17-acetoxy-	880	59, 147, 383
13α-hydroxy-3α-(3-methylvaleryloxy)-		
Villanovane, 17-acetoxy-13α-hydroxy-	879	59, 147, 383
3α-isovaleryloxy-		
Villanovane, 3α,15α-dihydroxy-	894	59, 147, 383
Wedelia-seco-kaurenolide, 15β-hydroxy-	763	57, 141, 374
Wedelia-seco-kaurenolide,	760	56, 141, 374
9-oxo-[wedelia-seco-kaurenolide]		
Wedelia-seco-kaurenolide, 9α-hydroxy-9-desoxo-	761	57, 141, 375
Wedelia-seco-kaurenolide, 9β-hydroxy-9-desoxo-	762	57, 141, 375
Wedeloside	679	53, 137, 370
Wedeloside, L-rhamnopyranosyl-	680	53, 137, 370
Wederegiolide, 15α-hydroxy-	866	56, 141, 374
Wederegiolide, 15β-acetoxy-	868	56, 141, 374
Wederegiolide, 15β-hydroxy-	867	56, 141, 374
Wyethic acid	6	7, 60, 219
Zoapatanol	55	10, 66, 230
Zoapatanol, pre-	1095	7, 61, 221
Zoapatlin,15-oxo-	847	59, 146, 382

Species Index

Alphabetized Listing of Species Names with Authorities

Abrotanella nivigena Muell., 178, 180
Acanthospermum australe (Loefl.)
    Ktze., 152, 213
Acanthospermum hispidum DC., 154
Achillea filipendula Lam., 165, 177
Achyrocline alata (H. B. K.) DC., 167
Achyrocline vargasiana, 153
Acritopappus confertus (Gardn.) K. &
    R., 168, 171, 172, 175, 178, 212
Acritopappus hagei K. & R., 169, 175,
    176
Acritopappus longifolius (Gardn.) K. &
    R., 167, 168, 169, 175, 204
Acritopappus morii K. & R., 168, 169,
    215
Acritopappus teixeirae K. & R., 175,
    176
Adenostemma caffrum DC., 191, 197
Adenostemma lavenia (L.) O. Kuntze,
    191, 198
Ageratina dendroides (Spreng.) K. &
    R., 167
Ageratina tristis (DC.) K. & R., 208
Ageratum fastigiatum (Gardn.) K. &
    R., 168, 169, 199
Ambrosia hispida Pursh, 193
Amphiachyris dracumculoides, 191
Arctotheca prostata, 180
Arctotis revoluta Jacq., 178, 179, 180,
    186, 196
Aristeguietia buddleaefolia (Benth.) K.
    & R., 157, 160
Aristeguietia pseudoarborea (Hieron.)

K. & R., 204
Aspilia floribunda (Gardn.) Baker,
    181, 190, 204
Aspilia foliacea, 178, 181, 190
Aspilia jolvana, 178, 179, 181, 188
Aspilia mossambicensis, 181, 205
Aspilia ovalifolia, 177, 179, 181, 188
Aspilia parvifolia Mattf., 171, 178,
    179, 181, 188, 190, 195, 196,
    213
Aspilia pluriseta Schweinf. subsp.
    pluriseta, 179, 181, 185, 188,
    189, 190, 195
Aspilia riddellii, 181
Aster alpinus L., 208
Aster bakeranus Burtt. Davy ex C. A.
    Smith, 179, 181
Aster pleiocephalus, 181, 188
Aster tanacetifolius, 185, 187
Athrixia arachnoidea Wood, 198
Athrixia elata Sond., 172, 181, 191,
    198
Athrixia phylicoides DC., 198, 199
Athrixia pinifolia N. E. Br., 178
Atractylis gummifera L., 191, 199
Austrobrickellia patens (D. Don. ex H.
    & A.) K. & R., 209
Austroeupatorium chaparense (B. L.
    Robins.) K. & R., 166, 169
Austroeupatorium inulaefolium (H. B.
    K.) K. & R., 169, 170, 178,
    179, 181, 185, 187, 189, 194,
    196, 202, 216

Ayapana amygdalina (Lam.) K. & R., 167, 172, 216

Baccharis alaternoides, 172, 176

Baccharis articulata Lam (Persoon), 173, 174, 217

Baccharis calvescens DC., 174

Baccharis cassinaefolia DC., 174, 175

Baccharis chilco H. B. K., 174

Baccharis concinna Barrosa, 185, 187, 190

Baccharis concinnia Barrosa, 181

Baccharis conferta H. B. K., 174

Baccharis crispa, 174

Baccharis eggersii Hieron, 207

Baccharis elliptica, 181, 188, 195

Baccharis genistelloides (Lam.) Per., 175

Baccharis gilliesii, 174

Baccharis grandicapitulata Hieron, 207, 217

Baccharis hutchisonii Cuatr., 173, 207

Baccharis incanum, 173, 203

Baccharis incarum Wedd., 216, 217, 218

Baccharis intermixta Gardn., 167, 181, 185

Baccharis kingii Cuatr., 203

Baccharis macraei, 173, 174

Baccharis magellanica (Lam.) Pers., 174, 217

Baccharis microcephala (Less.) DC., 175

Baccharis minutiflora Mart., 178, 179, 181, 194, 195, 196, 197

Baccharis nitida (R. & R.) Pers., 207

Baccharis oxydonta DC., 171

Baccharis pingraea DC., 214, 215, 216

Baccharis pithieri, 188

Baccharis polifera Griseb., 216

Baccharis polyphylla Gardn., 176, 181, 185, 187, 190

Baccharis pylicoides, 181

Baccharis quitensis H. B. K., 178, 179, 181, 194, 197

Baccharis ramosissima Gardn., 179, 181, 190

Baccharis rhetinodes Meyen. et Walp., 174

Baccharis rhetinoides Meyen. et Walp., 173, 217

Baccharis rhomboidalis, 217

Baccharis salicifolia Blake, 155, 174, 205, 210, 211, 212, 215, 216, 217

Baccharis salzmannii DC., 181, 188

Baccharis sarothroides A. Gray, 174, 210

Baccharis scoparia (L.) Sw., 154, 207, 211

Baccharis sternbergiana Steud., 166, 167

Baccharis subdentata DC., 174

Baccharis tola, 160, 171, 199, 200

Baccharis tricuneata (L.f.) Pers. var. tricuneata, 173, 174

Baccharis tricuneata var. lineata Cuatr., 173, 174

Baccharis trimera (Less.) DC., 175, 176

Baccharis trinervis Pers., 153, 215

Baccharis truncata Gardn., 176, 179, 181, 186, 188

Baccharis tucumanensis H. & A., 155, 207

Baccharis vaccinioides H. B. K., 210

Bahianthus viscidus (Baker) K. & R., 176

Balsamorhiza sagittata (Push.) Nutt., 178, 207

Bedfordia salicinia DC., 162, 184

Bejaranoa balansae (Hieron.) King et Robins., 213

Bejaranoa semistriata (Baker) K. & R., 150

Bishovia boliviensis K. & R., 157

Brickellia annulosa (DC.) A. Gray, 173

Brickellia argyrolepis B. L. Robins., 154, 155

Brickellia corymbosa (DC.) A. Gray, 155

Brickellia diffusa A. Gray, 154, 161

Brickellia eupatoriedes (L.) Shinner, 154, 155, 177

Brickellia paniculata (Mill.) B. L. Robins., 155

Brickellia squarrosa (Cav.) B. L. Robins., 155

Brickellia vernicosa B. L. Robins.,

155, 208

Brickellia veronicaefolia (H. B. K.) A. Gray, 155, 156

Brickellia veronicaefolia A. Gray var. typica, 156, 161

Cacalia bulbifera, 178, 180, 181, 196

Campovassouria bupleurifolia (DC.) K. & R., 178, 179, 180, 181, 189, 190, 191

Caramboa pithieri, 178, 181, 187

Caramboa trugillensis, 181

Carterothamnus anomalochaeta R. M. King, 154, 185

Centaurea cineraria, 153

Centipeda orbicularis Lour., 153, 162

Chiliotrichium rosmarinifolium Less., 213, 216, 217

Chromolaena collina (DC.) K. & R., 156

Chromolaena cryptantha, 181

Chromolaena laevigata (Lam.) K. & R., 210

Chrysanthemoides incana T. Norl., 165

Chrysanthemoides monolifera (L.) T. Norl., 164, 165

Chrysanthemoides monolifera (L.) T. Norl. subsp. canescens (DC.) T. Norl., 163, 164, 165

Chrysothamnus nauseusus (Pall.) Britt., 155, 156, 159, 214

Chrysothamnus paniculatus (Gray) Hall, 157, 158, 159, 160, 161, 163

Chrysothamnus viscidiflorus, 166

Coespeletia lutescens (Cuatr. & Arist.) Cuatr., 171, 178, 180, 181, 185, 186, 191, 193, 194, 198

Coespeletia marcana Cuatr., 178, 180, 181, 185, 186, 191, 193, 194, 196, 198

Coespeletia moritziana (Sch. Bip.) Cuatr., 171, 178, 180, 181, 186, 188, 189, 192, 194, 198

Conyza canadensis, 153

Conyza ivaefolia, 174

Conyza japonica (Thundb.) Less., 173, 176

Conyza podocephala DC., 153, 174

Conyza scabrida DC., 174, 176, 203, 205, 206

Conyza stricta Willd., 162, 169, 176

Conyza ulmifolia, 153

Critonia daleioides DC., 178, 181, 202

Critonia canescens

Critonia hebebotrya, 181, 202

Cronquistianthus bishopii K. et R., 212, 213

Croptilon divaricatum, 165

Denekia capensis Thunb., 156, 171, 208, 212

Desmanthodium fruticosum, 169

Dimerostemma asperatum Blake, 153

Dimerostemma brasilianum Cass., 153

Dimorphotheca aurantiaca Hort., 199, 200

Dimorphotheca pluvialis Moench., 165, 166

Dimorphotheca pseudoaurantiaca Sch. & Thell., 199, 200

Disynaphia halimifolia (DC.) K. & R., 150

Disynaphia multicrenulata, 178, 180, 181, 185, 187, 189, 195, 208

Doronicum hungaricum, 192

Doronicum macrophyllum Fisch., 207

Doronicum microphyllum

Echinacea purpurea Moench., 160

Enhydra fluctuans Lour., 181, 185, 187, 189, 196

Erigeron philadelphicus L., 171

Espeletia curialensis, 186

Espeletia figueirasii, 181, 194

Espeletia floccosa, 181, 194

Espeletia garciae Humb. & Bonpl., 178, 180, 181

Espeletia grandiflora, 178, 180, 181, 185

Espeletia hartwegiana, 171, 180, 186, 192, 198

Espeletia humbertii, 186, 196

Espeletia littlei, 186, 192, 196

Espeletia moritziana, 181, 194

Espeletia neriifolia, 180, 192

Espeletia schultzii Wedd., 178, 180, 181, 186, 192, 196

Espeletia tenor, 186

Espeletia timotensis, 192

Espeletia uribei Cuatr., 178, 180, 181, 186, 192, 193, 194, 195

Espeletia weddellii Sch. Bip. ex Wed.,

186, 192

Espeletia wedellii Sch. Bip., 178, 180, 207

Espeletiopsis garciae Cuatr., 178, 181, 186, 188, 189, 192

Espeletiopsis glandulosa Cuatr., 178, 181, 186, 188, 189, 192, 193, 198

Espeletiopsis guacharaca (Diaz) Cuatr., 171, 178, 179, 180, 181, 186, 189, 191, 192, 193, 194, 196, 198

Espeletiopsis purpurascens Cuatr., 178, 181, 186, 188, 189, 192, 193, 194, 196, 198

Espeletiopsis tachirensis (Aristeg.) Cuatr., 178, 181, 186, 188, 190, 192, 193, 194, 196, 198

Eupatorium album L., 191, 201

Eupatorium deltoideum, 192

Eupatorium inulaefolium H. B. K., 201, 209

Eupatorium jhanii B. L. Robins., 160, 206

Eupatorium maretiana

Eupatorium petaloideum Britt., 167, 202, 218

Eupatorium petiolare Moc. & Ses. ex DC., 181

Eupatorium salvia, 214

Eupatorium turbinatum Gray, 175, 214, 216

Eupatorium villosum Sweet, 206

Felicia erigeroides, 153

Fleischmannia deborabellae K. & R., 156

Fleischmannia pycnocephaloides (B. L. Robins.) K. & R., 156

Fleischmannia sinclairii (Benth. in Oerst) K. et R., 173

Fleischmannia viscidipes (B. L. Robins.) K. & R., 203, 204

Garuleum bipinnatum Less., 163, 164, 165

Garuleum pinnatifidum DC., 163, 164, 165

Garuleum sonchifolium (DC.) T. Norl., 163, 165, 176, 200

Geigeria burkei Harv. subsp. burkei var. burkei, 150, 151, 152

Geigeria burkei Harv. subsp. burkei var. elata Merxm., 151

Geigeria burkei Harv. subsp. burkei var. intermedia (S. Moore) Merxm., 151, 156

Geigeria burkei Harv. subsp. burkei var. zeyheri (Harv.) Merxm., 151, 152, 156

Geigeria burkei Harv. subsp. diffusa (Harv.) Merxm., 151

Geigeria burkei Harv. subsp. fruticulosa Merxm., 150, 151

Geigeria glabra var. glabra

Gnaphalium guadichaudianum, 171

Gnaphalium microcephalum, 153

Gnaphalium oligandrum (DC.) Hilliard & Burtt, 181, 192

Gnaphalium undulatum L., 171, 176, 178, 180, 181, 186, 199

Gochnatia glutinosa Don., 217

Gochnatia paniculata (Less.) Cabrera, 172, 173, 174

Goyazianthus tetrastichus (B. L. Robins.) K. & R., 172, 178, 180, 181, 187, 190, 192, 195, 217

Grangea maderaspatana Poir, 162, 178

Grazielia dimorpholepsis (Baker) K. & R., 170, 181, 188

Grazielia intermedia (DC.) K. & R., 150, 180, 181, 188, 189, 190

Grazielia serrata (Spreng.) K. & R., 181, 188, 191

Greenmaniella resinosa, 153

Grindelia acutifolia, 158, 159, 161

Grindelia aegialitis Gabr., 166, 214

Grindelia aphanactis Rydb., 158, 159

Grindelia boliviana Rusby, 157, 158, 210, 211, 215

Grindelia camporum Greene, 157, 158, 159, 160

Grindelia chilcoensis (Corub.) Cabrera

Grindelia chiloensis, 157, 158, 159, 160

Grindelia discoidea Hook. & Arn., 155, 214

Grindelia humilis (Hook. & Arn.), 158

Grindelia nauseosus, 158

Grindelia paludosa Greene, 157, 158, 159, 161

Grindelia perennis A. Nels., 158, 159
Grindelia pulchella, 157, 158, 159, 160
Grindelia robusta, 158, 159, 160
Grindelia squarrosa (Pursh.) Dunal, 157, 158, 159, 160, 207
Grindelia stricta DC., 157, 158, 159, 160, 161
Gutierrezia dracunculoides (DC.) Blake, 157, 163
Gutierrezia gilliesii Griseb., 208, 210
Gutierrezia grandis S. F. Blake, 203, 208
Gutierrezia lucida Greene, 156, 157
Gutierrezia mandonii (Sch. Bip.) Solbrig, 157
Gutierrezia sarothrae (Pursh.) Britt. & Rusby, 202, 203
Gutierrezia solbrigii, 208, 210, 213, 215
Gutierrezia spaerocephala Gray, 154, 214
Gutierrezia spathulata (Phil.) Kurtz, 210, 216
Gutierrezia texana (DC) T. & G., 211, 212
Gymnosperma glutinosa, 154
Haplopappus angustifolius, 176
Haplopappus ciliatus (Nutt.) DC., 173
Haplopappus foliosus, 176
Haplopappus glutinosus Cass., 210
Haplopappus paucidentatus Phil., 211, 216
Haplopappus pectinatus Phil., 167, 210, 215
Haplopappus tenuisectus, 158
Haplopappus venetus Blake, 158
Hartwrightia floridana Gray, 167, 172, 173
Hebeclinium macrophyllum (L.) DC., 209
Helianthopsis bishopii H. Robins., 181, 199, 200, 207
Helianthopsis microphylla (H. B. K.) H. Robins., 181, 199, 200
Helianthopsis utcubambensis H. Robins., 181, 199, 200
Helianthus angustifolius L., 170, 182, 187, 196
Helianthus annuus L., 180, 182, 186, 187, 189, 194, 196, 199, 200, 201, 202, 203
Helianthus argophyllus T. & G., 196, 201
Helianthus californicus DC., 201
Helianthus ciliaris DC., 201
Helianthus debilis Nutt. subsp. cucumerifolius (T. & G.) Heiser, 182, 186, 187, 189, 194, 195, 196, 197, 200
Helianthus debilis Nutt. subsp. debilis, 182, 187, 201
Helianthus decapetalus L., 170, 182, 185, 200, 212
Helianthus decapetalus var. multiflorus A. Gray, 212
Helianthus giganteus L., 179, 182, 192, 200
Helianthus grosseseratus, 194, 201, 218
Helianthus grosseserratus Martens, 186, 192
Helianthus heterophyllus Nutt., 182, 192, 193, 194
Helianthus hirsutus Raf., 182, 200, 202, 212
Helianthus laciniatus, 218
Helianthus maximiliani Schrader, 170, 192, 193
Helianthus niveus subsp. canescens (A. Gray) Heiser, 182, 186, 201
Helianthus nuttallii T. & G., 182
Helianthus nuttallii T. & G. subsp. nuttallii, 186, 194, 201
Helianthus occidentalis Riddell, 170, 182, 186, 187, 190, 194, 195, 196, 197, 200, 201
Helianthus occidentalis var. dowellianus, 182, 188, 194, 195, 200
Helianthus petiolaris Nutt., 182, 186, 194, 197, 201, 202
Helianthus radula (Pursh) T. & G., 182, 185, 187, 194, 195, 196, 197, 198, 201
Helianthus rigidus (Cas.) Desf., 182, 196, 201, 212, 218
Helianthus salicifolius A. Dietr., 196, 201

Helianthus simulans E. E. Wats., 182, 185, 186, 187, 195, 197
Helianthus strumosus L., 177, 202
Helianthus tomentosus Elliot, 201
Helianthus tuberosus L., 170, 182, 185, 188
Helichrysum acutatum DC., 153, 212
Helichrysum albirosulatum Kill., 171, 212
Helichrysum argentissimum Wood, 182
Helichrysum argyrolepis, 182, 216
Helichrysum aureum (Houtt.) Merile var. monocephalum (DC.) Hilliard, 182, 184, 185, 186, 192
Helichrysum bellum Hilliard, 182
Helichrysum calliconum Harv., 153
Helichrysum chionosphaerum DC., 177, 182, 200, 201
Helichrysum confertum N. E. Br., 156, 167, 170, 182, 208, 212, 214
Helichrysum cooperi Harv., 182, 184, 185
Helichrysum coriaceum, 157
Helichrysum davyi, 178, 179, 180, 182, 185, 196
Helichrysum dendroideum N. H. Wakefield, 179, 184, 194, 199
Helichrysum diosmifolium (Vent.) Sweet, 165, 197
Helichrysum fulvum N. E. Br., 182, 199, 201
Helichrysum heterolasium Hilliard, 179, 180, 182, 184, 185, 186
Helichrysum Krebsianum Less., 151, 153
Helichrysum miconiifolium DC., 179, 180, 182, 186
Helichrysum mimetes S. Moore, 150
Helichrysum mundii Harv., 150
Helichrysum nanum, 151
Helichrysum nudifolium (L.) Less., 153
Helichrysum nudifolium (L.) Less. var. nudifolium, 157, 214
Helichrysum odoratissimum Sweet, 153
Helichrysum oreophilum Klatt., 151
Helichrysum oxyphyllum Klatt., 151
Helichrysum pallidum DC., 180, 182, 185, 186, 196
Helichrysum pilosellum (L.f.) Less., 179, 180, 182, 185, 192
Helichrysum pinifolium (Lam.) Schrank, 157, 171
Helichrysum platypterum DC., 182, 192
Helichrysum refluxum N. E. Br., 182, 192, 196, 199, 201, 208
Helichrysum ruderale Hilliard & Burth., 185, 192
Helichrysum setosum Harv., 217, 218
Helichrysum subfacultatum Hilliard, 153
Helichrysum subfalcatum Hilliard, 151
Helichrysum subulifolium Hary, 182
Helichrysum suterlandii Harv., 156, 170
Helichrysum vernum Hilliard, 157, 171, 179, 182
Hemizonia congesta DC., 154
Hemizonia lutescens (Greene) Keck, 153, 154
Heterochrysum oreophilum, 213
Heterocondylus grandis, 182
Heteromma simplicifolium, 153
Heteropappus altaicus (Willd.) Novopokrov, 173, 174, 207
Heterotheca subaxillaris, 158, 160
Hinterhubera imbricata Cuatr., 173, 174
Hofmeisteria fasciculata (Benth.) Walp., 156
Hymenopappus newberryi (Gray) T. M. Johnst., 197
Ichthyothere connata Blake, 169, 170, 182, 188, 189
Ichthyothere cunabi, 182
Ichthyothere latifolia (Benth.) Gardn., 170, 182, 188, 189
Ichthyothere rufa Gardn., 179, 182, 188, 189
Ichthyothere terminalis (Spreng.) Malme, 170, 179, 180, 182, 186, 187, 188, 189, 190, 191, 192, 193, 194, 196, 197, 199
Ichthyothere ulei Thunb., 152, 180, 182, 187, 188, 190
Inula crithmoides, 156

Inula royleana DC., 166, 201
Isocoma coronopifolia Greene, 158,
    159, 214
Isostephane heterophylla, 182
Kingianthus paradoxus H. Robins., 150
Koanophyllon admantium (Gardn.) K.
    & R., 150, 162, 173
Koanophyllon conglobatum (DC.) K. &
    R., 150, 156, 157, 161, 162
Lagascea rigida (Cav.) Stuessy, 182,
    188, 192
Lasianthaea podocephala (A. Gray) K.
    Becker, 182, 192, 206
Lasiolaena morii K. & R., 150
Lasiolaena santosii K. & R., 150, 167
Liatris elegans (Walt.) Michx., 153
Liatris laevigata Nutt., 177
Liatris scariosa Willd., 173
Liatris spicata (L.) Willd., 174, 211
Libanothamnus granatesianus Cuatr.,
    178, 179, 182, 186, 189, 198
Libanothamnus ncriifolia (B. cx H.)
    Ernst, 179, 180, 182, 187, 192
Libanothamnus occultus (Blake) Cuatr.,
    171, 182, 187, 192
Libanothamnus spectabilis (Cuatr.)
    Cuatr., 171, 182, 187, 192
Libanothamnus tamanus (Cuatr.)
    Cuatr., 178, 182, 187, 192, 198
Libanothamnus wurdackii (R. T. & L.
    F.) Cuatr., 171, 178, 182, 187,
    192, 198
Lindheimera connata, 160
Lindheimera texana Gray et Engelm.,
    160, 214
Loxothysanus sinuatus (Less.) B. L.
    Robinson, 182, 189
Lychnophora sellowii Sch. Bip., 178
Macowania glandulosa N. E. Br., 155,
    162, 207
Madia sativa Mol., 154
Melampodium diffusum Cass., 152
Melampodium divaricatum (Rich. in
    Pers.) DC., 172, 206
Melampodium leucanthum Torr. &
    Gray, 209
Melampodium longipilum Robins., 152
Melampodium paludosum, 206
Melampodium paniculatum, 182

Melampodium perfoliatum (Carv.) A.
    Gray, 182, 206
Mikania alvimii K. & R., 170, 172,
    177, 215
Mikania arrojadoi Mattf., 179, 180,
    182, 184, 185, 186, 187, 189,
    194, 195, 201
Mikania banisteriae DC., 170, 178,
    184, 185, 202, 206
Mikania belemii K. & R., 179, 180,
    182, 189
Mikania congesta DC., 209
Mikania cordata (Burm. f.) B. L.
    Robins., 183, 189
Mikania goyazensis (B. L. Robins.) K.
    & R., 150
Mikania hirsutissima DC. var.
    hirsutissima, 183
Mikania luetzelburgii Mattf., 179, 180,
    183, 186, 189, 194, 195
Mikania micrantha, 180, 183
Mikania monagasensis, 183
Mikania mongenansis
Mikania oblongifolia DC., 189
Mikania officinalis Mart, 150, 151
Mikania periplocifolia Hook. et Arn.,
    213
Mikania pyramidata Donn. Smith, 165,
    170, 171, 177
Mikania schenkii, 183
Mikania sessilifolia DC., 150, 152,
    179, 180, 183, 184, 189, 191,
    195
Mikania triangularis Baker, 202, 217
Montanoa atriplicifolia (Pers.) Sch.
    Bip. in Seeman, 183
Montanoa frutescens, 183, 189
Montanoa leucantha (Lag.) Blake
    subsp. arborescens (DC.) Funk,
    152
Montanoa leucantha (Lag.) Blake
    subsp. leucantha, 152
Montanoa mollissima Brongn. ex
    Groenl., 152
Montanoa pteropoda Blake, 178, 179,
    180, 183, 187, 189, 192, 193
Montanoa tomentosa Cerv. in Llave &
    Lex. subsp. tomentosa, 152, 178,
    180, 212

Montanoa tomentosa Cerv. subsp.
microcephala (Sch. Bip.) Funk,
152
Montanoa tomentosa Cerv. subsp.
xanthiifolia (Schultz Bip. in C.
Koch) Funk, 152, 183, 187, 189,
192, 199, 201, 206, 207
Morithamnus crassus K. & R., 166,
167, 172, 175, 196
Nardophyllum lanatum (Meyen.)
Cabrera, 216
Neomiranda angularis, 173
Nidorella agria Hilliard, 162
Nidorella anomala Steetz, 199, 200
Nidorella auriculata DC. subsp.
polycephala (DC.) Willd., 153,
156
Nidorella hottentotica, 161
Nidorella resedifolia DC., 162
Nordophyllum lanatum, 213
Olearia heterocarpa S. T. Blake, 175
Olearia muelleri (sond.) Benth., 174
Olearia paniculata, 171
Ophryosporus chilca (H. B. K.)
Hieron., 203
Osteospermum auriculatum (S. Moore)
Norl., 163, 164, 165
Osteospermum barberiae (Harv.) Norl.,
153, 163, 164
Osteospermum ciliatum Berg., 163,
164, 165
Osteospermum corymbosum L., 163,
164, 165
Osteospermum fruticosum (L.) Norl.,
163, 164
Osteospermum incanum, 164
Osteospermum jucundum, 163, 164,
165
Osteospermum junceum Berg, 163,
164, 165
Osteospermum muricatum E. Mey ex
DC. subsp. muricatum, 163,
164, 166
Osteospermum oppositifolium (Ait.)
Norl., 164
Osteospermum polygalioides L., 164,
165
Osteospermum rotundifolium (DC.) T.
Norl., 163, 164, 165

Osteospermum scariosum (DC.) Norl.
subsp. scariosum, 164
Osteospermum subulatum DC., 165
Osteospermum thodei Maikotter, 164,
165
Othonna cylindrica DC., 177, 183, 192
Othonna floribunda Schltr., 177, 183
Othonna sedifolia, 163, 171, 183, 194,
195
Oxylobus adscendens, 207
Oxylobus arbutifolius (H. B. K.) A.
Gray, 154, 207
Oxylobus canescens, 203
Oxylobus glanduliferus (Sch. Bip.)
Gray, 154, 214
Oyedaea boliviana Britton, 183, 190,
192
Oyedaea buphthalmoides, 180, 183,
187, 192, 196
Oyedaea lanceolata (Russby) Blake,
179, 180, 183, 190, 192
Oyedaea rusbyi, 179, 180, 183, 187,
190, 191, 192
Oyedaea verbesinoides, 179, 183, 186,
188
Palafoxia arida B. L. Turner & M. I.
Morris, 177
Palafoxia rosea (Bush.) Cory, 156,
160, 163, 165, 166, 215, 216
Palafoxia texana DC., 214
Palafoxia texana var. ambigua, 156
Pegolettia senegalensis Cass, 150
Perymeniopsis ovalifolia, 183, 189
Perymenium discolor, 183, 196
Perymenium ecuadoricum Blake, 179,
180, 183, 187, 195, 196
Perymenium featherstoni, 179, 183,
187
Perymenium klattianum Fay, 180, 183,
192, 193, 200, 206
Perymenium ovalifolia (A. Gray) H.
Robins., 187
Perymenium serratum, 179, 183, 196
Peteravenia malvaefolia, 197, 200
Peteravenia schultzii, 200
Piptothrix jaliscensis B. L. Robins.,
209
Piptothrix sinaloae Blake, 209
Piqueria trinervia Cav., 202

Plagiocheilus prostratus Benth., 153
Planaltoa lychnophoroides Barroso, 169
Pleurocoronis pluriseta (Gray)
    K. & R., 154, 167
Podachaenium eminens, 156
Polymnia canadensis L., 195
Polymnia fruticosa H. B. K., 179
Polymnia fruticosum Benth., 179, 183,
    187
Polymnia maculata Cav. var. maculata,
    183, 192
Polymnia pyramidalis Triana, 179,
    183, 187
Printzia laxa N. E. Br., 171, 173, 174,
    175, 176
Psiadia altissima, 169
Pulicaria angustifolia DC., 206, 208
Pulicaria gnaphaloides (Vent.) Boiss.,
    174
Pulicaria salviifolia Bgl. in Mem., 216,
    217
Relhania acerosa (DC.) Bremer, 161,
    166, 167
Riencourtia oblongifolia, 192
Robinsonia gayana, 196
Robinsonia thurifera, 183
Rudbeckia fulgida Ait. var. fulgida,
    208
Ruilopezia bromeloides, 179, 183, 188,
    192
Ruilopezia figueirasii, 179, 180, 185,
    187, 192, 198
Ruilopezia jahnii (St.) Cuatr., 178,
    179, 183, 184, 186, 187, 189,
    192, 194, 196, 198
Ruilopezia lindenii (Sch. Bip.) Cuatr.,
    178, 179, 180, 183, 186, 187,
    188, 190, 192, 193, 194, 196,
    198
Ruilopezia marcescens, 178, 180, 187,
    190, 192, 193, 194, 198
Ruilopezia margarita Cuatr., 192, 196,
    218
Ruilopezia marcescens
Ruilopezia paltonioides, 178, 179, 180,
    183, 188, 192, 194, 195, 198
Ruilopezia ruizii, 178, 183, 186, 188,
    190, 192, 193, 194, 195, 198
Schkuhria multiflora H. & A., 171,

172
Sciadocephala schultze-rhonhofiae
    Matt., 185
Senecio asperulus, 200
Senecio erosus Wedd., 155
Senecio hypochoerideus DC., 217
Senecio pseudoorientalis, 153
Senecio sandersonii Harv., 165
Senecio subrubriflorus O. Hoffm., 163,
    164, 165
Senecio trichopterygius, 153
Senecio variabilis, 163
Sigesbeckia orientalis L., 177
Sigesbeckia pubescens Makino, 177,
    197, 206, 207
Silphium asteriscus L., 160, 161
Silphium compositum, 160
Silphium connatum, 156
Silphium integrifolium Michx., 160,
    161
Silphium perfoliatum L., 160, 161
Silphium terebinthinaceum Jacq., 160,
    161
Simsia dombeyana, 183
Simsia holwayi, 183
Smallanthus fruticosus (Benth.) H.
    Robins., 157, 179, 180, 183,
    185, 187, 192, 194, 195
Smallanthus glabratus (DC.) H.
    Robins., 180, 183, 185, 206, 208
Smallanthus reparius (H. B. K.) H.
    Robins., 183, 185, 187
Smallanthus siegesbeckia, 183, 185,
    187
Smallanthus uvedalia (L.) Mackenzie,
    179, 180, 183, 184, 185, 187,
    195
Solidago altissima L., 162, 163, 173,
    175, 207, 208
Solidago arguta Ait., 173, 176
Solidago canadensis L., 157, 160, 167,
    174
Solidago drummondii Torr. et Gray,
    213
Solidago elongata Nutt., 172, 173, 175
Solidago flexicaulis L., 213
Solidago gigantea, 157
Solidago gigantea Ait. var. serotina,
    162, 211

Solidago itatiayensis (Hieron.) K. & R., 176

Solidago juncea Ait., 173, 174, 183, 211

Solidago missouriensis Nutt., 166, 171, 177, 183, 209

Solidago multiradiata, 183

Solidago multiradiata var. multiradiata, 180

Solidago nemoralis Ait., 178

Solidago petradoria Blake, 158, 210

Solidago racemosa Greene, 213

Solidago rigida L., 183

Solidago rugosa Mill., 163, 174, 175, 176, 178, 180, 183, 190

Solidago sempervirens, 155, 157

Solidago serotina Ait., 162, 211

Solidago Shortii Torr. & Gray, 175

Solidago virgaurea L., 162, 163, 208

Steiractinia mollis Blake, 179, 180, 183, 187, 190, 192, 196, 197

Steiractinia sodiroi (Hieron.) Blake, 179, 183, 192, 196

Stevia andina, 156, 170, 183, 209

Stevia aristata, 170, 200, 202

Stevia berlandiera A. Gray, 156, 163, 170, 209

Stevia bupthalmiflora, 188, 189

Stevia eupatoria Will, 185, 209

Stevia jaliscensis B. L. Robins., 155

Stevia lemmonia A. Gray, 209

Stevia lucida Lag. var. Bipontini, 169

Stevia monardaefolia H. B. K., 156, 157, 183, 187

Stevia myriadenia Sch. Bip. ex Baker, 150, 151, 169, 173, 183, 188, 200

Stevia origanoides H. B. K., 170, 202

Stevia ovata Lag., 205

Stevia paniculata Lag., 205

Stevia polycephala Bertol, 176

Stevia rebaudiana Bertoni, 160, 170, 191, 214

Stevia salicifolia Cav., 155, 157, 163, 170, 209

Stevia setifera Rusby, 183

Stylotrichium rotundifolium Mattf., 150

Symphyopappus compressus (Gardn.) B. L. Robins., 172, 175

Symphyopappus reticulatus Baker, 172, 173, 175, 209

Tamania chardonii, 178, 179, 187, 188, 189, 198

Tetrachyron manicata, 183, 187, 192, 193

Tetrachyron orizabaensis Sch. Bip. ex Klatt. var. websteri Wussow & Urbatsch, 183, 187, 201

Tithonia longiradiata, 183, 193

Trichogonia salvaiaefolia Gardn., 204

Trichogonia villosa (DC.) Sch. Bip., 204, 205

Trichogoniopsis morii K. & R., 150, 155, 160, 197

Tridax peruviensis Powell, 208

Verbesina angustifolia (Benth.) Blake, 183, 193

Verbesina oncophora Rob. & Seat, 184, 193, 196

Vernonia divaricata, 184

Vernonia glabra (Steetz) Vatke var. glabra, 213

Vernonia venosissima, 184

Viguiera bishopii H. Robins., 169, 171, 184, 193, 196, 200, 201

Viguiera cordata, 184

Viguiera cordifolia Gray, 184, 200

Viguiera dentata (Cav.) Spreng., 184, 186, 187, 193, 201, 208, 209

Viguiera excelsa (Willd.) B. & H., 184, 193, 194, 197, 209

Viguiera excelsea (Willd.) B. & H., 193

Viguiera grammatoglossa DC., 179, 180, 184, 193, 199, 200

Viguiera hypargyrea, 184, 190, 194

Viguiera incana, 184

Viguiera insignis Miranda, 184, 193, 199, 200, 202

Viguiera lanceolata Britt., 201

Viguiera latibracteata, 193, 194, 195, 197, 201

Viguiera linearifolia, 184

Viguiera linearis, 184, 193, 196

Viguiera maculata Blake, 184, 186, 203

Viguiera oaxacana (Greenm.) Blake,

184, 193
Viguiera pazensis Rusby, 184, 187,
    188, 189, 199, 200, 201, 202,
    205
Viguiera porteri (A. Gray) Blake, 184,
    186, 189, 190, 194
Viguiera potosina Blake, 184, 186, 208
Viguiera procumbens (Pers.) Blake,
    184, 193, 201
Viguiera quinqueradiata (Cav.) A.
    Gray, 187, 188
Viguiera stenoloba, 170, 185, 196, 209
Viguiera stenoloba var. chihuahuense,
    193
Viguiera trichophylla, 184, 190, 201
Villanova titicaensis Meyer & Walp.,
    204
Wedelia asperrima Benth., 191
Wedelia buphthalmiflora Lorentz, 184,
    187, 188, 189, 193
Wedelia calycina L. C. Rich in Pers.,
    184, 186, 187, 189, 190, 193,
    197, 202
Wedelia glauca (Ort.) Hoffmann ex
    Hicken, 184, 189, 191
Wedelia grandiflora Benth., 179, 180,
    184, 185, 187, 188, 189, 190,
    193, 196
Wedelia helianthoides H. B. K., 184,
    188, 190, 193, 194, 195
Wedelia hispida H. B. K., 184, 193
Wedelia hookeriana Gardn., 179, 184,

187, 188, 189, 190, 193, 199,
    200
Wedelia pinetorum, 188, 190
Wedelia regis H. Rob., 204
Wedelia scaberrima Benth., 184, 186,
    188
Wedelia speciosa, 184
Wedelia trilobata (L.) Hitch., 184,
    185, 189, 190, 193, 199
Wedelia villosa, 184, 188
Wulffia maculata, 184
Wyethia helenioides (DC.) Nutt., 150
Xanthium strumarium L., 191
Xanthium strumarium L. var.
    strumarium, 191
Xanthocephalum linearifolium
    Greenm., 168, 169
Zaluzania angusta, 180, 184
Zaluzania angustifolia, 196
Zaluzania cinarescens, 184, 196
Zaluzania discoidea, 184, 193
Zaluzania subcordata, 180, 184, 193
Zaluzania triloba, 184, 193
Zexmenia gnaphaloides Gray, 150,
    153, 166
Zexmenia hispida A. Gray, 184, 193
Zexmenia phyllocephala (Hemsl.)
    Staad. & Steyer, 206
Zexmenia pinetorum, 188
Zinnia tenuiflora Jacq., 150
Zinnia verticillata Andr., 150

General Index

Anthemideae, 432
Astereae, 433
Asteroideae, 432
benzofuranes, 432
Biogenesis,
Ayapana, 386, 417
Brickellia, 389
Chrysothamnus, 387
Conyza, 388, 418
Helichrysum, 389
Koanophyllon, 387
Nidorella, 386, 418
Printzia, 388, 423
Relhania, 386, 420
Trichogonia, 426
cassane, 388, 425
chrysolane, 387
cleistanthane, 389, 427
cleistanthane, iso-, 389, 427
conyscabrane, 388, 423
ent-abietane, 389, 427
ent-primarane, 427
erythroxane, 389
normal-abietane, 388, 425
normal-pimarane, 388
printzianes, 388
relhaniane, 386
rosane, 426
sandaracopimarane, 425
solidagonane, 388
Biological Activity
Candida albicans, 487
Homeosoma electellum, 487
Nicotiana glutinosa, 486
Pectinophora gossypiella, 486

Schizaphis graminum, 485
Staphylococcus aureus, 487
Trirhabda spp., 488
abortifacient, 485
cocklebur poisoning, 488
Colorado potato beetle, 485
hypoglycemia, 487
miracidal activity, 487
molluscidal activity, 487
mutagenesis, 488
sheep and cattle toxins, 487
surface resin, 486
sweetener, 488
^{13}C-NMR Spectrometry
beyeranes, 506
clerodanes, 504
labdanes, 505
pimaranes, 506
kauranes, 506
Calenduleae, 433
Chromenes, 432
Cichorioideae, 432
Circular Dichroism, 495
Cistaceae, 495
Clerodanes
cis-ent-clerodane, 495
cis-normal-clerodane, 494
trans-ent clerodane, 493
trans-normal clerodane, 496
Compartmentation, 391
Cronquist, A., 432, 435
Cyclization, 385
Distribution
Baccharis, 435, 443, 470
Brickellia, 434, 441, 457

Helianthus, 436, 452, 482
Helichrysum, 435, 449, 480
Montanoa, 436, 456, 484
Stevia, 434, 442, 462
Viguiera, 436, 452, 482
Elongatolides, 494
Eupatorieae, 432, 433
Ericaceae, 485
Euphorbiaceae, 389, 485
Geranylgeranyl pyrophosphate, 385, 437
Gibberellin, 386
Grindelanes, 386
^{1}H-NMR Spectrometry
  absolute stereochemistry, 497
  3,4-epoxides, 501
  C-6 substituents, 499
  C-7 substituents, 504
  C-12 substituents, 504
  C-17 substituents, 504
  C-19 substituents, 504
  cis-ent-clerodanes, 504
  kolavanes, 494

  methyl chemical shift assignments, 498
  sandaracopimaranes, 506
Hardwickia pinnata, 493
Heliantheae, 433
Kaurene synthetase (AB activity), 389
Lactuceae, 432
Lamiaceae, 485
Misra, R., 493
Phytol Rule, 385
Plathyterpenones, 494
"Rules" of Conformation, 500
Sesquiterpene Lactones, 432
Solidagolactones, 494
Structure Description Codes, 3
Tori Equation, 503
Trichomes, 391
Velloziaceae, 389
Vernonieae, 432, 436
Wagenitz, G., 432, 438
X-Ray Analysis, 505
Zürcher Rules, 498